Forensic Engineering

Forensic Engineering
Damage Assessments for Residential and Commercial Structures

Second Edition

Edited by
Stephen E. Petty

CRC Press
Taylor & Francis Group
Boca Raton London New York

CRC Press is an imprint of the
Taylor & Francis Group, an **informa** business

Second edition published 2022
by CRC Press
6000 Broken Sound Parkway NW, Suite 300, Boca Raton, FL 33487-2742

and by CRC Press
2 Park Square, Milton Park, Abingdon, Oxon, OX14 4RN

© 2022 selection and editorial matter, Stephen E. Petty; individual chapters, the contributors

First edition published by CRC Press 2013

CRC Press is an imprint of Taylor & Francis Group, LLC

Library of Congress Cataloging-in-Publication Data
Names: Petty, Stephen E., editor.
Title: Forensic engineering : damage assessments for residential and commercial structures / edited by Stephen E. Petty.
Description: Second edition. | Boca Raton : CRC Press, 2021. | Includes index.
Identifiers: LCCN 2021006024 (print) | LCCN 2021006025 (ebook) | ISBN 9780367758134 (hardback) |
  ISBN 9781003189305 (ebook)
Subjects: LCSH: Building failures—Investigation. | Buildings—Natural disaster effects. |
  Weathering of buildings. | Forensic engineering. | Building inspection.
Classification: LCC TH441 .F67 2021 (print) | LCC TH441 (ebook) | DDC 690/.21—dc23
LC record available at https://lccn.loc.gov/2021006024
LC ebook record available at https://lccn.loc.gov/2021006025

ISBN: 978-0-367-75813-4 (hbk)
ISBN: 978-1-032-18061-8 (pbk)
ISBN: 978-1-003-18930-5 (ebk)

Typeset in Times
by codeMantra

INTERNATIONAL CODE COUNCIL®

About the International Code Council®

The International Code Council is a nonprofit association that provides a wide range of building safety solutions including product evaluation, accreditation, certification, codification and training. It develops model codes and standards used worldwide to construct safe, sustainable, affordable and resilient structures. ICC Evaluation Service (ICC-ES) is the industry leader in performing technical evaluations for code compliance, fostering safe and sustainable design and construction.

The second edition of this book is again dedicated in loving memory to my son, Mark Anthony Petty, who passed away at the early age of 21 on September 3, 2011, in New York City, due to complications of Crohn's disease. Mark had remarkable promise, excelling as one of the top students in his honors engineering and business administration programs during his senior year at the University of Cincinnati, along with an equally remarkable personality. He was a large part of EES Group, Inc., helping out on projects since he was 13 years old. Ten years later, Mark is tremendously missed by me, his colleagues here at EES, and everyone who knew him.

Contents

Preface to the Second Edition

As I wrote in the Preface to the First Edition, this book is intended to serve as a comprehensive resource to bridge engineering disciplines with the building sciences and trades (e.g., carpentry, masonry, and plumbing) disciplines. This is necessary to excel in the field of forensic engineering, particularly for those working for, or in, the insurance industry assessing claims. Oftentimes, those entering the field are engineers or those with a science background who lack the knowledge in building sciences, trades, and codes and standards associated with roofing systems, building envelope systems, carpentry, plumbing, wiring, and masonry.

As with most textbooks, and the same is true for this one, a book cannot realistically cover an entire field – in this case the field of forensic engineering. Broadly, forensic engineering is a subset of the field of forensic sciences and is defined as the field that applies engineering practices and principles to determine and interpret the causes of damage to, or failure of, equipment, machines, or structures. We expand that definition here to include the general causes of illnesses or poor health to those in such buildings or structures based on industrial hygiene and safety sciences.

This second edition updates all insurance statistics and codes and standards cited in the first edition. In addition, Chapter 22 on the Appraisal Process has been deleted and replaced with two new chapters: Chapter 22 on Plumbing, Piping, and Tubing Failures and Chapter 23 on Equipment Failures and Investigations. These are two areas often encountered by the forensic engineer that were not covered in the first edition. The old Chapter 23 on serving as an expert witness has been updated and renamed as Chapter 24. Significant enhancements have been made to Chapter 11 – Indoor Environmental Quality regarding current worldwide levels of exposure to substances and to the write-ups on bacteria and *Legionella*.

Acknowledgments

To the staff at EES Group, Inc. (both at the Ohio and Florida companies) who have invested tens of thousands of man-hours developing the methodologies outlined in this book based on experiences in the field and who have encouraged me to write this book.

To the administrative staff at EES Group, Inc. who contributed an immeasurable amount of time proofing and formatting the chapters, figures, and tables. Special thanks goes out to my wife Alexandra (Lexa) V Petty for her time proofing the manuscripts, obtaining permissions and all the other tasks needed to support the publisher.

To Taylor & Francis, CRC Press publishers for agreeing to issue this book, and in particular to Lisa Wilford, Editorial Assistant and Joseph Clements, Editor – Civil & Environmental Engineering, for their support, advice, and dedication to ensure that a quality textbook was issued.

To Margi M. Leddin, Vice President, Intellectual Property Contracts & Initiatives, International Code Council, and the International Code Council for their help and permissions in utilizing text and figures from the 2018 International Residential Code for One- and Two-Family Dwellings, the 2018 International Building Code, and the 2018 International Plumbing Code throughout this book. The materials used in this book contain information that is proprietary to and copyrighted by International Code Council, Inc. Portions of the information copyrighted by International Code Council, Inc. have been obtained and reproduced with permission by International Code Council, Inc., Washington, D.C. The acronym "ICC" and the ICC logo are trademarks and service marks of ICC. All rights reserved www.iccsafe.org.

To my colleagues and coauthors, who made major contributions to this book as chapter authors, chapter coauthors, or with detailed comments to ensure that information contained in the chapters was based on the best information we had available. A special thanks to Herbert D. Layman, Founder of U.S. Micro-Solutions, for his contributions and review of Chapter 13 covering mold and bacteria; Mr. Thomas E. Schwartz, Esq., Holloran White & Schwartz LLP, for his contributions and review of Chapter 24 on the topic of servicing as an expert witness.

To the National Fire Protection Association, Dennis J. Berry, Secretary of the Corporation and Director of Licensing. The materials used were reproduced with permission from NFPA 921–2011, *Guide for Fire and Explosion Investigations, Copyright© 2010, National Fire Protection Association*.

To APA – The Engineered Wood Association, Asphalt Roofing Manufacturer Association (ARMA), Brick Industry Association (BIA), CertainTeed, Insurance Institute of Business and Home Safety (IBHS), National Roofing Contractors Association (NRCA), New York State Department of Health, National Oceanic Atmospheric Administration (NOAA), RCI Foundation, RISA Technologies, Spray Polyurethane Foam Alliance (SPFA), Weather Decision Technologies, and Weyerhaeuser NR Company for all of their contributions.

To all who contributed and not mentioned, thank you for providing the requested materials for this book. Without your contributions, this book would not have been possible.

Editor

Stephen E. Petty, PE, CIH, CSP, is president and owner of EES Group, Inc. (www.eesgroup.us), a forensic engineering and environmental health and safety (EHS) corporation. EES Group, Inc. was founded in 1996; since that time, he has conducted or supervised thousands of forensic investigations and served as an expert witness in over 400 legal cases. Most recently, he has served as an exposure, Personal Protective Equipment (PPE), and warnings expert for the Monsanto Roundup, DuPont C8 and 3-M PFAS cases. He practices in the fields of chemical, civil, environmental, and mechanical engineering as well as the field of industrial health and safety. Mr. Petty is also a recognized expert in the field of heating, ventilation, and air conditioning (HVAC) and a holder of nine U.S. Patents, eight of which are in the HVAC field:

U.S. 6,649,062. November 18, 2003. Fluid-Membrane Separation. Petty.
U.S. 6,109,339. August 29, 2000. Heating System. Talbert, Ball, Yates, Petty, and Grimes.
U.S. 5,769,033. June 23, 1998. Hot Water Storage. Petty and Jones.
U.S. 5,636,527. June 10, 1997. Enhanced Fluid-Liquid Contact. Christensen and Petty.
U.S. 5,546,760. August 20, 1996. Generator Package for Absorption Heat Pumps. Cook, Petty, Meacham, Christensen, and McGahey.
U.S. 5,533,362. July 9, 1996. Heat Transfer Apparatus for Heat Pumps. Cook, Petty, Meacham, Christensen, and McGahey.
U.S. 5,339,654. August 23, 1994. Heat Transfer Apparatus for Heat Pumps. Cook, Petty, Meacham, Christensen, and McGahey.
U.S. 5,067,330. November 26, 1991. Heat Transfer Apparatus for Heat Pumps. Cook, Petty, Meacham, Christensen, and McGahey.
U.S. 4,972,679. November 27, 1990. Absorption Refrigeration and Heat Pump System with Defrost. Petty and Cook.

Mr. Petty began his working career in 1979 as a Scientist for the Battelle Memorial Institute (BMI) in Richland, Washington, where they operated the Pacific Northwest National Laboratory (PNNL), later transferring to their corporate headquarters in Columbus, Ohio, where he eventually worked as a Senior Research Engineer on a variety of private and public research projects. In 1987, he moved to the Columbia Gas System where he worked for approximately 10 years as a Senior Research Engineer and Section Manager in the areas of residential, commercial, transportation, and fuel cell research and product development. During this time, he was nominated to, and served on, 11 Industry Advisory Bodies [including Gas Research Institute (GRI), American Gas Association (AGA), U.S. Department of Energy Funding Initiative – $2 billion (USDOE-FI), and Gas Utilization Research Forum (GURF)]. He also served as a U.S. DOE expert reviewer in the areas of cooling, heat pumps, desiccants, and power generation. In 1996, he formed EES Group, Inc. located in Dublin, Ohio, to serve the insurance industry in claims assessments and the legal community in litigation matters as an expert witness in the areas of insurance claims disputes and toxic tort (primarily in the areas of exposure, PPE and warnings). In 2013, EES Group, Inc. was expanded into Florida and in 2015, the Ohio operations were sold to Mr. Ronald L. Lucy.

Mr. Petty earned a BS and an MS in chemical engineering at the University of Washington (1979 and 1982, respectively) where he graduated with honors and an MBA from the University of Dayton (1987) where he received the Father Raymond A. Roesch Award of Excellence for Outstanding Academic Achievement in the MBA Program for finishing first in his class. He is a Registered Professional Engineer in the states of Florida, Kentucky, Ohio, Pennsylvania, Texas, and West Virginia. He is also a Certified Industrial Hygienist (CIH) from the American Board of Industrial

Hygiene, a Certified Safety Professional (CSP) from the Board of Certified Safety Professionals, and an Asbestos Hazard Evaluation Specialist in the State of Ohio. He is also a working member of ASHARE Standard 2.3 – Gaseous Air Contaminants and Gas Contaminant Removal Equipment.

Mr. Petty is a member of the American Industrial Hygiene Association (AIHA), a member of the American Conference of Governmental Industrial Hygienists (ACGIH), and a member of the American Society of Heating, Refrigeration, and Air Conditioning Engineers (ASHRAE), the American Institute of Chemical Engineers (AIChE), the Society of Automotive Engineers (SAE), and Sigma Xi.

Mr. Petty currently resides in Pompano Beach, Florida, having made the move from Dublin, Ohio, toward retirement!

Contributors

George W. Fels
Curtiss-Wright Corporation
Cleveland, Ohio

Bryan E. Knepper
EES Group, Inc.
Waterville, Ohio

Herbert D. Layman
U.S. Micro Solutions
Latrobe, Pennsylvania

Ronald L. Lucy
EES Group, Inc.
Hilliard, Ohio

Noah Monhemius
EES Group, Inc.
Grove City, Ohio

Stephen E. Petty
EES Group, Inc.
Pompano Beach, Florida

Thomas E. Schwartz
Holloran White & Schwartz LLP
St. Louis, Missouri

1 Introduction

Stephen E. Petty
EES Group, Inc.

CONTENTS

PURPOSE/OBJECTIVES

The purpose of this chapter is to:

- Define the term "Forensic Engineering" and explain why forensic engineering is needed.
- Define areas within the insurance industry where forensic engineering services are often required.
- Define a standard forensic engineering inspection protocol.

- Explain why written reports are needed along with key elements of the basic forensic report.
- Define how to use the terms "not possible," "possible," "probable," "likely," and "certain."

Following the completion of this chapter, you should be able to:

- Understand where forensic engineering services are likely to be needed, especially by the insurance industry.
- Be able to conduct a forensic inspection using a standard protocol.
- Recognize the value and key components of a written forensic inspection report.
- Know and understand when and how to use the terms "possible," "probable," "likely," and "certain."

1.1   DEFINITION OF FORENSIC ENGINEERING/SCIENCES

A detailed discussion of the definition of forensic engineering follows:

"Forensic engineering is the application of engineering principals and methodologies to answer questions of fact. These questions of fact are usually associated with accidents, crimes, catastrophic events, degradation of property, and various types of failures,"[1] and further, "Forensic engineering is the application of engineering principles, knowledge, skills, and methodologies to answer questions of fact that may have legal ramifications."[1]

While this definition is applied to forensic engineering, it should be acknowledged that this field is practiced by not only engineers, but also other specialists involved with areas such as roofing system sciences, building envelope sciences, accident reconstruction, industrial hygiene (e.g., mold, bacteria, asbestos, and indoor air quality), and meteorology (rain, wind, snow, ice, hail, tornados, and hurricanes). Thus, the term forensic engineering in this book has been expanded to forensic engineering/sciences. The fundamental questions of fact to be addressed are:

- What are the failure(s) or condition(s) of concern?
- What are the magnitude and extent of the failure(s)?
- When did it occur (if this determination is needed and desired)?
- Why did it occur?

As noted in the preface, this last question, "Why did the failure(s) or concern(s) occur?" is complex, and this causation question must often be answered at multiple levels. For example, if a high wind caused failure of the roof, the failure may be due to high winds, but may have occurred at lower-than-design wind speeds due to improper design and/or installation. This example touches on the issue of the ultimate "root cause" of the failure, which requires analysis based on detailed site inspection information and subsequent analysis and review of the literature, pertinent codes and standards, and other information such as that obtained from interviews. It is common to arrive at a topical conclusion regarding the cause of a failure (e.g., wind) that is not the root cause of failure (e.g., faulty installation). Often, whether in claims resolution discussions or in litigation, this differentiation between a topical cause and a root cause of failure is the core of the arguments between opposing parties involved in a dispute.

What makes forensic engineering/sciences different from other fields of science is that it couples the academic fields of engineering and science with the practical fields, such as building/construction sciences and the trades such as those associated with carpentry, masonry, and plumbing. Building and construction sciences consist of knowing terminology, practices, and methodologies of trades such as carpentry, heating, ventilation and air conditioning (HVAC), plumbing, and electrical. Knowledge of residential and commercial codes and standards is also a must and bridges across all these areas. New engineers and other science professionals are rarely adequately trained in the trades or codes and standards; disciplines, which must be learned by trained forensic investigation professionals through experience. The training of engineers and scientists in this field requires considerable training beyond academics since much of the information needed to make forensics causation opinions lies in the practical fields, areas typically not covered in colleges and universities. Interestingly, often those growing up in rural environments have better entry-level skills in this field than those being raised in urban environments; most likely this is because they must be creative problem-solvers (i.e., cause versus effect) with limited resources given the environment in which they live.

Regardless of experience, the forensic investigator must be able to recognize when they may not have the skill sets to solve a given situation and must feel comfortable to rely on other, more experienced professionals for their help. Extending into areas beyond their education, training, and experience could be problematic should their report be challenged in litigation.

1.2   WHY FORENSIC ENGINEERING/SCIENCES?

The reason why forensic professionals are needed is typically distilled down to the desire by one or more parties to determine why a failure or issue occurred. The desire to seek this information usually involves determining responsible parties so costs associated with the failure can be properly allocated. The two categories of parties most likely interested in employing forensic professionals to make these determinations are associated with the insurance industry and the legal community. Other parties, such as building owners, may have an interest in determining these answers, but are generally unwilling to incur the costs or do not have the resources to employ such professionals.

Insurance companies are interested in making failure cause determinations for these primary reasons:

- Determine root cause failures and resulting responsible parties.
- Determine if they have coverage of a submitted claim based on root cause failures and timing of failure.
- Quantify the extent of damages.
- Demonstrate/illustrate how the failed system should be repaired/replaced according to best practices and/or codes and standards.
- Determine if other parties may have coverage for a submitted claim (i.e., concept of subrogation).

The insurance community will typically use this information to determine if they have coverage and to determine the costs of damages if they have coverage. The interest of the legal community in using forensic professionals is typically associated with the need to:

- Determine root cause failures and resulting responsible parties.
- Quantify the extent of damages.
- Qualify necessary repairs.
- Provide expert witness services for pending/actual litigation.

The legal community will typically use this information to help determine responsible parties for damage/injured parties and/or determine if claims have been appropriately addressed by insurance providers.

1.3   INSURANCE INDUSTRY CLAIMS STATISTICS

This book focuses on forensic investigations associated with insurance industry property claims; therefore, it is helpful to review what types of claims occur by topic and severity (i.e., cost of claims by type of claim).

   Data on property claims are typically organized by type and severity of the claim on an annual and regional or state basis. Thus, existing data on the claims are limited by the fact that claims vary by differences in year-by-year storm histories and by geographical locations within the United States. For example, hurricane-related claims would be more prevalent in the southeastern states, tornado claims in the south central and mid-western states, and earthquake claims in the western states. With these limitations in mind, national data from the Insurance Information Institute (III)[2] are presented for claims by type and severity in Table 1.1 and illustrated graphically in Figures 1.1 and 1.2.

TABLE 1.1

Claim Types by Frequency and Severity (Cost)

Type of Claim or Peril	Frequency[a]	Severity	Frequency (%)
Fire, lightning, and debris	0.28	$68,322	5.19
Other, including mischief and vandalism	0.66	$5,823	12.22
Theft	0.31	$4,264	5.74
Water and freezing	2.05	$10,234	37.96
Wind and hail	2.10	$10,182	38.89
Total	**5.40**		**100.00**

Source: Adapted from the Insurance Information Institute (2020).

[a]   Claims per 100 house years.

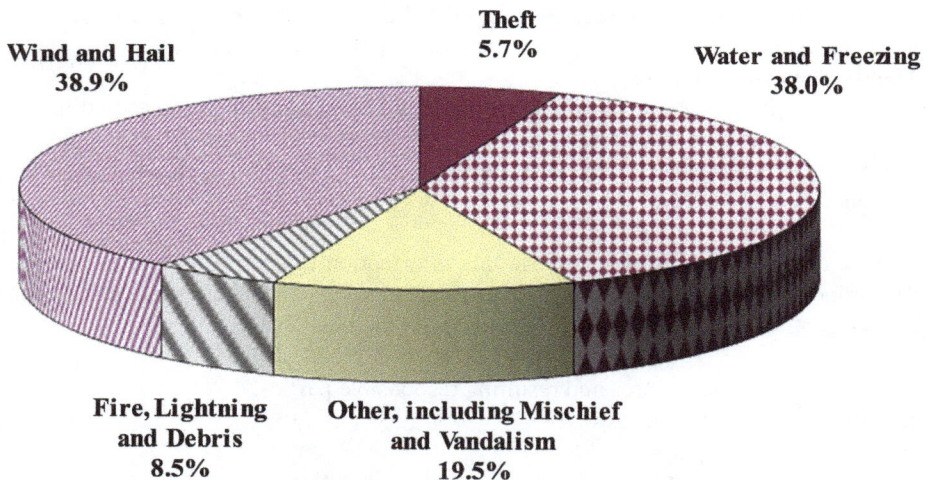

FIGURE 1.1   Claim frequency percentages by type of claim (2013–2017).

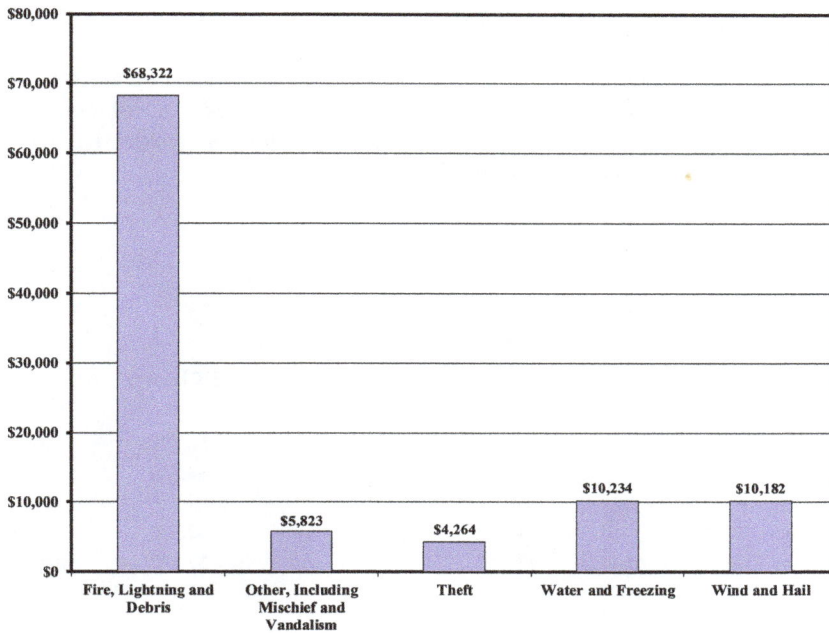

FIGURE 1.2 Claim severity (cost) by type of claim (2012–2017).

For the five categories of property claims listed over the time interval of 2012–2017, the most frequent claims are for wind and hail (2.10/100 house years – 38.9%) followed by those for water and freezing (2.05/100 house years – 38.0%). However, as illustrated in Figure 1.2, the claim severity is highest, by far, for fire claims (average of $68,322/fire, lightning, and debris claim). The claim frequency can vary from year to year due primarily to differences in severe weather as illustrated in Figure 1.3.

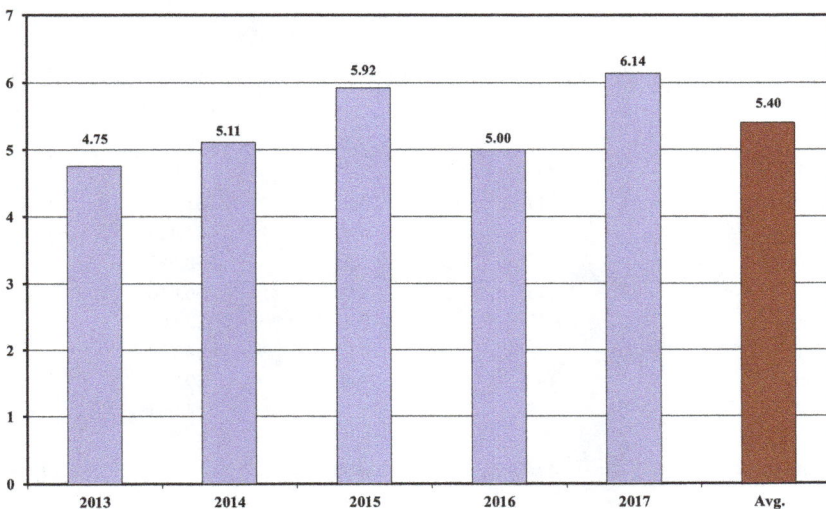

FIGURE 1.3 Annual claims by type of claim (2013–2017 and average of 5 years).

While the 2013–2017 average claim frequency was 5.40 claims/100 house years, the rate varied over this 5-year period from 4.75 to 6.14 (–12.0% to +13.7%).

Often regional or state claims data are more available for a given location since most insurance companies are authorized to provide insurance in specific states. An example of regional claims frequency and severity data for a mid-western insurance company for both residential and commercial claims is shown in Tables 1.2 and 1.3 and illustrated in Figures 1.4–1.7.

TABLE 1.2
Residential Claim Types by Frequency and Severity

Claim Type or Peril	Claims (#)	Claims (%)	$/Claim	$ (%)
Mold	13	0.57	$53,850	28.32
Hail	344	15.09	$8,729	16.45
PD liability	40	1.75	$4,720	1.03
Fire	96	4.21	$4,726	1.99
Mischief	4	0.18	$2,550	0.06
Ice/snow	77	3.38	$7,922	0.30
Collapse	7	0.31	$6,088	1.97
Other	59	2.59	$5,742	0.41
Theft	39	1.71	$1,406	0.30
Water	929	40.75	$5,258	26.76
Animal	5	0.22	$6,177	21.29
Wind	629	27.59	$5,157	0.14
Vehicle	38	1.67	$4,664	0.97
Total	**2,267**	**100.00**		**100.00**

Source: Adapted from the Midwestern Insurance Company (2015).

TABLE 1.3
Commercial Claim Types by Frequency and Severity

Claim Type or Peril	Claims (#)	Claims (%)	$/Claim	$ (%)
Fire	32	7.19	$91,874	35.81
Hail	24	5.39	$10,161	2.97
PD liability	124	27.87	$6,985	10.55
Ice/snow	17	3.82	$81,224	16.82
Mischief	2	0.45	$1,423	0.03
Collapse	6	1.35	$78,532	5.74
Other	18	4.04	$6,818	1.49
Mold	2	0.45	$2,164	0.05
Theft	19	4.27	$4,779	1.11
Water	109	24.49	$9,445	12.54
Wind	63	14.16	$14,057	10.79
Vehicle	29	6.52	$5,968	2.11
Total	**445**	**100.00**		**100.00**

Source: Adapted from the Midwestern Insurance Company (2015).

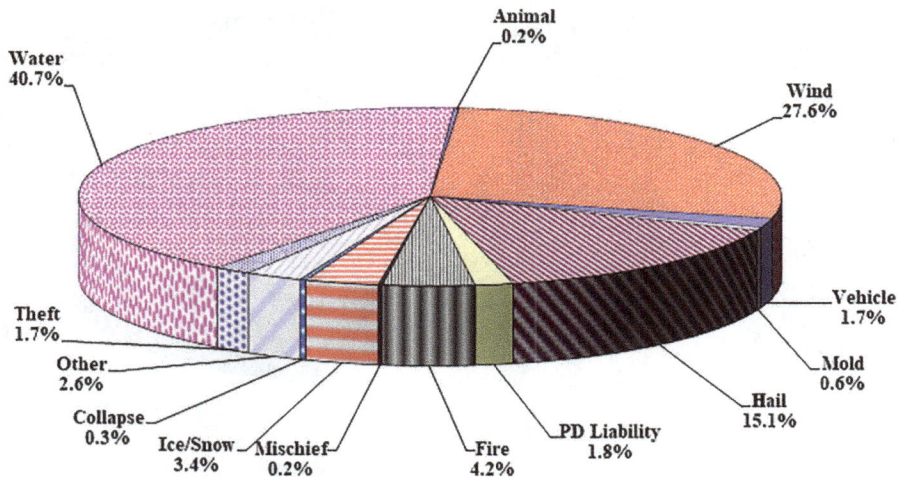

FIGURE 1.4 Residential claims frequency – Midwestern Insurance Company (2015).

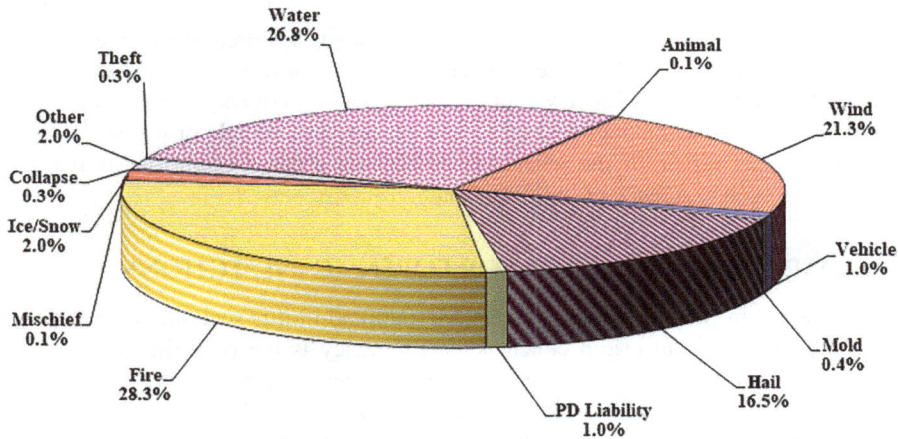

FIGURE 1.5 Residential claims severity – Midwestern Insurance Company (2015).

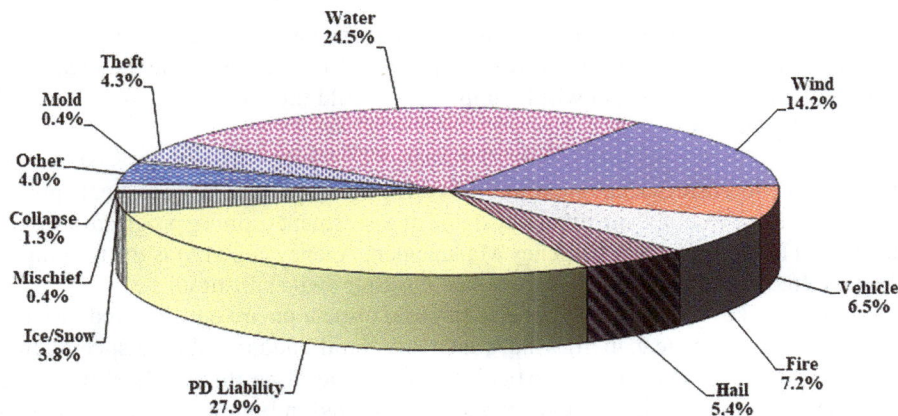

FIGURE 1.6 Commercial claims frequency – Midwestern Insurance Company (2015).

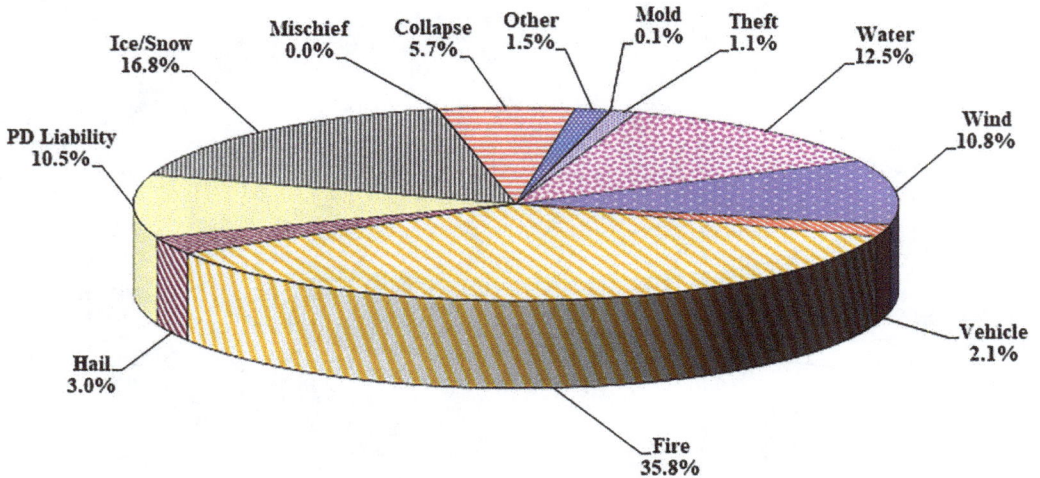

FIGURE 1.7   Commercial claims severity – Midwestern Insurance Company (2015).

Note that regional differences in claims will affect the needs for forensic engineers and scientists (i.e., the overall severity of this Midwestern Insurance Company was ~4.12 below overall national severity values for total claims). Property Damage (PD) Liability coverage is part of a car insurance policy. It helps pay to repair damage caused to another person's vehicle or property and typically helps cover the cost of repairs if the insured is at fault for a car accident that damages another vehicle or property such as a fence or building front.

1.4   STANDARD METHODOLOGY FOR FORENSIC INSPECTIONS

A consistent forensic inspection process, as the one outlined in this section, is very important for efficiently reaching cause and origin conclusions. Efficiency is not only the desire to minimize costs, but a desire to minimize the time needed from the owner or owner's representative so they are not inconvenienced. This need to be effective includes not only the desire to efficiently use money and time resources, but to do so in a way that most likely will result in the ability to actually make a determination of the cause and origin of a specific claim. We believe that the approaches outlined in this book:

- Are the most effective method for training new forensic engineers/scientists.
- Will result in a consistency of the inspection process. Consistency of the methodology is a tenant of the legal process and will be important should the conclusions reached be challenged in litigation.

As is the case in any forensic evaluation, training is critically important for consistent application of rules and methodology. While there are many damage assessment training programs available, one that is recognized by the Federal Emergency Management Agency (FEMA) is the training institute provided by the International Code Council, When Disaster Strikes Institute.

In this section, the key elements of a generic forensic inspection are outlined and discussed. For specific types of inspections (e.g., hail, wind, water, structural), additional case-specific inspection recommendations are provided in that particular chapter. Based on thousands of completed field inspections, the recommended inspection process is illustrated in Figure 1.8; elements in the flow-chart are discussed in the text that follows.

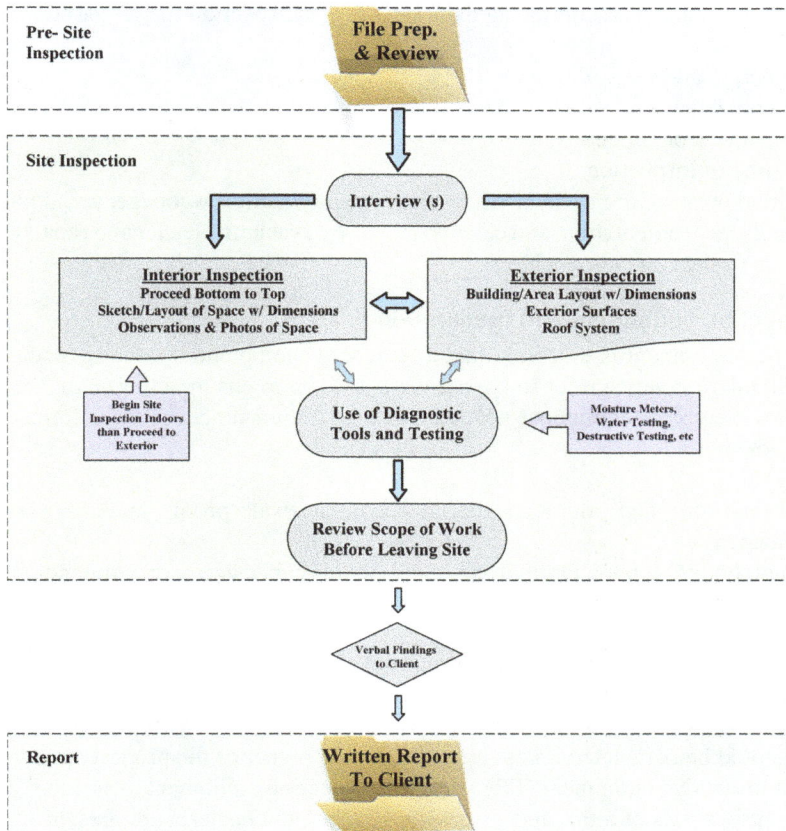

FIGURE 1.8 Generic forensic site inspection process flowchart.

The key steps to the inspection process are Pre-Inspection File Preparation, Site Inspection, and Post Inspection Written Report. These key steps are described in detail below.

1.4.1 PRE-INSPECTION FILE PREPARATION

The pre-inspection file preparation consists of obtaining and organizing the information, personnel, and equipment needed to conduct the on-site inspection accurately and in a time-efficient manner.

1.4.1.1 Gathering Information and Allocating Resources

It is good practice to gather as much detailed information as you can pertaining to the extent of the scope of work and the issues regarding a particular inspection from the client and property owner prior to the initial project setup. This information gathering effort often leads to effective management of time and resources including: improving the ability to estimate the time needed for the inspection accurately and allowing for a better identification of expertise and equipment needed to perform the inspection.

Examples of the elements recommended during the pre-inspection file preparation include:

- Site inspection address.
- Point of contact(s).
- Scope of work from client.
- Agreed time of inspection.
- Information on structure to be inspected (e.g., county auditor webpage, aerial photos).
- List of tools, instrumentation, and cleanup items (e.g., vacuum cleaner and repair materials).
- Directions to site inspection address.

1.4.1.2 Inspection Equipment and Preparation

Using the proper equipment is essential, and a review of the potential issues expected during the inspection is standard practice prior to the arrival at the site to ensure a thorough, accurate assessment. Examples of typical equipment needed and preparation practices for a forensic inspection include the following:

- Weather-resistant and durable digital camera (waterproof and shockproof are recommended).
- A field notebook/job book (a relatively small, manageable size is recommended).
- Ladders of different lengths.
- Flashlights.
- Tape measure and/or electronic measuring tool.
- Level and roof slope gauge.
- Specialized equipment if specific destructive testing is approved. Note that no destructive testing should be completed unless approved by the owner of the property and the client.
- Personal protective equipment (PPE) and fall protection equipment as needed (e.g., respirator for inspections of attics and spaces where mold and bacteria are present).
- Sampling media and collection equipment if mold, bacteria, or other indoor air quality (IAQ) issues are part of the scope of work.
- A cellular phone.
- A coworker for assistance on difficult, time-consuming, and/or potentially dangerous inspections (i.e., high, steep roof surfaces, surveying levels, etc.).
- A review of pertinent weather records for hail, wind, and water leak inspections.
- A review of pertinent building code/residential code requirements and industry best practice documents for the potential issue(s) expected during the inspection.

1.4.2 BASIC SITE INSPECTION METHODOLOGY

A consistent methodology based on industry best practices for forensic investigations is critical to ensuring consistency and completeness of the investigation process. Based on thousands of completed field inspections, this book provides recommended site inspection methodologies for property forensic investigations. The baseline recommended methodology is provided in this chapter; investigation-specific factors to add to this basic methodology can be found in the corresponding chapters that follow.

1.4.2.1 Basic Methodology – Site Arrival Best Practice

The protocol upon arrival to the inspection site typically includes photographing the front of the structure as a starting point on the camera for the subject inspection and introducing yourself to the property owner or point of contact if present. During this time, a business card should be left with the owner or point of contact to identify you as the inspector.

1.4.2.2 Basic Methodology – Property Owner/Point of Contact Interview

An interview should be conducted to obtain background information about the structure and specifics regarding the basis for the inspection. A well-done interview can often be the source of information that leads to the basis of the causation and origin of the claim for the inspection. Sometimes, based on inspection information, further follow-up questions may also be asked. The typical interview should include questions regarding the structure (e.g., age, length of time owned, square footage), the damage that has occurred (e.g., for wind damage to a roof, ask questions regarding the age of the roof, date of storm, direction from which the storm arrived, etc.), and if any improvements have been made to the structure. In addition to being a good listener, the key to a good interview is to ask questions that quantify information with respect to magnitude and time. For example, if a question is asked whether or not the shingles on the home have been replaced and the answer is "yes," then follow-up questions regarding the estimated date/year of the replacement should be asked along with information regarding whether or not the old layer was removed, what type of shingles were installed, and who was the installation contractor. A vital part of the interview is to always ask whether or not something occurred and then follow up with "what," "when," and "why" questions to flush out information on the topic. At the end of the interview, the inspector should let the owner or point of contact know how long the inspection will likely take, who will be receiving the inspection report, and when the report is likely to be issued.

1.4.2.3 Basic Methodology – Interior Inspection

After the interview has been completed, the formal on-site inspection begins. Best practices are to conduct the indoor portion of the inspection first (if the issue requires being indoors) followed by completion of the outdoor portion of the inspection. By conducting the indoor portion of the inspection first, the forensic inspector limits soiling of the indoor spaces from outdoor dirt, mud, and/or debris, and if the owner or point of contact needs to leave, this process minimizes the time they need to be present on the site. It should be noted that not every inspection requires interior observations such as for hail and wind damage claims.

Once indoors, best practices are to conduct the inspection from the lowest floor (i.e., basement and/or crawlspace) working upward to the attic spaces. This is particularly important in water and structural causation and origin inspections. During water leak inspections, since the law of gravity still works, water constantly attempts to find its level and flows downhill. By beginning the inspection on the lowest level, one begins to dial in on the source of the water intrusion as they move upward through the structure, noting the locations and patterning of the water damages as they go. Similarly, for structural damage inspections, many structural issues can be tied to basement or crawlspace foundation issues. In these situations, by beginning the interior inspection process at the basement/crawlspace level, the damages observed afterward in upper levels can be evaluated and any correlations/ties to the foundation can either be confirmed or denied based on their locations and patterning. Further, experience has shown that on numerous occasions, structural discrepancies go hand in hand with water intrusion causes/origins.

For each indoor level (e.g., basement/crawlspace, first floor, second floor, attic, etc.), experience has also shown that the following indoor inspection process leads to the highest success rate in determining problem cause(s) and effect(s):

1. Sketch the floor level and measure dimensions of each of the spaces on the level.
2. Conduct space inspections for each level (take inspection observations and photographs).
3. Repeat Steps 1 and 2 for each level (as needed).
4. Utilize relevant instrumentation or diagnostic tools; note that some methods should be conducted after the exterior inspection such as destructive and water testing (e.g., moisture meter measurements using conductance and/or capacitance probes, wall, floor, and ceiling level measurements, IAQ measurements such as temperature, carbon dioxide, carbon

monoxide, and humidity levels by space, FLIR© infrared photography for areas of possible water intrusion, specialized testing such as water infiltration testing, mold and bacteria sampling, formaldehyde testing, etc.).

5. Complete destructive testing as needed (if approved by client and/or property owner).

The key to this overall inspection process methodology is that the forensic inspector will visit each space at each level a minimum of three times (i.e., drawing floor plans and taking measurements, detailed inspection and photography by space, measurements and instrumentation by space), thereby maximizing the probability that the cause and effect of the issue(s) responsible for the forensic inspection will be recognized and determined.

1.4.2.4 Basic Methodology – Exterior (Non-Roof) Inspection

Once the indoor portion of the inspection has been completed, a complete and thorough investigation of pertinent exterior areas by elevation should begin. However, some inspections, like wind and hail damage claims, do not require an interior inspection in which case proceed to the exterior inspection after conducting the interview. Experience has shown that the following outdoor inspection processes lead to the highest success rate in determining problem cause(s) and effect(s):

1. Sketch the plan view of the structure, including roof features (note that the plan view is the vantage point from directly above and looking down on the structure).
2. Measure plan view dimensions.
3. Complete the exterior inspection one elevation at a time (i.e., take overview photograph and conduct elevation inspection with specific photographs of inspection findings).
4. If needed, complete a roof inspection – see below (i.e., take an overview photograph and conduct elevation inspection(s) with specific photographs of inspection findings).
5. Take outdoor measurements of spaces as needed (e.g., level measurements, moisture meter measurements, FLIR© infrared photography, specialized testing such as water infiltration testing using ASTM-based processes and equipment or equivalent, etc.).
6. Complete destructive testing as needed (if approved).

Typical exterior elevation inspections will include a description of the exterior finishes, damages, and/or failures of exterior finishes, pertinent details regarding installation for the damaged/failed finishes, local and overall ground slope directions (i.e., grading), condition of downspouts and gutters, staining patterns on finished surfaces, etc. It is effective to move consistently, either clockwise or counterclockwise around the structure while completing the inspection. This allows for an organized recollection of information that may have not been fully recorded in field notes and to recall where photographs were taken. If inadequate attic ventilation is possibly considered to be contributory to a failure, eave and gable ventilation opening locations, numbers, and dimensions should be recorded.

1.4.2.5 Basic Methodology – Roof Inspection

The roof inspection, if needed, should include the following:

- An overview of roof features (sketch and record dimensions, slopes, key elements, and any discrepancies. Note today that some of this information can also be obtained from third-party services using computer-imaging tools. In addition, it is typically helpful to take several photos at different angles).
- The roof construction as viewed from the eave (e.g., type of roof finishes, layers of roof finishes, presence/absence of and type of underlayment, presence/absence of drip edge molding, etc.).
- A roof inspection specific for the issue(s) of interest (e.g., hail damage, wind damage, roof water leaks).

Based on experience, special attention should be paid to flashing details at roof/chimney, roof/skylight, roof/furnace vent, attic vent or soil stack penetration interfaces, and wall/roof interfaces. Roof leaks often occur at these locations, especially when proper water management details were not installed or installed properly.

1.4.2.6 Basic Methodology – Collection of Evidence

Finally, any items collected as forensic evidence should be documented by photographs before and after removal. Prior to removal, physical dimensions of the item and its proximity to other structural elements should be recorded in the field notebook. Often a sketch is helpful to record this information. The time of evidence removal should also be recorded in the field notebook. If possible, the evidence should be placed in a plastic bag and labeled with a permanent marker. The bag and label should be photographed at the site location where the evidence was removed. Once back at the office, the evidence should be logged in using a chain of custody form. A copy of the form should be inserted with each evidence item and into the project file. The evidence should be stored in a climate-controlled space and stored until destructively tested or disposed of based on established holding times with the client.

1.5 WRITTEN INSPECTION REPORTS – WHY NECESSARY AND STANDARD COMPONENTS

This section provides a methodology for a written forensic investigation report and addresses the basis/rationale for preparing it. It also provides a listing and explanation of standard components that should be included in the written inspection report.

1.5.1 NEED FOR A WRITTEN INSPECTION REPORT

The reasons for mandating that a written report be prepared are that:

- Not all information committed to memory during a field investigation is documented initially in field logs or field notebooks.
- A written report prepared in a timely fashion (within a week) allows for documenting information not explicitly written in logs or field notebooks.
- Analysis of findings can be completed in a timely fashion resulting in better conclusions.
- It lessens the likelihood that verbal findings will be challenged by others.
- It provides a more professional view of the preparer to the client.
- It lessens and/or prevents inspection-related information from being lost if the inspection work/findings/conclusions should be questioned in the future.

In addition, it should be mandatory that all photographs are downloaded and labeled the day they were taken. Experience suggests that if photo files are not immediately labeled, this information may be lost. If the project is known to be likely associated with litigation, the raw (unmodified or unlabeled) photograph files should be copied to a separate file directory.

1.5.2 BASIC METHODOLOGY – ELEMENTS OF A WRITTEN INSPECTION REPORT

A methodology for a generic forensic inspection report, including key report elements, is outlined and discussed in this section. Report recommendations for case-specific inspections (e.g., hail, wind, water, and structural investigations) are provided in their representative chapters later in this book.

The elements of a well-written forensic inspection report should include a clear and thorough understanding of the reason(s) for the inspection and the scope of work to be performed, key elements/observations from the inspection, a review/discussion of findings, and conclusions reached as a result of the work completed. Reports consisting primarily of text are not as effective; well-written and effective reports should include visual elements such as sketches and photographs to illustrate findings.

A recommended report outline flowchart with key elements is illustrated in Figure 1.9; these elements are discussed in the text that follows.

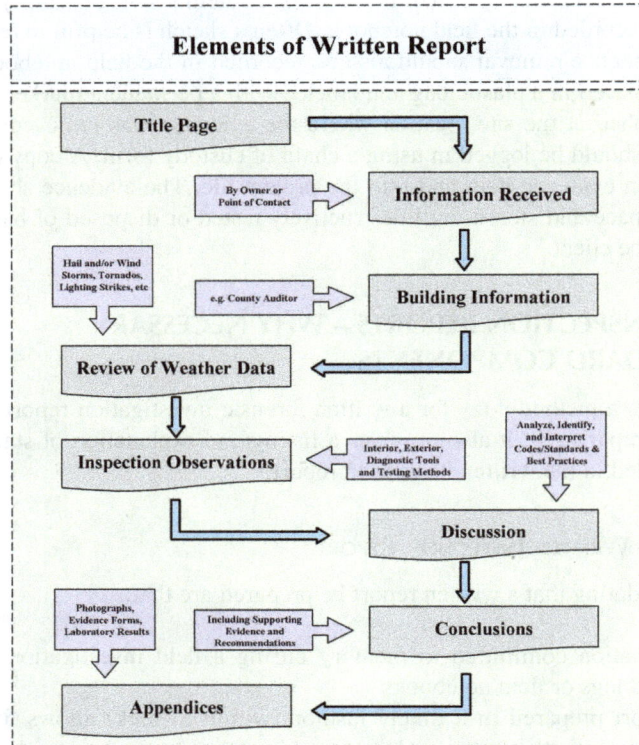

FIGURE 1.9   Report outline flowchart – key elements.

Another example of how to prepare a written forensic report is provided in Chapter 11 of Dickenson and Thronton's textbook on Cracking and Building Movement.[3]

1.5.2.1   Title Page, Cause for Claim/Inspection, and Scope of Work

The initial element of a well-written report should include a title page that identifies key elements pertaining to the inspection, such as the owner, address of the inspection location, client, date and time of inspection, and the name of the inspector. An example of a title page is shown in Figure 1.10.

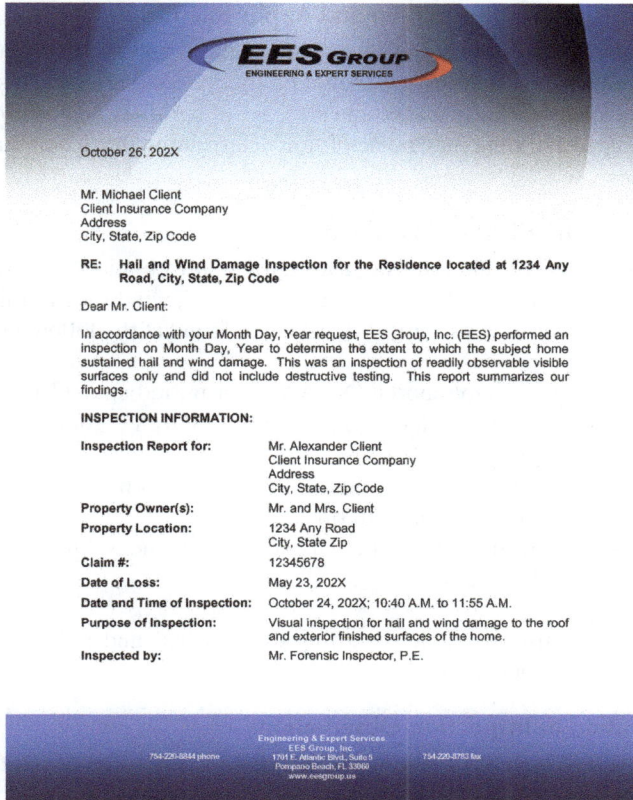

FIGURE 1.10   Written report – example title page.

It is important that the report clearly indicates the purpose for the inspection and that it was based on "readily observable visible surfaces" since it is nearly impossible to see all surfaces during an inspection due to clutter found in many buildings. Setting the limitations of the inspection in the Scope of Work is critical in today's litigious environment.

1.5.2.2   Documentation of Information Obtained During the Interview

A summary of information obtained during the interview with the property owner or point of contact should be included in the forensic report. Oftentimes information obtained from the interview provides insights into the cause of the issue being inspected and/or helps to focus the investigation itself. By documenting the results of the interview, the following can be achieved:

- A better understanding of the events/conditions that led up to the cause for the claim/ inspection is gained.
- Address the concern of the owner or point of contact and ensure they will be taken into consideration when determining the final outcome of the inspection.
- Provide a basis for specific areas to inspect/focus on during the inspection.

1.5.2.3   Review of Structure Information and Applicable Data

Information regarding the age of the structure, square footage, and finished materials (this can be obtained from most county auditor websites) should be summarized briefly. This provides information on ownership time history, building type and size, age of building, and time of improvements in some cases. The age of the building may be helpful in determining the age of building components such as roof finishes.

1.5.2.4   Review of Weather Data (If Pertinent)

Should it be pertinent to the specific forensic investigation, information regarding weather data is important to verify local weather conditions such as size of hail, wind speeds, rainfall amounts, and snow accumulation. Excellent sources for this data can be found at the following websites:

- http://www.spc.noaa.gov/climo/reports/ (NOAA's Storm Prediction Center).
- http://www.nws.noaa.gov/climate/index.php?wfo=iln (NOAA National Weather Center – Ohio Example Page).
- http://www4.ncdc.noaa.gov/cgi-win/wwcgi.dll?wwEvent~Storms (NOAA NCDC database on hail and wind storms by location with time).
- http://www.wunderground.com/ (Weather Underground home webpage – excellent source for local historical weather data).

Commercial resources are also available regarding data on hail and lightning (e.g., http://www.weatherforensics.com/2010/products/products.php).

1.5.2.5   Summarization of Inspection Observations

The bulk of the report will consist of summarizing observations taken at the time of the inspection. The best practice is to divide the written report into sections corresponding with the levels and areas inspected (i.e., first floor – living room) and to provide a format that will lead the client (or any other persons reading the report) through the inspection process and provide the supporting evidence needed for later analysis and accurate determinations/conclusions. Again, sketches and photographs illustrating key findings should be included to convey the results of the inspection to the reader effectively.

1.5.2.6   Discussion Section Including Pertinent Analysis

Experience has shown that it is important to add a discussion section to the report to provide a bridge between inspection observations and conclusions (i.e., how have the conclusions been reached based on observations) and to provide the owner or owner's representative with a basis for possibly resolving the issue(s) of concern. This section along with supporting evidence from the inspection observations, measurements, and instrumentation can be one of the most significant report elements that can answer the question "why?" Its general purpose is to bridge potential knowledge gaps between the inspection information and conclusions reached. Specifically the purpose of the discussion section will often:

- Analyze collected data during site inspection such as ventilation calculations, moisture output (e.g., vent-free space heaters), structural capacity of load bearing members, etc.
- Review pertinent building/residential codes, industry best practice documents, and manufacturer's installation instructions.
- Interpret, compare, and discuss the similarities and/or deviations of observed details with reviewed materials and experience.
- Offer professional/expert opinion on the cause and origin of the issue(s) at hand.

This section can also be used to provide the information needed in order to properly amend the current situation and possibly prevent it from happening again. In most cases, the damaged/failed components are likely due to improper design, installation, and/or construction.

For example, in the case of water leak causation and origin inspections, a discussion/review section pertaining to proper flashing details at various roof penetrations/interfaces versus what was actually observed during the inspection will help to explain why the roof leaked and how it should be repaired to prevent future leaks. Thus, it is important that the inspector explains not only why the situation occurred, but also how it can be fixed and/or repaired.

Another example might be the inclusion on the adequacy of attic ventilation if the roof shingles are prematurely thermally degraded.

1.5.2.7   Conclusions

Conclusions should follow the Discussion section and should be based on inspection observations, prioritized by information received by the client and found in the scope of work. New material should never be introduced in the conclusions section; such material supporting the conclusions should be located within the observation sections of the report. Conclusions reached should be readily apparent to the reader based on the observations and discussion sections presented earlier in the report.

The author should avoid speculation. The conclusions should be relevant to the scope of work and to what is known and can be proven. Engineers and scientists, in an effort to be helpful, often have a tendency to speculate beyond what is known or needed. Should the report be used in litigation, any speculation will be used by opposing council to denigrate any good work done and challenge the author on whether or not they typically provide conclusions based on speculation.

The report should then be signed (and stamped if required) by the professional(s) completing the inspection. Oftentimes, electronic signatures are used due to the desire to have reports provided electronically. Such reports should be sent in a format (e.g., .pdf) which can be protected so the file can be read (opened) and printed, but not modified to prevent doctoring.

1.5.2.8   Recommendations

Typically, recommendations are not included in the report unless either of the following situations arises: (1) they are requested by the client or (2) the conditions found during the inspection pose a hazard to those living or working at the site inspected. For example, if a structural beam supporting major portions of a roof has been found cracked and possesses the potential for collapse of the roof system, the owner and client should be informed verbally immediately, the conversations logged, and then the condition documented in the written report. Not only does this protect the inspector from potential litigation, it is also the proper thing to do, and such communication is called for in most professional ethics statements regarding the standards of practice for engineers and other professionals.

The primary reason why recommendations are typically avoided in forensic reports is that they tend to compel the client to complete recommended actions that may not be desired by the client.

1.5.2.9   Appendices

Appendices to the report should provide pertinent reference materials that were used and cited within the body of the report and provide a basis for the inspector's professional/expert opinion to the issue(s) for the particular structure. Common appendices to forensic inspection reports include:

- A sample of additional photographs taken during the inspection, which were not included within the body of the report.
- Chain of custody documents for evidence collected from the inspection site.

- Laboratory results.
- Other applicable reference material (e.g., industry best practice documents).

For documents covering code or best practices or standards, care should be used in providing documents dated after the installation. In litigation, such documents can result in the report and its conclusions being thrown out and rendering summary judgment against your client because they will argue (correctly so) that the information was not available to their client at that time. If post-dated (from time of installation) documents are used, and best practices have not changed over time, it should be noted that these documents simply illustrate known best practices.

1.6   TERMINOLOGY – USE OF THE TERMS "NOT POSSIBLE," "POSSIBLE," "PROBABLE," "LIKELY," AND "CERTAIN"

The way in which the relative certainty of a finding or conclusion is worded is important, not only to reflect actual knowledge in an oral or written report, but also may be critical should the inspection report output be part of, or subject to, litigation. Based on litigation experience, in addition to their legal skills, a lawyer's two advantages over individuals with technical/nonlegal backgrounds are: an understanding of the behavioral nature of individuals with technical backgrounds and their desire to please (i.e., provide an answer or solution); and an ability to sense weakness or lack of confidence (i.e., smell blood in the water).

In the former, engineers often extrapolate limited findings based on the inspection process to speculative broader conclusions. This trap is predictable and often set by lawyers, whereby, they attack the speculative conclusion(s) to discredit the other findings or conclusions. For example, an engineer testifying on fraud damage to a roof can readily use investigative techniques outlined in this book to prove that the damage is probably or likely fraud, but rarely can an engineer say with absolute certainty the damage was caused by fraud. Under this same hypothetical, the homeowner reports in the interview that Roofer X was on the roof the previous week. It would be speculative to state that this particular roofer caused the damage since the inspector did not see the roofer actually commit the fraudulent activity, nor does it rule out others who may have been on the roof. The lesson learned here is to limit the certainty reached about a situation to what is actually known, or can be supported technically, and to avoid extended, broader conclusions that are often simply speculation.

The second point to be made is that lawyers may not understand much of the technical information in an expert's report or testimony, but typically are excellent reads of defensive body language and quick to notice defensive language. By nature, all people will be nervous when their testimony is being taken (e.g., in depositions); thus, it is important to avoid situations where it becomes apparent that the technical person has extended/overstated their findings or conclusions beyond those that can be readily supported.

Thus, the important terms used to establish the level of the certainty in a finding or conclusion are "not possible," "possible," "probable," "likely," and "certain," which are defined as follows:

- *Not possible*: 0% probability.
- *Possible*: >0% probability.
- *Probable*: >50% probability.
- *Likely*: >75% probability.
- *Certain*: 100% probability.

Arguments are sometimes made that the terms "probable" and "likely" are interchangeable, but the term "likely" is defined as very probable or suggesting a higher degree of certainty than just probable. In findings and/or conclusions, this distinction is recommended.

For forensic projects, the client typically desires that the conclusions be reached "within a reasonable degree of scientific (or engineering) certainty" since this is the threshold of certainty needed in possible future litigation (see also Chapter 24 "Serving as an Expert Witness"). This term implies that the findings are "probable," or as more commonly stated, "more likely than not." In other words, the probability is greater that the conclusion *is valid*, rather than *not valid*. Note that the terms "likely" and "certain" would also meet this definition, but as discussed, they imply a higher level of certainty. The term "possible" may be true, but is of less value in reaching conclusions, since technically almost anything is possible.

One mistake commonly made by forensic engineers is to overstate or overreach when stating their conclusions by using the term "likely," or implying certainty when they state in a conclusion that the situation "was caused by X" rather than "was probably caused by X" or "was likely caused by X." In general, it is more accurate and less risky to use the lowest level of certainty that reflects the facts of the situation (i.e., use the word "probable" rather than the word "likely" or implicitly imply certainty).

Of course, the key to any conclusion is that it is supported by facts obtained using a rigorous investigation protocol and careful scientific/engineering analysis, including cross-checking conclusions to be reached from alternative pathways or from the literature and to base them on established methodologies accepted in the technical community.

IMPORTANT POINTS TO REMEMBER

- Forensic professionals include both engineering and other science professionals.
- Forensic professionals are defined by a combination of education, training, and experience. The training and experience are key elements for forensic professionals.
- Forensic professionals typically work for the insurance and legal communities. The insurance industry most often uses forensic professionals for hail, wind, water, fire, and structural claims.
- A consistent forensic inspection methodology, such as that outlined in this chapter, is important for both efficiently completing the inspection and possible legal challenges to results emanating from the inspection conclusions.
- A written report should be prepared following forensic inspections.
- Understanding the definitions of not possible, possible, probable, likely, and certain is critical in forming opinions in this field. Conclusions should be categorized as probable, likely, or certain to have value in terms of a "reasonable degree of scientific (or engineering) certainty."

REFERENCES

1. Noon, Randall K. *Forensic Engineering Investigation*. New York: CRC Press, 2001.
2. Insurance Information Institute. "Insurance Industries Statistics," Accessed May 20, 2020, http://www.iii.org/facts_statistics/homeowners-insurance.html.
3. Dickinson, Peter R. and Nigel Thornton. *Cracking and Building Movement*. London: RICS Business Services, Limited, 2004; Reprinted 2006.

2 Hail Fundamentals and General Hail-Strike Damage Assessment Methodology

Stephen E. Petty and Noah Monhemius
EES Group, Inc.

CONTENTS

PURPOSE/OBJECTIVES

The purpose of this chapter is to:

- Provide information on the formation and characteristics of hailstones, hailstorms, and the importance of these characteristics when performing hail damage assessments.
- Demonstrate the relationships between the size of hail and the dents they produce on metal surfaces.
- Provide a methodology for the determination of functional/cosmetic damage to roof and exterior finished surfaces and components.
- Document a general inspection methodology/protocol for performing hail damage assessments, including determining directionality and relative dating of hailstorms.

Following the completion of this chapter, you should be able to:

- Understand the characteristics and formation processes of hailstones.
- Understand that hailstorms produce a hail swath with various-sized hailstones.
- Understand the importance that hailstone size and hailstorm direction have on hail damage assessments.
- Understand how to determine the direction and approximate date of a hailstorm.
- Be able to determine the approximate size of hail that impacted a building by inspecting the soft metal surfaces.
- Be able to define clearly the difference between functional and cosmetic hail-strike damage.
- Understand the methodology for completing a visual hail damage inspection of a building and hail damage inspection report.

2.1   INTRODUCTION

The need to understand the full impact of hail damage to buildings and property has been an ongoing topic of concern to both owners and the insurance industry for many years. Annual damage from hail events in the United States for 2018 was $15 billion with estimates for 2019 at $22 billion.[1] Data from

The National Insurance Crime Bureau (NICB)[2–4] reported an increase in US total hail claims between 2013 and 2019 from 456,258 (2013) claims to 1,139,616 (2017) claims (Table 2.1 and Figure 2.1).

TABLE 2.1

Hail Insurance Claims – 2013 to 2019 – Total and by Top Ten States[2–4]

Location	2013	2014	2015	2016	2017	2018	2019
Texas	122,005	134,028	138,539	378,652	258,319	186,670	192,988
Colorado	32,741	99,565	50,285	90,002	118,645	191,679	69,742
Nebraska	45,860	80,293	22,193	58,142	81,053	23,424	56,897
Minnesota	45,860	16,688	30,641	20,759	75,835	24,865	49,973
Illinois	24,002	63,723	32,788	26,502	68,831	33,787	47,798
Oklahoma	60,131	11,760	26,302	0	0	0	0
Indiana	30,733	21,996	12,552	21,750	30,636	14,842	18,404
Kansas	52,404	39,222	36,337	49,864	58,939	38,117	50,737
Missouri	14,703	45,264	34,953	56,736	71,938	27,790	33,976
Iowa	0	0	0	10,845	67,731	25,664	19,744
North Carolina	0	0	0	0	22,850	10,466	25,026
South Dakota	27,819	30,634	16,428	0	0	0	0
New Mexico	0	0	0	11,529	0	0	0
Top ten states	456,258	543,173	401,018	724,781	854,777	577,304	565,285
All states	720,475	824,325	572,182	965,153	1,139,616	844,932	784,814
Increase/year (%)	−16.38	14.41	−30.59	68.68	18.08	−25.86	−7.12

Source: From NICB – https://www.nicb.org/news/blog/hail-claims-fluctuate-over-past-three-years.

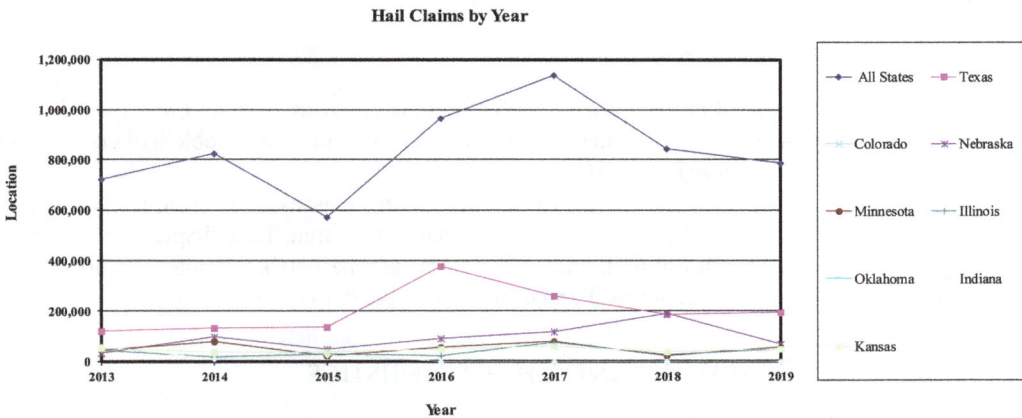

FIGURE 2.1 Total hail insurance claims with time – 2013 to 2019 – total & top ten states.

The top ten states with hail claims over this 7-year period were: Texas (1,411,201 – 24.12%), Colorado (652,659 – 11.15%), Nebraska (367,862 – 6.29%), Kansas (325,620 – 5.56%), Illinois (297,431 – 5.08%), Missouri (285,360 – 4.88%), Minnesota (264,621 – 4.52%), Indiana (150,913 – 2.58), Iowa (123,984 – 2.12%), and Oklahoma (98,193 – 1.68%). The top ten states reported 70.45% of all hail claims submitted over this timeframe.

Of the 5,634,380 hail claims over the 2013–2019 timeframe, with the five main policy types (Personal Property – Homeowners, Personal Auto, Personal Property – Farm, Commercial Multi-Peril and Commercial – Auto), which cover 96.3% of all claims over this timeframe (5,851,497 total claims), 57.6% (3,372,204 claims) were associated with Personal Property – Homeowners claims; 30.3% (1,771,670 claims) with Personal Auto; 5.2% (303,573 claims) with Personal Property – Farm; 2.08% (121,574 claims) with Commercial Multi-Peril; and 1.12% (65,359 claims) with Commercial – Auto claims. Thus, residential properties account for nearly 60% of all hail claims. Since 2006, the trend for hail claims has increased (Figure 2.2).

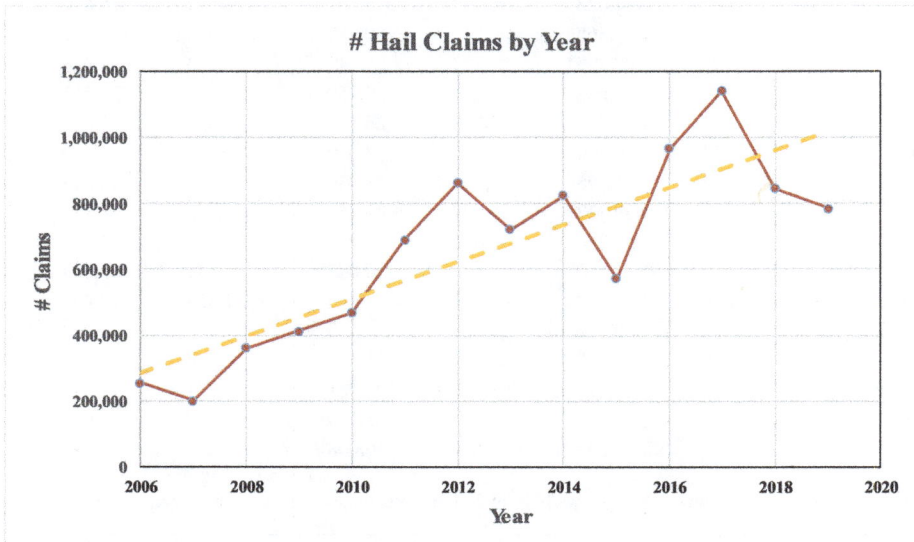

FIGURE 2.2   Total hail insurance claims with time – 2006 to 2019.

Finally, the NICB[5] reported an increase in US hail claims between 2006 and first quarter 2010 from 256,000 claims to 413,178 claims (increase of 61%) and a rise in questionable hail claims from 301 in 2006 to 711 in 2009 (increase of 136%).

Hail-strike damage claims are one of the areas where forensic assessments frequently occur as a result of this large and increasing number of claims filed in this area. This chapter is intended to provide an overview on the basics of hail, hail terminology, and hail-strike damage methodologies utilized in subsequent chapters (Chapters 3 and 4) on hail-strike damage assessments.

2.2 HAILSTONE FORMATION AND CHARACTERISTICS

To better understand the potential for damage that hailstones can create, it is important to gain more insight into the hailstones themselves. Information regarding the formation of hailstones within severe thunderstorms and their typical physical characteristics (i.e., hardness, shape, and size) is necessary when determining if the hailstones produced by a particular storm have the potential to functionally damage common exterior building and roof components.

2.2.1 HAILSTONE FORMATION

In order for a thunderstorm to have sufficient intensity to produce damaging hailstones, the following five conditions must be present[6]:

1. Air aloft cooler than normal.
2. Warm and moist air near the surface of the earth.
3. Strong winds aloft to assist in developing vertical motion.
4. Means of lifting the warm air to cause updrafts (typically in the form of a cold front).
5. Suitably cool air temperatures below the cloud formation so the hail does not melt before reaching the earth.

These conditions create a rising column of air, or updraft, which can exceed 100 feet/second (68 MPH). These types of conditions often all come together in the central plains and mid-western regions of the United States where warm, moist, unstable air from the Gulf area frequently meets cold fronts from the northwest part of the country. This results in frequent reports of large hail in these regions more so than in any other part of the country. The following national map (Figure 2.3) from Figure 2.1 in Allen et al.[7] graphically shows the maximum size of hail to fall by location both from 1979 to 2013 and from 1955 to 2013.

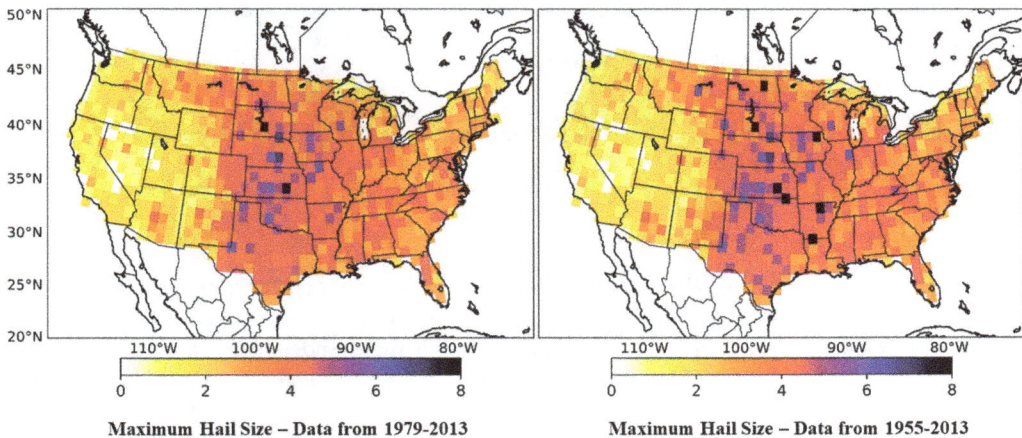

FIGURE 2.3   Maximum size of hail by location from 1979 to 2013 and from 1955 to 2013.

The formation of hailstones first begins with a hailstone "embryo" or "kernel," which typically consists of supercooled water freezing on contact with frozen raindrops, ice crystals, dust, or other types of nuclei, which have been drawn into the colder regions of the cloud and serve as a core for the initial growth of the hailstones[8,9] Then, due to the storm's updraft, the newly formed hailstones are lifted and cycled through different elevations (and temperatures) within the cloud, which allows more supercooled water to freeze on the surface of the hailstones, creating layers of accumulating ice and increasing the size. The hail then falls to the ground by gravity when the updraft is unable to support the weight of the hail or the updraft intensity weakens (Figure 2.4).

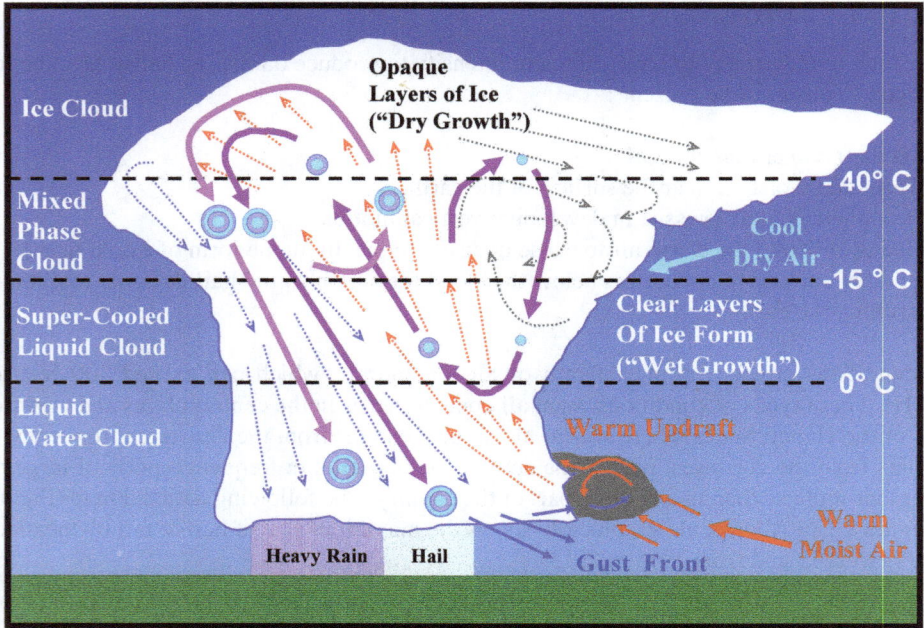

FIGURE 2.4 Hailstone formation within a thunderstorm cloud.

Note that hail differs from graupel (snowflakes coated with a layer of ice) where the latter is smaller and less dense than hail.

2.2.2 HAILSTONE CHARACTERISTICS

The physical characteristics of falling hailstones are dependent on their formation processes within the thunderstorm cloud. As mentioned in the preceding section, the formation of hailstones occurs as frozen particles are cycled through different regions of the thunderstorm cloud, thus, creating layers of ice. Taking the cross section of a hailstone (Figure 2.5) typically reveals onion-like layers of clear and opaque ice.

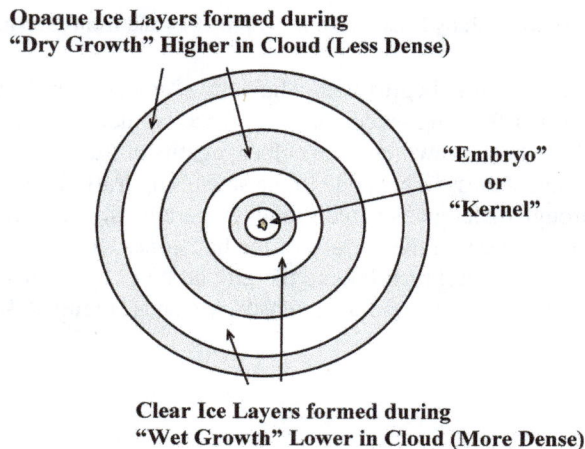

FIGURE 2.5 Cross-sectional view of hailstone – clear and opaque layers indicating location of ice formation.

The ice layers form on a hailstone in two ways, by "dry" growth and by "wet" growth. These layers indicate where in the thunderstorm cloud the ice formed.[8,9] The opaque layers of ice are formed during "dry growth" processes high in the storm cloud where the air temperatures are well below freezing, causing water droplets to freeze instantly upon impact with the forming hailstone, trapping air bubbles and giving it a "cloudy" or "milky" appearance.

The clear layers of ice are formed during "wet growth" processes lower in the storm cloud where the air temperatures are below freezing ($\leq 32°F$) but are not super cold. The slightly warmer temperatures cause the water droplets to surround and freeze slower on the forming hailstone, allowing for the air bubbles to escape and creating a more or less spherical shape.

The amount of times the hailstones are cycled by updrafts, how long they are in the thunderstorm cloud, and how high the thunderstorm cloud reaches into the atmosphere determine the shape and size of the hailstones that ultimately fall to the ground. The hardness, or density, of the individual hailstones varies depending on the layers of clear (no air bubbles – denser) and opaque (air bubbles – less dense) ice created during the formation of the hailstone. The densities of hail have been reported to range from 0.5 to 0.91 g/cm^3, with the latter-most value being the density of pure ice.[10–12] It is possible, however, for hailstones to have few or even no layers of differing ice. This condition indicates that the hailstones were "balanced" in an updraft causing them to form at a certain elevation above ground.[8]

2.2.3 SIZE AND SHAPES OF HAILSTONES

In general, the shape of smaller hailstones is more or less spherical and becomes more irregular with increasing size, with giant hailstones being very irregular and "jagged" in shape. As the size of the hailstones increases, the irregularities, which may not be noticeable in smaller hailstones, become more pronounced, making them appear less spherical.

As might be expected, giant hailstones form in high-intensity thunderstorms, with very high cloud tops, capable of generating relatively large uplift forces. The greater uplift allows for larger hailstones to be cycled through the higher regions of the cloud more frequently and as the wet hailstones from the lower region are pushed to the higher, colder regions at the top of the storm cloud, the conditions exist for the differing size hailstones to fuse together and form onto larger hailstones, creating the very "jagged" appearance of giant hailstones (see Figure 2.6).

When the updrafts associated with the storm become incapable of supporting the weight of the hailstones, they fall to the ground in various sizes. Large hailstones can be quite dangerous to humans. Interestingly, in November 2004, scientists commissioned by the National Geographic television channel examined the skeletal remains of more than 200 nomadic people from the ninth century, which were discovered in the remote Himalayan Garhwal region in 1942. After examination, the scientists concluded that the cause of death for these 200 people was most likely the result of one of the most lethal hailstorms in history in which hailstones "the size of cricket balls" fell and resulted in blunt force trauma.[13]

To date, the largest hailstone to have fallen in the United States on record (Figure 2.6) fell in Vivian, South Dakota on July 23, 2010 and was 8.0 inches in diameter, had a circumference of 18.62 inches, and weighed in at 1.94 pounds! The man who discovered the massive stone reported that, due to a 6-hour power outage, the stone had even melted slightly![14]

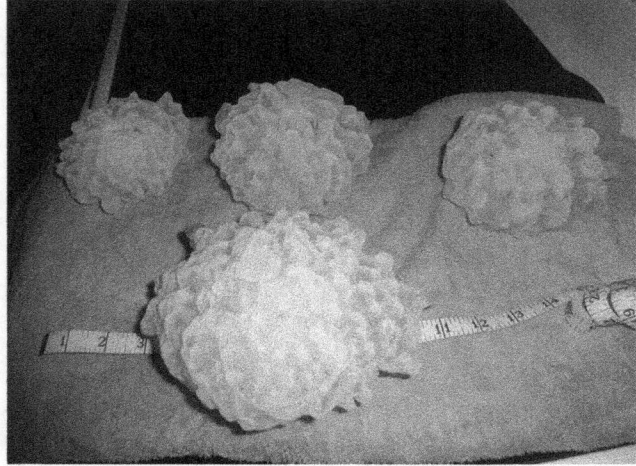

FIGURE 2.6   The largest hailstone recorded in the United States (to date) –Vivian, South Dakota Hailstorm –
July 23, 2010. (Photo Courtesy of the NOAA, Aberdeen, SD, Weather Forecast Office and the Department of
Commerce.)

Since hailstones are generally spherical in shape, a guide for reporting hail sizes is done by
comparing them to common circular objects of the same diameter. The smallest and most commonly
formed hailstones are approximately pea-sized (~0.25 inches in diameter) with giant hailstones
measuring upward to the size of grapefruits, softballs, and DVDs (+4.00 inches in diameter). The
following table (Table 2.2), given by the National Weather Service, provides the size of many
familiar circular objects that hailstones are commonly reported as being compared with.[15]

TABLE 2.2

**Chart for Estimating the Size of Hailstones
from the National Weather Service (NWS)**

Hail Size (inches)	NWS[a] Classification
1/4	Pea
1/2	Plain M&M
3/4	Penny
7/8	Nickel
1	Quarter
1-1/4	Half dollar
1-1/2	Walnut/ping-pong Ball
1-3/4	Golf ball
2	Hen egg/lime
2-1/2	Tennis ball
2-3/4	Baseball
3	Tea cup/large apple
4	Grapefruit
4-1/2	Softball
4-3/4 to 5	Computer CD-DVD

[a] Courtesy of the National Weather Service (NWS).

2.3 HAILSTORM CHARACTERISTICS

Knowing the formation processes of hail and their typical characteristics sets the basic groundwork for evaluating hail damage assessments; however, this does not tell the complete story of why hailstorm events have the potential to cause property damage. It is also important to understand the characteristics of the storm that produced the hailstones. The path, and/or total area, in which hailstones have fallen to the ground from a hail-producing thunderstorm is referred to as the hail swath. Knowing the distribution of the various-sized hailstones produced within the hail swath and the direction from which the hailstorm approaches the building is also crucial information when evaluating hail-caused damage to the roof system of a building or the exterior building envelope.

2.3.1 DISTRIBUTION OF HAILSTONES IN A HAIL SWATH

Hail swaths vary in sizes and can range from just a few acres to 10 miles across and hundreds of miles long.[16,17] Various-sized hailstones are produced from the thunderstorm as the hail swath is created (Figure 2.7).

FIGURE 2.7 Distribution of various sizes of hailstones produced in hail swath.

The smallest hailstones are produced throughout the entire hail swath, whereas the largest hailstones are produced by the storm fall in a smaller region of the path of the hailstorm. This is illustrated in Figure 2.8.[18]

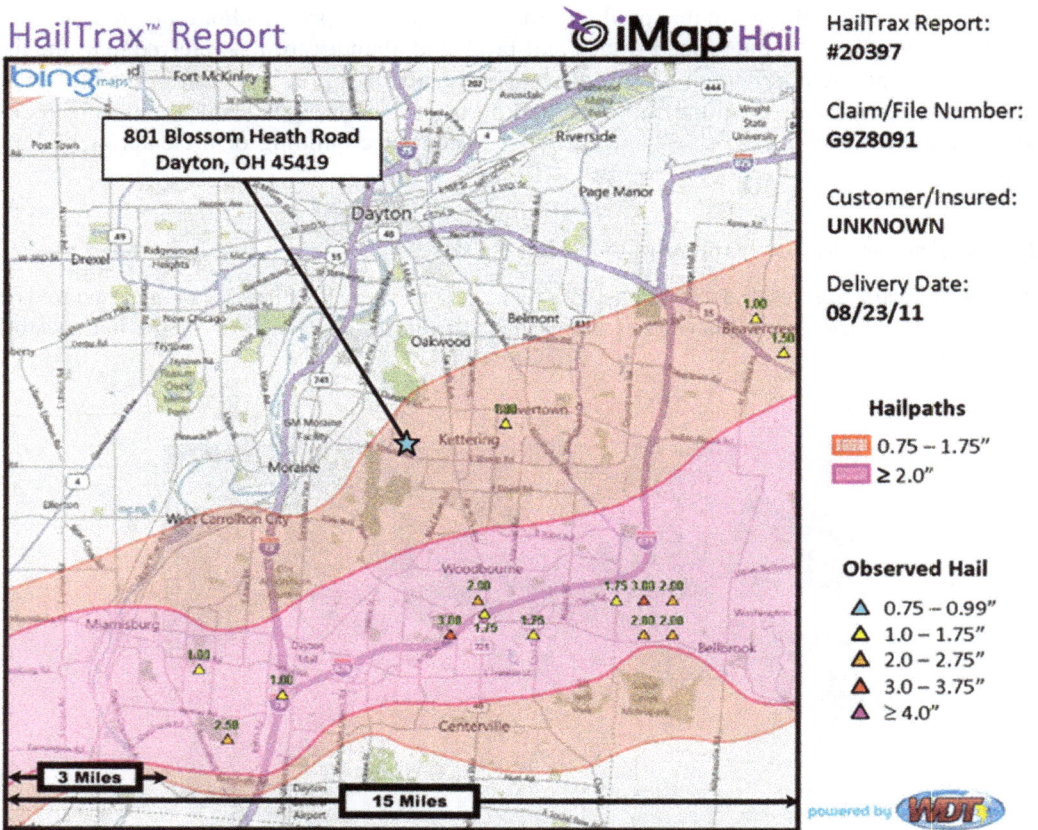

FIGURE 2.8 Hailstorm passing through Southwest-Central Ohio – May 25, 2011. (Courtesy of Weather Decision Technologies, Inc.)

As illustrated in Figure 2.8, the largest hail tends to fall in the center of a storm path with smaller hail on the edges (and often the beginning and end of the hailstorm swath). Whether or not a residence or structure is in the center of the path, the edge, the beginning, the end, or outside of the storm path, along with the size of the hail at various locations within the swath, will determine the possibility and/or extent of hail-strike damage to surfaces in its path. However, Verhulst et al. investigated actual hail sizes (229 buildings) vs those sizes projected by the National Oceanic and Atmospheric Administration (NOAA) database and fee-based commercially available weather data products from April 2016 hailstorms in the San Antonio, TX area, and concluded that an actual site inspection is warranted to determine the actual hail size that likely struck a specific location. The authors noted discrepancies and stated, "NOAA database weather reports of hail size within 3 miles (closest report) were generally within 0.5" of the field estimates at 74% of the sites/buildings evaluated, while hail swaths available from NWS were generally within 0.5" of the field estimates at 70% of the sites/buildings evaluated. Considering fee-based reports from meteorological services offering estimates of hailstone sizes, these estimates were within 0.5" of the field estimates for only 51% to 61% of the sites/buildings evaluated for the sources considered. Therefore, the data showed that spotter-based reports from NOAA databases within approximately 3 miles of the site/building provided the best correlation with field estimates of hail size."[19]

2.3.2 RANDOM FALL PATTERNS OF HAILSTONES

One important concept to keep in mind for later discussions on evaluating a building for evidence of hail-caused damages and determining the storm's directionality is that hail falls randomly (in no discernable pattern) from thunderstorm clouds and hits almost everything uniformly. Some building elevations may see more damage than others due to wind-driven hail, but the damage per unit area for elevation should be uniform. Damages to exterior building surfaces that are found in nonuniform patterns and not randomly distributed are probably man-made and/or not hail-strike damage. For example, dents found uniformly on aluminum siding on an elevation opposite from the direction the storm arrived may also suggest that this damage was either mechanical in nature or inflicted intentionally rather than from the result of hailstone impacts. However, dents on opposite elevations could also indicate multiple storm events, and therefore determining the directionality and fall patterns of the storm is an important aspect of the onsite investigation.

2.3.3 HAILSTORM DIRECTIONALITY AND DETERMINING FALL PATTERNS

The directionality of a hailstorm refers to the direction from which the hailstones (and hailstorm) arrived and struck the building. Similarly, the fall pattern of a hailstone refers to the way in which the hailstone fell to the ground, including whether it fell relatively straight down (i.e., little to no horizontal wind component) or was heavily influenced by the wind (i.e., wind-driven). Smaller hailstones produced by the storm may be swirled around by wind and impact parts of the building envelope even on the leeward side (i.e., opposite side of approaching storm) of the building, but a hailstorm predominantly travels in one direction, and the hail-caused damage to the building is more likely to be evident on the building elevation surfaces that faced the approaching hailstorm. Therefore, the directionality of the storm as well as the fall pattern of a hailstone may be determined by an inspection of the location, frequency, and orientations of hail-strike damage (i.e., spatter/burnish marks and dents) to the finished exterior surfaces of the building.

Directionality can often be determined by on-site inspections of various surfaces of the building envelope that exhibit spatter marks. Spatter marks (areas similar in appearance to the marks created by insects or falling bird droppings after they have impacted the windshield of a moving automobile) correspond to the area where the impacting hailstone has removed the thin layer of oxidation or biological growth (Figures 2.9–2.11). Based on an inspection of the orientation and geometry of the spatter marks, the direction of a hailstorm and the fall patterns of hailstones can be determined (Figures 2.11 and 2.12). Hail-strike spatter marks, similar to those shown in Figures 2.9–2.12, indicate hailstones that fell at an angle or with a certain degree of wind assistance. Spatter marks that are more circular in appearance with "tails" extending from nearly all sides, as shown in Figure 2.10, suggest a hailstone that fell relatively straight down and shattered upon impact with the surface. Additionally, by measuring the width of the initial impact area of the spatter mark (Figure 2.12), an estimation of the size of the impacting hailstone can be determined; however, other size indicators are more reliable (see Section 2.4.6 "Determining Hail Size: Correlation between Size of Dents in Metal versus Size of Hail"). It should be noted that spatter marks, otherwise known as burnish or skid marks, only affect the surface and will disappear with time.[20]

FIGURE 2.9   Spatter marks on PVC-covered awning.

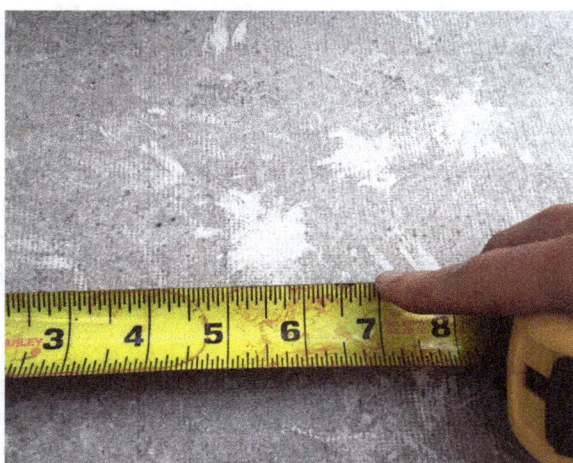

FIGURE 2.10   Spatter marks on PVC-covered awning.

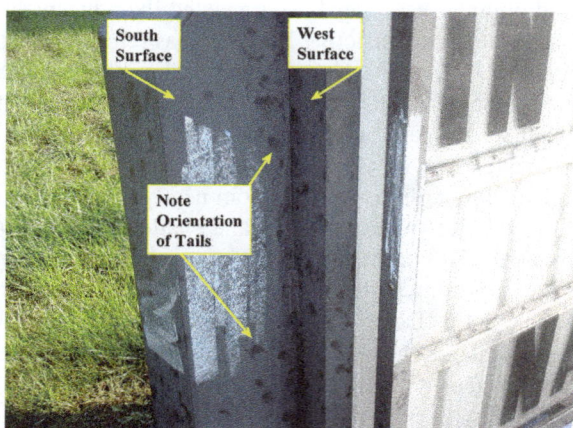

FIGURE 2.11   Spatter marks on west and south metal surfaces on home used to determine directionality.

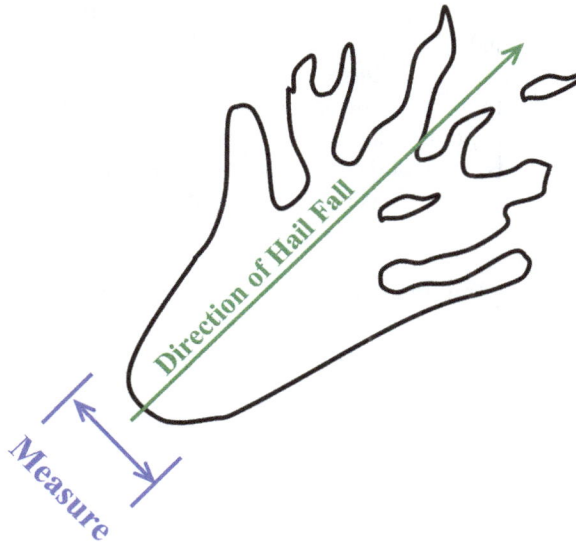

FIGURE 2.12 Proper measurement of a spatter mark.

Directionality information obtained through on-site observations is an important component to the hail damage assessment process/methodology since it defines the direction from which the hailstorm arrived. This information can be used to define which roof surfaces had the highest probability of hail-strike damage. In some cases, the storms can be dated, should multiple storms have struck the property.

2.4 HAIL–STRIKE DAMAGE ASSESSMENT FUNDAMENTALS

To become accurately familiarized with the conditions that play an important role for the determination of damage, a fundamental understanding of energy and impacts, as they relate to hail, must be gained and applied. That information, along with the material properties of the object's surface, as well as the determination of the size of hail that impacted the site, provides the framework needed for accurate analyses.

2.4.1 PHYSICS LESSON – BASICS OF ENERGY TRANSFER AND COEFFICIENT OF RESTITUTION

In order to evaluate the conditions that could lead to the formation of functional damage to any material, roofing or otherwise, one must have a basic understanding of the transfer and conservation of energy.

The law of conservation of energy states that the total energy associated with a particular system, or an object (i.e., hailstone), stays constant over time, thus is neither created nor destroyed.[21] In regard to a falling hailstone at a particular point in the sky (after falling from the cloud) at a height above ground, the total energy is a summation of both its potential (nonmoving energy due to height) and its kinetic energy (energy associated with motion) (Figure 2.12). The sum of these two components (potential and kinetic) will always be the same value unless the energy is transferred to another object. For a hailstone falling to the ground, the falling hailstone has a certain amount of total energy as it is falling (Eq. 2.1):

$$E_T = E_p + E_k = \text{Constant}\left(\text{Law of conservation of energy}\right) \qquad (2.1)$$

Potential energy, E_p, otherwise thought of as "position energy," is a function of the mass, m, of the hailstone, its height above ground, h, and gravity, g. This is like the water above a dam held in a lake before it begins to fall down over a dam. Once the water begins to move, or is moving, some of the potential energy associated with the water being held in the lake at a height above the bottom of the dam (like a hailstone above the ground) is converted to kinetic (moving) energy at the expense of potential energy. In other words, some of the potential energy is converted to kinetic energy (potential energy is reduced and kinetic energy is increased) even though the sum of both terms remains constant. Just before the water falling over the dam hits the river below, or just before a hailstone hits the ground, all the potential energy has been converted to kinetic energy (kinetic energy at a maximum and potential energy is at a minimum). Now, let's apply potential and kinetic energy equations to a hailstone to consider what it means with respect to the energy released when a hailstone strikes an object on the ground such as a roof shingle. The potential energy of the hailstone while falling is given by Eq. 2.2:

$$E_p = mgh \left(\text{Potential energy equation} \right) \tag{2.2}$$

Where:
E_p = potential energy of hailstone at height h
m = mass of hailstone
h = height of hailstone above the ground
g = gravitational constant (i.e. $9.8\,\text{m/s}^2$, $32.2\ \text{feet/s}^2$)

Note that the potential energy constantly decreases as the height decreases or becomes smaller.
   On the other hand, the kinetic (moving) energy of the hailstone is defined by Eq. 2.3:

$$E_k = \tfrac{1}{2} mv^2 \left(\text{Kinetic energy equation} \right) \tag{2.3}$$

where:
E_k = kinetic energy of hailstone at velocity v
m = mass of hailstone
v = velocity of hailstone at any given time

Remember that the total energy of the hailstone is the sum of the potential and kinetic energy and that when the height is zero, the potential energy is zero, or when the velocity is zero, the kinetic energy is zero.
   The constant force of gravity acting upon the stone causes the hailstone to accelerate at a constant rate toward the ground, thus increasing its speed or velocity, v. Since the velocity of the hailstone is *increasing*, this causes the kinetic energy of the hailstone to *increase* (Eq. 2.3). However, at some point, it reaches a terminal (constant) velocity when the drag force from the air stops the increase in velocity (i.e., acceleration is zero). The terminal velocity of an object is dependent on its weight and exposed surface area.
   Taking a look again at the equation for the law of conservation of energy given above, and knowing that the total energy of the system stays the same throughout the descent of the hailstone to the roof/ground, the following statements hold true (Figure 2.13):

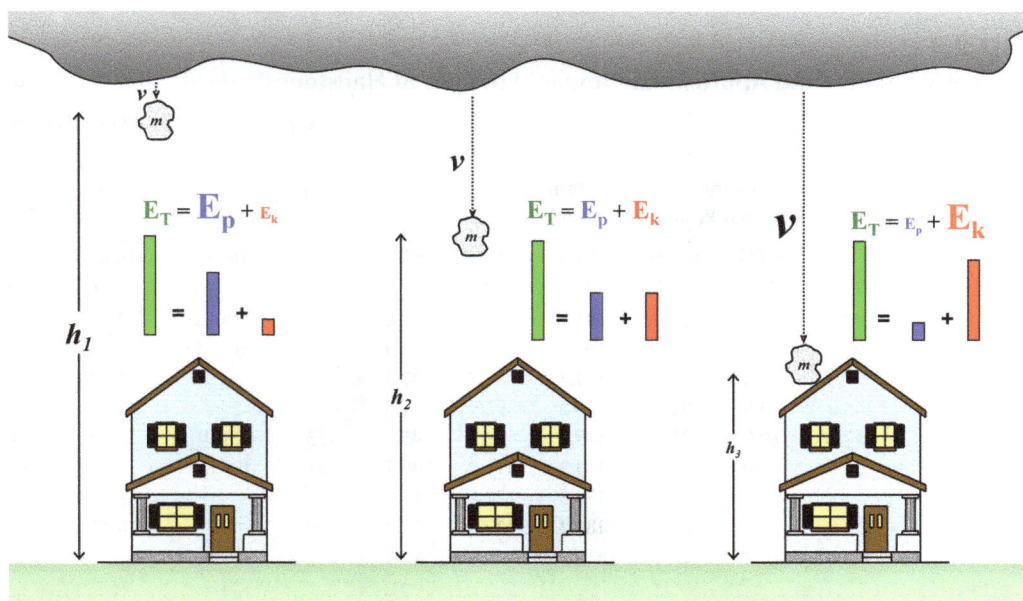

FIGURE 2.13 Law of conservation of energy for a falling hailstone – total energy stays the same while potential energy decreases and kinetic energy increases.

1. Figure 2.13, Position #1 (left): The total energy of the falling hailstone is predominantly due to its potential energy with a slight component of kinetic energy since it is moving.
2. Figure 2.13, Position #2 (middle): The total energy is somewhat split between both energy forms.
3. Figure 2.13, Position #3 (right): Just prior to impact with the roof system of a building, the total energy associated with the falling hailstone is predominantly due to its kinetic energy generated during its descent with a slight component of potential energy since it is still at a height, h_3, above ground.

Now that the total energy of the hailstone, just prior to impact with a roof surface, has been determined to be essentially kinetic energy, further analysis of different-sized hailstones can be performed.

In the 1960s, an early hail researcher named J. A. P. Laurie[22] analyzed data collected and published by Bilham and Relf in 1937 and calculated the terminal velocities and approximate impact energies associated with hailstones of differing sizes.[10,23] More recently, Heymsfield et al. have re-evaluated this earlier work to determine terminal hailstone velocities and impact energy.[11,12] The information gathered from this analysis is summarized in Table 2.3 and Figure 2.14.

TABLE 2.3

Terminal Velocities and Approximate Impact Energies of Hailstones[10–12,22,23]

Diameter		Terminal Velocity[10,22,23] (1960s Research Results)			Terminal Velocity[11,12] (2014 Research Results)			Approximate Impact Energy[10,22,22] (1960s)		Approximate Impact Energy[11,12] (2014)	
Inches	cm	ft/sec	MPH	m/sec	ft/sec	MPH	m/sec	ft-lbs	Joules	ft-lbs	Joules
1.00	2.54	73.0	50.0	22.3	77.7	53.0	23.7	<1	<1.36	1.11	1.51
1.25	3.18	82.0	56.0	25.0	93.1	63.5	28.4	4	5.42	2.91	3.94
1.50	3.81	90.0	61.0	27.4	107.9	73.6	32.9	8	10.85	6.38	8.64
1.75	4.45	97.0	66.0	29.6	122.3	83.4	37.3	14	18.96	12.39	16.80
2.00	5.08	105.0	72.0	32.0	136.2	92.9	41.5	22	29.8	22.03	29.87
2.25	5.72	117.0	80.0	35.7	149.9	102.2	45.7	53	71.9	36.60	49.63
2.50	6.35	124.0	85.0	37.8	163.2	111.3	49.7	81	109.8	57.64	78.15
2.75	6.99				176.3	120.2	53.7			86.92	117.85
3.00	7.62	130.0	88.0	39.6	189.2	129.0	57.7	120	162.7	126.47	171.47

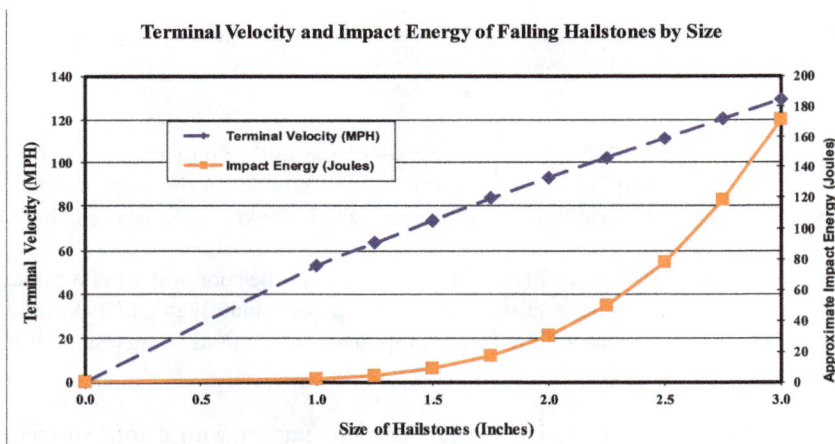

FIGURE 2.14 Approximate impact energies and terminal velocities of falling hailstones.

From Table 2.3 and Figure 2.14, one can conclude that:

- The terminal velocity and approximate impact energy of the hailstone summarized from Heymsfield et al.,[12] Laurie,[22] and Greenfield[23] appear to well approximate impact energy results with later experimental results by Heymsfield et al.,[11,12] but underestimated terminal velocity results as the size of the hailstones increases in diameter.
- For an increase of a quarter inch (0.25 inches) in hailstone diameter (above 1.0 inches), the terminal velocity (measured in MPH) increased steadily by an average of ~18%.
- For an increase in hailstone diameter from 1.0 inches (2.54 cm) to 3.0 inches (7.62 cm), the impact energy (measured in Joules) for hailstones increases by a factor of 113.6. The rate increases by a factor of 19.85 when the hailstone diameter increases from 1.0 inches (2.54 cm) to 2.0 inches (5.08 cm).

The energy calculations performed by these researchers used the terminal velocity of the hailstone as its vertical free-fall velocity; however, hail generally does not fall straight down but has a horizontal wind component, thus the energies associated with actual falling hailstones (i.e., diagonal fall trajectory) should be approximated by the combination of the free-fall (vertical) and wind velocity (horizontal) components of the stone.[24] In order to allow for the effect of wind on the impact energies of a hailstone, the following two conditions were evaluated: hailstones of 1.0 and 1.25 inches in diameter subjected to (1) 40 MPH and (2) 57.5 MPH (i.e., 50 knots used in the "severe thunderstorm" classification by the NOAA NSSL[25]) wind speeds. The increases in impact energy were calculated using the method of Koontz[24] and compared with the impact energies associated with the free-fall terminal velocity of 1.0- and 1.5-inch diameter hail (typically the size threshold for functional damage to most common roofing materials). The results from this analysis are summarized in Figure 2.15:

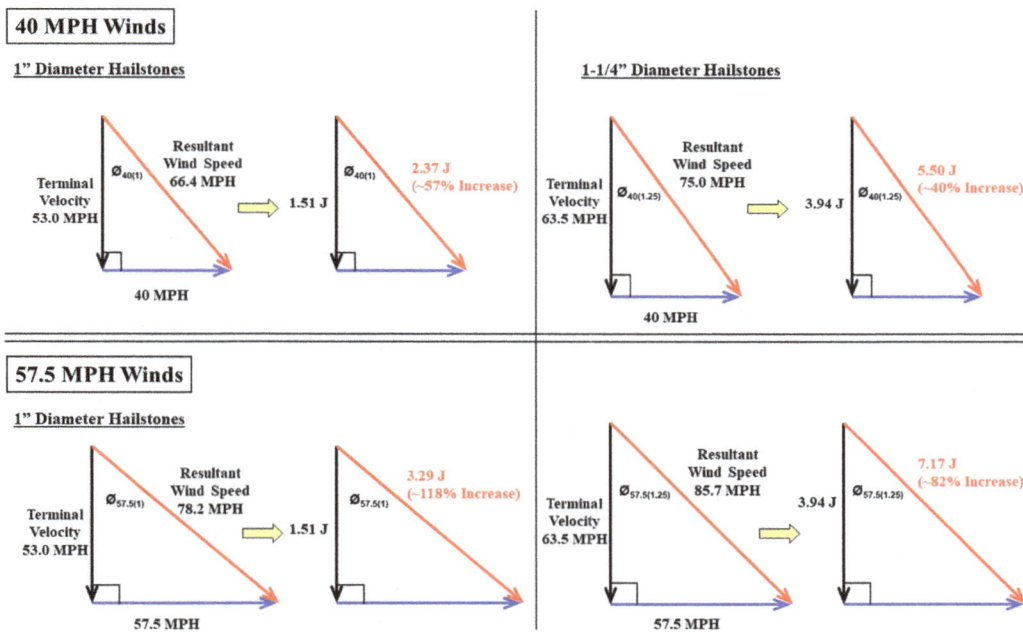

FIGURE 2.15 Wind effects on hail impact energies.

- For a 40 MPH horizontal wind component acting on hailstones 1.0 and 1.25 inches in diameter, the increases in approximate impact energies with the wind component added were ~57% and ~40%, respectively. Note that the calculated impact energies 2.37 and 5.50 Joules) are less than the approximate impact energy for the free-fall terminal velocity of 1.5 inches hail (8.64 Joules), which corresponds to typical-size hail for roof material damage.
- For a 57.5 MPH horizontal wind component (i.e., condition for a "severe thunderstorm") acting on hailstones 1.0- and 1.25-inch in diameter, the increases in approximate impact energies were ~118% and ~82%, respectively. Note, again, that the calculated impact energies (3.29 and 7.17 Joules) accounting for wind forces are less than the approximate impact energy for the free-fall terminal velocity of 1.5-inch hail (8.64 Joules), which corresponds to typical-size hail for roof material damage.

Based on this analysis of the effect of significant horizontal winds blowing during hailstorm events, hailstones below 1.5 inches in diameter would increase the impact energy of the hail but would most likely still not contain sufficient amounts of energy to reach typical-size threshold values for functional damage to common roofing materials.

For comparison, the wind speeds needed for 1.0-inch hail to reach the free-fall terminal velocity impact energy of 1.5-inch hail (8.64 Joules) would be ~127 MPH (comparable to an EF1 tornado or Category 2 hurricane)!

The forces acting on a hailstone can also be calculated using wind force equations:

$$F = 1/2 C_d \rho v^2 A$$

Where:

F = Force (N)
C_d = coefficient of drag, which is about 0.5 for a sphere
ρ = air density – 1.2 kg/m^3
v = wind speed (m/s)
A = area (m^2)

The results from this analysis are summarized in Table 2.4:

TABLE 2.4
Force of Wind on Hailstones

Wind Speed	Wind Force by Hail Diameter and Wind Speed (N or J/m)				
	Hail Diameter				
(MPH)	1.0	1.5	2.0	2.5	3.0
40	0.0012	0.0026	0.0047	0.0073	0.0105
50	0.0018	0.0041	0.0073	0.0114	0.0164
60	0.0026	0.0059	0.0105	0.0164	0.0236
70	0.0036	0.0080	0.0143	0.0224	0.0322
80	0.0047	0.0105	0.0187	0.0292	0.0420

The force (N) imparted by winds on hailstones illustrated in Table 2.4 ranged from 0.0012 to 0.042 N; the energy imparted on the hailstone (J or N-m) depends on the distance the hailstone is pushed by the wind.

The severity of hail damage to an impacted object is dependent on two main factors[20,23,24,26,27]:

1. The amount of kinetic energy the falling hailstone possesses at impact.
2. The fraction of energy that is actually transferred to the impacted object.

In order for an object to be functionally damaged by a hailstone impact, a sufficient amount of the kinetic energy from the falling hailstone has to be transferred to the object, which is at or above its threshold for damage. The measure of the energy transfer upon impact is called the *coefficient of restitution*, which can be defined as the ratio of the relative velocity of the hailstone after impact and just prior to impact.[20] Consider the following equation (Eq. 2.4):

$$\varepsilon = v_2 / v_1 \left(\text{Coefficient of restitution} \right) \tag{2.4}$$

Where:
    ε = coefficient of restitution (values range between 0 and 1)
    v_2 = velocity after impact
    v_1 = velocity before impact

A coefficient of restitution of "1" indicates an elastic collision (i.e., bounced) in that, the velocity of the hailstone before and after the impact is the same, only the direction of the hailstone reverses. In this case, no energy would be transferred to the item being struck (i.e., like a roof shingle). This condition would not be expected as the hailstone generally breaks apart and transfers most of its energy to the item being struck. A coefficient of "0" indicates the relative velocity of the stone after impact is "0," suggesting that all of the kinetic energy from the hailstone has either been transferred into the impacted object and/or been absorbed by the hailstone. This condition indicates that maximum damage was inflicted to the object or the hailstone or a fraction of each.[20]

Generally, the material or object the hailstone impacts will, to some degree, affect the coefficient of restitution. Obviously, a roofing material that has a higher value is more desirable concerning the potential for hail damage; however, this is typically constrained by economical factors/concerns.

This same coefficient of restitution factor must be accounted for when comparing laboratory results using solid ice balls (or even steel balls) to simulate actual hail-strike damage to various materials because actual hail is typically more of a composite material (see Figure 2.4) and more likely to break apart and/or absorb more energy on impact. This factor accounts for why hailstone size thresholds for actual hail-strike damage to materials tend to be greater than the results from laboratory tests using solid ice balls.

The Insurance Institute for Business & Home Safety, beginning in 2010, began to evaluate hail characteristics and the impact of hail strikes on materials with an emphasis on asphalt roofing shingles.[28] They have developed an impact-resistant protocol and rate shingles on their ability to resist damage from hail-strike impacts[29]; newer impact resistant shingles do show better resistance to hail strikes.

2.4.2 MATERIAL IMPACT RESISTANCE

The principle for the coefficient of restitution, in its very simplest terms, refers to the material's resistance to impact damage. Generally, for a given material, the three following principles affect the material's coefficient of restitution and apply in regard to its resistance to impacts, such as those associated with hail strikes and the potential for subsequent damage[30]:

1. The thicker the material, the greater the impact resistance.
2. The stiffer the underlying support, the greater the impact resistance.
3. The more worn/deteriorated the material, the less the impact resistance.

A potential fourth principle for a given material, temperature, should also be considered as affecting the impact resistance of a given material. Certain roofing materials, such as asphalt shingles, which are typically on residential and light commercial structures, become increasingly brittle at lower temperatures causing them to be less flexible and unable to absorb the impact energy associated with a falling hailstone.[23,24,27,30]

2.4.3 DAMAGE CLASSIFICATION: FUNCTIONAL OR COSMETIC?

Perhaps one of the greatest debates in determining what is and what is not hail-strike damage has to do with whether or not the strikes have caused *cosmetic* or *functional* damage. It is important to distinguish and clearly define what constitutes *functional damage* and how it differs from *cosmetic damage.*

The (US) Department of Defense Dictionary of Military and Associated Terms[31] provides the following definition for "functional damage assessment":

> The estimate of the effect of military force to *degrade or destroy the functional or operational capability* of the target to perform its intended mission and on the level of success in achieving operational objectives established against the target…

Much in the same way our military defines and assesses "functional damage" in regard to US military actions against foreign enemies and targets, forensic investigators must evaluate and determine if the impacted component has been functionally compromised, meaning that the event (in this case hail) negatively affected the functional or operational capabilities or capacity of the component. Within the roofing and insurance markets, the industry standard definition for "functional damage" to a building's roofing system is defined as: a reduction or diminishment of its water-shedding capabilities and/or a reduction in its expected long-term service life.[26,32,33]

The issue of functional damage vs cosmetic damage has been fought in the courts between insurance companies and their insured.[34] Classically, the argument is between the terms "direct physical loss" and "cosmetic damage" wherein the insured argues that cosmetic damage is direct physical loss (i.e., "distinct, demonstrable, physical alteration of the property").

From an engineering perspective, cosmetic damage typically only affects the appearance or aesthetic appeal of the component and does not significantly affect its service life.

Functionally damaged items from hail-strike damage are typically replaced unless the material is degraded beyond its life (e.g., severely degraded asphalt shingles or rotted wood shakes or shingles). When the debate occurs whether or not to replace cosmetically damaged items, the following questions should be answered:

1. Is the cosmetic damage readily visible?
2. Would the damage cause a reduction in property value when viewed by the casual observer?

Based on years of umpiring experience, examples of cosmetically damaged surfaces and whether or not to recommend replacement follow:

- Dented copper roof over a bay window or door entry (replace; readily visible to the casual observer).
- Dented metal garage door (replace; readily visible to the casual observer).
- Metal siding, fascia, gutters, and downspouts (replace; readily visible to the casual observer).
- Dented metal commercial roof – two stories up with parapet wall (do not replace; not readily visible to the casual observer).
- Steep-slope metal roof, surface dents readily visible from the ground (replace; readily visible to the casual observer).
- Granules in gutters from asphalt-shingled roof surfaces are generally associated with the degradation of shingle with time (do not replace; damage to shingles not readily visible and not considered functional damage).

Note in these last set of examples, the hail-strike damage did not cause functional damage, but in some cases would clearly reduce the value of the property since the hail-strike dents are readily visible. In the case of an appraisal, an umpire and appraisers would likely recommend that readily visible cosmetically damaged items be replaced even though the item (e.g., hail-strike damaged metal garage door) was still functionally adequate.

2.4.4 HAILSTONE SIZE THRESHOLDS FOR FUNCTIONAL DAMAGE TO ROOFING MATERIALS

Hailstones must be of a sufficient size before the impact energy generated is sufficient to cause functional damage to roofing materials. The more impact-resistant the material is, the larger the hailstone size must be to inflict damage. The point where the hailstone is of sufficient size to cause damage to a given material is known as the threshold size of hail for that material. The hailstone size threshold for functional damage to any roofing material is defined as[25,30,32]:

> The minimum, or smallest size of natural hail at which functional damage typically begins to occur and refers to hailstones which strike perpendicular to the surface of the roofing material which is in relatively good, mid-life conditions.

These threshold values presented here, and covered at length in Chapters 3 and 4, are based on a thorough review of past studies, literature, and from extensive field experience regarding hail damage assessments to residential and commercial properties.

When "Serving as an Expert Witness" (Chapter 24) or other venues where explicit opinions are required, considerable debate can occur regarding whether the damage (i.e., abrasions, dents, granule loss, spatter marks, etc.) to a surface was, or was not, caused by hail-strike impacts. The determination of the maximum size of hail to have struck a building, coupled with knowledge of hail-strike thresholds for a given roof covering or other material, provides a much more definitive measure of whether or not the hailstones to strike the building were of sufficient size to cause the reported damage as opposed to more subjective means, such as bruise count analysis (as on asphalt shingles). Thus, knowing hailstone size thresholds to building materials is also important to making such determinations.

Extensive experimental research regarding the effects of hail on roofing materials, proposed methodologies for damage assessment, and hail damage replication has been done dating back to 1952 when Rigby and Steyn first published test results of ice ball impacts on roofing materials in South Africa.[10,23,24,26,27,32,33,35-38]

However, as noted by Marshall et al.[26] and Petty et al.,[39] the "synthetic hail" produced for the impact testing procedures was made from frozen, molded water and was harder and less brittle than actual hailstones, which consist of layers of alternating density of ice (see Sections 2.2 "Hailstone Formation and Characteristics"). Thus, the actual size of real hailstones needed to cause threshold functional damage to surfaces would be expected to be somewhat larger than those associated with synthetic hailstones in the laboratory.

2.4.5 CONDITIONS LEADING TO INCREASED LIKELIHOOD OF FUNCTIONAL DAMAGE TO ROOFING MATERIALS

The size of a hailstone that impacts a building surface is the first and foremost factor to be established when determining the probability of hail-strike damage. If the hailstone size is determined to have been below the threshold for that particular roofing material (discussed above and at length in their respective chapters), then functional damage to that roof surface is unlikely.

Other variables that may affect the extent of damage to a roofing material include, but are not limited to, the following conditions (in no particular order of significance), which are discussed at length in the following paragraphs:

• Directionality (i.e., Hailstorm Direction).
• Angle of Hailstone Impact Relative to Roof Surface (i.e., Fall Pattern).

- Hailstones' Velocity Vector Perpendicular to Roof Surface.
- Density/Hardness of the Hailstones.
- Age and Condition of the Roofing Material.
- Impact Resistance of the Roofing Material.
- Number of Layers of Roofing Material.
- Condition of the Underlying Substrate (i.e., Roof Decking).
- Attic Ventilation Conditions.

2.4.5.1 Hailstone Directionality, Angle of Impact, and Perpendicularity

These three conditions pertaining to hailstorm events are closely related and dependent on one another and as such, are grouped in this section together. The directionality of the hailstorm, as mentioned before, is a significant factor to take into consideration when determining damage because it has a direct influence on how the falling hailstones approach and impact the surfaces of a building. The surfaces facing the direction of the arriving hailstorm would be expected to suffer more hail-strike damage than those surfaces in the opposite direction of the arriving storm.

For roof surfaces, two factors regarding directionality are windward vs leeward slopes and low-slope surfaces vs steep-slope surfaces.

Windward Slopes vs Leeward Slopes: Assuming the hail reaches threshold sizes, the windward slopes will see more hail-caused functional damage than the leeward slopes. This is due in large part to the fact that those roof elevations facing the oncoming storm have roof surfaces that are likely more perpendicular to the trajectory of the falling hailstones. The greater degree of perpendicularity, the greater resultant force vector acting on the roofing material (see Figure 2.16). This is consistent with the findings from a detailed analysis published by Petty et al.,[39] which showed that the number of bruise counts (i.e., functional damage to asphalt shingles) on the windward roof slopes were ~2.5 times greater than the number on the leeward slopes (see Chapter 3 for information regarding this study).

Low-Slope Surfaces vs Steep-Slope Surfaces: The amount of hail-caused functional damage to steep-slope and low-slope roof surfaces is dependent on the horizontal velocity vector for the prevailing winds and the pitch angles of the roof surfaces being evaluated. The combination of these parameters can lead to more perpendicular hail-strike impacts. Those elevations will be the elevations that sustain the most evidence of functional damage. Again, this correlation was included in the study performed by Petty et al.,[39] which showed that the greatest number of hail-strike bruise counts were located on the shallow (<4:12 pitch) and steep (>9:12) roof slopes (see Chapter 3 for information regarding this study) rather than intermediately sloped roof surfaces. Again, this condition is dependent on the fall pattern of the hailstone. If, from evidence gathered at the site, the hail appears to have been highly wind-driven (best determined by spatter marks and dents to vertical surfaces), then there would be an expectation that the steeper-sloped roof surfaces would have an increased likelihood for damage. Conversely, if the hail were not wind-driven, the greater damage would be observed on lower-sloped roof surfaces.

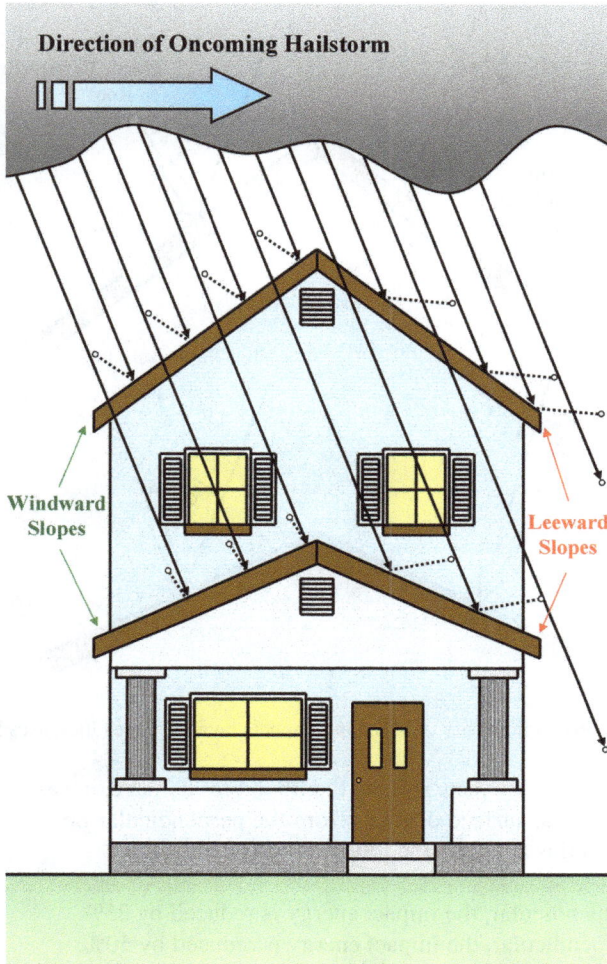

FIGURE 2.16 Hailstorm directionality and perpendicularity of impacts on roof slopes.

As further illustrated in Figure 2.17, greater perpendicularity allows for a greater resultant force, which transfers more of the kinetic energy associated with the falling hailstone into the roofing material, thus increasing its likelihood for potential damage.

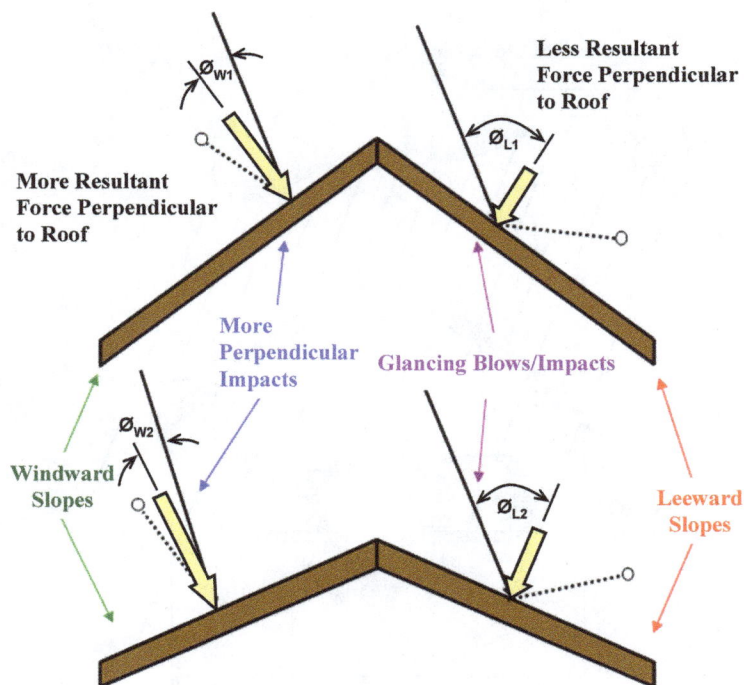

FIGURE 2.17 Greater perpendicularity of hailstone impacts on roof slopes increases likelihood for damage.

In an analysis of hailstone impacts and the likelihood for damage, it was determined "that when the angle of impact with the surface deviates from the perpendicular position, the impact severity diminishes."[20] Based on this analysis, the following was determined:

- At 30° from perpendicular, the impact energy is reduced by 25%.
- At 45° from perpendicular, the impact energy is reduced by 50%.
- At 60° from perpendicular, the impact energy is reduced by 75%.

Hailstones that strike the leeward slopes of a home result in more glancing blows to the roof surface, thus creating less resultant, perpendicular forces and decreasing the likelihood for damage. In glancing blows, energy transfer from a falling hailstone is dissipated somewhat when it strikes a surface at an angle <90°.

Referring to the examples of hailstone impacts in Figure 2.17, the degree of impact energies associated with them, in order from greatest to least impact severity, would be: $\emptyset_{W2} > \emptyset_{W1} > \emptyset_{L2} > \emptyset_{L1}$. This is based on the angle of impact, $\emptyset$, from the perpendicular position.

2.4.5.2 Hailstone Density/Hardness

All things being equal (e.g., hail is of sufficient size, strikes the roof surface at an angle that imparts a sufficient amount of energy transfer to cause damage, and the impact resistance of the roofing material is below that of the energy from the impacting hailstone), common sense suggests that the harder or denser the hailstones, the greater the chance for functional hail damage. The density of hailstones is dependent on the nature of their formation within the storm cloud and ranges from 0.7 to 0.91 g/cm³ with the latter being the density of pure ice (i.e., no bubbles).[10–12] Hailstones, which contain more "cloudy" or opaque layers of ice ("dry growth" formations high in the cloud), contain trapped air bubbles upon formation and are likely "softer" and will tend to absorb more of the energy and/or shatter upon impact. These characteristics reduce the amount of the kinetic energy of the hailstone transferred to the roofing surface and decrease the chance for functional damage to these surfaces.

2.4.5.3   Conditions Decreasing the Impact Resistance of Materials

The impact resistance of a material is dependent on the material itself, the age of the material, and the construction of the material and its substrate. Recall from Sections 2.4.1 and 2.4.2 that the degree of energy transfer from a hailstone to the impacted material depends on the coefficient of restitution; therefore, reducing the material's coefficient of restitution (i.e., impact resistance) would create conditions for the material to become more susceptible to potential hail-strike damage. Conditions that negatively affect impact resistance of a material are as follows:

- *Age*: an older, more deteriorated roof is less resistant than a newer roof surface to hail-strike impacts.
- *Number of layers*: a roof system that consists of multiple layers of roofing material will flex more, exacerbating hail-strike damage.
- *Substrate*: a softer, degraded substrate (i.e., roof decking) will flex more, exacerbating hail-strike damage.
- *Attic ventilation*: inadequate attic ventilation allows excess buildup of heat and moisture within the attic spaces, which can accelerate/expedite degradation to the roof decking and roofing materials. If permitted to persist for prolonged periods of time, inadequate attic ventilation can adversely affect the roofing material, causing it to become more brittle, thus reducing its impact resistance and increasing its susceptibility to functional hail damage.

2.4.6   DETERMINING HAIL SIZE: CORRELATION BETWEEN SIZE OF DENTS IN METAL VERSUS SIZE OF HAIL

The significance of the size of hailstones in relation to the formation of functional damage to roofing materials cannot be overstated. If the hailstones were not of sufficient size to generate sufficient energy upon impact with a given surface, the probability for hail-strike damage is very low. Should possible hail-strike damage be observed when hailstones are smaller than would be expected to cause such damage, other causes are likely responsible for damage to these surfaces. Therefore, determining the size of the hail, which likely impacted the area of the building, is of critical importance in order to proceed with the damage evaluation process.

Hailstones of sufficient size to potentially cause functional damage to roofing systems typically occur in rather small areas and can vary in size drastically from one location to the next due to the characteristics of hailstorms (see Section 2.3). Reports of hail size from weather agencies associated with the NOAA such as the National Climatic Data Center (NCDC) and the Storm Prediction Center (SPC) can provide insight into the possible size of hail in a given area. Also, in recent years, companies such as iMap®Weather Forensics (a division of Weather Decision Technologies, Inc. and producers of HailTrax™ now owned by DTN) have developed more sophisticated hailstorm/hail size information by combining radar imagery, computer algorithms, and human reports to help define boundaries for sizes of hailstones within a hail swath of a storm. However, since the size of hailstones can vary greatly by location within a hailstorm, these reports may not be completely indicative of the size that has impacted all areas. Therefore, methods to define the maximum size of hailstones to strike a specific location must still be used to help determine whether or not functional damage could have occurred at a given site from hail-strike impacts (note RICOWI[38] also supported this position stating "the inspection-based data would be considered more accurate" than estimates from algorithms).

A methodology used to determine the maximum size of hailstones to strike a given location is based on the size of dents that the hail strikes caused in metal surfaces.

Although knowledge on the interpretation of dent size versus actual hail size has not been defined by many, some research in this area exists and is reviewed in the following paragraphs.[10,20,33,35]

Historically, methods used to evaluate dents in metal and accurately correlate them to the actual size of the impacting hailstones were at times qualitative rather than quantitative. For example, statements such as the "dents in softer metals are close to the diameter of the hailstone"[33] were somewhat vague since the term "softer metals" was not defined. More authors recently made rather conservative estimates reporting, "Generally, the width of the ding in the sheet metal is about one-half the diameter of the hailstone that made it."[20] Although this statement is somewhat of a generalization, it provides a rule-of-thumb multiplier (i.e., 2.0) that is used by many inspectors within the industry. However, as summarized in the next section, two recent papers have provided a quantitative basis for determining the hailstone size to dent multiplier.

2.4.6.1   Ice Ball Impact Studies of Metal Surfaces to Determine Dent Multiplier

Studies have been performed[10,35] (Tables 2.5 and 2.6) to determine hailstone size to dent size multipliers using ice balls to simulate hailstones. Each study impacted several common metal roof appurtenances with this "synthetic hail." The hail was fired at respective terminal velocities in an attempt to determine information regarding dent multipliers.

TABLE 2.5

Crenshaw/Koontz Ice Ball Impact Study – Size of Hail vs Size of Dent for Various Metal Appurtenance Surfaces and Impact Angles[10]

Scenario	Hail Size (Inch)	Dent Size (Inch)	Ratio – Dent to Hail	Dent Multiplier – Dent to Hailstone Size (1/Ratio)	Average Dent to Hailstone Size Multiplier
Parapet caps – galvanized	1	0.4	0.40	2.5	2.0
steel 24 gauge – 90° impact	2	1.25	0.63	1.6	
	3	1.65	0.55	1.8	
Parapet caps – galvanized	1	0.55	0.55	1.8	1.7
steel 26 gauge – 90° impact	2	1.30	0.65	1.5	
	3	1.75	0.58	1.7	
Parapet caps – copper – 16 oz.	1	0.60	0.60	1.7	1.5
– 90° impact	2	1.30	0.65	1.5	
	3	2.10	0.70	1.4	
Parapet caps – aluminum	1	0.75	0.75	1.3	1.4
0.040″ thick –90° impact	2	1.30	0.65	1.5	
	3	2.25	0.75	1.3	
Mechanical unit[a] cabinets – 20	1	0.44	0.44	2.3	1.5
gauge thick – 45° impact	2	1.78	0.89	1.1	
	3	2.88	0.96	1.0	
Mechanical unit[a] cabinets – 20	1	0.67	0.67	1.5	1.1
gauge thick – 90° impact	2	1.78	0.89	1.1	
	3	4.22	1.41	0.7	
Aluminum coil fins[b] – HVAC	1	0.94	0.94	1.1	1.2
unit – 45° impact	2	1.56	0.78	1.3	
	3	2.31	0.77	1.3	
Aluminum flue vent caps	1.0	3.00	3.00	0.3	0.4
– 0.018″ thick – 90° impact	1.5	3.88	2.59	0.4	
	2.0	4.38	2.19	0.5	
	2.5	7.25	4.83	0.2	
	3.0	7.63	2.54	0.4	

[a] Splash (i.e. spatter) mark diameter.
[b] Width or diameter.

Findings from of this study reported the following:

- The maximum average dent multiplier for all materials was 2.0 (Galvanized Steel Parapet Caps – 24 gauge), and the minimum average dent multiplier was 0.4 (Aluminum Flue Vent Caps). For flue vent caps, the size of the dent always exceeded the size of the hailstone by a factor as high as 4.83!
- Excluding aluminum furnace vent cap results, the average dent multiplier for 1.0-inch diameter hail was 1.7. Furnace vent caps were found to be very soft; if the value were included in the average, it would unfairly increase the average of all the surfaces evaluated.
- Excluding aluminum furnace vent cap results, the average dent multiplier for 2.0-inch diameter hail was 1.4.
- Excluding aluminum furnace vent cap results, the average dent multiplier for 3.0-inch diameter hail was 1.3.
- Excluding aluminum furnace vent cap results, the average dent multiplier for all hail sizes across all materials tested was 1.5.
- Note that for HVAC coil fins (width), the dent width is close to the size of the impacting hail (i.e., dent multiplier of 1.2 or the size of the dent was approximately 83% the size of the hail)
- The effect of the angle of hail (45° vs 90°), based on HVAC cabinet data, implies that the width of the splash (or spatter or burnish) mark better approximates the size of the hail at 90° hail-strike angle vs 45° hail-strike angle.

Ice ball impact testing studies offer strikingly similar results for roof appurtenances[35] and are summarized and interpreted in Table 2.6.

TABLE 2.6

Ice Ball Impact Study – General Rules of Thumb for Inner Dent Width/Ice Ball Diameter for Common Roof Appurtenances[25]

| | Ice Ball Diameter | | Inner Dent Width | Dent to Hailstone |
| | Minimum | Maximum | as % of Ice Ball | Size Multiplier |
Roof Appurtenance	inch/(mm)	inch/(mm)	Diameter (%)	[1/(%/100)]
Lead soil stack flashing (nominal 3″ diameter – ~0.045″ thick base)	0.75/(19.1)	1.75/(44.5)	80	1.3
Galvanized steel turbine ventilator (nominal 12″ – blade thickness = ~0.032″)	0.75/(19.1)	1.25/(31.8)	90	1.1
Aluminum flue vent cap/cover (8″ diameter, ~0.020″ thick)	0.50/(12.7)	0.75/(19.1)	200	0.5
Aluminum static vent (9″×9″; ~0.024″ thick)	0.50/(12.7)	1.25/(31.8)	50	2.0
Aluminum air-conditioning unit fins (1″ wide×28″ long×0.007″ thick)	0.75/(19.1)	2.25/(57.2)	80	1.2

Findings from an analysis of this study are as follows:

- The maximum dent multiplier for all materials was 2.0 (aluminum static vent) while the minimum dent multiplier was 0.5 (aluminum flue vent cap/cover).
- The average dent multiplier for 1.0-inch hail was 1.4, excluding the aluminum flue vent cap/cover.

- The average hail between 1.0 and 3.0 inches was 1.5 with values ranging from about 1.1 to 2.0. This value indicates that the impact dents in the metal surfaces measured was approximately two-thirds the size of the hailstone.
- The aluminum coil fins for HVAC (i.e., air conditioner) provided identical dent multipliers of ~1.2. This indicated that the size of the impact dent in the coil fins was ~83% the size of the hailstone.

The ice ball impact studies provided invaluable information regarding accurate values for dent multipliers since they are based on results from simulated hail-strike dents to specific types of metals found on/near buildings rather than undefined generic metals. Moreover, given the hardness of ice balls (synthetic hail) versus actual hailstones, results should be conservative.

Based on a review and analysis of results from these two studies, it appears that a better typical hailstone size to dent size multiplier for hailstones sizes of interest to be utilized should be 1.5 rather than 2.0.

2.4.6.2 Hailstone Impact Dents in Metal – How to Measure Dent Diameters

While seemingly simple, the process of measuring a metal dent diameter is actually more complex than it would initially appear. Thus, dents must be defined and characterized so that they can be correctly measured in order to provide an accurate estimation of the hail size.

Hailstone impact dent in metal surfaces is typically circular to oval in shape and consists of two regions of deformations within the dent: an *inner dent* and an *outer dent*.[10,35,40] The inner dent is located about the center of the impact area and has well-defined slopes, whereas the outer dent has shallow slopes and surrounds the inner dent (Figure 2.18).

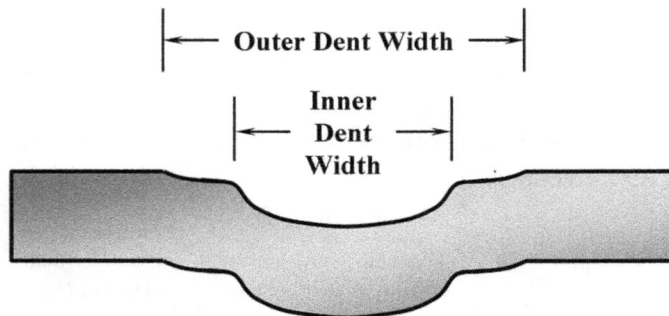

FIGURE 2.18 Cross-sectional view of dent to thin gauge metal.

The strongest correlation between the dent size and the diameter of the impacting ice ball was determined to be the smallest width of the *inner dent*.[35] According to Morrison, "the most effective method involved examining the deformation, visually determining the demarcation between slopes of inner and outer dents, and measuring the least dimension of the inner dent."[35]

Thus, one would naturally overstate the diameter of the dent if they used the outer ring or dent diameter and consequently would overestimate the size of the hailstone creating the dent.

2.4.6.3 Summary of Using Hail-Strike Dents to Estimate Maximum Hailstone Size

To best estimate the maximum size of hail to have struck surfaces at a building to be inspected, the methodology of measuring dents to metal surfaces coupled with a dent multiplier should be utilized. Dents are best highlighted in the field by rubbing a piece of chalk (experience suggests children's sidewalk variety works best) over the impact zone. The inner ring of the illuminated dent should be measured and recorded. This overall methodology for estimating the maximum size of hailstones to strike a given building is summarized in the following list of steps:

1. Locate impact zone on metal surface/roof appurtenance.
2. Rub a length of chalk over the impact zone.
3. Identify the transition between the inner and outer dent rings or diameters.
4. Measure the dimension, or diameter, of the inner dent ring.
5. Multiply the least measurement of the inner dent by its respective dent multiplier. If an average multiplier is used, a conservative value of 1.5 should be utilized. For more exact estimates, impact marks on specific metals should use specific multipliers for those metals (e.g., 1.2 for air conditioner outdoor unit coil fins).

As an example, Figure 2.19 depicts the measurement of an impact dent on the coil fins of a south-facing window air-conditioning unit. Two characteristics of the hailstorm can be identified in the photo: (1) based on the orientations of the marks, the impacting hailstones likely arrived from the left side (i.e., west side) of the photo indicating that the hailstones arrived from the west/southwest and (2) the width of the impact mark measures ~0.75 inches, suggesting that the diameter of the impacting hailstone was ~0.90 inches (0.75 inches × 1.2 dent multiplier).

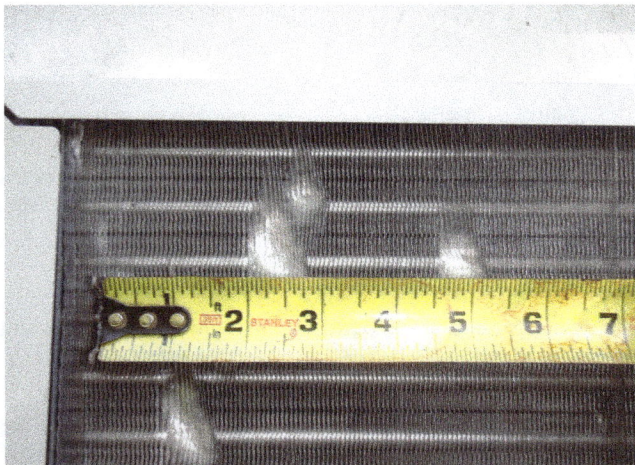

FIGURE 2.19   Measurement of impact dent in AC unit coil fins.

2.5   HAIL DAMAGE INSPECTION METHODOLOGY/PROTOCOL

The general methodology for completing hail damage inspections is outlined in Chapter 1, with some modifications for this specific type of inspection. It should be noted that the methodology for completing a hail inspection has also been outlined by others, especially for roof surfaces,[24,27,33] but the processes found to be most effective in completing thousands of inspections involve the following steps:

1. Property Owner/Occupant/Owner's Representative Interview.
2. Create/Obtain Plan View Sketch of Roof.
3. Take Overview Photographs of all Exterior Elevations.
4. Conduct Inspection of Exterior Surfaces of Structure for Each Elevation.
5. Conduct Inspection of Roof Surfaces.
6. Complete Inspection Report.

The processes used in completing each of these steps are detailed below.

2.5.1 Obtain Pertinent Information from the Property Owner, Occupant, or Representative through an Interview Prior to Inspection

After taking overview pictures of the residence/structure and noting the facing direction of the residence/structure, the first step in the process of inspecting and assessing for hail-strike damage to a particular structure is to gain insights on the storm and hail-strike damage from the property owner and/or occupant. Many times, the information gathered can be helpful when used in context with the observations documented during the inspection. Inquire and collect background information of the building, the local storm history, and reported storm damage (i.e., components damaged by the hailstorm). The following list of example questions asked to the property owner, occupant, and/or representative is very brief, but detailed in the information it can possibly generate.

2.5.1.1 General Building Information

- What is the approximate interior square footage?
- When was the building built and how long has it been under its current ownership?
- Have there been any modifications to the exterior or roof surfaces (i.e., new roof, siding, windows, etc.)? If so, when?

2.5.1.2 Roof Information

- What type of roof system is in place?
- What is its approximate age?
- Is the owner or owner's representative aware of any historic roof leaks into the interior of the residence/structure? If so, what are the interior areas affected and when did they occur or when were they discovered?

2.5.1.3 Storm History Information

- Was the property owner present at the building during the hailstorm event(s)?
- When did the hailstorm event(s) occur (i.e., date and time)?
- From which direction did the hailstorm arrive?
- What was the approximate size of the observed hailstones produced by the storm, if known?
- What were the damages to the building as a result of the hailstorm event(s)?

An example of typical information received from a property owner during a hail inspection interview follows:

- The two and one-half story home was built in 1911. The homeowner purchased the home about 25 years ago.
- About 7 years ago, tree branches fell onto the front part of the home causing damage. The asphalt shingles within the area of impact damage were subsequently replaced.
- The shingles on the back portion of the home and garage were more than 15 years old.
- No roof leaks were reported.
- A hailstorm passed through the area on May 7, 2010. The homeowner was home during the storm and reported seeing hailstones the size of golf balls.

- New roof surfaces were being installed on several of the homes in the neighborhood as a result of the hailstorm, so the decision was made to hire an independent roofing contractor to inspect the roof for damage.
- The roofing contractor inspected the home and informed the homeowner that there was evidence of hail damage to the shingles, gutters, and downspouts on the home and garage.

2.5.2 CREATE OR OBTAIN A BASIC PLAN-VIEW SKETCH OF THE ROOF

After all pertinent information regarding the building and the storm history has been gathered, the next step in the hail damage assessment process is to sketch out the roof surfaces in an inspection field book (Figure 2.20). The roof layout sketch is completed by drawing all hips and ridges as solid lines and all valleys as dashed lines. The plan view is a recreation of the roof surfaces as if viewed from directly overhead. It is extremely useful to have this roof layout on hand during the inspection so that key observations, damaged areas, and/or other areas of concern can be easily documented and identified. Once the inspection has been completed, the rough sketch can be converted to a cleaner-looking, more formal computer-aided layout (Figure 2.21).

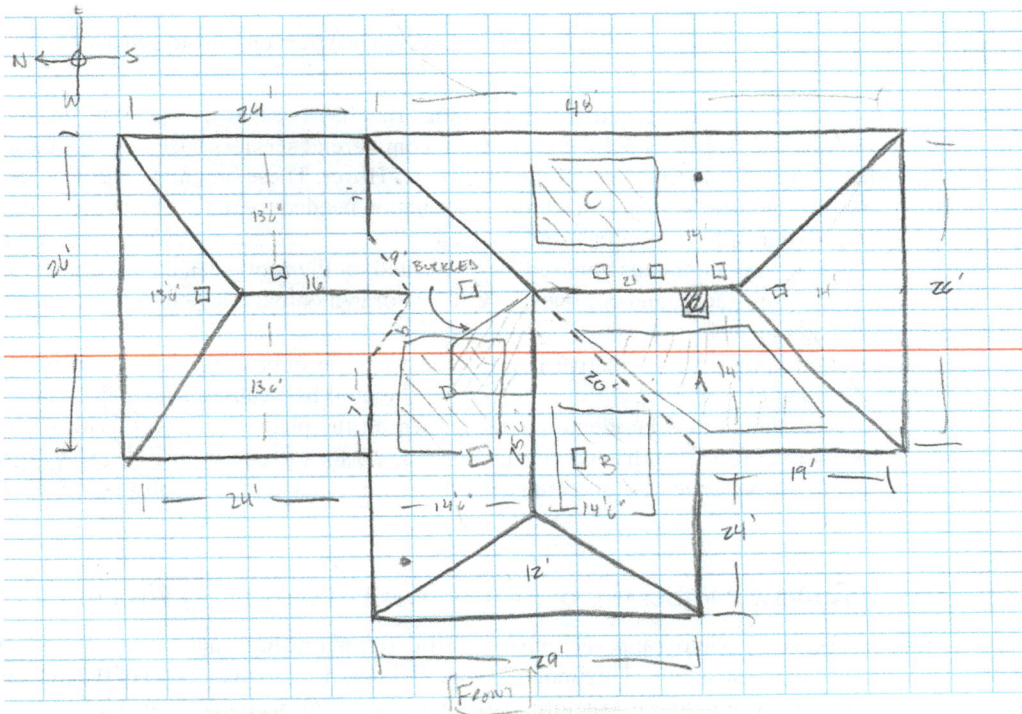

FIGURE 2.20 Example field book – plan-view sketch of roof.

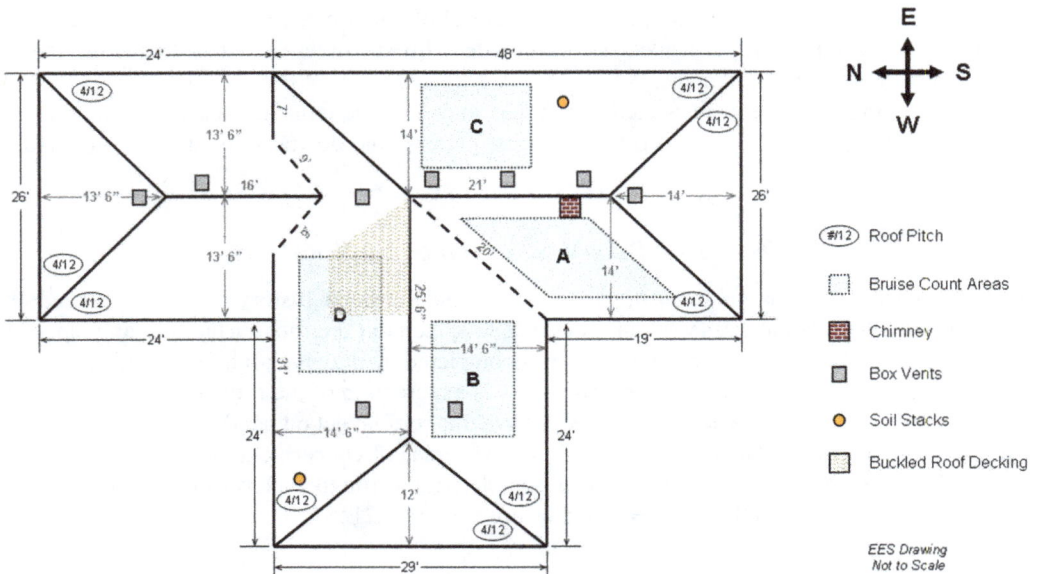

FIGURE 2.21 Typical computer-aided roof layout schematic (developed from on-site measurements).

Recently, the trend in the roofing and insurance industries has been to use satellite imagery technology for roof layouts, slopes, and dimensions from commercial services such as EagleView™ (formerly Pictometry® and GeoEstimator®) Technologies and Hover. These organizations provide a complete roof drawing with nearly all pertinent roof information needed for replacement (i.e., ridge/valley lengths, rake/eave lengths, roof areas, and pitches).

While in the field, all miscellaneous appurtenances located on the roof, such as box vents, furnace vents, ridge vents, soil stacks, and chimneys, should be drawn onto the layout drawing in case they are needed for later analysis.

Entire roof elevations and/or representative areas (i.e., test squares or bruise count areas) for each directional roof elevation chosen for a close inspection of potential hail-caused functional damage to the roofing material should also be identified and drawn on the plan view sketch/layout. This allows for documentation of the lengths and areas damaged and for the engineer/roof inspector or insurance company to estimate the cost of damaged surfaces without having to return to the site to obtain such information.

2.5.3 Take Broad-View Photographs of Each Exterior Elevation of the Structure

A broad-view, or overview, photograph encompasses all exterior finishes and shows the general condition of each side of the structure. This should be the first step prior to inspection of the finishes on a particular exterior elevation of the structure. Any photographs taken should be labeled that day for identification purposes. If the project is subject to probable litigation, one set of original, file-dated, unmodified digital photos should also be saved for future, possible reference by other parties.

2.5.4 Conduct a Systematic Inspection of the Exterior Building Envelope and Document any Damages

To perform a complete and comprehensive assessment of the extent to which the subject building has been damaged by hail, each exterior side of the building should be inspected and all observations documented both in writing in an on-site field book and visually with digital photography.

An inspection of all surfaces on each exterior side of the building must be performed, as hail falls randomly (in no discernable pattern) and hits almost everything uniformly from the direction it arrives. Any inconsistencies should be described in detail, focusing special attention to any signs of hail-caused damage.

Thorough documentation of these damages aids in future determination of hailstorm directionality and hailstone fall patterns (see Section 2.4.5) and helps "paint the hailstorm picture." Recall that the exterior sides of the building that exhibit the highest frequency of hail-strike damage help to define the direction from which the hailstorm likely arrived. Further, measurement of the diameter of the hail-strike dents and/or the width of spatter marks can be used to define the largest size of hail that likely impacted the building and provides a strong indication whether or not the hail was of sufficient size to damage various roof surfaces. When hail damage is present on the roof surface of a structure, there is typically extensive collateral hail-strike damage to the exterior surfaces/finishes, correlating to the same storm-facing elevations.

Typical exterior building envelope components that are inspected and typically exhibit evidence of hail impacts are noted below:

- *Exterior finish of the building*: Inspect and describe the exterior finishes and cladding of each side of the building (brick masonry, CMU, vinyl siding, painted wood siding, etc.).
- *Windows and doors*: Inspect and document the conditions of window/door frames, wraps, trim, screens, and shutters.
- *Gutters and downspouts*: Inspect, label, and describe each gutter and downspout on each side of the building.
- *Miscellaneous*: Inspect and document the condition of any additional items that are a part of the exterior of the building. Some of these surfaces include the metal surfaces of gas, electric, and water meters, heating, ventilation, and air-conditioning (HVAC) units, including the sheet metal and coil fins of air-conditioning units (both ground and window-mounted).

In the cases where probable hail-strike damage has appeared to have occurred to a particular surface, measure and document the diameter of any dents, shatter marks, and/or spatter marks, and take accurate representative or pertinent photographs of each instance of damage. Measure and document the dimensions of any damaged component or the area of any damaged portion of siding. This can aid in future estimations during the repair/replacement process.

Special care must be taken when inspecting metal siding, trim, and fascia for hail-strike damage as it is often difficult to see when the hail size is small, the metal is darker (especially gray) in color, and/or the day is cloudy. One method commonly used to make these determinations is to spray water on the surface; this causes the dents to stand out visually. Vinyl siding and plastic components will typically shatter when struck by hail of sufficient size. Shatter marks in vinyl siding most often occur on the lower protruding edges of the siding panels. The overall dimensions of the shatter marks are often greater than the diameter of the hail that caused the damage.

Note that much of the hail-strike damage documented to the exterior surfaces of a building is considered cosmetic and will likely not affect the functionality of the damaged component (see Section 2.4.3 "Damage Classification: Functional or Cosmetic?"); however, due to the visibility and possible reduction of property value, these items are often replaced to bring the property back to a pre-storm saleable condition. A good example would be impact dents in a standing-seam copper panel roof over an exterior bay window of a residential home. Unless the hailstone impact pierced the metal surface coating and/or disengaged a lapped seam element, the copper panel roof would not be considered functionally damaged but cosmetically damaged. However, as discussed earlier, since the dents are readily visible, the copper roof would likely be recommended for replacement since it is readily observable and would have negative impact on property value.

2.5.4.1 Mechanical Damage to Exterior Building Envelope Components

Once hail reaches sufficient size, it will begin to damage exterior surfaces depending on the size of the hail, the hailstone density/hardness, the type of surface, and the angle of impact. Discretion must be used and special attention should be paid, particularly for exterior siding components, to the locations of the damage since not all observed damage was likely the result of a hailstone impact and many times is most often misdiagnosed as such. Absence of similar damage higher on the exterior surfaces could indicate possible mechanical damage, or damage created by some other means than from hailstone impacts, which was more susceptibly lower near the ground. Two such examples would include objects inadvertently impacted against the side of the building from children and debris ejected by lawnmowers. This type of mechanical damage is often observed during field investigations. Sometimes, a brief general observation of the surrounding ground surfaces can add insight to possible sources for the observed apparent mechanical damages.

Dents and damage to some of the metal building envelope components, such as siding and downspouts, around the building are often debated and believed to be caused by hail strikes. In some of those cases, the observed damage is most likely attributed to mechanical damage. Clear indicators of mechanically caused damage are punctures or creases to the metal surface within the dent or scratches to the finished painted surface (Figures 2.22–2.24).

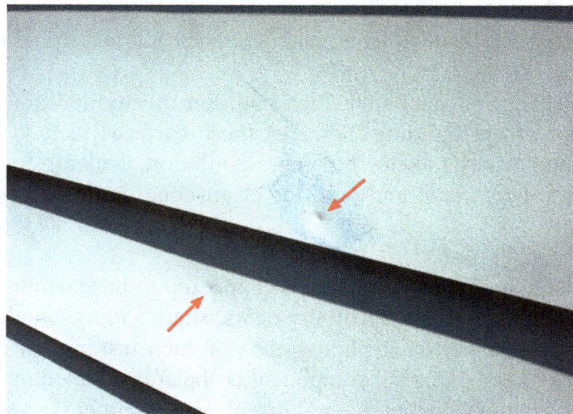

FIGURE 2.22   Punctured/creased mechanical dents in exterior metal siding.

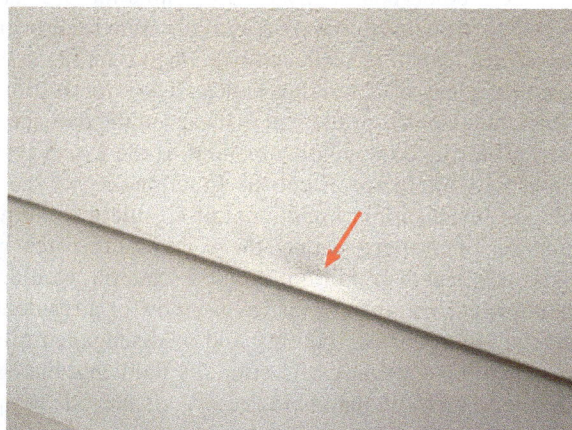

FIGURE 2.23   Punctured/creased mechanical dents in exterior metal siding.

FIGURE 2.24 Creased mechanical dents in downspout.

When hailstones strike a metal surface, such as exterior siding or a downspout, they may or may not create a dent, depending on the size of the hail produced, the hailstone density/hardness, the thickness of the metal, and the angle at which the hailstone strikes the metal surface. However, when a dent is created from a hailstone impact, it may leave behind a mark on the oxidized metal surface (i.e., spatter or burnish mark), but it will not create a puncture or crease in the metal, and it will not scratch the finished paint.

2.5.4.2 Common Exterior Damage Claims – Air-Conditioning (HVAC) Units

Recall from Section 2.4.6 ("Determining Hail Size: Correlation between Size of Dents in Metal versus Size of Hail"), the coil fins for outdoor air-conditioning units provide an excellent source of information for determining hailstorm/hailstone characteristics, assuming that one or more of the sides of the unit face the direction from which the hailstorm arrived. Such information includes:

- The direction from which the storm arrived (i.e., hailstorm directionality).
- The maximum width of the hail-strike dents to have struck the building (recall that dents to air-conditioner coil fins are a consistent indicator for hail size – dent multiplier for damage is 1.2).
- The locations and geometries of spatter/burnish marks to the sheet metal surfaces and dents to the coil fins can be used individually or in combination to determine hailstorm directionality and hailstone fall patterns (i.e., angle at which the hailstones fell).
- If a building has multiple air-conditioning units with varying manufacturing dates (determined from serial numbers that provide the date of manufacturing; the installation date is typically within months of manufacture), and given historic hailstorms arrived from different directions, the damage associated with storms of various dates can be relatively established.

Oftentimes with regard to hail damage, two issues often arise when hail strikes an outdoor air-conditioning unit and damages the coil fins: (1) the compressor in the unit fails due to hail-strike dents to the coil and/or (2) the unit or the coil must be replaced due to hail-strike dents. It would be highly unusual for a compressor to fail due to hail-strike damage. Often, the unit and its compressor may be approaching the end to its normal life (~12 to ~17 years) so it has failed (often short out) simply due to "old age." This can be checked in the field by determining the age of the unit from the interview and/or the serial number on the unit.

Regarding the issue of damage to air-conditioning outdoor unit coil fins, Sitzman et al.[41] studied this issue at length and concluded that even if the coil fins were completely flattened and combed out, the capacity of the unit was restored to within 1%–4% and its efficiency restored within 4%–6%. Moreover, capacity and efficiency of test units were not impacted when more than 50% or more of the fins were flattened. Thus, data suggests that coil fins struck by hail stones can be combed out without significantly affecting the performance of the system and that a large percentage of the coils must be flattened before the performance of the unit is impacted significantly. In addition, only one or two sides of the coil are typically impacted by hail-strike damage. If the unit is at/near the end of its service life, the issue of repair versus replacement of the coil may need to be considered if a majority of the coil fins have been impacted.

2.5.4.3 Common Exterior Damage Claims – Windows and Seal Failures

Sometimes, following a particular hailstorm event, owners believe that the window seals on the building's windows were damaged by hailstone impacts. The rule of thumb is that the hail must be of sufficient size to crack the window glass in order to break the window seal. For double-paned windows (known as insulated glazing units – IGU) installed between 1982 and 1991, the typical life of a window (and seal) ranges between ~10 and 25 years, depending on seal type and climactic conditions.[42] Earlier windows using single seal designs have shorter lives. During the inspection, one should determine if the window seals have failed on windows opposite the direction of the storm; if so, it is likely that the window seals have reached the end of their expected life. The typical causes for window seal failures are as follows:

- *Seals breaking down from exposure to water*: Windows without the proper safeguards to keep water from puddling around the perimeter seals will fail sooner.
- *Excess heat*: Most often failure occurs on windows with direct sun exposure (i.e., southern exposure). Heat causes the panes to expand and contract (termed "solar/thermal pumping"), and it softens and weakens the seals until they develop a crack and allow moist air in.
- *Old age*: Even the most elastic, flexible seal cannot last forever; eventually a seal will allow moisture to enter the window. Once moisture enters between the glass panes, the desiccants (whose sole purpose is to maintain dryness and absorb moisture) within the spacer bars separating the two panes become saturated and condensation between the panes forms. Therefore, evidence of condensation suggests that the seals have most likely failed.

2.5.4.4 Common Exterior Damage Claims – Potential for Hail-Strike Damaged Masonry and Concrete Surfaces

Owners sometimes will claim that hail has damaged brick masonry or concrete surfaces after a hailstorm has impacted the area. Masonry and concrete surfaces are much harder than hail and should not be affected by hail-strike impacts. In fact, the hail will shatter when striking these surfaces. The most likely cause of damage to these surfaces is freeze–thaw damage known as spalling. During inspections of chimneys and other masonry surfaces, one can count the number of spalled masonry units by elevation. If the rate of spall damage is similar or higher on the elevations not facing the direction of the incoming hailstorm, this will help to demonstrate that the damage to the masonry was not primarily related to the hailstorm hail-strike impacts.

2.5.5 Conduct a Systematic Inspection of the Roof Surfaces and Appurtenances

The majority of hail damage claims reported by property owners are for possible hail-strike damage to the roof surfaces of their home or business. Often, following a hailstorm, local roofing contractors will canvas the area, inspect an owner's roof, and inform them that it was damaged by the hailstorm. Oftentimes, this is based on showing the owner photographs of dents in a metal box vent or other roof appurtenance and/or surface defects to the shingles that can be perceived as hail damage. Hail-strike dents to metal roof surfaces may, or may not, be indicative of damage to finished roof surfaces, depending on the size of the hail that struck the building. As is often the case, there is a common misconception that if hail had the ability to create dents to metal, which can be perceived as a strong material, then the roofing material surely must be damaged too. In other cases, a homeowner knows that their neighbor received a new roof, so their roof too must be damaged. However, sometimes the neighbor's roof was not inspected and simply "bought" for any number of reasons that do not necessarily reflect whether or not the area was struck with hail of sufficient size to damage roof finishes. An inspection by a knowledgeable inspector is the only method for determining the extent of hail damage to a building.

Much in the same way as the exterior building envelope was inspected, a thorough and methodical hail damage assessment is performed on the roof surfaces of the subject structure. All observations are documented both in writing in an on-site field book and visually with digital photography. Included in the roof inspection are the following procedures:

- Adherence to proper safety equipment, protocol, and procedures.
- A description of the roof construction at a roof eave or access point.
- Measurements of all the necessary roof dimensions, if not already provided from an out-sourced service.
- Inspection for hail-caused dents in the metal surfaces and/or roof appurtenances.
- Inspection for hail-caused functional damage to the roofing material on each of the directional roof elevations, either in a representative test area (i.e., for asphalt shingles and wood shakes/shingles) or in entire roof elevations (i.e., for tile roof surfaces).
- Assessment of the overall condition of the roof surfaces.
- Documentation of areas of concern including inherent and imminent safety concerns and/or maintenance issues.

The processes used in completing each of these steps are discussed in further detail below.

2.5.5.1 Safety

Roof inspection workers are not covered under the new OSHA (Occupational Safety and Health Administration) Construction Fall Protection Standard – CFR 1926.500(a)(1)[43]:

> This subpart sets forth requirements and criteria for fall protection in construction workplaces covered under 29 CFR part 1926. Exception: *The provisions of this subpart do not apply when employees are making an inspection, investigation, or assessment of workplace conditions prior to the actual start of construction work or after all construction work has been completed.*

Regardless, it is good practice for companies who perform roof inspections to have a Health and Safety Plan in place and to train inspectors on the plan and company expectations.

The most important step in the inspection process, from a safety standpoint, is to follow company protocol(s) outlined in their Health and Safety Plan and to utilize sound equipment in order to ensure safety during any roof inspection. A ladder, man-lift, or other mechanical means should be used to access the roof, when necessary. Typical access points are along roof eaves, rakes, and peaks and at the lower ends of valleys. Appropriate footwear with optimal traction should be worn for safety and to protect the roof coverings from scuffs and footfalls. For roof surfaces below a 6:12 pitch, soft-soled shoes are typically adequate; on steeper roof surfaces, specialized roofing boots or shoes (e.g., CougarPaws™) can aide greatly in traction and mobility on residential and light commercial shingled surfaces. Regardless of footwear, the following techniques and advice should be considered with walking roof surfaces:

- In the case of steep-slope roofs in residential and light commercial construction (≥9:12 pitch), attempts should be made to walk along valleys, hips, and ridges where balance is enhanced and/or use safety harnesses.

- Attempts should be made to avoid stepping on cupped, curled, or otherwise fragile wood shingles, wood shakes, or mineral-based tiles, and discretion must be used with badly damaged areas.
- On steep-sloped roof surfaces, a harness or safety rope may be necessary.
- Never attempt to walk on wet, ice- or snow-covered wood, slate, tile, or metal roof surfaces.
- Lower-sloped (≤6:12) asphalt surfaces can often be walked when wet, although the ability to observe hail-strike damage to wet asphalt shingled surfaces is somewhat reduced.
- It is a good rule of thumb that higher-sloped roof surfaces (>9:12) be inspected by a team of at least two people equipped with harnesses and fall protection equipment.

Regardless of the circumstances, inspectors should take whatever means necessary to be safe during the inspection and should never put themselves in potentially dangerous and/or life-threatening situations.

2.5.5.2   Roof Construction at Eave or Access Point

Once at the eave or access point to the roof, observe and describe the roof construction. During this portion of the inspection, the inspector should document:

- Whether drip edge molding is installed along the eave or rake.
- Whether felt underlayment is installed over the roof decking.
- Whether ice guard is present (depending on location).
- The type of roofing surface material(s) installed.
- The number of layers of roofing surface material(s) installed.

The number of layers of roofing materials is an important inspection observation to take into account because it can directly affect the susceptibility for hail-caused functional damage to the top layer of roofing by decreasing the material's coefficient of restitution and, therefore, allowing more absorption of energy from an impacting hailstone (see Section 2.4.1 "Physics Lesson – Basics of Energy Transfer and Coefficient of Restitution" and Section 2.4.2 "Material Impact Resistance").

2.5.5.3   Roof Measurements

If not already provided by an outsourced service, all ridge lengths, eave lengths, valley lengths, elevation slopes, elevation dimensions, and elevation pitches should be measured and documented in the plan-view sketch in the field book. Even if the measurements are provided by an external source, these measurements should be spot checked to verify their accuracy. Any individual portions of the roof that may require replacement should also be measured and noted in the field book.

2.5.5.4   Inspection of Metal Surfaces and Roof Appurtenances
            for Hail-Caused Damages

Hailstones fall in a random pattern and hit almost everything on the roof surfaces of a building (see Section 2.3.2 "Random Fall Patterns of Hailstones"), including all of the miscellaneous metal surfaces and/or appurtenances, such as box vents, furnace flue vents, ridge vents, soil stack flashings, and chimney caps and flashings. When hailstones of sufficient size strike one of these soft metal surfaces, a dent is created, which can be readily observed and measured helping to determine the characteristics of the hailstorm, which are vital to the inspection process and damage assessment (see Section 2.4.5.1 "Hailstone Directionality, Angle of Impact, and Perpendicularity" and Section 2.4.6 "Determining Hail Size: Correlation between Size of Dents in Metal versus Size of Hail"). Recall from earlier that the average rule-of-thumb dent multiplier for most common metal roof appurtenances was 1.5. Note that this rule of thumb does not apply to aluminum flue vent caps (see Section 2.4.6) because of the relative softness and inconsistency of hail-strike dents to flue vent caps.

All metal surfaces and roof appurtenances should be inspected for evidence of hailstone impacts, documenting the measurements and dimensions of each damaged roof component both in the field book and by taking representative photographs of each damaged condition. Examples of hail-caused damage to typical roof components for residential, light commercial, and commercial buildings are provided in subsequent chapters pertaining to each structure type so an exhaustive list will be not be provided at this point.

Most of the hail impact damage to metal roof surfaces such as vents and flashings, whether residential/light commercial or commercial buildings, will be considered cosmetic in nature and not necessarily functional damage.[44] Thus the dents and/or spatter/burnish marks will not likely affect the functionality of a roof component or expected service life, but nevertheless are often considered damaged by insurance companies and removed/replaced.

2.5.5.5 Inspection for Hail-Caused Functional Damage to the Roofing Material

Recall from earlier in Sections 2.4.3 and 2.4.4 ("Damage Classification: Functional or Cosmetic?" and "Hailstone Size Thresholds for Functional Damage to Roofing Materials") that the industry standard definition for functional damage to any roof covering is either: (1) a reduction of its water-shedding capabilities or (2) a reduction in the expected long-term service life of the roof material.[26,32,33] The minimum size thresholds for hailstone impacts to start to cause functional damage for each of the individual roofing materials are discussed at length in later chapters devoted to those particular roof coverings common to residential, light commercial, and commercial buildings. Note that these size thresholds are the smallest size of natural hail at which functional damage typically begins to occur and refers to hail that strikes perpendicular to the surface of the roofing material, which is in relatively good, mid-life conditions.

Using the indirect hail-sizing information from Section 2.4.6 for metal surfaces and common roof appurtenances and the minimum size thresholds to cause functional hail damage to common roofing materials, it can be reasonably determined if hail-caused functional damage was more or less likely to occur to the roofing material on a particular building.

Equipped with this broad base of knowledge, an inspection of the roofing material is conducted for the presence of potential hail-caused functional damage. In general, the methods used within the forensic engineering industry typically include a visual inspection of either a representative area on each directional elevation of the roof, which is indicative of the conditions of the roofing system as a whole, or the entirety of each roof elevation on the subject building. The method used is dependent on the type of structure (i.e., residential/light commercial) and the type of roofing material (i.e., asphalt shingles, slate/concrete/clay tiles, roll roofing, Ethylene Propylene Diene Monomer (EPDM), Built-Up Roofing (BUR), Polyvinyl Chloride (PVC), etc.). Regardless of which method is employed, a thorough and detailed inspection should be performed, taking into account all conditions and factors that could potentially lead to the formation of hail-caused functional damage.

A more thorough and detailed discussion of characteristics of functional hail damage to residential, light commercial, and commercial roofing materials is provided in Chapters 3 and 4 based on building type and specific roofing material.

2.5.5.6 General Observations on the Overall Condition of the Roof Surfaces

To assess hail damage accurately to the roof covering on a particular building, general observations on the overall condition of the roof surfaces should be documented. Typical observations of general roof conditions would include, but are not limited to, the following:

- The appearance of the roof decking (i.e., wavy/buckled/subsided) and if it was soft to walk on.
- Missing areas of roofing material.

- Degraded roof areas.
- The overall visible condition of the roofing material, noting locations and patterns of deterioration (i.e., roof slopes exhibiting more deterioration).
- Areas of apparent mechanically caused damage (i.e., holes, scrapes, gouges, footfalls/foot traffic damage).
- The presence of biological growth (i.e., moss, algae, and/or lichens) on the surfaces of the roofing material.
- Areas of accumulated or ponded water.
- Areas of debris accumulations including fallen material from nearby trees.
- The presence of overhanging trees/bushes and other nearby buildings.

The reasons for taking general observations on the overall condition of the roof surfaces are two-fold: (1) first, it provides a thorough report of the roof and makes the property owner aware of any areas that may need additional attention and/or necessary maintenance because more often than not, they have not personally observed the condition of their roof; and (2) if observations of the roofing material show that it is older in appearance and in a deteriorated state, it can become more susceptible to hail-caused damage or other weather-related damaging factors such as high winds and heavy rainfall.

2.6   METHODS TO DETERMINE RELATIVE DATES OF HAILSTORMS

Oftentimes, situations arise where multiple hailstorms have been known to pass through specific areas at considerable lengths of time apart from each other. These situations become more complicated with regard to the home insurance industry where policy language on coverage limits the timeframe in which property owners have to submit a damage claim, or when multiple carriers had coverage for the property at different times. If a particular hail damage claim is date-specific, a forensic engineer or inspector is often called on to: (1) assess the property and determine if any evidence of hail damage is present and (2) determine, within a certain degree of engineering/scientific certainty, which of multiple storms likely caused the damages (if present).

The inspection should follow the methodology outlined in this chapter, but additional analysis of the collected information will typically be needed. Only when all of the available information has been gathered and analyzed can a determination be made on the relative date of the damaging hailstorm. Experience has shown that the following steps facilitate claims where dating of a specific storm is desired:

- Obtain as much information by interviewing all parties involved with the claim.
- Thoroughly review all hail weather data for the area in question, including reports from local and national weather agencies (i.e., newspapers, local news organizations, NOAA NCDC, and SPC hail reports, etc.) and possibly from companies such as iMap®Weather Forensics (a division of Weather Decision Technologies, Inc. and producers of HailTrax™), CoreLogic, and Verisk, which create more detailed information on local hail sizes and directions.
- Gathering visual evidence of the locations and patterning of hail-caused damage to the building envelope and roof appurtenances.
- Observe and document the exterior sides of the building, which exhibit the greater frequency of hail damage in order to determine the hailstorm direction.
- Make special note to the orientation of surfaces that were impacted by hail in order to attempt to determine the fall patterns of the impacting hailstones (i.e., straight down with no damage to vertical surfaces or wind-driven in which vertical surfaces are heavily impacted).

- Be sure to document the pitches of each roof elevation of the building.
- Measure the sizes of dents in metal and spatter/burnish marks on other available surface in order to determine the size of the hailstones that impacted the property. Note that the presence of spatter/burnish marks is generally clear indicator that a hailstorm event has passed through and impacted the area rather recently since they tend to wear away with time.
- Evidence of hail damage on opposite sides of a building typically indicates that a minimum of two hailstorms have impacted the building and possibly more. Special attention should be paid on the maximum size of the dents created in the metal surfaces and where exactly they are located and oriented. Attempt to document hailstorm directions by elevation and maximum sizes of dents/hailstones by elevation/direction.
- Record the serial numbers of air-conditioning units and other HVAC equipment (newer units that do not show evidence of damage may indicate that the damage to the other surfaces is older; however, special attention should be paid to the fall patterns of the hail-strike damage (coil fin elevations damaged and size of dents by elevation). Ages of air-conditioning outdoor units and locations of hail-strike dents to coil fins, coupled with weather data, can often provide the basis for identification of a specific hailstorm to have struck the building.

Using all this information, an analysis can be completed to determine whether or not the hailstorm of interest struck during a given carrier's coverage. It should be noted, however, even after all attempts have been made to gather all of the necessary information, an accurate determination of the date may not be able to be concluded.

2.7   HAIL DAMAGE INSPECTION REPORT

Recall from Chapter 1 ("Introduction"), the written report should summarize inspection findings, pertinent explanations of literature and conclusions reached as a result of the inspection. This should be completed within a timely fashion and before inspection, recollections fade. Typically, recommendations are not included since they have a tendency to be viewed as requirements for the client; the exception to this is any condition where the inherent safety of individuals is threatened, in which case these situations should be explicitly called out. A typical report outline for a hail damage inspection should include the following elements:

- Introduction (Information on Inspection Location and Client).
- Scope of Work (What is the scope of the inspection?).
- Information Regarding the Property (e.g., age, square footage information from a County Auditor Webpage).
- Hail Weather History at/near the Inspection Location.
- Summary of Interview(s).
- Summary of Exterior Observations.
- Summary of Roof Observations.
- Discussion/Analysis of Observations.
- Conclusions Regarding Hail-Strike Damaged Surfaces (i.e., surface and area/length).
- Photographs and Figures.
- Evidence and/or Supporting Documents.

Experience in the field of forensic engineering indicates that the use of drawings (e.g., Figure 2.20) and photographs within the inspection report rather than a report simply containing text is more effective in conveying inspection findings to interested parties.

IMPORTANT POINTS TO REMEMBER

- Hailstones consist of alternating layers of clear and opaque ice, which have differing densities, and thus the overall density of the hailstone is a function of how and where the hailstones were formed within the storm cloud.
- Sufficiently sized hailstones capable of causing functional damage to common roofing materials typically occur in rather small areas when compared with the size of hail created throughout the swath of a hailstorm and can change drastically in rather short distances.
- Hailstones fall randomly in no discernable pattern and should cause uniformly or evenly spaced damage patterns.
- The exterior sides of a building that exhibit the greatest frequency of hail damage marks help to determine the direction from which the hailstorm arrived (i.e., hailstorm directionality).
- The orientations of hail damage marks (spatter/burnish marks and dents) help determine the hailstorm directionality and the general fall pattern of the hailstones (i.e., straight down or wind-driven).
- The impact energies associated with falling hailstones are dependent on their size or mass, velocity, and brittleness.
- The horizontal wind component applied to heavily wind-driven hail increases the resultant velocity and subsequently the impact energy of the hail; however, it does not significantly increase it to levels at which functional damage to common roofing materials typically occurs and therefore is not a primary contributing factor when evaluating roof damage.
- The greatest likelihood for functional hail damage to a roofing material is from a hailstone that strikes perpendicular to the roof surface.
- Functional damage typically does not occur to common roofing materials until the *hailstone size threshold* has been reached.
- Conditions that decrease the coefficient of restitution of the roofing material cause the material to absorb more energy from an impacting hailstone and therefore increase its susceptibility to functional hail damage below typical size threshold values.
- Basing hail-strike damage on metal dent sizes and size of hail is more effective for determining damage to finished roof surfaces than more subjective parameters such as hail-strike bruise-counts (i.e., for common asphalt shingles).
- The hailstone size can best be determined by measuring the inner dents created in common metal surfaces and roof appurtenances.
- Excluding soft aluminum flue vent caps (roof appurtenance); the average dent multiplier (ratio of maximum hailstone size to inner dent diameter) applied to measured inner dents in metal surfaces is 1.5.
- Dents to the coil fins of air-conditioning units are one of the most accurate and consistent indicators of size of hail with a dent multiplier of 1.2.

REFERENCES

1. E. Leefeldt. "Are Hailstorms Getting Worse in the U.S.? Why 2019 Could Produce Record Damage," 2019. Accessed July 31, 2020, https://www.cbsnews.com/news/hail-damage-costs-this-year-could-hit-new-annual-high-in-u-s/.
2. D. Fennig. "National Insurance Crime Bureau (NICB) ForeCAST™ Report Regarding 2013–2015 United States Hail Loss Claims," May 2nd 2016.

3. T. Manasek. "National Insurance Crime Bureau (NICB) ForeCAST™ Report Regarding 2016–2018 United States Hail Loss Claims," July 25th 2019.

4. T. Manasek. "National Insurance Crime Bureau (NICB) ForeCAST™ Report Regarding 2017–2019 United States Hail Loss Claims," April 14th 2020.

5. Ohio Insurance Institute. June 9, 2010, Press Release – May 7–8 Northern Ohio Storm Losses Top $31 Million – Source of National Insurance Crime Bureau Hail Claims Data. Accessed March 2011, http://www.ohioinsurance.org/newsroom/newsroom_ full.asp?id=602.

6. N. F. Somes, R. D. Dikkers, and T. H. Boone. "Lubbock Tornado: A Survey of Building Damage in an Urban Area." U.S. Department of Commerce, National Bureau of Standards (NBS). Technical Note 558, March 1971.

7. J. T. Allen, M. K. Tippett, Y. Kaheil, A. H. Sobel, C. LePore, S. Nong, and A. Muehlbauer. "An extreme value model for U.S. hail size." *Monthly Weather Review*, American Meteorological Society, vol. 145, pp. 4501–4519, 2017.

8. National Oceanic and Atmospheric Administration (NOAA) National Severe Storms Laboratory (NSSL). "Hail Basics." Accessed January 26, 2012, http://www.nssl.noaa.gov/primer/hail/hail_basics.html.

9. E. Gaviola, and F. Alsina Fuertes. "Hail formation, vertical currents, and icing of aircraft." *Journal of Meteorology*, vol.4, pp. 116–120, 1947.

10. V. Crenshaw, and J. D. Koontz. "Hail: Sizing it Up!" 2010. Accessed March 2011, http://www.hailtrax.com/hail_size_it_up.pdf.

11. A. Heymsfield, and R. Wright. "Graupel and hail terminal velocities: Does a "supercritical" Reynolds number apply?" *Journal of the Atmospheric Sciences*, vol. 71, pp. 3392–34013, 2014.

12. A. Heymsfield, I. M. Giammanco, and R. Wright. "Terminal velocities and kinetic energies of natural hailstones, geophysical research letters, American Geophysical Union," pp. 8666–8672, 2014.

13. D. Orr. "Giant Hail Killed More than 200 in Himalayas." *The Telegraph*, November 7, 2004. Accessed August 3, 2020, https://www.telegraph.co.uk/news/worldnews/asia/india/1476074/Giant-hail-killed-more-than-200-in-Himalayas.html.

14. National Oceanic and Atmospheric Administration (NOAA). "'Volleyball' from the Sky: South Dakota Storm Produces Record Hailstone." United States Department of Commerce. Accessed August 3, 2020, https://www.weather.gov/abr/vivianhailstone. http://www.noaa.gov/features/02_monitoring/hailstone.html.

15. National Weather Service Forecast Office. "Hail Size Chart." Albany, NY. Accessed January 26, 2012, http://www.erh.noaa.gov/aly/Severe/HailSize_Chart.htm.

16. National Oceanic and Atmospheric Administration (NOAA) National Severe Storms Laboratory (NSSL). "Hail Climatology." Accessed January 26, 2012, http://www.nssl.noaa.gov/primer/hail/hail_climatology.html.

17. The National Severe Storms Laboratory (NSSL). "Severe Weather 101 – Hail Basics." Accessed August 3, 2020, https://www.nssl.noaa.gov/education/svrwx101/hail/.

18. May 25, 2011 HailTrax Report #20397 to EESGroup, Inc., www.weatherforensics.com.

19. S. M. Verhulst, J. D. Bosley, and A. K. Talbott. "Hail sizing: A comparison of on-site data with weather data," *Published at the Forensic Engineering 8th Congress*, held November 29 to December 2, 2018 in Austin, TX. Accessed August 3, 2020, https://www.nelsonforensics.com/wp-content/uploads/2018/12/Hail-Sizing-A-Comparison-of-On-Site-Data-with-Weather-Data.pdf.

20. R. K. Noon. *Forensic Engineering Investigation*. New York: CRC Press, 2001.

21. J. M. Smith, and H. C. Van Ness. Introduction to Chemical Engineering Thermodynamics, 3rd ed. New York: McGraw-Hill Book Company, 1975.

22. J. A. P. Laurie. "Hail and Its Effects on Buildings," Research Report 176, NBRI, Pretoria, South Africa, 1960.

23. S. H. Greenfeld. "Hail Resistance of Roofing Products." Building Science Series #23, National Bureau of Standards, August 1969.

24. J. D. Koontz. "The effects of hail on residential roofing products." *Third International Symposium on Roofing Technology*, April 1991, Montreal, CA, pp. 206–215.

25. National Oceanic and Atmospheric Administration (NOAA) National Severe Storms Laboratory (NSSL), "Thunderstorm Basics." Accessed January 29, 2012, http://www.nssl.noaa.gov/primer/tstorm/tst_basics.html.

26. T. P. Marshall, R. F. Herzog, S. J. Morrison, and S. R. Smith. *Hail Damage Threshold Sizes for Common Roofing Materials*. Dallas, TX: Haag Engineering, Co., 2002.

27. W. C. Cullen, "Hail damage to roofing: Assessment and classification", *Proceedings of the Fourth International Symposium on Roofing Technology*, NRCA/NIST, 1997, Gaithersburg, MD, pp. 211–216.

28. T. Brown-Giammanco, and I. M. Giammanco. "An Overview of the IBHS Hail Research Program, Insurance Institute for Business and Home Safety, 11.1". Accessed September 10, 2020, https://ams.confex.com/ams/29SLS/webprogram/Manuscript/Paper348661/Brown-Giammanco_SLS_Overview%20of%20the%20IBHS%20Hail%20Research%20Program.pdf.

29. Insurance Institute for Business and Home Safety, 11.1. "IBHS Impact Resistance Test Protocol for Asphalt Shingles." Accessed September 10, 2020, https://ibhs.org/wp-content/uploads/2019/06/ibhs-impact-resistance-test-protocol-for-asphalt-shingles.pdf.

30. K. Shephard, and N. Gromicko. "Mastering Roof Inspections: Hail Damage, Part 6." InterNACHI. Accessed January 26, 2012, http://www.nachi.org/hail-damage-part6-33.htm.

31. B. Leonard. "Department of Defense Dictionary of Military and Associated Terms." Joint Publication 1–02, April 12, 2001, amended Through April 2010, p. 194.

32. S. L. Morrison. *Hail and Composition Shingles.* Dallas, TX: Haag Engineering Co., 1993, revised March 1995.

33. T. P. Marshall, and R. F. Herzog. "Protocol for assessment of hail-damaged roofing", *Proceedings of the North American Conference on Roofing Technology*, Sept. 16-17, 1999, Toronto, Ontario, CA, pp. 40–46.

34. Advance Cable Company, LLC, et al., vs Cincinnati Insurance Company, LLC, et al. Case No. 14–02620-Consolidated with Cross Appeal Case No. 14–2748, United States Court of Appeals for the Seventh Circuit, Document 16-1, filed October 20, 2014.

35. S. J. Morrison. "Dents in metal roof appurtenances caused by ice ball impacts," *Proceedings 12th International Roofing & Waterproofing Conference*, February 2002, Orlando, FL, pp. 1–19.

36. C. A. Rigby, and A. K. Steyn. The hail resistance of South African roofing materials, South African Architectural Record, vol. 37, no. 4, pp. 101–107, 1952.

37. T. P. Marshall, R. F. Herzog, S. J. Morrison, and S. R. Smith. *Hail Damage to Tile Roofing.* Dallas, TX: Haag Engineering, June 2004. Accessed September 10, 2020, https://www.researchgate.net/publication/237557281.

38. Roofing Industry Committee on Weather Issues, Inc. (RICOWI, Inc.). "Hailstorm Investigation – Dallas", Fort Worth, TX, May 24, 2011. Accessed September 10, 2020, https://docplayer.net/10564289-Hailstorm-investigation-dallas-fort-worth-tx-may-24-2011.html.

39. S. A. Petty, M. A. Petty, and T. Kasberg. "Evaluation of Hail-Strike Damage to Asphalt Shingles Based on Hailstone Size, Roof Pitch, Direction of Incoming Storm, and Facing Roof Direction." Interface Magazine, May/June 2009, pp. 4–10.

40. R. C. Mathey. "Hail Resistance Tests of Aluminum Skin Honeycomb Panels for the Relocatable Lewis Building, Phase II," 1970, p. 19.

41. M. J. Sitzmann, F. K. Lu, and S. R. Smith. "Hail effects on air-conditioning performance." *Proceedings of IMECE2007, SME International Mechanical Engineering Congress and Exposition, IMECE2007–41518*, Seattle, WA, November 11–15, 2007.

42. B. C. Tandy, and A. G. J. Way. "R&D to improve site practices for collection and clean separation of composite (glass) materials in the construction and demolition industry." Project code: GLA2–011, Final Report, Published by The Waste & Resources Action Programme, March 2004, www.wrap.org.uk.

43. Occupational Safety and Health Administration, Construction Fall Protection Standard – 29 CFR 1926.500(a)(1). Accessed September 10, 2020, https://www.osha.gov/laws-regs/regulations/standardnumber/1926/1926.500.

44. S. Cochran. "Is It "Really" Hail Damage?" The Investigative Engineer Magazine, January 2003.

3 Hail Damage Assessments for Residential and Light Commercial Exterior Finished Surfaces and Steep-Slope Roof Systems

Stephen E. Petty
EES Group, Inc.

CONTENTS

PURPOSE/OBJECTIVES

The purpose of this chapter is to:

- Define the characteristics of hail and hail-caused functional damage to surfaces associated with residential and light commercial structures.
- Document a methodology for assessing hail damage claims to residential and light commercial structures.

Following the completion of this chapter, you should be able to:

- Understand the impacts of hail on residential and light commercial exterior surfaces.
- Understand the differences between functional and nonfunctional, or cosmetic damage to common residential and light commercial roofing materials.
- Be able to distinguish between functional hail-caused damage and other defects such as natural degradation, manufacturing-related anomalies, installation defects, and mechanical damage to common residential and light commercial roof coverings.
- Be able to perform a thorough visual inspection for evidence of hail-caused damages to the exterior cladding system (siding, trim, windows, doors, screens, gutters, etc.) and roof coverings (asphalt shingles, wood shingles/shakes, slate/clay tiles, etc.) of a structure.
- Know and understand the size thresholds for functional hail damage to common residential and light commercial roofing materials.

3.1   INTRODUCTION

Hailstones, when of sufficient size, can damage building surfaces, including exterior finished components and roofing materials. Examples of hailstones between 0.75 and 1.0 inches are shown in Figure 3.1.

FIGURE 3.1   Hailstones 0.75–1.0 inches collected in a cup by a resident following a hailstorm in Akron, OH.

This chapter covers hail-strike damage to exterior finished surfaces and steep-slope roofing systems typically associated with residential and light commercial buildings. Light commercial buildings are defined as those typically >5,000 to 10,000 square feet in area used for commercial purposes, but with construction methods and systems typically found in residential structures. Hail-strike damage to low-sloped, typically commercial, roofing systems is covered in Chapter 4.

3.1.1 NEED FOR HAIL DAMAGE INSPECTIONS FOR THE RESIDENTIAL AND LIGHT COMMERCIAL MARKET

Recall from Chapter 2 that damage from hail events in the United States approaches $22 billion each year.[1] That being said, a great potential for profit can be gained by proactive roofing companies/contractors who canvas residential and light commercial neighborhoods that have recently been struck by hail. Their perceived exterior finish and roofing expertise, combined with the lack of knowledge by property owners and some insurance representatives, provides them a distinct advantage in allowing for replacement of exterior finishes, outdoor air-conditioning units, roof surfaces, and other building components, which may not have been functionally damaged by hail.

Hail-strike damage claims, particularly concerning residential and light commercial properties, are one of the areas where forensic investigations frequently occur as a result of the actions by entrepreneurial roofing "experts" and the increasing number of claims reported by insurance companies (see Section 2.1). Therefore, a proper inspection methodology for accurately evaluating these properties must be outlined to produce consistent assessment methods that can be relied upon.

3.2 COMMON EXTERIOR BUILDING ENVELOPE COMPONENTS DAMAGED BY HAILSTONE STRIKES

In Chapter 2, the inspection methodology for conducting hail damage assessments was introduced, including the information that can be obtained by performing a systematic inspection of the exterior building envelope. Important information such as the directionality, fall patterns (i.e., straight down or wind-driven), and the size of the hail can be identified during a close inspection of the exterior components.

Recall, also from Chapter 2, that all surfaces on each exterior side of the structure should be inspected, documenting field observations both in writing in an on-site field book and visually with a digital camera. Special care should be taken to look for any signs of potential hail-caused damage and any inconsistencies should also be documented.

Examples of typical hail-strike damage to exterior surfaces of a residential or light commercial building are illustrated in Table 3.1.

TABLE 3.1

Examples of Hail-Strike Damage to Typical Residential/Light Commercial Exterior Surfaces

Exterior Surface	Photograph
Air-conditioner outdoor unit Burnish/spatter marks to sheet metal andbending of coil fins	

(Continued)

TABLE 3.1 (*Continued*)
Examples of Hail-Strike Damage to Typical Residential/Light Commercial Exterior Surfaces

Exterior Surface	Photograph
Electrical box metal cover Typically hail-strike burnish/spatter marks	
Gas metal cover Typically hail-strike burnish/spatter marks	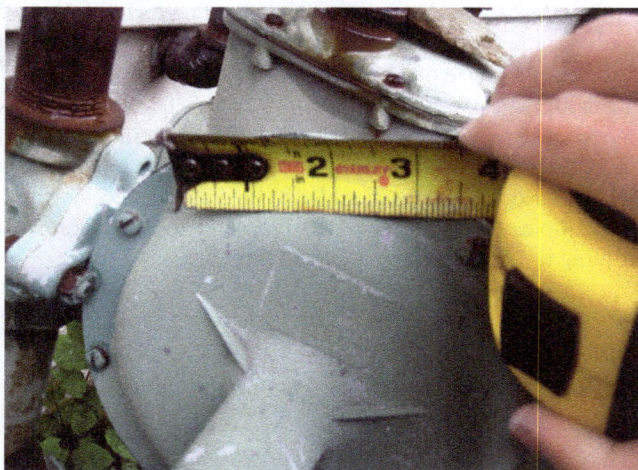
Metal window wrap – sill Hail-strike dent – illustrated using chalk	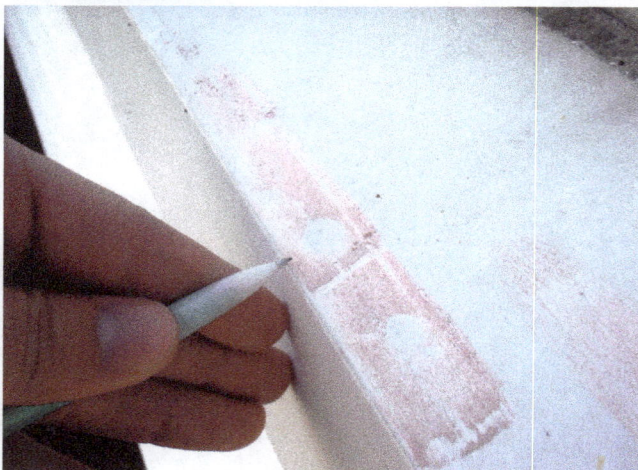

(*Continued*)

TABLE 3.1 (*Continued*)
Examples of Hail-Strike Damage to Typical Residential/Light Commercial Exterior Surfaces

Exterior Surface	Photograph

Metal window wrap – sill

Hail-strike dent – measured and illustrated using chalk

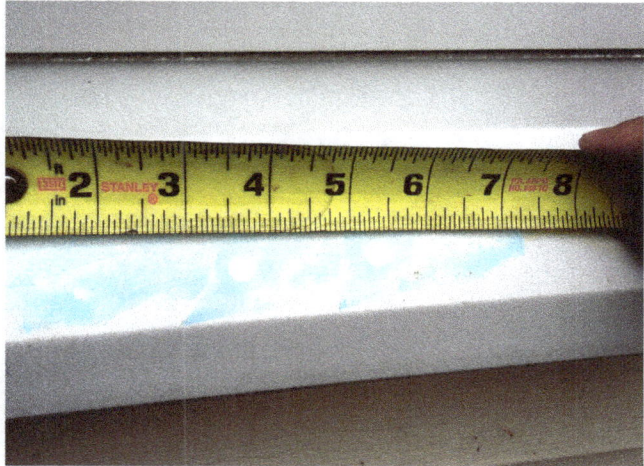

Painted wooden window sill

Hail-strike damage – note freshness of wood at impact points

Window screen

Hail-strike tear – measured

(Continued)

TABLE 3.1 (*Continued*)
Examples of Hail-Strike Damage to Typical Residential/Light Commercial Exterior Surfaces

Exterior Surface	Photograph
Downspout Hail-strike dent – measured	
Gutter Hail-strike dent – measured	
Metal gutter guard Hail-strike dents	

(*Continued*)

TABLE 3.1 (*Continued*)
Examples of Hail-Strike Damage to Typical Residential/Light Commercial Exterior Surfaces

Exterior Surface	Photograph
Vinyl gutter guard Hail-strike shatter marks	
Plastic basement window well cover Hail-strike shatter marks	
Metal siding Hail-strike dent	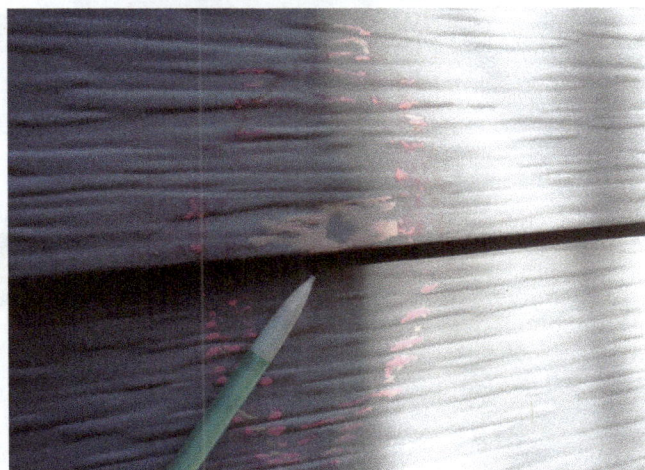

(*Continued*)

TABLE 3.1 (*Continued*)
Examples of Hail-Strike Damage to Typical Residential/Light Commercial Exterior Surfaces

Exterior Surface	Photograph
Metal siding Hail-strike dent with large burnish/spatter mark	
Vinyl siding Hail-strike shatter mark	
Vinyl siding Hail-strike shatter mark to lower protruding edge	

(*Continued*)

TABLE 3.1 (*Continued*)

Examples of Hail-Strike Damage to Typical Residential/Light Commercial Exterior Surfaces

Exterior Surface	Photograph
Painted wood siding Hail-strike chip	

3.3 COMMON METAL SURFACES AND ROOF APPURTENANCES WITH HAIL-CAUSED DAMAGES

When deliberating the probability of hail-strike damage to roofing material on a residential or light commercial structure, the size of the hailstones should be the first and foremost factor to determine. If the hailstone size is below the hailstone size threshold for a roofing material (asphalt shingles, wood roof systems, tile roof systems, etc.), the hail likely does not contain sufficient amounts of energy at impact to functionally damage the roof surface.

Recall from Section 2.4.6 in Chapter 2 that the size of the hail at a particular site correlates consistently to the size of the dents in the metal surfaces it has impacted, particularly roof appurtenances. Therefore, by identifying impacted components and accurately measuring the *inner dent* diameters, a consistent and reliable method for determining the maximum hailstone size can be achieved, and the evaluation and likelihood of functional damage to the roofing material can be more accurately determined.

Examples of typical hail-strike dents to exterior metal surfaces and roof appurtenances of a residential or light commercial building are illustrated in Table 3.2.

TABLE 3.2
Examples of Hail-Caused Damage to Residential/Light Commercial Roof Metal Surfaces and/or Roof Appurtenances

Hail-Caused Damage	Photograph
Hail-strike dents in metal box vent (12″×14″ slant-back)	
Hail-strike dents in furnace (flue) vent cap	
Hail-strike dents in metal ridge vent	

(Continued)

TABLE 3.2 (*Continued*)
Examples of Hail-Caused Damage to Residential/Light Commercial Roof Metal Surfaces and/or Roof Appurtenances

Hail-Caused Damage   **Photograph**

Hail-strike dents in metal soil stack flashing

Hail-strike dents in south- and west-facing chimney counter flashings (helps with determining hailstorm directionality)

Hail-strike dents in metal valley flashing

(Continued)

TABLE 3.2 (*Continued*)
Examples of Hail-Caused Damage to Residential/Light Commercial Roof Metal Surfaces and/or Roof Appurtenances

Hail-Caused Damage	Photograph
Hail-strike dents in skylight wrap/cladding	

3.4 DETAILED INSPECTION METHODOLOGIES FOR DIFFERENT ROOF FINISHES

Detailed inspection methodologies for various roof finishes of residential and light commercial structures are discussed in the following sections.

3.4.1 ASPHALT ROOF SYSTEMS/SHINGLES

Asphalt or composition shingles are the most common residential roofing material in the United States and cover approximately four out of five residential homes. They are also used extensively on light commercial structures. Asphalt shingles are durable, versatile, and more importantly, afford-able when compared with other common roofing materials such as those discussed in later sections. To further increase value and appeal, shingle manufacturers are designing and manufacturing shingles that mimic natural materials such as tile, wood, cedar shakes, or slate.[2]

While asphalt shingles are manufactured in a wide range of styles, there are generally four different types of asphalt shingles: dimensional, laminated, or architectural shingles; strip shingles, including multi-tab shingles (usually three- and four-tab); interlocking shingles; and large individual shingles (generally rectangular or hexagonal in shape and do not contain cutouts or tabs).[2] The most common asphalt shingles observed in the field are either three-tab or dimensional shingles with the trend toward the use of dimensional or laminated shingles.

Although the specific construction of asphalt shingles varies with differing styles, the basic materials and manufacturer processes used are relatively consistent. The base of most all shingles is formed from a reinforcement layer (or base mat) typically composed of organic felt or fiberglass. A binder layer, consisting of asphaltic bitumen, is then sprayed onto the mat. This serves as a waterproofing material and to bond granules to the reinforcement layer. The granules (crushed stone) are then bonded to the binder layer to help shield sublayers from ultraviolet radiation, reflect heat that could permeate into the building, add weight to the shingle, and provide color and aesthetic appeal. Dimensional shingles are manufactured by adding a partial second layer in spots to provide a more appealing roof surface.

Since asphalt shingles are the most common residential roofing material in the United States and are also commonly used for light commercial roofing, the number of claims to replace "hail-damaged" asphalt roof surfaces far outweighs that of any other roof covering. For roofing contractors, potential and actual hail-damaged roof systems provide very significant business opportunities. As a result of these opportunities and in some cases a lack of knowledge as to what constitutes hail damage, forensic engineers and scientists are often employed to help resolve conflicts between insurance companies and owners. The following paragraphs provide definitions on functional damage to asphalt shingles and then give example photographs, taken from site investigations, first of functionally damaged shingles that were caused by hail and then of non-hail defects that are commonly mistaken for hail damage.

Damage to asphalt shingles, as it pertains to hailstone impacts, can be classified as functional or cosmetic. Hailstones may leave impact marks and/or displace granules without affecting the functionality of the shingle. The two conditions of the industry standard definition for functional damage to asphalt shingles (as defined previously in Section 2.4.4) refer to impact-caused damage that:

- Ruptures the shingle reinforcement (i.e., organic or fiberglass mat).
- Causes a significant displacement of granules sufficient to expose the underlying asphalt to ultraviolet radiation.

Each of the functional damage conditions stated above is discussed in further detail below.

3.4.1.1 Functional Damage Condition #1: Ruptured Shingle Reinforcement

When sufficiently sized hailstones impact an asphalt shingled roof surface and cause rupturing of the organic or fiberglass mat, the water-shedding capability has been compromised and the shingle is functionally damaged as a result. These impact spots on an asphalt shingle are commonly referred to in the industry as a "hail-strike bruise" and are typically discovered by visual means and/or by pressing downward on the shingle surface in damaged areas to test the firmness of the shingle mat. The threshold size for damage to asphalt shingles from hail impacts begins with a hailstone measuring 1.5–2.0 inches in diameter[3]; others have reported thresholds of damage to light asphalt shingles (e.g., builder's grade – 20 to 30-year warranty) from hailstones measuring down to 1.0 inches in diameter,[4] but field experience[5,6] suggests that the higher threshold is more representative for functional damage from actual hailstones. Typically, the number of hail-strike bruises will be minimal (i.e., 0, 1, or 2) as the size of the impacting hailstones reaches the functional damage threshold, then as size increases, the number of bruises increases exponentially and jumps to higher intensities (i.e., >10). These lower counts are questionable given the size of hail vs damage arguments, but cannot be ruled out because of the shape and size of the defect(s). Experience suggests that asphalt shingle bruise counts for shingles not beyond their normal life have thresholds associated with hail size; that is, the bruise counts are zero or very low, until the hail reaches a sufficient size at which time the count jumps well above ten bruises per square.

Examples of hail-caused rupturing of asphalt shingle mats are given in Figures 3.2 and 3.3.

FIGURE 3.2   Hail-strike bruise to asphalt shingle that ruptured fiberglass reinforcement – top side (a) and fractured mat (b).

FIGURE 3.3   (a) Hail-strike bruise to an older asphalt shingle that ruptured the shingle reinforcement. (b) Another example of a hail-strike bruise to an older asphalt shingle that ruptured the shingle reinforcement.

Functional damage from a hailstone impact, or a "hail-strike bruise," to asphalt shingles that have not reached the end or their normal useful service life typically bears the following physical characteristics:

- Diffuse looking pattern of granule loss with a circular to oblong shape.
- Relatively smooth edges at the impact zone (not sharp edges often seen by blister defects).
- Residual granules embedded into the asphalt mat near the center of the damage area (rarely does the damage area or bruise result in loss of all the granules in the bruise area).
- The impact area is soft to the touch in comparison to unaffected areas.
- Rupturing of the shingle mat as evidenced by looking at the bottom side of the shingle (see above).

Despite arguments often heard in the field, experimental work suggests that any hail-strike impact bruise damage will not appear later or worsen with time.[7]

Hail-strike bruises are reported as "intensities," which are calculated per square of roof surface area (100 square feet) for each directional roof elevation of the residential or light commercial structure and are, generally, easier to report in a table format such as Table 3.3.

TABLE 3.3

Example Table for Hail-Strike Bruises and Intensities

Bruise Count I.D.	Elevation	Dimensions[a]	Area[a] (square feet)	Total Defects	Hail Bruise(s)	Bruise Intensity[b]
A	North	12′×10′	120	32	5	4.2
B	West	12′×10′	120	20	5	4.2
C	East	15′×8′	120	17	0	0.0
D	South	12′×10′	120	18	0	0.0

[a] All dimensions are approximate.

[b] Per 100 square feet of roof area.

Shingles that have exceeded their natural life (i.e., are degraded, brittle, and weathered) will demonstrate a different type of damage than shingles not beyond their normal service life, as discussed above. In these situations, the shingle will be shattered with the damage reflected by chunks of granules being displaced (Figure 3.4).

(a)   (b)

FIGURE 3.4 Hail-strike bruise to older asphalt shingles – overview (a) and displaced chunks of granules (b),

Note that shingles more poorly supported (e.g., cap shingles and shingles covering ridge vents) will be damaged by smaller-sized hail than more supported shingle. Referring to Chapter 2 (Section 2.4.1 "Physics Lesson – Basics of Energy Transfer and Coefficient of Restitution") and discussions earlier in this chapter, the less-supported shingles reduce the coefficient of restitution of the shingle and allow for more transfer of energy into the shingle, thus making it more susceptible to functional hail damage below typical-size thresholds (Figure 3.5):[3]

FIGURE 3.5   Hail-strike bruise to less-supportive ridge vent shingle.

3.4.1.2 Functional Damage Condition #2: Significant Granular Loss Exposing the Underlying Asphalt

The loss of granules from asphalt shingles is a common and normal characteristic of shingles. Over the lifetime of a shingle, granule loss occurs gradually as a result of aging, rainfall, ice, snow, wind, and hail. Loss of granules, often seen in the gutters, will lead a roofing contractor or property owner to believe the asphalt roof surface was damaged by hailstone strikes. Granule loss by itself is not generally considered functional damage or to have an impact on the life of a roof surface[7] and is considered to be cosmetic damage because neither the water-shedding capability of the roof system nor the service life of the shingle has been compromised.

Though significant granule loss from hailstone impacts would leave the underlying asphalt binder exposed to ultraviolet radiation and potentially cause further deterioration and reduction of service life, research indicates that the long-term effects of the exposure from granule loss do not compromise the functionality of the shingle. In a study by Haag Engineering Company spanning 15 years, new and weathered asphalt shingles were impacted with simulated hailstones in order to analyze the long-term effects of hailstone impacts. Dents caused by impacts that did not initially rupture the shingle mat did not change measurably over the 15 years of the study and therefore were not considered functional damage.[7] Specifically, for three-tab shingles, the research suggested that hail-strike impacts that dislodged granules did not expose the asphalt mat or affect the service life of the material and therefore were considered not functional damage. Similarly, for laminated shingles tested, hail-strike impacts associated with the simulated hailstones also dislodged granules but did not expose the asphalt binder. With exposure to natural weathering over time, the impact areas for the laminated shingles did shed additional granules, but on closer inspection, a bed of granules was immersed in the asphalt and continued to protect the asphalt and reinforcement. Thus, this too was not considered functional damage.

The results of the study appeared clear: (1) functional hail damage to asphalt shingles (new or weathered), which ruptured the shingle reinforcement, was immediate and identifiable and did not appear to worsen with time, and (2) granular loss as a result of hailstone impacts did not affect the functionality of the shingles.

It should be noted that some confusion exists in the area despite available research findings because shingle manufacturers sometimes indicate that granule loss is damage. The nature of this damage is not stated (i.e., cosmetic or functional) nor is a basis for the statement provided.

3.4.1.3 Correlation of Hail Damage to Asphalt Shingles to Hail Size, Pitch of Roof, and Directions of Roof Slopes Relative to an Oncoming Hailstorm

As previously discussed in Chapter 2 and in earlier sections of this chapter, functional hail damage to roofing materials encompasses several variables (see Section 2.4.5). Of particular importance when evaluating a particular roof system is: (1) the size of the hail, (2) the direction from which the hailstorm arrived, and (3) the angle of hailstone impact/roof pitch with variables (2) and (3) being closely related (see Section 2.4.5.1).

Shingles will sustain differing levels of damage from hailstone impacts (assuming the hail that impacted the site was of sufficient size to begin to cause functional hail damage), depending upon the slope or steepness/pitch of the roof elevation and its relation to the directionality of the oncoming storm.

EES Group, Inc. conducted a detailed analysis using data from 729 hail damage inspections over a 5-year period, which resulted in the following correlations pertaining to the variables listed above[5]:

- *Size of hail vs bruise counts to asphalt shingles*: The estimated maximum size of hail must be between 1.75 and 2.0 inches in diameter to cause increased levels of hail-strike bruises per roof elevation (i.e., greater than ten hail-strike bruises per 100 square feet).
- *Bruise count vs slope of roof*: The hail-strike bruise count vs roof pitch (steepness) suggests that shallow and steep roofs (defined as those measuring below a 4:12 pitch and steeper than 9:12, respectively) had more hail-strike bruises than moderately sloped roofs (roofs with 5:12 to 9:12 pitches).
- *Bruise count on slope facing the hailstorm vs bruise count on the opposite face*: The shingles on roof elevations facing toward an incoming storm contained almost 2.5 times as many hail-strike bruises as roof elevations facing away from the incoming storm.
- *Bruise count on slope facing the hailstorm vs bruise count on the perpendicular faces*: Roof elevations facing toward an incoming storm contained over 2.0 times as many hail-strike bruises as roof elevations perpendicular to the incoming storm.

Of particular interest were the results from the second bulleted item: more hail-strike bruises were discovered on low-sloped (<4:12 pitch) and steeper-sloped (>9:12 pitch) than moderately sloped roof elevations. One explanation for this result may be that hailstorms typically have either hail falling essentially vertically (not influenced by horizontal wind vector components) or heavily wind-driven hail (heavily influenced horizontally by wind – Figure 3.6). Note the results from this analysis are consistent with earlier sections discussing the importance of hailstorm directionality, angle of impact, and perpendicularity of hailstone impacts (see Section 2.4.5.1).

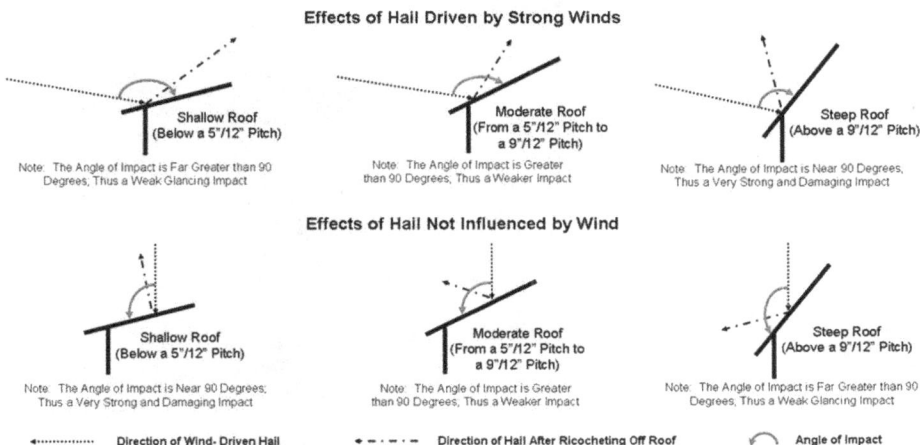

FIGURE 3.6 Effects of hail driven by strong winds and uninfluenced by winds on roof slopes of differing pitches.[5]

3.4.1.4   Non-Hail-Strike Damage to Asphalt Shingle Roof Surfaces

Various internal and external forces acting on a roof covering such as asphalt shingles may be mistaken for damage from hailstone impacts. As stated in the introductory paragraphs of this section, there are discrepancies in opinions on what constitutes a hail-strike bruise or hail damage to an asphalt shingled roof system. It is common for roofing contractors to either incorrectly attribute such damage to hail or sometimes fraudulently (e.g., State Farm Fire & Casualty vs Radcliff)[8] create simulated hail damage defects (see also Chapter 5 – Synthetic Storm Damage (Fraud) to Roof Surfaces) since they stand to benefit monetarily if the building owner's insurance provider approves the claim for roof replacement.

The ability to differentiate between damage caused to surfaces by hailstone impacts and that caused by other forces (internal or external) is critical in performing accurate hail damage assessments. Oftentimes, non-hail surface defects present on a roof surface are the result of one or more of the following factors/conditions and not attributable to hailstone impacts: age, thermal degradation, weathering, shingle quality, inadequate attic ventilation, and synthetic hail damage (fraud). Examples of common defects to asphalt shingles not likely attributable to hailstone hail-strike impacts are provided in Table 3.4.

TABLE 3.4

Non-Hail Defects to Asphalt Shingles

Non-Hail Defect	Photograph
Blister	
Flaking (lighter amounts)	

(*Continued*)

TABLE 3.4 (*Continued*)
Non-Hail Defects to Asphalt Shingles

Non-Hail Defect	Photograph
Flaking (heavy amounts) Typically indicative of insufficient attic ventilation and/or aging of shingles (at or near end of effective service life)	
Aged and heavily degraded Typically indicative of insufficient attic ventilation and/or aging of shingles (at or near end of effective service life)	
Craze cracking Typically indicative of insufficient attic ventilation	

(*Continued*)

TABLE 3.4 (*Continued*)
Non-Hail Defects to Asphalt Shingles

Non-Hail Defect   **Photograph**

Clawing

A downward curling at the corners of the
shingles. Clawing commonly occurs when
the bottom of the shingle shrinks relative to
the top of the shingle

Curling/cupping

Nail pop

Result of raised nail from decking movement
during thermal cycling rubbing against
underside of shingle

(*Continued*)

TABLE 3.4 (*Continued*)
Non-Hail Defects to Asphalt Shingles

Non-Hail Defect	Photograph
Vertical/diagonal cracking Created by internal forces due to thermal cycling at locations above adjoining shingles	
Defects caused by lichen growth Created when lichen embeds and subsequently dislodges from shingle surface. Differentiated from percussive forces in that the underlying mat is not fractured	

(*Continued*)

TABLE 3.4 (*Continued*)
Non-Hail Defects to Asphalt Shingles

Non-Hail Defect	Photograph
Mechanical cuts	
Artificial/synthetic hail damage (fraud) See Chapter 5	

Some of the examples in Table 3.4 have been argued by roofing contractors and property owners to be caused by hail, but the lack of fracturing to the shingle reinforcement beneath the defect suggests that the defects were caused by internal forces and not by external forces such as hail impacts. Descriptions of several of the non-hail damage defects shown in Table 3.4 and common to asphalt shingles follow.

3.4.1.4.1 Blistering

Blistering on the surface of a shingle is commonly mistaken for hailstone impact damage. These surface defects are caused by volatiles in the asphalt binder or moisture from the shingle mat being vaporized due to high temperatures. Though these may appear to be the result of a hailstone impact (to the untrained eye), the following physical characteristics deviate from those of a functional hail-strike bruise:

- There is a surface void in the shingle with missing granules and asphalt binder.
- The edges of the blisters are sharp to the touch and not smooth as with hail-strike bruises.
- The underlying mat will not contain a fracture indicative of an impact.

Blistering is typically more pronounced on southern-facing roof surfaces due to increased direct sun exposure and, therefore, higher shingle temperatures (note this is only true for homes in the northern hemisphere; the opposite would occur in the southern hemisphere). Blistering defects can also commonly occur to homes with inadequate attic ventilation, which does not allow for the proper removal of excess heat and moisture that builds up in the attic spaces. Regardless of cause and location, blistering can be of two basic types: a small rash type 0.75 inches or less in size or a larger tent type. Larger tent-type blisters may cause premature failure of the roofing material. The smaller rash-type blisters affect the appearance of the roof only and do not necessarily shorten the life of the roofing material.[9]

3.4.1.4.2 Buckling

Buckling is usually caused by improper spacing of shingles. If the ambient temperature is hot enough, shingles will expand, and if the spacing is too tight to allow for the expansion, the shingles will have no place to go but up. Buckled shingles will have a rippled or wavy appearance and will not usually subside with time.

3.4.1.4.3 Cracking

Cracking can be caused by hailstone impacts or exposure to ultraviolet light and resultant shrinkage and should not be attributed to hail when caused by the latter. Ultraviolet exposure causes the lighter hydrocarbons in the bitumen mix to break down, volatilize, and outgas. The loss of this material then causes the affected material to shrink creating the cracked appearance.

These cracks tend to be relatively uniform in distribution across elevations that receive exposure to sunlight. They generally initiate at the top surface and diminish in width with depth into the material, where ultraviolet light cannot penetrate.

The interior of the cracks will appear to be weathered, oxidized, and/or may contain wind-borne debris. This is because the cracks open slowly, as opposed to those caused by hail, which open immediately upon impact. Unlike those caused by hailstone impacts, cracks caused by weathering will be more prevalent on elevations that receive more exposure to sunlight and/or with poor attic ventilation.

3.4.1.4.4 Curling and Cupping

Curling and cupping constitute the drying out of the topmost layer of the shingle, resulting in an upward curling, which is exacerbated by exposure to ultraviolet light. The top layer dries, resulting in a reduction of mass that subsequently causes shrinkage. The differential shrinkage between the top and bottom layers creates tension that causes the shingle to curl or cup. The susceptibility of a shingled roof to cupping and curling is increased if the shingles are nailed too high, too far from the edge, or if too few shingle fasteners were installed. Shingles that are cupped or curled are more susceptible to hailstone impacts because the curled portion of the shingle is unsupported.

3.4.1.4.5 Clawing

Clawing is similar to curling/cupping; however, the bottom layer of the shingle dries out causing differential shrinkage between the top and bottom layers, creating tension, but in the opposite direction causing the shingles to claw or the edges of the shingle to curl downward toward the roof surface.

3.4.1.4.6   Granule Loss

Granule loss is another condition that can be caused by hail or other forces. Granules can shed over time due to expansion and contraction of a shingle along with other factors, including, but not limited to, rainfall. As discussed at length above, granule loss should not be attributed to hailstone impact if the shingle mat/reinforcement is free of a fracture.

3.4.1.4.7   Defects/Degradation the Result of Insufficient Attic Ventilation

The incorporation of a properly ventilated attic space is one of the most important design considerations as it pertains to ensuring the maximum service life of roof coverings and roof assemblies, not to mention the reduction in energy costs and the prevention of ice damming. If premature degradation of the roof due to insufficient or inadequate attic ventilation is suspected, the number and size of soffit, gable, and roof vents should be measured for potential future use in attic ventilation calculations. For a more detailed explanation of attic ventilation, refer to Chapter 12.

3.4.1.4.8   Artificial/Synthetic Hail Damage (Fraud)

On occasion, when inspecting an asphalt-shingled roof surface for functional damage attributed to hail impacts, damaged areas are discovered that appear to have been artificial or man-made. These defects are made in an attempt to mimic hail-caused bruises in the hopes that they will affect the outcome when the determination for roof replacement is warranted. Generally, artificial or man-made damage to asphalt shingles has similar characteristics and exhibits patterning atypical of randomly distributed hailstorm events. Refer to Chapter 5 for further information regarding artificial or synthetic hail damage to asphalt shingles.

3.4.2   Roll Roofing Roof Systems

Roll roofing (i.e., strips of granule covered asphalt felt) is sometimes encountered by an inspector during residential and light commercial inspections. These systems are commonly found on low-sloped roof areas below which point shingles are not allowed to be installed.

These roof systems often leak due to improper installation of underlayment. The materials may be confused with modified bitumen (mod-bit) roof surfaces (see Chapter 4); however, roll-roofing material can be distinguished from mod-bit finished surfaces in that the roll roofing will typically rip easily when a corner is torn, whereas mod-bit roofing typically will not tear easily. The inspection of hail-strike functional damage to roll-roofing surfaces should be completed in the same fashion as asphalt shingle roof finishes, due to their similar construction characteristics (i.e. use of representative test squares), and they typically have similar-size thresholds for functional damage (i.e., 1.5–2.0 inches in diameter hailstones).

3.4.3   Wood Shake and Shingle Systems

Wood roof surfaces provide residential and light commercial property owners with an alternative roofing material that helps give their property a rustic or earthly appearance. Wood roof surfaces can last from 30 to 40 years dependent on the grade and quality of the roofing product, when it is properly installed and maintained, and the climate in which it is located. The most common type of wood used in the industry is Western Red Cedar. Wood roof coverings are designated as wood shakes or wood shingles.

The following discussions provide background information on definitions and guidelines for wood shakes and shingles then goes into detail on differentiating between hail-caused functional damage and defects not created by hail impacts.

3.4.3.1 Definitions and Guidelines

The Cedar Shake and Shingle Bureau (CSSB) is the trade organization for the wood roofing industry, and it offers numerous resources regarding general information, installation, care and maintenance, and quality control. The two major distinctions between shakes and shingles are the way they are manufactured and installed[10,11]:

- Shakes are typically created from splitting a cedar block (on one or both sides) although some shakes are taper-sawn and sawn on both sides. As a result of splitting shakes, one end of the shake is typically thicker than the rest of the shake, which gives a measurable distinction between shakes and shingles. Shakes typically range from 0.5 to 0.75 inches at the exposed end. On the other hand, shingles are always sawn on both sides. This gives shingles a uniform thickness throughout the length of the wood.
- Shakes are often installed with felt between each layer and are often interlaid to provide two-ply thickness (i.e., interlayment), which results in greater exposure (greater than one-third of the shake is exposed) and decreases the number of shakes required for the roof surface. Shingles, however, require only one felt layer over the decking beneath the shingles for the entire elevation of the roof. Shingles also must overlap each other to produce a three-ply roof, which decreases exposure (typically less than one-third the length of the shingle) and increases the number of shingles required for the roof.

Cedar shakes and shingles are also standardized into different types of classifications, depending on the quality and grade of the wood. Grading of the wood shakes and shingles is determined by the type of cut (i.e., edge grain, slash grain, or flat grain) and the locations of defects such as knots, sapwood, width, etc. Industry best practices recommend both cedar shakes and shingles to be a minimum of No. 1, or premium grade wood with limitations on face defects, edge grain/flat-grain percentages, and dimensions, although lower grades are available.[10,11]

3.4.3.2 Inspection for Functional Hail-Strike Damage to Wood Roof Surfaces

Functional hail damage to wood shakes and shingles is characterized by a distinct impact mark (known as a "peck mark") coincident with a fresh split in the wood. The fresh appearance of the impact mark and the internal surface within of the split indicate that it was caused by a recent hailstone impact (Figure 3.7).

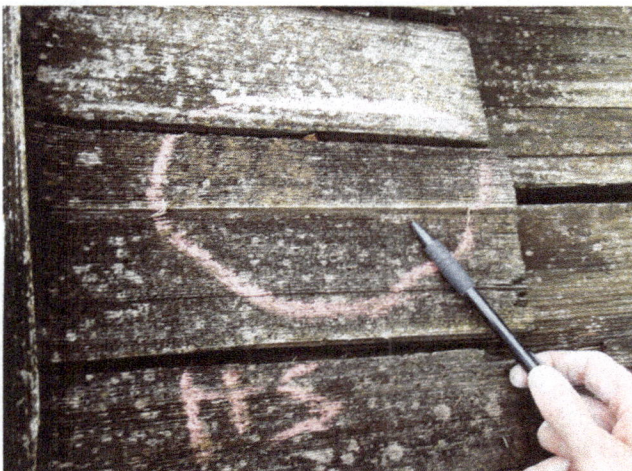

FIGURE 3.7   Hail-strike split to wood shake – impact damage.

Laboratory studies conducted by Haag Engineering Company[12] and the National Bureau of Standards[13] have concluded that hailstones must reach a diameter of ~1.0 inches to split a wood shingle in poor condition and 1.25 inches to split a wood shake in good condition. The Roofing Industry Committee on Weather Issues, Inc. (RICOWI) reported splitting or puncturing of wood shingles and shakes that occurred when hailstones exceeded 1.5 inches in diameter.[6] The lower threshold is for shakes or shingles affected by fungal rot, erosion, cupping, or curling.

In order to determine the extent of hail-caused damage to the wood shakes or shingles accurately, there must be a close examination of each shake/shingle within each representative test square. Each individual shake/shingle should be inspected closely and hand-manipulated to detect splits. Any areas of hail damage should be circled with chalk and a designated letter is written nearby to mark it. Non-hail damages may also be marked, but with a different letter or designation, differentiating it from hail damage defects.

The number of split shakes or shingles associated with hail-strike peck marks noted by the inspector will often total more than those caused by hail-strike impacts due to splits caused by foot traffic from roofing contractor employees, insurance inspectors, and sometimes the property owner themselves. Thus, the number of splits observed to wood shingles and shakes during a hail-strike damage inspection will likely be a maximum since some of the splits were likely caused by foot traffic after the storm rather than from the hail-strike impacts.

Peck marks that do not split the wood initially will not cause delayed cracking, are not considered functional damage, and will not shorten the expected service life of a wood shake/shingle (Figure 3.8).

FIGURE 3.8   Hail impact mark with no split to wood shake.

In fact, these impact marks, or dents, are often undetectable after normal weathering has taken its course, allowing the fibers that were compressed during the impact to recover due to normal moisture absorption. In some instances, where the hail is minor, no mark may be left on the wood roof at all.[11]

Within each representative test square, if there are numerous defects and splits in the shakes/shingles, there must be an emphasis on a close examination of each to determine whether or not they were caused by hail impacts. Two common splits observed that are not the result of hail impacts are "foot falls" and "weathered chips/splits," each of which is discussed further below.

"Foot Falls" are defined as fresh splits in the shakes/shingles, which are caused by foot traffic and not by hail (Figure 3.9).

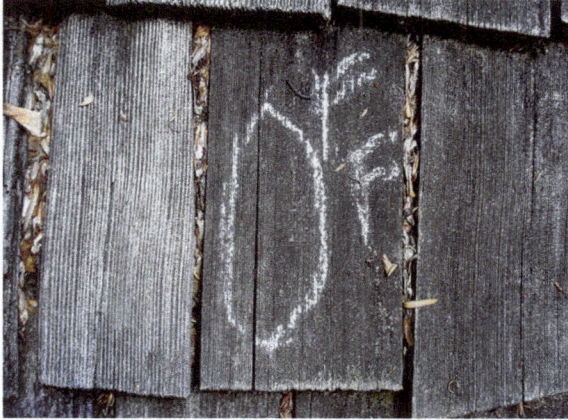

FIGURE 3.9 Fresh foot fall to wood shake (no evidence of impact mark).

Foot falls are differentiated from hail-strike splits and characterized by the following:

- The wood surfaces within the split are fresher in appearance than the exposed surfaces; however, the shake/shingle can be pieced back together when manipulated by hand.
- There is no evidence of hail impacts (i.e., dents or impact marks) along the split.
- The affected shake/shingle is located on the roof in a typically trafficked area (e.g., along ridges and valleys).

Foot falls are commonly observed when a roof is undergoing damage investigations and is associated with several parties walking on the roof surface.

"Weathered Chips/Splits" are defined as defects, which were not, to a reasonable degree of certainty, the result of hail impacts. Weathered chips/splits can be differentiated from hail-strike splits and are characterized by the following:

- The weathered appearance (i.e., grayer color) of the wood surfaces within the split, which will be similar in appearance to the exposure of the shake/shingle.
- The chip/split will have rounded and weathered edges.
- There is no evidence of hail impacts (i.e., dents or impact marks) along the split.*

*During the course of the assessment, there may be evidence of a fresher hail impact mark along a weathered split in a shake/shingle. The rounded edges of the split and the weathered appearance to the wood within the split indicate that the split was not the result of the recent hail impact (Figure 3.10).

FIGURE 3.10 (a) Weathered split wood shake with fresh impact mark. (b) Weathered split wood shake without fresh impact mark.

Hail-strike splits are reported as "intensities," which are calculated per square of roof surface area (100 square feet) for each directional roof elevation of the structure and are generally easier to report in a table format such as Table 3.5.

TABLE 3.5
Example of Reported Wood Shake Split Counts and Intensities

ID	Elevation	Dimensions[a]	Area[b] (ft²)	o al Splits	Foot Falls	Weathered Splits	Probable Hail Splits	Hail Split Intensity[b]
A	SW	15′×8′	120	>50	2	>50	1	0.83
B	NE	15′×8′	120	36	0	35	1	0.83
C	SE	(14′6″+6′)/2 x 12'	123	>100	4	>100	1	0.81
D	NW	15′×8′	120	>100	2	>100	0	0.00

[a] All dimensions are approximate.
[b] Per 100 square feet of roof surface area.

3.4.3.3   Importance of Care and Maintenance

Proper care and maintenance are important to ensure the functionality and increase the longevity of any roofing material, but are of particular importance for wood roof systems. Cedar wood roof surfaces need to breathe and, therefore, need to be kept clean of any accumulations of debris that will affect the life span by not allowing the wood to properly dry out.[10,14] This includes buildup of tree debris (i.e., leaves, branches, etc.) from overhanging branches of nearby trees and biological growth, in particular moss growth. Moss retains moisture, which can harm the wood over time and cause rot.

Fungal rot softens the material (i.e., reducing the wood's coefficient of restitution and impact resistance), which increases the susceptibility of a wood shingle to damage from a hailstone impact. Fungal rot is typically located on northern roof elevations, roof surfaces beneath overhanging tree branches, and along the lower butt edges of the shakes and shingles.

Of particular importance is the fungal growth along the lower, "butt" edges of shakes/shingles due to their nearly constant shaded condition (Figure 3.11). If left in place and not maintained, the fungal growth will soften the lower edges of the wood and cause it to become eroded, split, and brittle over time, creating a condition known as "butt rot" (Figure 3.12).

FIGURE 3.11   Heavy fungal growth along lower butt edges of wood shake.

FIGURE 3.12   Severe "butt rot" to wood shakes.

Additional defects common to wood shakes and shingles, which are not the result of hail impacts, are given in Table 3.6.

TABLE 3.6
Examples of Non-Hail Defects to Wood Shakes/Shingles

Non-Hail Defect	Photograph
Displaced and missing wood shingles	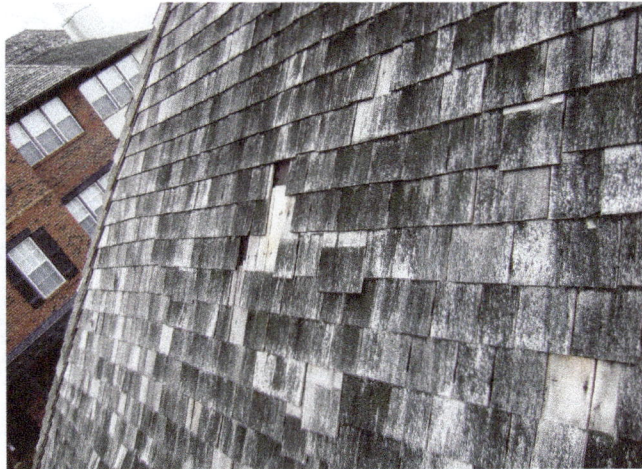

(Continued)

TABLE 3.6 (*Continued*)
Examples of Non-Hail Defects to Wood Shakes/Shingles

Non-Hail Defect	Photograph
Curled wood shingles	
Curled and cupped wood shingles	
Rotted and eroded wood shakes	

(Continued)

TABLE 3.6 (*Continued*)
Examples of Non-Hail Defects to Wood Shakes/Shingles

Non-Hail Defect	Photograph
Eroded and missing wood shingles beneath an overhanging tree	
Heavy lichen and/or biological growth	

3.4.3.4 Repairing Wood Roof Surfaces

One of the beauties of wood roof surfaces is the fact that they can typically be repaired rather than completely replaced, given the nature of the material. However, considerations should be made to replace the entire roof system if the costs of repair exceed 80% of the total replacement cost.[10] However, the cost of total roof replacement must take into account all factors, including the possible need for re-decking or even substrate reinforcement for load bearing capacity. The repair vs replacement decision can be correctly provided with proper assessment protocols.[11,15]

Individual wood shakes or shingles that have been functionally compromised by hail-strike splits can be replaced using the proper tools and methods. Refer to repair and replacement procedures outlined by the CSSB.[11,14] In the course of repairing hail-damaged wood shakes, the rule of thumb is that an additional shingle or shake will be damaged during the repair activities. This additional damage should be accounted for in repair estimates.

3.4.4   SLATE, CLAY, CONCRETE, AND ASBESTOS TILE ROOF SYSTEMS

These particular roof systems (hereinafter referred to as tile systems) are often not seen due to geographical preferences. Nevertheless, the forensic investigator will encounter them from time to time and must be prepared to determine the extent of hail-strike damage to these roof systems. The most common tile roof systems are:

- Slate.
- Clay.
- Concrete.
- Fiber Cement.
- Asbestos.

These roof systems have been utilized throughout history for their impressive water-shedding capabilities and the ability to withstand years of continuous weathering. For example, it is believed that good-quality clay roof tiles have a typical service life of 70 years or longer while slate tiles can have a life span of up to 150 years, depending on where it was quarried. In the case of slate tiles, Vermont green slate tile is more durable, longer-lived, and more resistant to hailstone impacts than Pennsylvania black slate tile, which is a relatively softer slate with a somewhat shorter life span.

Aside from the basic strength and durability of the base tile materials, the effectiveness and longevity of these systems as a water-shedding assembly and its functionality depend upon whether proper installation details were followed, as well as care and maintenance. For example, the life of a clay tile roof system is dependent on the life of the felt below the tile; the tiles simply protect the felt. These felts are designed to last between 75 and 175 years, depending on the slope of the roof (steeper is better). When replacing the tile, it is critical that the underlying felt be inspected and repaired, when necessary.

3.4.4.1   Tile Roof System Inspection Methodology, Definitions, and Guidelines

Unlike the methodology used with asphalt shingles and wood shakes/shingles, in order to determine the extent of hail-caused damage to tile roof systems accurately, there must be a close examination of each tile on each roof elevation of the structure and not just within a representative test square. Then, from area measurements of the roof and the tiles (exposures and widths), the number of tiles covering the roof surfaces can be estimated and an approximate percentage of hail-damaged tiles can be calculated, which can aid in further repair protocol.

Functional damage to a roof system, as originally given in Chapter 2 (see Section 2.4.4) and reiterated throughout this chapter, is (1) a reduction of its water-shedding capabilities and/or (2) a reduction in the expected long-term service life of the roof material. The size threshold at which functional damage typically begins to occur to tile roof systems is when hailstones are ~1.5–2.0 inches in diameter, depending on the area of the tile impacted, which have differing levels of vulnerability.[13] Examples of the hail diameter needed to cause threshold damage to various tile roofing materials are summarized in Table 3.7.

TABLE 3.7

Threshold Hail Size for Hail-Strike Damage to Tile Roofing Materials

Tile Material	Threshold Hail Size (Inches)	References
Typical tile product (13 types of tile products)	1.5–2.0	[6,16]
Asbestos cement	1.5–2.0 (edge)	[13]
	2.0 (center)	
Asbestos cement	1.5 (no fractures)	[16]
	1.75 (corners began breaking)	
	2.0 (fractures)	
Clay	1.25–1.5 (some breaking of corners of tile)	[16]
	1.5 (shatter)	
Clay	1.5 (threshold)	[4,12]
Clay (red)	1.75 (unsupported)	[13]
	2.0 (center)	
Wood-fiber cement	1.5	[16]
Concrete	1.0 (none damaged)	[16]
	1.25 (4 of 13 had corners damaged	
	1.50 (7 of 13 damaged)	
	2.50 (all tiles broken)	
Concrete (most)	1.5 (threshold)	[12]
Concrete (most)	1.75 (threshold)	[13]
Concrete (most)	2.0 (threshold)	[4]
Concrete (red, gray)	2.5 (threshold)	[17]
Slate	1.5 (threshold)	[4]
Slate	1.5–2.0 (crack thresholds)	[13]

For roof tile to be considered functionally damaged by hail impacts, there must be evidence of any of the three following conditions:

- Penetration or puncture through the tile.
- Split in the face of the tile or a significant chip on the edge with evidence of a hailstone impact mark.
- Discernable impression left behind that broke through significant surface layers without piercing the tile.

Each tile roof system listed above has unique characteristics that make it more desirable, whether it is functionality or aesthetic appeal. Examples of hail-caused functional damage to tile roof systems are shown in Table 3.8.

TABLE 3.8

Examples of Functional Hail-Strike Damage to Tile Roof Systems

Hail Defect	Photograph
Hail-strike penetrations and impressions to slate tiles Note heavier deterioration of damaged tiles	
Hail-strike chips to slate tile edges Repaired with metal bibs	
Hail-strike split/fracture to slate tile Note impact point at top of split/fracture	

(*Continued*)

TABLE 3.8 (*Continued*)
Examples of Functional Hail-Strike Damage to Tile Roof Systems

Hail Defect   **Photograph**

Hail-strike split to slate tile
Note sharp/fresh edges and impact mark

Hail-strike penetration to asbestos tile
Note fresher (i.e., lighter) edges

Hail-strike split to clay tile
Note location of fresh impact mark along split

Oftentimes, upon examination of the tiles, there are holes observed in the face of the tiles, particularly with slate. A common misconception would be to consider these holes as hail-caused penetrations since a hailstorm might have passed through the area recently; however, in order to determine whether they were created by an external impact, such as hail, or from beneath the tile, the hole must be closely examined. When slate is perforated or punctured, the impact leaves a hole with clean and sharp edges on the impact side of the tile and a cratered hole on the side opposite the impact[18,19] (Figure 3.13).

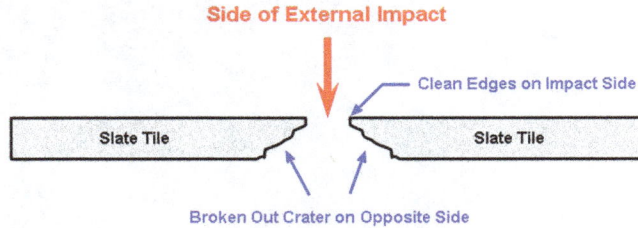

FIGURE 3.13   Punctured slate tile – clean hole on impact side and cratered hole on opposite side.

An example of this phenomenon is when slate tile manufacturers create fastener holes in the tiles. During this process, the back side of the tile is punctured, creating a hole with clean edges on the back side and a cratered or concave hole on the front side, which effectively creates a countersink for the nail fasteners.[19]

Oftentimes, when hailstones are not of sufficient size to cause functional damage to tile roof systems, spatter marks to the tile exposures will be observed, but neither penetrations, splits, chips, nor impressions will be present (Figure 3.14). This is *not* functional damage and will not likely shorten the expected service life of the tiles.

FIGURE 3.14   Hail impact spatter marks on slate tiles.

In some instances, when evidence of hailstones has not reached the size threshold to cause functional damage to tiles, penetrations, splits, and/or impressions can occur, but are most likely present to tiles that exhibit higher levels of deterioration. The case in which this scenario appears to be most common is with the long-term delamination of slate tiles. Delamination is the process by which the surface silicate layers of the slate tile separate from the tile and are shed from the roof over time, thus slowly thinning the tile, decreasing its impact resistance, and causing it to become more susceptible to functional hail damage from hailstone impacts. Typically, delamination is more prominent on southern-facing roof surfaces due to the increased sun exposure (Figure 3.15).

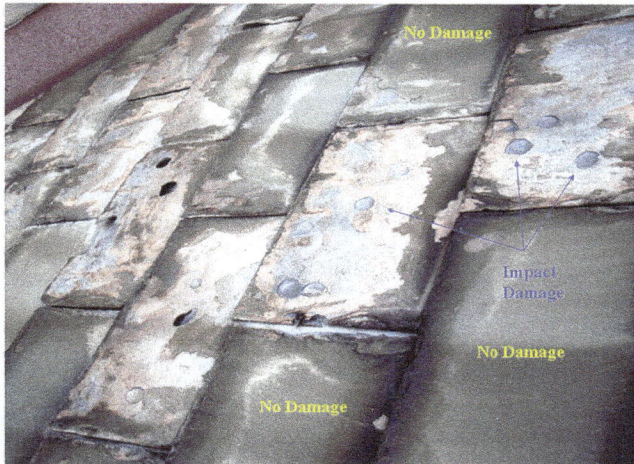

FIGURE 3.15   Hail damage to more deteriorated slate tiles.

On each roof elevation, there are potentially numerous defects (holes, cracks, chips, etc.) to the tiles and once again there must be an emphasis on a close examination of each to determine whether or not they were caused by hail impacts.

A somewhat unique tile system sometimes encountered is an asbestos tiled roof surface. Asbestos roof tiles often show cosmetic patterning from hail-strike impacts that remove a surficial layer or mold/algae/fungal growth and a fine layer of asbestos fibers. While readily visible, this appearance is not functional damage and will return to its original appearance with time as the roof surface re-weathers.

Since the methodology of evaluating tile roof systems differs slightly from that of other residential or light commercial roofing materials, such as asphalt shingles and wood roof systems, the level of functional hail-strike damage to the roof of a particular structure is not reported in terms of "intensities" but rather a percentage of roof surface area for each elevation inspected. It is often easier to report such levels of damage, as well as numbers and location of addition non-hail defects, in a table format such as Table 3.9. Note that preparing a table prior to the inspection can greatly facilitate the inspection process.

TABLE 3.9

Example Table of Summary of Roof Defects to Tiles by Elevation

ID	Elevation	Area (square feet)	Est. # Tiles	Weathered Chips	Non-Storm Cracks	Eroded Holes	Penetrations/ Splits/ Impressions	Missing/ Displaced Tiles	Total Defects	Tiles with Hail Damage (%)
1	South	48	171	6	3	3	1	1	14	0.58
2	North	48	171	3	0	0	0	0	3	0.00
3	East	453	1,606	14	12	6	0	3	35	0.00
4	West	76	270	17	4	5	58	0	84	21.48
5	North	300	1,064	11	9	3	1	2	26	0.09
6	East	65	231	3	1	0	0	0	4	0.00
7	West	65	231	9	3	2	4	1	19	1.73
8	North	94	334	4	3	0	0	1	8	0.00
9	East	65	231	4	2	0	0	0	6	0.00
10	West	65	231	7	3	0	24	0	34	10.39
11	North	293	1,039	11	9	2	3	2	27	0.29
12	East	69	245	4	3	4	3	1	15	1.22
13	West	433	1,535	28	14	12	215	9	278	14.01
14	North	55	195	3	1	0	0	0	4	0.00
15	South	55	195	3	4	2	4	3	16	2.05
Totals		**2,184**	**7,749**	**127**	**71**	**39**	**313**	**23**	**573**	**4.04**

3.4.4.2   Tile Roof System – Examples of Non-Hail Damage

Exposure to water, repeated freezing and thawing of moisture in the tile, biological growth, and other factors could cause surface deterioration or weathering defects to tiles. Table 3.10 provides examples of common defects encountered during hail damage evaluations, which were not, to a reasonable degree of engineering/scientific certainty, the result of hailstone impacts.

Some of the examples in Table 3.10 have been argued by roofing contractors and property owners to have been caused by hail, but the lack of sharp edges and fresher-in-appearance (i.e., lighter) surfaces of the split or defect suggests that the defects were not likely the result of hail impacts. Descriptions of several of the non-hail damage defects shown in Table 3.9 and common to tile roof systems follow.

TABLE 3.10

Examples of Non-Hail-Related Damage to Tile Roof Systems

Non-Hail Defect	Photograph
Non-hail crack to slate tile Note no evidence of impact mark along crack	
Displaced and missing slate tiles	
Eroded hole to slate tile Note location over nail head	

(*Continued*)

TABLE 3.10 (*Continued*)

Examples of Non-Hail-Related Damage to Tile Roof Systems

Non-Hail Defect	Photograph
Eroded hole to slate tile Note weathered appearance and no sharp edges	
Weathered chip to slate tile	
Chip to clay tile Note weathered appearance	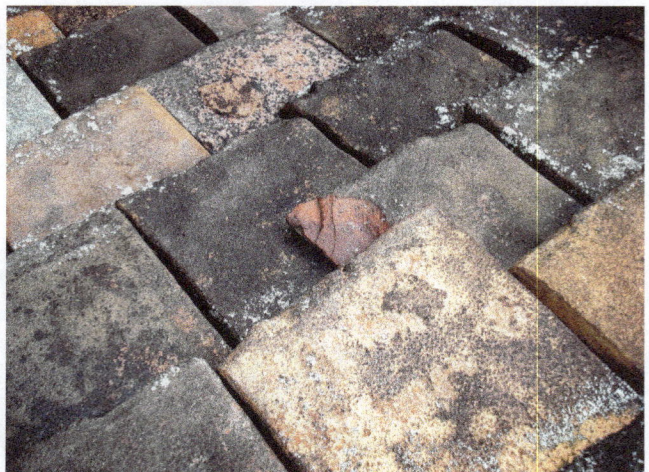

(*Continued*)

TABLE 3.10 (*Continued*)

Examples of Non-Hail-Related Damage to Tile Roof Systems

Non-Hail Defect	Photograph
Surface pitting and spalling to clay tile Attributable to repeated freeze/thaw cycling	
Foot fall to clay tiles Fresh cracks likely a result of foot traffic	
Eroded hole with weathered crack in clay tile	

(*Continued*)

TABLE 3.10 (*Continued*)
Examples of Non-Hail-Related Damage to Tile Roof Systems

Non-Hail Defect	Photograph
Displaced Spanish-style concrete tile	
Weathered crack to concrete tile Note duller, rounded edges	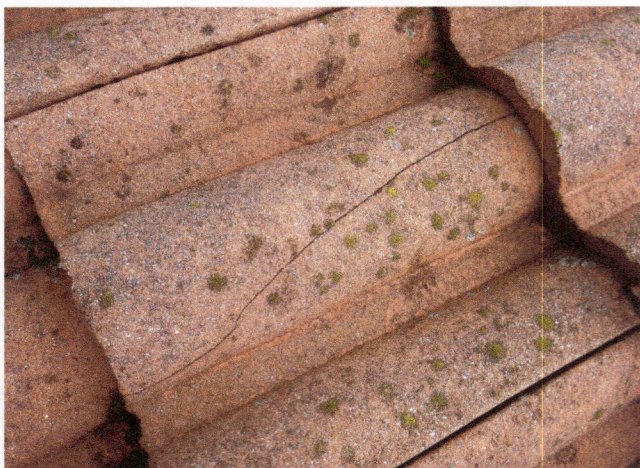
Weathered chips to asbestos tiles Note weathered, rounded edges	

(Continued)

TABLE 3.10 (*Continued*)

Examples of Non-Hail-Related Damage to Tile Roof Systems

Non-Hail Defect	Photograph
Non-hail cracks to asbestos tiles	
Hail removed biological growth from surface of asbestos tiles Not functional damage	

3.4.4.2.1 Eroded Holes

Eroded holes are holes in the tile with eroded edges that were not, to a reasonable degree of probability, the result of hail. Oftentimes, eroded holes will be located directly over a nail fastener of an underlying tile. Over time, the nail head rubs against the underside of the tile, wearing a hole into it.

3.4.4.2.2 Cracks

Cracks are splits or cracks in the tiles that contain no evidence of hail impact marks (i.e., spatter marks or impressions) along their length and were not, to a reasonable degree of probability, the result of a hail impact.

3.4.4.2.3 Chips

Chips are defects to the corners or edges of a tile that could possibly have been caused by hail and is dependent on the size of hail, the condition of the tiles, and the angle of impacts. Chipped edges do not typically lessen the water-shedding capabilities of the tile and therefore, are generally not considered functional damage. Chips, which are not likely the result of hail impacts, typically will have a rounded, weathered appearance and will not contain evidence of a hail impact along the fractured/chipped edge.

3.4.4.3 Repairing Tile Roofs

Much like wood roof systems, an advantage of tile roofs is that in most cases, the roof can be repaired by replacing individual damaged tiles. There is often a push for full replacement of tile roof surfaces, either knowingly or not, by roofing contractors who lack knowledge regarding the reparability of these roof systems and/or see a chance for large profit. Often, typical roofing contractors have little "true" experience working with tile roofs and likely do not know the proper methods of repair. The same lack of knowledge also exists with many insurance adjusters who often agree to a total roof replacement.[18,19] Typical costs for replacement of slate tiles and slate tile metal valleys are: (1) $75–$100/tile and (2) $200–$250 per foot of valley. An additional repair factor, breakage associated with the repairs themselves (0%–30%) should be accounted for in any estimates.

One of the most knowledgeable and foremost experts in the field of slate roofing is Joseph Jenkins, who has several published works regarding the subject. His book, titled "The Slate Roof Bible (Second Edition),"[19] is probably the preeminent work in the field of slate roofing and provides nearly everything a professional would need during installation or an evaluation process. He too advocates slate roof repairs rather than replacement. He also describes appropriate repair methods at length (i.e., Nail and Bib Repair and Slate Hook Repair).[19]

Another leading publication within the tile roofing industry is the "Concrete and Clay Roof Tile: Installation Manual for Moderate Climate Regions" authored by the Tile Roof Institute (TRI) and the Western States Roofing Contractors Association (WSRCA).[20] In it, the proper repair/replacement methods for individual clay and concrete tiles, which have been functionally damaged, are given.

3.5 USEFUL EXPERIENCE AND RULES OF THUMB FOR HAIL-STRIKE DAMAGE TO RESIDENTIAL AND LIGHT COMMERCIAL ROOF SYSTEMS

Based on completing thousands of hail damage assessments of residential and light commercial structures, several "rules of thumb" can be used when evaluating the roof system of a particular structure:

- Hail falls randomly (in no discernable pattern) and hits nearly everything rather uniformly; therefore, a systematic inspection of the exterior components of the structure provides insight into important hailstorm information.
- For asphalt shingles, the typical threshold hailstone size for functional damage to occur begins when hailstones are ~1.5–2.0 inches in diameter. This is also the size where hail-strike bruise intensities become significant (>1–3 bruises per square).

- For asphalt shingles, the number of hail-strike bruises on windward slopes (i.e., those facing the incoming hailstorm) are ~2–2.5 times greater than the number on leeward (opposite) slopes or slopes perpendicular to the incoming storm.
- Weathered and deteriorated asphalt shingles are less impact-resistant and are more susceptible to hail-caused functional damage; therefore, damage can sometimes occur when hail is below typical-size thresholds.
- For wood shingles and shakes, the typical threshold hailstone size when significant numbers of hail-strike splits begin to occur when hailstones are ~1.25–2.0 inches in diameter.
- For slate tiles, the typical threshold hailstone size when significant numbers of hail-strike penetrations and splits begin to occur when hailstones are ~1.5–2.0 inches in diameter, depending on the quality of the tiles.
- For clay tiles, the typical threshold hailstone size when significant numbers of hail-strike penetrations and splits begin to occur when hailstones are ~1.25–2.0 inches in diameter, depending on the quality of the tiles.
- For concrete tiles, the typical threshold hailstone size when significant numbers of hail-strike penetrations and splits begin to occur when hailstones are ~1.5–2.5 inches in diameter.
- For asbestos tiles, the typical threshold hailstone size when significant numbers of hail-strike penetrations and splits begin to occur when hailstones are ~1.5–2.0 inches in diameter.

IMPORTANT POINTS TO REMEMBER

- Functional hail damage to a residential or light commercial roofing material is defined by: (1) a reduction of the water-shedding capabilities and/or (2) a reduction in the expected long-term service life of the roof material.
- Functional hail damage to a particular residential or light commercial roofing material typically begins when the impacting hailstones reach the material's threshold size for damage, where the level of impact energy has reached the point at which damage can occur. This hailstone size threshold is ~1.5–2.0 inches in diameter for typical asphalt shingles.
- The greatest likelihood for functional hail damage to a roof system is from a hailstone that strikes perpendicular to the roof surface.
- The directionality of the hailstorm and the size of the hailstones, which have impacted a particular residential or light commercial structure, can be determined from a thorough visual inspection of the exterior and roof surfaces.
- Basing hail-strike damage on metal dent sizes and size of hail is more effective for determining damage to roof finished surfaces than more subjective parameters such as hail-strike bruises.
- Functional damage to asphalt shingles, in the form of a hail-strike bruise, is immediate, identifiable, and does not appear to worsen with time. Granule loss by itself is not generally considered functional damage or to have an impact on the service life of an asphalt-shingled roof surface.
- Functional damage to wood shakes and shingles is characterized by a fresh split in the wood coincident with evidence of a hailstone impact mark on the wooden surface.
- Functional damage to slate and other tile roof systems is typically characterized by either (1) a fresh puncture or penetration in the tile, (2) a fresh fracture or split, or (3) an impression that has caused significant surface layers of the tile material to become compromised.

- Oftentimes, the relative time in which a hailstorm occurred and caused functional damage to a roofing material can be determined by its appearance. Damage impact points on asphalt shingles and wood roof systems weather and appear grayer with time, while the damage to tile roof surfaces often appears duller with rounded, weathered edges of penetrations, splits, fractures, and chips.
- In most cases where functional damage has occurred, individual replacements/repairs can be made to wood and tile roof systems.

REFERENCES

1. Leefeldt, Ed "Are Hailstorms Getting Worse in the U.S.? Why 2019 Could Produce Record Damage," 2019. Accessed July 31, 2020, https://www.cbsnews.com/news/hail-damage-costs-this-year-could-hit-new-annual-high-in-u-s/.
2. Asphalt Roofing Manufacturing Association (ARMA). Accessed March 2011, www.asphaltroofing.org/.
3. Noon, Randall K. *Forensic Engineering Investigation*. New York: CRC Press, 2001.
4. Morrison, Scott L. *"Hail and Composition Shingles."* Dallas, TX: Haag Engineering Co., 1993, revised March 1995.
5. Petty, Stephen E., Mark A. Petty, and Tim Kasberg. "Evaluation of Hail-Strike Damage to Asphalt Shingles Based on Hailstone Size, Roof Pitch, Direction of Incoming Storm, and Facing Roof Direction." *Interface Magazine*, May/June 2009, pp. 4–10.
6. Roofing Industry Committee on Weather Issues, Inc. (RICOWI, Inc.). "Hailstorm Investigation – Dallas" Fort Worth, TX, May 24, 2011. Accessed September 10, 2020, https://docplayer.net/10564289-Hailstorm-investigation-dallas-fort-worth-tx-may-24-2011.html.
7. Morrison, Scott J. "Long-Term Effects of Hail Impact on Asphalt Shingles – An Interim Report." *American Conference on Roofing Technology 2002*, 30–39. Accessed September 11, 2020, https://haagglobal.com/am-site/media/long-term-effects-of-hail-impact.pdf.
8. State Farm Fire & Casualty Company vs J.M. Radcliff and Coastal Property Management, LLC. State of Indiana, Hamilton Superior Court, No. 1, Cause No. 29D01-0810-CT-1281, 2009.
9. Canadian Asphalt Shingle Manufacturers Association (CASMA). "Blistering of Asphalt Shingles." Technical Bulletin No. 21, June 2019. Accessed September 11, 2020, https://www.casma.ca/technical-bulletins/blistering-of-organic-asphalt-shingles.
10. New Roof Construction Manual, Cedar Shake and Shingle Bureau (CSSB), September 2014. Accessed September 11, 2020, http://www.cedarbureau.org/cms-assets/documents/roof-manual-full-09-2014.pdf.
11. Cedar Shake and Shingle Bureau (CSSB). "Certi-label™ Cedar Roofing: A Claim Adjuster's Guide to Handling Hail-Related Claims", March 2006. Accessed September 11, 2020, http://www.cedarbureau.org/cms-assets/documents/24826-485192.cssb-adjusters-guide-to-hail.pdf.
12. Haag Engineering. "Hailstorm Characteristics." 2006. http://www.haagengineering.com/ehail/chas/eHail/hailstorm.html.
13. Greenfeld, Sidney H. "Hail Resistance of Roofing Products." Building Science Series #23. National Bureau of Standards, August 1969, p. 9.
14. Cedar Shake and Shingle Bureau (CSSB). "Cedar Roof Care and Maintenance", 2004.
15. Marshall, Timothy P. and Richard F. Herzog. "Protocol for Assessment of Hail-Damaged Roofing." *Proceedings of the North American Conference on Roofing Technology*, Sept. 16-17, 1999, Toronto, Ontario, CA, pp. 40–46. Accessed September 11, 2020, https://haagglobal.com/am-site/media/protocol-for-assessment-of-hail-damaged-roofing.pdf.
16. Marshal, T. P., Richard F. Herzog, Scott J. Morrison, and Steven R. Smith. "Hail Damage to Tile Roofing." Haag Engineering. Accessed September 11, 2020, https://ams.confex.com/ams/pdfpapers/81093.pdf.
17. Koontz, Jim D. "The Effects of Hail on Residential Roofing Products." *Third International Symposium on Roofing Technology*, April 17–19, 1991, Gaithersburg, MD, pp. 206–15. Accessed September 11, 2020, http://docserver.nrca.net/technical/374.pdf.
18. Jenkins, Joseph. "What the Hail?" *Traditional Roofing Magazine*, 2007. Accessed September 11, 2020, http://www.theslateroofexperts.com/downloads/What_The_Hail.pdf.
19. Jenkins, Joseph. *The Slate Roof Bible (Second Edition)*. Pennsylvania: Joseph Jenkins, Inc., 2003.
20. Tile Roof Institute (TRI) and the Western States Roofing Contractors Association (WSRCA). "Concrete and Clay Roof Tile: Installation Manual for Moderate Climate Regions", July 2006.

4 Hail Damage Assessments to Low-Sloped Roof Systems

Ronald L. Lucy and Stephen E. Petty
EES Group, Inc.

CONTENTS

PURPOSE/OBJECTIVES

The purpose of this chapter is to:

- Introduce the types of commercial low-sloped roof systems.
- Provide a methodology for assessing hail damage to low-sloped roof systems.
- Provide hailstone size thresholds required to cause damage to low-sloped roof systems.

Following the completion of this chapter, you should be able to:

- Have a general understanding regarding the types of commercial low-sloped roof systems.
- Be able to perform a hail damage assessment for low-sloped roof systems.
- Be able to identify hail-strike damage on commercial low-sloped roof systems.
- Know hail size thresholds for functional hail damage to common commercial low-sloped roof systems.

4.1   INTRODUCTION

By definition, low-sloped roof systems consist of a category of roofs that are installed on slopes at a 4:12 or less pitch. Many commercial buildings (and portions of some residential buildings) throughout the United States have flat or low-sloped roof systems. A couple of major differences between commercial low-sloped roof systems and residential steep-sloped roof systems are: (1) the higher cost of removal and replacement and (2) commercial low-sloped roof surfaces typically have a much greater surface area.

As with steep-sloped roof surfaces, commercial low-sloped roof systems are susceptible to hail-strike damage. It is not uncommon for the replacement cost of a commercial low-sloped roof system to be ten times greater (i.e., $100,000 vs $10,000) or more than the replacement cost of a steep-sloped roof system.

To conduct a proper hail-strike damage inspection of a commercial low-sloped roof system, the same basic inspection guidelines discussed in Chapters 1 and 2 are followed; major elements of the inspection should include the following steps:

- Interview with building owner, tenant, maintenance personnel, and/or the owner's representative. Note that in many cases, the tenant or maintenance staff will have a better knowledge on the history of the roof than the owner who may not often be at the property.
- Complete an inspection of the exterior surfaces to identify and document hail-strike damages. All observations should be recorded in a field notebook with key findings photographed. This allows one to identify the direction from which the hailstorm arrived, estimate the maximum size of the hail that struck the building, and determine if the hail that struck the building fell vertically or was wind-driven.
- Sketch out or obtain an aerial report from a commercial service and confirm building and roof dimensions. Vents, HVAC units, and other appurtenances should be marked on the sketch or commercial service schematic.

- Inspect the metal surfaces and equipment for likely hail-stone strikes and other dents. A best practice is to record the manufacturer, serial numbers, model numbers, and date manufactured (if listed) from the mechanical units such as Heating, Ventilation and Air-Conditioning (HVAC) outdoor units. The manufacturing date is typically encoded in the serial number of the unit and is usually a good indicator of the date the unit was installed. This information can assist the forensic investigator in dating a hailstorm (if needed).
- Visually inspect the roof surfaces for hail-strike damage; illustrations of typical hail-strike damage and hailstone size thresholds by type of commercial roof system are provided later in this chapter.
- Destructive test the roof system (if necessary) to verify functional hail-strike damage. It is a best practice to photograph the test cuts prior to the cut, after the test cut is made, and after the area has been repaired.
- Analyze findings and prepare a written report. The form of the report would follow the outline previously discussed in Chapters 1 and 2.

One of the key differences in a commercial low-sloped roof inspection and a steep-sloped roof inspection is that destructive testing is often needed to be performed on the commercial roof surface. A properly trained forensic investigator should be prepared to perform destructive testing (i.e., test cuts) to the roof surface and make necessary repairs afterwards. Such test cuts can be self-performed or aided by a roofing contractor. In all cases, this should be completed with permission of the owner and possibly the insurance company (if applicable) to avoid future liability. The test cut can provide validation of the roof construction details and knowledge regarding the presence of moisture or damage to the underlying materials, i.e., number of roof layers for a built-up roof system (BUR). Additional costs may also be incurred to properly repair the roof system and/or possibly bring the roof system up to current code. An example of a test cut revealing four layers of roof covering and insulation (three layers of BUR with one layer of a single-ply membrane) is shown in Figure 4.1.

FIGURE 4.1   Multiple layers of a BUR roof system.

The type of insulation present as part of a membrane roof system can also be determined by test cuts. This is important when hail strikes are of sufficient energy to damage the membrane, allowing water intrusion to damage the insulation and decking below. If this occurs, portions or all of the insulation and/or decking may have to be removed and replaced. Test cuts can also provide an excellent means to view the back side of the membrane to identify whether or not a hail stone actually struck the surface with enough force to cause a fracture or cracking to the material.

4.2 DESCRIPTIONS OF LOW-SLOPED ROOF SYSTEMS

In order to identify and assess whether or not a low-sloped roof system has sustained functional hail-strike damage, it is important to know the different finished surfaces. Low-sloped roof systems can consist of the following finishes:

- Asphaltic bitumen built up in layers (BUR).
- Bitumen modified with polymers (Mod-Bit).
- Single-ply synthetic materials.
- Metal panels.
- Sprayed-on systems.

Forty years ago, asphaltic-based BUR systems (commonly referred to as hot-mopped tar) dominated the marketplace. In the 1970s, during the time of the oil embargos, the base cost of petroleum increased, raising the cost of BUR and modified bituminous (Mod-Bit) roof finished surfaces. This resulted in a declining use of BUR systems and provided a demand for alternate types of roof systems such as synthetically made single-ply materials. Consequently, BUR systems currently represent 20% of the roofing market, which is down from 45.7% in 2000. Single-ply roof systems are projected to represent 59.7% of the re-roofing market and 66% of the new construction market in 2011.[1] The rise in use of single-ply roofing systems is primarily due to several factors including the following:

- They are relatively easy to install.
- Labor costs for installation are lower.
- Lower or more competitive pricing of single-ply roof systems.
- A declining labor force of experienced installers for BUR systems.
- Energy code requirements that promote more reflective roof systems.

The balance of this chapter will provide a brief description of the common low-sloped roof system finished surfaces, various types of defects found on these surfaces, and the threshold size of hail needed to cause significant damage to low-sloped roof systems.

4.3 BUILT-UP (BUR) ROOF SYSTEMS

4.3.1 DESCRIPTION

BUR roof systems consist of layers of bitumen and reinforcement fabrics that are applied in the field. BUR can be applied in 2–5 layers with hot mopping, cold process asphalt, or self-adhesive materials. The roofing industry traditionally has assigned 5 years of anticipated service life to each felt ply; hence, a 25-year service life could be expected on a five-ply BUR.[2] Bitumen serves as the glue that holds the plies together and provides the overall weatherproofing to the roof system. The bitumen used can be either asphaltic or coal tar pitch with a majority of newer roofs using the asphaltic bitumen. Asphaltic bitumen comes from the bottom of the distillation processes used during the refining of crude oil. The reinforcement fabrics or plies stabilize the membrane, bridge gaps, aid in controlling bitumen thickness, provide impact resistance, and in some cases, provide fire resistance. The application of BUR membranes requires experienced and skilled laborers. The BUR plies must be installed in void-free layers to ensure long-term performance. Poor application can result in various types of defects that can shorten the life of the membrane. BUR membranes are installed as either smooth or gravel-covered surfaces. Aluminum- or zinc-based coatings are often found on BUR membrane surfaces to increase the reflectivity and provide UV protection of the roof surface.

4.3.2 BUR ROOF SURFACE LIFE AND COMMONLY ENCOUNTERED DEFECTS

According to some sources in the roofing industry, the mean service life of a fiberglass-reinforced membrane can range between 15 and 25 years.[3,4] Learning the age of the roof system during the interview process, if possible, can be important to determine if the roof system is at or beyond its useful life. It is not uncommon to investigate a hail-damage claim and find the roof contains numerous age-related defects and is simply beyond its effective service life. Some types of defects can be mistaken for hail-strike damage or claimed to be hail-strike damage. Defects in BUR membranes can be associated with installation anomalies, normal aging, and exposure to the elements. Common age and installation defects observed on BUR roof systems include blisters, ridging, bare spots in gravel, flashing failures, and alligator cracking. Examples of common types of defects to BUR roof surfaces, along with their probable causes, are illustrated in Table 4.1.

TABLE 4.1
Examples of BUR Defects and Probable Causes

Defect Type	Probable Cause	Photograph
Blisters	Expansion of volatile fractions of bitumen or air or water, in warm/sunny weather conditions. Can occur due to voids in the substrate or adhesive application and from applying roofs over wet substrates	
Ridging or buckling	Movement of the reinforcement, deck, or substrates. Can be a result of thermal movement or where the reinforcements are not properly bonded to the roof deck	
Deterioration from ponding water	Improper design with inadequate slopes for drainage or obstructed drains	

(Continued)

TABLE 4.1 (*Continued*)
Examples of BUR Defects and Probable Causes

Defect Type	Probable Cause	Photograph
Bare spots from loss of gravel	Gravel applied in adverse weather. Too thin a layer of too fine gravel at edges and corners. Inadequate adhesion of gravel at edges, corners, or through the field of the roof	
Flashing failures	Inadequate allowance for movement. Poor adhesion or inadequate protection of flashing felts. Damage to capping at parapets and expansion joints	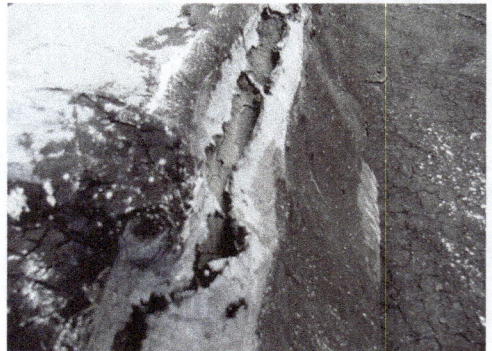
Alligator cracking	Occurs on smooth surface bitumen on bare spots in the gravel from thermal exposure	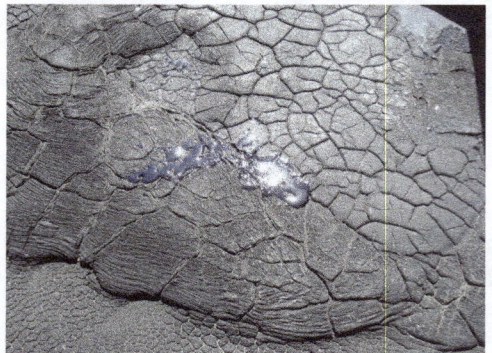

4.3.3 BUR – THRESHOLDS FOR HAIL-STRIKE DAMAGE

Research has been completed on the resistance of BUR to hail-strike damage since at least the 1960s. In 1969, one of the earliest studies was performed to evaluate hail-strike damage to roofing finishes, Sidney Greenfield evaluated the resistance of BUR membranes and other surfaces to synthetic hail (ice balls).[5] Robert G. Mathey and William C. Cullen also performed similar testing for the National Bureau of Standards in 1974.[6] Additional studies were conducted by Haag Engineering in 1988 and 1993[7] and by Vickie Crenshaw and Jim D. Koontz in 2000.[8]

The net results of these studies suggested that damage to smooth-surface BUR began with hailstones >1.5 inches in diameter and elevated damage occurred with hailstone 2.0 inches in diameter or greater.[9] The study also suggested that damage occurred to gravel-covered BUR finishes with hailstones >2.0 inches in diameter. The Roofing Industry Committee on Weather Issues, Inc. (RICOWI)[10] reported impact marks without fractures or minimal damage from hailstones ranging in size from 1.75″ to 2.25″ or larger. Similar to composition shingles (see Chapter 3), research has found that the BUR was more prone to damage at softer, less supported roof areas (i.e., flashings along parapet wall) than supported areas of the membrane over denser substrates. Gravel-surfaced BUR membranes were also much more resistant to hail damage.

When hailstones are of sufficient size, readily observable visible damage occurs to BUR membranes. The damage can consist of spalling of the surface coating, divots into gravel surfaces, concentric ring-type fractures, and penetrations into the reinforcement plies. Based on dozens of field inspections and experience, hailstones must reach at least 2.0 inches or greater in diameter before hail damage occurs to smooth BUR roof surfaces.

4.3.4 BUR – Inspection Case Studies

The following case studies illustrate lessons learned from actual hail-strike damage inspections to BUR roof surfaces.

4.3.4.1 Small Hailstones to a Smooth-Surfaced BUR

In this case study, a property was located in the path of a 2010 hailstorm. This particular inspection was conducted 7 months after the hailstorm passed through the area. Inspections of the metal surfaces indicated the hail that struck the building and roof surfaces was up to 1.0 inch in diameter. The roof was covered with two layers of three-ply smooth-surfaced BUR with a reflective coating and was about 20 years old. The reflective coating was chipped and contained circular impact marks (Figure 4.2).

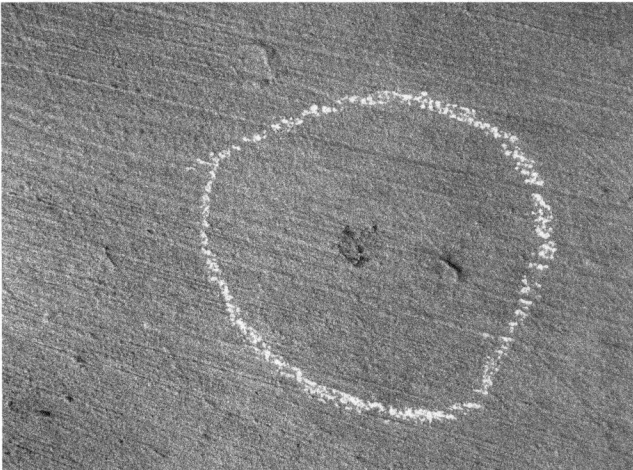

FIGURE 4.2 Hail chips in BUR reflective coating.

Test cuts were made into the membrane at locations where the hailstones had impacted (i.e., circular marks) the roof surface; no cracks, penetrations, or punctures were present (Figures 4.3 and 4.4).

FIGURE 4.3   Test cut into BUR membrane.

FIGURE 4.4   Lack of hail-strike impact damage to BUR plies.

Further, no evidence of moisture infiltration was present in the lower layers of the roof surface. In this case, the hail was not of sufficient size to compromise the membrane. Recommendations were made to clean the roof surfaces and reapply the reflective coating. Visually, the impact markings on the BUR surface suggested that the membrane might have been compromised by hail-strike damage; however, the test cuts revealed that this damage was surficial and cosmetic. Without performing test cuts to examine the bottom side of the membrane and the underlying substrates, it is likely the BUR roof surface would have been removed and replaced based simply on the cosmetic damage that had been observed.

4.3.4.2   Large Hailstones to a Smooth-Surfaced BUR

A church building, with a roof surface that was partially covered with a BUR roof membrane, was located near the center of the hailstorm path. The hailstorm reportedly dropped hailstones up to 3.0 inches in diameter in the area. The inspection of the metal surfaces indicated that the maximum size of hail that struck the church was ~2.0 inches in diameter. Extensive hail damage, up to ten hail-strikes per 100 square feet, was present on the smooth-surface BUR. Photographs of the hailstone impacts are shown in Figures 4.5 and 4.6.

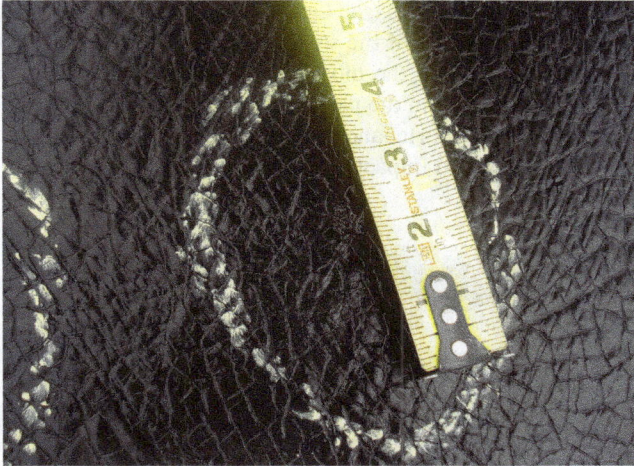

FIGURE 4.5   Hail damage to BUR flood coat and top ply.

FIGURE 4.6   Hail damage to BUR flood coat and top ply.

In this case, the hail was of sufficient size to cause damage and the BUR membrane was removed and replaced.

4.3.4.3 Old Hailstone Damage to a Smooth-Surfaced BUR

In this case study, a large commercial building contained reported hail-strike damage to the BUR roof surfaces. The building was covered with two layers of three-ply BUR that was reported to be <30 years old. The membrane was heavily degraded and the roof had been leaking for at least 10 years. Temporary patch repairs were made by staff in unsuccessful attempts to stop the roof leaks. A roofing contractor was contacted and reported that hail damage was present on the roof surfaces. The forensic investigator was asked to assess the extent of hail damage and determine when the storm occurred, if possible. Within the past year (2010), a new insurance carrier had picked up coverage for the building and the date of the storm event was critical to assignment of the damage claim to the proper insurance carrier.

In reviewing many weather data sources (see Chapter 2), several hailstorms had passed through the area during the past 4–5 years. Most of the storms contained hail that ranged from 0.75 to 1.0 inches in diameter. Data confirmed that the most severe hailstorm to strike the area occurred in 2007 and the building was located within the path of that specific hailstorm.

Numerous metal roof vents and appurtenances contained evidence of dents, indicating that hailstones up to 2.5 inches in diameter had impacted the building (Figure 4.7).

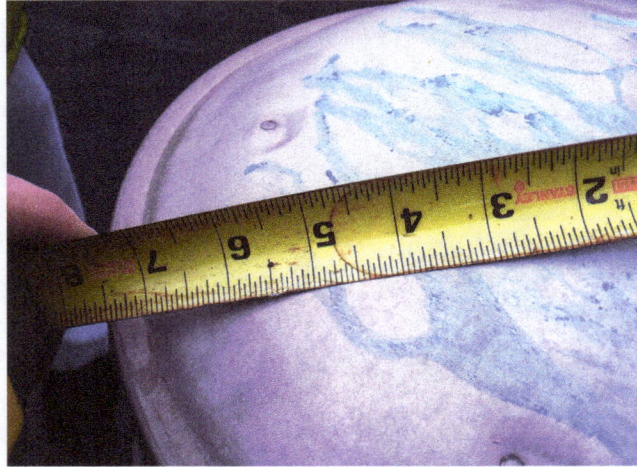

FIGURE 4.7   Hail dent in roof vent.

The finished surface of the BUR also contained circular indentations and penetrations consistent with hailstone impacts (Figure 4.8).

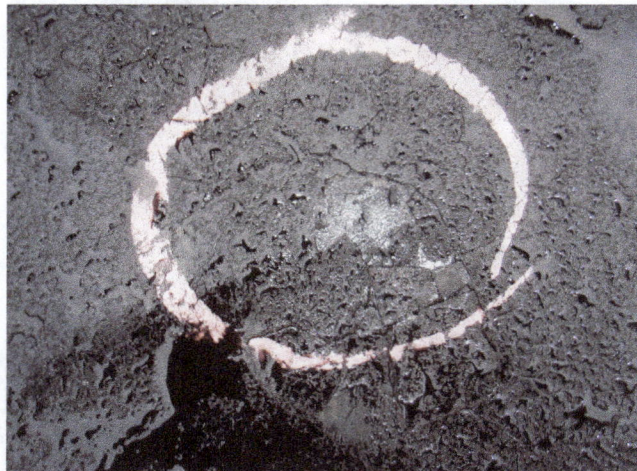

FIGURE 4.8   Hail penetration into BUR membrane roof surface.

A test cut indicated that there were two layers of BUR on the roof surface. The hailstone caused impact damage that had penetrated almost entirely through the top BUR layer (Figure 4.9).

FIGURE 4.9   Hail penetration into BUR membrane layer.

The hail strikes clearly caused damage to the membrane, warranting roof replacement. The review of weather data and correlation with the size of the dents to the metal surfaces confirmed that the hailstorm that caused the damage to the membrane occurred in 2007. Therefore, the forensic investigator was able to date the hailstorm likely responsible for the hail-strike damage, which fell outside of the policy coverage period for the current insurance carrier.

4.4   MODIFIED BITUMEN (MOD-BIT) ROOF SYSTEMS

4.4.1   Description

Mod-Bit roof systems, sometimes called Polymer Modified Bitumen roof systems, have become an important segment of the commercial roofing markets over the past 20 years. Mod-Bit systems are premanufactured asphaltic bitumen sheet membranes that are modified with either atactic poly-propylene (APP) or styrene-butadiene-styrene (SBS) polymers. The thickness of modified bitumen membrane is typically between 120 and 180 mils (1 mil = 1/1,000 inch), which is much greater than single-ply membranes that can range up to 90 mils. These polymers are relatively more resistant to strain (stretching forces), provide greater flexibility, and allow the membrane to withstand greater temperature extremes. These attributes make it possible to reduce the number of plies (lowering the material and labor costs) to the roof system and still provide the same types of water-proofing features as a BUR roof system.

Mod-Bit roof systems can be adhered by torch, cold-adhesive or self-adhesive methods, or mopped into place. The membrane is commonly covered (surfaced) with granules or foils, most commonly when manufactured. Field surfacing of reflective coatings, emulsions, and flood coats with aggregate are also observed.

4.4.2   Mod-Bit Membrane Roof Life and Commonly Encountered Surface Defects

The mean service life of Mod-Bit membrane roof surfaces is about 15 years.[3,4]

Similar to BUR membranes, Mod-Bit membrane roof defects occur as a result of manufacturing, installation, weathering, and age-related issues. Mod-Bit roof failure modes, in order of occurrence, are: (1) defective lap seams, (2) shrinkage, (3) checking, (4) blistering, (5) delamination, (6) slippage, and (7) splitting.[11] Examples of common types of defects to Mod-Bit roof surfaces, along with their probable causes, are illustrated in Table 4.2.

TABLE 4.2

Examples of Mod-Bit Defects and Probable Causes

Defect Type	Probable Cause	Photograph
Blisters	Expansion of volatile fractions of bitumen or air or water, in warm/sunny weather conditions. Can occur due to voids in the substrate or adhesive application and from applying roofs over wet substrates	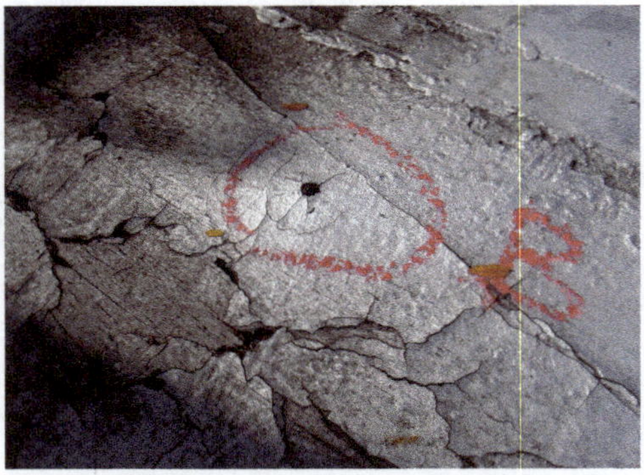
Protruding fasteners causing spalling of coating	Movement of the structure, or moisture intrusion from preexisting conditions, roof leaks, or ponding water	
Deterioration from ponding water	Improper design with inadequate slopes for drainage or obstructed drains	
Defective lap seam	Adhesive failure sometimes due to application temperature of installation issues	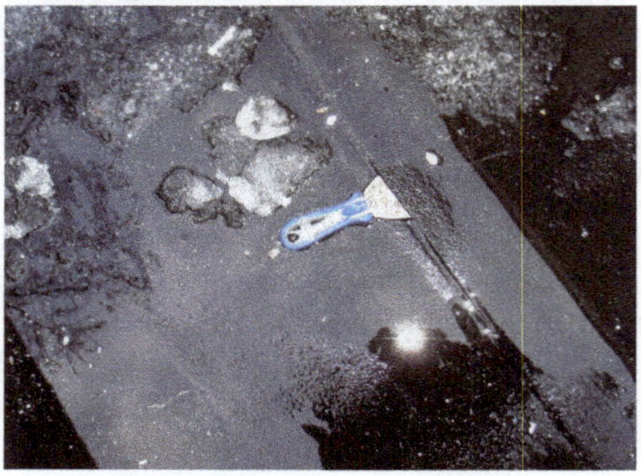

(*Continued*)

TABLE 4.2 (*Continued*)
Examples of Mod-Bit Defects and Probable Causes

Defect Type	Probable Cause	Photograph
Checking	Aging and long-term exposure to ultraviolet radiation	

4.4.3   Mod-Bit – Thresholds for Hail-Strike Damage

Only a single laboratory study on the hail size thresholds for hail-strike damage to Mod-Bit membranes was identified in a review of the literature. In an investigative report dated April 21, 2004 from the RICOWI,[12] several modified bitumen membrane roof systems were inspected for possible hail-strike damage. The study noted that bruising/fracturing was indicated by circles where granule loss was visible and by an indentation that was either visible or could be felt with finger pressure. With respect to age/condition, roof slope, and support conditions of the material, the study concluded that fracturing to the Mod-Bit membrane was observed on all roof surfaces where the hail size was 2.0 inches or larger in diameter. Further, no hail-strike damage was observed to Mod-Bit membrane roof surfaces when the hail size was <1.5 inches in diameter to 2 inches in diameter.[9] Other work suggests that hail damage from hailstones begins at ~1.75 inches in diameter.[10] Actual hail-strike damage inspection results on Mod-Bit roof systems completed by EES concur with these threshold findings.

4.4.4   Mod-Bit – Inspection Case Studies

The following case studies illustrate lessons learned from actual hail-strike damage inspections to Mod-Bit roof surfaces.

4.4.4.1   Small Hail to Reflective-Coated Mod-Bit Membrane

This case study was associated with a building struck by a hailstorm in 2008. The Mod-Bit roof surface investigated as part of this inspection consisted of an ~30-year-old smooth surface APP Mod-Bit roof system that was surfaced with a reflective coating. The purpose of the investigation was to determine if hail-strike damage to the Mod-Bit roof membrane was present since the roof reportedly had begun to leak following the hailstorm. Observations indicated that the roof was poorly drained and contained deep cracks and open seams; however, the locations of these defects did not correlate with the reported leak areas.

A survey of the metal roof surfaces indicated the hail that struck the building was up to 1.0 inch in diameter. The exact locations of where the hailstones struck the roof surface were readily visible in the form of burnish marks, which had removed the weathered oxidation layer from the Mod-Bit membrane surface (Figure 4.10).

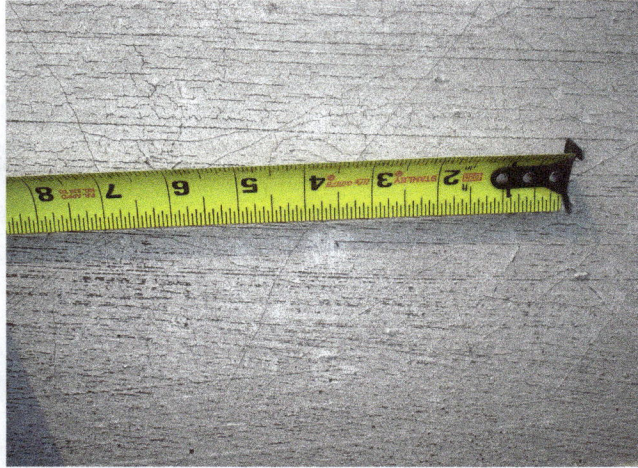

FIGURE 4.10   Burnish marks on Mod-Bit membrane surface.

Other than cosmetic damage to the surfacing, no additional damage from hail-strike impacts was observed.

4.4.4.2   Large Hail to Gravel-Surfaced Membrane

In this case, a commercial warehouse building was covered with a fairly new (within 5 years) granular surface SBS Mod-Bit membrane. The roof of the building reportedly had been leaking for the past few years in multiple areas. The leaks had caused little damage to the building contents so the building owners were not overly concerned. When business slowed down, the building was put up for sale and the owners needed to address the roof leaks. After walking the roof surfaces, their roofing contractor reported widespread hail damage to the Mod-Bit membrane roof surfaces. The owners then filed a hail-damage claim with their insurance carrier. In reviewing the hail-storm history, several hailstorms reportedly passed through the area in the 2007–2008 timeframe. Which storm caused this reported hail-strike damage to the roof system posed issues for both the insurance company and the building owner. Similar to a BUR case study mentioned earlier in this chapter, the building was insured by one insurance carrier up until the end of 2007 and another from 2008 forward. For coverage reasons, it was important to both insurance companies that a determination be made of when the hailstorms struck this roof system and which one(s) caused damage, if any, by storm and date.

A survey of the metal surfaces suggested that large hail, upward of 2.5 inches in diameter, had struck the building. The Mod-Bit membrane contained up to ten circular fracture marks per 100 square feet of roof area that had punctured through the membrane (Figure 4.11).

FIGURE 4.11 Hail fracture mark in Mod-Bit membrane roof surface.

In reviewing weather data reports for the area over the timeframe of interest, it was determined that four significant hailstorms had passed through the area. Three of the four hailstorms were reported to produce hailstones ranging from 0.75 to 1.0 inches in diameter. The fourth hailstorm, which occurred in mid-2007, produced hailstones up to 4.0 inches in diameter and the subject building was located directly within the path. In this case, the date of the hailstorm that likely caused damage to the Mod-Bit membrane roof surface could be determined by the size of the hail produced. Since only one of the four hailstorms contained hail large enough to damage Mod-Bit membrane roof surfaces, the date of the damage could be determined by the forensic investigator.

This investigation showed that hail up to 2.5 inches in diameter could cause elevated damage to Mod-Bit roof surfaces. The study again showed the importance of using the size of denting to determine maximum hail size to strike a subject building and the importance of storm weather data to provide a reasonable timeframe on when the damage occurred.

4.4.4.3 Large Hail to Smooth-Surfaced Reflective-Coated Mod-Bit

The third Mod-Bid roof case study involved a high-rise condominium building struck by a fairly significant hailstorm in 2010. The roof surface was covered with an APP Mod-Bit roof membrane that was ~25 years old. The roof deck was poorly sloped and contained widespread areas of ponded water. The membrane contained heavy surface deterioration, including alligator cracking and heavy blistering. The reflective coating was worn or eroded away due to the heavy ponding. Following the hailstorm, the roofing contractor found moisture beneath the membrane that reportedly had not been there the previous year. Neither the roofing contractor nor the owner was able to provide any previous maintenance/inspection records of the roof.

Based on hail-strike dents in the metal surfaces on the building, the largest size hail to strike appeared to be upward of 2.0 inches in diameter. The membrane surfaces showed no visible evidence of hailstone fractures or indentations. In a few areas, some of the reflective coating was chipped away by probable hail-strike impacts (Figure 4.12).

FIGURE 4.12 Chipped coating from probable hail strikes to Mod-Bit roof surface.

Since the surface coating was damaged, the question was whether or not this damage extended through the membrane. To answer this question, six test cuts were made into the roof surface to investigate the back side of the membrane and substrates. In each case, the test panels were taken where evidence of surface damage was present. Neither the back side of the membrane nor the fiberboard insulation showed signs of hail-strike impacts, indentations, or fracturing (Figure 4.13). The samples were viewed off site using a high-powered microscope. No evidence of fracturing in the back sides of the membrane samples was observed (Figure 4.14). This indicated that the hail-strike damage was limited to surficial damage.

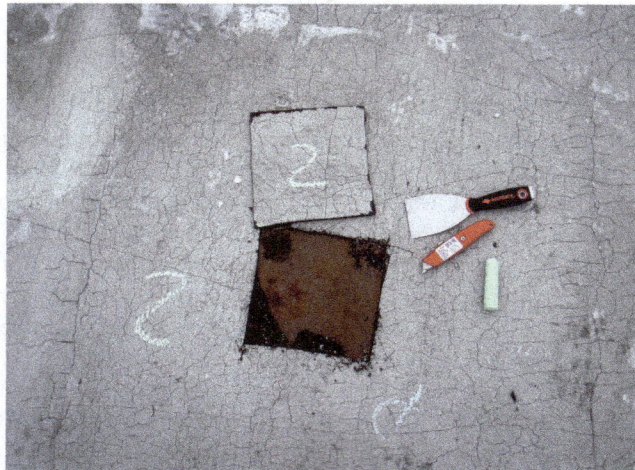

FIGURE 4.13 Overview of test cut to Mod-Bit roof surface.

FIGURE 4.14   Back side of Mod-Bit membrane (high-power microscope).

Again, the light damage to the reflective coating could have been easily mistaken for functional hail damage to the membrane had test cuts not been performed.

4.5   ETHYLENE PROPYLENE DIENE MONOMER (EPDM) ROOF SYSTEMS

4.5.1   DESCRIPTION

Ethylene Propylene Diene Monomer (EPDM) roofing, often referred to as a "rubber" roof membrane, is a single-ply roof system that falls into the category of thermoset or non-weldable materials. Because of lower costs and greater durability, EPDM has become a mainstay in the commercial roof market, replacing butyl, chlorobutyl, and neoprene as the elastomer single-ply membrane of choice. The EPDM membrane can be installed fully adhered, mechanically attached, loose-laid, or installed with a combination of these three methods.

Fully adhered systems are installed by applying a proprietary adhesive to bond the membrane to the roof surface. This method is effective, but often the adhesive is not uniformly applied. In addition, if the approved substrate/underlayment utilized has a peel strength less than that of the adhesive, separation failure may occur below the adhered surface. For example, sometimes the ISO insulation board attachment to a metal decking may be too weak, allowing the entire roof system to lift during high wind events even though the adhesive bonding between the membrane and ISO board remains intact.

Mechanically attached systems use intermittent attachment of the membrane throughout the perimeter and field of the roof surface. Mechanical fasteners consist of decking screws and disks (stress plates) that aid in the distribution of roof loads over a greater surface area. The tops of the fasteners are then covered by adjacent membrane sheets.

Loose-laid or ballasted systems are only attached to the roof surface at the perimeters with penetrations that allow the membrane to expand and contract under normal thermal conditions or wind load. The ballast consists of round smooth (river) stones intended to hold the roof in place. Stone is typically applied at perimeters and at ten pounds per square foot (PSF) for lower wind uplift areas and 20 PSF for higher wind load areas. Ballasted roof systems are generally limited to buildings that range in height from about 35 to 45 feet up to 105 feet.

4.5.2   EPDM MEMBRANE ROOF LIFE AND COMMONLY ENCOUNTERED SURFACE DEFECTS

EPDM membranes have a mean service life of between 15 and 20 years, depending on the method of installation.[3,4] Based on conversations with EPDM manufacturers' technical departments at Firestone Building Products, Carlisle Syntec, and KENDA StaFast Roofing Systems, a distinct code is stamped on their products that provides information on when the material was manufactured. The manufacturing date is typically a good indicator of when the EPDM roof system was installed. Examples of these codes are provided in Figures 4.15–4.17.

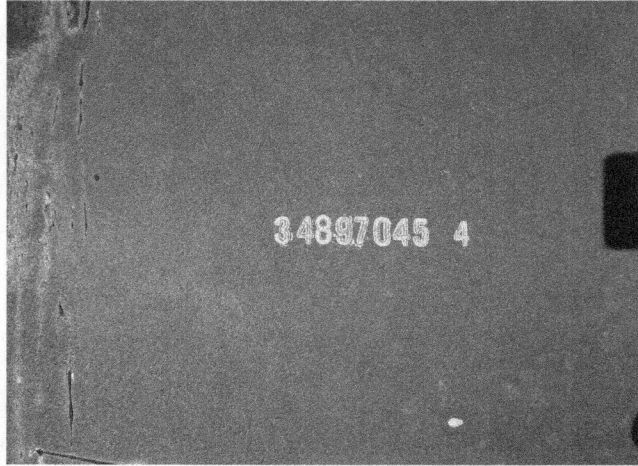

FIGURE 4.15   Example of a firestone EPDM membrane stamp.

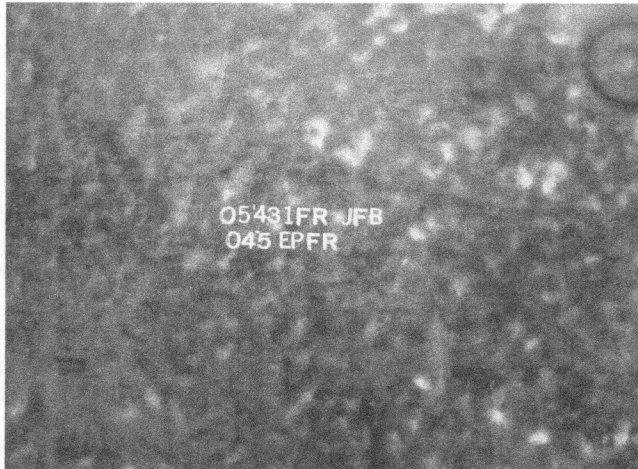

FIGURE 4.16   Example of a Carlisle EPDM membrane stamp.

FIGURE 4.17 Example of a KENDA EPDM membrane stamp.

In Figure 4.15, the white linear stamp for a Firestone EPDM membrane is shown. The first three digits of the code represent the calendar day of manufacturing (348); the next two digits represent the year (97); the next three digits represent the material thickness (045 mils); and the last digit represents the shift in which the material was manufactured (4). Thus, this stamp indicates that this membrane was manufactured in mid-December of 1997 during the first shift.

In Figure 4.16, the Carlisle Syntec code is typically stamped with blue lettering, although the coloring fades to a lighter tint. The first two digits represent the month (05); the next digit represents the Plant ID (4); the next two digits represent the calendar day (31); the first two letters (FR) represent the product specification; the next two letters present the calendar year alphanumerically, whereas $J = 10$ and $F = 6$ or 2006; the next letter represents the shift (B); and the remaining numbers represent of thickness (45 mils). This stamp indicated that the membrane was manufactured on May 31 of 2006.

In Figure 4.17, the KENDA StaFast Roof systems stamp is much simpler to decipher. The first two numbers represent the year (04); the next two numbers represent the month (10); and the last two numbers represent the day (11). Hence, the membrane was manufactured on October 11, 2004.

The most common defects to EPDM membranes are shrinkage and defective seam laps. The defects can occur as a result of manufacturing, installation, weathering, and/or age-related issues. Examples of common types of defects to EPDM roof surfaces, along with their probable causes, are illustrated in Table 4.3.

TABLE 4.3

Examples of EPDM Defects and Probable Causes

Defect Type	Probable Cause	Photograph
Membrane separation at perimeters	Expansion and contraction of membrane due to lack of perimeter attachment	
Membrane separation at perimeter	Shrinkage and improper perimeter fastening	
Protruding fasteners causing membrane tears	Movement of the structure, or moisture intrusion from preexisting conditions, roof leaks, or ponding water	
Loose seam	Shrinkage or improper installation	

4.5.3 EPDM – Thresholds for Hail-Strike Damage

Most roofing professionals have found that EPDM membranes are the most resistant to hail-strike damage of all the single-ply membranes. Research on the hail resistance to EPDM membranes conducted by Jim D. Koontz in 2009 stated that the hailstone size threshold to cause damage to EPDM membranes was 2.5–3 inches in diameter.[13] RICOWI[10] reported field examples of single-ply membrane systems typically with no visible damage to hailstones up to 2 inches in diameter, but did observe one system with multiple repairs that had damage with hailstones of 1.75 inches in diameter. Typically, moderate damage (fractures and punctures) was observed with hailstones at 2.5 inches in diameter. As with any type of roofing materials, the more rigid or supported the membrane, the more resistant the membrane will be to hail-strike damage.

4.5.4 EPDM – Inspection Case Studies

The following case study illustrates lessons learned from an actual hail-strike damage inspection to an EPDM membrane roof.

A commercial building was inspected for hail-strike damage after a hailstorm in 2010 struck the building. The hailstorm reportedly dropped hailstones measuring up to 1.5 inches in diameter. The roof surface was finished with a mechanically attached EPDM membrane that was reported to be >20 years old and was heavily weathered. Roof leaks had been occurring intermittently for the past couple of years. The owner reported that the leaks appeared to worsen following the hailstorm and that a hail-strike damage claim was filed. Hundreds of pinholes were found across the surface by the owner's contractor that was thought to have been caused by hailstone impacts. The reinforcement scrim was also visible through the top of the membrane (Figure 4.18).

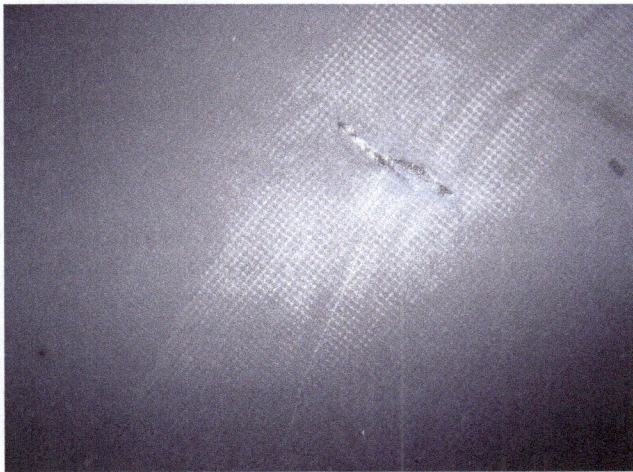

FIGURE 4.18 EPDM scrim visible through top side of EPDM membrane.

The pinholes were heaviest where the scrim was more exposed. In many cases, the holes were filled with ice, suggesting that moisture infiltrated and was beneath the membrane (Figures 4.19 and 4.20).

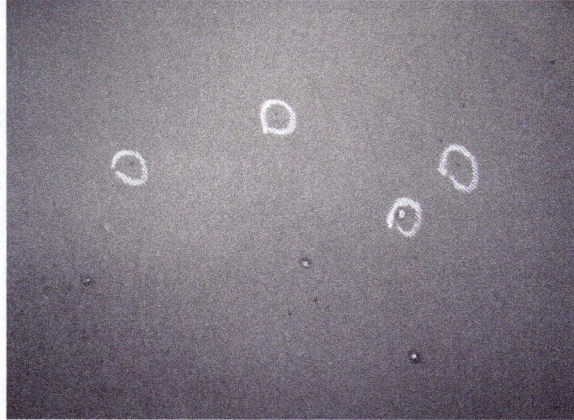

FIGURE 4.19   Pinholes in EPDM membrane.

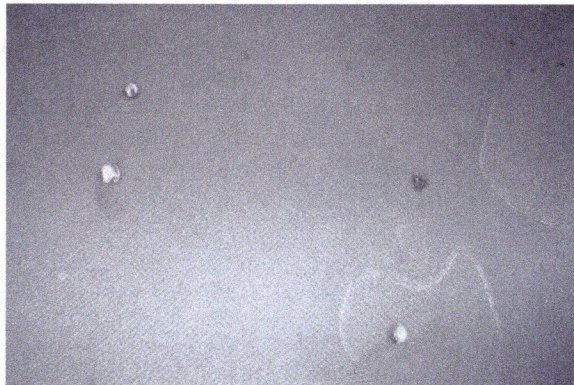

FIGURE 4.20   Pinholes in EPDM membrane filled with ice.

Destructive testing illustrated that the membrane was not fractured around the pinhole defects and indentations were not present in the insulation board below the defects. This supported the hypothesis that the pinholes observed in the EPDM surface were not a result of hail-strike impacts but were the result of long-term aging or chalking. Further, it was observed that the maximum size of hail to have struck the roof surface was not likely of sufficient size to cause damage to an EPDM roof surface. Dozens of EPDM hail inspections performed by EES have also shown that hailstones ranging from 1.0 up to 2.5 inches in diameter did not result in hail-strike damage.

4.6 THERMOPLASTIC ROOF SYSTEMS

4.6.1 DESCRIPTION

Thermoplastic membranes are single-ply membranes that are weldable at the seams and openings. The weldable properties allow for relatively easy installation, especially with roof surfaces that contain multiple penetrations. The seams and terminations are heat fused with a hot air gun to form watertight seams. Polyvinyl chloride (PVC) and thermoplastic polyolefin (TPO) are the two major categories of thermoplastic membranes.

PVC roofing was first introduced in Europe in the 1960s with significant usage beginning in the United States in the 1970s. In the late 1970s and early 1980s, PVC roof systems were found to perform poorly. The reasons for poor performance were associated with the low thickness of the membrane, poor formulation characteristics, and the lack of reinforcement needed to help support the polymer. While PVC roof surfaces were successful in the milder European climates, they did not perform well with the thermal temperature swings that occur in much of North America. Improvements to the formulations of the PVC polymer (e.g., better plasticizers or flexible non-chlorine polymers) and the addition of reinforcements (e.g., utilization of woven or nonwoven fabrics) have overcome many of these early performance issues.

Like EPDM roof membranes, PVC roof membranes can be installed using fully adhered, mechanically attached, or ballasted securement systems. In lieu of field splicing with proprietary adhesives, as is done with EPDM membranes, the thermoplastic properties of the PVC membranes allow the splices to be mated through field welding with a hot air gun.

TPO membranes, the second type of thermoplastic roof membranes, are based on polypropylene and ethylene polypropylene rubber. These materials are polymerized together during the manufacturing process. TPO has been used in various applications, including the automobile industry since the 1980s. In 1989, unreinforced TPO moved into the single-ply roofing industry. By 1993, the non-reinforced membrane was replaced with membranes containing reinforcement fabrics.

The advantages of these thermoplastic products are that the polymerization process can be completed at low temperatures and the TPO polymer does not contain chlorine. This latter advantage has allowed marketers to tout this product as more environmentally safe or a "green" building product.

TPO membranes are typically installed with mechanical attachment or are fully adhered to the roof surface. The combination of the built-in reinforcement fabric and TPO plies provides this membrane type with relatively high breaking and tearing strength along with a resistance to punctures. As with other thermoplastic roofing materials (i.e., PVC), TPO can be field-welded with a hot air gun during application, eliminating the need for adhesives at splice and perimeter terminations.

4.6.2 THERMOPLASTIC MEMBRANE ROOF LIFE AND COMMONLY ENCOUNTERED SURFACE DEFECTS

The mean service life of PVC membranes is ~13.8 years, and the mean service life of a TPO membrane is 12.7 years.[3] The types of defects found on thermoplastic roof membrane surfaces vary based on age and workmanship. Defects tend to increase with the age of the membrane and accelerate as the roof surface reaches its effective service life. Examples of common types of defects to TPO membrane roof surfaces, along with their probable causes, are illustrated in Table 4.4.

TABLE 4.4

Examples of Thermoplastic (TPO) Membrane Defects and Probable Causes

Defect Type	Probable Cause	Photograph
Membrane separation at perimeters	Expansion and contraction of membrane due to lack of perimeter attachment	

(Continued)

TABLE 4.4 (*Continued*)
Examples of Thermoplastic (TPO) Membrane Defects and Probable Causes

Defect Type	Probable Cause	Photograph
Protruding fasteners causing cracks to membrane	Movement of the structure, or moisture intrusion from preexisting conditions	
Separated seams	Expansion and contraction of membrane due to lack of perimeter attachment	
Fractures around fastener plates	Age-related stress fracturing due to a loss of plasticizer	

4.6.3 THERMOPLASTIC MEMBRANE ROOF SURFACES – THRESHOLDS FOR HAIL-STRIKE DAMAGE

Research has been completed on hail size thresholds for hail-strike damage to PVC membranes; however, no information was available for hail-strike damage to TPO membrane surfaces, but it would be expected to be above 1.0 inch in diameter hailstones. For PVC roof surfaces, research found that hail-strike damage occurred to aged PVC membranes when struck by hailstones measuring 1.0 inch in diameter[14] and that older PVC membranes with higher measured plasticizer loss were even more susceptible to hail-strike damage even though the threshold was still 1.0-inch diameter hailstones. It should be noted that this threshold of 1.0 inch in diameter hailstones appears to be the lowest for single-ply membrane systems. Dozens of PVC hail inspections performed by EES have also shown that the hailstone threshold for damage is 1.0 inch in diameter.

4.6.4 THERMOPLASTIC – INSPECTION CASE STUDIES

4.6.4.1 Hail-Strike Damage to Aged PVC Membrane

In this case study, a roof was inspected at a car dealership that was finished with a PVC membrane and was reportedly 17 years old. Numerous age-related defects were observed throughout the membrane surface. These observations, coupled with age life data, suggested that the PVC membrane was nearing the end of its effective service life. The roof had been leaking for some time and many patch repairs had been made to the roof surface. Reportedly, the leaks worsened over time, which resulted in the building owner contacting their insurance carrier and filing a hail-strike damage claim. Hail damage was reportedly found throughout the membrane roof surfaces during subsequent inspections by an insurance adjuster and a roof inspector.

During the subsequent forensic inspection, small dents up to 0.5 inch in diameter were found on the metal surfaces of the building that were consistent with hail-strike impacts. These dents suggested that hail up to 1.0 inch in diameter had struck the building. A review of past weather data from the area indicated that at least six hailstorms, spanning a 4-year period, had dropped hailstones up to 1.0 inch in diameter in the vicinity of the building. This weather data confirmed site-specific hail sizes based on hail-strike dent observations. The PVC roof surface contained hundreds of circular fractures consistent with hailstone impacts (Figure 4.21).

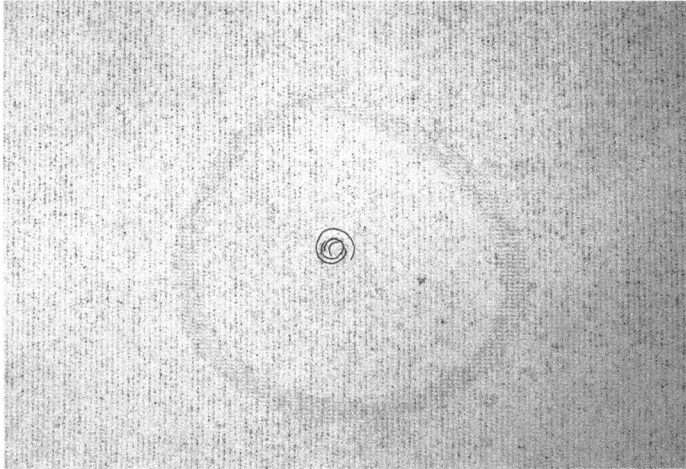

FIGURE 4.21 Hail-strike fracture in PVC membrane roof surface.

The number of fractures ranged from 5 to 13 per 100 square feet of roof surface area. Four test cuts were made into the membrane to verify that the fractures were caused by exterior impacts (i.e., hailstones). At each test cut location, the membrane was fractured, the indentations penetrated through the PVC membrane, and the insulation below the membrane was wet. This suggested that the fracturing of the membrane surface was likely a result of hail-strike damage, which allowed water to enter the roof system substrates (Figure 4.22).

FIGURE 4.22   Test cut into hail-struck damaged PVC membrane – moisture on insulation below.

This investigation demonstrated that hail-strike damage from hailstones as small as 1.0 inch in diameter was capable of causing widespread damage to aged PVC membranes.

4.6.4.2   Hail-Strike Damage to Newer PVC Membrane

A large hotel facility, located in the path of a hailstorm, was inspected for hail-strike damage. The low-sloped roof surface was finished with a PVC membrane; the steep-sloped roof surfaces were covered with standing seam metal panels and dimensional asphalt shingles. The PVC membrane roof surface on the main building was reported to be ~10 years old; another area of PVC roof membrane, installed over a pool building, was reported to be 4–5 years old. Following the hailstorm, the main hotel and pool areas covered with PVC roof membranes were showing evidence of roof leaks (e.g., ceiling staining).

During the forensic inspection, surveys of the metal surfaces indicated the size of the hail that struck the building was up to 1.0 inch in diameter. This maximum size of hailstone was consistent with weather data records from hailstorms reported in the area. Distinct circular fracture marks and indentations were present throughout the surface of the membrane (Figure 4.23). These hail-strike marks penetrated through the membrane.

FIGURE 4.23   Hail-strike fractures in PVC membrane roof surface.

Of specific interest, the impact mark intensities varied greatly between the main hotel (10 years old) and pool (4 years old) roof surfaces, both of which appeared to have been subjected to the same hailstorms. The older PVC membrane surface contained up to 36 hail-strike fracture marks per 100 square feet; the newer PVC membrane over the pool contained ten hail-strike fracture marks per 100 square feet of roof surface area. This investigation confirmed that a hailstone threshold of 1.0 inch in diameter was capable of causing functional damage to PVC membrane roof surfaces and confirmed research by Foley et al.[14] that the degree of hail-strike damage increases with roof age for PVC membrane roof surfaces.

4.7 METAL ROOF SYSTEMS

4.7.1 Description

While the initial cost of a metal roof installation is typically greater than single-ply or BUR systems, they are one of the lowest long-term maintenance cost commercial roof systems. In the past, traditional metal roofs consisted of lead and copper metals, providing an aesthetically appealing finish. However, due to the cost and craftsmanship required to install lead and copper roof systems, they were increasingly less utilized. With the advent of coated steel and aluminum metals, this changed and these roof systems have gained more popularity.

The modern metal roof systems, based primarily on coated steel panels, are divided into two classes: structural standing seam metal roofs and architectural metal roofing panels with flashings. A structural standing seam roof is a panel that spans more than 3 feet and has the ability to resist gravity and wind-uplift loading. Architectural metal panels provide an aesthetic function and rely on steep slopes (6:12 or greater) to shed water from the roof surfaces. Architectural metal panels must be supported by a solid roof deck. Both of these metal roof systems can also be defined as fixed or floating metal roof systems. A fixed metal roof system is one where the metal is through-fastened along the longitudinal seams and requires expansion joints at intervals of 30 feet or less. A floating metal roof system is typically one where the panel is fixed at one location along the panel length and movement of the panels is allowed through the use of sliding clip fasteners.

One of the main limitations of metal roof systems is the propensity of the material for thermal expansion. A metal roof system can reach temperatures up to 150°F on summer days and as cool as −30°F on colder winter days. This implies that the metal roofing system can see temperatures that range by up to 180°F. Technically, the change in temperature (i.e., differential thermal expansion) causes movement in the panels (shorten with decreasing temperature and lengthen with increasing temperature), which creates force on the fasteners and/or crimped seams since portions of the system are tied to a fixed position.

Panel clips are used to secure the panels to the purlins (members spanning across the tops of rafters). The two types of clips used in the metal roofing industry are single (fixed) or dual (sliding or articulating). Single-component clips are typically used for hydrokinetic (steep slope) roof systems and dual-component clips are typically used for hydrostatic (low slope) roof systems. Single-component clips are generally not used for most low-sloped roof systems because of the need to allow the metal roof panels to move as the roof temperature varies.

Temperature-induced thermal movement can cause clips, fasteners, and/or metal panels to move, which in turn provides the opportunity for roof leaks and resulting water intrusion into a building.

4.7.2 Metal Roof System Life and Commonly Encountered Surface Defects

Most defects with metal panel roof systems occur due to installation, thermal movement, or natural aging. Typical modern metal panel roof systems have a mean service life of 26.5 years; one of the highest service lives among low-sloped roof systems. Examples of common types of defects to metal roof surfaces, along with their probable causes, are summarized in Table 4.5.

TABLE 4.5
Metal Panel Defects and Probable Causes

Defect Type	Probable Cause	Photograph
Gaps forming at crimp seams	Differential movement causing abrasion through the seam	
Fastener pullout	Thermal movement elongating holes	
Fastener pullout and separated end lap seams	Aging and degradation of the metal panels	

4.7.3 MODERN METAL PANEL ROOF SYSTEMS – THRESHOLDS FOR HAIL-STRIKE DAMAGE

Damage to metal roof systems is defined as either functional damage or cosmetic damage. In a technical bulletin issued by the United States Steel Company (USS),[15] hailstone damage is characterized as aesthetic damage or functional damage. Functional damage is defined as damage that diminishes the water shedding capabilities and reduces the expected surface life of the roof. For a metal roof panel to be considered functionally damaged by hail impacts, there must be evidence that hailstones have reduced the water-shedding capability or reduced the expected long-term service life of the metal roof panel. Hail-caused functional damage can occur in the following ways:

- *Rupturing the metal*: Penetration or puncture in the metal roof panel with enough impact energy to cause cracks or splits in the surface or protective coating.
- *Disengagement of lapped elements*: Hail-caused dent or "impression" at a location along the seam of the metal roof panels that would create a gap or disengagement of the lapped elements, which would disrupt the water-shedding capabilities of the metal roof system at that particular location.
- *Disengagement of a fastener*: Hail-caused dent or "impression" at a fastener location for a metal roof panel that would create a gap or disengagement of the fastener, which would disrupt the water-shedding capabilities of the metal roof system at that particular location.

Literature reports that functional hail damage to metal roofing panels will not occur for metal roofing until hail reaches 2.5 inches in diameter or greater.[16] Typically, hail with a diameter greater than this threshold is required to cause penetrations that cause functional damage to metal roofing panels. EES has inspected modern metal panel roof systems buildings that had been impacted by hailstones from up to 2.5 to 3.0 inches in diameter and has not observed functional damage to these systems.

Cosmetic hail damage reported as dents or impressions (see Figure 4.24) will have an adverse effect on the appearance but does not affect the water-shedding performance of the roof system or the expected service life.

FIGURE 4.24 Hail-strike dents in standing seam metal panels.

4.7.4 MODERN METAL PANEL ROOF SYSTEM – INSPECTION CASE STUDY

4.7.4.1 Hail-Strike Damage to Metal Roof on a Warehouse Building

A 20-year-old warehouse building was reportedly struck by large hailstones in the early summer of 2011. The roof was covered with standing seam metal panels. The roof system was in relatively good condition, although some ongoing roof leaks were present at/near piping penetrations through the metal roof panels.

During repairs around one of the pipes, a contractor noticed dents in some of the metal surfaces and the owner subsequently filed a claim with their insurance carrier for hail-strike damage to the roof system. During a subsequent forensic investigation, the size of the probable hail-strike dents in the exterior and roof surfaces of the building suggested that hail up to 2.5 inches in diameter had struck the building. Numerous dents were observed across the metal panels (Figure 4.24).

The dents did not harm the protective coating nor cause seam separation when the integrity of the seams was tested. Thus, the hail-strike damage was concluded to be cosmetic since the dents would not cause loss of the water-shedding capability or shorten the life of the roof system.

4.8   SPRAYED POLYURETHANE FOAM (SPF) ROOF SYSTEMS

4.8.1   DESCRIPTION

Sprayed Polyurethane Foam (SPF) roof systems were introduced into the market in the late 1960s. The advantages of these systems are their insulating and solar reflectivity characteristics.

Proper application of this system is somewhat complex and requires qualified workmanship, training, and attention to sensitive weather conditions, lift requirements, curing, and surface finish. After a number of failures related to workmanship and the improper application of these types of roof systems, proper installation guidelines were developed by the Spray Polyurethane Foam Alliance (SPFA). The guidelines were developed for proper application, specifications, and detailing of the foam and coatings.

SPF is a spray-applied liquid mixture that forms a waterproof membrane over the roof deck during application. The liquid mixture combines a Part A isocyanurate with a Part B hydroxyl resin or polyol. When combined during installation, the mixture will expand from 20 to 30 times its original volume within a few seconds. After this foam mixture sets, a protective coating is then applied to the surface of the foam.

Weather conditions during the installation and application practices are critical to the proper installation of SPF roof systems. The temperature should be at least 40°F and wind speeds should not exceed 12 MPH. Prior to applying SPF, the roof deck must be properly inspected and cleaned. Applying SPF over dust, rust, dirt, or other contaminants will restrict the ability of the foam to adhere to the roof deck. Any moisture present on the roof deck can limit the adhesion and has the potential to disrupt the chemical reaction between the two-part foaming agents. Further, the foam can only be applied in a minimum of 1/2″ per pass and the full foam thicknesses should be applied in a single day. If the size of the roof surface does not permit a one-day application, the area should be divided into segments. The surface finish must resemble smooth orange peel, coarse orange peel, or on the verge of popcorn prior to applying the protective coating. The protective coating should be applied on the same day as the foam to protect it from ultraviolet degradation.

The SPF must be coated for protection from exposure to sunlight and to provide resistive properties for foot traffic and abrasive forces. The coatings can consist of polyurethane elastomers, acrylics, and silicones. Mineral surface aggregate can be applied to the top coat to provide ultraviolet protection and to increase fire resistance to the roof surface.

4.8.2   SPF ROOF SURFACE LIFE AND COMMONLY ENCOUNTERED DEFECTS

The SPFA recommends that the coating be frequently inspected and reapplied every 8–15 years. Assuming that the integrity of the surface coating is maintained, the roof systems should have a life in excess of 20 years.

Defects or issues with SPF roof systems are typically caused by installation and aging. If the foam is applied onto wet substrates, eventually the foam will blister, causing holes, which will cause voids in the foam over time. The surface texture following application is also important in order to provide an acceptable substrate for the application of the protective coating. If the substrate finish is too rough, the dry film coating will have varying thicknesses and coverage. This condition can make the roof surface more prone to voids or pinholes, which can exacerbate age-related cracking. Examples of common types of defects to SPF roof surfaces, along with their probable causes, are summarized in Table 4.6.

TABLE 4.6

Examples of SPF Defects and Probable Causes

Defect Type	Probable Cause	Photograph
Severe degradation	Heavy ponding water causing coating and foam degradation	
Surface cracking	Moisture entrapment and degraded coating. Lack of proper maintenance to coating system	
Blisters and bubbling	Moisture entrapment due to wet substrates or water intrusion	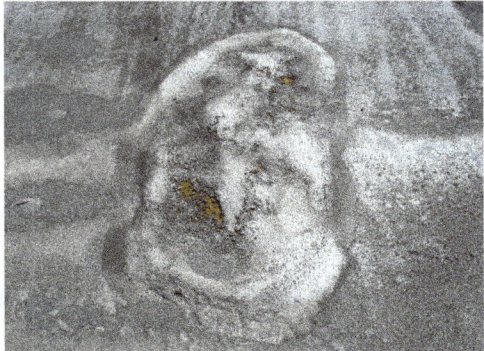
Splitting	Improperly fastened/adhered insulation causing movement of the substrate	

4.8.3 SPF Roof Finishes — Thresholds for Hail-Strike Damage

No single threshold exists for SPF roof systems. Experience suggests that hailstones as small as 0.75 inch in diameter can cause surface damage to SPF membrane surfaces.

The thresholds for hail-strike damage to SPF roof finishes are defined by various categories, with the effective damage to the system increasing with increasing hailstone size. Hail-strike damage to SPF roof coverings is classified from minor to severe. Minor damage (caused by hailstones from 0.75 up to 2.0 inches in diameter) is associated with a bruise or fracture to the protective coating with no penetration into the foam. Since the closed-cell foam structure of the membrane tends to repel water, this minor damage is typically not susceptible to water intrusion and can be repaired by caulking and recoating the blemishes. More severe hail damage (2.0 inches in diameter and greater) occurs when the hailstones puncture through the foam leading to immediate water intrusion. This damage would result in removal and reapplication of the foam.

Since the levels of damage can vary based on hailstone size and age/deterioration of the SPF, the SPFA issued a paper titled "Recommendations for Repair of Spray Polyurethane Foam (SPF) Roof Systems Due to Hail and Wind Driven Damage."[17] This paper provides a breakdown on repair methods for various hail sizes and the number of defects. These repair versus replacement recommendations and methods are summarized in Table 4.7.[17]

TABLE 4.7

SPFA Extent of Hail-Strike Damage to SPF Roof Systems by Hailstone Size and Recommended Repairs

Degree of Damage	Size and Severity	Extent per 100 square feet	Recommended Repair
Light	1/2″ diameter or less and <1/8″ deep	Less than ten cracks, cuts and/or dents	Caulk and coat dents, cuts, and cracks. Note: Recoat should be considered based on remaining service life.
		More than 20 cracks, cuts, or dents	Recoat as required to fill in cracks. Note: Some caulking may be required to fill in cracks
Moderate	1/2″ to 3/4″ diameter and <1/4″ deep	Less than ten cracks/dents	Coat/caulk cracks
		More than 20 cracks, cuts, or dents	Recoat as required to seal cracks. Note: Some caulking may be required to seal deeper cracks
Heavy	3/4″ to 1-1/2″ and 1/4″ to 1/2″ deep	Less than ten cracks/dents	Removed damaged SPF: Caulk holes and recoat as required
		More than 20 cracks/dents	Scarify 1/2″ of roof surface: Re-foam and coat
Severe	1-1/2″ or larger and 1/2″ or deeper	Less than ten cracks/dents	Removed damaged SPF: Caulk holes and recoat as required
		More than 20 cracks/dents	Scarify 3/4″ of roof surface: Re-foam and coat

Source: Courtesy of spray polyurethane foam alliance.

Repair recommendations are based on a thorough and detailed inspection to evaluate the degree, size, and severity of the damage caused by hail-strike impacts. Test cuts will also be required to determine the depth of the impacts and to determine if water has entered the roof system. If water or high levels of moisture have been found in the roof system below the coating, at least that impacted section must be removed and replaced and cannot simply be repaired by recoating the surface.

4.8.4 SPF – INSPECTION CASE STUDIES

4.8.4.1 Hail-Strike Damage to SPF Roof – No Reported Roof Leaks

In this case, the SPF roof surface on a commercial building was determined to have been impacted with hail measuring up to 1.5 inches in diameter. The roof was covered with a SPF roof system that was ~2 inches thick. The roof system was ~20 years old, was not drained well, and the surface coating was deteriorated in appearance. Numerous apparent hail-strike dents and cuts, measuring up to 0.75 inch in diameter, were present in the roof surface (Figures 4.25 and 4.26).

FIGURE 4.25 Probable hail-strike damage to SPF roof surface.

FIGURE 4.26 Probable hail-strike damage to SPF roof surface.

The average hail-strike damage equaled to 23 impacts per 100 square feet of roof surface area. Two tests cuts made at the largest impact marks indicated that the hail penetrated no more than 0.5 inch into the foam (Figures 4.27 and 4.28).

FIGURE 4.27   Test cut into SPF roof system.

FIGURE 4.28   Hail-strike impact depth into SPF roof system.

Based on destructive testing and moisture meter test results, it was determined that the closed-cell foam structure prevented the lower foam layers from becoming saturated with water. No major leaks were occurring into the interior of the building.

According to the recommendations outlined in Table 4.7,[17] the damage was characterized as heavy. Based on field findings and the SPFA guidance, recommendations were made by the forensic investigator to scarify the top 1/2″ of the SPF surface, replace this 1/2″ of foam, and then reapply the surface coating. The information gained from the test cuts, coupled with SPFA repair recommendations, allowed this SPF roof to be repaired rather than removing and replacing the entire roof system.

4.8.4.2   Hail-Strike Damage to SPF Roof – Reported Roof Leaks

In this case, the SPF roof surface on a large commercial building was determined to have been struck with hail and roof leaks were causing water damage within the building. A contractor investigating the cause of the leaks found dents in the metal surfaces and many holes and dents in the SPF roof system. A claim was filed with the owner's insurance carrier for hail-strike damage to the roof system.

During the subsequent forensic investigation, an interview was performed with a long-term maintenance employee. It was determined that the SPF was 15 years old and that it had been applied over a BUR roof system. A few leak areas began occurring 4 years prior to the reported hailstorm event. According to the maintenance person, the leakage became more widespread during the past

3 years. Some roof repairs had been completed by the maintenance person, but the roof-leak situation continued to worsen.

A review of the hailstorm history in the area indicated that the last hailstorm that passed over the area occurred 4 years prior, at/near the timeframe when the roof began leaking. The SPF roof surface was in very poor condition. It was poorly drained and contained large areas of water that was ponded across the roof surface (Figure 4.29).

FIGURE 4.29   Water ponded on SPF roof surface.

The SPF roof surface was heavily blistered and contained numerous surface splits. In four areas, the SPF was observed to be floating on the roof. The feel when walking on the roof surface was like that felt when walking on a waterbed (Figure 4.30).

FIGURE 4.30   SPF floating on ponded water.

The size of the hail that struck the building was about 1″ in diameter. The membrane surface contained numerous circular impact marks that were consistent with hailstone impacts (Figure 4.31).

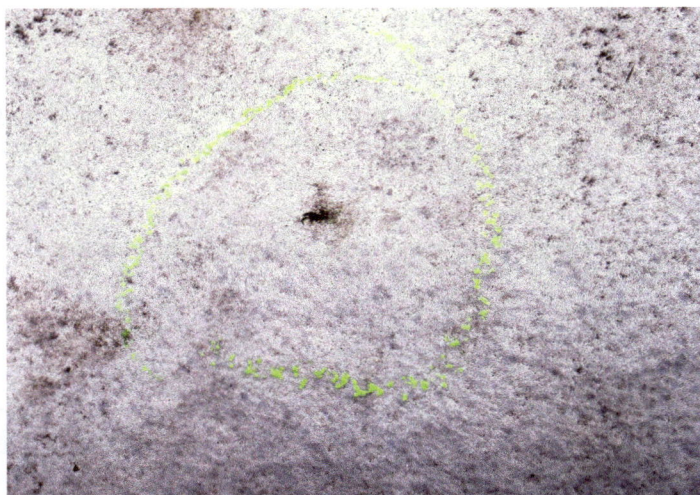

FIGURE 4.31   Probable hail-strike damage to SPF roof surface.

The intensity of probable hail-strike impacts was measured and determined to be 4–5 hits per 100 square feet of roof surface area. Using the SPFA criteria in Table 4.7[17], this level of damage would be characterized as light to moderate. The repair recommendations would have been to simply caulk and seal the defects. However, because of the known roof leaks, the poor condition of the roof surface, the lack of immediate repairs following the hailstorm, and the resulting water penetration into the roof system, this recommendation could not be followed. The poorly drained roof surface had exacerbated water entry into the roof system and degraded the SPF roof finish. Thus, due to passage of time (~4 years) since the hailstorm and the resulting entry of water into the roof system over this period of time, the SPFA recommendations could not be followed. Given the delay in repairs, the only option available was a more costly removal and replacement of the roof system.

IMPORTANT POINTS TO REMEMBER

- Hail damage to low-sloped roof systems can be difficult to identify and quantify and will vary based on the type of roof system. Blistering to bitumen-based roof systems can be easily misinterpreted as a hailstone impact. Fastener protrusions and age-related cracking can also be easily misinterpreted as hail damage on single-ply and SPF roof membranes. In all cases, the forensic inspector should be prepared to make test cuts (i.e., destructive testing) into a roof system. Test cuts are often the only way to determine whether or not hail damage is, or is not, present to commercial roof systems within a reasonable degree of engineering or scientific certainty.
- A detailed interview with the person(s) having knowledge of the building and/or roof history is important. Experience has shown that most building owners are not aware of the age or condition of their roof surfaces. Maintenance staff can often be an excellent source of providing information.
- With exception of PVC and SPF roof systems, it is extremely rare that hail damage will cause immediate leakage through a roof system.

- For commercial low-sloped roof systems, the following thresholds can be used regarding hail-strike damage:
  - Hailstones must be 2.0 inches in diameter or greater to cause functional hail-strike damage to BUR membrane roof systems. Gravel BUR roof surfaces will have even higher hail size thresholds. The intensity of hail-strike damage to BUR roofs will increase with the age of the roof surface.
  - Hailstones must be 2.0 inches in diameter or greater to cause functional hail-strike damage to Mod-Bit membrane roof systems.
  - Of all the single-ply roof systems, EPDM membranes are the most resistant to hail. Hailstones must be 2.5 inches in diameter or greater to cause functional hail-strike damage to these roof surfaces.
  - Thermoplastic PVC membranes have been found to be the least resistant membrane surface to hail-strike damage. Widespread functional hail-strike damage can occur with hailstones measuring 1.0 inch in diameter or greater to PVC membrane roof systems.
  - Modern metal roof systems can easily be dented, creating a less pleasing aesthetic, but are very resistant to functional hail-strike damage. Hailstones must be 2.5 inches in diameter or greater to cause functional damage to modern metal roof systems.
  - The damage and reparability of SPF roof systems depend on the size of the hailstones to strike the roof system, the roof age and condition, and the timeframe between the storm and repairs.

REFERENCES

1. Ashley, St. John. "In with the new." Professional Roofing, July 2011, p. 8.
2. Griffin, C. W. and Richard L. Fricklas. *Manual of Low-Sloped Roof Systems: Fourth Edition*. New York: McGraw Hill, 2006, p. 209.
3. D'Annunzio, John. "Service Life Predictability." Roofing Contractor, May 2004, p. 14.
4. Hoff, James. "A New Approach to Roof Cycle Analysis." Interface Magazine, January 2007, pp. 5–12.
5. Greenfeld, Sidney H. "Hail Resistance of Roofing Products." Building Science Series #23, National Bureau of Standards, August 1969, p. 9.
6. Mathey, Robert G. and William C. Cullen. "Preliminary Performance Requirements for Bituminous Membrane Roofing." National Building Standards, National Bureau of Standards, November 1974.
7. Marshall, Timothy P. and Scott J. Morrison. "Hail damage to built-up roofing." Haag Engineering. *Presented at the American Meteorological Society Conference*, October 4–8, 2004, Seattle, WA.
8. Crenshaw, Vickie and Jim D. Koontz. "Simulated hail damage and impact test resistance test procedures for roof coverings and membranes." *Presented at the Third International Symposium on Roofing Technology*, April 17–19, 1991, Gaithersburg, MD.
9. Haag Engineering. "Hailstorm Characteristics." 2006. http://www.haagengineering.com/ehail/chas/eHail/hailstorm.html.
10. Roofing Industry Committee on Weather Issues, Inc. (RICOWI, Inc.). "Hailstorm Investigation – Dallas", Fort Worth, TX, May 24, 2011. Accessed September 10, 2020, https://docplayer.net/10564289-Hailstorm-investigation-dallas-fort-worth-tx-may-24-2011.html.
11. Griffin, C. W. and Richard L. Fricklas. *Manual of Low Slope Roof Systems, Fourth Edition*. New York: McGraw-Hill Companies, January 2006, p. 282.
12. Roofing Industry Committee on Weather Issues, Inc. "Hailstorm Investigation Oklahoma City, Oklahoma", April 21, 2004. Oklahoma: RICOWI, Inc., November 1, 2005.
13. Koontz, Jim D. and Thomas W. Hutchinson. "Hail impact testing of EPDM roof assemblies." *Presented at the RCI 24th International Convention and Trade Show*, Dallas, TX, March 12–19, 2009.

14. Foley, Frank J., Jim D. Koontz, and Joseph K. Valaitis. "Aging and Hail Research of PVC Membranes." Accessed September 11, 2020, http://www.heartlandap.com/Websites/AndreasDaughtersite/Blog/3767668/Aging%20and%20Hail%20Research%20of%20PVC%20Membranes.pdf.
15. United States Steel Corporation. "Hail Damage on Coated Sheet Steel Roofing." Technical Bulletin TBP 2005, p. 17.
16. e-Hail™ Report Generator Provided by Haag Engineering. "Hailstorm Characteristics." June 3, 2009. http://www.haagengineering.com/ehail/chas/eHail/hailstorm.html.
17. Spray Polyurethane Foam Alliance. "Recommendations for Repair of Spray Polyurethane Foam (SPF) Roof Systems Due to Hail and Wind Driven Damage." January 2003.

5 Synthetic Storm Damage (Fraud) to Roof Surfaces

Stephen E. Petty
EES Group, Inc.

CONTENTS

PURPOSE/OBJECTIVES

The purpose of this chapter is to:

- Demonstrate how often synthetic (fraud) hail damage may be observed during hail/wind damage inspections.
- Identify/illustrate where hail and wind fraud typically occurs.
- Provide methodology to determine whether or not artificial/fraudulent hail damage has occurred to inspected surfaces.
- Provide methodology to determine whether or not artificial/fraudulent wind damage has occurred to inspected surfaces.

Following the completion of this chapter, you should be able to:

- Understand where typical fraudulent hail and wind damage occurs on structures.
- Be able to distinguish differences between real and artificial hail/wind damage (i.e., coin scrapes, ball-peen hammer, and golf ball in a sock).

5.1   INTRODUCTION

The terms "artificial," "questionable," "synthetic," and "fraud" can be used somewhat interchangeably to describe intentional man-made damage to building finished surfaces. Heretofore, the term "fraud" or "fraudulent" will be used to represent all these terms.

As a consequence of either difficult times or the proverbial need to "get something for nothing," fraudulent storm damage attributed to the actions of hail and wind has been increasing. Data from The National Insurance Crime Bureau (NICB)[1] reported that during the timeframe of 2006–2009, hail claims increased 61.7% while questionable hail claims have more than doubled to 136.1%. Overall, from 2006 to 2009, total questionable hail insurance claims rose from 301 to 711 (Table 5.1 and Figure 5.1).

TABLE 5.1
Questionable Hail Insurance Claims – 2006–2009 – Total and by State

Location	2006	2007	2008	2009	Totals	Totals (%)
Texas	52	71	84	310	517	29.03
Illinois	37	25	114	81	257	14.43
Minnesota	17	39	55	17	128	7.19
Indiana	96	14	4	9	123	6.91
Colorado	8	18	15	76	117	6.57
Georgia	3	5	40	47	95	5.34
Ohio	9	43	13	16	81	4.55
Louisiana	8	8	13	9	38	2.13
Missouri	15	8	21	9	53	2.98
Kansas	5	3	20	24	52	2.92
Top ten states	**250**	**234**	**379**	**598**	**1,461**	**82.09**
All states	**301**	**275**	**492**	**711**	**1,780**	**100.00**
% Increase/year	**N/A**	**−8.60%**	**78.79%**	**44.45%**	**Overall:**	**136.05**

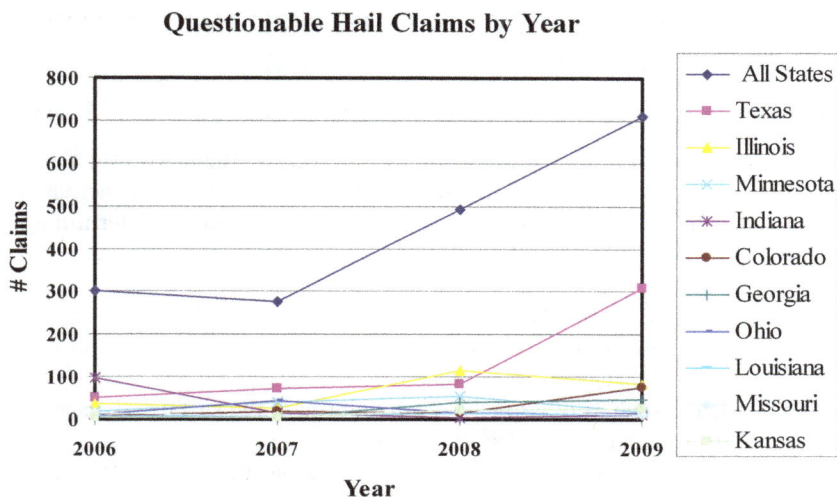

FIGURE 5.1   Questionable hail insurance claims with time – 2006–2009 – total and by state.

The top ten states for questionable hail claims over this 4-year period were Texas (517 – 29.05%), Illinois (257 – 14.44%), Minnesota (128 – 7.19%), Indiana (123 – 6.91%), Colorado (117 – 6.57%), Georgia (95 – 5.34%), Ohio (81 – 4.55%), Louisiana (38 – 2.14%), Missouri (53 – 2.98%), and Kansas (52 – 2.92%). These states reported 82.1% of all the questionable hail claims reported during this time frame.

Comparing data in Table 5.1 and Figure 5.1, a disproportionate number of questionable hail claims were filed in the states of Illinois and Indiana. The result of these questionable claims has been media reports of hail claim fraud to roof and exterior finished surfaces[2,3] and legal action filed by insurance companies against those reported to be committing such fraud.[4] Interestingly, the disproportionately high number of questionable hail damage claims filed in Indiana has decreased with time after State Farm filed legal action[4] against a firm and individuals reportedly performing hail damage fraud to asphalt-shingled roof surfaces.

Based on experience, most of the hail and wind fraud appears to be concentrated on finished roof surfaces (e.g., asphalt shingles) with some hail fraud to metal siding, downspouts, and gutters. Moreover, since asphalt shingles are used to cover four out of five residential roof surfaces in the United States,[5] much of the fraud found on finished roof surfaces is to asphalt shingles. In fact, given the extent of suspected fraud to asphalt-shingled roof surfaces by a major insurance carrier, EES Group, Inc. was asked by the carrier to prepare a Technical Bulletin on this topic. This bulletin was based on original research on this topic, including simulating such roof damage experimentally.[6]

Therefore, based on insurance industry statistics, press reports, and experience, it is apparent that fraud claims are on the rise. The focus of this chapter will be on the recognition and evaluation of hail fraud damage to asphalt-shingled roof surfaces with some discussion of fraud damage associated with wind to roof surfaces and hail to other surfaces such as metal siding.

5.2   RECOGNITION AND EVALUATION OF POSSIBLE HAIL DAMAGE FRAUD TO ASPHALT-SHINGLED ROOF SURFACES

5.2.1   INTRODUCTION TO HAIL DAMAGE FRAUD TO ASPHALT-SHINGLED ROOF SURFACES

Without firsthand observations of the damage being inflicted, it is difficult to absolutely prove or dis-prove fraud claims, particularly the individual or company responsible for the activity since rarely does one actually view the fraudulent event.[6,7] The difficulty in definitively determining whether or not fraud has occurred lies within the following factors:

- No actual observation(s) of the damage being inflicted to the roof surface (i.e., insufficient evidence to tie the damages to a specific person or company, or more than one company has inspected the roof surface for storm damage).
- Confounding factors such as other mechanical damage to the surface, the age, condition, or type of the surface, and/or actual storm damage.
- Lack of information on what instrument, object, or tool may have caused the damage along with the original condition of the component prior to the inflicted damage (e.g., date of damage – age, temperature of shingles when damage occurred, type of surface damaged – roof vs other materials, and foot traffic history – homeowner and/or other contractors).
- Lack of information regarding damage caused by various instruments used to create fraudulent damage.

However, using available inspection evidence, knowledge of the physical tendencies of actual hail defects (e.g., hail-strike bruise) versus those committed during a fraudulent event and experi-ence, one can determine whether the damage observed is, or is not, consistent with fraud damage. Determining the responsible party may be more difficult, especially if more than one contractor has inspected the roof surfaces.

5.2.2   CHARACTERISTICS OF FRAUDULENT HAIL DAMAGE TO ASPHALT-SHINGLED ROOF SURFACES

Upon inspection of many roof surfaces, including asphalt, factors that may guide the inspector to suspect that some, if not all, of the reported/inspected hailstorm roof damage may be fraudulent are as follows:

- Damage is concentrated in or limited to easily accessible roof areas.
- Unusual damage patterns.
- A storm event was either (1) not present in the area, or (2) the size of the hail was not con-sistent with hail known to have fallen from the storm event.
- Presence/patterning of simulated hail-strike defects not consistent with the size of the hail that fell (or did not fall) and/or patterning of the defects.

Further discussion of these factors (red flags) follows. Bear in mind that natural acts such as hail-storms should result in hail-strike damage that is random, relatively indiscriminate, consistent with the size of the hail that struck the area, and consistent with the direction from which the storm arrived.

5.2.2.1 Damage Is Concentrated in, or Limited to, Easily Accessible Roof Areas

The following overall location factors are often seen with artificial damage on roof surfaces:

- The damage is concentrated on low-sloped, less-steep roof elevations.
- The damage is concentrated near roof valleys, ridges, or rakes (i.e., easily walked).
- The damage is concentrated in areas that restrict the view of onlookers (i.e., back side of home).
- The damage is found on roof surfaces opposite the side of where the weather or storm(s) arrived.

The concentration of fraudulent hail damage in more easily accessible roof areas is an indication of possible fraud (Figure 5.2).

FIGURE 5.2 Probable hail fraud damage to asphalt shingle roof – damage near accessible location(s) and back of home.

Fraudulent roof hail damage is often concentrated on the back of a home or where the perpetrator cannot be seen intentionally damaging the roof. These location factors should be distinguishable from legitimate damages such as heaviest damage found on roof elevations facing the incoming storm or on susceptible shingles.

5.2.2.2 Unusual Damage Patterns

The following patterning factors often are seen with artificial damage on roof surfaces:

1. The defects occur or are more numerous on roof elevations opposite the direction from which the hailstorm arrived (Figures 5.3 and 5.4). Figure 5.3 is a schematic of the example roof with bruise counts (defects/100 square feet of roof area), and Figure 5.4 is a photograph of the actual roof area.

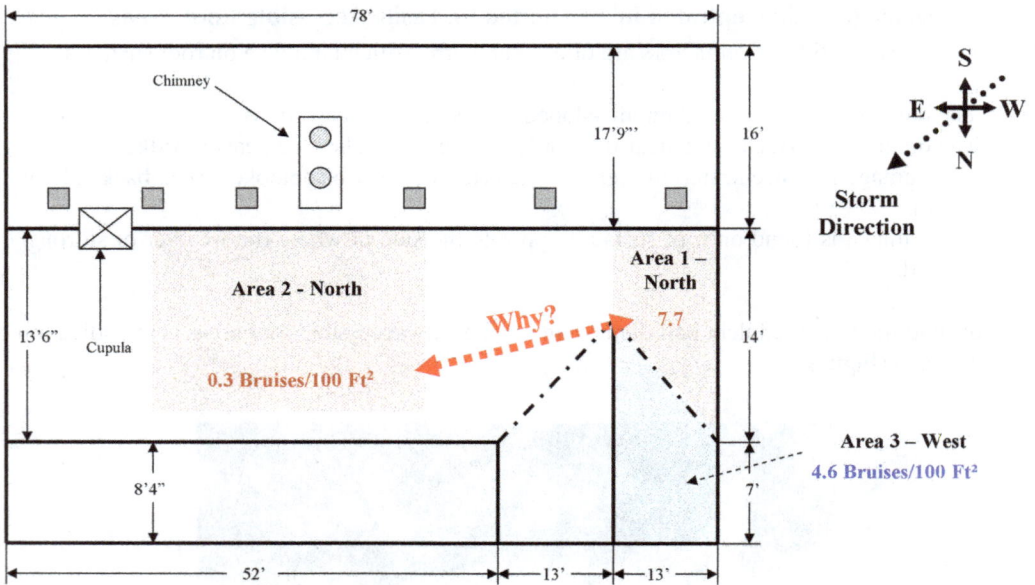

FIGURE 5.3   Inconsistent location of extent of hail damage to roof shingle surfaces.

FIGURE 5.4   West end of north roof elevation – high number of shingle defects.

As illustrated in Figure 5.3, in a situation where the hailstorm arrived from the south-west, one would expect the greatest defects to occur on the south and west roof elevations as was observed to exterior finished surfaces, gutters, and downspouts. However, in this case, the defect counts on two north-facing elevations were inconsistent (0.3 and 7.7) and with those on the west elevation (4.6; elevation facing the storm should be the highest). The suspect Area 1 is also readily accessible from a nearby valley.

2. The damages or defects are oriented in linear, wavy, or zigzag patterns rather than random (Figure 5.5).

FIGURE 5.5   Zigzag defect pattern.

3. The defects are centered on shingle exposures (Figure 5.6).

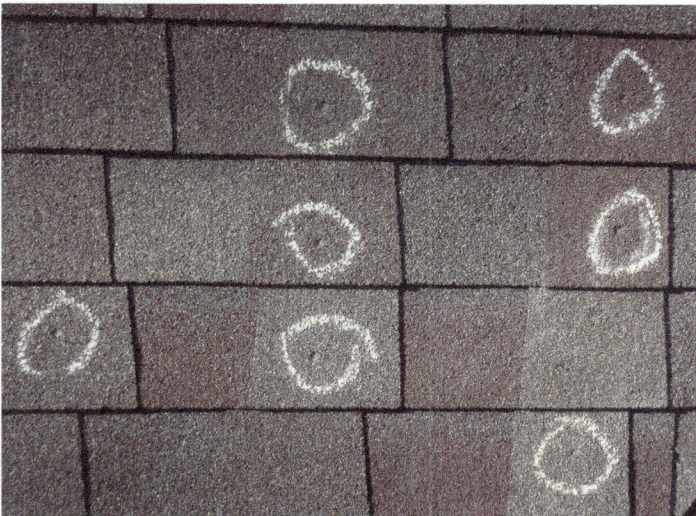

FIGURE 5.6   Defects centered on shingle tabs – ball peen hammer.

4. The defects are consistent in size, orientation, or appearance (Figure 5.7); see also Section 5.2.2.4.

FIGURE 5.7 Consistent and localized mechanical damage.

Fraud is normally associated with abnormal patterning that is: (1) inconsistent with respect to storm and slope directions, (2) very consistent or uniform in patterning within a limited area or areas, or (3) oddly mechanical in appearance and within limited areas. These are consistent "tells" regarding hail fraud damage to asphalt shingles.

5.2.2.3 Storm Event(s) Either Not Present in the Area or Size of Hail Not Consistent with Hail-Stone Sizes from Known Storm Events

Numerous research sources are available to determine if storm events occurred in the area of interest and/or the maximum size of hail reported from given storms in an area. Weather data sources include:

- The National Oceanic & Atmospheric Administration (NOAA) National Climatic Data Center (NCDC).
- The NOAA Storm Prediction Center (SPC).
- The National Weather Service (NWS).
- Private Companies (e.g., HailTrax™, CompuWeather and HailStrike™).

If significant hail-strike roof damage is to be observed, it is reasonable to expect to find records of a storm in the area with hail strikes large enough to cause such damage (Chapters 3 and 4). Experience with these services for hailstone size information, versus actual site visit hail investigations, suggests that they are conservative (i.e., maximum size of hail estimated to have fallen in an area slightly greater than actually observed or occurred) regarding maximum size of hail for a given area. Therefore, if the weather data does not indicate that the size of the hail that fell onto the residence/structure being inspected was large enough to cause the observed damage, the claim of hail-strike damage should be questioned. Also, the bruise-count damage to shingles should be consistent with size of the hailstones causing hail-strike dents in metal surfaces on the roof, gutters, downspouts, window screens, siding (both metal and vinyl), and outdoor air-conditioner coil fins (Figure 5.8).

FIGURE 5.8 Heavy damage to shingles inconsistent with denting in top of box vent.

If not, then claims regarding hail-strike damages to such surfaces should be questioned.

Finally, the lack of significant hail-strike damage to other surfaces on one or more elevations that should have been struck should call into question claims of significant hail damage to the roof shingles or surfaces.

5.2.2.4 Presence or Size of Simulated Hail Defects

The size of hail that strikes a building can be reasonably determined by analyzing the size of the dent and the storm history (Chapter 2). If the size of hail determined to have struck a residence was not likely capable of damaging asphalt shingles, one should not find significant numbers of hail-strike bruises to the shingles. Similarly, if the shingle defects are larger than those likely to be caused by the estimated hail size (from dent data), they probably did not result from hailstone hail-strike impacts. Nevertheless, some individuals have been very clever, and others not so clever, in attempting to synthesize hail-strike impact damage to various surfaces, including asphalt shingles.[2-4] Examples of methods reportedly used to create hail-strike bruises fraudulently on asphalt shingles include the following:

- Spinning coins such as dimes and quarters (i.e., using the edge of the coin) – Figure 5.9.
- Ball-peen hammer – Figures 5.6 and 5.10 – direct impact.
- Handle end of a screwdriver – twist and spin.
- Golf ball in a sock – direct impact.
- Scrape from spinning the end of a key or a utility knife blade.
- Using the end of a wooden cane.
- Spikes from golf shoes.

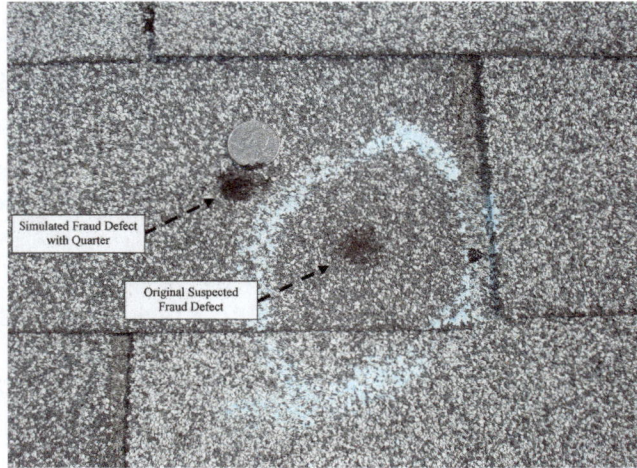

FIGURE 5.9   Field fraud damage vs simulated damage using a spun quarter.

FIGURE 5.10   Simulated damage using a ball peen hammer.

Petty[6] completed experiments in an attempt to duplicate simulated defects on cold and hot asphalt-shingled roof surfaces; results of these experiments are summarized in Table 5.2.

TABLE 5.2

Simulated Hail Damage Characteristics

Simulated Damage	Defect Dimensions Warm Shingles (~77°F)			Defect Dimensions Hot Shingles (~115°F)			Comments
	Min. Dia. in Inches	Max. Dia. in Inches	Depth in Inches	Min. Dia. in Inches	Max. Dia. in Inches	Depth in Inches	
Ball-Peen Hammer							
Light impact	0.3260	0.3500	0.0040	0.4355	0.4855	0.0665	Granules crushed and pressed
Heavy impact	0.3975	0.4800	0.0265	0.5150	0.5935	0.1183	into shingle; uniform circular diameter defect with granules missing in center of defect
Golf Ball in a Sock				Very little damage observed; a few granules lost			
Spinning a coin – 25¢	0.5305	0.5985	0.0651	0.3525	0.4060	0.02925	Uniform circular diameter defect with granules missing in center of defect. Much easier to create simulated defect with hot shingles. Best simulation of hail-strike bruise; especially with hot shingles

As illustrated in Table 5.2 and Figure 5.9, spinning coins provided the most realistic simulated hail-strike bruises to asphalt shingles. The key to identifying potential fraud, provided that the asphalt shingles are not past their service life, is the observation that actual hail-strike bruises tend to leave granules in the center of the defect (Chapter 3), whereas with spun coins, the granules are missing from the center of the defect. In addition, the spun coin defect is less diffuse looking than actual hail-strike bruises.

Ball-peen hammer damage is distinctive, with circular defects and crushed granules. With these defects, nearly all the granules are present in the defect, albeit crushed, unlike an actual hail-strike bruise.

5.3 CHARACTERISTICS OF FRAUD (MAN-MADE) HAIL DAMAGE TO OTHER ROOF SURFACES

While less commonly observed, fraud damage to other finished roof surfaces such as wood shake and shingle, slate, and tile surfaces does occur. These are likely less observed simply because less roof surfaces are finished with these materials. The most common fraudulent defects are additional cracks in those individual components (e.g., a shake or tile) due to foot traffic. The difficulty in this area is that some incremental foot damage will legitimately occur as the result of normal inspections conducted by roofing contractors and insurance/engineering inspectors. While more difficult to spot, suspected fraud on these surfaces tends to follow similar trends discussed regarding asphalt shingle fraud:

- Nonuniform patterning of defects.
- Defects are present in greater concentrations near readily accessible surfaces and/or on elevations opposite the direction(s) from which the storm arrived.
- Defect characteristics are inconsistent with the size of hail that fell (i.e., hail that fell was too small to cause such damage – especially slate or tile), or the shape of the defect is inconsistent with hail-strike damage.

Fortunately, like most fraud-related activities, the individual committing the fraud may not produce a consistent and realistically appearing defect pattern on the surface fraudulently damaged due to a lack of knowledge or laziness. In addition, these individuals typically target a neighborhood or small geographical area, so nearby roof surfaces will also be characterized by similar fraud defects and patterning. Thus, while it is more difficult to determine likely fraud on these surfaces, the lack of overall consistent patterning of defects on impacted surfaces generally exposes fraudulent activity on such roof surfaces.

5.4 CHARACTERISTICS OF FRAUD (MAN-MADE) HAIL DAMAGE TO EXTERIOR BUILDING ENVELOPE COMPONENTS

Fraud to other finished surfaces on the exterior around the building envelope occurs most frequently to metal (e.g., aluminum) siding, metal downspouts, and/or HVAC (air-conditioning) outdoor unit coil fins. Most of the same characteristics of fraud discussed in the previous section on asphalt-shingled roof surfaces apply to these surfaces as well. Factors that may make the inspector suspect that some or all of the storm damage may be fraudulent, based on experience and the literature,[8] are as follows:

- Damage is concentrated in or limited to easily accessible areas.
- Unusual damage patterns; locations of damage are not consistent with the direction from which the storm arrived.
- Either the storm event was not in the area or the size of hail-strike dents was not consistent with the size of hail from known storm events.
- Presence or size of simulated hail defects (inconsistent with damage created by falling hailstones).
- Damage to other building items (e.g., metal siding, metal gutters, or downspouts, air-conditioning coil fins) not consistent with damage on roof surfaces.

Typically the fraud damage to metal siding, downspouts, and sometimes gutters is accomplished by using a pressing action with a thumb, finger, or golf ball.[4] Damage to HVAC outdoor unit coil fins (Figure 5.11) is typically accomplished using either a thumb, or a hammer, or back of a tool like a screwdriver and is normally quite obvious due to patterning.

FIGURE 5.11   Likely fraud damage to an HVAC outdoor unit coil fins.

5.5 RECOGNITION AND EVALUATION OF POSSIBLE WIND DAMAGE FRAUD TO ASPHALT-SHINGLED ROOF SURFACES

5.5.1 INTRODUCTION TO WIND DAMAGE FRAUD TO ASPHALT-SHINGLED ROOF SURFACES

As with hail damage, without a firsthand observation of the damage being inflicted, it is difficult to absolutely prove or disprove fraud claims, especially the specific individual or company responsible for the fraud. Determining the responsible party may be more difficult, especially if more than one contractor has inspected the roof surfaces.

The most important factor for identifying probable wind damage fraud is being familiar with the characteristics of legitimate wind damage. This knowledge should enable the inspection professional to distinguish between likely legitimate damage from wind forces and fraudulent wind damages within a reasonable degree of certainty. Using available inspection evidence such as knowledge of the physical tendencies of actual wind damage (e.g., typical locations and wind speeds) versus those that occurred during a fraudulent event, one could determine whether the damage observed is, or is not, consistent with fraud damage.

As will be discussed in Chapters 6 and 7, wind damage to asphalt shingle roofs has the following typical characteristics:

- Occurs when the wind reaches at least design wind speeds (e.g., 90 MPH) or greater, assuming proper installation (i.e., correct fastener installation and not a cold-weather installation where adhesive strips did not seal).
- Damage first occurs at peaks, along rakes, and at the eaves.
- Damage is greatest in the direction from which the storm arrived and associated with collateral damage to other building systems such as gutters, downspouts, and window screens.
- Tabs are creased or partially missing, but mainly remain present. Wind speeds sufficient to blow off shingle tabs typically result in the tab being blown away. Lifting tabs still present with debris (e.g., leaf and dust) under the tabs is either a sign of improper installation or possible fraud damage.

Modern building codes require that all building systems, including asphalt shingle roof systems, be designed to meet specific minimum wind speeds, depending on their location and other factors. Confusion sometimes exists because the warranty of the shingle manufacturer will only cover the roof to 70 or 75 MPH, for example. This discrepancy reflects the shingle manufacturer's concern about installation practices; for their warranty derates the actual ability of the system to meet modern codes simply due to recognized limitations in the actual installation of shingles by contractors.

Large numbers of easily lifted shingles, which contain adhesive strips that never initially sealed (i.e., no adhesive residue suggesting contact or debris present along the strip), are typically evidence that the shingles were installed in weather that was colder than recommended and never properly seated. This condition can result in wind-caused damage, but the damage was not associated with how the roof shingle system should have performed had the shingles been installed during proper temperature conditions (see Chapter 6).

Another installation condition that leads to failure of the shingles to adhere properly is the installation of fasteners into the adhesive strip. This not only reduces the surface area for adequate adhesion, but if the fastener is tipped at an angle, it can prevent contact of the overlying shingle at this location. Shingle manufacturers explicitly warn roofing contractors on each bundle wrapper to not drive fasteners into or above the adhesive strip and that fastening into the strip interferes with sealing and contributes to blow-offs.

Roof observations that may be indicators of fraudulent wind damage to asphalt-shingled roof surfaces include:

- Lack of a severe storm event(s) associated with the reported date of loss.
- Unusual patterning of missing shingle tabs such as:
  - On elevations opposite the direction from which the storm arrived.
  - Lack of damage to susceptible shingles. Damage at locations within the field of the elevation not typically associated with wind blow-off (as opposed to ridges, rakes, and eaves).
  - Damage to shingle locations easily accessed.
- Unusual mechanical damage to missing/displaced shingle tabs and/or near-by tabs:
  - Tool marks and/or scratches to the shingle mat (underside).
  - Delaminated adhesive/sealant strips.
  - Tear from side edge of shingle, no crease beyond tear.
  - Shingle creased in two different directions.
  - Creased dimensional shingles.
  - Scratched asphalt or paint.
  - Debris under tabs.
- Heavy or unusual placement of tarps or patterns of tarps.
- Homeowner reported that contractor spent excessive amount of time on roof.

Additional details regarding each of these types of wind fraud to asphalt-shingled roof surfaces follow.

5.5.1.1   Lack of a Severe Storm Event Associated with the Reported Date of Loss

Assuming that the date of loss is consistent with when the storm damage occurred, weather records typically are researched to determine whether or not windstorms struck the area, and if so, the maximum wind speeds associated with such storms. If no windstorm was in the area, or if the wind speeds were below design wind speeds (Figure 5.12), damages attributable to wind should not have occurred.

STATION: AKRON CANTON OH
MONTH: JUNE
YEAR: 2010
LATITUDE: 40 55 N
LONGITUDE: 81 26 W

	TEMPERATURE IN F:						:PCPN:	SNOW:	WIND				:SUNSHINE:	SKY	:PK WND	
1	2	3	4	5	6A	6B	7	8	9	10	11	12	13	14	15	16	17	18
										12Z	AVG	MX 2MIN						
DY	MAX	MIN	AVG	DEP	HDD	CDD	WTR	SNW	DPTH	SPD	SPD	DIR	MIN	PSBL	S-S	WX	SPD	DR
1	79	61	70	6	0	5	0.00	0.0	0	6.1	16	300	M	M	8	18	21	310
2	80	58	69	5	0	4	0.33	0.0	0	6.9	18	220	M	M	7	138	22	210
3	78	63	71	7	0	6	0.14	0.0	0	6.8	12	50	M	M	9	1238	13	70
4	78	61	70	5	0	5	0.60	0.0	0	6.5	23	180	M	M	7	138	26	180
5	76	65	71	6	0	6	0.73	0.0	0	8.1	21	200	M	M	10	138	28	200
6	75	58	67	2	0	2	0.52	0.0	0	12.8	21	280	M	M	9	138	35	220
7	70	52	61	-4	4	0	0.00	0.0	0	6.6	3	190	M	M	7		7	10
8	73	48	61	-5	4	0	T	0.0	0	3.7	8	170	M	M	5		15	250
9	76	56	66	0	0	1	1.29	0.0	0	8.7	15	210	M	M	9	138	18	190
10	79	61	70	4	0	5	0.00	0.0	0	7.8	17	290	M	M	5		26	270
11	79	58	69	3	0	4	T	0.0	0	7.0	14	180	M	M	7		18	140
12	83	68	76	9	0	11	0.04	0.0	0	8.9	24	290	M	M	7	13	30	290
13	82	64	73	6	0	8	0.10	0.0	0	7.0	15	270	M	M	8	1	22	220
14	78	62	70	3	0	5	T	0.0	0	4.0	10	80	M	M	8	18	13	190
15	79	65	72	5	0	7	0.10	0.0	0	7.6	13	120	M	M	9	18	17	110
16	83	66	75	7	0	10	0.00	0.0	0	12.0	25	260	M	M	8	1	35	280
17	73	57	65	-4	0	0	0.00	0.0	0	5.1	12	360	M	M	6		14	350
18	83	53	68	-1	0	3	0.00	0.0	0	5.4	14	200	M	M	3	18	20	200
19	85	68	77	8	0	12	0.00	0.0	0	8.3	16	240	M	M	5		22	220
20	83	61	72	3	0	7	0.00	0.0	0	6.0	15	330	M	M	4		22	270
21	84	61	73	4	0	8	0.00	0.0	0	3.4	8	180	M	M	6		12	270
22	87	68	78	9	0	13	0.34	0.0	0	9.0	22	270	M	M	7	138	30	280
23	88	68	78	8	0	13	0.54	0.0	0	6.9	15	220	M	M	6	138	20	220
24	81	63	72	2	0	7	0.19	0.0	0	9.2	24	330	M	M	6	13	30	320
25	83	61	72	2	0	7	0.00	0.0	0	4.8	14	310	M	M	3	18	17	300
26	84	63	74	4	0	9	0.14	0.0	0	7.8	17	240	M	M	7	18	25	230
27	89	69	79	9	0	14	0.50	0.0	0	10.1	28	240	M	M	7	138	38	240
28	83	67	75	5	0	10	0.37	0.0	0	9.1	20	220	M	M	8	1	26	260
29	72	55	64	-6	1	0	T	0.0	0	6.7	16	320	M	M	5		26	10
30	71	52	62	-8	3	0	0.00	0.0	0	7.2	18	310	M	M	4		24	10

FIGURE 5.12 Weather data used to determine likely local maximum wind speeds. (Courtesy of NOAA – http://www.nws.noaa.gov/climate/getclimate.php?wfo=cle.)

In this example, on the dates near the reported date of wind damage, the maximum wind speeds reported by NOAA were 28 MPH (2-minute maximum) and 38 MPH (maximum gust), much less than shingle warranty (70–75 MPH) or design wind speeds (90 MPH) for this specific area.

5.5.1.2 Unusual Patterning of Damaged Shingles

The first factors that may signify fraudulent wind damage to asphalt shingles are situations where the wind-damaged shingle tabs are greater on elevations opposite the direction from where the storm arrived or are more prevalent on areas other than the peak, rake, and eaves for a given location. Like with hail, wind fraud to asphalt shingles will often occur at more accessible locations (e.g., near eaves or valleys) and not be present to the same extent at the peaks, rakes, and eaves. Simply counting the number of missing tabs by elevation, and then within certain areas of a given elevation, will begin to reveal oddities in the patterning of reported wind damage.

Once fraud is suspected based on general patterning and/or the lack of known weather events that could cause such damage, an examination of the damaged or lifting shingles, as well as adjacent shingles, will provide information on whether or not mechanical tools have been used to fraudulently remove portions of shingle tabs or complete shingle tabs. Examples of potential situations and possible causes of the damage situation follow:

1. *Partially missing shingle tabs*: If the shingles are generally well seated, look for evidence of scrapes, scratches, and partially delaminated adhesive strips at/near the partially missing tab(s). This often suggests fraudulent activities by the mechanical destruction of the seal between the tabs.

2. *Creased shingle tabs – multiple crease lines and directions*: If the shingles are generally well seated, evidence of multiple crease lines to the overlying tab (Figure 5.13) and scrapes, scratches, and partially delaminated adhesive strips to the underlying/overlying shingle tabs (Figure 5.14) suggests fraudulent mechanical activities intended to dislodge the shingle tab.

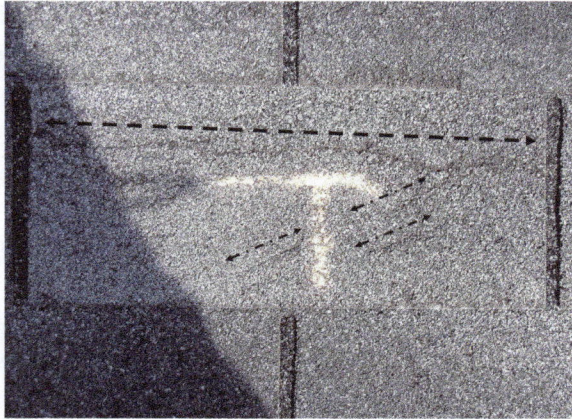

FIGURE 5.13 Likely wind fraud damage to asphalt shingle tab – multiple crease lines (tab mechanically lifted from bottom right).

FIGURE 5.14 Likely wind fraud damage to asphalt shingle tab – partially detached adhesive strip.

Note, based on the crease-line pattern, the shingle tab appears to have been lifted mechanically from the bottom right corner. Debris under the shingle tab may also be an indicator of fraudulent activities depending on when they may have occurred.

3. *Creased dimensional shingles:* Due to the stiffness, design, and dynamics associated with the creasing of a dimensional shingle (typically ~36″ or wide), any visible creasing, especially creasing of the entire shingle, suggests fraudulent mechanical activities (Figure 5.15).

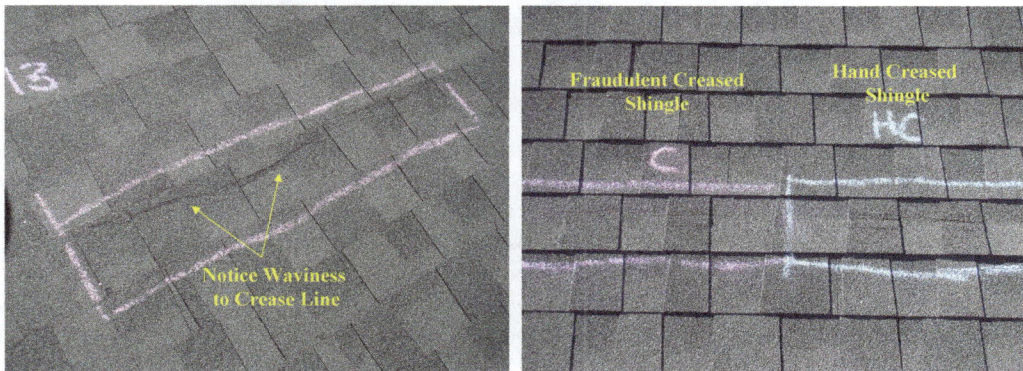

FIGURE 5.15   Likely wind fraud damage to dimensional shingle (a) and hand creased shingle adjacent to fraudulent creased shingle – notice similarities in creasing pattern (b).

A simple test to replicate the suspected fraudulent damage consists of manually creasing the shingle by hand to an undamaged shingle and comparing the two forms of damage. Similarities between the two could suggest intentional mechanical damage.

4. *Missing creased shingle tabs:* Scrapes, scratches, and delamination of the adhesive strips – Again, if the shingles are generally well seated and the missing tabs are outside of the typical wind damage areas, evidence of scrapes, scratches, and partially delaminated adhesive strips to the underlying/overlying shingle tabs (Figure 5.14) suggests fraudulent mechanical activities to dislodge the shingle tab.

Note that delaminated adhesive strips are a common observation on fraudulent wind damage investigations. Delamination of a sealant strip implies that the adhesive bonds were stronger than those between the granules and its asphalt shingle adhesive bonds; just like a weld being stronger than the adjoining metals. Delamination is apparent when portions of the granular base or asphalt mat of the adjoining shingles are torn from either shingle and are attached to the sealant strip. This type of failure rarely occurs unless encouraged mechanically.

5.5.1.3   Excessive Use of Tarps

The following factors have been encountered on fraudulent wind investigations:

- Excessive use of tarps.
- False use of tarps.

On one particular wind damage assessment associated with ~88 building apartment complex (Figure 5.16), nearly every roof elevation within the complex was blanketed with tarps following Hurricane Ike.

FIGURE 5.16   Apartment complex – aerial view of tarps on buildings.

While the installation of tarps is a common practice to prevent water intrusion, it may also be used to suggest damage that does not actually exist or exists to a lesser extent than suggested by the tarped area. Regardless of intention, in the process of placing a tarp on a roof surface, the roof is damaged by the penetration of fasteners (i.e., nails or staples) to secure the tarp in place.

In this case, apparently a contractor at some point in time figured out that some insurance adjusters may buy damaged elevations or entire roofs without looking underneath the tarps. This can be due to the difficulty in removing the large numbers of fasteners securing the tarp in place or the presumption that the tarps were placed on the roof surface to prevent potential water intrusion or further damage. During the removal of smaller tarps throughout the complex, some legitimate wind damage was uncovered. However, when several other larger blue tarps were removed, little to no visible evidence of wind damage was found (see Figure 5.17). This fraudulent activity suggests a greater amount of damage than what actually existed.

FIGURE 5.17   Before and after removal of large tarp – notice limited amount of wind damage beneath large and excessive tarp.

5.5.2   EXAMPLE FRAUD WIND DAMAGE – ASPHALT ROOF – REPORTING INFORMATION

Using the tools outlined in the previous sections, one can perform a wind damage inspection of damaged shingle tabs to determine whether or not fraud wind damage is present, and if present, the extent to which it is present. An example scenario for illustration purposes follows:

Wind speeds for the times near the date of loss were documented to be below 90 MPH. For this example, Figure 5.18 illustrates a plan-view schematic of an asphalt-covered roof evaluated for potential wind and wind fraud damage.

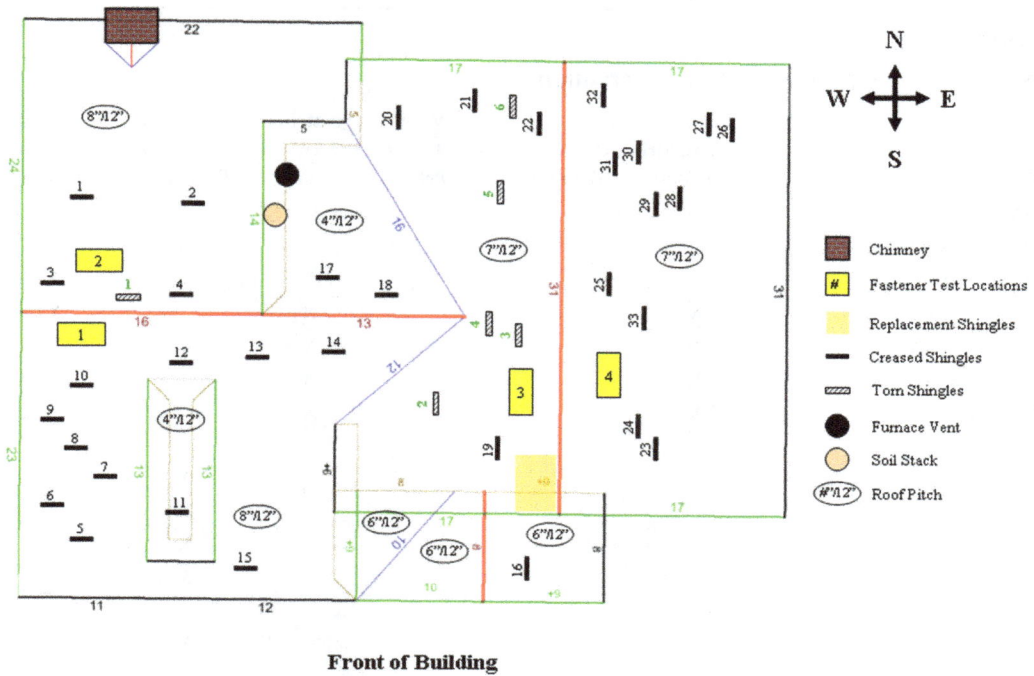

Front of Building

FIGURE 5.18   Example roof schematic with creased and torn asphalt shingle locations.

Creased and torn shingles are illustrated and damaged shingle tab observations by tab ID number are summarized in Table 5.3.

TABLE 5.3
Example Damaged Shingle Tab Observations

ID	Fresh Damage	Historic Damage	Adhesive Strip Delaminated	Defects Present Below Shingle	Within Linear Pattern	Scratches Present on Mat	Likely Wind Damaged	Likely Mechanically Damaged
C1	X		X			X		X
C2	X		X			X		X
C3	X		X			X		X
C4	X		X			X		X
C5	X				X			X
C6	X				X			X
C7	X			X	X	X		X
C8	X				X			X
C9	X				X			X
C10	X			X	X			X
C11	X							X
C12	X				X			X
C13	X				X	X		X
C14	X			X	X			X

(Continued)

TABLE 5.3 (*Continued*)
Example Damaged Shingle Tab Observations

ID	Fresh Damage	Historic Damage	Adhesive Strip Delaminated	Defects Present Below Shingle	Within Linear Pattern	Scratches Present on Mat	Likely Wind Damaged	Likely Mechanically Damaged
C15	X							X
C16	X							X
C17	X							X
C18	X		X			X		X
C19	X		X			X		X
C20	X		X			X		X
C21	X		X			X		X
C22		X					X	
C23		X					X	
C24		X					X	
C25		X		X			X	
C26		X					X	
C27		X					X	
C28		X		X			X	
C29		X					X	
C30		X					X	
C31		X		X			X	
C32		X					X	
C33		X					X	
Total:							12	21

Observation categories (i.e., questions raised and answered) used in Table 5.3 during the inspection of the shingle tabs were as follows:

- Do surfaces show evidence of fresh damage (i.e., lighter/brighter coloration on surfaces)?
- Do surfaces show evidence of historic damage (i.e., duller coloration on contact surfaces)?
- Is the adhesive strip delaminated?
- Are mechanical defects present below the damaged shingle tab?
- Do the damaged shingle tabs follow a liner defect pattern?
- Are scratches present on the mat?

Of 33 damaged shingle tabs, and based on an analysis of these six characteristics, 21 shingle tabs were determined to be likely damaged by fraud and 12 by forces attributable to wind action.

Overall conclusions reached in this specific example, including example language used, were as follows:

- The reported wind speeds reported were below the speed of 90 MPH used for designing building systems in the mid-west.
- Up to ~15% of the shingle tabs on the roof elevations lifted easily. Based on experience, shingles that lift but are not creased or torn off are typically not lifting due to wind action. This is often attributable to installation deficiencies (see Chapter 6 for information on lift tests).

- The roof elevations contained 12 creased shingle tabs that were likely damaged some time ago (based on criteria above) by forces from wind action. The primary cause of these damages was likely installation deficiencies exacerbated by wind forces. Evidence to support this conclusion follows:
  - The wind speeds reported were below design speeds.
  - Fasteners were installed through and above the adhesive strip of the overlying shingle.
  - Fasteners were underdriven, overdriven, and driven at angles not flush to the shingle surface.
  - The shingle damage was determined to be historic (see above criteria).
- The roof elevations contained 21 creased shingle tabs that were likely damaged recently (based on above criteria) by mechanical manipulation. These damages appear consistent with simulated wind damage. Evidence to support this conclusion follows:
  - The wind speeds reported were below design speeds.
  - The observed mechanical damages were fresh in appearance.
  - Shingles on the roof surfaces that were susceptible to damage due to fastener "pops" were observed to be free of damage.
  - Linear patterning was present on some of the freshly damaged shingles.
  - Many of the freshly damaged shingles contained concentrations of mechanical scratches at the bottom edge of the mat.
  - Many of the freshly damaged shingles had been delaminated from the adhesive strip.
  - Most of these damaged shingles were located within the field of the roof surface as opposed to areas more susceptible to wind-related damages (i.e., ridges, rakes, eaves).

IMPORTANT POINTS TO REMEMBER

- Natural acts such as hailstorms and high wind events should result in damage patterns consisted with the direction from which the storm arrived.
- Fraud damage most often occurs to finished roof surfaces, metal siding, and metal downspouts. Asphalt shingles are damaged by fraudulent activities more frequently because they are more commonly installed.
- Fraudulent hail and wind storm damages are typically located on easily accessed roof areas and away from the view of onlookers.
- Patterning and defect analysis, along with knowledge of the characteristics of legitimate hail or wind damages, can often allow one to distinguish between legitimate storm damage and fraud damage.
- Roof surfaces that contain tarps should always be uncovered, or portions uncovered, to validate wind damage.

REFERENCES

1. Ohio Insurance Institute. "May 7–8 Northern Ohio Storm Losses Top $31 Million." June 9, 2010. http://www.ohioinsurance.org/newsroom/newsroom_ full.asp?id=602.
2. Coughlin, Matt. "Pair Charged in Lehigh Hail-Damage Fraud." April 29, 2011. http://www.phillyburbs.com/news/crime/pair-charged-in-lehigh-hail-damage-fraud/article_4efcbfb7-7029-5e3a-b8f8-53f5396075a6.html.
3. Graham, Bob. "Fraudulent Hail Loss Claim Spike Blamed on 'Fly-by-Night' Contractors." National Insurance Crime Bureau (NICB), June 22, 2010. http://ifawebnews.com/2010/06/22/fraudulent-hail-loss-claim-spike-blamed-on-%e2%80%98fly-by-night%e2%80%99-contractors/.

4. State Farm Fire & Casualty Company vs Joseph Martin Radcliff and Coastal Property Management, LLC a/k/a CPM Construction of Indiana, Plaintiff. Statement of Material Facts Not in Dispute, Designation of Evidence and Memorandum in Support of its Motion for Partial Summary Judgment, Cause No. 29D01–0810-CT-1281, Hamilton Superior Court No. 1, Indiana, 2009.
5. Asphalt Roofing Manufacturers Association (ARMA). "Frequently Asked Questions." 2012. http://www.asphaltroofing.org/resources_faq. html#ss1.
6. Petty, Stephen. "EES Group, Inc. Technical Bulletin #5: Fraud and Artificial Damage to Asphalt Shingles," July 2007.
7. *"Haag Certified Roof Inspector Program – Residential Edition."* Dallas, TX: Haag Course Workbook, 2008.
8. Manasek, Thomas and National Insurance Crime Bureau (NICB). "2017–2019 United States Hail Losses Claims". ForeCAST Report, April 14th 2020.

6 Wind Damage Assessments for Residential and Light Commercial Roofing Systems and Finished Surfaces

Stephen E. Petty
EES Group, Inc.

George W. Fels
Curtiss-Wright Corporation

CONTENTS

PURPOSE/OBJECTIVES

The purpose of this chapter is to:

- Provide information and examples on how to identify wind damage to residential and light commercial roofing surfaces.
- Discuss factors that can either decrease or increase the resistance of wind uplift to common residential and light commercial roofing surfaces.
- Document a methodology for assessing wind damage claims to residential and light commercial roofing systems and finished surfaces.

Following the completion of this chapter, you should be able to:

- Understand the underlying theory of wind forces acting on buildings.
- Be able to identify wind-related damages to various residential and light commercial roofing materials.
- Recognize the differences between non-wind-related damages (e.g., natural degradation from conditions such as inadequate attic ventilation, manufacturing-related anomalies, damages associated with improper installation, and mechanical damage) and wind-related damages.
- Be able to perform a thorough visual inspection for wind damage to components on the exterior surfaces of a home and light commercial building (siding, downspouts, gutters, etc.) and roof surfaces (asphalt shingles, wood shingles/shakes, slate/clay tiles, etc.).

6.1 INTRODUCTION

High winds have the potential to cause considerable damage to residential and light commercial roof systems and exterior finishes. High winds passing over a roof can create uplifting forces on the entire roof system and potentially remove individual pieces or, in more severe cases, remove entire sections of the roofing system. A photograph of wind damage to an asphalt shingle roof is illustrated in Figure 6.1.

National consensus and state building codes require buildings and other structures to be able to withstand forces generated from certain minimum wind speeds without damage occurring to the roof or structure. Unfortunately, some building components may not be constructed or installed so as to comply with industry standards and common practices for wind resistance. For example, based on experience, many residential and light commercial asphalt-shingled roof surfaces are not installed (e.g., fastener location) in accordance with manufacturers' instructions or industry best practices and are prone to wind blow-offs. In this case, the improper fastener locations can alter the resistance of the shingle to wind uplift and allow for it to be more susceptible to wind damage.

FIGURE 6.1   Wind damage to asphalt shingles.

6.1.1 WIND BASICS

As wind gusts pass over the roof surface of a residential home or light commercial building, aerodynamic forces are formed that cause uplifting forces and pressures on the various roofing and building components. Wind damage to roof systems and building components typically occur at the points on the home or building where the greatest uplifting forces are present. For roof systems and finished surfaces, these most vulnerable locations include the windward roof edges as well as the leeward side of ridges (e.g., peaks, rakes, and eaves). Figure 6.2 illustrates the dynamics of wind forces acting on a pitched roof surface.

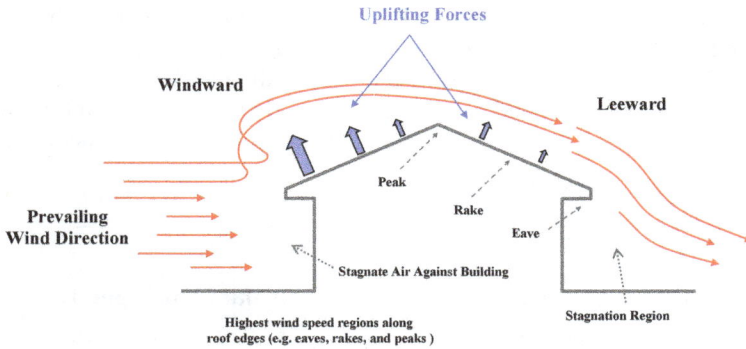

FIGURE 6.2 Wind-generated forces acting on a pitched roof.

For roof systems, the primary cause of wind damage is the pressure differential acting on the roofing component. These positive and negative forces can "push" or "pull" on a component, creating a moment of force that the component is not able to resist. When wind flows across the top surface of the component, it creates a negative or uplifting pressure (similar to the uplift on an airplane wing); this continual pressure can weaken the method of securement or adhesion holding the component in place (i.e., chemical adhesion such as a sealant strip or mechanical adhesion such as nail fasteners).

In order to understand the effects of wind gusts on a structure fully, the topic of these forces and pressures associated with these wind gusts must be briefly discussed. In Figure 6.2, the prevailing wind imposes a force, or load, on the structure, and this wind load is directly related to the wind velocity. The relationship between load or force and pressure is given in Eq. 6.1:

$$F = P/A \tag{6.1}$$

where:
F = lbs force
P = lbs force/square feet
A = Area square feet

The relationship between the wind speed and the velocity pressure can be expressed using a simplified version of the Bernoulli equation (Eq. 6.2)[1]:

$$P = 0.00256v^2 \tag{6.2}$$

where:
P = lbs force/square feet
v = wind speed in miles/hour (MPH)

Combining these two equations, one can derive the force associated with a given wind speed (Eq. 6.3):

$$F = P/A = 0.00256v^2 * A \tag{6.3}$$

where:

$F =$ lbs force

$v =$ wind speed/velocity in miles/hour (MPH)

$A =$ Area square feet

This simplified form of the Bernoulli equation does not take into account the height of the building, the local geographical terrain, the importance (i.e., usage) of the building, or the directionality of the wind gusts. These factors have a greater relevance with low-sloped roof systems rather than steep-sloped roof surfaces and are addressed at length in Chapter 7. The resulting pressure from a given wind gust impinging on the windward perpendicular surface of a structure is illustrated in Table 6.1.

TABLE 6.1

Pressures Associated with Perpendicular Wind Speed

Wind Speed (MPH)	Resulting Pressure (psf)
10	0.3
20	1.0
30	2.3
40	4.1
50	6.4
60	9.2
70	12.5
80	16.4
90	20.7
100	25.6
120	36.9
150	57.6

Attention is directed to wind gusts blowing over a structure due to the creation of a pressure differential between a vertical sidewall and a sloped roof surface. For example, if a 30 MPH wind gust blows against the side of a building or residence, similar to Figure 6.2 and according to the equation above, an average pressure of 2.3 psf would be acting against the sidewall. As wind flows across the eave of the building, the wind speed could increase to 50 MPH. This increase of wind speed will create a change, or difference, in the air pressure of 4.1 psf, from a total of 6.4 psf. Similar to the uplift imposing on an airplane wing, this pressure differential can potentially pull up components from the roof system (i.e., decking, shingles, etc.). If the roof area of a home or light commercial building measures 1,925 square feet (35 by 55 feet), this amounts to a total lifting force of 6,353 lb$_f$ acting on the roof surface. This illustrates the point that when small pressure differences are applied over relatively small areas, the forces become very large.

For example, consider another example where in one case (e.g., a Midwest location) the wind speeds were gusting up to 70 MPH compared with wind gusts of 90 MPH at another location (e.g., near coastal area). What would be the relative differences in lifting forces associated with these two different wind speeds? In this case, assuming a 2,000-square-foot roof surface, the relative forces and pressures would be 12.5 psf (25,088 lb$_f$) and 20.7 psf (41,472 lb$_f$), respectively. Thus, in this simplified example, for a 29% increase in wind speed, the wind forces increased by 65%.

These forces, which can be observed on most residential and light commercial roof systems (e.g., asphalt shingles, clay, concrete, slate, and wood shingles/shakes) and building components (e.g., siding, fascia, soffits, gutters, and downspouts), can completely displace sections of these surfaces or components. For asphalt shingles, once cohesive failure allows a shingle tab to become lifted and unadhered, the wind load acting on the tab can increase almost exponentially, resulting in the creasing and eventual tearing of the shingle.[2]

6.1.2 WIND-ASSOCIATED FAILURE MODES

Possible wind-related failures of finished roof surfaces and other building components can be attributed to one or more of the following three causes: (1) wind forces in excess of the design speed; (2) wind coupled with installation deficiencies; and/or (3) wind coupled with aging and weathering or thermal degradation. Occasionally, manufacturing-related issues or defects to finished surfaces are encountered, but this is not common and is not covered in this chapter.

6.1.2.1 Design Wind Speeds

Modern building codes such as the International Building Code (IBC)[3] require that roof systems be able to withstand the uplift pressures associated with minimum basic straight-line wind speeds. The ultimate design wind speeds vary in magnitude and are dependent on the geographic location of the structure and other factors such as the *risk category* of the building. These minimum requirements are ultimately based on the American Society of Civil Engineers (ASCE) Standard 7, "Minimum Design Loads of Buildings and Other Structures."[4] Code requirements for minimum wind loads for typical residences and light commercial buildings (defined as Risk Category II in the IBC) are shown in Figure 6.3.

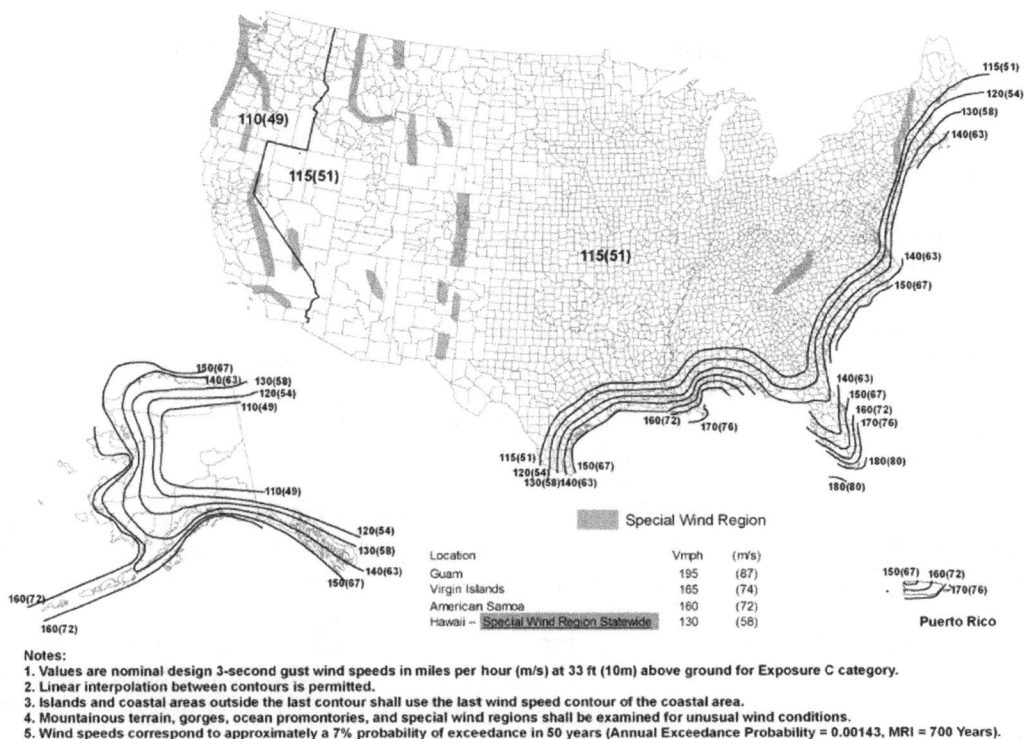

Location	Vmph	(m/s)
Guam	195	(87)
Virgin Islands	185	(74)
American Samoa	160	(72)
Hawaii – Special Wind Region Statewide	130	(58)

Notes:
1. Values are nominal design 3-second gust wind speeds in miles per hour (m/s) at 33 ft (10m) above ground for Exposure C category.
2. Linear interpolation between contours is permitted.
3. Islands and coastal areas outside the last contour shall use the last wind speed contour of the coastal area.
4. Mountainous terrain, gorges, ocean promontories, and special wind regions shall be examined for unusual wind conditions.
5. Wind speeds correspond to approximately a 7% probability of exceedance in 50 years (Annual Exceedance Probability = 0.00143, MRI = 700 Years).

FIGURE 6.3 Ultimate design wind speeds – risk category II. (Courtesy of International Code Council.)

For much of the Midwest, the ultimate or strength design wind speed is 115 MPH, whereas the design wind speeds for coastal areas can be up to 180 MPH. When necessary, these ultimate design wind speeds presented in the IBC can be converted into nominal wind speeds (90 and 140 MPH, respectively) used for cladding and building components by Equation 16.33 in Section 1609.3.1 of the code. Additional information of design wind speeds can be found in Chapter 7.

Since components of roofing systems and buildings are required to withstand or perform against minimum wind speeds in accordance with their geographical building codes and industry standards, by design, the roofing system and building components should remain intact up to the code-required minimum wind speeds. In some cases, these design wind speeds lead to confusion when, for example, shingles installations are warranted at lower wind speeds (e.g., code wind speed of 90 MPH; warranty wind speed of 70 MPH). This discrepancy does not mean that the manufacturer's shingles, when properly installed, do not meet code requirements; it simply reflects their concerns that the shingles may not be properly installed. Thus, the warranty simply represents a derating of the design wind speed where they will cover blow-offs under warrantee based on recognized poor installation practices.

6.1.2.2   Factors Contributing to the Resistance of Wind Uplift or Blow-Off

Various manufacturers, trade associations, government agencies, and modern building codes have developed several requirements and recommendations for the application of roof systems to increase the resistance of a material to wind forces. For the most part, these factors rely on proper installation procedures and continued proper maintenance. These factors for the three most common roofing materials used in residential homes and light commercial buildings are discussed below (i.e., asphalt shingles, wood shakes and shingles, and tile roof systems).

6.1.2.2.1   Asphalt Shingles

Factors that contribute to the amount of resistance asphalt shingles have to the forces of wind are as follows:

- The effectiveness of the sealant or adhesive strip that bonds the courses of the shingles together and helps prevent lifting.
- The mechanical properties of the asphalt shingles.
- The orientation of the roof surface with respect to wind direction.
- The pitch of the roof surface (the steeper the pitch, the greater the effect of gravitational forces).
- Installation methods of attachment (nails, correct number of fasteners, etc.).
- Sufficient attachment schedule of the roof sheathing and truss members to ensure appropriate wind uplift resistance of the decking and framing members.[5–8]
- More than two layers of shingles installed. Codes and best practices only allow for two layers of shingles to be installed at one time.[9]

Based on experience from completing hundreds of residential and light commercial wind damage inspections, improper workmanship, premature deterioration due to inadequate attic ventilation, and aging are often the primary causes for reported wind damage to roof finished surfaces that are not actually wind damage.

Effectiveness of the Sealant/Adhesive Strip: Most modern asphalt shingles are manufactured with a heat-sensitive adhesive commonly referred to as a sealant or adhesive strip and can be composed of asphalt, a polymer or elastomer, a cross-linker/resin, and/or petroleum oil.[10] Properly sealed adhesive strips are the primary defense of a shingle against uplifting wind forces. Therefore, the failure of an adhesive strip to keep the shingles sealed to one another will render the shingles more susceptible to wind damage.

The design level of adhesion expected by a manufacturer can only be achieved when the heat-sensitive strip fully adheres to the adjacent shingle and the shingle is installed in a manner to promote this adhesion. Based on the experience, the sealant strip will fail either adhesively or cohesively.

When the sealant strip fails to properly adhere (i.e., adhesive failure), the shingles are more vulnerable to blow-offs and other wind-related damages simply because the shingle can lift. This failure is apparent when a lack of adhesive residue transfer is observed on the underlying shingle or overlying shingle mat except in cases of fraud where the adhesive bond has been mechanically broken (Chapter 5). This condition can often be attributed to one or more of the following causes: (1) cold weather installation, (2) fastener positions prohibiting full adherence, and/or (3) windborne debris on the sealant strip.[2,11–15] Each of these factors is discussed below:

- Cold weather installation (typically below 40 F) limits/prevents the activation of the thermally activated sealant strip. Strong winds that impact the roof surface prior to sufficient warming and activation can damage shingles (e.g., causing them to curl up) and/or blow debris between the shingles, which prevents future adhesion once the weather warms.

The Asphalt Roofing Manufacturers Association[16,17] (ARMA) provides a technical bulletin with additional recommendations during cold weather installation of asphalt shingles for improved efficiency and performance. This bulletin requires following the manufacturers' recommendations and the use of adhesive during cold weather installation of shingles, but experience suggests that this is rarely/never done. Manufacturers recommend asphalt shingles be installed with ambient temperatures between 40°F (4°C) and 85°F (26°C); typically, installations are not recommended below 40°F (4°C).[18,19]

- The position of the fasteners can limit the contact area between the sealant strip and the overlying or underlying shingle. This includes fasteners driven through the adhesive strip as well as fasteners overdriven, underdriven, or driven at angles not flush to the surface of the shingle. Any fasteners that are underdriven or driven at angles can not only limit the contact area for the adhesive strip but also prevent the shingle from lying flush against the roof surface, which is required to achieve adequate adhesion. This explains why best practices recommend not installing the nails within the adhesive area and manufacturers warn the installer that doing so can result in shingle blow-off.
- Wind-borne debris on the sealant strip such as pollen, dirt, and leaves can also limit or prevent proper adhesion from occurring.

Cohesive failure of the adhesive strip is defined as a failure of the adhesion properties of the sealant once proper adhesion has already taken place. This failure mode is apparent when sufficient adhesive residue is observed on both the underlying and overlying shingle mats and no apparent mechanical damage (e.g., fraud – Chapter 5) is present. The cohesive failure of a shingle adhesive strip can be caused by: (1) fatigue stresses resulting from movement of the shingle and/or roof decking, (2) thermal degradation and oxidation or age hardening, (3) manufacturer defect, and/or (4) forces from wind uplift. Each of these factors is discussed below:

- Expansion and contraction (i.e., movement) of a shingle or of the decking occurs with every temperature change. This constant repetitive force acting upon the sealant strip of a shingle causes fatigue. Fatigue resulting from cyclic loading is known to cause many materials to fail below the yield stress.
- Thermal degradation including the effects of oxidation, ultraviolet radiation, and excessive heat are only a few stresses that can deteriorate the asphaltic-based material within the adhesive strip of a shingle.[2,20]
- A manufacturing defect would be associated with either an improper formulation or lack of sufficient adhesive strip area. The former is difficult to identify without considerable analysis and cost. It might also be manifested as premature aging of the adhesive strip. Most likely, this situation would be discovered only as part of a product recall where the

manufacturer self-identified the issue. The later situation, while rare, is readily observable due to a lack of adhesive strip materials on the shingle surfaces.

- *Forces of wind uplift*: Once winds reach design wind speeds, the shingles, even if properly installed, will begin to blow off in the normally susceptible areas. If wind speeds are below the design threshold(s), the primary cause of blow-off is likely other than wind.

Shingles that have experienced cohesive failure (i.e., currently unadhered) and are not physically damaged (i.e., tears and creases) can be resealed to the roof surface by hand sealing or hand dabbing with roofing cement.

In reviewing several shingle manufacturers' warranties pertaining to wind, it appears that their warranties range from 5 to 10 years, sometimes even 15 years. This time period reflects the life expectancy of the shingle to remain adhered to the roof surface and thus, how long the adhesive strip will resist wind uplift (10 years based on communication with a shingle manufacturer). EES mined data from 310 wind-damage inspections of asphalt-shingled roof surfaces between 2008 and 2010, which revealed that the percentage of shingles lifting (i.e., not sealed) increased with age up to a timeframe ranging from 8 to 12 years after which time it was relatively constant. This again suggests that shingles have an adhesive strip life span of ~10 years as stated by the manufacturer. After this timeframe, the shingle is more susceptible to blow-offs and wind-related damages.

When "delamination" of a sealant strip occurs, it implies that the adhesive and cohesive bonds were greater in strength than the attached shingles. This is like a weld being stronger than the adjoining metals. Delamination is apparent when portions of the granular base or asphaltic mat of the adjoining shingles are torn from either shingle and are attached to the sealant strip. This type of failure is typically not considered a sealant strip failure. In some instances, supporting evidence can suggest that this delamination can be a result of artificial or simulated wind damage.

Application and Installation Methods: Deviation from shingle manufacturers' installation instructions can also increase the susceptibility of damage to shingles by wind forces. Based on experience, the most common mistakes regarding the fastening of shingles are as follows:

- Positioning fasteners through or above the adhesive strip.
- Driving fasteners too far into the shingle, not far enough into the shingle, or at an angle not flush with the surface of the shingle.
- Not installing fasteners in the designated "nail zone" (applicable to most modern dimensional shingles).
- Installing fasteners too close or too far away from the side edge of the shingle.
- Installing too few fasteners per shingle (e.g., three as opposed to four). Note that for high wind regions and steep-slope applications, six nails per shingles are recommended and are typically required by building code.
- Installing fasteners that are too short to penetrate the roof sheathing fully.
- Using staples rather than nails (i.e., best practices strongly encourage the use of nails) as fasteners.

Proper techniques and installation methods must meet the requirements of local building codes and the recommendations of various manufacturers (e.g., CertainTeed and Elk-GAF) and trade associations [e.g., ARMA[9,12], The National Roofing Contractors Association (NRCA), APA – The Engineered Wood Association (APA)[5-8] and FEMA[11]]. The proper method for installing fasteners for asphalt shingles is illustrated in Figure 6.4.

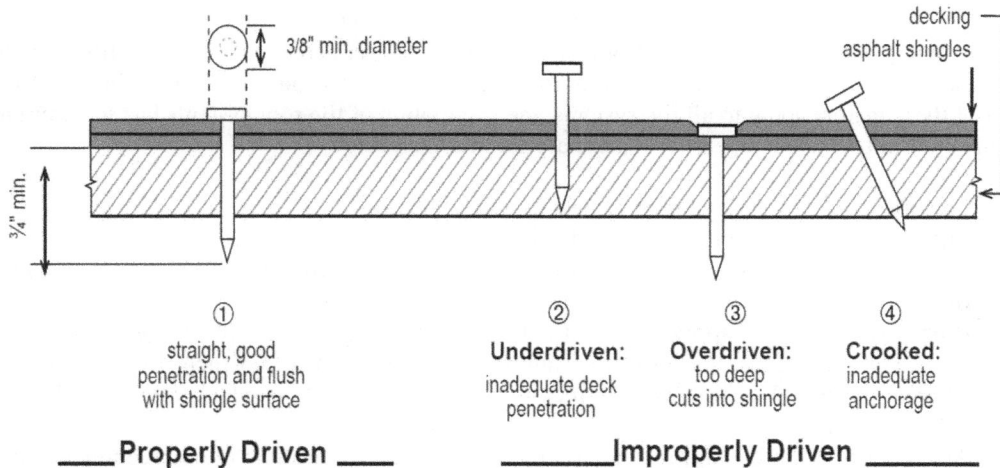

FIGURE 6.4 Fastener application for asphalt shingles. (Courtesy of Asphalt Roofing Manufacturers Association, Inc.)

It should be noted that staple fasteners were allowed both by codes and by manufacturers until about the year 2000 after which time only nail fasteners were allowed/recommended. Experience suggests that staple fasteners are less effective than nail fasteners and make shingles more prone to wind blow-off. The head size of the nail provides more surface area, which holds the shingle better in place, particularly during high winds, whereas staples can tear through the shingle more easily.

6.1.2.2.2 Wood Shakes and/or Shingles

Factors that contribute to the amount of resistance wood shakes and/or shingles have to the forces of wind are[13,21–23] as follows:

- The orientation of the roof surface to the wind direction.
- The slope or pitch of the roof.
- The age and quality of the wood shingles/shakes.
- Method of attachment and installation workmanship (i.e., a minimum of two fasteners driven flush and firmly into the roof sheathing). The Cedar Shake and Shingle Bureau (CSSB) recommended the use of ring-shank nails in high wind areas to resist fastener pullouts.
- Selection of the roof deck; proper selection provides greater uplift resistance to wind.
- Care and maintenance of the roof.

The primary factors that can diminish performance and resistance to wind uplift are: (1) improper installation (e.g., installations failing to follow local code requirements and/or recommendations set forth by the CSSB, NRCA, and other associations), (2) aging of the wood shingle or shake, and (3) quality (i.e., grade) of the wood shake and/or shingle. As expected, as the age and resulting deterioration of the shingle or shake increase, the susceptibility of wind-related damage also increases.

The CSSB and NRCA can be referenced for proper installation of wood shingles and shakes. Comparison of actual best practices with actual installation practices observed will provide the basis for determining whether or not the wood shingles or shakes were properly installed.

6.1.2.2.3 Slate, Clay, Concrete, and Asbestos Tile Roof Surfaces

Factors that contribute to the amount of resistance slate, clay, concrete, and asbestos tile roof surfaces have for the forces of wind are similar to those outlined for wood shakes and shingles. While most of these factors apply to all tile systems, the installation of tile roof systems has some unique features to limit wind damage.[13,24–27] These are summarized below:

- The orientation of the roof with respect to wind direction.
- The slope of the roof will affect wind resistance to tiles.
- The age and quality of tiles and their resistance to both freeze/thaw cycles and breakage from wind-borne debris.
- Proper selection and application of the roof sheathing (i.e., solid panels or wood battens). For slate roofs, the NRCA recommends a minimum of 5/8-inch-thick decking, whereas Joseph Jenkins Inc.[24-26] recommends using solid wood sheathing at least 3/4 inch thick.
- Application method (e.g., for slate, a minimum of a 3-inch headlap often recommended; however, Joseph Jenkins, Inc.[24-26] recommends the use of a 4-inch headlap in high wind regions).
- Attachment method (i.e., number, type, and length of fasteners, wire anchors, and/or clips). The following factors are listed for slate, clay, and concrete tiles:
  - For slate tile, fastener best practices are summarized in *The Slate Roof Bible* publications by Jenkins.[24-26] and the FEMA.[27] Figures 6.5a and b illustrate nose clips used to help fasten replacement tiles.
  - For clay and concrete tiles, the *Concrete and Clay Roof Tile Installation Manual* specifically designates the various methods and/or combinations for either mechanical or adhesive attachment, depending on the desired performance and geographic conditions.[28]
- A minimum of two fasteners per tile.
- The fasteners should penetrate firmly into the sheathing. Experience has shown that the nails should penetrate 3/4 of an inch into the sheathing.
- Utilization of appropriate type of nails (preferably ring-shank) or screws.
- For additional support, ensure the use of clips or hangers along eaves, ridges, and hip zones for they receive the highest forces of wind action.
- Ensure that fasteners are driven perpendicular to the roof sheathing and not driven tightly against the tile. Tiles are intended to "hang" from the fastener. Driving the fastener flush against the tile will increase the risk of breakage.

(a)   (b)

FIGURE 6.5   Use of nose clips on slate tiles - Vermont (a) and Pennsylvania (b).

6.1.2.3 Age of the Roof System and Components

All roofing materials can and will ultimately age and deteriorate over time. The extent of this natural deterioration is dependent upon several factors, including the quality of the material, slope direction, attic ventilation, and quality of the attachment materials. It is well recognized and documented that roof systems that are deteriorated beyond the age of their intended service life are more susceptible to damage not only from hailstorms as discussed in previous chapters, but also from wind gusts below the design resistance. These conditions are more prevalent with asphalt shingles, such as organic matted shingles, which are more prone to thermal defects related to natural aging and weathering and effects of improper attic ventilation. While these situations are common, wind gusts cannot be identified with sole responsibility for the failure of the roofing components.

Further, roof surfaces on the south and west elevations are more prone to thermal degradation and damage since they receive more exposure to the sun. This condition of apparent premature aging of the roof surface can be significantly exacerbated by poor attic (or plenum space above cathedral ceilings) ventilation. Such roof surfaces are prone to wind damage, but the primary cause of the blow-off often is inadequate ventilation, causing premature aging of the roof surface(s).

Finally, not only do all roofing materials age, but so do the attachment materials (e.g., nails, screws, and/or clips). These can corrode and/or move or loosen with time. For example, as often observed on older homes with a tile roof system, the mechanical fasteners and attachment devices have corroded and are in a much poorer condition than the tiles themselves.

6.2 WIND FAILURE FORENSIC INVESTIGATIONS – OVERVIEW OF METHODOLOGY

The vast majority of forensic investigations, regarding straight-line wind damage to residential and light commercial buildings, center on damage to the roof system and its finished surface(s). Thus, this section focuses on the methodology associated with completing roof and finished surface field investigations. Collateral damage to other building components such as siding, fascia, soffits, gutters, and downspouts is typically apparent, as these components are either missing or bent. Wind damage from winds associated with tornados is typically associated with twisted debris and debris fields and is discussed at length in Chapter 19.

6.2.1 WIND DAMAGE INSPECTION METHODOLOGY

Following the basic forensic inspection methodology outlined in Chapter 1 as well as the detailed and extensive methodologies of completing a hail damage inspection outlined in Chapters 3 and 4, the methodology for completion of wind damage inspections to residential and light commercial roof finished surfaces consists of the following elements:

1. Interview with property owner, occupant, and/or the owner's representative.
2. Obtain or create a plan view sketch of the roof.
3. Take overview photographs of all exterior elevations.
4. Conduct an inspection of the exterior surfaces of the structure for each elevation.
5. Conduct an inspection of the roof surfaces.
6. Complete a written inspection report.

These elements are discussed in detail below. In addition, the Applied Technology Council (ATC) 45 Field Manual: Safety Evaluation of Buildings after Windstorms and Floods, may also be used as a guideline in forensic investigations.

6.2.1.1   Wind Damage Inspection Methodology – On-Site Interview

When first arriving at the site, the owner, occupant, or the owner's representative should be interviewed regarding the local storm history and the apparent wind damage to the building. If they are not present, the interview should be completed by phone. Questions to be asked during the interview should include:

- When did the wind event(s) occur (i.e., date and time)?
- From which direction did the windstorm arrive?
- Is the wind speed associated with the storm event known? If so, what was it?
- What is the age of the roof surface? When was the last time it was replaced?
- Have any roof repairs been completed? If so, when, what, where, and by whom?
- What wind-associated damages occurred to the residence or structure as a result of the storm event(s), if known:
  - Damages on the exterior surfaces (e.g., missing/displaced gutters, downspouts, window screens, etc.).
  - Damages on the roof surface(s) (e.g., missing/displaced/torn asphalt shingles, tiles, or wood shakes).
- What was the date when the damages were first noticed?
- Were any damages present to the structure before the storm event had occurred?

6.2.1.2   Wind Damage Inspection Methodology – Create or Obtain Plan View Sketch of Roof

A plan view sketch of the residence or structure should then be completed or validated. Like outlined in Chapter 2, a recent trend in the roofing and insurance industries has been the use of satellite imagery technology for roof layouts, slopes, and dimensions from commercial services such as EagleView™ Technologies and Pictometry® (formerly GeoEstimator®). These organizations provide drawing schematics of the roof surfaces with pertinent information needed for its replacement (i.e., ridge/valley lengths, rake/eave lengths, roof areas, and pitches). If a satellite-based plan view of the roof was not previously obtained, a layout of the roof surface should be sketched in a field book as detailed in Chapter 2. If an image was obtained from a commercial service, verify measurements and slopes on roof surfaces. Insert information on types of finished roof surfaces, locations of box vents, plumbing vent stacks, furnace stacks, chimney(s), and other prominent roof features. Also record key elements such as locations and numbers of shingles or tiles that have been replaced/repaired and/or missing/creased/torn, areas of heavily deteriorated surfaces, areas covered with tarps, and other information that may be helpful regarding the storm history (e.g., missing gutters, downspouts, detached siding, fascia, or soffit, etc.). An example of a typical roof layout with roof appurtenances and key details is provided in Figure 6.6.

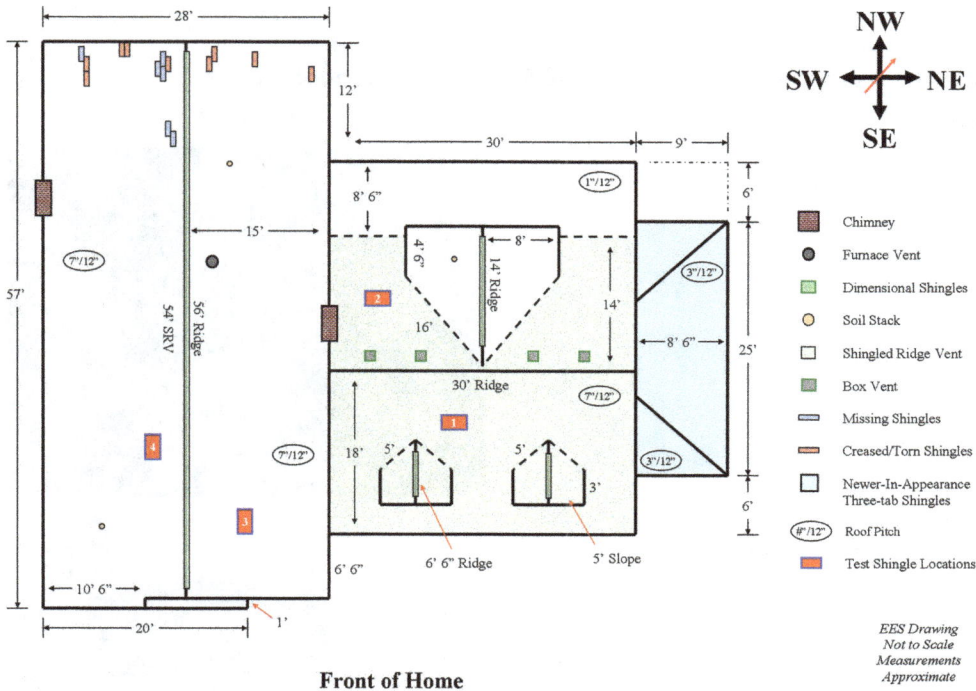

Front of Home

FIGURE 6.6   Typical roof schematic (developed from on-site measurements).

6.2.1.3   Wind Damage Inspection Methodology – Complete Exterior Inspection(s)

To perform a complete and comprehensive assessment to which the subject structure has been damaged by wind forces, begin with a walk-around and inspection of surfaces and features by elevation. This portion of the inspection provides a determination of exterior surfaces and components that were probably damaged by wind forces and provides a basis for determining the direction from which the storm arrived and which roof surfaces should have been most impacted.

This portion of the inspection should begin by starting on one elevation and then moving either clockwise or counterclockwise around the residence or building. High winds will displace exterior finished surfaces (e.g., siding and fascias), water management components (e.g., gutters and downspouts), and window components (e.g., screens and shutters). It should be noted, damages to a home could also be the result of wind-borne debris such as tree branches, slate tiles, and/or asphalt shingles. Such wind-borne debris has been observed to shatter vinyl siding, scratch, and possibly dent/crack other exterior components such as windows and doors. All areas/objects probably damaged by forces of wind action should be documented in the field notebook (i.e., location, area, length, or other dimensions). Items not damaged by forces of wind should also be documented to help prevent later disputes. Observations should be recorded in writing in a field notebook and photo documented. An initial elevation overview photograph taken at the beginning of an inspection for a particular elevation, followed by detailed photographs taken on that elevation, makes identification/labeling of photographs easier. Photographs of typical wind damage to residential exterior surfaces are illustrated in Table 6.2.

TABLE 6.2
Examples of Wind Damage to Typical Residential Exterior Surfaces

Exterior Component   **Photograph**

Downspout
Detached from upper elbow

Metal fascia
Missing sections of fascia on rake

Gutters
Displaced gutter along eave

Vinyl siding
Multiple missing sections of siding

(Continued)

TABLE 6.2 (*Continued*)
Examples of Wind Damage to Typical Residential Exterior Surfaces

Exterior Component	Photograph
Undereave soffits Missing pieces of vinyl soffits of gable	
Vinyl siding Partially displaced piece of siding	

Improper installation practices and settlement can be responsible for damage to siding attributed to forces of wind. For example, the installation manual provided by the Vinyl Siding Institute (VSI) puts great emphasis on proper attachment and nailing patterns, as well as orienting the overlap joints of the siding away from prevailing winds, to provide the maximum wind resistance.[29] Aside from settlement and unhinging of overlapping sections of siding (e.g., metal and vinyl siding), one of the most common discrepancies found on residential and light commercial structures regarding the exterior siding (e.g., metal, wood, or vinyl) is that the fasteners were: (1) not driven firmly into a solid exterior surface such as wood studs or block/concrete walls, (2) fasteners were of insufficient length, and (3) too few fasteners were used and/or the location/placement of the fasteners was incorrect. When possible, document the fastener type, dimensions, spacing, and placement in the field notebook. This information can later be compared with manufacturers' recommended instructions and/or code requirements to determine if installation deficiencies contributed to the damage attributed to forces from wind.

6.2.1.4   Wind Damage Inspection Methodology – Complete Roof Inspection

The next step in assessing a residence or structure for wind damage is to complete a detailed inspection of the roof surface. Care should be taken to use proper safety equipment and adhere to safety protocol and procedures during this phase of the inspection process.

Similar to the general inspection methodology outlined in Chapter 2 for the inspection of the roof surfaces, the inspection should include, but not be limited to, the following elements:

1. A description of the roof construction. At the roof eave or access point, determine the roof construction (e.g., one vs multiple layers, application of underlayment, presence of drip

edge molding, etc.). If multiple roof finishes or differences in appearances for the same roof finish exist, this should be completed for each roof type or finish.

2. Measure or verify measurements of roof dimensions.
3. Inspect for, document in writing, and photograph wind and other damage to roofing materials and appurtenances (e.g., creased asphalt shingles, missing tiles/shakes/shingles, damaged roof appurtenances, etc.). These should be documented in writing by sketching them on the roof plan view and then recording them in the field notebook.
4. Inspect for, document in writing, and photograph roof items apparently replaced along with any other evidence of past repairs such as tar/caulking repairs.
5. Inspect for, document in writing, and photograph the overall condition of the roof surfaces and any non-wind anomalies (e.g., drainage issues, degraded surfaces, missing or damaged flashing, or other factors associated with premature aging of or damage to roof surfaces).

Additional discussions regarding each of these five elements follow.

6.2.1.4.1   Roof Construction at Eave or Access Point

The roof construction should be observed at the roof eave or access point for each roof surface finish as described in Chapter 2. A determination should be made for the presence or absence of drip edge molding, felt underlayment, ice guard, and the type of roofing materials, including the number of layers of finished surface materials. This information provides a basis for whether the roof system conforms to state or local residential building codes. For example, 2018 International Residential Code for One- and Two-Family Dwellings[30] (2018 IRC) as well as the IBC[3] requires no more than two layers of any type of roofing material and no additional layers of material over wood shake/shingles or a tile roof system (e.g., clay/slate/concrete/asbestos). If, for example, it was observed during the inspection that a roof surface has a layer of asphalt shingles installed over a layer of slate tiles, this roof system would violate the 2018 IRC. Note, when determining the layers of asphalt shingles at the roof eave, a starter strip at the eave is not considered a layer for code and best practices purposes.

The roof construction is also important in addressing the issue of resistance to wind uplift. For most residential and light commercial roofing finished surface materials, the most common method of mechanical attachment to the roof surface is by nail fasteners. In order to provide the needed resistance to strong wind forces, the nail fastener must penetrate firmly into the roof decking material. Best practices are that nail fasteners should penetrate ~3/4 of an inch into or through the roof sheathing in order to ensure they remain in place and provide required wind resistance for most roof surfaces. With the presence of additional layers, it has been found that standard nail fasteners may not be long enough to penetrate into the sheathing fully.

6.2.1.4.2   Inspection of Metal Surfaces and Roof Appurtenances

Metal surfaces and roof appurtenances including box vents, soil stacks, furnace vents and vent caps, gutters, and ridge vents, which can be dislodged or displaced by high winds, should be inspected. Identification (type), location, material(s) of construction, and physical measurements of the damaged component(s) should be documented in the field notebook with photographs.

Installation details of a component can provide an explanation of why the roofing appurtenance became susceptible to forces from wind action. For example, a loose and/or slightly displaced section of a metal ridge vent could possibly be the result of forces from high wind action; however, it should also be checked for proper installation (e.g., the type, number, and type of fasteners should be recorded) for later analysis. If the nail shanks are not long enough to penetrate into the decking as directed by the manufacturer, this discrepancy can reduce the resistance of the component to wind-related forces. Wind gusts might have contributed to the damaged component, but in a particular instance, the nail fasteners may not have been properly secured or attached. Table 6.3 illustrates examples of damage to roof appurtenances.

TABLE 6.3

Examples of Damage to Roof Appurtenances

Roof Appurtenance	Photograph
Metal ridge vent Displaced/loose end	
Satellite dish Damaged/displaced from roof	
Furnace vent Missing cap	
Furnace vent Missing cap	

6.2.1.4.3   Identification and Quantification of Wind Damage on Each Roof Elevation, or Facet

All potentially wind damaged (i.e., missing, displaced or dislodged, ripped, torn, cracked, penetrated) elements of a finished surface should be quantified by elevation. In addition, for each elevation, test sections should be inspected to document the method of application and attachment of the shingles, shakes, tiles, or other finished surfaces.

The highest concentration of damage from forces of wind action should be observed along the peak, rake, and eave areas. If the damage is more extensive outside of these areas, than installation, the time of year of the installation, or fraud may be partially, or totally responsible for the damages observed. Under these circumstances, particular attention should be paid to installation practices (e.g., the method of attachment) and for evidence of fraud (Chapter 5).

In addition, during the roof surface inspection, information should be collected to distinguish between wind-related damages and non-wind-related damages such as those associated with natural or premature aging/deterioration (e.g., blistering, thermal splits, small mechanical tears, chips, etc.). This may include documenting information related to ventilation of the attic (soffit, gable, box, or ridge vents) to determine if the attic space was adequately ventilated since the lack of adequate ventilation is a common cause of premature degradation of the roof surfaces (Chapter 12).

If wind damage is present on a roof surface, the quantity or extent of damage, the ease of reparability, and the cost-effectiveness should be taken into consideration to determine whether a full or partial roof replacement is ultimately recommended. For example, a recently installed layer of shingles with only minor levels of wind damage would typically warrant spot repairs. In this case, replacement shingles with a comparable match should be easily attainable with little to no noticeable color variation. On the other hand, a comparable match might not be obtainable for older homes that possess shingles or other roofing components near or past the end of their effective service life. The ease of reparability would be too difficult due to their overall condition (i.e., brittleness).

Additional information on the methodologies to be followed for roof inspections of the most commonly encountered roof finished surfaces follows.

6.2.1.4.4   Roof Wind Damage Inspections – Detailed Methodology for Inspection of Asphalt Shingles Finished Roof Surfaces

As noted in Chapter 2, asphalt shingled surfaces are those most often utilized on residential and light commercial structures and thus are the surfaces most commonly encountered during hail and wind forensic roof investigations. Failures associated with wind damage to asphalt-shingled roof surfaces include creased shingles, torn shingles, slipped or displaced shingles, and completely missing sections of shingles. Examples and descriptions of damage to asphalt shingles are illustrated in Table 6.4.

TABLE 6.4
Examples of Damage to Asphalt Shingles

Description	Photograph
Creased three-tab shingles The result of continuous "flapping" of the shingle tab from wind uplift. Unsealed or unadhered tabs are more susceptible. The crease is typically linear and along the top portion of the tab. This mechanism is typically associated with three-tab shingles	
Torn three-tab shingle The eventual tearing of a creased shingle	
Missing three-tab shingles Exposed nail fasteners driven through the gray and weathered adhesive strip	

(Continued)

TABLE 6.4 (*Continued*)
Examples of Damage to Asphalt Shingles

Description	Photograph
Missing, torn, and creased three-tab shingles A group of damaged shingles near the roof ridge. Notice the exposed felt underlayment and the potential source of water intrusion	
Missing dimensional shingle	
"Slipped" dimensional shingles Shingle "slippage" likely the result of improper installation (i.e., spacing) coupled with gravitational forces (i.e., nails likely driven above recommended location) allowing the shingles to be more susceptible	 

During the initial portion of the inspection of an asphalt-shingled roof, the methodology should consist of documenting and photographing the overall condition of the roof surfaces as well as any modifications or abnormalities. This should include:

- A description of the overall condition of the roof system (i.e., good, fair, or poor).
- Quantification of the area(s) and number of replacement shingles, if any.
- Locations and the condition of historic caulking or sealant repairs.
- The presence and location of biological growth (e.g., moss, algae, lichen) if present.

- Whether or not the shingles typically are well adhered to the roof surface. Inspecting multiple shingles on each cardinal elevation can provide a general understanding of how well the shingles are adhered. For example, a "lift test" of 20 random shingles should be completed on each cardinal elevation. This can provide a percentage of the number of unadhered (or easily lifted) shingles present for that cardinal direction.
- The condition of the roof sheathing. Is it wavy in appearance, firm, or soft to walk on?
- Documentation of areas and possible causes for water intrusion where likely present. This might include areas with exposed wood sheathing below missing shingles.

After documenting and photographing the roof construction, general roof conditions, and areas of shingle damage(s) on the roof surfaces, one should further document the installation method and fastening patterns for each roof elevation. As illustrated in Table 6.5, asphalt shingles are often improperly fastened.

TABLE 6.5

Examples of Improper Asphalt Shingle Installation Methods and Nailing Patterns

Description	Photograph
Three-tab shingle Staple fastener (~1″ in crown width) driven flush into surface of shingle and driven through the adhesive strip. Staples should be driven parallel to the long edge of the shingle (i.e., horizontally)	
Three-tab shingles The end nail fastener is spaced too far from the right edge of the shingle. Nails should be driven ~1 inch from the side edge	
Three-tab shingle Evidence of insufficient adhesion between shingles demonstrated by sealant residue – staple in right adhesive strip	

(Continued)

TABLE 6.5 (*Continued*)

Examples of Improper Asphalt Shingle Installation Methods and Nailing Patterns

Description	Photograph

Three-tab shingle

Nails driven through and above shiny and glossy sealant strip. Notice little adhesive residue on underside of overlying shingle likely indicating poor contact adhesion between shingles

Dimensional shingle

Nail fastener driven flush into surface of shingle yet driven above the sealant strip. In addition, the nail is driven well above the recommended "nailing zone" as indicated by the set of horizontal white lines beneath the sealant strip

Dimensional shingle

Little to no adhesive residue on underside of overlying shingle suggesting adhesive failure

Dimensional shingle

Multiple improperly driven nails located above and through the sealant strip. Notice the top underlying surface of the lifted shingle, little contact residue is present indicating that the nails prevented full adhesion at those locations

(Continued)

TABLE 6.5 (*Continued*)
Examples of Improper Asphalt Shingle Installation Methods and Nailing Patterns

Description	Photograph
Dimensional shingle Nail fastener overdriven through the surface of shingle. In this instance, the nail completely penetrated through the shingle creating a clean edged hole. These situations, often encountered with weighted dimensional shingles, provide less support from the nails and can lead to shingle "slippage"	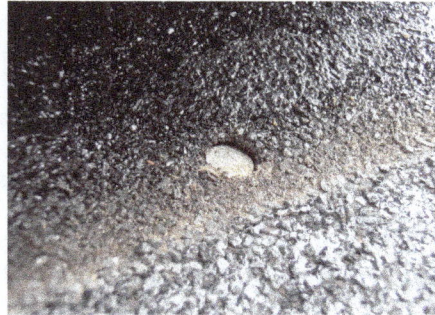
Fastener test shingle #1 – southeast elevation (see Figure 6.4) – overview	
Fastener test shingle #1 – southeast elevation Nail driven flush through the adhesive strip yet driven too far from side edge. The nails should be driven 1 inch from the side edges	
Fastener test shingle #1 – southeast elevation The bottom nail is overdriven above the adhesive strip. This condition is called high nailing and results in shingles that are not secured through the double-ply area (often denoted as the nail zone) and leaves the shingles bound to a roof with only half the recommended number of fasteners (four nails instead of eight)	

The recommended method for inspecting the fastening characteristics of asphalt shingles is to examine random test shingles, with at least one per elevation (remember to document these locations on the plan roof sketch as shown in Figure 6.6). Based on experience, these limited random test shingle examinations likely represent the method of shingle attachment employed throughout the roof surface. Roofing contractors tend to be people of habit and consistency, and this level of examination has been found to describe the quality of workmanship adequately throughout the roof surface. For each test location, the following information should be recorded in the field notebook and by photographs:

- *Condition of Adhesive Strip*: Inspect and document the condition of the shingle adhesive or sealant strip, if present, of the damaged shingle. Questions to be answered are:
  - Is the adhesive strip shiny or glossy in appearance, suggesting failure of activation to the underlying or overlying shingle?
  - Is there any evidence or presence of wind-borne debris such as dirt and pollen that might have prevented proper adhesion? This may be an indication that the shingles were installed in cold weather and never properly sealed.
  - Is there the presence of adhesive residue on the underlying or overlying shingle to indicate a cohesive failure?
  - Is there any evidence of tool marks/scrapes at/near the adhesive strips or has the adhesive strip separated with part of the shingle felt/granules? This suggests possible fraud (Chapter 5).
- *Fastener Type and Patterns*: Inspect and document the fastening patterns near failure locations and at least once on each elevation. Questions to be answered are:
  - Is the shingle fastened with nails or staples? Staples were allowed until about the year 2000 in many codes, but typically are no longer allowed due to fastening issues.
  - How many fasteners were used to secure the shingle? Typically, this should be four, but under some circumstances (e.g., high pitch and high wind areas) there should be six.
  - Are the fasteners driven above, through, or below the adhesive strip? All shingle manufacturers specify that fasteners be installed below the adhesive strip and typically note that installation of fasteners through the adhesive strip will lead to wind blow-off.
  - Are the fasteners driven flush and perpendicular into the shingle or are the fasteners overdriven, underdriven, and/or driven at angles not flush to the surface of the shingle?
  - For some dimensional or laminated shingles, are the fasteners driven within the indicated "nail line" or "nail zone" located below the sealant strip?
  - Document the location and spacing of the fasteners from the bottom edge of the shingle as well as from the side edges to ensure they meet manufacturer specifications for installation.
  - Document any rusting or corrosion to the fasteners. Rusting conditions often indicate that the damaged shingles have been exposed to weathering events for an extended period of time (months/years vs days/weeks). This may be helpful when trying to attribute damages to specific event timeframes.
- *Overall Installation Method*: The installation method of the shingles should be inspected and documented with written field notes and photographs. Questions to be answered are:
  - Are the shingles installed in a vertical or "racked" installation or are they installed in a diagonal or stair-step method?
  - A "racked" installation method is when the initial course of shingles is installed directly up the slope of the roof elevation in a vertical column. Subsequent courses must be

slid beneath the overlying first course to "butt" the shingles together. This "racking" pattern of installation is not the suggested method of installation according to some shingle manufacturers and best practices for optimal wind performance.

- Is any evidence present, including scratches, tears, or punctures that may be consistent with wind-borne debris or falling objects from higher roof elevations?

6.2.1.4.5 *Roof Wind Damage Inspections – Detailed Methodology for Inspection of Wood Shakes/Shingles and Slate, Clay, Concrete, and Asbestos Tile Finished Roof Surfaces*

Wood shakes/shingles and slate, clay, concrete, and asbestos tile roof finished surfaces can provide owners with an alternative roofing material that helps give the residence or light commercial building a sometimes rustic or earthy appearance. The characteristics and the longevity of these systems as a water-shedding assembly depend on several factors, including: (1) the quality of the material employed, (2) extent of care and maintenance, and (3) installation practices used. Various resources such as the NRCA,[13] CSSB,[21] Tile Roof Institute,[28] Haag Engineering,[31] and articles and books written by Joseph Jenkins[24-26] can provide a distinction between the various manufacturing processes, types, and grades of each of these finished surfaces. Regardless of finished surface, the following detailed methodology is recommended for wood shingle and shake and clay, concrete, and asbestos tile finished roof surfaces.

During the initial portion of the inspection of a non-asphalt-shingled roof finish, one should document and photograph the overall condition of the finished roof surfaces as well as any modifications or abnormalities. This should include:

- A description of the overall condition of the roof system (i.e., good, fair, or poor).
- Quantification of the area(s) and number of replacement shakes, tiles, etc., if any.
- Locations and the condition of historic caulking or sealant repairs.
- The presence and location of biological growth (e.g., moss, algae, lichen) if present.
- Whether or not the finished surfaces were well secured to the roof sheathing/substrate.
- The type, construction, and condition of the roof sheathing/substrate.
- Is the roof and decking wavy in appearance, firm, or soft to walk on?
- Documentation of areas and possible causes for water intrusion, where likely present. This might include areas with exposed wood sheathing below missing shingles, shakes, or tiles.

Unlike asphalt shingles, which can exhibit several forms of wind damage, wood shakes/shingles and the various components of tile roof systems are typically either displaced or completely removed by forces of wind action. In some occasions, these materials can also be damaged from wind-borne debris or other roofing components. Generic terms related to these surfaces are listed and defined below:

- *Displaced* materials are defined as components that may be slightly loose or be partially displaced from the original fastened position.
- *Missing* defines a material that is completely displaced from the roof. These types of defects are more likely to be attributable to wind forces, as a portion and/or the entire shake may be missing from the roof elevation.
- *Debris-Caused Damage* defines damages from wind-borne debris such as tree branches or other roofing materials falling from higher elevations.

Examples and descriptions of damage to wood, slate, and clay roof systems are illustrated in Table 6.6.

TABLE 6.6

Examples of Damage to Wood Shingles, Slate, and Clay Tile Roof Systems

Description	Photograph
Wood shingle Missing shingles exposing bright underlying surface	
Slate tile Missing tile exposing bright underlying surface	
Clay tile Displaced tile	
Slate tile Displaced tile	

As with asphalt shingles, these roofing systems eventually degrade with time. However, the degradation process is somewhat different, depending on the finished surface. The effects of erosion, repeated freezing and thawing, and other factors can cause surface pitting, delamination, or other forms of natural degradation over time. Therefore, the effectiveness and ability of the material to resist forces from wind action are proportionate to the age of the material, implying that as the surface ages and naturally deteriorates, the material becomes less resistant to wind damages and other stresses.

This is also true for the mechanical attachments such as nails, screws, clips, and other metallic components, especially for tile, which may not outlive the finished materials themselves causing displacement and/or loss of shingles, shakes, or tiles not fundamentally associated with wind. In order to assure these components are capable of lasting as long as the system employed, experience and common institutions such as the NRCA recommend the type and thickness of the material should have been used that correspond to the service life of the system (e.g., hot-dipped galvanized, stainless steel, copper, bronze, or cut-brass materials).

Due to these complexities, experience in the determination of whether or not damages are wind-related on a 75-year-old slate or clay tile finished roof surface, for example, is essentially a requirement. For instance, does the presence of missing or displaced tiles on the roof surface indicate a direct result of wind forces, or would these damages also be attributable to the age of the fasteners and tiles? Thus, the age and overall condition of the roof system cannot be overlooked and must be considered a factor in the damages present. A combination of ambient wind gusts, which are likely below the design threshold for wind resistance, coupled with the age and overall condition of the roof system would contribute to the presence of missing and displaced tiles. Additionally, metal valley and flashing components may become degraded to the point where water infiltrates and degrades the underlying roof system, allowing the tiles to be dislodged or missing. The wind may have finally caused the tiles to move, but the underlying cause of damage was a result of water intrusion from the deterioration of the metal components. Figure 6.7a and b shows photographs of missing slate tiles attributable to age and natural deterioration (e.g., delamination) as well as forces of wind action.

(a)   (b)

FIGURE 6.7   Missing (a) and degraded (b) slate tiles.

For newer roof finished surfaces, where the age of the roof system is less of an issue, one must examine installation and attachment methods as well as damage associated with forces from wind. This would include information on nail penetration length, the addition of nose clips, etc.

For damage reported as having occurred recently, a helpful tool in providing a timeframe of when the damage(s) could have occurred is visually inspecting the underlying shake/shingle surface of the missing or displaced component for color patterns or presence of debris. For example, an exposed underlying or damaged surface that is weathered, dull, or darker in appearance would indicate that the wood or tile component has been missing/displaced for quite some time (months/years vs days/weeks). On the other hand, a missing wood shake/shingle or tile that exposes a bright colored or fresher-in-appearance underlying surface would suggest that the damage occurred recently (days/weeks). These types of inspection details could also help the forensic investigator date damage from specific storms. Examples of missing and detached wood, slate, and clay tiles attributable to forces of wind action are shown in Figures 6.8–6.10.

FIGURE 6.8   Missing wood shake exposing weathered and dull underlying surface.

FIGURE 6.9   Missing slate tile exposing bright, underlying surface.

FIGURE 6.10   Recently displaced and/or loose wood shake.

Natural weathering defects and anomalies common to wood shingles and shakes, like those mentioned in Chapter 2 (e.g., curling and cupping), may be thought to exhibit signs of wind-related damage, but are not considered damages attributable to wind action unless they are slightly loose, displaced, or completely missing from the roof surface.

Unlike wood roofing materials, slate, clay, and other tile systems are somewhat brittle and more susceptible to cracking, chipping, and shattering. These types of defects are attributable to weathering anomalies, inspection practices (i.e., walking on the surface), severe weather, or even wind-borne debris. Even clay and concrete tile roof systems that are well attached to a roof surface can be easily broken by windborne debris, or "missiles."[27,28] A single failed tile can initiate a cascading failure of other tiles on the roof surface. Figure 6.11 illustrates a cascading failure as viewed from above, and Figure 6.12 depicts two clay tiles that were likely damaged from windborne debris.

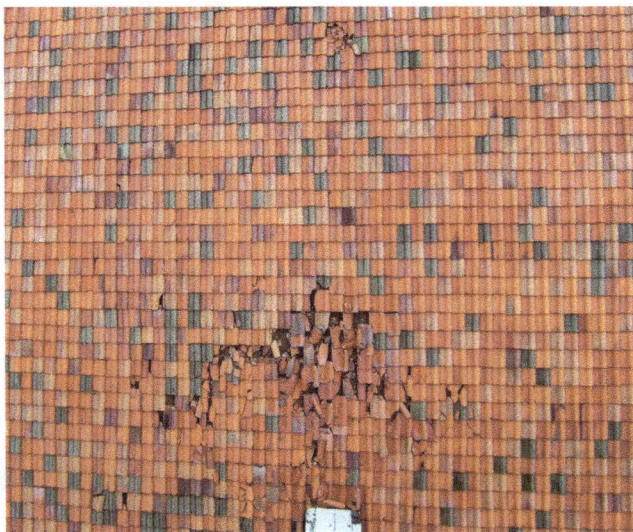

FIGURE 6.11   Cascading failure of clay tiles from debris (looking down from higher elevation).

FIGURE 6.12   Damaged clay tiles from falling debris (tree limbs/branches).

It should be noted that any wind-related or even non-wind-related damages to wood shakes/ shingles or to the various components of a tile roof system can nearly always be repaired without replacement of the entire roof elevation or system.

FEMA also provides guidance on metal roof systems in high-wind regions.[32]

6.2.1.5   Wind Damage Inspection Methodology – Inspection Report

A typical wind-damage inspection report should follow the general outline presented in Chapter 2. General elements and details unique to wind-damage inspections are discussed below.

A residential and light commercial wind damage report should include the following elements:

- Introduction (information on the inspection location and client).
- Scope of work (what is the scope of the inspection?).
- Information regarding the property (e.g., age of the structure and roof surface, recent repairs or modifications).
- Storm history (dates of storm(s) in the area, recent and past).
- Wind weather history at/near the inspection location. For example, the National Weather Service (NWS) provided by the National Oceanic & Atmospheric Administration (NOAA) offers monthly/daily climatological data for various geographical locations. This allows the inspector to analyze measured daily wind speeds to determine whether the wind gusts exceeded the resistance threshold prescribed under modern building codes.
- Summary of interview(s).
- Summary of exterior observations (i.e., wind-related damages).
- Summary of roof observations:
  - Roof construction.
  - General observations of the overall condition of the roof surface.
  - Location and quantity of wind-related damages (e.g., missing, displaced, torn, creased, slipped materials).
  - Specific observations of the installation patterns (e.g., nail placement for asphalt shingles).
- Discussion/analysis of observations:
  - Explain, if possible, why the damages were present on the exterior and roof surfaces of the residence.

- Did the overall poor condition of the roof surface contribute to the damages?
- Was the roofing system installed to provide the greatest resistance to wind blow-offs or related damage, or were deficiencies present, which could have contributed to the damages (i.e., improper fastening patterns and methods)?
- Tables quantifying the extent of wind damage to the residence (see below).
- Conclusions.
- Photographs and figures not included within the report.
- Chain of custody for any evidence and/or supporting documents.

Experience has shown that the use of figures, photographs, and tables greatly increases the effectiveness of the report. For example, a table providing information from the fastening methodology outlined above for four random test shingles (two for three-tab and two for metric-sized dimensional shingles) is illustrated in Table 6.7.

TABLE 6.7
Test Shingle Observations for Three-Tab Shingles (see Figure 6.6)

I.D.	Roof Elevation	Shingle Width	Shingle Exposure	Fastener Positions[a]				Fastener Height[b]
				1	2	3	4	
1[c]	Southeast	39-3/8"	5-1/2"	4-1/4"	14"	25-3/4"	34-5/8"	9"
2[c]	Northwest	39-3/8"	5-3/4"	2-1/4"	15-1/2"	23-3/4"	35"	8-1/4"
3	Northeast	36"	5"	3/4"	11-5/8"	23-5/8"	**Missing**	6-3/4"
4	Southwest	36"	4-1/2"	1-1/2"	13-1/4"	26"	35-1/4"	7-1/2"

[a] Measured in from left side edge of shingle.
[b] Measured up from bottom edge of shingle to highest fastener.
[c] Dimensional asphalt shingles.

Similarly, Table 6.8 illustrates an example summary table of wind-related damages present on each roof facet of a slate tile roof surface. A similar approach for tabulated data can be applied to wood shakes/shingles, clay, and other tile roof systems.

TABLE 6.8
Summary of Slate Roof Defects by Roof Facet

Facet I.D.	Elevation	Area (square feet)	Est. # Tiles[a]	Weathered Chips	Weathered Cracks	Eroded Holes	Missing Tiles	Displaced Tiles	Wind Damaged (%)[b]
A	East	333	546	6	3	1	0	1	0.2
B	North	70	115	0	0	2	0	0	0.0
C	South	70	115	1	0	0	1	2	3.2
D	South	197	323	3	5	0	4	0	1.5
E	West	90	148	0	0	0	2	0	1.7
F	South	180	296	2	1	1	6	0	2.5
G	West	145	238	0	0	0	1	2	1.5
H	North	123	202	2	4	1	0	0	0.0
	Total	**1,208**	**1,625**	**14**	**13**	**5**	**14**	**5**	**1.2**

[a] The calculated exposed area per tile was ~0.7.4 square feet.
[b] Included both displaced and missing tiles.

IMPORTANT POINTS TO REMEMBER

- Weathered and deteriorated materials near and/or past their expected and designed service life are more susceptible to weather-related damages, including forces from wind uplift.
- For asphalt shingles, the effectiveness of the sealant strip will decrease over a period of time. Experience has shown that the life span of the sealant strip to remain fully bonded as expected by manufacturers typically ranges from 8 to 12 years. After this timeframe, the shingle becomes more susceptible to cohesive failure and weather-related damages.
- Wind-related damages to asphalt shingles are classified/determined as creased/torn, displaced, or missing. Shingles that are lifting, but contain no physical damage, are typically not damaged from forces from wind action. If shingles are merely unadhered and can be easily lifted, then they are usually not the direct result of forces from wind action. These shingles can be resealed by a method called "hand dabbing."
- Wood, slate, clay, or other types of tiles often outlive fastening and flashing components. Experience has shown that wind-related damages to these systems are typically not only a result of weather and age deterioration to the roofing material, but also age and corrosion of the fasteners.

REFERENCES

1. Patterson, Stephen and Madan Mehta. "Wind Pressures on Low Slope Roofs." Roof Consultants Institute Foundation. Publication No. 0.01, Revised 2005.
2. Smith, Thomas L. "Improving Wind Performance of Asphalt Shingles: Lessons from Hurricane Andrew." *Proceedings of the 11th Conference on Roofing Technology*, September 21–22, 1995.
3. International Code Council. *2018 International Building Code (IBC)*, Accessed July 27, 2020, https://codes.iccsafe.org/content/IBC2018.
4. American Society of Civil Engineers (ASCE). "Standard 7–16: Minimum Design Loads for Buildings and Other Structures." 2017.
5. APA – The Engineered Wood Association. "Retrofitting a Roof for High Wind Uplift." Data File Number A410, July 2007.
6. American Plywood Association. "Roof Sheathing Fastening Schedules for Wind Uplift." APA Publication T325, Tacoma, WA, March 2006.
7. American Plywood Association. "Wind-Rated Roofs – Designing Commercial Roofs to Withstand Wind Uplift Forces". APA Publication G310, Tacoma, WA, January 2009.
8. American Plywood Association. "Design/Construction Guide – Non-Residential Roof Systems." APA Publication A310, Tacoma, WA, May 1996.
9. Asphalt Roofing Manufacturing Association (ARMA). "Nail Application of Asphalt Strip Shingles for New and Recover Roofing." 2017. Accessed July 27 2020, https://www.asphaltroofing.org/nail-application-of-asphalt-strip-shingles-for-new-and-recover-roofing/.
10. Nichols, Tom. "A Stronger Bond." *Professional Roofing*, 2010. Accessed July 27, 2020, https://www.professionalroofing.net/Articles/A-stronger-bond--12-01-2010/1809.
11. Federal Emergency Management Agency (FEMA). "Asphalt Shingle Roofing for High-Wind Regions." *Home Builder's Guide to Coastal Construction*, Technical Fact Sheet No. 7.3. Accessed July 27, 2020, https://www.fema.gov/media-library-data/20130726-1536-20490-8282/fema499_7_3.pdf.
12. Asphalt Roofing Manufacturers Association (ARMA). "Nail Application of Asphalt Shingles for New and Re-cover Roofing." Technical Bulletin Form No. 221-RR-94, June 1996.
13. The National Roofing Contractors Association (NRCA). "The NRCA Roofing Manual: Steep-Slope Roof Systems." 2017.
14. Koontz, Jim D. "Performance Attributes of Fiberglass Shingles." *Interface Magazine*, July 2007, pp. 23–30.
15. Corbin, Raymond L. "Mechanics for Properly Sealing Fiberglass Reinforced Asphalt Roofing Shingles." *Interface Magazine*, April 2001, pp. 10–14.

16. Asphalt Roofing Manufacturers Association (ARMA). "Recommendations for Installation of Asphalt Roofing Shingles in Cold Weather." May 2007.

17. Asphalt Roofing Manufacturers Association (ARMA). "Recommendations for Installation of Asphalt Roofing Shingles in Cold Weather." May 2015. Accessed July 27, 2020, https://www.asphaltroofing.org/media/2017/05/Recommendations-for-Installation-of-Asphalt-Roofing-Shingles-in-Cold.pdf.

18. Owens Corning. "Cold Weather Shingle Installation". Technical Bulletin TB-05, September 20 2017. Accessed July 27, 2020, http://www.owenscorning.com/NetworkShare/Roofing/10022497-Cold-Weather-Shingle-Installation-Technical-Bulletin---TB-05.pdf.

19. IKO. "IKO Equipment and Materials – Asphalt Shingles." 2020. Accessed July 27, 2020, https://www.iko.com/na/pro/building-professional-tools/learn-about-roofing/cold-weather-roofing/#:~:text=According%20to%20Roofing%20Construction%20%26%20Estimating,and%20more%20prone%20to%20breakage.

20. Dutt, Om "Adhesion of Granules on Asphalt Shingles: Assessing the Effects of Weather Factors." National Research Council of Canada (NRCC). Institute for Research in Construction (IRC), 1987. Accessed July 27, 2020, https://nrc-publications.canada.ca/eng/view/ft/?id=896e48c6-7f4e-4762-b509-1d3644e8e40a.

21. Cedar Shake and Shingle Bureau (CSSB). "New Roof Construction Manual". June 2014. Accessed July 27, 2020. http://www.cedarbureau.org/cms-assets/documents/roof-manual-full-09-2014.pdf.

22. Cedar Shake and Shingle Bureau (CSSB). "Cedar Roofing Fasteners." Technical Bulletin, March 2007. Accessed July 27, 2020, file:///C:/Users/SEP/Downloads/CSSB+Cedar+Roofing+Fasteners%20(4).pdf.

23. Federal Emergency Management Agency (FEMA). "P-762: Design Considerations, Regulatory Guidance, and Best Practices for Coastal Communities." Local Officials Guide for Coastal Construction, February 2009. Accessed July 27, 2020, https://www.fema.gov/media-library-data/20130726-1706-25045-9843/logcc_rev1.pdf.

24. Jenkins, Joseph. "Hurricane Force Winds and How They Affect Slate Roofs." *Traditional Roofing Magazine*, Issue #5, 2006. Accessed July 27, 2020, http://www.traditionalroofing.com/downloads/TR5_hurricane_article.pdf.

25. Jenkins, Joseph. *The Slate Roof Bible*, Third Edition. Joseph Jenkins, Inc.: Grove City, OH, June, 2016.

26. Jenkins, Joseph. "Slate Roofs: Why Learn the Hard Way?" *Interface Magazine*, October 2008, pp. 14–30. Accessed July 27, 2020, http://www.theslateroofexperts.com/downloads/RCI_slate_mistakes.pdf.

27. Federal Emergency Management Agency (FEMA). "Tile Roofing for High-Wind Regions." *Home Builder's Guide to Coastal Construction*, Technical Fact Sheet No. 7.4. Accessed July 27, 2020, https://www.fema.gov/media-library-data/20130726-1536-20490-8282/fema499_7_4.pdf.

28. Tile Roof Institute (TRI) and the Western States Roofing Contractors Association (WSRCA). "Concrete and Clay Roof Tile: Installation Manual for Moderate Climate Regions." July 2015. Accessed July 27, 2020, https://tileroofing.org/wp-content/uploads/TRI-Installation-Guide-2015-1.pdf.

29. Vinyl Siding Institute (VSI). "Vinyl Siding Installation Manual." 2017. Accessed July 27, 2020, https://www.vinylsiding.org/wp-content/uploads/2016/03/2017-Vinyl-Siding-Installation-Manual.pdf.

30. International Code Council. "2018 International Residential Code for One- and Two-Family Dwellings (IRC)." 1st Printing August 2017. Accessed July 27, 2020, https://codes.iccsafe.org/content/IRC2018.

31. *Haag Certified Roof Inspector Program Workbook,* Residential Edition. Haag Engineering Co.: Dallas, TX, 2008.

32. Federal Emergency Management Agency (FEMA). "Metal Roofing in High-Wind Regions." Home Builder's Guide to Coastal Construction. Technical Fact Sheet No. 7.6. Accessed July 27, 2020, https://www.fema.gov/media-library-data/20130726-1536-20490-8282/fema499_7_6.pdf.

7 Wind Damage Assessments for Low-Sloped Roof Systems

Ronald L. Lucy and Stephen E. Petty
EES Group, Inc.

CONTENTS

PURPOSE/OBJECTIVES

The purpose of this chapter is to:

- Provide an understanding on how buildings are designed to resist wind uplift pressures.
- Discuss factors that can either decrease or increase the wind uplift resistance of low-sloped roof systems.
- Document a methodology for assessing wind damage claims to low-sloped roof systems.

Following the completion of this chapter, you should be able to:

- Have a basic understanding of wind uplift design principals.
- Be cognizant of installation issues that can make a roof susceptible to wind damage.
- Perform a visual inspection and identify/document wind damage to low-sloped roof systems.

7.1 INTRODUCTION

As discussed in Chapter 4, low-sloped roof systems contain a pitch of 4:12 or less. They are commonly found on light commercial, commercial, industrial buildings, and occasionally on portions of residential roofs. This chapter focuses on wind damage claims to low-sloped roof systems.

It is not uncommon for wind forces to damage the membrane, insulation, and decking of low-sloped roof systems. Buildings impacted by wind damage can also be impacted by damage attributable to water intrusion since storms producing high winds are frequently accompanied by rains.

National consensus standards, building codes, and manufactures provide requirements and details to ensure that roof systems perform under the design wind speeds to prevent catastrophic damage from occurring. With the exception of storms producing hurricane or tornado force winds, in theory, damage should not occur if the basic installation requirements of low-sloped roof systems are followed. However, improper installation, lack of maintenance, and deterioration can result in the failure of roof system components at wind speeds well below the design speeds. This chapter will provide some basic information on how wind uplift pressures act on a roof surface, discuss some of the common causes of roof failures as a result of wind forces, provide a methodology for low-sloped roof wind damage inspections, and provide case study examples to illustrate wind damage to various low-sloped roof systems.

7.2 WIND FORCES

The following discussion on wind uplift on low-sloped roof systems is restricted to a slope of $\leq 7°$ or a pitch of ~1.5:12 or less. Positive (+) and negative (−) signs within this section signify pressures acting toward and away from the surface of the roof, respectively.

7.2.1 Wind Pressure Interaction on Buildings

Wind forces can exert both positive and negative pressures on buildings. As previously discussed in Chapter 6, the pressure acting on an area generates a force (i.e., force = pressure × area), which in turn can create a force acting on the surface or side walls of a building. The variation in wind speed and direction (i.e., gustiness or turbulence) results in pressures/forces that do not remain constant and can exert both positive and negative pressures on the surface of a structure over time. Usually there will be a dominance of pressure exerted on a surface that will be either positive or negative pressure. For example, when wind strikes a rectangular building, positive pressure will be exerted on the windward side of the building while negative pressure will be exerted on the leeward side and across the roof surface. A schematic illustrating wind pressure interaction on a rectangular building is shown in Figure 7.1.

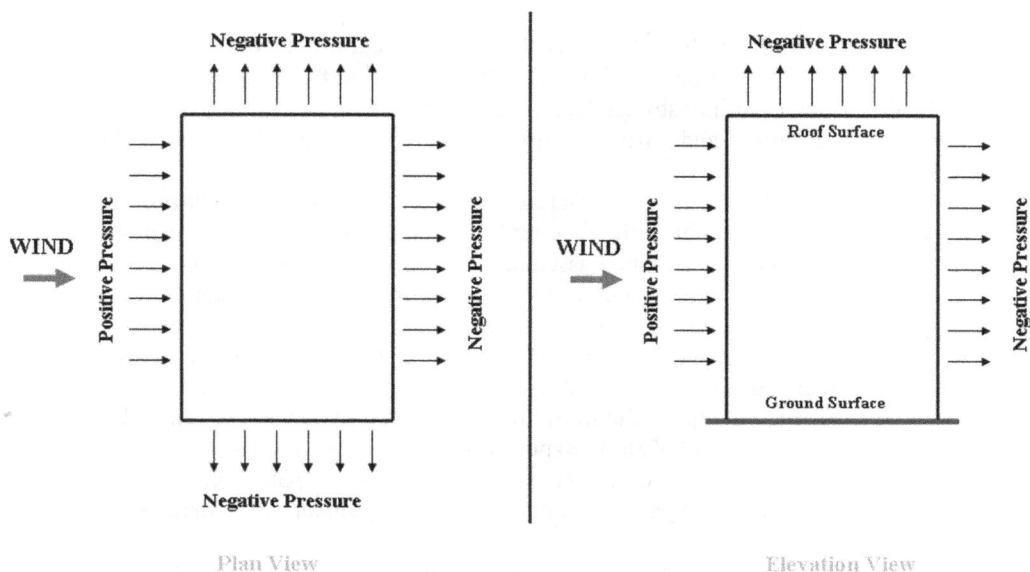

FIGURE 7.1   Pressure interaction on a building.

On the windward wall, positive pressure has been found to increase with the height (eave height) of the building; however, the negative pressure has shown no appreciable change in relation to the building height on the leeward walls.

The interaction of wind pressure on a building can also be affected by openings in the building. Buildings with openings can cause a cupping effect on the wind, resulting in a ballooning effect to the interior. Consequently, from a wind load standpoint, buildings are classified as either enclosed, partially enclosed, or open.[1]

Enclosed buildings have no effective openings except for cracks/gaps in the perimeter walls, windows, and doors. If these openings were subjected to infiltration and exfiltration of air, the internal pressures generated are typically much smaller than buildings having partial openings. Partially enclosed buildings consist of structures such as aircraft hangars or dock areas with roll-up doors. These openings tend to create the highest interior pressure, resulting in the greatest ballooning effect. Buildings where all of the walls, from the floor to the roof, are open are defined as open buildings. In this case, the wind is able to pass through the structure creating no adverse effects to the interior pressure.

Similar to wind interaction on wall surfaces, the wind pressures on roof surfaces vary based on the roof effective wind areas. From a design standpoint, the roof surface areas are divided into corners, perimeters, and field (middle portions) areas. The highest pressures will occur at the corners, followed by the perimeters, and then the field of the roof surface. The averaging of pressures over these different areas defines the correct number of fasteners or securing methods that will be needed for the average given areas.

7.2.2 DESIGN WIND SPEEDS

Model building codes such as the 2018 International Building Code (IBC) require that roof systems are able to withstand the uplift pressures associated with 3-second gust wind speeds at 33 feet above the ground surface. The ultimate design wind speeds vary in magnitude and are dependent on the geographic location of the structure and other factors such as the risk category of the building. The risk category is based on who may occupy the structure; risk categories from the 2018 IBC, Section 1604.5 are shown below[2]:

Risk category I: Building or other structures that represent a low hazard to human life in the event of failure, including but not limited to: (1) agricultural facilities, (2) certain temporary facilities, and (3) minor storage facilities.

Risk category II: Buildings and other structures except those listed in Risk Categories I, III, and IV.

Risk category III: Buildings and other structures that represent a substantial hazard to human life in the event of failure, including but not limited to: (1) buildings and other structures whose primary occupancy is public assembly with an occupant load >300, (2) buildings and other structures containing Group E occupancies with an occupant load >250, (3) buildings and other structures containing educational occupancies for students above the 12th grade with an occupant load >500, (4) Group 1–2, Condition 1 occupancies with 50 or more care recipients, (5) Group 1–2, Condition 2 occupancies not having emergency surgery or emergency treatment facilities, (6) Group 1–3 occupancies, (7) any other occupancy with an occupant load >5,000, (8) power-generating stations, water treatment facilities for potable water, wastewater treatment facilities, and other public utility facilities not included in Risk Category IV, and (9) buildings containing quantities of toxic or explosive materials not included in category IV.

Risk category IV: Buildings or other structures designated as essential facilities, including but not limited to: (1) Group 1–2, Condition 2 occupancies having emergency surgery or emergency treatment facilities, (2) ambulatory care facilities having emergency surgery or emergency treatment facilities, (3) fire, rescue, and police stations and emergency vehicles garages, (4) designated earthquake, hurricane, or other emergency shelters, (5) designated emergency preparedness, communications and operations centers, and other facilities required for emergency response, (6) power-generating stations and other public utility facilities required as emergency backup facilities for Risk Category IV structures, (7) buildings and other structures containing quantities of highly toxic materials that exceed maximum allowable quantities per control area as given in Table 307.1(2) or per outdoor control area in accordance with the International Fire Code and are sufficient to pose a threat to the public if released, (8) aviation control towers, air traffic control centers, and emergency aircraft hangers, (9) buildings and other structures having critical national defense functions, and (10) water storage facilities and pump structures required to maintain water pressure for fire suppression.

Using these risk categories, design wind speeds were specified by the American Society of Civil Engineers (ASCE) in their ASCE 7-16 standard "Minimum Design Loads for Buildings and Other Structures"[3] and is utilized by most code bodies. The design wind speeds were then prepared on three maps (for risk categories I, II, III, and IV) and incorporated as the Ultimate Design Wind Speeds Maps into Section 1609 of the 2018 IBC. An example of the Risk Category II map was shown in Figure 6.3 in Chapter 6. For example, a Risk Category II building located in Chicago, IL, would have an ultimate design speed of 115 miles/hour while a Risk Category IV building at this same location would have an ultimate design speed of 120 miles/hour.

7.2.3 CALCULATING WIND UPLIFT PRESSURES

Once one has the ultimate design wind speed, the next step is to determine the associated wind and uplift pressures. The velocity pressure exerted on a roof surface from wind forces is based on the simplified Bernoulli's velocity pressure equation, P (velocity pressure) $= 0.00256 * v^2$ where v equals the wind speed in MPH (see Chapter 6, Eq. 6.2). For example, assuming a wind speed of 115 MPH, the resultant velocity pressure force on the roof surface or building would equate to 33.85 psf (pounds per square foot). This simplified velocity pressure force does not reflect other factors such as velocity pressure coefficients related to the height of the building (K_h), surface roughness or topographical factor (K_{zt}), directional factor (K_d), and the importance factor (I). Each of these factors is used to adjust the simplified Bernoulli equation for wind pressure to reflect local conditions (Eq. 7.1):

$$q_h = 0.00256 * (K_h) * (K_{zt}) * (K_d) * (v^2) * I \tag{7.1}$$

Where:
   q_h = velocity pressure at calculated mean roof height
   K_h = the velocity pressure coefficient at a mean roof height
   K_{zt} = topographic or roughness factor
   K_d = directionality factor
   v = wind speed at 33 feet above ground level
   I = importance factor.

Each of these adjustment factors is briefly discussed below:
   The velocity pressure coefficient (K_h) factor accounts for the gradient height that can change the profile based upon the different exposure categories. This coefficient increases or decreases based upon the height of the building. The values of K_h from *Wind Pressures on Low-Slope Roofs*[1] are provided in Table 7.1.

TABLE 7.1
Values of K_h

Mean Roof Height		Exposure Category		
Feet	Meter	B	C	D
0–15	0–4.6	0.70	0.85	1.03
20	6.1	0.70	0.90	1.08
25	7.6	0.70	0.94	1.12
30	9.1	0.70	0.98	1.16
40	12.2	0.76	1.04	1.22
50	15.2	0.81	1.09	1.27
60	18.0	0.85	1.13	1.31
70	21.3	0.89	1.17	1.34
80	24.4	0.93	1.21	1.38
90	27.4	0.96	1.24	1.40
100	30.5	0.99	1.26	1.43
120	36.6	1.04	1.31	1.48
140	42.7	1.09	1.36	1.52
160	48.8	1.13	1.39	1.55
180	54.9	1.17	1.43	1.58
200	61.0	1.20	1.46	1.61

Source: Courtesy of Roof Consultants Institute Foundation.

The surface roughness, or topographical factor K_{zt}, accounts for the wind pressure of the local ground conditions in the direction from which the wind arrives and is divided into the following three categories[1] (the surface roughness factor "A" is not used and was deleted after the 1998 version of ASCE 7[3]):

Surface roughness B: Represents urban/suburban areas with closely spaced buildings of the size of single-family dwellings or taller. Highly wooded areas, city centers, and downtown areas are included in this roughness category.

Surface roughness C: Generally refers to a relatively open terrain with scattered obstructions, generally <30 feet tall. Large water bodies in hurricane prone regions also belong to this roughness category.

Surface roughness D: Refers to flat, unobstructed areas and large water bodies outside hurricane prone regions.

Wind speed is also affected by the topography of the site and is assigned a topographic factor. Hence, areas with relatively flat terrain would not be subject to this factor, but buildings located on an isolated hill or escarpment would be affected. Buildings on relatively flat terrain are assigned a topographical factor (K_{zt}) of 1.0.

Wind behavior will vary slightly based upon the type of structure and affect the velocity pressure. The directionality factor (K_d) is taken into account for the type of structure. The directionality factor ranges from 0.85 for most free-standing buildings to 0.95 for hexagonal or round chimneys, tanks, or similar structures.

The importance factor (I) is used to adjust the structural reliability of a building; in most cases, (I) is assigned a value of 1.0. For buildings that have higher performance requirements (e.g., hospitals, fire department buildings, or other emergency services buildings), (I) is typically assigned a value of 1.15.

Using Eq. 7.1 above, a 40′ tall risk category II (wind speed of 115 MPH) partially enclosed free-standing building located on flat terrain in exposure category B should be designed for a velocity pressure of 29.93 psf [q_h = (0.00256) * (1.04) * (1.0) * (0.85) * (115)^2 * 1.0]. Note that the net effect of the correction factors in this example decreased the wind velocity pressure from 33.85 psf (simplified Bernoulli equation) to 29.93 psf (ASCE design equation) for an actual design situation, a decrease of 11.6% [((33.85 − 29.93)/33.85) * (100)].

The velocity pressure equation must be further adjusted to determine roof wind uplift pressure for different areas on the roof surface. To determine the amount of uplift pressure exerted on the different areas of a low-sloped roof system, the velocity pressure is multiplied by the difference between the external and interior pressure coefficients (Eq. 7.2) derived from wind tunnel studies conducted by ASCE:

$$P_{ul} = q_h \left[\left(\text{GC}_{p(\text{external pressure})} \right) - \left(\text{GC}_{pi(\text{internal pressure})} \right) \right] \qquad (7.2)$$

Where
   P_{ul} = uplift pressure
   q_h = calculated velocity pressure at mean roof height
   GC_p = external pressure coefficient, product of gust factor
   GC_{pi} = internal pressure coefficient, ballooning effect

Values for the coefficients $\underline{\text{GC}}_p$ and GC_{pi}, derived by ASCE,[3] are listed in Tables 7.2 and 7.3 for various conditions:

TABLE 7.2
Values of External Pressure Coefficient GC$_p$ for Low-Sloped Roofs

	Mean Roof Height >60 feet		Mean Roof Height >60 feet	
	Zone	GC$_p$	Zone	GC$_p$
Effect of a parapet: If a parapet >3 feet is provided	1 (field)	−1.0	1 (field)	−1.4
around the entire roof, Zone 3 may be considered	2 (perimeter)	−1.8	2 (perimeter)	−2.3
as Zone 2.	3 (corner)	−2.8	3 (corner)	−3.2
Negative values of GC$_p$ indicate that the pressure is				
away from the roof surface, i.e., uplift pressure.				

Source: Courtesy of Roof Consultants Institute Foundation.
Restricted to roofs with a slope of ≤7° (1.5:12).

TABLE 7.3
Values of Internal Pressure Coefficient GCpi

Enclosure Classification	GCpi
Open buildings	0.00
Partially closed building	0.55
Enclosed buildings	0.18

Source: Courtesy of Roof Consultants Institute Foundation.

Using the velocity pressure of 29.93 psf presented earlier and multiplying it by the difference of the external and internal pressure coefficients, the roof uplift pressure in the field results in a pressure of $P_{ul} = [29.93 * (−1.0 − 0.55)] = −46.3$ psf. Inserting the appropriate factors for the perimeters and the corners, the uplift pressures at the perimeter and corners would equate to −70.3 and −100.2 psf, respectively (note uplift pressures by definition are negative values acting away from the roof surface).

In the next step of the design process, the designers of the roof system need to ensure that the method of securement will satisfy these uplift pressures. For instance, if one were installing 3′ × 4′ panels of insulation board using fasteners with a pullout strength of 200 lb and the design uplift pressures are calculated in the field, perimeter, and corners [46.3 (field), 70.3 (perimeter), and 100.2 corner], the number of fasteners can be calculated using Eq. 7.3:

$$F_n = (A_{ib} * P_{ul})/F_{ps} \tag{7.3}$$

Where:
F_n = # of fasteners required for a given area A_d
A_{ib} = area of insulation board
P_{ul} = absolute value of wind uplift pressure (use positive value in this case)
F_{ps} = fastener pullout strength

Using Eq. 7.3, the number of fasteners needed for the field, perimeter, and corners of the roof would be:

Field: 12 × 46.3 = 555.6/200 = 2.78 fasteners (round to 3)
Perimeter: 12 × 70.3 = 843.6/200 = 4.22 (round to 4)
Corners: 12 × 100.2 = 1,202/200 = 6.01 (round to 6)

This same method can be applied to mechanically attached single-ply membrane or metal roof systems.

7.3   LOW-SLOPED ROOF SYSTEM FAILURE MODES

In this section, common low-sloped roof types and their common methods of installation, along with their common failure modes from forces of wind, are discussed.

7.3.1   LOW-SLOPE INSTALLATION METHODS AND OVERVIEW OF FAILURE CAUSES

Commercial low-sloped roof systems are installed using a variety of methods, depending on the type of roof membrane and substrates. The most common installation methods include mechanically attached, fully adhered, or ballasted roof systems. Each of these methods is designed to ensure proper wind uplift resistance for the roof assembly.

Mechanically fastened systems require the use of fasteners for securing the insulation and/or membrane to the decking. As outlined in Section 7.2, the pullout strength and number of fasteners are dictated by the pullout strength and the wind uplift pressures. On a side note, the corrosion resistance of fasteners selected is critical in preventing low-sloped roof failures. For example, less expensive carbon-coated fasteners were used for a reroofing project and installed into a wet substrate can corrode across the entire cross section of the fastener.[4]

Fastener back-out is another situation that can lead to tenting, membrane punctures, and possible roof blow-offs. Fastener back-out is controlled by: (1) using the proper shank length, (2) the nature of the drill point, (3) the thread design, and (4) selection of the stress plate and fastener head.[4] With the addition of thicker insulation during reroofing applications, the fastener shank length is increased, creating a vertical cantilever, which results in a substantial loss of flexural stiffness against lateral movement. This allows for increased movement that occurs from cyclic wind cycles, resulting in fastener back-out. In addition, the wind forces can cause an oscillation effect on the fastener, causing the drill point to enlarge and thus loosening the fastener within its anchorage.

Fully adhered systems are installed by applying a proprietary adhesive to bond the membrane to the roof surface. The thickness of the adhesive and the application as well as the spacing of the beads affect the system's performance resisting wind uplift. In many cases, roof failures with fully adhered systems occur when either too little adhesive is applied or inadequate coverage is applied. Failures of these systems have also been found as a result of delamination between the membrane and insulation facer.

Ballasted roof systems rely on gravel/stone or pavers as the necessary securement to provide resistance to wind uplift pressures. The size and weight of the ballast materials vary depending on the severity of the wind loading. Ballasted systems are not recommended to be installed on buildings >150 feet in height and in areas with design wind speeds >100 MPH; however, this height may be less and can vary by building type and locality. Perhaps one of the most difficult issues related to ballasted roof systems is finding and repairing water leaks in the membrane; this requires removal of some of the ballast, which may not be properly redistributed following repairs causing the roof system to become susceptible to wind damage.

7.3.2   TYPES AND TYPICAL CAUSES OF FAILURE FOR LOW-SLOPED ROOF SYSTEMS

According to FM Global,[5] about 80% of roof blow-offs start at the perimeter flashing while the remaining 20% begin throughout the field of the roof. According to engineers surveying wind-related roof damage from hurricane Andrew, it was estimated that 90% of the damages occurred as a result of installation deficiencies that violated local, state, or national building codes.[6] Types of improper installation practices found included the following:

- Insufficient number of nails to secure the sheathing to the wood trusses
- Substitution of staples versus nails (typically in shingle applications)
- Insufficient mopping with adhesive or inadequate number of fasteners to secure insulation boards

- Use of defective or substandard fasteners
- A lack of proper securement of perimeter edge metal

When wind damage is found to occur at wind speeds below the design speeds, it is probable that the damage is associated with improper installation and that wind is not the primary cause of failure.

7.4 INSTALLATION PRACTICES FOR PROPER SECUREMENT OF LOW-SLOPED ROOF SYSTEMS

Regardless of the installation method, the roof system must be installed to resist wind uplift forces. In this section, industry best practices associated with the proper installation of low-sloped roof systems at the perimeter, components above the roof deck, and the membrane itself are reviewed.

7.4.1 PERIMETER FLASHING AND ATTACHMENTS FOR LOW-SLOPED ROOF SYSTEMS

Perimeter flashings on low-sloped roof systems are typically made of lighter gauge metals that form fascia, copings, or edge metal flashings. These flashings provide perimeter closure to prevent moisture and winds from entering into or underneath the roof system. Based on experience, wind damage normally propagates along the perimeter of these systems, resulting in failure(s) of the flashing components. When the perimeter metal (fascia, nailer, or copings) fails, it exposes the roof membrane to suction and peeling forces across the field of the roof and can result in the roof system failing at wind speeds well below the design speeds. Typical perimeter edge failures include the following:

- Loss of light gauge metals such as fascia and copings.
- Blow-off of inadequately strapped light gauge metal gutters.
- Uplift of nailers inadequately anchored to masonry walls.

Failures can also occur due to moisture intrusion facilitating corrosion of the fasteners, which can diminish their strength of attachment.

7.4.1.1 Perimeter Nailers

One of the key elements for roof perimeters is the installation of wood nailers. The wood nailer serves as the base tie-in for both the perimeter metal and the roof membrane. Improper or inadequate securement of the wood nailer can result in catastrophic low-sloped roof failures. Common installation deficiencies or failures have included: (1) anchoring wood nailers into masonry or concrete block walls using roofing nails or improper fasteners, (2) using less than the required number of fasteners, and (3) rot associated with moisture intrusion. FM Global Property Loss Prevention Date Sheet 1–49 for Perimeter Flashings provides the following recommendations for the installation of wood nailers[5]:

- The top surface of the nailer(s) should be level with the roof edge.
- The nailer should be wide enough to allow for two rows of fasteners.
- Masonry anchors should be used when securing a wood nailer into masonry walls.
- Fasteners and/or anchors should be spaced in accordance with their pullout strength and calculated wind uplift pressures.
- Wood nailers anchored to masonry or steel should be a minimum of $1\text{-}1/2'' \times 5\text{-}1/2''$ and should be Douglas Fir, Southern Yellow Pine, or wood having similar decay resistance properties.
- Nails used to secure wood such as fascia, cant strips, and top nailers to other wood members should be long enough to penetrate $1\text{-}1/4''$ into the wood.

7.4.1.2   Perimeter Fascia, Gravel Stops, and Copings

A second common perimeter low-sloped roof failure mode is the blow-off of edge metal such as fascia, gravel stops, and copings. This usually occurs as a result of insufficient anchorage or lack of continuous hook strips to provide additional reinforcement along the perimeters. Figure 7.2 illustrates recommended details for typical flat roof and parapet wall perimeter terminations.

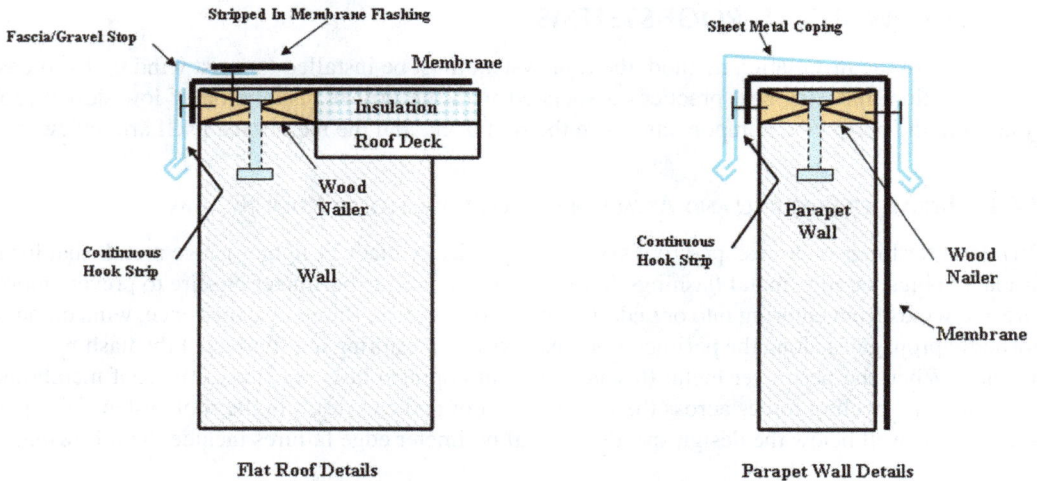

FIGURE 7.2   Recommended edge metal flashing details.

FM Global Property Loss Prevention Date Sheet 1–49 provides the following recommendations for perimeter edge metal flashing details[5]:

- Gravel guards and hook strips should not be heavier than 24–26 gauge. Hook strips should be one gauge heavier than the fascia.
- Gravel guards should be installed in lengths of 8′–10′, lapped 2″ at side joints, and covered with a 4″ wide cover plate. The guard should be set in cement and nailed 1″ back from the edge with fasteners spaced 4″ on centers.
- The horizontal part of the guard should be set in roofing cement and nailed 1″ from the back edge with fasteners spaced 4″ on centers.
- The top of the gravel guard should be stripped with membrane and lap the nail heads by at least 2″. A second 9″ wide membrane should be stripped in to provide 4″ overlap of the metal and provide at 4″ of overlap over the base membrane.
- Hook strips should be anchored with nails long enough to penetrate the wood nailer 1-1/4″. The nail head should be a minimum of 3/16″ in diameter. Screws can be used in place of nails but should be long enough to penetrate 3/4″ into the wood or 3/8″ into metal. Fasteners (nails or screws) should be corrosion-resistant or treated to resist corrosion. The hook strip should be fastened to the wall with 16″ on center spacing.
- Metal fascia and flashing should be secured to wood nailers at the bottom edge with a continuous hook strip. The metal sections should be secured at each end under the joist cover with slotted holes to permit expansion and contraction.
- Metal coping and counter flashing should be secured with hook strips attached to the wall exterior. The inside surfaces should be secured with galvanized screw fasteners with neoprene washers spaced 30″ on center. The screws should be long enough to penetrate

into the wood a minimum of 1″. For higher wind speeds, the fastener spacing should be decreased to 20″ on center spacing.

- Metal counter flashing should be attached to masonry walls with masonry anchors (no plastic materials) and spaced 30″ on center. Each anchor shall have a minimum pullout resistance of 200 pounds.

It has been found that fascia metal can fail even when hook strips are used. This can occur when the hook strips are fastened at the top of the strips, thus increasing the bending stress of both components. This type of failure can be eliminated by fastening the hook strip to the exterior wall surface.

7.4.2 ROOF DECK AND ABOVE-DECK COMPONENTS FOR LOW-SLOPED ROOF SYSTEMS

The roof deck and above-deck roof components are also susceptible to wind damage due to the negative/uplift wind pressure forces that can act on the field of the roof. This typically affects the membrane, insulation substrates, and in more severe cases, the roof decks and perimeter components (i.e., parapet walls). The ballooning or tenting effects can cause mechanically or fully adhered membranes and insulation to tear free from the decking, resulting in substantial damage. Field experience has shown that this can occur from the following causes:

- Not using enough fasteners with adequate pullout strength.
- Improper placement of fasteners.
- Not applying sufficient amounts of adhesive for fully adhered systems.
- Insulation facer board delamination.
- Reroofing over wet materials, increasing the propensity of fastener corrosion.

Proper attachment of the above-deck insulation and membrane roof covering is critical to ensure that the roof system performs under the rated wind pressure conditions.

7.4.2.1 Roof Insulation for Low-Sloped Roof Systems

In most commercial roof systems, a variable thickness of insulation will be added above the roof deck and below the membrane to increase the energy efficiency of the building. The insulation also serves as the primary substrate through/onto which the roof covering membrane is mechanically attached, fully adhered, or loose-laid with ballast. It is critical that the insulation boards are properly anchored to the roof deck in order to perform under the design wind speeds. FM Global Property Loss Prevention Data Sheet 1–29 – Roof Deck Securement and Above-Deck Roof Components – Section 2.2.10.6 provides the following requirements for the installation of insulation under low-slope roof surfaces[7]:

- No insulation boards >4′ × 8′ should be used.
- The maximum recommended insulation board size if adhered or hot mopped with adhesive is 4′ × 4′ in area, except for 1/2″ wood fiber or 5/8″ gypsum board insulation, which can be up to 4′ × 8′ in area.
- Insulation boards should be installed in a staggered pattern and provide a minimum bearing of 1″ over steel deck flanges.
- When insulation boards are cut, secure each piece with the appropriate number of fasteners or adhesive ribbons for the full board.
- Fasteners should only be installed in dry substrates. Wet substrates can cause deterioration of the fasteners.
- Fasteners must be driven perpendicular to the decking and only through the top flange of steel ribbed decking.

• Fasteners must be embedded 1″ to 1-1/2″ into structural concrete, 1″ into wood decks, or ~1/4″ of the nail head should be exposed on the underside of the deck. Screw-type fasteners for metal decks should be at least 3/4″ longer than the assembly being secured.

7.4.2.2   Roof Membrane Attachment

As briefly discussed in Section 7.3.1, roof membranes are secured to the deck in one of three ways: (1) with mechanical attachment, (2) full adherence, or (3) with ballast.

Mechanically attached single-ply roof coverings and metal panel roof systems rely on fasteners or anchors to provide attachment of the roof surface to the structure. In order to meet the required wind uplift pressures, the number, spacing, and pullout resistance of the fasteners must follow design requirements for wind uplift. Special attention should be paid to the perimeters and corners where the highest wind forces occur. Experience has found that inadequate perimeter attachment is the primary cause for roof failures.

For metal roof panels, the wind uplift resistance for the entire panel is transferred to the hold-down clips and fasteners anchoring the clips to the structural members. Lack of adequate clip design or spacing and improper or inadequate seaming lead to metal roof failures below design wind speeds. Uplift forces must be accounted for when designing clips and clip spacing. Seams should be properly sealed as required by the manufacturer and industry best practices. Improper or inadequate seeming is another failure mode observed in the field.

Fully adhered single-ply systems typically require a uniform coverage of urethane-based adhesives over the field of the roof. The adhesive is applied over the substrate (i.e., insulation boards) and the membrane is then rolled over the adhesive. Some observed failures (discussed later in this chapter) have occurred as a result of nonuniform applications of adhesives.

In addition, when adhering single-ply membranes to faced insulation boards such as polyisocyanurate boards (a common insulation board used in the single-ply industry), it has been found that the facer is subject to delamination (separation of the facer from the insulation), leading to failures at wind pressures well below the design. To account for this condition, most industry best practices recommend that a cover board be installed over the installation to prevent this failure mechanism.

Ballasted roof systems use loose-laid single-ply membranes. The ballast (stone or pavers) serves as the primary means of providing wind uplift resistance to the loose-laid membrane. Ballasted roof systems are divided into three categories (or systems) based on the severity of the wind loading:

System 1: Requires a nominal ballast of 1-1/2″ smooth, river bottom stone spread over a uniform weight of 10 psf over the entire roof or concrete pavers at 15 psf (10 psf for interlocking type).

System 2: Requires 2-1/2″ smooth river bottom stone applied at a uniform weight of 13 psf in the corners, and 10 psf for the perimeters and field or concrete pavers at 22 psf minimum weight.

System 3: Requires 2-1/2″ smooth river bottom stone in the field of the roof applied at a 13 psf weight. In lieu of ballast in the corners and perimeter, System 3 requires that a fully adhered or mechanically anchored membrane with a minimum of 90 psf uplift is used in the corners.

For all three systems, best practices limit the building height of ballasted systems to 150 feet or less to prevent ballast from spilling over the roof surface to the ground below. Experience has shown that most low-sloped roofs with ballasted systems failed as a result of scouring or removal of the ballast and not replacing it after maintenance activities. Of note, structural calculations should be made to ensure that the weight of the ballast does not exceed the designed loads of the building.

7.5   LOW-SLOPED ROOF SYSTEM INSPECTION METHODOLOGY

When performing a wind-damage inspection on a low-sloped roof system, the same methodology that is discussed in Chapters 1 and 6 should be used. In addition to a visual inspection, steps in the methodology of the investigation of a low-sloped roof system can often include destructive testing and wind uplift testing. Low-sloped roof inspections are sometimes more difficult and can cover a much larger surface area than residential and light commercial building inspections. These additional steps are discussed below.

7.5.1 Visual Inspection of Low-Sloped Roof Surfaces

In many cases, a visual inspection is all that is needed to perform a detailed wind damage assessment for low-sloped roof surfaces. The visual inspection should include both the exterior elevations and the roof surfaces of the structure.

Prior to accessing the roof, an exterior walk around should first be performed. During the walk-around, any damages to components or cladding should be photographed and documented. Attention should be given to the roof edges and perimeter flashings. Sometimes perimeter flashings that have been displaced as a result of wind uplift forces may not be readily visible from the roof surface and must be examined from the sides of the structure.

Once the inspection of the exterior elevations is completed, the roof inspection should be initiated. At times, the vast surface area of a commercial building can seem extremely challenging and daunting for any forensic investigator; however, using a systematic approach can reduce the inspection into a manageable task. This is done by breaking down the inspection into the following components: (1) perimeter of roof edge, (2) field of the roof, and (3) roof appurtenances (i.e., vents and mechanical units). At the beginning of the roof inspection, a sketch of the roof surface(s) should be drawn in a field notebook. It is important to document the locations of the vents, mechanical units, drains, etc., and record the roof measurements both in the field notebook and on the sketch.

Since most roof failures occur at the perimeters, this is one of the most important areas to investigate. The entire perimeter of the roof surface should be inspected, documented, and photographed. Key details to observe include, but are not limited to, the following:

- The securement method of the perimeter or corners.
- The types, widths, and lengths of the flashings.
- The type and spacing of the fasteners.
- The condition of the flashings and fasteners.
- The method by which the membrane is terminated at the edge.
- Loose or unsecured edges. The edges can be checked by hand or with a roof probe.
- Damages to the roof system.

The entire field of the roof should also be walked when performing the inspection; all information should be documented in writing in the field notebook and key details should be photographed. When walking the field, it is best to use a grid-type pattern. Within the field, damage is most likely to have occurred, and will be more noticeable, at the roof seams. Key details to document when inspecting the field of the roof should include the following:

- The method of securement (i.e., mechanically attached or fully adhered).
- The widths of the seam overlap.
- The widths of the roof system panels.
- The type and spacing of the fasteners for the seams (if mechanically attached) and for the insulation boards. In some single-ply membrane systems, such as ethylene propylene diene monomer (EPDM) or thermoplastic olefin (TPO), the seam and insulation fasteners can be observed through the membrane.
- Adherence of the seams. This can be checked with a small rod or seam probe.
- Insulation board displacement.
- Obvious signs of wind damage.

Finally, all roof appurtenances should be checked to ensure that they are secure and show no evidence of wind-related damage (i.e., displaced components). For example, it is not uncommon to find vents caps missing on older roof systems.

7.5.2 DESTRUCTIVE TESTING OF LOW-SLOPED ROOF SURFACES

Destructive testing is oftentimes the only way to render a solid conclusion on the cause(s) of why the roof system failed. Destructive test cuts may include removing and observing small or large sections of the membrane, underling insulation, and/or support systems. As discussed in Chapter 4, destructive testing should only be performed if the roof can be repaired immediately following the observations, permission is granted from the owner and/or insurance company, and the testing will not void the warranty of the roof system.

When a fully adhered single-ply membrane roof system fails, the underside of the membrane and substrates (i.e., insulation) must be observed. This may sometimes require cutting and opening a large swath of the membrane in order to observe the extent and quality of the applied adhesive.

7.5.3 UPLIFT TESTING OF LOW-SLOPED ROOF SURFACES

In situations where a roof system has been found to fail at wind speeds well below design values, an excellent way to determine whether or not the roof was installed to meet design wind loads is to perform uplift testing. The two main types of uplift tests are the negative pressure test and the bonded uplift test. The purpose of these tests is to determine if the roof systems will remain in place when uplift pressures at/or exceeding the design pressures are exerted on the roof system.

The FM Global Property Loss Prevention Data Sheet 1–52 – Field Verification of Roof Uplift Resistance – recommends that the number of uplift tests should be based on the area of the roof.[8] For example, if the roof area is 10,000 square feet or less, a minimum of three tests is recommended: one in the field, one at the perimeter, and one at the corner. For a roof area between 10,000 and 60,000 square feet, the recommended minimum number of tests is five: two in the field, two at the perimeters, and one at the corner. It is important to remember that the minimum number of wind uplift tests performed on a building is dependent on the roof area. Lastly, it is recommended that the roof system be tested at 1.5 times the design pressures in order to provide a safety factor of 1.5. Details on each of the two uplift test procedures follow in the next two sections.

7.5.3.1 Negative Pressure Test – Low-Sloped Roof System Surfaces

The negative pressure test is essentially a nondestructive method of testing that can be performed on fully adhered built-up roof (BUR), modified bitumen (Mod-Bit), or single-ply membrane roof systems. This test can also be performed on mechanically attached membranes as long as the fasteners are spaced no more than 2 feet in the center in either direction. This test cannot be performed on ballasted roof systems and is usually not cost-effective to perform on smaller roof surfaces.

The test apparatus consists of a vacuum pump within a 5-foot-by-5-foot dome or compartment, which is capable of creating negative pressure inside the test chamber. With the compartment sealed, a deflection bar is used to measure the degree that the membrane "lifts" up. A basic schematic of a negative pressure test apparatus is shown in Figure 7.3.

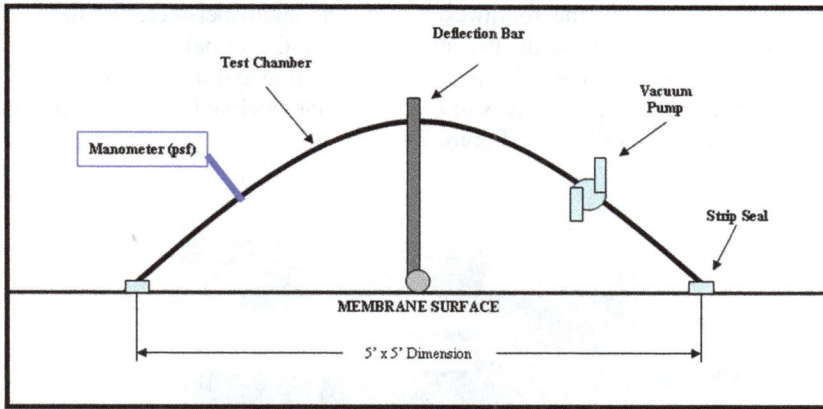

FIGURE 7.3 Negative pressure test apparatus.

To perform the test, the apparatus is placed and sealed onto the membrane surface. Negative pressure is then applied to the dome, starting at an initial pressure of 15 pounds per square foot (psf). The pressure is increased in increments of 7.5 psf with each increment held for 1 minute until the design test pressure is reached or failure occurs. Failure of the roof system is defined when the membrane balloons up or when the deflection of the bar is >0.25 inch.

7.5.3.2 Bonded Uplift Test – Low-Slope Roof System Surfaces

Unlike the negative pressure test, the bonded uplift test is a destructive test method used to determine the uplift pressure of a roof system. The fact that this method is a destructive test method makes it less attractive than the negative pressure test.

The test equipment can be purchased at any hardware store and generally consists of plywood sheets, screws, eyebolts, adhesive, and repair materials. The equipment includes a calibrated spring scale or an equivalent force measuring device, a hand chain or hydraulic hoist, and a tripod. The bonded uplift test utilizes two pieces of plywood measuring 2 feet by 2 feet that are fastened together. The plywood is then adhered to the smooth roof surface. After a curing period, the roofing membrane is cut at the perimeter of the plywood. The attached plywood/roof assembly is then attached to a scale/tripod assembly and upward force is applied in increments of 7.5 psf starting at 15 psf and held for 1 minute at each increment until failure occurs or the design threshold is reached. It is important to remember that this is a destructive test and immediate repairs will be required following the test.

7.6 WIND DAMAGE CASE STUDIES

The following case studies represent actual forensic damage investigations to various types of low-sloped roof systems. The case studies are intended to provide the reader with knowledge regarding the methods, findings, and interpretation of results associated with actual failures of low-sloped roof systems.

7.6.1 BLOWN-OFF TPO MEMBRANE

This forensic investigation involved the failure of a 2-year-old TPO roof membrane. The roof membrane was reportedly failed as a result of a windstorm, which produced 75 mph peak wind gust speeds. The membrane was mechanically fastened to 1″ foam board insulation over wood board sheathing members. The roof covered a brick masonry building reported to be over 120 years old.

The roof membrane failed at the southwest corner and south perimeter of the building. The wind forces had blown off sections of the perimeter fascia, and over half of the membrane that was covering the roof surface was torn off. Sections of the sheathing and trusses were also damaged as a result of the uplift exerted from the forces of the membrane peeling back across the roof surface. A photograph of the damage is shown in Figure 7.4.

FIGURE 7.4   South perimeter membrane failure.

During the forensic investigation, it was determined that the membrane and perimeter metal were fastened to a nominal 2×10 wood nailer. The wood nailer was quasi-fastened into the two-wythe masonry wall with screws. Many of the screws within the wood nailers were either not screwed/ anchored into the degraded brick or penetrated into/through the hollow cores of the brick masonry. In addition, the perimeter fascia was not fastened along the perimeter with a continuous cleat (Figure 7.5).

FIGURE 7.5   Overview of perimeter securement.

Forensically, it was concluded that the primary cause of the roof failure was the result of insufficient perimeter attachment of the wood nailer and fascia. This failure could have been avoided if the following best practices had been followed:

- The wood nailer should have been installed with anchor bolts 12″ into the brick masonry. Additionally, the hollow cores used for attachment should have been filled with concrete.
- The perimeter fascia should have been installed with continuous hook strips to prevent uplift along the bottom edges of the fascia.

7.6.2 Missing Ballast on EPDM Membrane Roof

A reported windstorm with 67 mph peak wind gusts passed over a building covered with a loose-laid ballasted EPDM membrane roof. The 50′ × 60′ EPDM roof surface area was covered with 1–3 inches of smooth stone ballast. The EPDM membrane was reportedly between 10 and 15 years old. A claim was filed by the owner of the building who reported that the center of the roof surface had ballooned up from the wind storm and damaged the membrane, which resulted in water intrusion around the skylights and parapet wall.

During an interview with the owner, it was reported that a large tree limb had fallen onto the roof ~1 year prior to the storm and caused a substantial hole/rupture into the membrane. The owner contracted the work with a local "mom and pop" firm (now out of business) to make the necessary repairs. Once these repairs to the roof were completed, the owner paid the bill, but did not inspect the repairs or the condition of the roof once the repairs were made. The owner also reported that a second contractor had later been engaged to make some temporary repairs to the membrane around the skylight.

During the forensic investigation of the roof surface, the ballast was observed to have been removed throughout the center of the roof system (Figure 7.6).

FIGURE 7.6   Ballast removed through center of roof.

The area of removed ballast exposed at least three repair patches, evidently, from where the fallen tree limb had punctured the membrane roof surface (Figure 7.7).

FIGURE 7.7 Membrane repairs.

More recent membrane repairs were observed at each corner of the skylight. The membrane in these areas was pulled away at the corners of the skylight, providing openings for potential water intrusion (Figure 7.8).

FIGURE 7.8 Repairs around skylight.

Lastly, the wood nailer along the perimeters was fastened on top of the brick masonry parapet wall with nails. However, the nails were not fully engaged into the brick wall, thus leaving an ~2-inch wide gap. This in turn allowed air from high winds to infiltrate into the space beneath the wood nailer and membrane (Figure 7.9).

FIGURE 7.9 Improperly installed nailer.

Forensically, it was concluded that the wind uplift damage was caused by several factors. First, the contractor making the initial tree-damage repairs did not replaced/re-level the ballast that was scoured away to make the repairs. This left a portion of the membrane loose-laid on the decking with essentially no means of wind uplift resistance. Second, the wood nailer installed into the brick was not properly secured. Had lag bolts or masonry screws of sufficient length been used, this would have provided closure and firm attachment at the edge of the membrane, thus preventing winds from entering beneath the membrane.

7.6.3 NEWER BUR MEMBRANE OVER OLDER BUR MEMBRANE

An ~10,000 square feet, barrel-shaped BUR roof membrane reportedly failed along the west windward perimeter as a result of a windstorm that produced peak wind gusts of 66 mph. The multilayer BUR membrane roof was reported to be at least 20 years old. A large strip of the BUR had peeled back from its support along the eave all the way to the ridge of the barreled roof. At the time of the inspection, a large blue tarp was secured over the top of this damage area (Figure 7.10).

FIGURE 7.10 Tarp on area of damaged BUR.

The top layer of the membrane had peeled back exposing a bottom layer of BUR, which in turn had allowed water to flow into the building. Interview information from the property owner revealed that the roof had been leaking for several years.

A test cut into the BUR revealed that the top layer consisted of a three-ply membrane applied over an older existing four-ply layer of BUR. Edge metal was present along the older original layer of BUR; however, the newer layer of BUR was only adhered to the roof edge (over the existing layer) with no perimeter securement (Figure 7.11).

FIGURE 7.11   Perimeter edge detail.

At the failure area, the newer and unsecured BUR layer was peeled away, exposing the heavily degraded older (bottom) BUR layer (Figure 7.12).

FIGURE 7.12   Degraded membrane.

Forensically, it was concluded that this BUR roof failed for two reasons. First, the top layer of membrane was essentially not secured to the roof edge. No edge metal was installed to provide closure to the edge of the membrane. Second, the heavy degradation and aging of the older layer of BUR membrane weakened the adhesion between the bottom (older) and top (newer) layers of the roof system.

7.6.4 Insufficient Adhesive for Fully Adhered EPDM Membrane

A fully adhered EPDM membrane roof installed on a two-story strip mall was reportedly wind damaged during a windstorm that produced peak wind gust speeds of 53 mph. The fully adhered membrane had been installed 2 years prior to the reported date of failure. The failure occurred within the field and along the southern edge of the roof (windward side of the building) at the south parapet wall (Figure 7.13).

FIGURE 7.13 Overview of failure area.

The membrane throughout the center of the roof was loose and oscillating during periods of light winds. Sections of the coping stones had been displaced along the south parapet wall. Temporary repairs appeared to have been made to secure the membrane to the remaining sections of the parapet wall. Linear sections parallel to the parapet wall on the roof were actually sections of the capstones (also covered with light snow), which were left on the roof surface to aid in securing the loose membrane onto the decking until permanent repairs could be completed. A close-up picture of the capstone section is shown in Figure 7.14.

FIGURE 7.14 Capstones securing roof membrane.

Through destructive testing, it was found that the initial failure likely occurred throughout the field of the roof due to lack of application of sufficient adhesive. Consequently, the field portion of the membrane loosened with time, worsening with each wind event and ultimately pulling the capstones off the top of the parapet wall and onto the decking. At this point, the roof system catastrophically failed. Test cuts revealed that the fully adhered membrane was not actually fully adhered. Spotty adhesive coverage had been applied between the membrane and insulation boards (Figure 7.15).

FIGURE 7.15 Back sides of membrane showing incomplete adhesive application.

In reviewing the membrane manufacturer's installation instructions, a uniform coverage of adhesive is required to be applied to the underside of the membrane to ensure proper adhesion and provide the required wind uplift resistance. It was clear from an examination of the adhesive patterning on the bottom side of the membrane from the test cuts that the adhesive had not been uniformly applied across the back side of the membrane.

This failure could have been avoided had the installation contractor followed the manufacturer's installation guidelines and applied a uniform application of the bonding adhesive.

7.6.5   Improperly Secured EPDM Membrane Seams

The roof surfaces of a bowling alley were covered with a fully adhered EPDM membrane and portions of standing seam metal roof panels; the age was unknown. The gable-style roof sloped east and west toward the perimeter gutter system and contained numerous penetrations (i.e., HVAC units and vent stacks). The membrane was installed over a plywood substrate and was heavily degraded. A wind storm with peak wind gust speeds of 54 mph reportedly passed over the subject building. The building owner claimed wind damage to the membrane leading to multiple water leaks and damage to the interior of the building. An ~90′ × 100′ section of the EPDM membrane was covered with tarps on the east (leeward) side of the building (Figure 7.16).

FIGURE 7.16   Overview of damaged EPDM roof.

Numerous membrane patches, along with sections of older and newer-in-appearance membrane panels, were observed throughout the roof surfaces, suggesting that several repairs had been made to the membrane over time. Protruding fasteners (securing the plywood substrate) were observed throughout the field of the roof. Many patches were observed over the fastener locations, and some of the fasteners had caused holes in the membrane. The membrane could be easily lifted from the substrate. A roofing-type tar was also observed along several seams throughout the field of the roof, suggesting problems with the membrane seaming.

Tests cuts into the membrane were completed in areas not containing a tarp and in other areas by removing sections of the tarps. Results from the test cuts revealed the following installation deficiencies: (1) the seam laps were adhered to the roof surface, but contained no splice flashing tape along the seams, and (2) there was little to no adhesive present to bond the membrane to the plywood substrate. These deficiencies are shown in Figures 7.17 and 7.18.

FIGURE 7.17   No flashing tape at seam lap.

FIGURE 7.18   Adhesive at seam, no adhesive under membrane.

Forensically, it was concluded that the installation provided little to no membrane attachment along the seams throughout the field of the roof. This allowed the roof system to fail at wind speeds well below the code criteria and design wind speeds for this location.

7.6.6 Improper Edge Details for EPDM Membrane

A one-story warehouse building was covered with a fully adhered EPDM membrane. Peak wind gust speeds of 54 mph reportedly passed through the area, and the EPDM membrane was torn away from the western (windward) perimeter of the building. The membrane was mechanically fastened over wood fiberboard sheathing, which was installed over steel decking. It was not until the building tenants reported water dripping into the building that the owner became aware of the damages.

Once inspected, it was observed that the membrane had torn from the west roof edge and most of the membrane was detached along the perimeter. Under this area, the exposed wood fiber insulation boards were degraded and contained heavy water damage. The membrane throughout most of the field of the roof was detached, loose, and rippled (Figure 7.19).

FIGURE 7.19 Overview of membrane damage along west perimeter.

The edge metal was detached, but still connected to the membrane (Figure 7.20).

FIGURE 7.20 Detached edge metal.

The edge metal was fastened with 5/8″ long nails spaced ~5″–7-1/2″ on center. The nails were fastened into the wood fiberboard sheathing, but were not of sufficient size to fully engage into the wood nailer. The edge metal was not fastened with continuous hook strips, and nails were found to have only been used along the horizontal edge of the metal (no vertical fasteners). The edge metal along the east perimeter revealed the same installation details and was also found to be lifted or pulled up from the insulation board (Figure 7.21).

FIGURE 7.21   Lifted edge along east perimeter.

Test cuts revealed that the fully adhered membrane was merely spot adhered. Spotty areas of adhesive were observed on the underside of the membrane; where tested, the membrane was almost entirely detached from the insulation boards.

Forensically, it was concluded that while forces of wind were a secondary cause for failure of this roof system, the primary sources were tied to several factors. The length of the nails fastening the edge metal into the insulation substrate was only 5/8″. Nails used to fasten materials into wood need to be long enough to penetrate a minimum of 1-1/4″ in the wood nailers. The nails used to fasten the membrane to the perimeter were not long enough to engage the insulation boards fully, let alone the wood nailer. No hook strips were present to secure the bottom edge of the metal against the exterior wall. Properly installed hook strips may have kept the edge metal in place. The improper and spotty application of the bonding adhesive rendered the membrane susceptible to uplift forces even though the failure began at the edge of the roof.

7.6.7   ROTTED PURLINS SUPPORTING A METAL PANEL ROOF SYSTEM

The metal roof panels and wood purlins on a three-sided metal panel shed were reportedly extensively damaged when peak wind gust speeds of 45 mph passed through the area. No wind damage was reported to the adjoining properties or structures near the building.

Upon investigation, it was found that a large section of the metal panels and associated wood purlins were broken and torn away from the structure. Figures 7.22 and 7.23 provide overview pictures of the roof damage.

FIGURE 7.22 Overview of wind roof damage to pole barn – from ground.

FIGURE 7.23 Overview of wind roof damage to pole barn – from roof.

The investigation indicated that the metal panels were fastened with 1-7/8 inches nails into nominal 2×8 wood purlins spaced on 24-inch centers. The nails were spaced ~5–6 inches on center horizontally and ~24 inches on center vertically. The nails appeared to be of sufficient length to penetrate fully through the metal panels and into the purlins. Inspection observations also suggested that the securement of the panels to the purlins was adequate. Further, there was no evidence of lifting or displacement of the intact sections of the metal panels to suggest that they were affected by wind uplift forces. Many portions of the remaining purlins were found to be heavily damaged, stained, and were rotted from water intrusion (Figure 7.24).

FIGURE 7.24 Water damaged and rotted purlins.

Additionally, rotted and broken purlins still attached to the metal panels were observed in a debris pile.

Forensically, the primary cause of this roof system failure was due to the lack of building maintenance. Long-term water intrusion caused significant damage to the purlins, which weakened the metal panel support members. This allowed for wind forces well below the code criteria and design wind speeds for this location to lift and blow off the metal panels. This failure could have been avoided had the roof system been properly maintained and any water-damaged structural members been promptly replaced.

7.6.8 OLDER MECHANICALLY ATTACHED EPDM MEMBRANE

A vacant manufacturing building was covered with a mechanically attached EPDM roof membrane over an existing BUR roof surface. Markings on the membrane and information from the owner indicated that the membrane was over 25 years old. The owner reported wind damage to several areas of the membrane.

During the subsequent investigation, it was found that the roof was in very poor condition. Numerous defects were present throughout the roof surface that were related to installation and aging versus damage from wind forces. Numerous repairs had been made to the roof over time, of which many were improper and simply "band-aids." Photographs and descriptions of defects observed to this roof surface are provided in Table 7.4.

TABLE 7.4

Membrane and/or System Defects

Description	Photograph	Deficiency
Lifted/ corroded edge metal		Water intrusion occurring near the roof edge caused corrosion to edge metal and fasteners. The membrane was not properly flashed at the corner termination
Protruding fasteners and batten strips		Heavy water infiltration was occurring at the membrane termination with the masonry wall. This led to severe degradation of the roof decking causing the fasteners to release. In many cases, the protruding mechanical attachment pieces produced tears to the membrane
Improper termination at parapet wall		The membrane was fastened to the vertical edge of the parapet wall, which was covered with bitumen. No copings or flashings were used to prevent water from entering between the wall and membrane. Instead, caulk (heavily degraded) was the only means of attachment. Heavy water damage and rotting were observed to the decking below this intersection (see Chapter 8 for proper details)

(Continued)

TABLE 7.4 (*Continued*)
Membrane and/or System Defects

Description	Photograph	Deficiency
Improper membrane attachment to masonry wall		The membrane was fastened to the wall with a batten bar, but non-counter flashing was installed to prevent water migration between the interface. Best≈practices recommend the installation of counterflashing tucked into the masonry for proper water management. Heavy water damage and rotting were observed to the decking
No base tie-ins at corners		The membrane was pulled away from the wall causing tears in the corner. The patch at this corner serves as a temporary band-aid. Best practices require base tie-ins to secure the membrane to the wall surface
Torn membrane on steep dormer roof		The membrane tore loose at the seams. EPDM membranes are limited to 4:12 slopes. The excess slope in this case (7:12) put stress along the membrane seams eventually tearing loose

The damages and defects were related to installation deficiencies, which negatively affected the water management details around perimeter attachments. This, in turn, led to long-term water intrusion. The membrane was also found to be relatively brittle, which indicated that the membrane was beyond its effective service life.

Forensically, the failure of this roof system was caused by installation deficiencies, water intrusion, and natural aging.

7.6.9 BALLOONED FULLY ADHERED EPDM

A fully adhered EPDM membrane was installed over the roof of a warehouse building. Six months after the installation, peak wind gust speeds of 69 mph reportedly passed through the area. The owner was present at the building during the storm and observed the membrane ballooning as much as 20′ high over the roof surface. During the interview portion of the inspection, it was reported that the membrane was installed by a local roofing contractor who was not certified and had very little experience with installation of the single-ply roof membrane.

During the subsequent forensic investigation, the failed roof membrane was being removed, which allowed for detailed observations of the original roof installation. It was observed that the EPDM was installed over the top of several layers of roofing and insulation materials. The cross section of the roof system is provided in Figure 7.25.

Roof Cross-Section

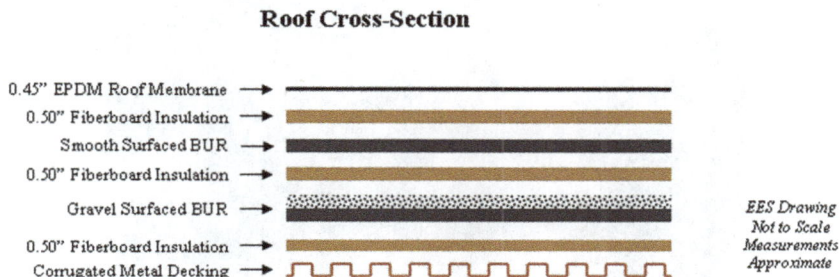

0.45" EPDM Roof Membrane →
0.50" Fiberboard Insulation →
Smooth Surfaced BUR →
0.50" Fiberboard Insulation →
Gravel Surfaced BUR →
0.50" Fiberboard Insulation →
Corrugated Metal Decking →

EES Drawing Not to Scale Measurements Approximate

FIGURE 7.25 Roof cross section.

The following deficiencies were found:

- There were many protruding fasteners throughout the field of the roof.
- The insulation boards were degraded, suggesting long-term water intrusion.
- The membrane was heavily blistered and degraded.
- Several sections of the steel decking were heavily corroded.

Portions of the recently removed sections of EPDM were lying in the worksite dumpster. It was found that little to no adhesive residue was present on the underside of the membrane surface (Figure 7.26). A section of the EPDM membrane that was still present on the roof surface was also pulled back and revealed similar conditions (Figure 7.27).

FIGURE 7.26 No adhesive to back side of membrane.

FIGURE 7.27   Nonuniform adhesive application along edge.

It appeared that the initial roof failure occurred in the center (field) portion of the roof. The ballooning effect ultimately applied forces along the perimeter and tore off (internally) sections of the metal coping. Forensically, the cause of this roof failure was that the membrane was not properly bonded to the insulation boards exacerbated by the wet conditions of the multiple layers of substrates.

7.6.10   TPO Membrane Covering an Aquatic Complex

A building covered with a fully adhered TPO roof system suffered extensive uplift damage as a result of wind speeds well below the design speeds. The TPO membrane was <2 years old. The building was an aquatic complex, which housed an Olympic-sized swimming pool and smaller dive pools used for Scuba training and certification. Since the roof system was expected to have high humidity levels, the architect had specified a continuous vapor barrier around the perimeter of the building.

Video evidence and photographs from the owner showed that the roof was ballooning up through the entire central portion. The initial failure appeared to have started at the southwest corner of the building where the highest wind forces were directed. An overview of the roof surface is provided in Figure 7.28.

FIGURE 7.28 Overview of roof area damage.

Several tests cuts were made within the field of the roof and along the perimeters. The membrane was mechanically fastened along the perimeters with a metal reinforcement strip with a sufficient number and length of fasteners. In almost all cases, the screws were almost entirely corroded (Figure 7.29). The perimeter test cut also revealed that the vapor barrier stopped at the deck and was not continuous at the interface with the parapet wall. Moisture staining was observed on the insulation boards along the perimeter and corner; a strong chlorine odor was detected at the deck/wall interface.

FIGURE 7.29 Corroded screw fastener.

It became apparent that a continuous vapor retarder was not installed as specified. This allowed for water vapor to travel between the roof deck and membrane and condense out as liquid once the vapor impacted the colder surfaces of the parapet wall. This moisture ultimately corroded and weakened the fasteners. Once the perimeter fasteners began to fail, the entire roof was susceptible to wind uplift forces.

The American Society of Heating, Refrigerating, and Air-Conditioning Engineers (ASHRAE) provides information regarding the proper design of natatorium buildings. ASHRAE notes that if the building envelope is not properly designed, condensation will occur, which can decay and degrade building materials, causing heavy corrosion that can result in roof collapses or failure in worst-case scenarios.[9] Forensically, the cause of this roof failure was a lack of proper vapor barrier design that allowed a corrosive environment to exist within the roof system, ultimately degrading the fasteners.

7.6.11 INSULATION FACER BOARD DELAMINATION

A 260-foot-tall building was covered with a fully adhered EPDM membrane over the top of iso-cyanurate insulation boards and a concrete deck. The membrane was found to have been properly installed and fastened along the perimeters. Plastic totes or containers were filled with water and placed on the roof to provide uplift resistance over the damaged area (Figure 7.30).

FIGURE 7.30   Overview of roof area.

The field portion of the membrane had failed or become detached from the insulation boards reportedly as a result of a storm with 66 MPH winds. Ultimately, it was discovered that the roof failure was due to detachment between the membrane and insulation board facer (Figures 7.31 and 7.32).

FIGURE 7.31   Separation of facer from insulation board.

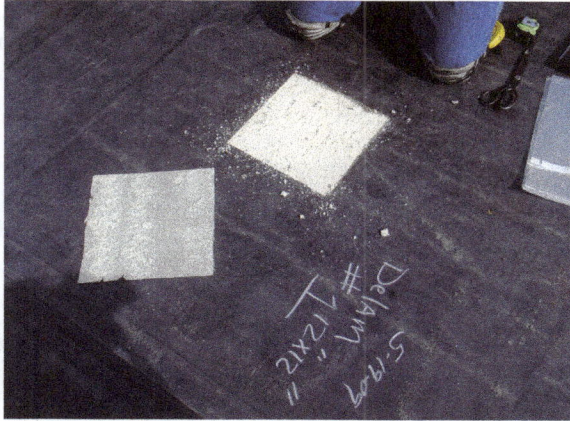

FIGURE 7.32   Separation of facer from insulation board.

Test cuts indicated that the membrane was bonded with a uniform coating of adhesive. The cuts also revealed moisture within the insulation and on top of the concrete deck, suggesting that water had penetrated the roof system. Five negative pressure wind uplift tests were performed on roof areas that contained no damage (i.e., separation). The membrane failed the negative pressure tests in four of the five locations. This indicated that the roof system was failing well below the design pressures.

Forensically, the cause of roof failure was the separation between the insulation facer and insulation. Facer delamination has been reported to be caused by several factors, including heavy traffic during construction, moisture intrusion, and sometimes, manufacturing defects. As a result of many similar failures through the industry, the National Roofing Contractors Association (NRCA) began recommending in 2000 that cover boards be installed over the isocyanurate insulation boards as a best practice for adhered single-ply membranes.[10]

7.6.12   Unlocked Standing Seam Metal Panels

A large warehouse distribution center was covered with a standing seam metal roof system (SSMR). A storm, reportedly producing 63 MPH winds, passed over the building and caused widespread displacement of the metal panels (Figure 7.33).

FIGURE 7.33   Overview of roof damage.

The panels were secured to the roof with Z-clips that appeared adequately sized and spaced. The panel seams were typically manually seamed or locked at the locations of the securement clips (Figure 7.34). The remaining portions of the seams were open and had not been manually or mechanically seamed (Figure 7.35).

FIGURE 7.34 Locked seam at securement clip.

FIGURE 7.35 Unlocked seam.

In the manufacturer's installation instructions, a critical step was to provide a continuous lock of the seams along the entire length of the panels. The NRCA recommended the following, "Once two or more panels are secured, the seaming process can begin. If the panels are of az mechanical interlock design, it is good practice to crimp the panels together with hand seamers before using a mechanical seamer. This should be done close to the location of the clips so that there will be less likelihood of panels disengaging from the clips and adjacent panels before being seamed. It will also keep the panels firmly nested together for seaming, which may prevent the seamer from disengaging from the damaging the seam."[11]

Forensically, it was concluded that the cause of this roof failure at less than design wind speeds was because the panels had not been fully seamed as required by industry best practices.

IMPORTANT POINTS TO REMEMBER

- Negative pressure is exerted by forces of wind across a roof surface creating uplift forces.
- Higher pressures will typically be exerted on partially enclosed buildings than on closed or open buildings.
- The highest wind pressures occur at the corners and perimeters of a roof surface; therefore, codes, standards, and best practices require enhanced securement in these areas.
- The uplift forces are dependent on the height, exposure, and type of building.
- In properly installed and maintained building roof surfaces, components and cladding should be able to withstand the ultimate design wind speeds and damage should not occur at lesser speeds.
- Roof systems can be installed by fully adhering with mechanical attachment and ballasted. In each case, the type of installation should ensure adequate wind uplift resistance.
- Roughly, 80% of low-sloped commercial roof failures occur at the corners or perimeters of structures. Further, ~90% of low-sloped roof failures have been reported to be caused by faulty installation practices.
- Perimeter failures are typically due to insufficient securement or aging/maintenance issues.
- Improper adhesive application in the field portion of the roof system is common in roof failures.
- A systematic inspection methodology for low-sloped roof system failures, as suggested in this chapter, will ensure that the cause of the roof failure is properly and efficiently determined.
- Destructive testing and/or uplift testing will likely be required to determine the cause of low-sloped roof failures.

REFERENCES

1. Patterson, Stephen and Madan Mehta. "Wind Pressures on Low Slope Roofs." Roof Consultants Institute Foundation, Publication No. 0.01, 4th Edition, 2018.
2. International Code Council. "2018 International Building Code (IBC)". Accessed July 22, 2020, https://codes.iccsafe.org/content/IBC2018/chapter-16-structural-design#IBC2018_Ch16_Sec1604.
3. American Society of Civil Engineers (ASCE). "Minimum Design Loads for Buildings and Other Structures." ASCE Standard 7–16, Reston, VA, 2011.
4. Griffin, C. W. and Richard L. Fricklas. *"Manual of Low-Slope Roof Systems"*, 4th Edition. New York: McGraw Hill, 2006, p. 179.
5. FM Global. "FM Global Property Loss Prevention Data Sheets 1–49". Perimeter Flashing, Revised September 2000.
6. Griffin, C. W. and Richard L. Fricklas. *"Manual of Low-Slope Roof Systems"*, 4th Edition. New York: McGraw Hill, 2006, pp. 131–132.
7. FM Global. "FM Global Property Loss Prevention Data Sheets 1–29". Roof Deck Securement and above Roof Deck Components, February 2020. Accessed July 22, 2020, https://www.fmglobal.com/research-and-resources/fm-global-data-sheets.
8. FM Global. "FM Global Property Loss Prevention Data Sheet 1–52". Field Verification of Roof Wind Uplift Resistance, February 2020. Accessed July 22, 2020, https://www.fmglobal.com/research-and-resources/fm-global-data-sheets.
9. American Society of Heating, Refrigerating and Air-Conditioning Engineers, Inc. *"2019 ASHRAE Handbook – Heating, Ventilating, and Air Conditioning Applications."* Atlanta, GA: ASHRAE, 2019, Chapter 6 – Section 6.2.

10. D'Annunzi, John. "Polyisocyanurate Insulation: Concerns Revisited." August 29, 2002. https://www.roofingcontractor.com/articles/83969-polyisocyanurate-insulation-concerns-revisited.
11. The National Roofing Contractors Association. *"The NRCA Roofing and Waterproofing Manual"*, 5th Edition (2006 Update). Rosemount, IL, p. 545. Accessed July 22, 2020, http://staticcontent.nrca.net/masterpages/technical/manual/06pdfs/06update.pdf.

8 Water Infiltration – Cause and Origin Assessments Steep- and Low-Sloped Roof Systems

George W. Fels
Curtiss-Wright Corporation

CONTENTS

PURPOSE/OBJECTIVES

The purpose of this chapter is to:

- Identify and address common areas on the roof surface (i.e., transitions, valleys, penetrations, etc.) that are prone to water intrusion.
- Provide an overview of valuable instruments and techniques to aid in the investigation of water leaks attributable to roofing issues.
- Document a systematic approach (i.e., methodology) for handling an investigation of water infiltration associated with steep and low-sloped roofing systems.

Following the completion of this chapter, you should be able to:

- Identify the basics of proper flashing and water management details associated with steep and low-sloped roof systems.
- Conduct a methodical and systematic inspection to locate and identify the source and cause for the water intrusion.
- Be able to document findings using this systematic methodology for a completed site inspection.

8.1   INTRODUCTION

When an owner of a property first observes water damage through staining patterns on surfaces, dripping of water, and/or ponding of water on the floor, their first thoughts lead to "why is the leak occurring" and "where is the water leak coming from"? Often the answers to these questions are not readily determined. Consequently, an owner will turn to their insurance company for help in determining the cause of water damage and to obtain compensation for items damaged by water.

The questions of where the water is coming from and what the cause is has led to the need for experts to determine the cause and origin of water damage claims. Experience suggests that the answer to these questions can be quite simple and visible in some cases yet very complex and not so visible in other cases. Water cause and origin investigations can be some of the most complex issues a forensic inspector can encounter, and often the complexity of the situation will not become apparent until a detailed analysis is undertaken.

Regardless of the complexity of the water infiltration investigation, a systematic approach or methodology, coupled with experience, offers the best opportunity for determining the cause(s) and origin(s) of water intrusion into building envelopes. Within this field of investigative engineering, the roles of these services are to identify the source of the water entry (i.e., the "where") and then determine the reason for the failure (i.e., the "why"). This chapter will address leaks associated with different types of roofing systems and the most common areas that are prone to water entry.

Steep-sloped roofing is generally defined as elevations with a pitch of 2:12 or greater, although some definitions of steep-sloped roofs set values at 4:12 or greater. This compares with low-sloped roof surfaces, which are defined as slopes <4:12. Steep-sloped roofing materials can include asphalt shingles (the most common), tile (slate, clay, concrete, etc.), and wood (shake or shingles). These types of roofing materials are considered hydrokinetic or water-shedding versus hydrostatic or water holding, which are typically employed with low-slope roof systems.[1] Therefore, these interlaced materials must be able to *shed* or allow water to run off the roof surface with the help of gravity to avoid water backups or issues where water can wick underneath these materials due to capillary action. On the other hand, because of the gradual grade on low-sloped roofs, these systems are typically designed with the application of a waterproof membrane intended to prevent slow-moving water from penetrating underneath the system.

It should be noted up front that in most instances, roof water infiltration is the result of, and possibly the combination of, age/deterioration, improper installation, and/or the lack of proper water management details. For most steep-sloped roofing applications, water entry is typically most prevalent at transitions, terminations, penetrations (i.e., pipes, stacks, chimneys), and along eaves. In very few cases (unless severely damaged) does water infiltrate through the actual roofing material as long as the roof is installed on an adequately sloped surface. Water intrusion associated with a low-sloped roof system is associated with penetrations (e.g., pipes, supply lines/ducts, etc.), terminations and wall interfaces (i.e., parapet walls), and issues related to water drainage or ponding on the roof surface.

8.2 COMMON LEAK AREAS ASSOCIATED WITH ROOF SYSTEMS

This portion of the chapter will focus on some of the most common causes of water intrusion associated with steep- and low-sloped roof systems.

8.2.1 APPLICATIONS FOR STEEP-SLOPED ROOF SYSTEMS

This section focuses on causes for water intrusion associated with steep-sloped roof surfaces based on experience, a review of industry best installation methods, and a review of modern building codes. Information from the National Roofing Contractors Association (NRCA),[1] APA – The Engineered Wood Association,[2] and other references[3–5] was also reviewed for this section. Many of the examples are associated with asphalt-shingled roof surfaces since this is the most prevalent surface for residential and light commercial structures.

The following are common areas where if installation requirements and/or best practices are not followed can result in conditions conductive to water entry:

- Underlayment and Ice Damming Protection.
- Eave and Rake Details.
- Gutters and Roof Drainage.
- Valleys (i.e., Open, Closed-Cut, Closed-Woven).
- Vertical Walls (Including Chimneys).
- Roof Penetrations (Plumbing Stacks, Skylights, Box Vents).

These will be reviewed in the following sections.

8.2.1.1 Underlayment and Ice Damming Protection

Prior to the application of the roofing material, an appropriate roofing underlayment must be selected first. There are several primary functions of underlayment, but the single underlying role is to provide additional defense to moisture intrusion. Underlayment is available in two types: (1) water-resistant and (2) waterproof and provides an additional weather barrier between the roofing materials and wood sheathing.

Asphalt-saturated felt is the most common form of water-resistant underlayment used throughout the roofing industry. This type of underlayment is not water-impermeable and relies more on its water-shedding abilities to divert water off the sloped roof surface before it can enter a structure. Typical language for water-resistant roofing materials found in building codes and best practices for steep-sloped roof systems, by slope, follows:

- For roof surfaces with a 4:12 pitch or greater, a minimum of one layer of water-resistant underlayment should be applied and should overlap the preceding or successive course by a minimum of 2 inches and contain an end lap of a minimum of 4 inches. Caution must be taken to ensure the underlayment is absent of any visible distortions.[3]

- For roof surfaces less than a 3:12, modern building code requires a double layer of appropriate underlayment, applying successive sheets with a minimum of 19-inch exposures. It should be noted that based on experience and the recommendations of other trade associations, including the NRCA, asphalt shingles should not be installed on roof surfaces with a pitch of 3:12 or less.[1,3]

The other type of underlayment consists of a waterproof or impermeable membrane, which consists of a self-adhering polymer-modified bituminous sheet. These self-adhering membranes are typically used as ice dam protection at roof edges (i.e., eaves), along valleys, and other penetrations where water entry is of concern since they are a true waterproof material. Two layers of standard asphaltic felt cemented together with either roofing cement or other adhesives can also qualify as ice dam protection. Typical language for waterproof roofing materials found in building codes and best practices, for steep-sloped roof systems, by slope, follows:

- For ice dam protection, modern code requires a form of waterproof membrane (two cemented layers of underlayment or ice guard/shield) be extended past the inside of the exterior wall by at least 24 inches.[3]
- For roof surfaces with a pitch less than a 4:12, the membrane should be extended a minimum of 36 inches upslope from inside of the exterior wall.[1]
- The application of ice/water shield should be applied along all roof edges, valleys, and most roof penetrations such as skylights and chimneys.

Additional information regarding the application of underlayment and the prevention of ice dams can be found in articles written by ARMA,[6,7] Fricklas,[8] and FEMA.[9]

8.2.1.2 Eave and Rake Details

In steep-sloped applications, the exterior wood members along roof eaves and rake lines are susceptible to exposure from the weather and damage from possible water entry. These wood members typically include the roof sheathing and wooden fascia board. Precautions should be made to ensure that these construction elements are shielded, particularly from water runoff from the roof surface.

Most modern residential homes are typically constructed with roof overhangs as part of the eave and rake construction and make water infiltration at these areas more difficult to reach the interior portions of the home. Nevertheless, conditions do arise where water can enter at these locations and penetrate into the building envelope. For example, many older cape cod-style homes do not have overhangs along the eaves and rakes so prevention details along these edges are critical for protecting the building.

Starting at the bottom edge of the roof system, the application of drip edge molding allows any water traveling down the steeped roof surface to be diverted away or collected into the gutter system. Figure 8.1 depicts a simple construction detail of the eave of a steep-slope application.

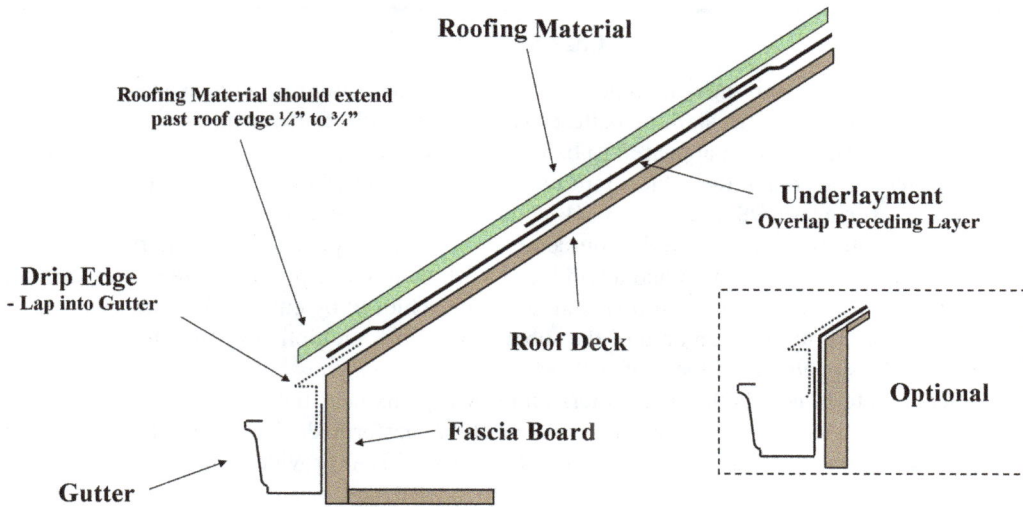

FIGURE 8.1 Eave construction details.

Based on experience, recommendations of various trade organizations and shingle manufacturers, and the requirements set forth by modern building code, the following guidance should be followed for eave construction with regard to water entry at the eaves:

- Drip edge molding should be present along all edges of the roof surface, including eaves, gables, and rakes.
- Drip edge molding should be installed underneath the bottom edge of the underlayment or ice guard, or install the edge molding on top of the underlayment. In the latter case, lap the underlayment behind the gutter system for additional protection.
- Metal drip edge along an eave should be fastened to roof decking and should overlap the back inside edge of the gutter. In addition, the drip edge should extend up the roof surface a minimum of 2 inches.
- Finished roofing surface materials should either be installed flush with or extend past the drip edge molding, to divert water runoff away from the roof components.

8.2.1.3   Gutters and Roof Drainage

The main purpose of a drainage system is to collect rainwater from the roof surface and direct it away from the structure. An improperly functioning drainage system can be the result of poor design, installation deficiencies, and/or lack of periodic maintenance. All of these factors can contribute to water overflowing the gutter, backing up, or wicking up underneath the roofing material resulting in significant water intrusion into the building. Similarly, poorly performing roof drainage systems can allow for soil erosion as well as water intrusion into the basement of a building.[10]

Proper gutter design, installation, and maintenance are necessary to minimize water damage from rain, ice melt, and melting snow from entering the roof system. The proper design capacity of the gutter/downspout drainage system is dependent on the slope of the roof surface, the surface area in question, and the average rainfall intensity for the geographical region of the building. To illustrate proper gutter and downspout design, the following case study serves as an actual example of the effects of the design of these systems regarding water intrusion.

Case Study 8.1

Interior water-damage staining was reported in a residential building with asphalt shingles in Cincinnati, Ohio. Based on the staining patterns and evidence collected during the inspection of the residence, the interior damage appeared to be the result of excessive water runoff overflowing the drainage system. The area of the subject building was being drained by one corrugated rectangular downspout measuring ~2-3/8 inches by 3-1/4 inches. The roof drainage area, which collected into the sections of gutter believed to be the source of water intrusion present in the interior, was ~900 square feet. The gutter in question had a width of ~5-1/8 inches, a depth of ~3-1/8 inches, and was ~33 feet in total length. (The roof drainage area can be achieved by either calculating the area of each elevation or taking the plan area of the subject elevations multiplied by an adjustment factor dependent on the slope/pitch of the roof surface.)

In order to determine whether the gutters and downspouts near the interior damage provided adequate drainage, simple sizing calculations must be performed. The Sheet Metal and Air Conditioning Contractor's National Association, Inc.[5] (SMACNA) provides guidance on gutter and downspout design.

Gutter design: For this case, the rainfall intensity data for a 5-minute duration within a 10-year period for the city of Cincinnati, Ohio, is given to be 6.8 inches of rainfall per hour. The minimum gutter width, W (ft), needed to optimally drain a roof area is found by Eq. 8.1 for rectangular gutters:

$$W = 0.0106 * M^{-4/7} * L^{3/28} * (I * A)^{5/14} \qquad (8.1)$$

where

  M = the ratio of the gutter depth to width,
  L = the length of the gutter (feet),
  I = the rainfall intensity (in/hour) of the subject area, and
  A = the area that is being drained (square feet).

Therefore, the $I * A$ (Intensity × Area) value from above is ~6,120. The depth-to-width ratio (M) of the gutter was 0.61, and the length (L) of the gutter section was 33 feet. Inserting these values into the calculation for W gives a needed gutter width of ~0.46 feet, or 5.52 inches. SMACNA recommends rounding up to the next largest gutter size, in this case 6 inches in width. Since, the gutters on the home measured ~5-1/8 inches in width, the gutter design for the roof area in question was not sufficiently large enough to meet the SMACNA standard design requirements for this roof area.

Downspout design: The section of roof surface near the location of interior staining was drained by a single downspout that extended down to a subgrade drainage system. According to SMACNA, the size of downspout in this particular situation had an area of ~7.73 square inches. Using the same drainage factors for Cincinnati, Ohio, given by SMACNA (data based on 2002 data from the U.S. Weather Bureau), the drain rate of a downspout is 130 square feet/square inch of downspout, meaning that the downspout can drain 1,005 square feet/hour (7.73 square inch × 130 square feet/square inch). As stated, the total roof area drained by this single downspout was ~900 square feet. Since the SMACNA design suggests that this downspout can drain a maximum of 1,005 square feet/hour, the downspout appeared to be of sufficient size to channel water runoff from the roof surface. It must be noted that these calculations were based on a level, clean, and unobstructed gutter.

Discussion: This case study simply illustrates the need for a properly designed roof drainage system. Even though the single downspout was sufficient enough to provide adequate draining capacity for the roof elevation in question, the section of gutter was not. This implies that this gutter would not handle the large volume of rainfall that would occur during a design rain event leading to water backing up beneath the shingles and entering into the subject home and/or overflowing the gutter onto the ground below and possibly flowing against the foundation wall and into the basement.

To prevent water-related problems, one must conduct periodic maintenance of the drainage system by clearing out all built-up debris and assuring the system is secured firmly in place. Tree debris and sediment are often observed clogging gutters and can create conditions where water can overflow the system and potentially damage the roof system at the eave and/or infiltrate into the structure at the eave/foundation.

Additional information and photographs associated with gutter/downspout drainage systems, as well as issues resulting in basement water intrusion, can be found in Chapter 10.

8.2.1.4 Valleys (i.e., Open, Closed-Cut, Closed-Woven)

Roof valleys are formed when two sloping elevations intersect at two down-slope planes. Since these areas experience a large amount of water runoff, they provide an ideal location for water intrusion. The primary purpose of valleys is to provide a pathway for rainwater to drain away from the roof surface. These intersections must be able to resist water infiltration from wicking of rainwater as well as from entry during periods of snow and ice melting in colder climates.

For asphalt shingles (three-tab and dimensional), the three different types of valleys employed for steep-sloped roof systems are open, closed-cut, and closed-woven:

- *Open valleys*: Typically, metal flashing is installed along the intersection of each adjoining elevation and then the roofing materials are installed over the top of the flashing edges, leaving a clear and unobstructed channel for the runoff of water.
- *Closed-cut valleys*: Roof materials from one elevation are laid over and installed onto the adjoining elevation. Materials from this adjoining elevation are then installed over and cut along the valley forming a mitered joint.
- *Closed-woven valleys*: Used only with strip shingles in which shingles from each adjoining elevation extend across the valley and successive courses of shingles are then interlaced or installed in a woven pattern. This type of valley is not as predominant as the other two types of valleys.

The most common causes for water intrusion at valleys for steep-sloped applications are discrepancies or deficiencies concerning the installation of these valleys. Various trade organizations (e.g., NRCA), manufacturers (e.g., CertainTeed), and modern building codes have provided the following general details regarding the proper installation of all types of valleys to limit water infiltration (note, these guidelines can apply to most strip shingles):

- All valleys should be lined with a layer of a water-resistant underlayment that is 36 inches in width and centered in the valley. The field underlayment should be applied to overlap the underlayment in the valley by at least 2 inches. In cold climates, a self-adhering membrane should be applied in the valley instead.
- The shingles cut along the valley must have their top corners "cropped" or "trimmed" (remove a small 2″ triangle). These trimmed corners help direct any water getting under those shingles back into the valley so it can be properly drained away.
- Shingles or metal valley flashing should extend slightly past the edge of the roof eave.
- When nailing the shingles to the roof surface, do not position the nail of the shingles closer than 6 inches from the center line of the valley. The closer the nail is to the center line, the greater the potential for water infiltration through that nail penetration.

- Optionally, apply and embed the cut shingles into a continuous strip of roofing sealant/ cement. Experience has shown that this would provide additional protection against the possibility of water runoff wicking back up underneath the shingles.

8.2.1.4.1   Open Valleys

Specific recommended best practices for installation of open valleys are as follows:

- Valleys should be lined with metal at least 24 inches wide. If the valley is lined with two layers of mineral-surfaced roll roofing, the bottom layer must be 18 inches wide and the top must be 36 inches wide.[3]
- Best practices and manufacture's installation guidance state that the valley should be tapered; meaning the valley is wider at the bottom (at the eave) than at the top (at the ridge). This allows for greater water runoff and limits the accumulation of ice/snow buildup.
- The metal valley liner should be secured with either metal clips/cleats or nails spaced anywhere from 8 to 24 inches. Best practices suggest not nailing directly into the valley liner.
- Shingles must overlap the metal lining a minimum of 4 inches. Some best practices also recommend that the shingles should be installed at least 3 inches away from the valley centerline.

Photographs of open valleys for three-tab and dimensional shingles are illustrated in Figure 8.2a and b.

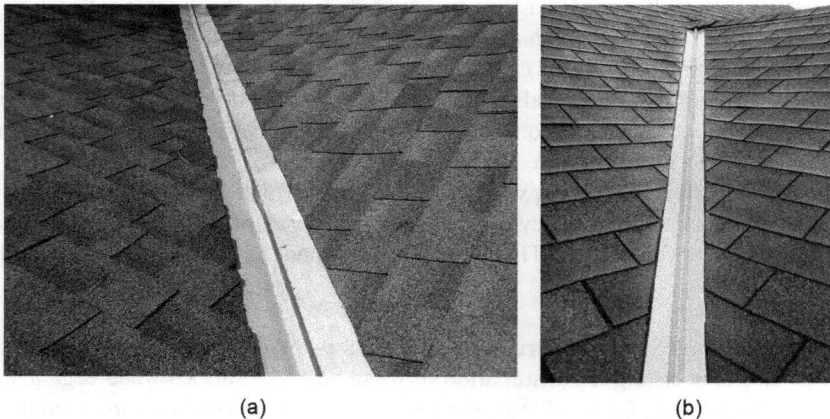

(a)   (b)

FIGURE 8.2   Example of open valley with dimensional shingles (a) and three-tab shingles (b).

Recommended flashing details from NRCA for an open valley installation are shown in Figure 8.3.[1]

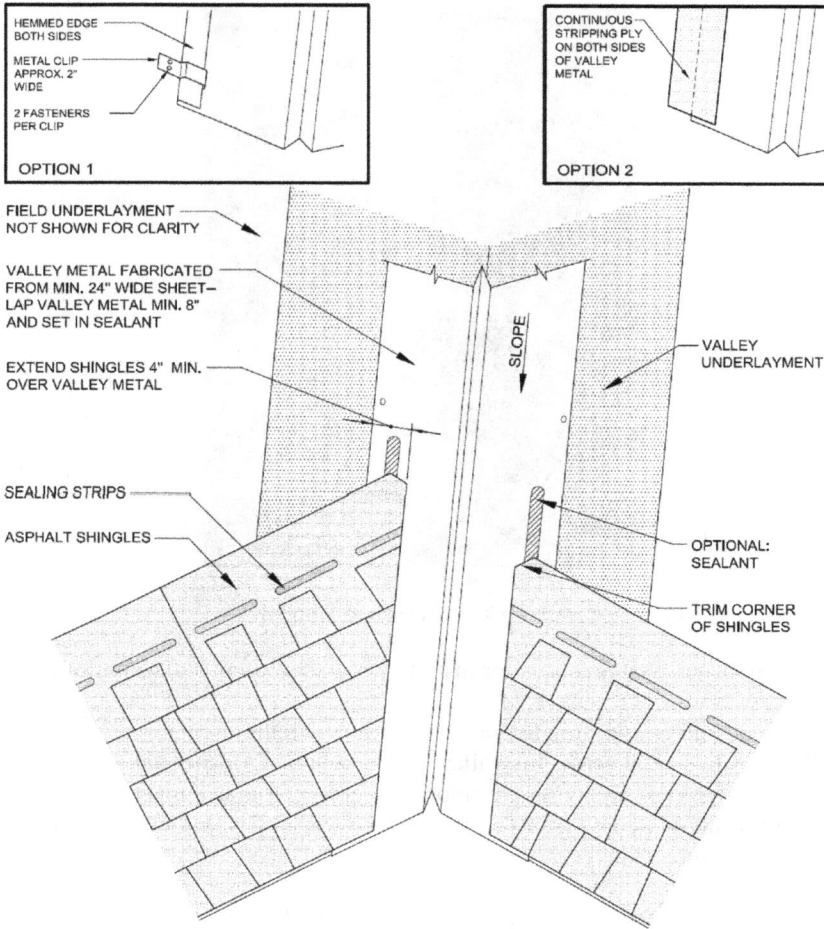

OPTION 1

HEMMED EDGE
BOTH SIDES

METAL CLIP
APPROX. 2"
WIDE

2 FASTENERS
PER CLIP

OPTION 2

CONTINUOUS
STRIPPING PLY
ON BOTH SIDES
OF VALLEY
METAL

FIELD UNDERLAYMENT
NOT SHOWN FOR CLARITY

VALLEY METAL FABRICATED
FROM MIN. 24" WIDE SHEET—
LAP VALLEY METAL MIN. 8"
AND SET IN SEALANT

EXTEND SHINGLES 4" MIN.
OVER VALLEY METAL

SEALING STRIPS

ASPHALT SHINGLES

SLOPE

VALLEY
UNDERLAYMENT

OPTIONAL:
SEALANT

TRIM CORNER
OF SHINGLES

FIGURE 8.3 Open valley details. (Reprinted with Permission of the National Roofing Contractors Association.)

Based on experience, this installation will provide additional protection from water intrusion than with more minimal installations often encountered.

8.2.1.4.2   Closed-Cut Valleys

An illustration of a closed-cut valley is shown in Figure 8.4.

FIGURE 8.4 Example of closed-cut valley with dimensional shingles.

Specific recommended best practices for installation of closed-cut valleys follow:

- For closed-cut valleys, the shingles on the elevation with the smaller roof area intersecting the valley should be laid across the valley a minimum of 12 inches and left uncut.
- The shingles from the adjoining elevation (the elevation that drains a larger volume of water runoff) should be laid over the centerline of the valley and be trimmed back from the valley centerline ~2 inches.

An alternative method for closed-cut valleys is to install an additional course of shingles vertically prior to applying the shingles on the larger area elevation. Proper installation details for a closed-cut valley with three-tab asphalt shingles provided by CertainTeed Corporation, along with an alternative approach for dimensional shingles, are illustrated in Figure 8.5a and b.

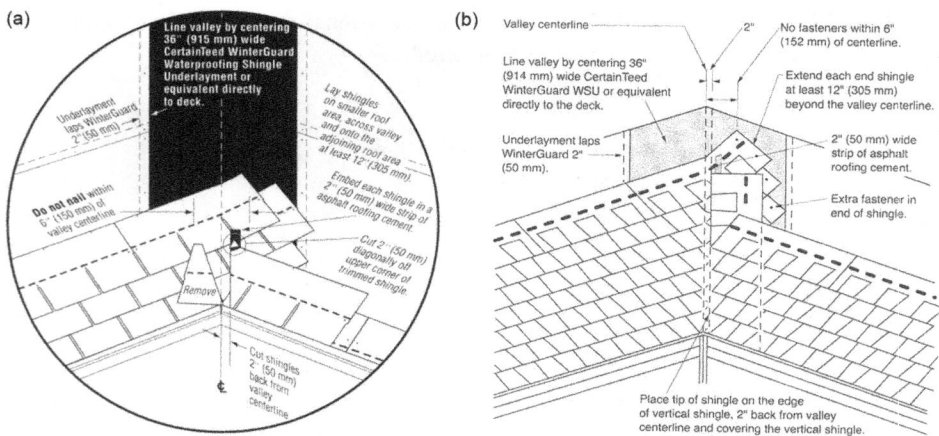

FIGURE 8.5 Proper installation of closed-cut valley with three-tab shingles (a) and alternative method, with dimensional shingles (b). (Courtesy of CertainTeed Corp.)

8.2.1.4.3   Closed-Woven Valleys

An illustration of a closed-woven valley is shown in Figure 8.6.

FIGURE 8.6   Example of closed-woven valley with dimensional shingles.

Specific recommended best practices for installation of closed-woven valleys follow:

- Starting from the eave, the first course should extend past the centerline and onto the adjoining elevation at least 12 inches and proceed in an interlaced pattern. Remember to start with the shingles on the elevation with the least area of water runoff.
- Remember, no nails should be installed within 6 inches of the valley's centerline. Oftentimes, an additional end nail might be needed for full securement.

8.2.1.5   Valley Deficiencies

Given a basic understanding of the different types of valleys employed on steep-sloped roof applications, common deficiencies observed during on-site field inspection activities along with how these deficiencies can contribute to water intrusion follow.

8.2.1.5.1   Untrimmed Corners

Untrimmed Corners are one of the most common installation deficiencies observed on residential properties. For the cases of closed-cut and open valleys, shingles along one or both adjoining roof elevations should be cut along the length of the valley. Oftentimes, the upper corners of the cut shingles are not trimmed. Trimming is intended to aid in the proper diversion of draining water back into the valley and not underneath the shingles. Consequently, visible evidence of dirt, debris, drip patterns, and even efflorescence deposits underneath the shingles is indicative of water traveling underneath the shingles and possibly into the home.

Table 8.1 depicts these untrimmed corners along an open valley.

TABLE 8.1

Untrimmed Shingle Corners Leading to Potential Water Intrusion

Description	Photograph
Efflorescence patterns and dirt on valley metal flashing indicating draining water traveling back beneath shingles at untrimmed corner location	
Untrimmed shingle corner protruding past the cut edge of the shingles and into the valley. This corner can "catch" water runoff and direct it underneath the shingles	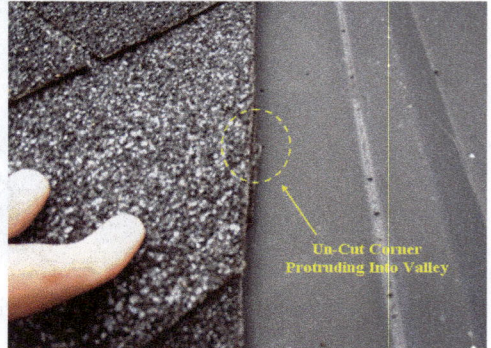

8.2.1.5.2   Shingles Not Embedded with Sealant/Caulking

Although not an installation requirement, experience has shown that the addition of roofing cement and/or sealant underneath the cut shingles for both metal and closed-cut valleys provides added protection against valley water runoff traveling underneath the shingles.

8.2.1.5.3   No Waterproof Membrane (Ice Guard)

Waterproof underlayment, especially in cold climates, serves as a last layer of protection in the defense against water intrusion from ice and snow buildup. When either rain falls on the ice/snow or the ice/snow melts during a warm-up, the ice guard/barrier will protect a potential and leak-susceptible roof intersection. Increasingly, manufacturers and local building authorities are advocating for the application of an ice guard/barrier rather than typical water-resistant underlayment as a base layer for the construction of valleys.

8.2.1.5.4   Nail Fasteners Driven Too Close to Valley Centerline (Open Valleys)

The metal flashings should only be secured along the outer edges and not pierced by nail fasteners. Driving nails through the valley flashing, especially close to the cut edge of the shingles along the valley centerline or the open valley metal, creates pathways for water to enter the structure. Figure 8.7a and b illustrates situations where an open valley flashing was secured by nails very close to the cut edge of the shingles.

(a)   (b)

FIGURE 8.7   (a) Evidence of water presence at nail depression and (b) nail driven too close to centerline of open valley.

Upon lifting the shingles along the valley, evidence of drip staining, efflorescence, and dirt/debris was found on the top side of the flashing at the location of this nail depression (Figure 8.7a). Also, note the lack of asphalt plastic cement sealing down the cut edges of the shingles.

8.2.1.5.5   *Nail Fasteners Driven Too Close to Valley Centerline (Closed-Cut Valleys)*

Similar to open valleys (above), shingle nail fasteners that are installed too close to the valley centerline, or too close to the cut edge of the shingles along the valley, provide potential pathways for water to enter the structure. The fasteners for these shingles should be placed away from the centerline a sufficient distance (no closer than 6 inches from the centerline of the valley) and then be manually sealed with asphalt plastic cement to provide additional resistance to water intrusion.

8.2.1.5.6   *Shingles Incorrectly Cut along Wrong Elevation of Valley (Closed-Cut Valleys)*

With closed-cut valleys, which adjoining elevation along the valley contains the larger and smaller roof surface area must first be identified. Special attention must be made to which shingles should be installed first (i.e., bottom of valley) and which should be installed second (i.e., top of valley). Since the adjoining elevation with the greater surface area will drain a higher volume of water runoff, the elevation with the smaller drainage area of water should be first laid across the valley.

Case Study 8.2

Based on interior measurements and observations, water intrusion was thought to be originating along a closed-cut valley on a residential building. As shown in Figure 8.8, a closed-cut valley was installed along the intersection of roof elevation A and roof area elevation B. The surface area of roof elevation A, which drained into the valley, was calculated to be ~148 square feet compared with elevation B, which was calculated to contain a drainage area of ~62 square feet. Considering that elevation A has the potential to drain a greater volume of water runoff, the shingles from elevation B should be installed first (lower layer) and the shingles from this elevation should be trimmed back from the centerline. This helps to ensure that the larger amount of water from elevation A cannot breech or infiltrate underneath the shingles from the adjoining valley.

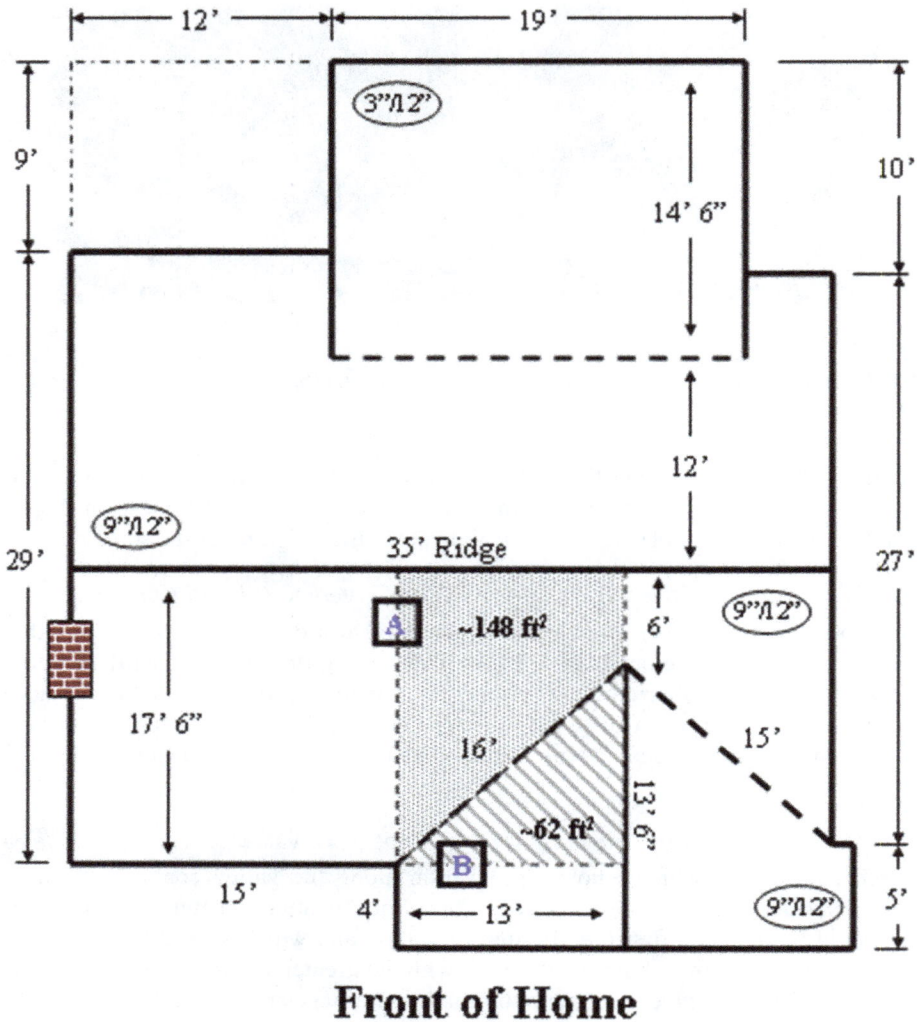

FIGURE 8.8 Drainage area calculated along a closed-cut valley.

In this example case, the installation was backward from the recommended installation best practice. The shingles along the valley were installed in the opposite and incorrect manner, thus allowing water to easily back up underneath the shingles.

8.2.1.5.7 Improper Termination of Valley at Home Interface

In order to function properly, valleys must be clear and free of obstructions so water can drain freely. According the NRCA,[1] "a clear, unobstructed drainage path is desired in valleys so the valley may carry water away quickly and perform successfully for the service life of a roof system." This simply means that if a valley is obstructed, even partially, the potential risk of water infiltration under the shingles designed to shed water can increase. Keeping valleys clear also means the removal of windborne debris such as leaves, sticks, etc., and proper design of the drainage system to ensure a clear and unhindered pathway for water runoff. Finally, this best practice also applies to the valleys associated with chimney crickets and saddles. These valleys should not drain water directly against the corners of chimneys; but should be slightly offset for an unobstructed flow past the corners of the chimney. Table 8.2 provides field examples of clogged and obstructed valleys.

TABLE 8.2
Clogged and Obstructed Valleys

Description	Photograph
Heavy tree and leaf debris at termination of valley	
Heavy debris at termination of valley	
Heavy amount of tree debris in valley	
Buildup of debris and vegetation at intersection of exterior wall and masonry chimney	

Case Study 8.3

Photographs in Table 8.3 depict two different situations associated with water intrusion that ultimately led to the discovery of an improperly designed and/or constructed roof valley. In both cases, the interior wall and ceiling surfaces directly below an exterior corner contained heavy water damage staining. In addition, water damage and standing water were found in the basement directly below these areas.

Upon investigation for both cases, a roof valley was found to terminate at the exterior corner above the interior staining locations. Making matters worse in one case, a few overhanging tree branches covered the lower elevations, causing a buildup of leaf debris (i.e., leaves, twigs, braches, etc.) at the valley termination point. This poor design of the roof drainage system, coupled with other factors, allowed water runoff from the valleys to drain against the corner of the exterior wall and enter the building in each scenario.

TABLE 8.3
Two Examples of Improper Termination of Valley at Home Interface

Description	Photograph
Case study 8.3.1: Overview of roof valley above interior water damage Case study 8.3.1: Centerline of valley improperly terminates directly into corner interface	 

(*Continued*)

TABLE 8.3 (*Continued*)

Two Examples of Improper Termination of Valley at Home Interface

Description	Photograph
Case study 8.3.2: Overview of valley draining water runoff directly against corner of exterior wall Case study 8.3.2: Potential water entry point if proper flashing details not followed	

Unless proper flashing and water management details are employed, these junction terminations to valleys form ideal points for water intrusion into the structure.

8.2.1.6 Vertical Walls and Chimneys

Proper flashing along the intersection of a roof and vertical wall provides protection from water entry. The flashing employed along a vertical wall and roof interface typically involves a two-piece system. Step, apron, or backer/cricket flashing is attached to the roof surface and extends up the vertical wall to provide a watertight system along the intersection of the roof surface and penetrating wall. Counter flashing then overlaps and covers the top leading edge of these types of flashing. Unlike a vertical wall, a chimney employs the use of not only step and counter flashing, but also flashing on the upslope and down slope wall and roof intersections (i.e., crickets and aprons).

In most problem situations encountered in the field at these locations, deficiencies were detected with the flashing system such that it was not providing a watertight seal either along the intersection with the wall or with the roof surface.

This section briefly provides common installation techniques of flashing associated with chimneys, front (or head) walls, and sidewalls. Common improper installation methods and deficiencies typically encountered are also discussed. Roof systems covered with asphalt shingles (either three-tab or dimensional) are emphasized due to their popularity along with interfaces with masonry walls, but similar details and basic concepts apply to all types of steep-sloped roofing materials and wall cladding.

For steep-sloped roofing, four common types of metal flashings that are used for the prevention of water intrusion at roof/vertical wall interfaces follow:

1) *Apron or base flashing:* Metal flashing material used at the transition of a front (or head) wall and sloped roof. Apron flashing is also employed on the down slope intersection of the roof and chimney wall.
2) *Step flashing*: Metal flashing engaged along the intersection of a sloped roof surface and a vertical wall. Step flashing should always be used at any junction of a roof surface and vertical wall component.
3) *Counter flashing*: Flashing used to overlap and protect the step and apron flashing. Counter flashing is found on both chimneys and vertical walls.
4) *Cricket or backer flashing:* Typically installed on the upslope side of a chimney along the interface of the sloped roof surface.

The proper installation of each type of metal flashing along with its appropriate application is discussed below.

8.2.1.6.1 Apron or Base Flashing

Two different flashing situations occur on typical steep-sloped roof systems: (1) apron flashing on chimneys and (2) flashing at roof/vertical front wall interfaces. A discussion of best practices for both types of flashing follow:

Chimney flashing: APA – The Engineered Wood Association[2] prepared a publication that provides excellent details and best practices for preventing water infiltration into residential homes. Within this publication (Figure 8.9), recommended details for apron flashing on a masonry chimney were provided.

Apron flashing for downslope portion of masonry chimney.
Underlayment shown pulled away from chimney.

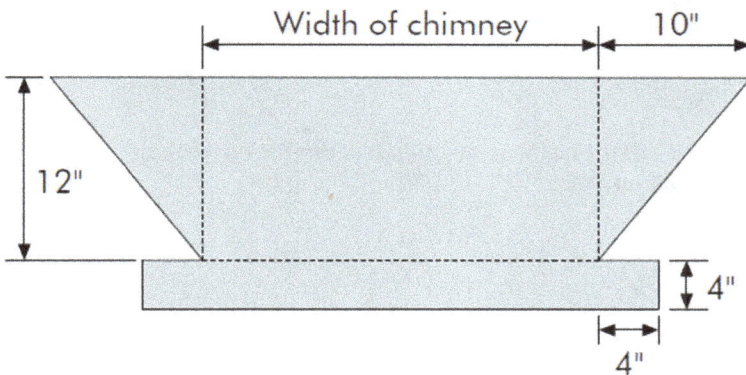

Underlayment

Coat of masonry
primer

Asphalt plastic
cement behind
flashing

Apron flashing
applied over
shingles and
set in asphalt
plastic cement

Width of chimney   10"

12"

4"

4"

FIGURE 8.9   Installation of apron flashing. (Courtesy of APA – The Engineered Wood Association.)

From experience, key installation deficiencies associated with apron flashing occur along the bottom, or down slope, side of a masonry chimney. These deficiencies do not meet the following criteria: (1) the flashing should extend at least 4 inches onto the roof surface and 6 inches up the wall surface, and (2) the flashing should be set in or sealed with roof cement. It is also common for the bottom edge of the apron flashing to be slightly hemmed to ensure water runoff does not wick underneath the flashing.

Front (or head) wall flashing: Again, the APA – The Engineered Wood Association[2] publication mentioned above provides excellent details and best practices for installation of flashing at vertical head walls. Figure 8.10 illustrates the fact that the flashing used at the horizontal wall-to-roof intersection should be a continuous piece of flashing sealed down with asphaltic cement. The flashing should be installed and nailed over the felt underlayment and cutouts in the penultimate course of shingles. The last course of shingles should be trimmed and adhered to the flashing.

A SHINGLE-TYPE ROOF AT A HORIZONTAL WALL-TO-ROOF INTERSECTION

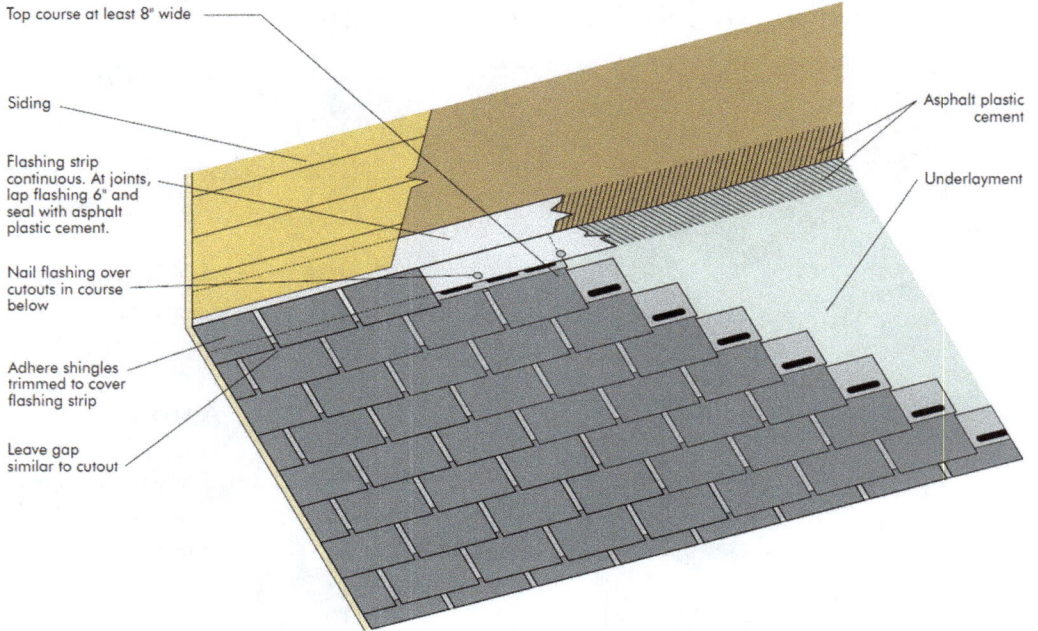

Top course at least 8" wide

Siding

Flashing strip
continuous. At joints,
lap flashing 6" and
seal with asphalt
plastic cement.

Nail flashing over
cutouts in course
below

Adhere shingles
trimmed to cover
flashing strip

Leave gap
similar to cutout

Asphalt plastic
cement

Underlayment

FIGURE 8.10   Installation of head flashing. (Courtesy of APA – The Engineered Wood Association.)

Figure 8.11 from CertainTeed also illustrates an industry-based installation technique for flashing along a front vertical wall/lower roof interface.[4]

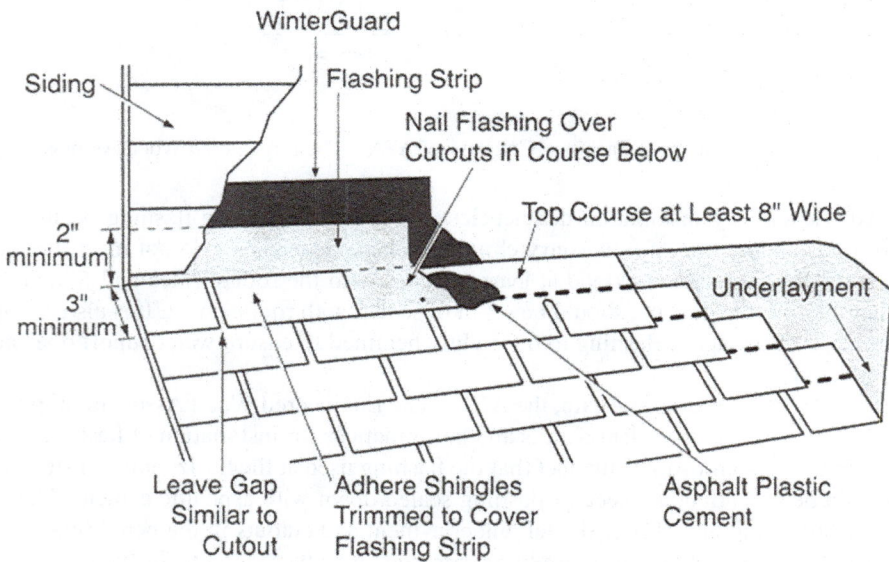

WinterGuard

Siding

Flashing Strip

Nail Flashing Over
Cutouts in Course Below

Top Course at Least 8" Wide

2"
minimum

3"
minimum

Underlayment

Leave Gap
Similar to
Cutout

Adhere Shingles
Trimmed to Cover
Flashing Strip

Asphalt Plastic
Cement

FIGURE 8.11   Installation of head flashing. (Courtesy of CertainTeed Corp.)

Similar to the recommendation set forth by APA – The Engineered Wood Association, CertainTeed first recommends that underlayment be lapped up the vertical wall (for colder climates, the use of waterproof ice guard underlayment is encouraged). Then the metal flashing should extend up the vertical wall at least 2 inches and overlap the last shingle by at least 3 inches. Note that this distance up the vertical wall, 2 inches, is much less than that recommended by APA (6 inches). Experience suggests that the 2-inch value is too low at many locations where ice and snow can build up at these interfaces and then melt, allowing water to get behind the flashing. The metal flashing should not be nailed to the vertical wall but secured by embedding it in asphalt roofing cement at the roof surface. However, it is recommended that the metal flashing in this situation be nailed above the cutouts of the underneath course of shingles. For both three-tab and dimensional shingle applications, if any nail heads are exposed, a dab of caulking/sealant must be applied to cover the nail head.

Experience and best practices have shown that these flashing details at lower roof/vertical wall interfaces, whether it is a chimney or exterior wall, are essential for preventing or limiting water entry into structures. Proper installation best practices, based on information gained from field assessments and industry best practice documents for this type of flashing, are as follows:

- Roofing underlayment or ice guard for severe climates and the flashing component should lap up the vertical wall and be installed underneath the building paper/wrap. It should extend twice the vertical flashing length.
- The vertical portion of the flashing should extend up the wall at least 3–4 inches.
- The apron flashing should extend out from the vertical wall at least 4–6 inches.
- Only one edge of the apron/base flashing should be secured to allow for potential expansion and contraction as temperatures vary. As illustrated in the references above, the flashing can be nailed to the roof surface coupled with sealant/caulking at the roof/wall interface. Best practices are to avoid exposed nail heads.
- While this situation is to be avoided, any face-driven nails should be covered with a dab of caulking/roofing cement. It should be noted that roofing cement and/or sealant would ultimately degrade over time, requiring periodic maintenance. Once this occurs, or in the absence of a sealant, moisture can penetrate into the structure around the nail fasteners.
- If the apron flashing is not overlapped by a form of counter flashing or wall cladding, the apron flashing must be tucked into the wall material. For example, if a piece of apron flashing is employed along a vertical masonry head wall, the flashing must be tucked into a mortar joint (see counter flashing section below).

Table 8.4 provides examples and descriptions of improper apron flashing installations encountered during on-site inspections.

TABLE 8.4

Examples of Improper Application of Apron Flashing

Description	Photograph
Apron flashing on the down slope side of a masonry chimney. Notice nail fasteners not covered with roofing cement/sealant	
Apron or head wall flashing secured with nails covered with daps of roofing cement. Notice the wall cladding in this situation is vinyl siding	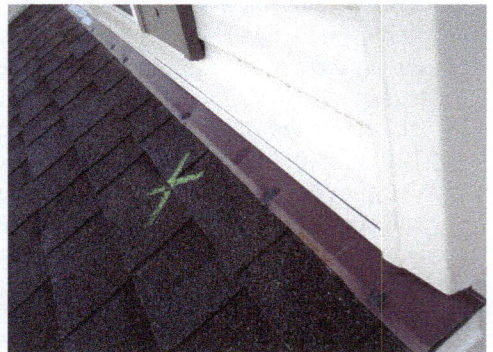
No apron flashing present along intersection. Notice face-driven nail heavily rusted	
Apron flashing along flat surface with use of no counter flashing or nail fasteners. Apron flashing is not secured and tucked into masonry wall	

8.2.1.6.2   *Step Flashing*

The step flashing applies to any vertical, or side, wall and sloped roof interface. In the process of flashing a chimney, step flashing in conjunction with the installation of asphalt shingles would follow the installation of the apron flashing and framing of the cricket. Examples of industry best practices for step flashing installation are provided by the APA – the Engineered Wood Association[2] and the NRCA.[1] Installation details from both references are shown in Figures 8.12 and 8.13.

FIGURE 11A

A SHINGLE-TYPE ROOF AT A SLOPED WALL-TO-ROOF INTERSECTION

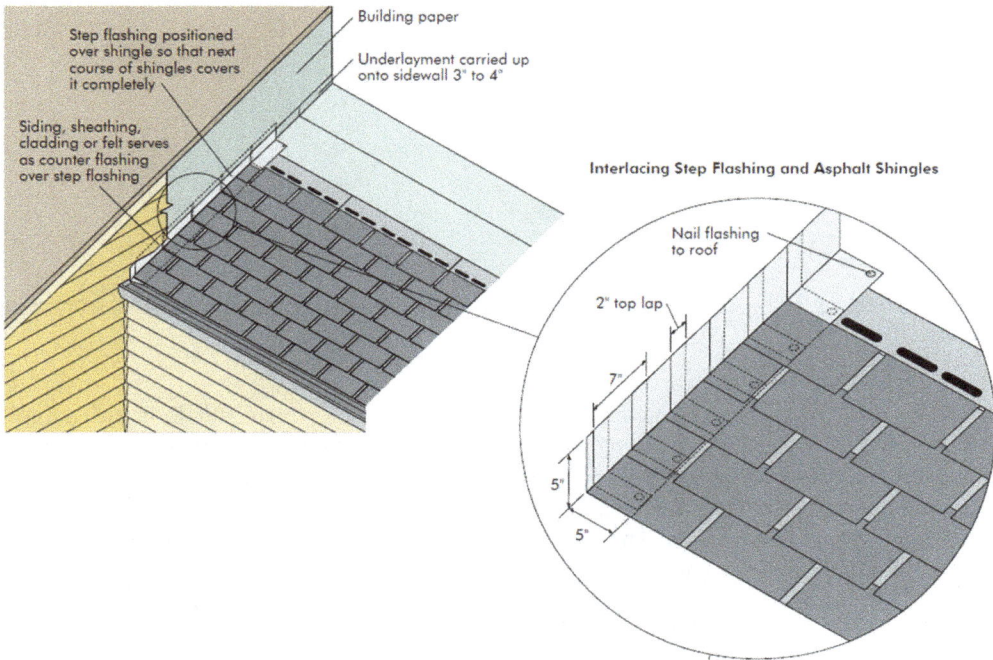

Step flashing positioned over shingle so that next course of shingles covers it completely

Siding, sheathing, cladding or felt serves as counter flashing over step flashing

Building paper

Underlayment carried up onto sidewall 3" to 4"

Interlacing Step Flashing and Asphalt Shingles

Nail flashing to roof

2" top lap

7"

5"

5"

FIGURE 11B

CLOSE UP OF FLASHING DETAIL

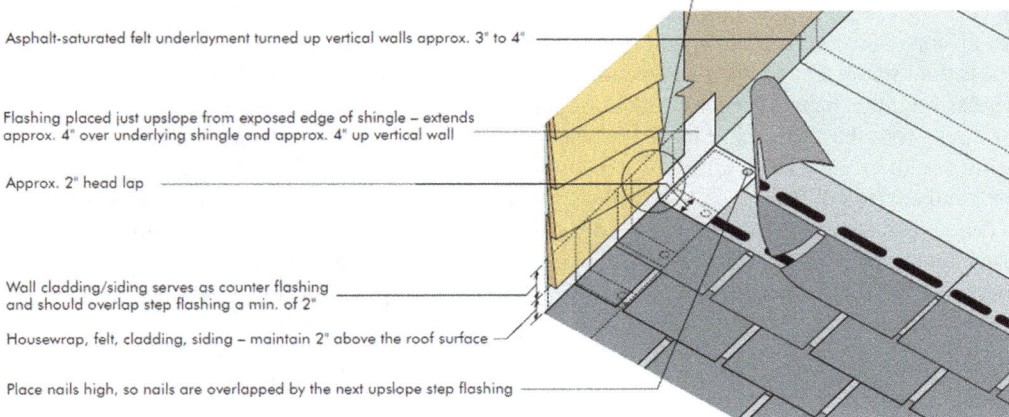

Asphalt-saturated felt underlayment turned up vertical walls approx. 3" to 4"

Flashing placed just upslope from exposed edge of shingle – extends approx. 4" over underlying shingle and approx. 4" up vertical wall

Approx. 2" head lap

Wall cladding/siding serves as counter flashing and should overlap step flashing a min. of 2"

Housewrap, felt, cladding, siding – maintain 2" above the roof surface

Place nails high, so nails are overlapped by the next upslope step flashing

FIGURE 8.12   Step flashing details along wall-to-roof interface. (Courtesy of APA – The Engineered Wood Association.)

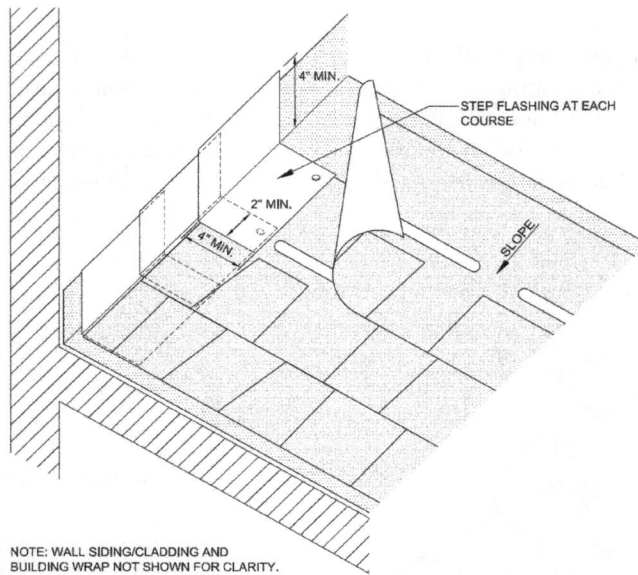

FIGURE 8.13 Step flashing at a vertical wall. (Reprinted with Permission of the National Roofing Contractors Association.)

Note that these best practices recommend that the roof underlayment should be turned up the wall 3–4 inches underneath the building wrap/paper. This is often not done. In addition, best practices suggest that step flashing be installed vertically up the wall 4–5 inches and horizontally along the roof and below the shingles 4–5 inches. Some debate exists on the distance step flashing should extend vertically and horizontally. This debate probably arises due to many suppliers and roofing contractors continuing to use narrow (L-shaped) step flashing pieces (5 inches wide), which would only allow 2-1/2 inches of coverage up the wall and out onto the underlying shingles. This type of coverage may work in mild climates, but experience has shown that to work successfully (keep water out), the step flashing should extend 4–5 inches underneath the intersecting steep-slope materials in areas of moderate and severe weather climates, areas of heavy rainfall, and/or areas with the potential for snow and ice accumulation. The flashing should be installed flush against both surfaces. Additionally, the nail fasteners should be placed high so the nails are overlapped by the next upslope piece of step flashing.

It is also important to install diversionary flashing, also known as kick-out flashing, where the step flashing intersects a vertical wall along the eave to ensure that the roof drainage is directed out and away from the wall. Experience has shown that a lack of, or improper installation of, kick-out flashing at these interfaces allows water to enter the wall system at these locations. The flashing joints and corners of this piece of flashing should be soldered or otherwise made watertight.

The step flashing details for the side walls of chimneys are installed in a similar manner as described above. Illustrations from APA – the Engineered Wood Association,[2] presenting general guidelines for the proper best practices installation of step flashing for a masonry chimney, are shown in Figure 8.14.

Interlace step flashing with shingles. Set step flashing in asphalt plastic cement.

FIGURE 8.14 Step flashing along a side masonry chimney wall. (Courtesy of APA – The Engineered Wood Association.)

Proper installation best practices, based on information gained from field assessments and industry best practice documents for this type of flashing, are as follows:

- The flashing component in conjunction with appropriate roofing underlayment should lap up the side wall and be installed underneath the building paper/wrap when applicable.
- For interlaced step flashing, the flashing should extend up the wall surface and onto the roof surface beneath the shingles at least 4–5 inches.
- If step flashing is employed along side walls, ensure that the step flashing is properly overlapped by a form of counter flashing. If means of counter flashing are not employed, then the step flashing must be tucked into the cladding material (applicable to masonry, stucco, or other solid finishes).
- Nail fasteners driven into the interlaced flashing on the roof surface should be located "high" to prevent water intrusion around the fastener penetration.
- Sealant, roofing cement, and caulking are considered a secondary means of waterproofing and should not be used as a primary sealing method in lieu of flashing.

Table 8.5 provides examples and descriptions of step flashing installations encountered, both proper and improper, during on-site inspections.

TABLE 8.5
Application of Step Flashing

Description	Photograph
Nail fasteners not driven flush against surface creating penetrations for potential water entry	
Interlaced step flashing extending ~3-1/2″ up vertical side wall covered with vinyl siding as counter flashing	
Step flashing extending underneath shingles with nail fastener driven "low." Nails should be driven high and overlapped by the preceding piece	
Apron flashing along flat surface with use of no counter flashing. Apron flashing not secured and tucked into masonry wall	

(Continued)

TABLE 8.5 (*Continued*)
Application of Step Flashing

Description	Photograph
Step flashing along side masonry wall of chimney not tucked into mortar joint exposing large gap	

8.2.1.6.3   Cricket or Backer Flashing (Typically Applicable to Chimneys Only)

A "cricket" or saddle is utilized to prevent the buildup of ice and snow at the upslope side of a chimney and to divert water around the chimney. Modern building code states a cricket or saddle is required when the upslope width measurement of the chimney parallel to the ridgeline exceeds 30 inches in width.[3] However, industry best practices and experience have shown that crickets should be employed when any of the following criteria are met[1]:

- If the roof surface could expect a large volume of water runoff including the accumulation of ice and snow. In addition, a form of ice guard should be applied to the roof deck around the base of the chimney as well as up the walls in areas of severe climate.
- The building/roof surface is susceptible to the accumulation of tree debris/leaves. The buildup of debris behind a chimney could allow water to back up beneath the roofing material.
- When the width of the chimney exceeds 24 inches in width and the pitch to the roof surface is measured at a 6:12 or greater.

These are intended as precautionary details to prevent water entry at these types of interfaces.

Based on a review of several manufacturers, trade organizations, and best practice documents, the application of either a cricket or piece of continuous backer flashing should extend up the wall of the chimney at least 4 inches and should extend up the roof surface at least 18–24 inches. All joints of the cricket or saddle flashing should be soldered together and the edge should be hemmed. Unless a counter flashing system is employed, the cricket or backer flashing should be tucked inside a reglet or embedded in a mortar joint. Remember, any valleys used in conjunction with a cricket should not terminate, or direct water runoff, against the vertical wall of the chimney. Should this poor design feature be employed, the probability of water entry at this location is greatly increased. Table 8.6 illustrates actual situations encountered where water intrusion occurred at the upslope side of a chimney.

TABLE 8.6
Application of Cricket and Backer Flashing

Description	Photograph
Metal valleys along shingled cricket direct water runoff against corners of masonry chimney. For this particular inspection, water entry was found to occur at both of these corners	
Large gaps along interface with stone masonry and backer flashing. Since counter flashing is not employed, backer flashing should be tucked into chimney	
The upslope width measurement of the chimney parallel to ridge far exceeds 30″. Therefore, a cricket should be installed rather than backer flashing	

8.2.1.6.4 *Counter Flashing*

All of the various types of flashing discussed above are employed to protect the building envelope from water intrusion. However, an additional piece of flashing is typically required to cover and overlap the top edge of the step, apron, or backer/cricket flashing. Counter flashing can be in the form of a wall covering (or cladding material) or as a separate piece of metal flashing tucked into or behind the wall. Cladding material such as stucco, wood, metal, or vinyl siding can act as a form of counter flashing and should extend past and cover the flashing along the vertical wall a minimum of 2–4 inches. Several options and variations are associated with counter flashing, but the underlying concept is to prevent water from getting behind the lower flashing.

In the case of masonry chimneys/walls and therefore the absence of a wall covering, a piece of counter flashing is typically inserted into the masonry to protect the exposed top edge of the apron, step, cricket and/or backer flashing. Based on experience, counter flashing installed along a vertical masonry wall should be tucked into or behind the masonry. Simply applying sealant/caulking along the leading edge is not a permanent method for water prevention. Remember, sealant/caulking should only be used as a secondary means of waterproofing because sealant/caulking will degrade over time, thus requiring a reapplication to provide adequate protection. This is often not done by owners out of lack of knowledge or other reasons leading to later water entry issues. Figure 8.15 illustrates an industry best practice method (NRCA) for installation of flashing at a side wall.[1]

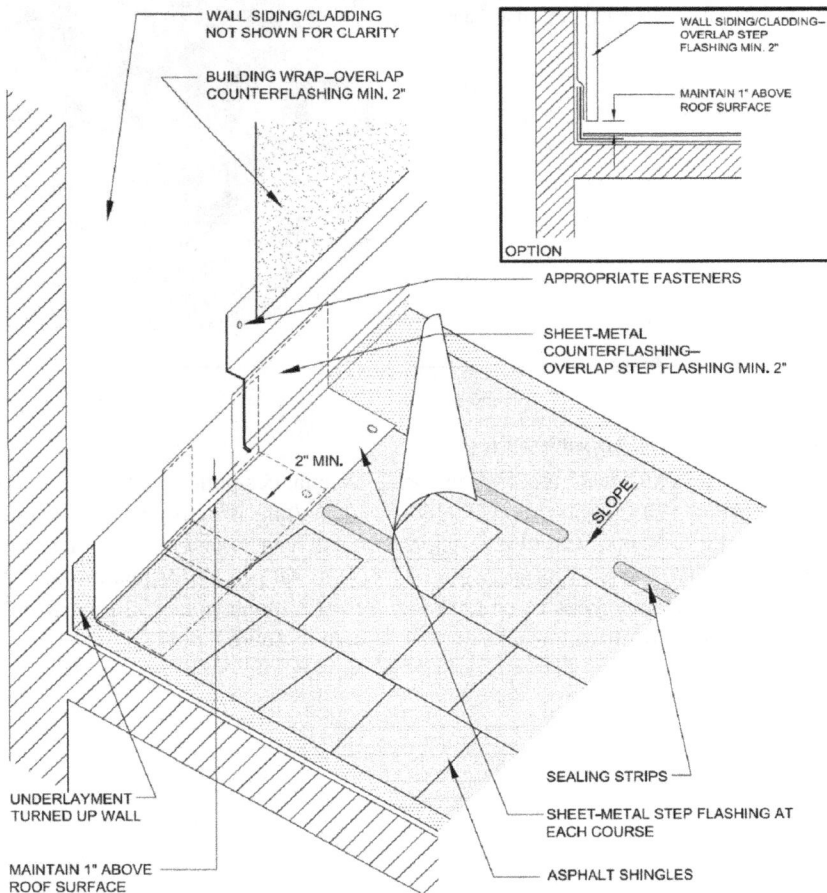

FIGURE 8.15 Sidewall flashing. (Reprinted with Permission of the National Roofing Contractors Association.)

Table 8.7 illustrates common deficiencies with counter flashing and masonry walls.

TABLE 8.7
Deficiencies in Counter Flashing

Description	Photograph
Heavy deterioration of sealant along edge of counter flashing resulting in large gaps for water entry	
Continuous flashing along side wall of chimney. However, leading edge of flashing only protected by sealant and not secured into the masonry	

8.2.1.7   Roof Penetrations/Appurtenances

On most steep-sloped roof systems, penetrations or appurtenances penetrate through the surface of a watertight roof system. These penetrations include (1) vents/fans providing exhaust ventilation for the building; (2) piping or vents needed to dissipate excess heat, moisture, or other contaminants from the attic space; (3) soil stack vent piping; and/or (4) metal piping for heating appliances (i.e., flue gases). The most common types of penetrations or roof appurtenances range from static and power vents to plumbing and furnace stacks and are shown in Table 8.8.

TABLE 8.8

Examples of Various Roof Penetrations

Description	Photograph
Slant-back metal box vent	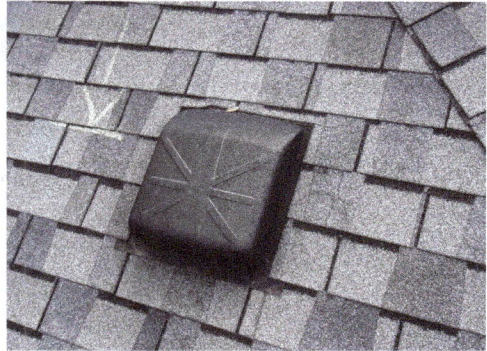
Soil stack pipe with rubber boot flange	
Turbine exhaust vent	
Power vent with plastic cover	

These locations are potential sources for water intrusion and must be properly sealed to prevent water from entering a structure. For adequate waterproofing around these roof penetrations, a flat metal or rubber-like material typically is installed underneath the steep-sloped roofing materials on the upslope side of the flange. The flange of the penetration also extends onto the top of the roofing materials down slope of the penetration. Some penetrations (i.e., furnace vents and plumbing stacks) use a collar or gasket attached or sealed around the intersection of the flange and penetration to provide sufficient waterproofing capabilities. As discussed throughout this chapter, the application of roofing cement and/or sealant underneath the flange and surrounding roofing materials is a secondary method of sealing and waterproofing and not intended for the primary means of water protection such as that provided by the flange or gasket.

The balance of this section will provide an overview of industry best practices for sealing roof penetrations around appurtenances and will also discuss a few of the most common discrepancies and deficiencies associated with these roofing penetrations that can lead to water intrusion into the structure.

8.2.1.7.1 Pipe Stacks

In the construction of newer buildings, vent stacks utilize an elastomeric, or rubber-based, gasket and/or boot flange at the base of the penetration or pipe stack. Unfortunately, these materials can undergo degradation caused from environmental factors such as heat and UV radiation. From experience, the first signs of cracking and splitting appear around the 8–12-year range. This aging and deterioration of the boot has the potential to allow water intrusion along this interface. In addition, furnace and plumbing vents on older homes oftentimes only contain a metal boot flange and/or collar near the base of the stack near the interface of the roof. During construction, these intersections or joints on the stacks can be either soldered or welded together or sealed with caulking or sealant. As with most items exposed to the elements, the materials along these intersections can weather, age, and weaken over time allowing for possible gaps at these interfaces. Table 8.9 illustrates examples of deteriorated sealing materials around plumbing and furnace stacks.

TABLE 8.9
Plumbing and Exhaust Vent Deficiencies

Description	Photograph
Cracked and degraded rubber gasket around boot flange leading to potential water intrusion	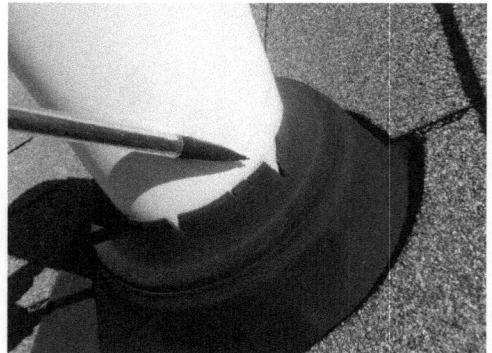

(Continued)

TABLE 8.9 (*Continued*)
Plumbing and Exhaust Vent Deficiencies

Description	Photograph
Cracked and deterioration to rubber gasket around older soil stack vent	
Large visible crack in roofing cement along intersection of furnace vent stack and roof surface	
Degraded caulking around intersection of rain collar of furnace vent stack	

Preventative roof maintenance should be conducted periodically by the property owner or their representatives to determine the conditions of these flanges, boots, and gaskets at roof and appurtenance interfaces. This inspection should include all caulking and sealant seams for weathering and degraded conditions. Degraded roof appurtenance seals should be replaced to ensure the integrity of the roof system against water intrusion.

8.2.1.7.2 Static Vents

As with all roof penetrations, proper installation of static vents (e.g., box, power, turbine, etc.), allows for water runoff to be diverted around the object and continue downward off the roof surface without entering the structure.

The bottom metal flange of a roof appurtenance must be installed underneath the roofing materials on the upslope side and on the top side the shingles on the down slope side of the vent. Table 8.9 illustrates the bottom flange of a vent stack properly overlapping the roofing shingles. However, if the bottom end of the flange is improperly installed underneath the roofing material, then some of the water runoff will travel beneath the shingles and enter into the structure. Table 8.10 shows the improper installation of box vents causing the roof sheathing to be water damaged by the resulting water intrusion.

TABLE 8.10
Improperly Installed Box Vents

Description	Photograph
Overview of improperly installed box vent; notice area of soft decking likely due to water intrusion. Bottom flange tucked underneath the shingles on down slope side of the vent directly water beneath shingles	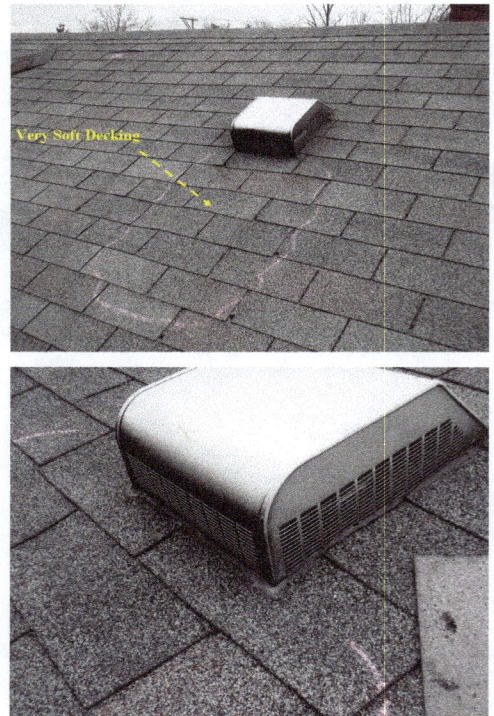

(Continued)

TABLE 8.10 (*Continued*)
Improperly Installed Box Vents

Description	Photograph
Plastic square box, or static, vent with bottom plastic flange sealed beneath bottom shingles. However, water runoff traveling down can penetrate degraded sealant and enter the building	

8.2.1.7.3 Skylights

Skylights are popular for increasing the natural lighting in room as well as the aesthetics and curb appeal of any building or residence. However, they are often not properly installed, leading to water intrusion. Essentially the same concept and general flashing guidelines for chimney flashing are employed for flashing a skylight. The only difference is that some skylights arrive with preassembled flashing kits that are to be assembled and installed according to the manufacturer's installation instructions. Proper installation best practices, based on information gained from field assessments and industry best practice documents for this type of flashing, are as follows:

- The apron flashing should extend up the vertical wall of the skylight and underneath the roofing material. In addition, the lower edge of the apron should contain a hemmed edge.
- The step flashing should extend underneath the intersecting steep roofing materials ~4–5 inches.
- The backer flashing should extend upslope a minimum of 18 inches underneath the roofing surface.
- The integrated skylight frame acts like counter flashing and should lap the step and likely apron flashing by ~2 inches.

If a skylight is found to be leaking, the reinstallation and/or re-flashing is probably necessary. The use of roofing tar or cement is a temporary fix (i.e., a secondary method of sealing and waterproofing) and should not be viewed as a permanent solution to the leak.

8.2.2 Low-Slope Applications

Chapter 7 discussed in detail the principles of high wind forces, their effects on various low-sloped roof systems, and the means to ensure wind resistance. In a similar manner, this portion of the chapter will discuss general means for water resistance for low-sloped systems.

The underlying principle for the prevention of water entry into low-sloped roof systems follows from recommendations and requirements set forth by trade organizations, manufacturers, and local building codes. When these recommendations are not followed, either due to improper design, lack of knowledge, lack of care and maintenance, or poor installation/workmanship, the likelihood of future and potential issues related to water infiltration increase.

The two most common issues found with water entry to low-sloped roof systems are associated with improper roof drainage and improper flashing details. Both topics are discussed further below.

8.2.2.1 Roof Drainage

In addition to the quality of materials and the manner of application, the durability and longevity of a roof system are also dependent on the resistance to weather-related issues such as rainfall and snow accumulation. Ponding water, for example, present for long periods of time on a low-sloped roof system, can prematurely degrade and deteriorate the roofing surface material and ultimately lead to water infiltration into the roof system and structure below. Too much ponding water on a low-sloped roof can even lead to the collapse of the roof system.

Experience has shown that many water and moisture intrusion problems related to roof drainage are related to two issues: (1) the ability of the roof surface to drain and (2) the design of the actual system intended to divert and drain away water, melting snow, and/or ice melt accumulation. Details regarding the proper slope and drainage system design are discussed below.

8.2.2.1.1 Slope

In order to prevent water intrusion into the structure, the roof surface must be quickly drained of any accumulated water. A well-drained roof system, therefore, must contain a properly sloped system. For a low-sloped roof surface (slope <4:12), modern building codes and best practices require the roof surface to contain a minimum slope as follows:

1) Metal roof surface ranging from 1/4:12 to 3:12 depending on metal roof used.
2) Mineral roof surfacing – 1:12.
3) Clay and concrete tile roof surfacing – 2-1/4:12.
4) Mineral surface roll roofing – 1:12.
5) Wood shingles – 3:12; wood shakes – 4:12.
6) BUR – ranges from 1:12 to 2:12 depending on type.
7) Thermoset single ply membrane – 1/4:12.
8) Sprayed polyurethane foam – 1/4:12.
9) Liquid applied coatings – 1/4:12.

Thus, depending on the low-sloped roof surfacing material encountered, one must be familiar with minimum slope requirements for that material.

As the buildings encountered get larger, increasingly, roof surfaces such as membrane roof surfaces with a minimum slope requirement of 1/4:12 or an ~2% slope will be encountered.[3] This 2% allows for some tolerance for many of the imperfections inherent in the building. However, buildings can experience significant deflection, which can negate this 2% design slope. Therefore, careful considerations must be made to ensure positive drainage conditions for all loading deflections.

Positive drainage is defined by the absence of ponding water on the roof surface within 48 hours of a rain event. Positive slope can be achieved by structural means or tapered insulation as well as the use of localized crickets and roof saddles.[11] The reasons for why positive drainage is important are outlined in *The Manual of Low-Sloped Roof Systems* as follows[12]:

- Periodically, structural roof collapses are caused by ponded water following heavy rains. This is typically the result of the progressive increase of ponding water exceeding the structural capacity of the roof deck due to increased deflection spans.
- Ponding water has the ability to infiltrate through the membrane by any imperfections from natural aging and weathering such as cracks and splits to unsealed lap seams. These sources of water entry can be a result of typical weathering defects or by poor workmanship issues. In areas of colder temperatures, these issues become more of a concern, because during colder months, the formation of ice can cause delamination and further damage to the membrane as the freezing water expands and contracts during freeze/thaw cycles.
- Stagnant water can promote the growth of vegetation, algae, or other biological organisms. If this growth is left untreated, the membrane can become damaged.
- The consistent exposure of water can accelerate premature deterioration and degradation to the membrane material, which can cause shrinkage, for example.

Asphalt Roofing Manufacture's Association (ARMA) also addressed in detail the negative effects of ponding water on low-sloped roof surfaces.[13] A clear and evident indication of long-term standing water on a low-sloped roof surface is the buildup of sediment, deposits, and even vegetation growth. These types of observations are illustrated in Table 8.11.

TABLE 8.11
Areas of Ponding Water

Description	Photograph
Standing water with heavy sediments and tree debris	
Large widespread ponding water throughout roof surface	

(Continued)

TABLE 8.11 (*Continued*)
Areas of Ponding Water

Description	Photograph

Ponding water with heavy sediment suggesting negative drainage

Ponding water around membrane penetrations

A roof collapse due to excessive ponding water. The ponding water was the result of clogged primary roof drains and absent emergency (or secondary) drains

8.2.2.1.2 Drainage System

In addition to meeting the minimum slope requirements, the proper sizing and the correct number of roof drains must be included in the roof system design to ensure the prompt removal of water from the surface of the system being drained. The two types of drainage systems typically encountered are external drainage systems and internal drainage systems.[11] An external drainage system consists of either: (1) a scupper penetration through a parapet wall discharging to a conductor head or (2) a simple gutter and downspout system. An internal drainage system describes a system where the roof drains are located within the field of the roof surface. Internal drains are connected to a plumbing system (i.e., leaders) beneath the roof deck that carries roof waters down through the interior of the building. Either system requires the drains to be located at the lowest points of the roof surface (since water flows downhill). Best practices state the locations of the interior drains should be installed away from load-bearing walls or columns that provide less deflection of the roof decking.[12]

Minimum low-sloped roof drain design requirements are addressed in detail in section 1502 of the 2018 International Business Code (IBC)[3] and Chapter 11 of the 2018 International Plumbing Code (IPC).[14] Key points from these documents for proper minimum drainage design requirements follow:

- Size requirements for drains, gutters, leaders, conductors, and secondary drains should be calculated from the 100-year, 1-hour rainfall (inches), and the maximum projected roof surface area.
- Industry best practices also recommend that all roof drains should have strainers to keep the drain area from clogging.
- Secondary emergency drainage is required where the roof perimeter construction is extended above the roof in such a manner that water will be entrapped should the primary drain be backed up.
- In the situation of scuppers as the emergency drain, the scupper must be sized properly to prevent ponding water from exceeding the design rain load of the roof and must not have an opening <4 inches (102 mm).

Secondary or overflow drains are intended to provide additional drainage to serve as emergency drains should the primary drains be blocked or clogged. Code requires secondary drains for roof systems employing interior drains as the primary drainage system. Secondary drains can consist of through-wall scuppers or additional slightly raised internal drains. Regardless of the method used, the secondary drains should be designed so that personnel can easily observe them to ensure the drains are accessible should blockage occur to the primary drains. Specific methods to properly design these roof drainage systems can be found in the references listed at the end of this chapter.

Scuppers and drains: The individual components of the drainage system such as scuppers and interior drains must not only be installed to ensure proper drainage but also to provide a watertight seal between these components and the roof surface to avoid water entry into the roof system and the structure below.

While there are numerous installation details for each type of roof drain and roofing material, the same general principle typically applies to ensure a watertight seal. The flashing should be continuous and absent of seams through or near the mouth or opening of the drain. This is typically accomplished by installing components in a "shingled" fashion based on the direction of the water flow. The application of appropriate sealants around the interface of either the wall or roof surface and drain flange is only intended for additional protection (i.e., secondary seal). An important point often neglected is positive drainage at the intersection of the scupper and membrane. The thickness of seams in junction with flashing components can create a small lip or hump, which can prevent positive drainage at the lower end of the scupper opening. The use of tapered insulation or roof saddles in these areas should be considered to provide additional slope at these locations. Figure 8.16 provides a cross section of a typical roof drain.

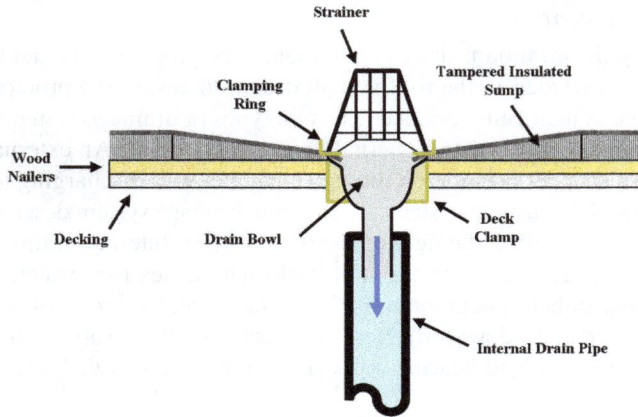

FIGURE 8.16   Side profile of an interior roof drain.

Photographs of scuppers, interior drains, and conductor pipes obtained during on-site field inspections where water intrusion was reported are displayed in Table 8.12.

TABLE 8.12
Field Inspection Examples of Drainage Systems

Description	Photograph
Through-wall scupper with heavy rusting conditions and a large rust hole in the bottom on the unit. Evidence of sediment and debris around the mouth of the opening coupled with rusting conditions suggests the presence of long-term ponding water	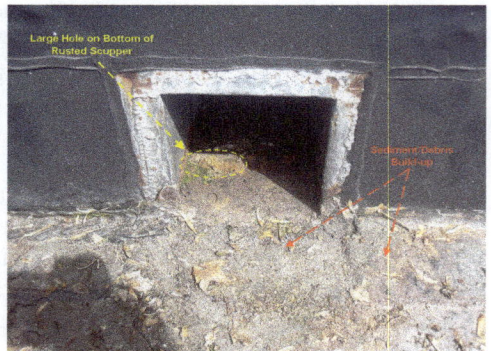
Heavy buildup of debris and ponding water inside cavity of through-wall scupper suggesting inadequate slope and positive drainage around unit	

(Continued)

TABLE 8.12 (*Continued*)
Field Inspection Examples of Drainage Systems

Description	Photograph
Ponding water sediment and dirt outlining elevated roof cricket, which appears to provide positive drainage toward interior roof drains	
Roof drain partially clogged and blocked due to accumulation of vegetation, sediment, and debris. This buildup can restrict the flow and drainage of the roof surface creating potential water entry issues	
Heavy rusting and repairs to interior surface of roof drain along with drip stains on horizontal drainage pipe suggesting long-term water entry issues with roof drain	
Rusting conditions to elbow of vertical conductor and horizontal pipe of interior drainage system	

Gutters/conductors and downspouts: When through-wall or open scuppers are employed on a building, the water is typically diverted and collected by a conductor head and downspout system that moves the collected water away from the building. However, built-in or externally attached gutters must be leak-free, able to support the weight of ice and snow accumulation, and be constructed from durable materials that resist long-term weather conditions and internal stresses.

Gutter systems utilized along the perimeter edge of a low-sloped roof surface must contain similar characteristics to those discussed in the section above for steep-sloped systems. Gutter design considerations should include the following design features[15]:

- Gutters must be designed and sized correctly to properly drain the roof surface and be able to withstand various environmental effects and conditions.
- Gutters must be watertight (welded, riveted, and soldered, or lapped with sealant then riveted).
- The perimeter edge metal should overlap the inside edge of the gutter.
- A periodic maintenance schedule should be employed to ensure a clear and efficient drainage system.

Further information regarding the drainage system can be found from the NRCA, *The Manual of Low-Sloped Roof Systems*, SMACNA, and from the RCI.

8.2.2.2 Flashing

Flashing seals the joints and seams along various junctures where the low-sloped material is interrupted. These interruptions range from curbs, HVAC units, pipe penetrations, and parapet or other vertical walls, to name a few. There are numerous, if not hundreds, of installation details on how to properly flash the various components of a low-sloped roof system given a specific roofing material. Most manufacturers and trade associations such as the NRCA and ARMA produce specific details of flashing techniques for a variety of interruption situations often encountered. The base and counter flashings are utilized throughout these roof system details to provide water protection into the building.

The most common areas of concern regarding water entry consist of horizontal and vertical roof surface intersections. Horizontal intersections entail perimeter roof edges, interior drains, pipe stacks, and HVAC units, whereas vertical terminations consist of parapet, curbs, or any type of vertical wall. To address these intersections, best practices for flashing details provided by the NRCA are summarized below[12,15,16]:

- Base and counter flashings must be anchored firmly to supports.
- The flashed joints must be located above the highest water level and ensure positive drainage to divert water away from these joints.
- Avoid sharp bends by creating contoured surfaces with the use of cant strips. Bituminous materials should be installed at angles <45° to circumvent damage.
- Allocate for the expansion and contraction (differential) movement of materials. Components along certain junctures can wrinkle, split, and/or delaminate causing large gaps for possible water intrusion. Accommodations to such excessive movement should be considered.

These principles should be considered during the initial design of the roof system to minimize the potential for future water entry into a given structure. Figure 8.17 illustrates the kind of flashing detail available for designers, in this case for roof flashing of a masonry parapet wall.

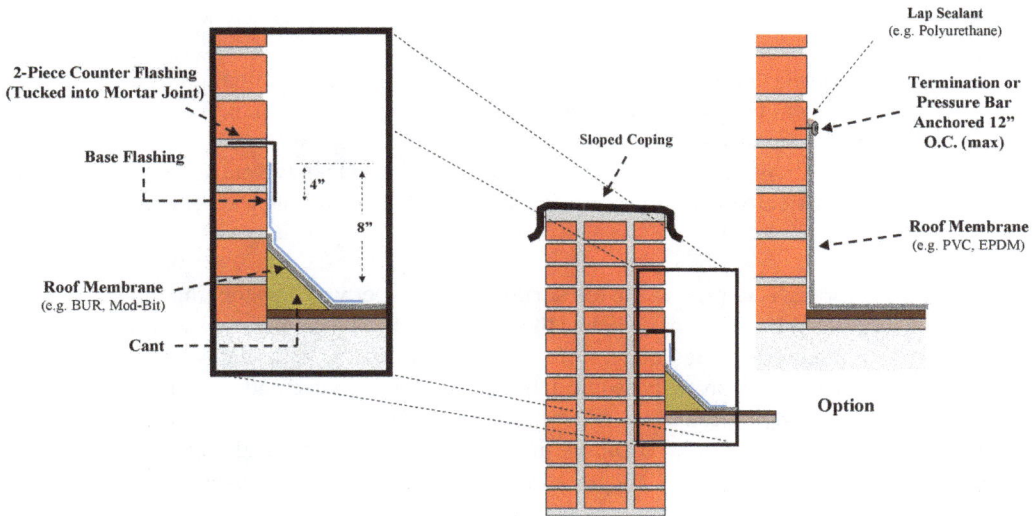

FIGURE 8.17 Flashing detail at masonry parapet wall.

8.2.2.2.1 Base Flashing

The function and purpose of base flashing are very similar to step flashing on steep-sloped applications. The base flashing material should be impermeable and nonporous to water entry, flexible and durable enough to withstand differential movement and varying weather conditions, firmly attached, and compatible to the roofing material itself. Base flashing materials are lapped vertically up the side of a component, curb, or vertical wall. Unlike steep-sloped roof situations where separate pieces of flashing are used, the base flashing is essentially a continuation of the same low-sloped roofing material lapped or turned up the side wall of the component. Best practice guidelines suggest that the base flashing should be applied up a vertical wall or curb a minimum of 8 inches above the roof surface.[12,15,16] These nonmetallic flashing materials are typically installed after the application of the roofing system and can be either mechanically fastened (i.e., termination bars or anchor fasteners) or fully adhered.

From experience, most water intrusion issues associated with failures of the base flashing are the result of heavy deterioration of the flashing material taking the form of cracking, splitting, and tearing along the intersection of the roof surface and vertical wall. Periodic inspections of the roofing system and occasional maintenance should be made to either replace or repair damaged or deteriorated flashing and to limit water entry to the structure at these locations.

8.2.2.2.2 Counter Flashing

Similar to the details discussed above for various steep-sloped applications, counter flashings are typically formed from pieces of sheet metal installed to cover the leading edge of the base membrane flashing. The sole function of the counter flashing is to protect and shield the exposed joint from water passing over the top edge of the flashing and then entering underneath the roof surface at these junctions.

There are numerous styles of counter flashing depending on the application, roof material, and ease of construction; the reader is referred to the NRCA[15,16] for best practice details in this area. Interestingly, the NRCA notes that a single piece of counter flashing may create difficulties for potential reroofing efforts causing unwanted and excessive repairs and recommends avoiding such counter flashing. A single counter flashing system also has limited movement and can be damaged due to expansion and contraction during temperature changes. A two-piece metal counter flashing system with a receiver or reglet is recommended, noting that it can provide several advantages from additional water protection as well as easing future issues associated with maintenance or reroofing.[12]

Finally, the counter flashing should be installed slightly above the termination of the base flashing to ensure independent movement of the counter flashing. However, it should overlap the leading edge of the flashing by at least 4 inches.

Some small penetrations such as those associated with pipe stacks typically use metal rain collars or pipe boots, which are designed to provide a watertight seal around the penetration.

8.2.2.2.3 Coping

The purpose of a coping is to protect the top surface of a parapet wall by stopping moisture or rain water from entering into the wall cavity below the parapet wall. The coping, typically metal, should: (1) be angled to avert ponding water on its surface, (2) contain an adequate and watertight seal along all joints or seams, and (3) be properly secured. If not adequately joined and sloped, the intersection of individual pieces of coping can allow water to penetrate the lap joint and enter the wall cavity.

The two most common options for joining the edges of sections of metal coping are: (1) soldering/welding the pieces together or (2) the use of mechanical fasteners (i.e., rivets, screws, and bolts) in conjunction with waterproofing materials such as sealant, solder, or gaskets. Similar to a two-piece counter flashing system, considerations must be made to account for differential movement of individual panels of coping such as the use of cleats and clips.[12,15,16] Oftentimes, a secondary waterproof membrane, lapping down both sides of the parapet wall, is utilized for additional water proofing protection of areas where coping is used. Examples of situations where failures of the coping system have occurred are illustrated in Table 8.13.

TABLE 8.13

Field Inspection Examples of Coping Failures Contributing to Water Entry

Description	Photograph
Large gaps between joint of clay tile copping allowing for potential water entry	
Heavy tenting or pullout of base flashing along parapet wall. Termination bar likely not fully anchored to wall	

(*Continued*)

TABLE 8.13 (*Continued*)

Field Inspection Examples of Coping Failures Contributing to Water Entry

Description	Photograph
Large gaps along base flashing lapped up masonry wall absent of either a termination bar or counter flashing	
Heavy degraded caulking/sealant along joint of metal coping. No means of mechanical attachment	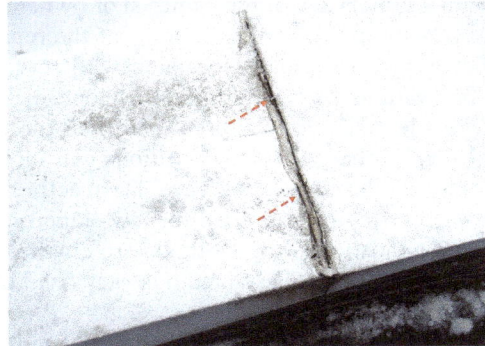

8.2.2.3 Care and Maintenance to Low-Sloped Roofing Systems

An important factor in water prevention is proper care and maintenance associated with all types of roof systems. A reduction in the effectiveness of the surfacing material will ultimately increase over time for all types of low- and steep-sloped roof systems leading to the potential for water entry into the structure. Weathering and age-related defects on low-sloped materials can include the loss of plasticizers causing splits, cracks, and seam separation of the roof surface. The service life span of most common low-sloped materials can vary and is dependent on climatic conditions (i.e., UV radiation), quality of the material (i.e., thickness and material characteristics), extent to which the installation matched best practices, and to the extent it is maintained. Without a periodic maintenance plan to repair and/or replace damaged roofing system components, water intrusion is possible if not expected. For instance, best practices recommend that an annual inspection be performed of the roof system, including assessing the condition of roofing cement/sealant/caulking around the building. This may include removing the old and degraded sealant and reapplying it where needed. However, repairs may be more extensive such as the need to replace damaged areas of a roof surface with small patches of replacement materials. *The Repair Manual for Low-Slope Roof Systems*[17] published jointly by ARMA, NRCA, and Single-Ply Roofing Industry (SPRI) provides excellent guidance for the identification and specific repair procedures for low-sloped membrane materials. As the manual notes, "The primary purpose of maintenance and repairs for roof systems is to extend the roof's service life so as to prolong and enhance the original investment made in the roof system." However, they caution that some of these minor and small repair methods should not be intended or expected to be permanent solutions. Continuous use of temporary roof repairs, while common, increases the risk of premature loss of the roof system life and collateral damage such as water entry degrading the structure below. Simply applying several "band-aids" may provide a temporary fix, but they do not provide any long-term solutions. At some point, the property owner needs to realize that permanent repairs are the best and most viable option.

8.3 METHODOLOGY FOR WATER CAUSE AND ORIGIN INSPECTIONS

When an insurance claim is made regarding interior damages resulting from water intrusion, a detailed inspection is conducted to determine the cause and source of the water entry into the subject property. Water inspections run the gauntlet from being very simple to being very complex; however, the inspector will often not know the complexity of the situation until arriving at the site. Therefore, a systematic inspection methodology is critical to efficiently determining the cause(s) of the reported roof water leak(s).

Similar to the baseline inspection methodology outlined in Chapter 1, the inspection begins with an interview of the owner(s) or owner's representative for background information regarding the property and the history of the reported damages associated with the water intrusion. For suspected issues associated with roof systems, the process then continues with detailed observations of the building envelope including the attic space (if applicable) and roof system, identifying the source, and then the causation of the intrusion. The process or method of performing visual observations can include both nondestructive and destructive testing to ascertain the cause(s) of leaks. Destructive test methods and repairs, such as those identified in Chapter 4, will be needed for low-sloped roof water cause and origin inspections. Frequently, the diagnostic testing methods are employed to aid in isolating and identifying the source of water intrusion. These methods vary from the use of moisture meters and/or infrared (thermal) cameras to water testing.

The following sections will provide details for completing on-site water cause and origin inspections associated with water entry into a building envelope associated with roofing systems.

8.3.1 Interview with the Property Owner(s) and/or Owner's Representative

The first step in the process of identifying water intrusion is to gain insight about the home/building from the owner or owner's representative. The information can be helpful when used in context with the observations documented during the inspection. The following list of specific water cause and origin questions, asked to the property owner or their representative, will assist the inspector in determining the cause and origin of the water entry along with areas that may need to receive prioritized attention during the inspection.

8.3.1.1 Home/Building Information

- What is the approximate square footage of the home/building?
- When was the home/building constructed and how long has the current property owner owned the home/building?
- Have there been any modifications to the exterior of the home/building (i.e., new roof, siding, windows, etc.)?
- What type of heating and cooling system is employed in the subject property?

8.3.1.2 Roof Information

- What type of roof system is in place?
- What is the approximate age of the roof system?
- Have any recent repairs or modifications been made to the roof system?

8.3.1.3 Water Intrusion History

- Where, in the subject property, has water damage been observed?
- When was the leak first noticed?
- Is the water leak currently active?
- Does the leak or damage propagate during particular times (i.e., heavy rains, snow, and ice)?
- What areas/surfaces have been observed to be water-damaged?
- Have any repairs been made to the damaged area or to the suspected source of the water leak? If so, when did they occur and to what extent?

8.3.2 INTERIOR INSPECTION

8.3.2.1 Plan View Sketch and Measurements of Interior

After pertinent information regarding the home/building has been obtained, continue with the assessment by sketching a schematic of the damaged area by floor/level. Depending on the extent of this area, it may only require a sketch of a single room, a specific corner of the building, or an entire floor. The inspector will have the discretion on which areas of the building to sketch, but for time management and cost efficiency reasons, the inspection of floors free of water damage is typically not performed unless later observations suggest a need for it to be done.

As noted in Chapter 1, the inspection typically begins on the lowest floor. Since water flows down-hill, by starting at the lowest level first, a pattern and history of where the water source may be coming from begin to develop as the inspection moves higher and higher elevations within the structure.

For each elevation inspected, a floor plan is sketched out to scale as closely as possible, and then measurements are made of each space on the elevation to be inspected. Throughout the inspection, specific areas of damage and other important observations should be identified within the sketch. An example of an interior floor plan sketch for a residential home, coupled with areas of observed water damage staining and the location of diagnostic tools used (i.e., moisture meter), is shown in Figure 8.18.

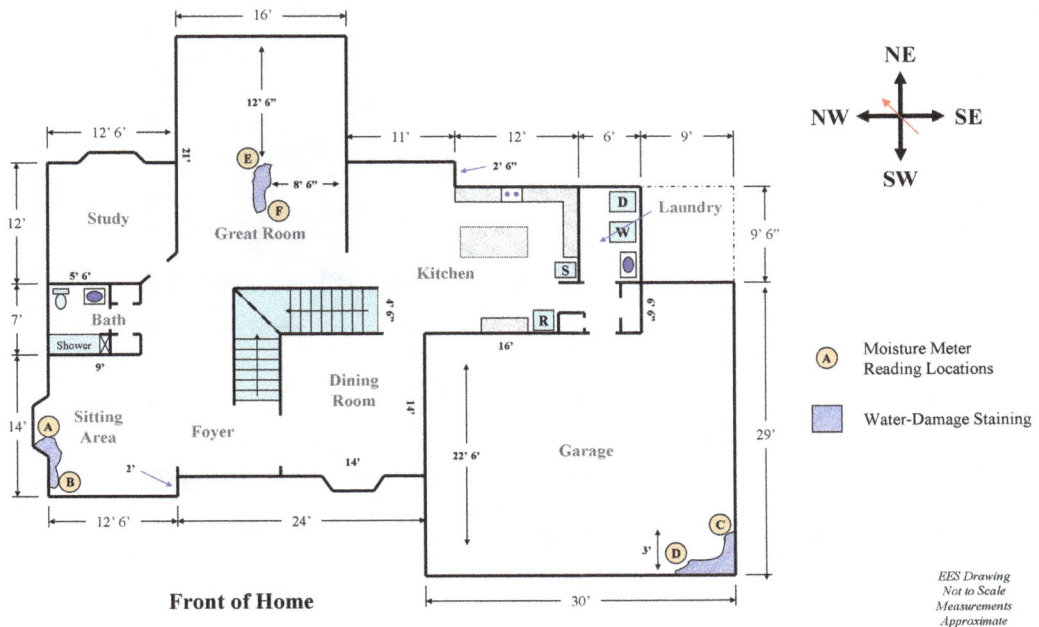

FIGURE 8.18 Computer-aided plan view of interior floor with key observations.

8.3.2.2 Interior Observations (by Floor and Room)

During this portion of the inspection, general observations of each room are made, including noting the finish of the visible surfaces (e.g., carpeting, laminate flooring, wallpaper, painted/textured drywall, etc.). Areas of water damage staining, possible visible mold, and active leaks are noted in writing in a field notebook, photographed, measured, and delineated with respect to the floor plan or sketch. This approach not only aids in pinpointing the likely source of the leak, but it also provides the client with the dimensions of the damaged area, which could be needed for estimating costs of repairs. Example photographs of interior water damage staining obtained during on-site inspections are shown in Table 8.14.

TABLE 8.14
Examples of Interior Water Staining

Description	Photograph
Several circular water damage stains on ceiling surface	
Water damage staining to acoustic ceiling tiles	

8.3.2.3 Attic Space Observations

A general idea of the location of the water entry should have been gained by inspecting the interior floors of a given structure. Inspecting the attic space (if applicable) is the next step to connecting interior damage to the roof system. Some low-sloped roof systems do not contain an attic space, but typically contain a plenum space that can be inspected, which will connect interior water damage to specific roof system leaks.

Information recorded should include the construction of the attic (i.e., trusses vs deck boards), the presence (and extent) of insulation/vapor barrier, the forms of ventilation, any appliances venting into the attic (e.g., a bathroom vent exhausting into the space), and evidence of water damage staining and possible visible mold patterns. For example, the presence of water damage staining along a valley beam is likely indicative of a water leak in the proximity of the valley, whereas staining patterns around a roof penetration (i.e., furnace vent, soil stack, or chimney) suggest that location as the likely source of water entry.

Remember, water follows with the force gravity and flows from higher elevations to lower elevations so the drip and staining patterns should always be followed back to their source(s). A sketch of the attic should be drawn to illustrate key dimensions and findings. For example, Figure 8.19 illustrates a situation where water intrusion was observed around a furnace vent pipe.

FIGURE 8.19   Evidence of water intrusion around furnace pipe.

Assessment of water-stained surface areas usually provides clear identification of the location of the water leak with regard to the roof system. Examples of water damage staining in an attic space are provided in Table 8.15.

TABLE 8.15

Examples of Water Staining Evidence in an Attic Space

Description	Photograph
Drip patterned staining originating from a valley	
Drip patterned staining on face of side wall originating from roof/wall intersection	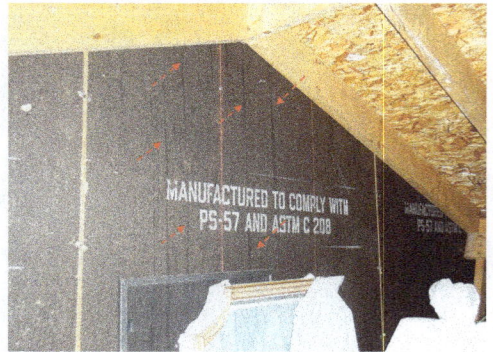
Drip stains down furnace vent pipe originating from metal collar on roof surface	
Heavy historic and active staining patterns along face of brick chimney	

8.3.3   EXTERIOR AND ROOF SYSTEM INSPECTION

Once the interior and attic portions of the inspection are completed, the exterior elevation and roof inspections should be completed.

8.3.3.1   Exterior Walk-Around

The exterior walk-around entails an inspection of each exterior surface for observations that might assist in helping to determine the cause and origin of the roof leak. All water damage observations should be recorded in the field notebook and photographed. Examples of conditions that should be documented follow:

- Rotting conditions on the exterior surfaces (i.e., wood siding/fascias)
- Water staining on exterior surfaces
- Overflow and drip stain patterning on gutters or the ground surface below
- Staining on the upper portion of the exterior wall cladding near intersection of the undereave soffit
- Major defects or inconsistencies (i.e., missing sections of siding, tree impact damage, tarps, etc.)

Water intrusion to exterior surfaces is further discussed in detail in Chapter 9.

8.3.3.2   Roof System Assessment

Next, the roof system is inspected for evidence of areas where water infiltration may be occurring. This portion of the inspection should also be initiated by drawing a schematic of the roof with measurements and key observations. An example of a roof schematic is shown in Figure 8.20.

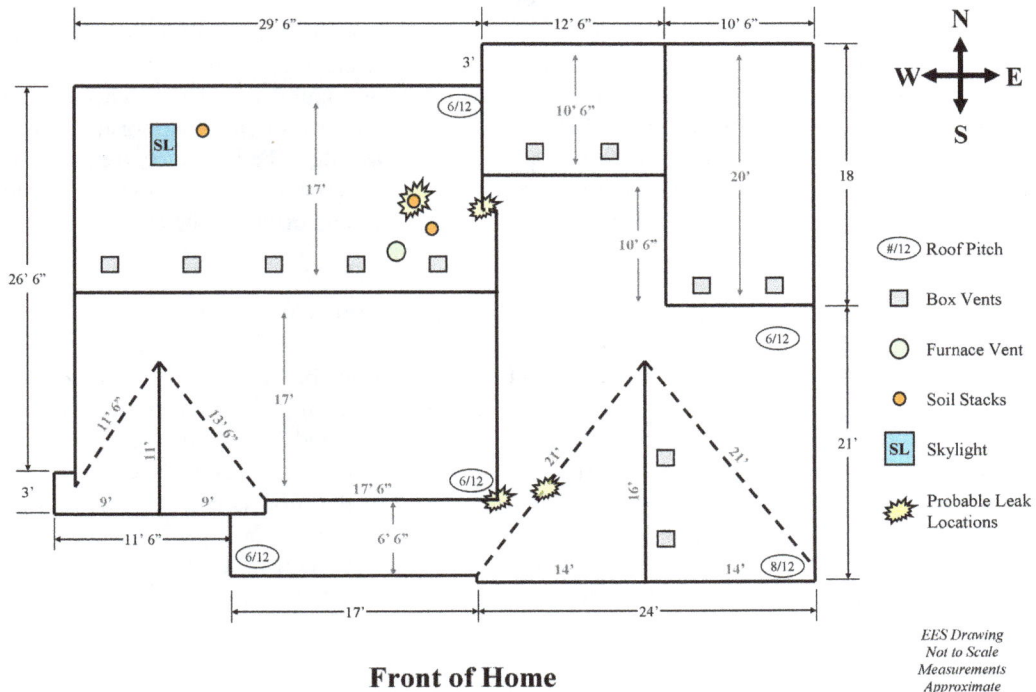

Front of Home

FIGURE 8.20   Roof schematic drawing.

Since the interior and attic inspections have been completed, the inspector should have a relatively good idea regarding the location(s) of the source(s) of the roof water leak(s) based on the interior observations and measurements (i.e., the collected evidence of water entry should lead the inspector to the source of the leak). Observations should be documented in writing in the field notebook and by photographs. The assessment of the roof system should include, but not be limited to, documentation of the following:

- Roof construction (i.e., asphalt, tile, slate, membrane, BUR, EPDM, Mod-Bit, etc.).
- Assessment of the overall condition of the roof surfaces.
- Location of the water leak(s) based on interior/attic observations and measurements. A few common leak location areas associated with steep- and low-sloped roof systems are listed below.
    - Steep-Sloped Roof Systems:
        - Eave and Rake Details
        - Underlayment and Ice Damming Issues
        - Valleys (i.e., Open, Closed-Cut, Closed-Woven)
        - Vertical Walls (Including Chimneys)
        - Roof Penetrations (Plumbing Stacks, Skylights, Box Vents)
    - Low-Sloped Roof Systems:
        - Perimeter Attachment of Membrane
        - Roof Penetrations (i.e., Pipe Stacks, HVAC units, etc.)
        - Membrane Attachment at Wall Terminations
        - Roof Drainage
        - Membrane Seam Laps
- Detailed description of the likely leak location(s) and cause(s) for leak(s) at that location.

8.3.3.3 Diagnostic Tools and Testing Methods

In some cases, the use of diagnostic tools and field tests can aid the inspector in correlating the path of water intrusion with the extent of interior damage. Some tools can directly measure the moisture content of a material (i.e., high, normal, low), whereas other tools can detect the presence of temperature differences in building materials, which may be an indicator of elevated moisture levels. While these tools can provide a "road map" or pattern that may allow the inspector to retrace the path of the water intrusion back to the source, actually simulating water entry conditions (i.e., water testing) is always the recommended approach to verify the cause and origin of the leak.

8.3.3.3.1 Moisture Meters

Moisture Meters are simple and easy to use devices that can determine the moisture content of most common building materials.

Modern moisture meters provide the user with a list of various building components (i.e., the moisture content in various species of wood, sheetrock, etc.) in which moisture levels can be measured. Two commonly used moisture meters are dielectric-based and conductance-based meters. Dielectric moisture meters are a nondestructive, noninvasive type of meter that sends out and receives back an alternating electrical field. The impedance of the field as it passes through the material in contact with the meter is related to the moisture in the material adjacent to the meter. The receiver in the meter senses the reduction in the relative strength of an electrical field and correlates it with moisture levels in the material. On the other hand, a conductance, or sometimes referred to as resistance-type meter, measures moisture by determining the relative conductivity of the media between the two metal probes. These electrodes, or pins, are physically inserted into the material in order to measure the moisture content.

Both meters have their pros and cons. The impedance meter is affected by changes in density of materials in the wall (e.g., areas near studs read higher), and the conductance meter can be fooled by foil-backed wallpaper. Neither meter can locate an intermediate leak location that dries out (e.g., from a rain event). Nevertheless, moisture meters are often helpful in locating/verifying a source of water intrusion related to steep- and low-sloped roof systems. Further, both meters can help to avoid destructive testing, which can further damage the roof system.

8.3.3.3.2 Forward-Looking Infrared Cameras

Forward-Looking Infrared Cameras use infrared radiation to detect the presence of temperature differences in building materials that may be an indicator of elevated moisture levels. Note that independent verification of the higher moisture level, using a moisture meter or other tools, is required to verify the presence of moisture since they are based on temperature differences that may or may not be associated with areas of higher moisture. This device is also useful when determining thermal defects and/or air leakage within the building envelope such as the effectiveness of the insulation or lack thereof.

8.3.3.3.3 Water Testing

Water Testing is a validation technique that can simulate water intrusion into the structure in a controlled manner in order to trace the pathway of intrusion. Validation of water leaks using actual water testing should be done whenever possible since it provides validation of all other evidence leading to the location as the source of the leak.

Two different types of nondestructive water testing methods are available and should be used during on-site inspections rather than a "garden hose" test. The problem with using a garden hose is that the amount and intensity of water applied to the possible leak location are not calibrated to any recognized test method and could invalidate conclusions reached using it if the matter were to be involved in litigation.

The first method utilizes a water spray rack built and calibrated to ASTM E1105.[18] The apparatus and method are designed to simulate wind-driven rain events on surfaces of a structures. The apparatus sprays water on the surface in question (such as windows and doors) at a rate of five gallons per hour per square foot of area. The only difficulty with this apparatus is that it is bulky and somewhat difficult to handle and position on steep-sloped roof surfaces.

The second test method is based on AAMA 501.2[19] and utilizes a calibrated spray nozzle with a pressure gauge attached to a hose (i.e., garden) to simulate rain events. Water from the nozzle is directed at the questionable surface or intersection to recreate a water intrusion event.

Simulating an active water leak using these methods can take anywhere from <1 to 20 minutes or more. Recreating water entry into a building depends on various factors such as the distance the water must travel and/or the thickness and type of material that water would need to penetrate. Oftentimes, the combination of water testing and the use of a diagnostic tool such as moisture meters can help in validating the source of the water leak for situations where visible proof does not immediately occur.

Confirming the source of the water entry by water testing, coupled with the interior and exterior observations, provides the inspector with a high probability of having identified the source or cause of the leak. Examples of water testing applications are shown and described within Table 8.16.

TABLE 8.16
Water Testing Applications

Description	Photograph
Spray nozzle with pressure gauge calibrated according to AAMA 501.2	
Water being sprayed at the interface of the shingled roof surface and vertical wall covered with siding	
Water being sprayed along the open metal valley to determine possible entry points	

8.3.3.3.4 Boroscope

This optical illuminated device consists of a flexible tube that can be inserted into tight spaces such as inside roof cavities or exterior wall systems. A boroscope with a camera should be used to photographically (or by video) capture findings of interest. This device also avoids the need for destructive testing.

8.3.3.3.5 Destructive Testing

Destructive Testing is a testing method in the field, which is also used to validate the source and/or the cause of water entry in a concealed space. This method requires the precise removal of a section of building material by destructive means in order for the inspector to get a better understanding of the situation or to confirm a causation opinion based on other evidence collected during the inspection. This could include destructively removing a small area of roofing surfacing materials (often done with low-sloped roof water entry inspections). Destructive test cuts performed on a low-sloped roof membrane can provide an evaluation of the condition of the substrate below the finished roof surface as well as the method of installation. Examples of destructive testing of roof surfaces are illustrated Table 8.17.

TABLE 8.17
Destructive Testing Applications

Description	Photograph
A precise test cut in the water damage drywall ceiling exposing leaky plumbing lines	
Test cut into spray polyurethane foam (SPF) material. A Tramex® survey encounter moisture meter was used to measure moisture content	
Test cut into roof surface covered with three-tab asphalt shingles	

8.4   WATER CAUSATION AND ORIGIN INSPECTION REPORT

Similar to the inspection report methodology outlined in Chapter 1, roof water causation and origin reports should include the following elements:

- Introduction (Information on Inspection Location and Client).
- Scope of Work (What is the scope of the inspection?):
  - Summary of interview(s).
  - Home and roof information.
  - History of water leak.
- Summary of Interior Observations (by floor):
  - Interior and attic space observations (if applicable).
  - Information on extent of water damage (i.e., use of diagnostic tools).
- Summary of Exterior Walk-Around and Roof Assessment Observations:
  - Provide general information of the exterior and roof surface.
  - Identify approximate location of water leak as determined by interior and roof measurements and observations.
  - Provide a detailed assessment for the reason of the water infiltration (i.e., degraded caulking, installation deficiencies, visible holes/gaps, etc.).
  - Validate source of water intrusion (i.e., water and/or destructive testing).
- Discussion/Analysis of Observations:
  - Explain the reason for the water intrusion (i.e., improper flashing details or installation practices). If so, provide the reader with the appropriate method.
  - Did the overall poor condition of the roof surface contribute to the damages?
- Conclusions.
- Photographs and figures.
- Evidence and/or Supporting Documents.

IMPORTANT POINTS TO REMEMBER

- Extending the life span and service life of a building roof system must involve periodic inspections and maintenance plans to address potential problems in a timely fashion.
- Weathered and deteriorated roofing components near and/or past their expected and designed service life can make a roof system more likely to leak.
- The proper design of roof drainage, installation methods and procedures, the quality of roofing materials, and occasional maintenance are critical aspects for the prevention of water intrusion associated with roof systems.
- Due to their complexity, water cause and origin inspections require a systematic approach to determine the cause(s) and origin(s) of the roof leak(s).
- Inspection tools such as infrared cameras, boroscopes, and moisture meters along with water testing are nearly always needed to determine the cause and origin of roof leaks.
- The most common locations for roof leaks are at roof junctions and intersections. The causes for leaks at these locations range from improper water management details like flashing and workmanship to inadequate design and maintenance (i.e., termination of valleys into exterior walls and clogged valleys/gutters from tree debris and sediment).
- These conditions, if not addressed, will lead to increased consequential exterior and interior damages and possibly failure (collapse) of the roof system.

REFERENCES

1. The National Roofing Contractors Association (NRCA). "The NRCA Roofing Manual: Steep-slope Roof Systems." 2017.
2. APA – The Engineered Wood Association. "Designing Roofs to Prevent Moisture Infiltration.". Form Number A535B, Revised August 2017.
3. International Code Council. "2018 International Building Code (IBC)." Accessed September 17, 2020, https://codes.iccsafe.org/content/IBC2018/chapter-15-roof-assemblies-and-rooftop-structures#IBC2018_Ch15_Sec1502.
4. *Shingle Applicator's Manual*, Thirteenth Edition. Malvern, PA: CertainTeed Corporation, 2019.
5. National Association, Inc. (SMACNA). *Architectural Sheet Metal Manual*, Seventh Edition. Chantilly, VA: Sheet Metal and Air Conditioning Contractors, September 2012.
6. Asphalt Roofing Manufacturers Association (ARMA). "Installation of Self-Adhering Membranes in Steep Slope Roofing." Accessed September 17, 2020, https://www.asphaltroofing.org/use-of-self-adhering-membranes-as-underlayments-in-steep-slope-roofing/.
7. Asphalt Roofing Manufacturers Association (ARMA). "Preventing Damage from Ice Dams." March 2019. Accessed September 17, 2020. https://www.asphaltroofing.org/media/2017/05/Protecting-Against-Damage-From-Ice-Dams-27March2019.pdf.
8. Fricklas, Richard, James D. Carlson, and Terrance Simmons. "Underlayments for Steep-Slope Roof Construction." *NRCA 11th Conference on Roofing Technology*, Rosemont, IL, September 21–22, 1995.
9. Federal Emergency Management Agency (FEMA). "Roofing Underlayment for Asphalt Shingle Roofs" *Home Builder's Guide to Coastal Construction*, September 2004. Technical Fact Sheet No. 19. Accessed September 17, 2020, https://www.fema.gov/media-library-data/20130726-1604-20490-7914/ra1_roof_underlayment.pdf.
10. Steven Winter Associates, Inc. "The Rehab Guide Volume 3: Roofs." U.S. Department of Housing and Urban Development, March 1999.
11. Patterson, Stephen, Madan Mehta, and J. Richard Wagner. "Roof Drainage". Roof Consultants Institute Foundation (RCIF), Publication No. 02.03. 2003.
12. Griffin, C. W. and Richard L. Fricklas. *Manual of Low-Sloped Roof Systems*, Fourth Edition. New York: McGraw Hill, 2006.
13. Asphalt Roofing Manufacturers Association (ARMA) "The Effects of Ponding Water." Form No. 115-BUR-93, September 2019. Accessed September 17, 2020, https://www.asphaltroofing.org/media/2019/09/Effects-of-Ponding-Water-on-Low-Slope-Roof-Systems-FINAL.pdf.
14. International Code Council. "2018 International Plumbing Code (IPC)." Accessed September 17, 2020, https://codes.iccsafe.org/content/IPC2018/chapter-11-storm-drainage#IPC2018_Ch11_Sec1110.
15. The National Roofing Contractors Association (NRCA). "The NRCA Roofing Manual: Architectural Metal Flashings, Condensation Control, Reroofing." 2018.
16. The National Roofing Contractors Association (NRCA). "The NRCA Roofing Manual: Membrane Roof Systems." 2019.
17. Asphalt Roofing Manufacturers Association (ARMA, National Roofing Contractors Association (NRCA), and the Single-Ply Roofing Industry (SPRI). "Repair Manual for Low-Slope Membrane Roof Systems." 1997.
18. ASTM International. "ASTM Standard E1105: Field Determination of Water Penetration of Installed Exterior Windows, Skylights, Doors, and Curtain Walls by Uniform or Cyclic Static Air Pressure Difference." 2015.
19. Intertek. "AAMA Standard 501.2-15: Quality Assurance and Diagnostic Water Leakage Field Check of Installed Storefronts, Curtain Walls, and Sloped Glazing Systems." 2015.

9 Water Infiltration – Cause and Origin Assessments Exterior Residential and Light Commercial Building Envelope

Stephen E. Petty
EES Group, Inc.

CONTENTS

PURPOSE/OBJECTIVES

The purpose of this chapter is to:

- Provide a best practices approach to investigating water infiltration issues with common exterior claddings for residential and light commercial structures.
- Document a systematic approach and methodology for handling water infiltration investigations associated with common exterior claddings.
- Provide best practices details of water management for exterior wall envelopes and components.

Following the completion of this chapter, you should be able to:

- Understand the importance of water management within conventional wood-frame construction.
- Identify the four principles of water management and how they relate to the exterior building wall envelope.
- Understand the key elements for proper water management in exterior claddings common to residential and light commercial structures.
- Identify common deficiencies in water management details for the common exterior claddings discussed.
- Be able to understand and identify best practices flashing and drainage details for the common exterior claddings discussed.
- Conduct a methodical and systematic visual inspection of the exterior cladding of a residential or light commercial structure as it pertains to water infiltration.
- Be able to create a formal written report of inspection findings in accordance with best practices.

9.1 INTRODUCTION

The exterior building envelope of a residential or light commercial structure is one of the key components to its line of defense against the infiltration and possible accumulation of damaging water and moisture. If allowed to go unnoticed and persist, either knowingly or not, excessive amounts of moisture within the wood-framed wall system can eventually lead to biodegradation and rot, which can then lead to potential compromises in the structural integrity of the building and possibly eventual failure. Therefore it is of utmost importance that exterior walls, along with their key exterior envelope components (i.e., windows, doors, wall penetrations, etc.), be provided with the proper water management details in order to prevent such water/moisture infiltration and accumulation.

The invariable high costs associated with the damaging effects (both damage and repairs) of water penetration to the finishes and supporting structure of a building are one of the leading causes for insurance claim frequency and severity (i.e., dollar amount). Recall from Chapter 1 (Section 1.3) of the example given for annual regional claim data for a mid-western insurance company that water losses were ~40.8% and 24.5% of the total number of residential and commercial claims filed and ~26.8% and 12.5% of the total amount spent for residential and commercial restoration services.[1] Water damage claims in California exceed $500 million/year.[2] Annual losses from deterioration of buildings in Canada range from $235 to $380 million dollars.[3] Further, in today's litigious environment, if proper water management details were not followed during building design and construction, which then eventually led or contributed to a water loss, then the general contractor, subcontractor, or both, can be held liable, and potentially heavy economical tolls could incur for all parties involved.

The importance of proper detailing in building construction when dealing with water is undeniable. This chapter, much like Chapter 8, which addressed low- and steep-sloped roof systems, addresses the determination of the cause and origin of water infiltration specifically through the exterior building envelope of conventional wood-frame construction (i.e., residential and typical light commercial structures). Through the following sections, the reader will be given the knowledge base needed to perform such investigations beginning first, with an introduction into design and water/moisture management, as they pertain to conventional wood-frame exterior wall construction, and then outlining the code required and recommended industry best practices details and common water/moisture management deficiencies encountered for common exterior cladding and building envelope assemblies.

For water/moisture to damage materials, the following four conditions must be satisfied[3]:

- A *moisture* source must be available.
- There must be a *route* or means for the moisture to travel.
- There must be some *driving force* to cause moisture movement.
- The material(s) involved must be *susceptible* to moisture damage.

To eliminate a moisture problem, one of the four conditions simply has to be eliminated. In forensic engineering projects, the objective is to determine which of these four conditions has impacted the building.

9.2 MOISTURE CONTROL DESIGN CONSIDERATIONS

The design of a weather-resistant exterior building envelope assembly requires careful consideration and a conscious awareness of the interaction between water/moisture and building components and an adequate plan of action to ensure the long-term durability of the exterior wall system.

Requirements for conventional wood-framed construction wall-coverings are primarily governed by the 2018 International Residential Code (IRC) for One- and Two-Family Dwellings. Current code language[4] gives the following general requirements for all exterior systems:

> *R703.1 General*: Exterior walls shall provide the building with a weather-resistant exterior wall envelope. The exterior wall envelope shall include flashing as described in Section R703.4.
> *Exception*: Log walls designed and constructed in accordance with the provisions of ICC 400.

It should be noted and fully understood that the building code provides only general, or minimal, requirements pertaining to protection from exterior water infiltration and does not address all of the issues needed to ensure long-term moisture resistance. Improved performance and protection of the building envelope warrant the use of best practices (sometimes referred to as "code-plus"). These best practices tend to be more prescriptive and provide more detailed requirements and guidelines than the minimal requirements found in codes.[3,5,6] Once they are recognized, best practices that produce consistent and desirable results are often adopted and incorporated into codes to ensure historical problems are not repeated in the future.[6,7]

How moisture in buildings is controlled depends on the location and weather characteristics of that location, which are typically associated with climate zones. In 2018, ASHRAE/IECC updated their climate zones (Figure 9.1).

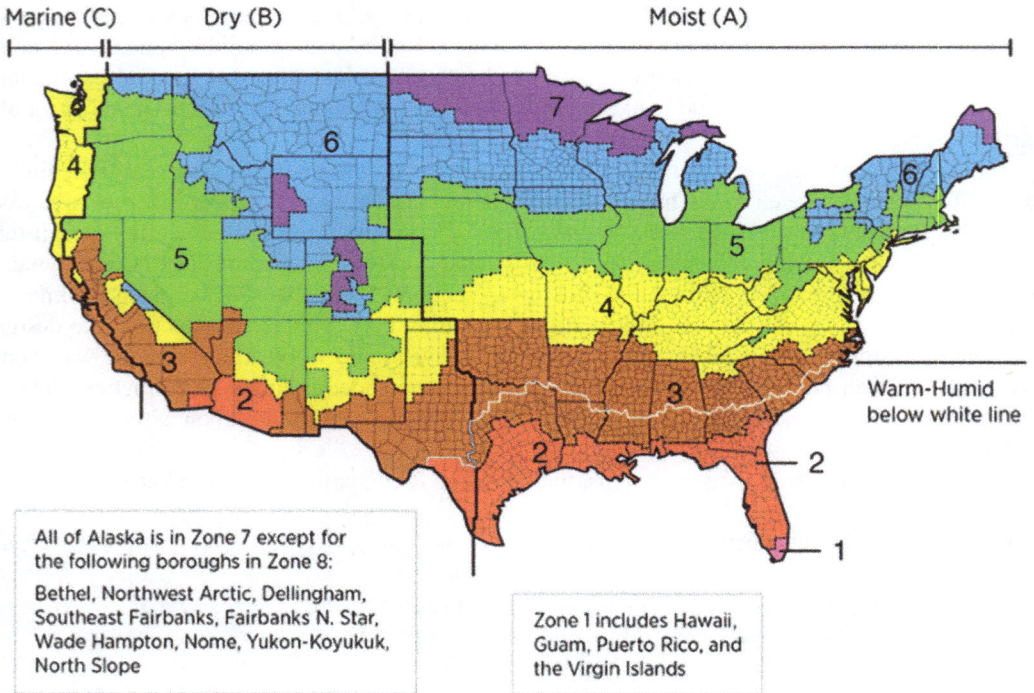

Marine (C)   Dry (B)   Moist (A)

Warm-Humid
below white line

All of Alaska is in Zone 7 except for
the following boroughs in Zone 8:

Bethel, Northwest Arctic, Dellingham,
Southeast Fairbanks, Fairbanks N. Star,
Wade Hampton, Nome, Yukon-Koyukuk,
North Slope

Zone 1 includes Hawaii,
Guam, Puerto Rico, and
the Virgin Islands

FIGURE 9.1 ASHRAE climate zones. (Permission ASHRAE.)

As detailed below, the best practices in controlling moisture/water infiltration into wall systems are the result of long-term historical experience by those practicing in their specific fields of expertise and which have been accepted and proven to be reliable methods.

9.2.1 WALL MOISTURE SOURCES AND TRANSPORT MECHANISMS

Water/moisture in the form of precipitation (i.e., rain and melting snow, sleet, and ice, etc.) is the primary source of water/moisture that can infiltrate the exterior building envelope and has the most significant potential for water damage to construction materials associated with building envelopes. Other moisture sources affecting the exterior walls include water-vapor-laden air, from either the exterior or the interior of the building, which can condense on wall cavity surfaces under conditions where these surfaces are at/below the dew point temperature of the moist air (Figure 9.2).[7,8]

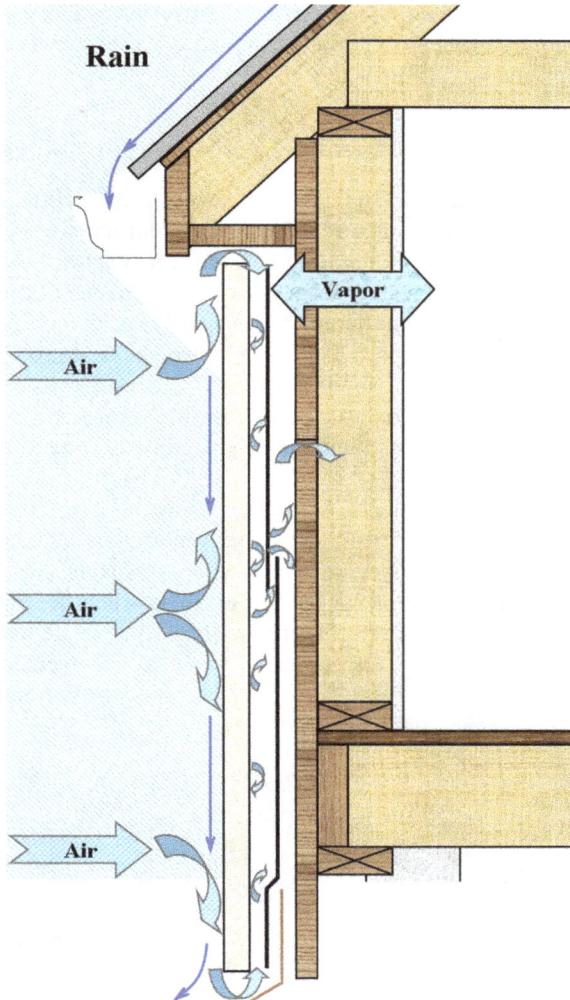

FIGURE 9.2 Sources of moisture and transport mechanisms.

The combined effect of moisture loading for a given structure is dependent on the geographic location and climatic conditions as well as site-specific factors such as building construction, exposures, and architectural details.[5]

The movement of water/moisture is dependent on its chemical state (i.e., liquid water or gaseous water vapor-laden air), which determines its method of transportation into the structure. Transport mechanisms[3,5] for water/moisture include the following:

- Liquid Flow and Capillarity (Liquid Water).
- Air Movement (Water Vapor).
- Diffusion (Water Vapor).

Liquid flow is the primary transport mechanism for water/moisture infiltration and accumulation in wall cavities. It should be noted that although the constant force of gravity acting on the water dictates its general movement (i.e., downward), the effects of wind and capillary action (i.e., movement of water due to surface tension and molecular attraction) can cause the water to move in nearly all directions.[3,5,6] Water vapor within the air (exterior and interior) is transported by the movement of air through leakage points in the building envelope and/or direct movement through building materials caused by vapor pressure differentials.[5]

9.2.2 Importance of Moisture Content Control for Wood and Other Products

Methods recommended by best practices, including design, construction, and maintenance, are critically important in managing the amount of water entering an exterior wall system and the moisture content of building materials. Maintaining the moisture content below particular threshold levels assures that deterioration and decay will not occur for the given building materials.[5] For conventional wood-framed construction, the moisture sensitivity, or the moisture content threshold at which wood decays, is typically around 22%–27% by weight, and rapid rotting conditions are present when the moisture content is above 30%–35%.[5,9–11] As a reference, wood that is protected by water and subjected to normal atmospheric conditions will generally equilibrate to a moisture content of ~8%–14%, depending on geographical location, climatic conditions, and relative humidity but typically will not exceed 15%.[5,10,11]

Conventional wood-frame construction can typically allow for a small amount of intermittent water infiltration for the life of the structure. Intruding water/moisture can be absorbed, distributed, and dissipated throughout a wooden structure without causing any structural deficiencies (Figure 9.3). Where water damage problems tend to occur is when design, construction, and/or maintenance issues allow water/moisture to enter the wall cavity at a rate that exceeds the capacity of the wood to dissipate or eliminate the water.[8,12] When wood is subjected to prolonged periods of continual wetting that raises the moisture content of the wooden framing levels at or above the decay threshold, damage ensues (Figure 9.4).

FIGURE 9.3 Intermittent light water staining to framing below window leak.

FIGURE 9.4 Prolonged wetting/rotting to lower framing of Tudor-style home.

Typically, before the source of water/moisture infiltration has caused the situation to reach this point, evidence of water damage staining and efflorescence (crystalline deposits that are left behind from evaporated water or moisture) are noticed by building owners on the finished portions of the home, but not in all cases. Due to the gravitational pull of the water and oftentimes the presence of a water vapor retarder or air infiltration barrier within the wall construction, water that enters an exterior wall cavity causes subsequent damage. Even historic and long-term decay may go unnoticed since the interior finished surfaces appear unaffected.

Steel present is concrete, masonry mortar, and grout tends to corrode at relative humidity between ~60% and ~100% with maximum rates between ~80% and ~85%. Corrosion of steel in these materials drops/stops as the relative humidity drops below ~50%.[3]

9.2.3 PRINCIPLES IN WATER MANAGEMENT

Ideally, the *prevention* of water infiltration through the exterior building envelope would be the standard that all structures are designed to achieve; however, designing strictly in this fashion is impractical. The ultimate goal is to keep the finished and structural materials of the building dry, to a certain extent (see moisture content discussion above). Therefore, one must design, construct, and provide periodic maintenance to ensure that water/moisture is properly *managed* and *controlled* in an expected fashion. That is, the exterior envelope should be expected, and designed, to allow a finite amount of moisture to enter the structure (but not too much) and then have a mechanism to capture and redirect the moisture back outdoors.

In 1996, a survey was conducted by the Canada Mortgage and Housing Corporation (CMHC) in the (eastern) coastal climate province of British Columbia, Canada,[13] in order to examine building wall envelope performance problems (i.e., water penetration, water damage, wood decay/rot, etc.), which appeared to plague numerous low-rise, multiunit, wood-framed residential buildings over a 10-year period. For the survey, both "problem" and "control" buildings, all no more than 8 years old, were selected based upon historic and reported water management performance. Results from the survey are summarized below.

- "Problem" buildings (a total of 37) were defined as buildings with moisture problems within the exterior wall, which resulted in damages equaling $10,000 or more to repair. Exterior wall claddings included stucco, wood, and vinyl.

- "Control" buildings (a total of nine) were defined as buildings that had not experienced wall moisture problems over a period of at least 5 years.

The study concluded that: (1) greater attention to detail was needed for water management principles, including moisture entry, drainage, and drying of the walls; and (2) local climate conditions should be considered when designing water management construction strategies. Some key differences and findings between "problem" and "control" buildings from the survey included the following[13]:

- Walls on the "problem" buildings had greater exposures (i.e., to wind and from smaller roof overhangs) than "control" buildings.
- "Control" buildings had fewer architectural details and more of the details were flashed compared with "problem" buildings.
- Construction details were often poorly designed in both buildings; however, the problems arose in the clarification and communication between designers and trade personnel.
- Almost all problems were associated with details such as windows, decks, walkways, balconies, and wall penetrations.
- All exterior cladding types experienced problems, although the buildings with higher reported problems occurred with stucco wall types.

Based on the information from this survey and from past experience, exterior wall assemblies that have experienced water infiltration problems lacked adequate water control and management construction details.

The principles of water control and management deal with building features and architectural design and construction details and are generally governed by the four D's (listed in order of general importance)[12]:

1. *Deflection:* Details limiting the exposure of the exterior envelope to precipitation events and the potential for liquid water to contact or infiltrate the wall envelope.
2. *Drainage:* Wall assembly details that redirect incidental infiltrating water out from the wall system and back to the exterior.
3. *Drying:* Conditions and details allowing for the drying of wet building materials.
4. *Durability:* Construction details and materials that provide adequate tolerance to moisture.

These principles of water control and management are generally considered to be the primary details for water management (as opposed to secondary details; see Section 9.2.4) and are detailed further, along with key exterior building envelope components, in the following sections.

9.2.3.1 Deflection

The first and foremost principle of water management involves the deflection of potential water or moisture from contacting and/or penetrating through the exterior wall envelope. Studies[13,14] (Straube[3,15] provides a simplified summary) and experience have shown that designing and constructing building features and details that limit the exposures of exterior walls to moisture sources can significantly aide in the other remaining principles, particularly drainage and drying as they deal with water that has incidentally infiltrated the wall envelope.[5,12] Note that ~70% of the water striking a wall stays on the wall and of this amount, ~1% penetrates the cladding.[16,17]

Common architectural and building design features that provide deflection include the following (also refer to Figure 9.5)[5,12]:

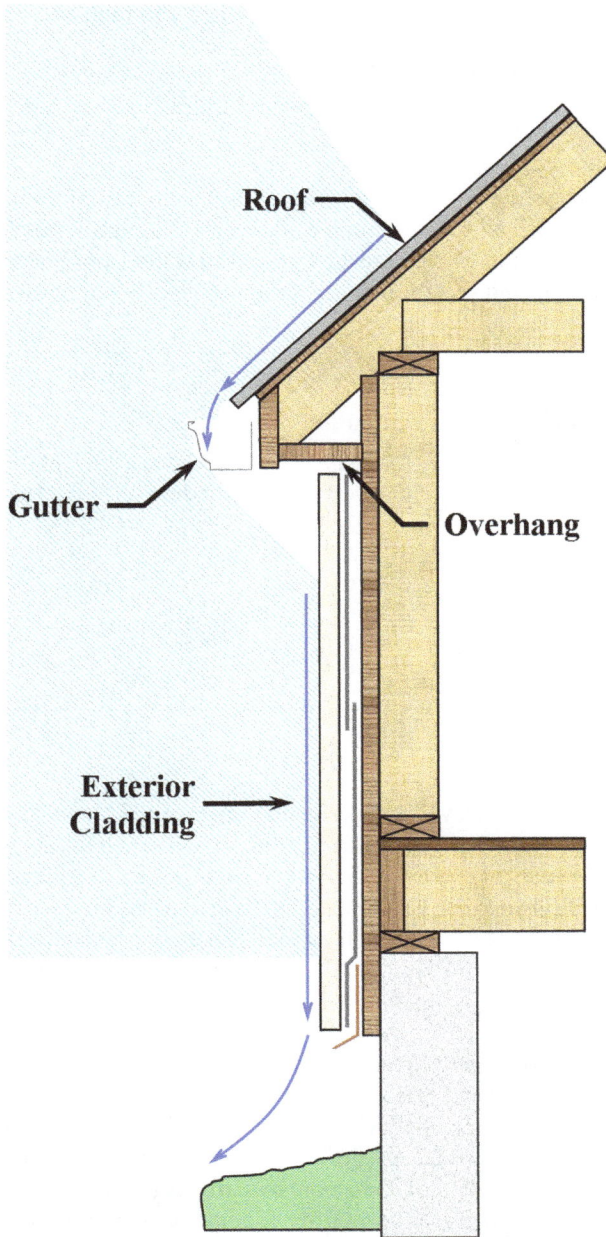

FIGURE 9.5 Water management principle – deflection.

- Exterior cladding.
- Sheltering of the building exposure from prevailing wind and weather patterns.
- Roof overhangs (i.e., soffits) and proper water runoff drainage systems (i.e., gutters and downspouts).
- Proper flashing and caulking details at interfaces susceptible to water infiltration.
- Water vapor retarders at required locations within wood-framed wall assemblies.
- Air infiltration barriers within the wall assembly to prevent/limit air leakage.

Interestingly, overhangs reduce wind-driven rain in low-rise and high-rise buildings by factors of 4 and 1.5, respectively.[18]

Site-specific conditions (e.g., local weather conditions) of the deflection mechanisms should be taken into account when conducting exterior wall water infiltration cause and origin investigations.

9.2.3.1.1 Exterior Cladding

Exterior cladding is the first line of defense for exterior water infiltration from precipitation events. Depending on the type of exterior wall system (i.e., barrier wall or drainage plane/cavity walls), the exterior cladding may be the only barrier intended to stop all water from entering the wall assembly. Ventilation through cladding and sheathing for several materials is summarized in Table 9.1.

TABLE 9.1
Cladding and Sheathing Ventilation[19]

Type of Cladding or Sheathing	Flow Rate (CFM/feet^2)	Flow Gap (inches)	Air Changes per Hour (ACH)
Wood siding	0.1	3/16	20
Vinyl siding	0.5	3/16	200
Brick veneer	0.15	1	10
Stucco (vented)	0.01	3/8	10
Stucco (directly applied)	None	None	0
Sheathing flanking flow	0.05	3/16	10

9.2.3.1.2 Sheltering and Overhangs

Sheltering and overhangs are conceived and designed during initial home development, but affects the deposition of water on the exterior walls from wind-driven precipitation for the lifetime of the structure. The wetting patterns of exterior walls due to wind-driven rain events in specific climates are dependent on the building's shape and orientation to prevailing weather events, aerodynamics, raindrop diameter, and wind speed[14,15] as well as local vegetation and surrounding obstructions.[5] Studies[14,15] of wind-driven rains and their wetting patterns on buildings concluded the following associations:

- The wettest locations on blunt-edged, rectangular buildings are on the upper, windward corner, followed by the top and side edges.
- The side walls remain "relatively dry" when wind-driven rain is impacted normally or perpendicular to the windward face of the building. The wetting patterns for the side walls increase as the angle of attack increases more toward a perpendicular angle to the walls.
- Cornices, or overhangs, decrease the wetting conditions along the top and side edges of the building face.
- A peaked roof reduces the rain impact on the windward face by redirecting the airflow more up and away from the building face.
- Balconies and canopies have a local sheltering effect of the building wall surfaces below their locations.

These can all affect the amount of water (either positively or negatively) that physically contacts the building envelope and subsequently contributes to the exterior moisture loading.

9.2.3.1.3 Flashing

Flashing details at wall penetrations and along component interfaces and projections, if properly designed and implemented into the exterior wall envelope, can provide sufficient means of deflecting any water or moisture from entering the wall system. Current IRC language[4] gives the following general requirements for flashing:

> *R703.4 Flashing*: *Approved* corrosion-resistant flashing shall be applied shingle-fashion in a manner to prevent entry of water into the wall cavity or penetration of water to the building structural framing components. Self-adhered membranes used as flashing shall comply with AAMA 711. The flashing shall extend to the surface of the exterior wall finish. *Approved* corrosion-resistant flashings shall be installed at all of the following locations:
>
> 1. Exterior window and door openings. Flashing at exterior window and door openings shall extend to the surface of the exterior finish or to the water-resistive barrier (WRB) complying with Section 703.2 for subsequent drainage. Mechanically attached flexible flashings shall comply with AAMA 712. Flashing at exterior window and door openings shall be installed in accordance with one or more of the following:
>
>   1.1. The fenestration manufacturer's installation and flashing instructions, or for applications not addressed in the fenestration manufacturer's instructions, in accordance with the flashing manufacturer's instructions. Where flashing instructions or details are not provided, pan flashing shall be installed at the sill of exterior window and door openings. Pan flashing shall be sealed or sloped in such a manner as to direct water to the surface of the exterior wall finish or to the WRB for subsequent drainage. Openings using pan flashing shall also incorporate flashing or protection at the head and sides.
>
>   1.2. In accordance with the flashing design or method of a registered design professional.
>
>   1.3. In accordance with other approved methods.
>
> 2. At the intersection of chimneys or other masonry construction with frame or stucco walls, with projecting lips on both sides under stucco openings.
> 3. Under and at the ends of masonry, wood, or metal copings and sills.
> 4. Continuously above all projecting wood trim.
> 5. Where exterior porches, decks, or stairs attach to a wall or floor assembly of wood-frame construction.
> 6. At wall and roof intersections.
> 7. At built-in gutters.

Again, it should be noted that the building code provides only general, or minimal, requirements pertaining to protection from exterior water infiltration, and the best practices are typically referred to when dealing with exterior building envelope water management details. Further details regarding best practices for installation of flashing are provided by the BIA[20,21] for masonry, HUD,[22] and others discussed in later sections.

9.2.3.1.4 Vapor Retarders and Air Barriers

Vapor retarders and air barriers are commonly installed within exterior wall assemblies to prevent the movement of unwanted vapor transmission and air leakage, respectively, both of which carry with them the possibility of condensation and potentially damaging amounts of water accumulation within the walls.[7,8,12] The IRC prescriptively specifies interior vapor barriers (Class I: <0.1 Perm; Class II: >0.1 to <1.0 Perm; and Class III: >1 to <10 Perm – in the United States, a perm is defined as 1 grain of water vapor per hour, per square foot, per inch of mercury – grain/hour-square feet-mm Hg – flowing through a wall system) for framed walls depending on climate zones. Exceptions to Class III barriers are allowed with minimum continuous exterior insulation.[23]

Vapor transmission is the molecular passage of water through building materials that is driven by a differential vapor pressure across the wall whose direction of transmission is dependent on geographic location and climate conditions. Vapor transmission only poses a problem in wall construction when there is a strong thermal bridge (i.e., drastic temperature drop) located within the wall assembly that allows the vapor to contact a surface that is at or below the air's specific dew point temperature, or the temperature at which the vapor will condense (i.e., turn from vapor to liquid), and form liquid water within the wood-framed cavity, thus causing the wood to absorb the moisture and increasing its moisture content and increasing its susceptibility to decay/rot if unmanaged for a prolonged period of time.[8,12] Typically, in colder regions, water vapor from the interior living spaces can pass through the interior wall finishes and condense on the cooler surfaces of the exterior wall sheathing and framing. The reverse pathway is also possible in areas with hot and humid climates where the vapor transmission is directed indoors. In this scenario, the water vapor associated with relatively humid outdoor air contacts the cooler exterior side surfaces of the air-conditioned interior finishes and condenses on these cooler surfaces.[8] This confusion on which side of the wall (interior vs exterior) has been debated for decades and depends mostly on the climate zone where the building is located.[16,23–25]

Moisture-laden air movement through the wall system is created by differential air pressure differences between the exterior and the interior portions of the building. Much like vapor transmission, the problems arise when condensation occurs within the wood framing of the wall. The difference lies in the difficultly of condensation formation. Vapor needs only to pass straight through building materials, whereas moisture-laden air must find a leak in order to enter the wall cavity. Once in the wall cavity, the length of the pathway the air is allowed to travel through determines whether condensation will form. The longer the path, the more time the air has to cool to the dew point temperature, the more condensation occurs, and the more water accumulates within the wall.[12]

Vapor barriers, as opposed to retarders, have the intent of stopping the flow and are increasingly being considered particularly in colder climates.[25]

In terms of forensic engineering inspections, due to the nature of the water formation within the walls, oftentimes, long-term and historic deterioration of the wooden wall framing goes unnoticed since no physical evidence of water damage is present on the finished surfaces of the home, and the issue is only then discovered when conditions that are more problematic develop.

The presence of vapor retarders and air infiltration barriers may or may not be able to be determined during site investigations unless destructive testing is approved and performed.

9.2.3.1.5   Sealants

Sealants such as caulks are used by contractors to fix leaks in the building envelop. However, caulks should be considered as a secondary water control measure and not a primary means of control since they lose their sealing capacity from weathering in a relatively short time period. For example, BIA recommends use of caulking, but only after backer rods have been installed. Aside from limited life (Table 9.2), caulks can chemically attack cements and are susceptible to failure where movement in the building system may occur.[20]

TABLE 9.2
Properties of Caulks[22]

Type of Caulk	Estimated Life (years)	Uses	Cost/Tube (~10 oz.)	Clean Up
Acrylic latex	5 to 20+	Most dry surfaces	$2.00 to $4.00	Water
Butyl	4–10	Masonry and metal	$2.50 to $3.00	Paint thinner
Kraton	10–15	Most dry surfaces	$5.00 to $7.50	Paint thinner
Oil	1–3	Most dry surfaces	$1.00 to $2.00	Paint thinner
Polyvinyl acetate	1–3	Indoor surfaces only	$1.50 to $2.00	Water
Polyurethane	15 to 20+	Masonry	$4.50 to $10.00	Acetone, MEK
Silicone	20+	Glass, aluminum, mostly dry surfaces – not for masonry	$4.00 to $7.00	Paint thinner, naphtha, toluene
Styrene rubber	3–10	Most dry surfaces	$2.00 to $2.50	Paint thinner

9.2.3.2 Drainage

Incidental moisture getting beyond the deflection components and infiltrating the exterior wall assembly (for drainage plane/cavity walls) must be adequately managed so as not to create a problem to the interior or structural components of the wall. Drainage typically is not much of an issue in buildings when annual rainfall is <20 inches (50.8 cm) per year[26,27] (see Figure 9.6).

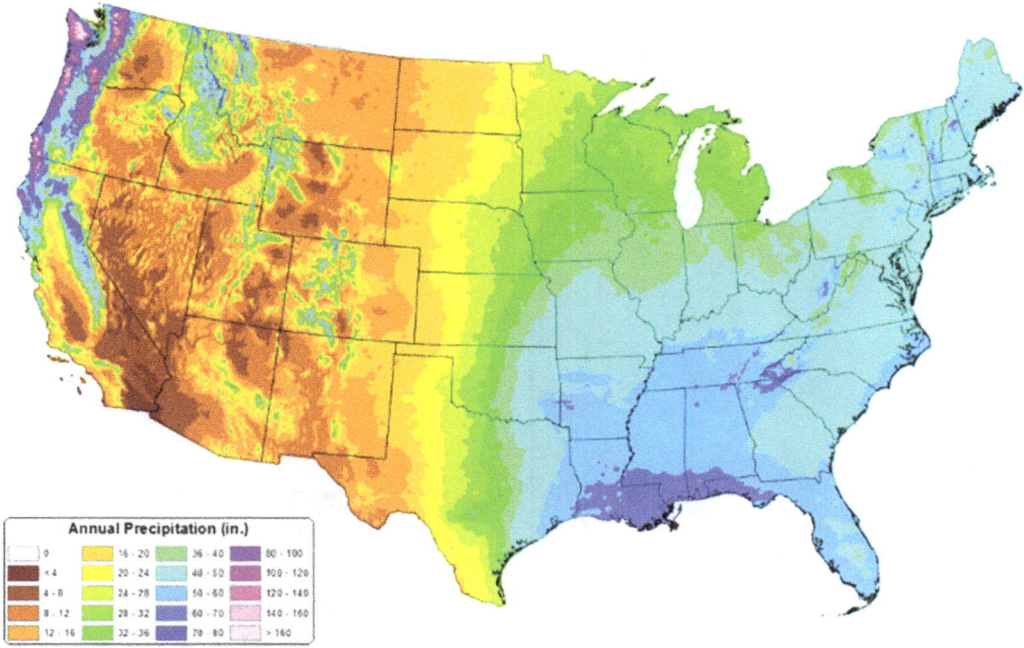

FIGURE 9.6 Average annual rainfall in the continental United States.

Second only to the principle of deflection, drainage ensures that the bulk incidental moisture is collected and then properly returned to the exterior side of the wall, via gravity, where it can then be carried away from the building by site drainage. In conventional wood-frame construction, drainage of the exterior wall envelope is generally created by the use of a drainage plane and/or an air cavity within the wall assembly (Figure 9.7).

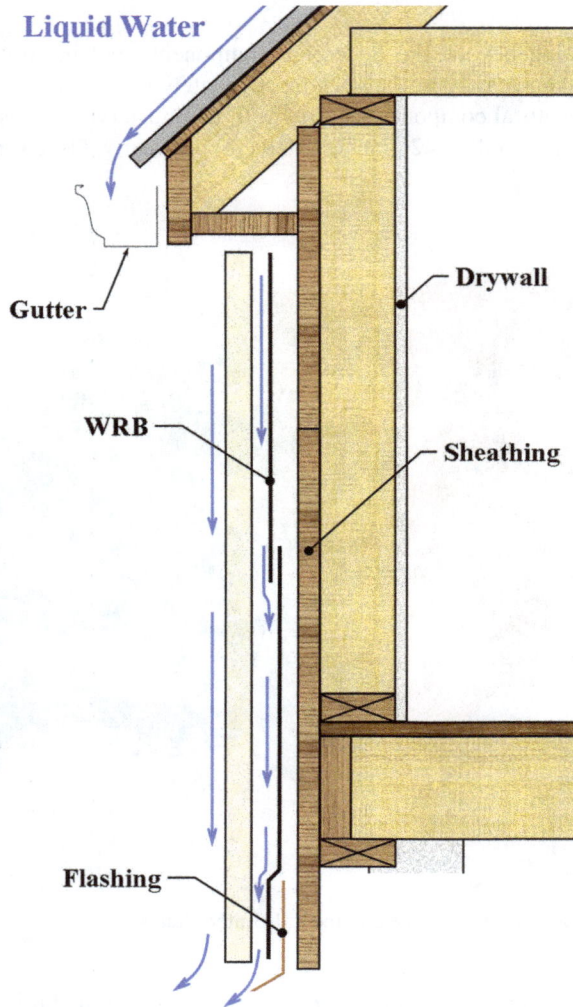

FIGURE 9.7   Water management principle – drainage.

9.2.3.2.1   Drainage Plane

A drainage plane is the component interface within the wall assembly at which the inward movement of the bulk moisture infiltrating past the exterior cladding is stopped and then redirected
downward along the exterior side of the interior wall cavity and back to the exterior. The drainage
plane in wood-frame construction typically consists of a WRB and properly designed and incorporated flashings along interfaces and around wall penetrations.

The WRB, typically in the form of house wrap or building paper, and when installed correctly,
primarily serves to shed the incidental ingress of liquid water from the wall sheathing and the interior portions of the building. It should also be permeable enough to allow for the passage of water
vapor through the material so as not to allow for the formation of potentially damaging condensation within the wall assembly. Depending on the type of WRB material selected for construction,
the WRB can also serve as an air infiltration barrier.[28,29]

In order to assure the proper functionality of the WRB, as with most other exterior wall envelope
components, it must be installed correctly in accordance with building code requirements and best

practices recommendations. Generally, the WRB must be lapped properly in a continuous, shingle-wise fashion (i.e., upper layer overlapping lower layer) the entire height of the wall in order to facilitate the downward flow of liquid water over its seams. Ensuring that each successive layer of the WRB is lapped in the correct fashion and the lap distance is adequate is the key to the prevention of water infiltration from such forces as gravity, wind, surface tension, and capillary action at locations commonly susceptible to water intrusion (i.e., windows, doors, wall penetrations, deck interfaces, roof-to-wall interfaces, etc.).[29] Current language in the IRC for One- and Two-Family Dwellings[4] gives the following general requirements for WRBs:

R703.1.1 Water resistance: The exterior wall envelop shall be designed and constructed in a manner that prevents the accumulation of water within the wall assembly by providing a water-resistant barrier behind the exterior cladding as required by Section R703.2 and a means of draining to the exterior water that penetrates the exterior cladding.
Exceptions:
1. A weather-resistant exterior wall envelope shall not be required over concrete or masonry walls designed in accordance with Chapter 6 and flashed in accordance with Section R703.4 or R703.8.
2. Compliance with the requirements for a means of drainage, and the requirements of Sections R703.2 and R703.4, shall not be required for an exterior wall envelop that has been demonstrated to resist wind-driven rain through testing of the exterior wall envelope, including joints, penetrations, and intersections with dissimilar materials, in accordance with American Society for Testing and Materials (ASTM) E331 under four (4) specific conditions.

Water-resistive barrier: One layer of No. 15 asphalt felt, free from holes and breaks, complying with ASTM D 226 for Type I felt or other approved WRB shall be applied over studs or sheathing of all exterior walls. No. 15 asphalt felt shall be applied horizontally, with the upper layer lapped over the lower layer not <2 inches (51 mm). Where joints occur, felt shall be lapped not <6 inches (152 mm). Other approved materials shall be installed in accordance with the WRB manufacturer's installation instructions. The No. 15 asphalt felt or other approved water-resistance barrier material shall be continuous to the top of walls and terminated at penetrations and building appendages in a manner to meet the requirements of the exterior wall envelope as described in Section R703.1.

It should be noted that actual inspections have found leakage into the wall system as a result of the felt being installed backward with the lower layer overlapping the upper layer, allowing water to enter the open seam.

Also, as discussed in previous sections within this chapter, the building code provides only general, or minimal, requirements pertaining to protection from exterior water infiltration, and the best practices are typically referred to when dealing with exterior building envelope water management details.

Manufacturers of WRBs typically recommend the following installation details for house wrap and building paper:

- When used as both a WRB and air infiltration barrier for residential and low-rise applications, all house wrap seams (horizontal and vertical) and terminations (roof-to-wall, sill plates, etc.) must be taped.[30]
- When used solely as a WRB for residential and low-rise applications, only the vertical seams of the house wrap need to be taped.[30]
- House wrap should have a minimum lap of 6 inches at all terminations, seams, penetrations, and transitions.[30]

- Building paper is recommended to have a 3-inch overlap along horizontal seams (minimum of 2 inches required) and a minimum of a 6-inch overlap along vertical seams.[31]
- All forms of WRB must be installed and properly integrated with wall penetrations (i.e., windows, doors, etc.) and flashing in order to form a comprehensive moisture control system.[30,31]

Flashings that have been properly designed and incorporated into the continuous drainage plain of the exterior wall aid in the redirecting of incidental water infiltration down and away from the building and back to the exterior of the wall. Due to its integrated installation into the wall assembly, flashings should be designed for durability and should serve to function as long as the exterior covering.[28] Flashing is required at locations susceptible to exterior water infiltration and at the lower terminations of the WRB as to maintain the continuity of the drainage plane behind the cladding. At each of these locations, it is important that the flashing extends to the exterior face of the cladding so that water that has infiltrated into the wall can have a proper means of exiting back to the exterior. Specific flashing details given by best practices are discussed in later sections pertinent to exterior cladding type and wall penetration. Generally, best practices installation details for metal flashings are given in the Architectural Sheet Metal Manual created by the Sheet Metal and Air Conditioning Contractor's National Association (SMACNA).[32]

9.2.3.2.2   Air Cavity

An air cavity between the interior side of the cladding and the drainage plane (i.e., WRB and flashings) serves to act as a capillary break, or drainage cavity, between the two vertical surfaces aiding in the drainage ability of the wall assembly. It also serves to provide increased air circulation leading to greater degrees of drying potential (see Section 9.2.3.3), and it can help to balance the differential air pressures on either side of the exterior cladding. Oftentimes, it is this air pressure difference that is a driving force for moisture to enter the wall assembly in the first place.[5,12,33] Of these functions, its role with respect to the drainage of infiltrated water is of primary importance as it allows for the greatest prevention of excess accumulation of water within the wall, thereby decreasing the chances for interior- or structural-related water issues. Minimum required air cavities for brick masonry are 1 inch (25.4 mm)[3,34,35] and recommended values for stucco and other cladding are 3/16 to 1/4 inches (4.76–6.35 mm),[26,27] but no minimum venting areas are specified.

9.2.3.3   Drying

The objective of water/moisture management within exterior wall envelopes is to maintain the delicate balance between the wetting and the drying of the wall construction materials (especially wood) and adequately controlling its moisture content as to maintain it below thresholds for decay and rot.[5] Straube notes that removal of moisture can occur four ways[3,26,36]:

- Drainage by gravity.
- Capillary transport of bound liquid water to, and evaporation from, the outer surfaces of the wall materials.
- Diffusion and/or convection of water vapor outward through the wall cavity and inward into the wall or building interior.
- Convective flow of exterior air through the air space (e.g., ventilation).

As mentioned previously, intermittent and incidental water that infiltrates the wall envelope can and will be absorbed, distributed, and dissipated by the wood and eventually eliminated without incident by the mechanisms listed above. However, conditions leading to decay and rot occur when the rate of water absorption exceeds the ability of the wood to dry, and the moisture content of the wood is raised for prolonged periods of time. The drying potential for the wooden wall sheathing and framing, following incidental water/moisture infiltration, is dependent on the local environmental conditions of the building and how they relate to air movement (i.e., ventilation) and vapor diffusion.[5,12,33,37]

The impact of air cavities on ventilation and other factors behind exterior claddings has been extensively researched over the past decades through field studies and theoretical analyses.[18] A review of such research is provided in the following conclusions regarding the use of air cavities with respect to moisture removal and other factors[33]:

- An air cavity can provide several important functions to the exterior wall envelope: (1) it can provide a capillary break, (2) it can provide a gravity drainage plane for incidental water infiltration (see Section 9.2.3.2), (3) it can serve as a ventilation channel, which can improve building material drying capabilities, and (4) it can act as a pressure equalizer for the cladding.
- Air cavity ventilation does not always improve the drying potential of a wall. The local climate conditions (i.e., temperature, humidity, solar radiation) and performance of the material layers adjacent to the air cavity (i.e., WRB, permeability) both play important roles in actual performance.
- Wall cavity ventilation is generally and primarily beneficial for most wall structures, allowing them to dry out from incidental water/moisture leakage into the wall cavity. However, it does have occasional minor drawbacks such as helping to bring moisture into the wall during certain conditions (i.e., water vapor from leaked water retention driven inward through more permeable WRB causing "summer condensation" within the wall cavity).
- Wall cavity ventilation is particularly important for masonry and stucco claddings (i.e., more water absorptive).

Although air cavity ventilation can help to improve the drying capability of the wall system in some instances, the principle of drying should not be relied upon as a primary control mechanism for water/moisture management since it is a much slower process.[9,12] The principles of deflection and drainage should still remain the primary means of water management within the exterior wall envelope, and more emphasis in drying should be paid when conditions warrant such emphasis.

9.2.3.4 Durability

"Durability is defined as the ability of a building or any of its components to perform the required functions in a service environment over a period of time without unforeseen cost for maintenance or repairs."[38] The proper design, construction, and maintenance of a durable exterior cladding and wall envelope can significantly impact the long-term sustainability and performance of a building.

In 2000, the U.S. Green Building Council (USGBC) developed the Leadership in Energy and Environmental Design (LEED) Green Building Rating System® in order to evaluate quantifiably the environmental impact and performance of a building by using a whole-building approach. The LEED® program is a voluntary, consensus-based, third-party rating system for new and existing buildings based on such key green building performance areas as sustainable sites, water efficiency, energy and atmosphere, and materials and resources, to name a few. The rating system, based on a maximum 69-point scale, is based on credits that are obtained by meeting/exceeding the criteria for each key performance area.[39,40] For more information regarding the LEED® program or the USGBC, visit www.usgbc.org.

The Canadian Green Building Council's (CaGBC) adaptation of the LEED® Green Building Rating System functions in much the same way, but is tailored more specifically toward Canadian climates, construction practices, and regulations.[41] The CaGBC's LEED® rating system is based on a maximum 70-point scale, with the one additional point available through the "Durable Building Credit" (number 8) in the "Materials and Resources" category (MRc8), which evaluates the structure's durable qualities.[40] Robert Marshall, one of the creators of Canada's LEED® MRc8 credit and author of the PCI Journal article "Delivering Durable Building Envelopes,"[40] stated that their motivation was to "prevent moisture and structural deterioration that can cause the collapse of a building envelope." Further, the intent of the durability credit was to minimize the amount of materials used and the constructive waste over the life of a building that results from premature failure of the building and its components and assemblies.[40] Marshall believed that a durable LEED credit would lead to a reduction of premature failures, lawsuits, insurance claims, and loss of reputation within the construction industry.[40] For more information regarding the LEED® Canada program, the CaGBC, or the MRc8 credit, visit www.cagbc.org.

The reference standard for durability that has been used extensively by architects and engineers is the Canadian Standards Association (CSA) publication number S478-95 (R2007) titled, "Guidelines on Durability in Buildings."[42] This publication summarizes the agents and mechanisms related with durability and gives advice and guidance to designers, builders, owners, and operators into the design, operation, and maintenance requirements for buildings and their associated components. In order to meet the requirements for the "Durable Building" credit in the LEED® Canada green building rating system (i.e., MRc8) mentioned previously, a building designer must develop and implement a "Building Durability Plan" in accordance with the principles of CSA S478-95. More specifically, the building must be designed and constructed where the predicted service life meets or exceeds its design service life, and where the design service life of a particular component or assembly is shorter than that of the building, those components or assemblies can be readily and easily replaced. Finally, the Building Durability Plan should document a quality assurance program, which helps ensure that the predicted service life is achieved.[38,40]

It is beyond the scope of this book to address all of the variables associated with durability, component design, and predicted service lives that would typically be incorporated into a building durability plan. However, during the conceptual design phase in the development of a building, considerations should be made for anticipated lives, maintenance, and possible future repairs. Durability design considerations for conventional wood-frame wall construction envelopes should generally include the following[38]:

- Develop a Building Durability Plan and review it often during construction.
- Make informed decisions and optimize the design of all building components early on in the design using life cycle assessment tools.
- Select design strategies that are appropriate to the geographic location.
- Specify realistic levels of workmanship that are based on practical construction methods.

For more detailed information on design considerations for building envelopes for wood-framed structures, please refer to the "Durability" website jointly owned and operated by the Canadian Wood Council (CWC) and FPInnovations (www.durable-wood.com).

9.2.4 SECONDARY DETAILS IN WATER MANAGEMENT

The principles of water management for exterior wall envelopes (i.e., the four D's) were discussed at length in the preceding sections. If these were considered to be the primary details for water management, then the secondary details would be those that contribute to the overall effectiveness to the water-resistive system of the building. However, these secondary details should not be relied upon to serve as the main sources for moisture control. Another perspective would be that the primary details are those that serve to provide long-term control of water for the approximate service life of the exterior finishes, while secondary details would need to be continually checked and maintained to ensure functionality for the life of the wall system.

Secondary water management details typically refer to the use of caulks and sealants along water- and air-susceptible gaps and joints. Modern construction techniques and repair activities rely rather heavily on caulking and sealants for resistance to water and air infiltration. While caulks and sealants have their place within the weather-resistive system of the building, they cannot be solely relied upon to serve as a long-term means of moisture control. This is due primarily to their propensity to fail prior to the predicted service life of the exterior finishes and/or the wall system from the combined effects of aging, weathering, building component movement, installation/application deficiencies, and lack of periodic maintenance.[43,44] Due to exposure to extreme weather conditions and differential amounts of movement with the building, caulks and sealants will deteriorate and eventually fail likely creating cracks or gaps through which water can potentially infiltrate. Oftentimes, even the implementation of a periodic and diligent maintenance program cannot keep the building free of cracks and gaps.[12] Therefore, it is in the best interest of the building, as a whole, that the four principles of water management be designed and strongly relied upon to keep the walls of the building dry and free of excess moisture.

9.3 GENERAL WATER MANAGEMENT DETAILS FOR COMMON EXTERIOR FINISHES

The primary mechanisms for controlling the infiltration and subsequent accumulation of water within an exterior wall assembly were outlined and explained in the preceding sections. To review, these primary details refer to the four D's: Deflection, Drainage, Drying, and Durability. The interrelationship between all four principles must be accounted for during the design, construction, and maintenance of the exterior wall envelope; however, for the purposes of ensuring a sufficiently dry wooden wall cavity, the principle of drainage should be given particular emphasis. This is based in large part to the likelihood that water, in some way or another, will find its way past the exterior cladding. Although specific details may change slightly, conventional wood-framed wall construction should be expected to have a method of draining any of this incidental moisture back to the exterior.

Experience and surveys of water-related building envelope failures[13] have indicated that the vast majority of problems have been related to the incidental infiltration of water between wall components and/or at penetrations and the lack of proper drainage behind the exterior cladding. The water enters the wall system and remains there for prolonged periods of time allowing for rot and decay of the wooden structural framing.

General water management details for common exterior wall finishes for conventional wood-frame construction are summarized in the following sections. In some instances, case studies from forensic site investigations will be given to help provide clarity in the wall construction/finish details and emphasize the importance of the primary details with respect to moisture control.

9.3.1 STUCCO

Conventional stucco consists of a mixture primarily composed of water, sand, and Portland cement. When Portland cement is combined with water a reaction occurs, which forms a paste, and with time causes the cement to harden and become rigid. Its aesthetic appeal and versatility, along with durability and cost-effectiveness, have made it an attractive choice in North American buildings for over 300 years. Due to its particular porous nature and propensity to crack, stucco wall assemblies tend to work well in many dryer climates and where location and architectural details help limit its exposure to wind-driven rains (i.e., deflection principles).[45]

There are two different methods that can typically be used to apply stucco to an exterior wall. The first method, known as traditional or "three-coat" stucco, involves the application of three separate layers of stucco: a scratch layer, a brown layer, and a finish or "color" coat. The other method is known as "two-coat" (sometimes referred to as "one-coat") stucco. It involves the application of a base coat and then the addition of a finish coat, which can utilize conventional cement color finish or synthetic acrylic color finish. Modern-day construction of a stucco-clad wall is illustrated in Figure 9.8.

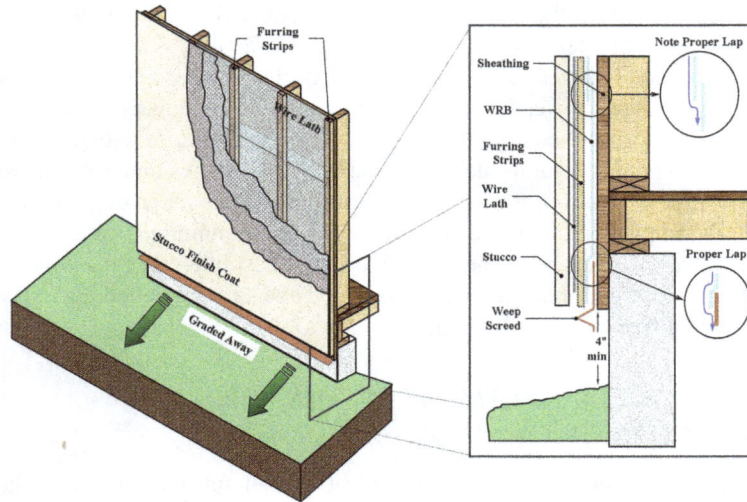

FIGURE 9.8 General construction and water management details – typical stucco-clad wall.

A modern stucco-clad wall consists of the following sequence of construction activities:

- A WRB is fastened to the exterior wall sheathing. Current IRC requires that the WRB for "exterior plaster" (i.e., stucco) over wood-based sheathing and in compliance with ASTM C 926 and ASTM C 1063 be water-resistive and vapor-permeable with "a performance at least equivalent to two layers of Grade D paper." The WRB is allowed to be a single layer when its water resistance is "equal to or greater than that of 60-minute Grade D paper and is separated from the stucco by an intervening, substantially non-water-absorbing layer of designed drainage space."[4,43]
- A metal/wire lath (which may or may not contain a water-resistive paper backing) is secured to the sheathing directly over the WRB and holds the stucco coats in place on the wall.
- The subsequent coats of stucco are applied over the metal/wire lath and each other.
- The lower edge of the stucco wall system is terminated above finished grade with either a weep screed or a casing bead/flashing combination along its lower edge. Current IRC[4] requires that a corrosion-resistant weep screed with a vertical flange of at least 3-1/2 inches be provided and requires that it be placed a minimum of 4 inches above the earth or 2 inches above paved areas.

Current IRC[4] requires that Portland cement stucco applied over metal or wire lath shall be not less than three coats. When applied over masonry, concrete, pressure-preservative treated wood, or decay-resistant wood, or gypsum backing, it shall be not less than two coats.

Typical water-related problems associated with this general form of stucco-clad wall construction are threefold:

1. The WRB must be installed correctly (i.e., properly lapped) and sufficiently to stop and redirect infiltrated water from reaching the wooden sheathing. Oftentimes, the WRB used beneath the stucco cladding is improperly lapped along horizontal/vertical seams and/or insufficiently to properly resist the accumulation of water, which can build up in the wall assembly. This results in long-term water-related deterioration of the sheathing and the potential for the growth of molds within the wall cavity.

Case Study 9.1

In the example case study depicted in Figure 9.9, a site investigation was performed to determine the cause of vertical cracks in the exterior stucco finish and a small, light area of water damage staining to an interior wall at the location of an electrical outlet. Upon inspection into the subject area (including approved destructive testing), the staining was associated primarily with exterior water that had infiltrated the stucco finish and permeated through the inadequate WRB, which in this case appeared to be one layer of building paper. The vertical cracks in the stucco were equally spaced on ~16-inch centers, corresponding with the locations of the wooden wall studs, and were caused by dimensional variations in the wood as it absorbed moisture and swelled, creating excess stress in the stucco.

FIGURE 9.9   Inadequate water-resistant barrier leading to mold.

2. Due to the inherent construction/application design, the lath is fastened directly to the sheathing and the stucco base/scratch coat is then applied directly over the sheathing/lath interface. This method causes the stucco and lath to be in very close proximity, and even contact, or bond to, the WRB. The lack of a defined drainage cavity (i.e., gap) between the two surfaces creates increased surface tension and the capillary action of water to allow for it to remain within the wall assembly and not drain properly down toward the bottom termination of the wall or other outlet locations. This nearly constant interaction or contact between water and the WRB effectively causes the WRB to lose its water repellency.[12,46]

There are a few ways to combat this issue. One way is to use two layers of building paper over the sheathing behind the stucco, as required by building code and best practices. The air space, or drainage cavity, is created between the two layers. The outer-most layer (closer to stucco) serves as a bond break, which allows for drainage and the innermost layer (closer to sheathing) to be free to repel water as intended.[46]

Another way to aid in the drainage of water within a stucco-clad wall is with specialized WRBs. One product on the market specifically designed for stucco walls is DuPont™ Tyvek® StuccoWrap™. This product is manufactured with drainage grooves, which, when installed properly (i.e., grooves oriented vertically), is intended to facilitate drainage behind the stucco. However, research experiments appear to indicate that this product, by itself, does not provide adequate drainage. The experiment did show that this product "worked perfectly" when a layer of "cheap felt paper" was added over the DuPont™ Tyvek® StuccoWrap™ before stucco application.[46]

The final construction method is the utilization of a rainscreen wall (Figure 9.10). Essentially, this type of construction introduces a well-defined cavity between the back side of the stucco and the drainage plane (WRB) by installing preservative-treated wood furring strips vertically (coinciding with the studs) between the metal lath (typically paper-backed) and the WRB. This allows for gravity-induced drainage of any incidental water that may infiltrate the stucco-clad finish.

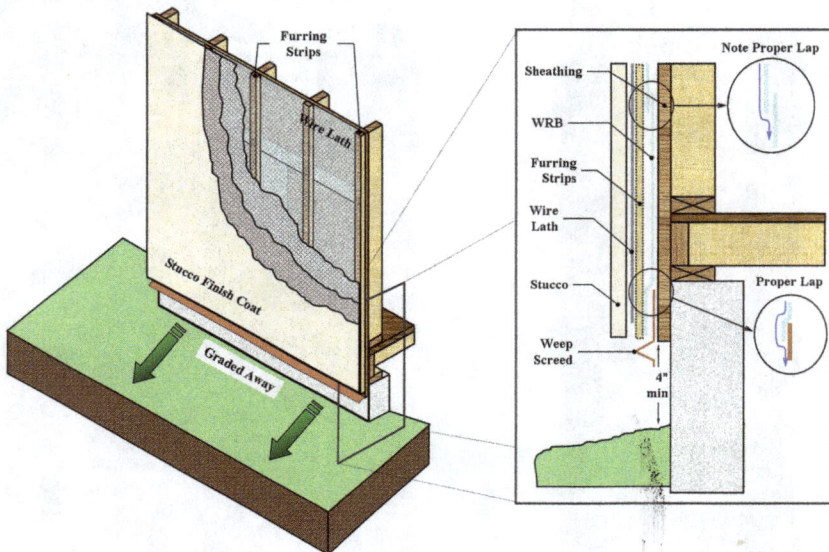

FIGURE 9.10 General construction and water management details – stucco-clad wall (rainscreen construction detail).

3. In order for the principle of drainage to function as it was intended, the lower edge of the stucco *must* be terminated *above* the level of the surrounding finished grade (which also needs to slope away from the foundation). Infiltrated water has to have an unobstructed pathway back to the exterior of the building.

Case Study 9.2

In the case study shown in Figure 9.11, the owners of this home had noticed a small amount of what appeared to be mold on the lower baseboard of their family room wall after moving a bookshelf. Subsequent remediation activities involved removing the interior drywall, which then led to the discovery of widespread water damage staining, mold, and deterioration of the framing members. A site investigation was completed in order to find the source of the water/mold.

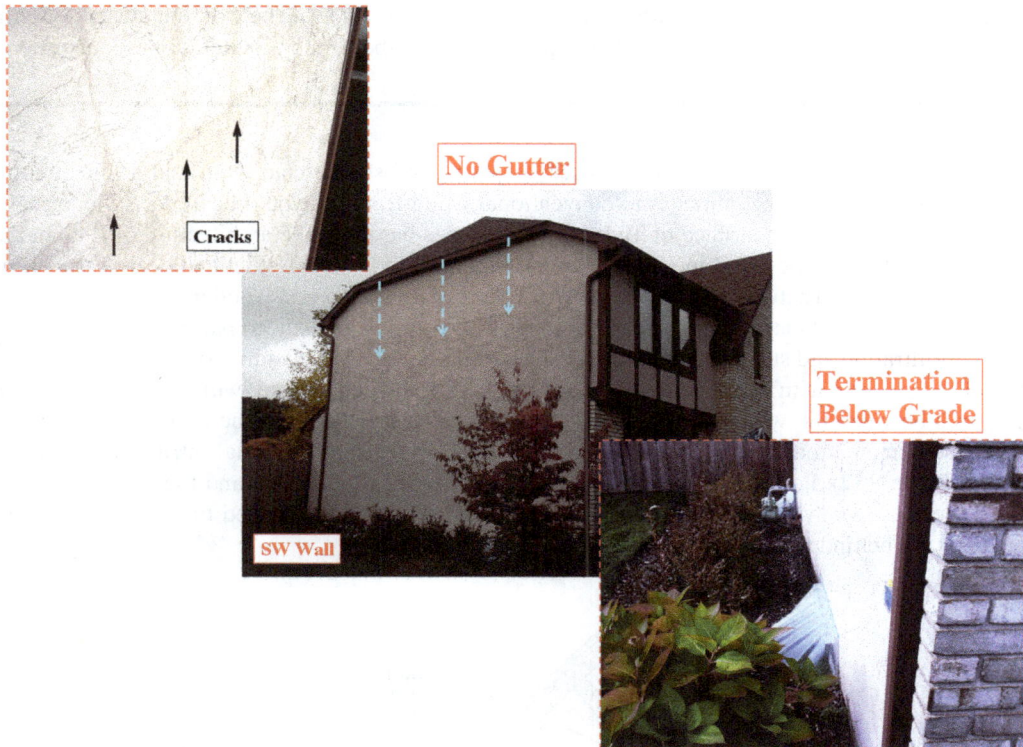

FIGURE 9.11 Stucco wall (SW elevation) – no gutter along roof eave, stucco termination below grade, and crack patterning to stucco finish coat.

Following the initial interior inspection of each floor, it was evident that the condition noticed on the first floor was widespread and present throughout the entire southwest exterior wall. An exterior inspection revealed the following:

- The affected wall faced southwest (i.e., direction of predominant wind and weather patterns).
- No gutter was attached to the roof eave and drip lines were visible in the local landscaping along the foundation.
- The lower edge of the stucco was terminated below the local landscaping.
- Distinct and discernable cracks were present in the finish coat. The patterning of the cracks corresponded with the sill plate and wall studs of the wood-framed construction. Evidence of moisture staining was present along the cracks.
- The lower portion of the wall, finished just above grade, appeared to bulge outward.

In the case of this home, the lack of adequate deflection mechanisms (i.e., wall's exposure to prevailing wind and weather patterns and lack of gutter) likely exacerbated exterior water to infiltrate the stucco finish where it was obstructed from draining back to the exterior due to the termination of the stucco below finished grade. This led to a lack of a properly functioning drainage mechanism, which allowed for the accumulation of water within the wall assembly, leading to cracks in the stucco and widespread water staining, mold, and deterioration to the wooden wall framing of the home. A complete lack of a WRB between the sheathing and stucco also contributed significantly to the conditions. The entire situation caused major headaches and worries for the homeowners who were distraught at the time of inspection and had to pay for the repairs since they were associated with poor workmanship by their builder. The solution was to remove the existing stucco system, including damaged wall members and then replace wall members, add a gutter/downspout system, and properly reinstall the stucco cladding.

In addition to the three most common causes, a specialized stucco-clad wall, and a fourth common cause for water-related damages to conventional wood-framing would be Tudor-style, or half-timbered, homes. The appearance of Tudor-style homes is based on the architecture of England at the end of medieval times, and modern-day construction mimics the aesthetic appeal of the original post-and-beam structural framing look by incorporating stucco and wooden trim boards. This method of stucco application can leave the exterior wall framing and stucco particularly vulnerable to water infiltration and subsequent damage.[44] The design of the wall cladding typically has the surfaces of the wooden trim boards slightly above that of the adjacent stucco, which is formed within the spaces between the wood. These raised surfaces and seams between the two dissimilar components can create areas that are susceptible to water collection and potential intrusion behind the stucco (Figure 9.12). For this reason, the intersections between the stucco and the wooden boards (which there are many) must be properly sealed and continually maintained to help deflect water from entering behind the stucco.

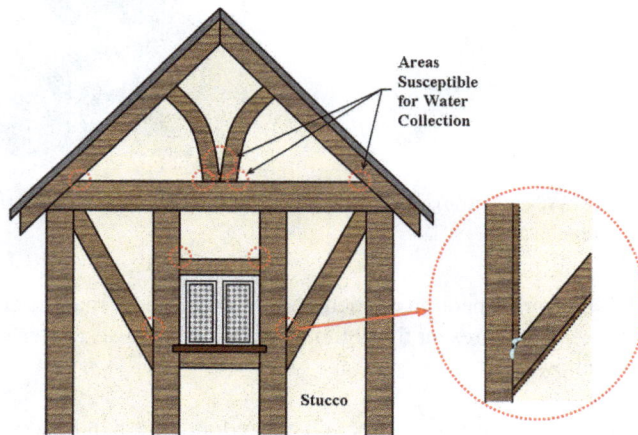

FIGURE 9.12   General construction and water management details – points of water collection on Tudor-style stucco.

Case Study 9.3

In the case study, illustrated in Figure 9.13, the owner of this home was in the process of replacing the wooden trim boards on the front of his Tudor-style home when he discovered heavy water damage and rot to the lower portions of the exterior wall framing. However, no evidence of water damage was noticed on interior portions of the home.

FIGURE 9.13 Tudor-style home – water damage and rot to wall framing below stucco/wood intersection and untaped foam sheathing seam.

A site investigation was completed to determine the cause of the damage. At the time of the site inspection, the exterior wall on the front of the home (west elevation) was in the process of being replaced with new framing, stucco, wood trim, and the appropriate WRB and flashings, so the original conditions for that wall were unable to be determined at the time. However, inspecting other elevations around the home revealed that the sealant along the intersections between the stucco and the wooden trim boards was heavily weathered, degraded, and cracked and provided potential pathways for water to infiltrate behind the stucco.

Photographs were provided by the insurance adjuster who was present when the original west elevation wall was removed. Further analysis of the photographs exposed the following to the investigator:

- Double-sided, foil-faced foam sheathing was used as the WRB behind the stucco and wooden boards. However, the wood panels were butted up against one another and the resulting seams were not taped or covered in any way.
- Evidence of water staining was present on the foil-faced surfaces of the foam sheathing at the vertical seams between the stucco and wooden trim boards.
- The wall stud directly below the vertical, untaped seam in the foam sheathing contained heavy water damage staining and light rotting conditions.
- The water damage to the stud continued down toward the bottom of the wall where it was not properly flashed, causing long-term rot to the lower portions of the studs, the sill plate, and even the band board.

The improper water management details for this Tudor-style stucco-clad wall, particularly those associated with the principles of deflection (i.e., sealant along stucco/wood interfaces) and drainage (i.e., untaped drainage plane/foam insulation), allowed for water infiltration directly into the wood-framed wall. The conditions were further exacerbated by the fact that the affected west wall faced predominant wind and weather patterns, thus allowing for it to be subjected periodically to wind-driven precipitation.

9.3.2   BRICK AND STONE MASONRY VENEER

Masonry walls have been around for centuries and are an extremely durable construction material. However, as with most other exterior cladding systems, most problems encountered by masonry walls are directly related to the unintended consequences of water infiltration. Typical, water-related problems to masonry include: (1) water penetration, (2) damage from freeze–thaw actions (i.e., cracking, spalling, disintegration, etc.), (3) dimensional changes, and (4) the appearance of efflorescence (refer to Chapter 16 for more information on efflorescence).[47]

Historic construction of masonry walls (typically brick) intended for them to serve as both the structural system for the building and its primary water resistor. Due to the masonry walls' monolithic and large size, this typically was not a problem; water would infiltrate to some degree but not enough to reach the interior portions of the building.

Modern-day masonry veneer walls are generally constructed of a single wythe of brick or stone and only provide aesthetic appeal and *some* water deflection. In fact, under normal service life conditions for many masonry veneer walls, it is nearly impossible to provide the necessary deflection components needed to keep a heavy wind-driven rain from penetrating masonry to some degree.[27] Figure 9.14 shows water penetration through a brick veneer wall (at a mortar joint) during a water test.

FIGURE 9.14   Water penetration through brick veneer during water test.

Water was visible on the interior side of the wall after just a few minutes of spraying.

Due to the inherent porosity of brick and mortar, the best approach to ensure the wall will perform well is to assume that some amount of water will infiltrate behind the masonry and then to provide proper detailing in order to redirect this water back out to the exterior.

Effective water management for anchored brick veneer walls is typically obtained by using the rainscreen method, thereby utilizing a drainage plane and an air space between the brick and the wooden wall sheathing (Figure 9.15).

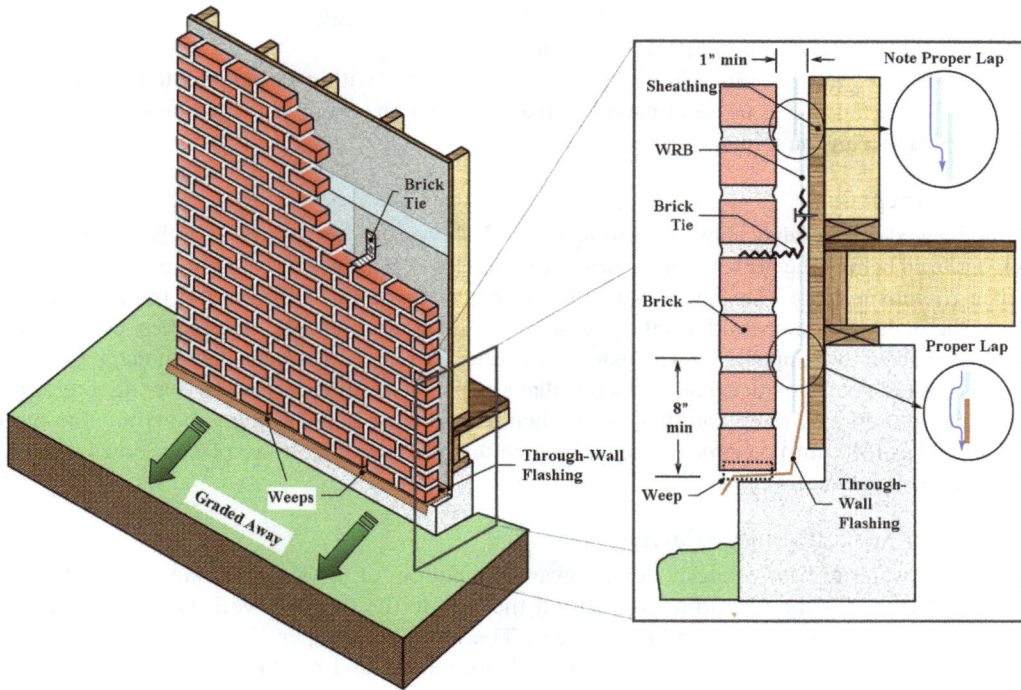

FIGURE 9.15 General construction and water management details – typical brick masonry veneer wall.

Water that penetrates through the single wythe of brick reaches the drainage cavity (i.e., air space) where it then flows down the back face of the brick toward the bottom of the wall until it encounters the through-wall flashing, which then redirects it back to the exterior through the weeps.

Current IRC requirements and best practices recommendations for anchored brick veneer construction, as they pertain to water management with conventional wood-frame construction, are detailed below.

9.3.2.1 Through-Wall Flashing

Through-wall flashing typically refers to a membrane installed beneath the first course of brick located at the base of the wall above finished grade and at locations of support in the exterior brick veneer. It serves to collect any incidental water infiltration and facilitates its drainage back to the exterior of the wall. Current IRC[4] requires only that flashing be placed at these locations within the masonry veneer wall and does not give any prescriptive details regarding proper installation or dimensions. Long-accepted trade practices recommendations, such as those given by the Brick Industry Association (BIA), state that proper design requires that flashing be placed at wall bases, window sills, heads of openings, shelf angles, projections, recesses, bay windows, chimneys, tops of walls, and at roofs.[34,35,47] Further, best practices state the following:

- The flashing should extend beyond the face of the brick wall in order to form a drip edge.
- Flashing should not be terminated short of the face of the brickwork.
- The flashing should extend a minimum of 8 inches vertically up the backing wall where it is then lapped by the WRB.
- Flashing sections should be lapped at least 6 inches with each other and sealed with mastic or flashing-compatible adhesive.

- Preformed flashing corner pieces or field cut, lapped and sealed sections should be incorporated to achieve continuity around corners.
- The flashing must be turned up at least 1 inch at each end within a head joint to form a dam where the flashing is not continuous such as over and under openings and on each side of vertical expansion joints.

9.3.2.2 Weeps

Weeps, located immediately above the through-wall flashing, serve as pathways for the water to drain back out from behind the brick veneer. Current IRC[4] and best practices[3,21,34,35] require that weep holes be spaced a maximum of 33 inches (825 mm) on center and not be <3/16 of an inch (5 mm) in diameter. Again, the building code provides only general or minimal requirements, and the best practices are typically referred to when dealing with exterior building envelope water management details.

The BIA (i.e., best practices) recommends that an open head joint, formed by leaving mortar out of a joint, be used as a weep and that weeps should be at least 2 inches high and spaced no more than 24 inches (610 mm) on center. They also state that the metal, mesh, or plastic screens may be placed in the head joint weeps.

9.3.2.3 Air Space/Drainage Cavity

In order for water to flow properly down toward the bottom of the air space (and drained out of the wall), there must be a continuous path all the way to the through-wall flashing and weeps located at the base of the wall and other areas. Therefore, it is imperative that the air space be of sufficient width and kept clean, particularly from mortar and mortar droppings, which tend to fall into the space as the wall is being constructed. Mortar droppings may clog the weeps, preventing proper drainage, and can even create a direct, continuous pathway for water/moisture to span the air space from brick to sheathing oftentimes at protrusions such as brick ties. In these instances, if an inadequate, highly permeable WRB is present over the wood-based sheathing, liquid water could pass from the water-saturated mortar spanning the air space to and through the WRB and into the interior portions of the wall assembly. This method of water/moisture transport is due to capillary continuity and occurs when a porous building material, such as mortar or concrete, is in direct contact with the WRB, thus reducing or eliminating its water repellency.[46,48] Building code and best practices both require at least a 1-inch wide air space between the sheathing and the back side of the brick. Best practices may also include the utilization of a drainage mat between the sheathing and brick to help prevent mortar from entering the air space. These mats are typically made of a plastic mesh or other porous material to allow for the proper drainage of gravity-fed water.[47]

9.3.2.4 Common Deficiencies with Brick Veneer Contributing to Water Infiltration

Water (and mold) cause and origin investigations dealing with brick veneer, more often than not, reveal a deficiency in the drainage principle of water management in the walls, particularly through-wall flashing, weeps, and/or air/drainage spaces.

9.3.2.4.1 Through-Wall Flashings and Weeps

This particular field investigation has, in numerous instances (particularly in residential construction), been omitted altogether by builders and bricklayers. These are the primary means of redirecting any form of water within the wall assembly back to the exterior and should not be omitted.

Although through-wall flashing and weeps have not typically been used in residential construction in the past, best practices would warrant their presence. Situations such as face sealing the exterior of the brick veneer could theoretically prevent/deflect water from leaking through the brick, but research and experience suggest that this approach is unreliable. The coating weathers and ages quickly and is often not replaced or is improperly installed.[12] Face sealants are secondary water management details, not unlike caulks and sealants, and cannot be relied upon as primary water management systems.

Another solution might be the construction of a massive, monolithic, multi-wythe barrier wall system that would, like historical construction, provide such a large cross-sectional width for the water to penetrate, that it would likely dry before reaching the interior. However, this solution is not practical in modern construction.

Figure 9.16 illustrates good examples of proper installation and incorporation of through-wall flashing and weeps at the lower termination of brick veneer above finished grade.

Unfortunately, the proper construction details shown in Figure 9.16 are rarely encountered when problems with water infiltration are discovered with brick veneer in forensic investigations.

FIGURE 9.16 Example of proper installation and incorporation of through-wall flashing and weeps in a brick veneer wall.

Case Study 9.4

In the case study shown in Figure 9.17, the owners of this brick home had discovered that water was present within the wall cavity of the first floor dining room and in the basement area below the dining room. Further, it was determined through a comprehensive interview that the brick veneer adjacent to and below a gutter opposite the affected interior areas was wet or saturated with water after rain events.

During the course of the investigation, the following observations were documented:

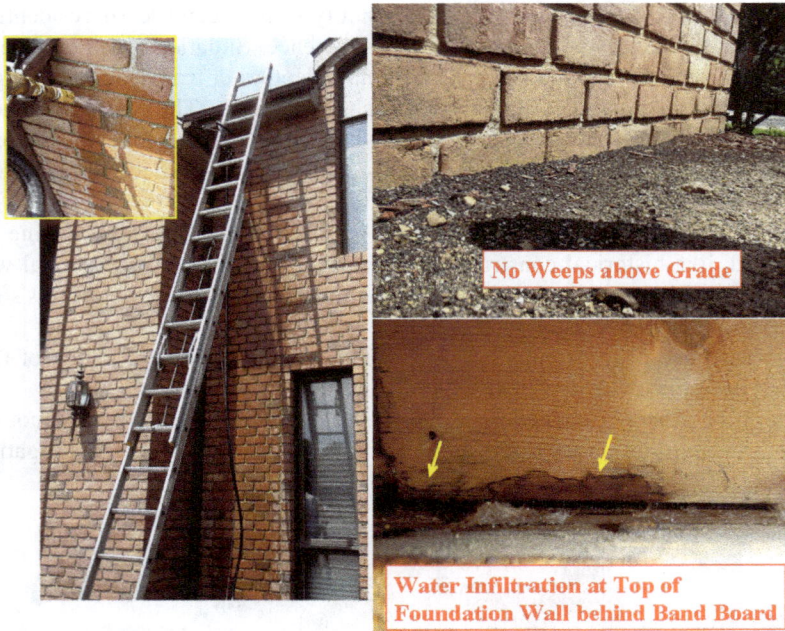

FIGURE 9.17   Water test of brick veneer – clogged gutter – no weeps – water infiltration.

- The gutter along the brick wall was clogged with debris, particularly around the inlet for the downspout.
- Evidence of soil washout was present in the ground surface directly below the gutter, suggesting past overflows of water.
- No weeps were observed along the lower courses of brick above the local finished grade surfaces. It was probable that through-wall flashing was not present; however, destructive testing of the wall would be needed in order to verify if that was the case.

As depicted in Figure 9.17, a water test was conducted on the exterior and set up in a fashion to recreate conditions when the gutter would overflow, spilling water onto the brick wall. It took a fair amount of time, but evidence of water infiltration was discovered in the basement at the top of the CMU foundation wall. Water began to seep out from under the wooden band board in an area where patterns of water staining were observed during the inspection of the basement.

From this investigation, it was readily apparent that the lack of a proper drainage system, including weeps for the brick veneer, leads to the interior areas of concern. Of course, proper maintenance of the gutter (i.e., removing debris) would have lessened the severity of the issue, but nevertheless an adequate drainage system for the wall would have prevented water from entering the home.

Case Study 9.5

In the case study shown in Figure 9.18, a site investigation was conducted after this homeowner noticed water on the floor of the kitchen pantry. To determine the cause of water infiltration, the interior side of the wall in the pantry was removed and subsequent mold growth and water damage were observed.

An inspection of the exterior finishes opposite the wall in question revealed the following observations:

FIGURE 9.18   Water test of through-wall flashing with an end dam needed beneath concrete cap at corner interface with brick veneer and vinyl siding – lack of flashing.

- The exterior finish consisted of a lower brick veneer wall and upper vinyl siding separated by a precast concrete cap. Metal apron flashing was present along the interface between the cap and the siding and extended up behind the siding.
- The slope of the concrete cap was measured and found to slope toward a corner interface between the cap, brick veneer, and vinyl siding.
- No through-wall flashing was present beneath the concrete cap along its intersection with the lower brick veneer.
- Through-wall flashing and weeps were present in the lower course of brick just above finished grade.
- The exposed wall cavity from the interior revealed a rather narrow air space between the back face of the brick and the exterior sheathing board. Further, mortar droppings were present along the bottom of the air space and partially obstructed the functioning drainage system of the wall.

A water test was conducted at the corner interface between the concrete cap, brick veneer, and vinyl siding opposite the interior kitchen pantry. Immediately upon commencement of the test, water began pouring into the pantry.

According to best practices, through-wall flashing should have been installed beneath the concrete cap projection atop the lower brick veneer. Additionally, due to the nature of the construction at the corner interface between the cap, brick veneer, and vinyl siding, an end dam, formed in the through-wall flashing, was needed to redirect water away from the interior of this home.

9.3.2.4.2 Air Spaces

Air Spaces immediately behind brick veneer walls are essential to the drainage capacity of the wall assembly. Further, as stated in preceding sections, the air space must be of sufficient width and largely free of potentially obstructing debris. This primarily includes mortar droppings, which tend to fall into this space during construction of the wall. In the case study given in Figure 9.16, through-wall flashing and weeps were present at the bottom of the wall; however, mortar droppings were observed within the air space that appeared to obstruct the ability of the wall to drain incoming water sufficiently to prevent interior intrusion and damage. Given enough time and *only intermittent* water infiltration, the wall drainage may have functioned adequately, but in the case of this particular home, large amounts of water seemed to be entering due to the lack of through-wall flashing and end dam between the brick veneer and the concrete cap.

9.3.2.5 Deficiencies with Stone Veneer Contributing to Water Infiltration

Stone veneer can consist of either: (1) natural stone such as sandstone, limestone, marble, or granite, which is durable and weather-resistant, or (2) manufactured stone, which is comprised of cement mixed with lightweight aggregates and color pigments and used to simulate natural stone at a fraction of the cost of natural stone.

The method of effective water management for stone veneer can be similar to that of brick or can even closely mimic stucco, depending on the type of stone veneer being used and its specific application conditions.

For example, for natural stone, best practice recommendations such as those from the Building Stone Institute (BSI)[49–51] state, much like brick veneer installation, an air space should be present behind the natural stone veneer for drainage and for air circulation, along with vent holes near the bottom and top of the wall to promote ventilation.

For manufactured stone and adhered natural stone veneer over wood-frame construction (Figure 9.19), best practices recommend that the stone veneer be set into a mortar scratch coat and setting bed that has been applied to a metal lath over a suitable WRB. This method of application is much like stucco and, as such, can pose similar problems in regard to proper drainage of any unintended water infiltration. Therefore, it is recommended that a defined drainage plane or weep system be utilized behind the cladding such as that created with the use of two layers of building paper or an optional drainage mat.[52]

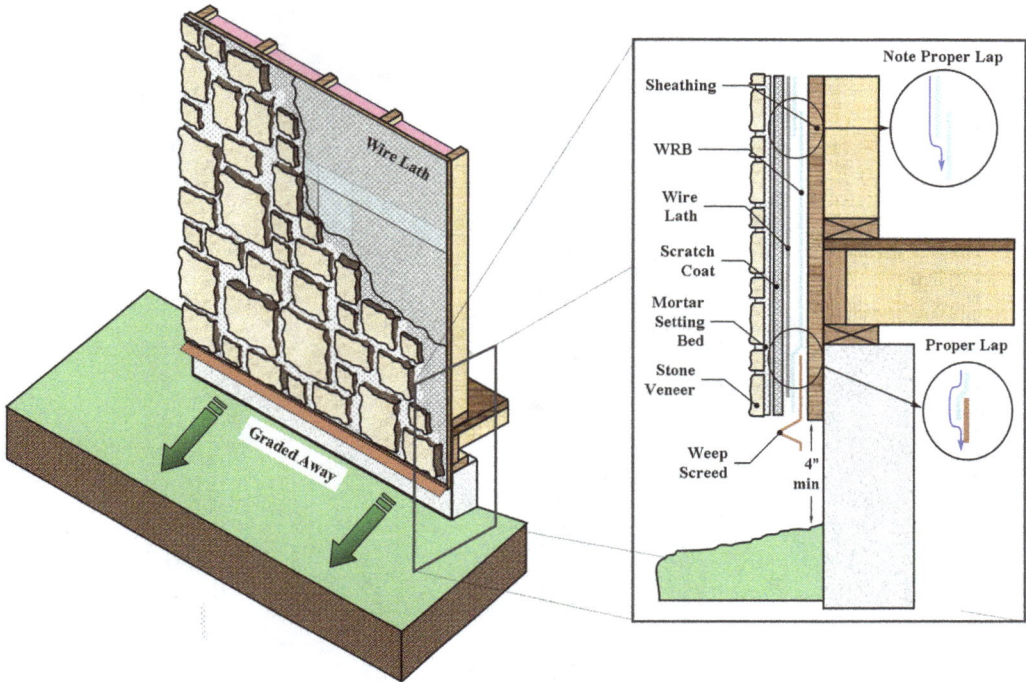

FIGURE 9.19 General construction and water management details – typical manufactured stone or adhered natural stone veneer wall.

Water infiltration problems with stone veneer, just like all of the exterior coverings discussed so far, are the result of poor detailing with respect to water management principles, particularly deflection and drainage, which are generally the most important.

Case Study 9.6

One of the disadvantages of manufactured stone veneer is its higher absorption of water (by weight) since it is made using rather porous cement.[52] Thus, the drainage plane behind the veneer is heavily relied upon to protect the wall sheathing and interior portions of the home. In the case study shown in Figure 9.20, the owners of this home noticed dampness to the carpeting along the north wall of their master bedroom, opposite a manufactured stone veneer exterior wall. When the interior drywall was removed, wet conditions were discovered to the insulation and Oriented Strand Board (OSB) sheathing.

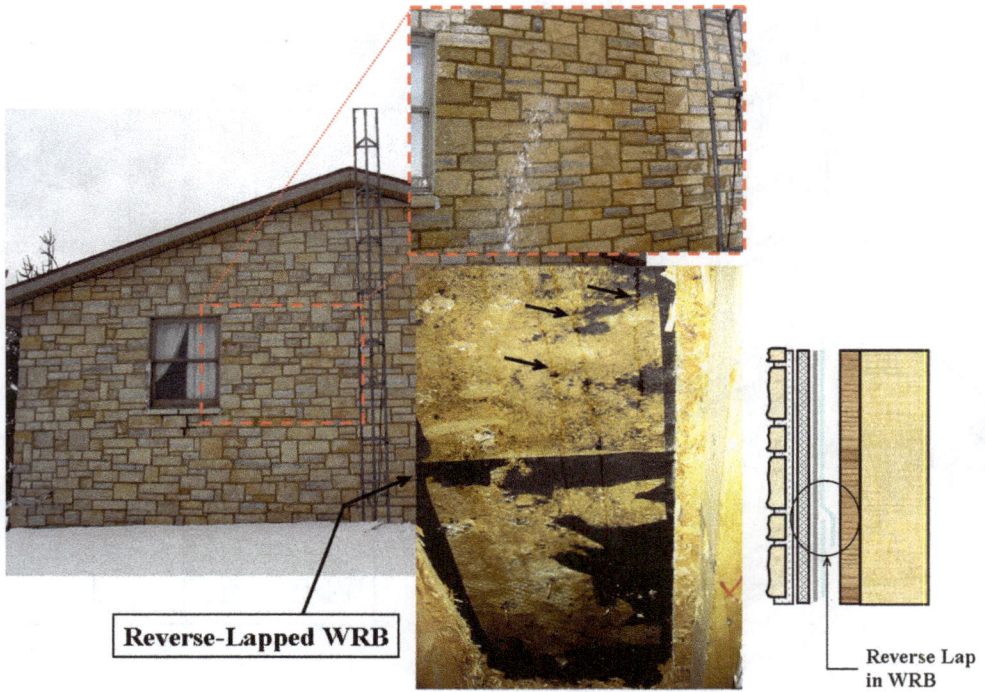

FIGURE 9.20   Improper lapping of WRB causing interior water damage through manufactured stone veneer wall.

During the site investigation, destructive testing was conducted from the interior side of the subject wall and removal of a portion of the exterior OSB sheathing, thereby exposing the WRB and the back side of the stone veneer. A critical deficiency in the installation of the WRB was revealed where the upper course of the black felt paper was reverse-lapped behind the lower course of paper, allowing for water to travel down the drainage plane until it encountered this improper lap causing it to drain through the lapped seam and onto the OSB sheathing. Recall from earlier discussions of WRB s that they must be shingle-lapped with the upper course overlapping the lower course.

Case Study 9.7

In the case study shown in Figure 9.21, an entire condominium complex was experiencing water infiltration problems, opposite the exterior manufactured stone veneer. A contractor for the condominium association took it upon himself to remove portions of the stone veneer. When he did, he discovered heavy water damage, rot, and mold to the OSB sheathing. The patterning of the damages traced back to where the roof eaves intersected the stone veneer.

This particular instance is discovered quite often with stone veneer and stucco walls and is a prime example for the need of kick-out diversionary (KOD) flashing at the roofline.

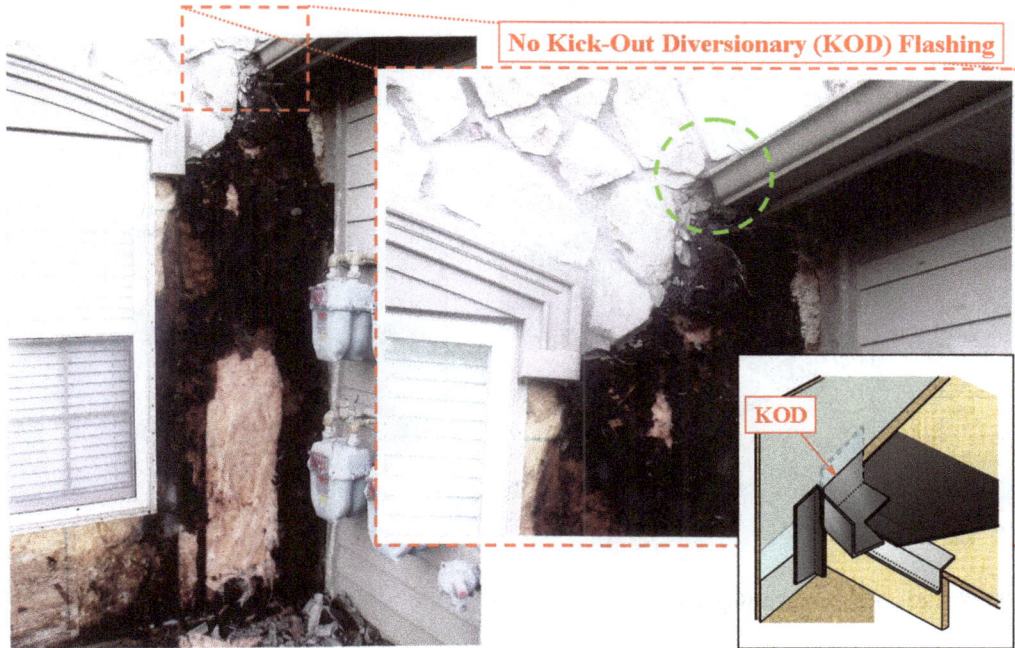

FIGURE 9.21 Heavy water damage, rot, and mold to sheathing behind manufactured stone veneer.

9.3.2.5.1 Kick-Out Diversionary (KOD) Flashing

KOD flashing is installed at the roof eave intersection to help divert draining water into the gutter instead of into the building, when properly installed. During a precipitation event, water draining from the roof travels alongside the sloped roof-to-wall interface. Recall from Chapter 8 that this water is prevented from entering along this interface with step flashing that has been installed behind the exterior cladding and properly integrated with the water-resistive system of the wall and the roofing material. At the eave where this step flashing terminates, the bend in the "L"-shaped flashing is actually on the interior side of the stone veneer (or stucco); therefore, water traveling along the step flashing is then channeled directly behind the cladding and onto the exterior sheathing.

Best practices and manufacturers of stone veneer, such as Owens Corning[®53] and Centurion[®] Stone,[54] realize the need for KOD flashing and provide necessary requirements for its dimensions and incorporation into the water-resistive system of a building (i.e., adjacent sections of flashing and WRB). Experience gained from field assessments and these best practices[53–55] is illustrated in Figure 9.22.

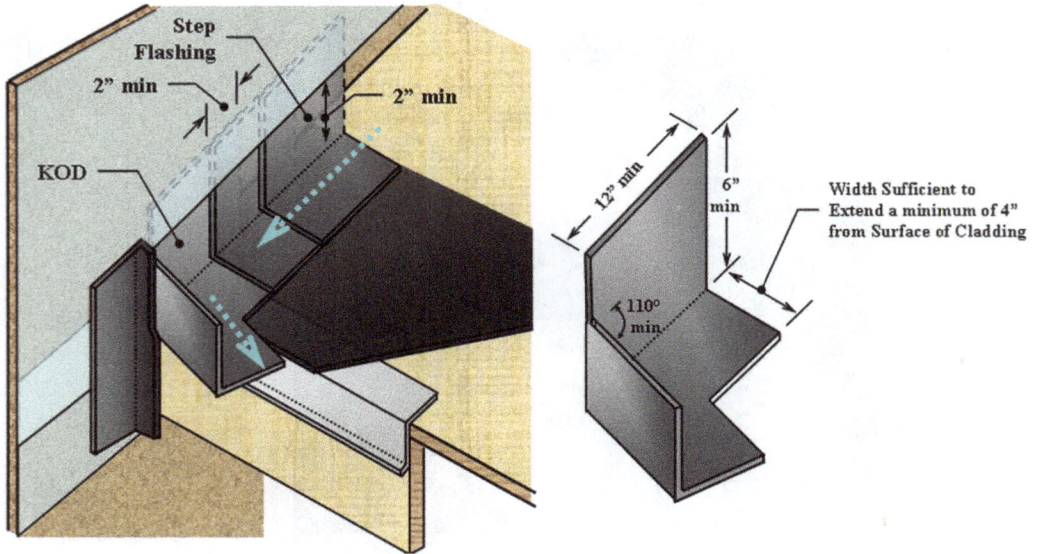

FIGURE 9.22   Kick-out flashing details.

Best practices consist of the following details:

- KOD flashing should be installed in a manner similar to step flashing (Chapter 8).
- The KOD flashing should consist of either a preformed, seamless flashing component or a one-piece flashing section with watertight seams.
- The vertical leg of the flashing should extend a minimum of 6 inches up the wall behind the WRB, which is lapped over to facilitate drainage.
- The horizontal leg of the flashing should extend a minimum of 6 inches out onto the roof *over* the drip edge and roof underlayment.
- The length of the flashing should extend a minimum of 6 inches up slope from the lower edge of the roof decking.
- When installing the first course of step flashing, make sure it overlaps the up slope edge of the kick-out flashing by a minimum of 2 inches.
- The bottom edge of the WRB and exterior cladding should be terminated a minimum of 2 inches above the level of the roofline to prevent water from wicking up the wall as it drains.
- The angle of the diverter should be bent to provide a minimum of 110° from the vertical leg of the flashing and the exterior wall to prevent negative drainage and debris accumulation at the kick-out.

Case Study 9.8

In order for the kick-out diversionary flashing to function as it was intended, it is essential that it be the right type and installed correctly.

In the case study depicted in Figure 9.23, the owners of this stucco-clad home noticed water and buckled hardwood flooring on the first floor of their home and water in the basement below this area, following heavy snow accumulation and subsequent warmer periods, causing the snow and ice to melt. During the course of the inspection, it was apparent that the source of the moisture was originating near the intersection between the stucco and the roofline where KOD flashing *was* present.

FIGURE 9.23 Improper KOD flashing causing draining water to funnel into the exterior wall assembly behind the stucco cladding.

Water testing of the subject area revealed the following information:

- The KOD flashing was bent at an angle slightly past perpendicular with the exterior wall. A small amount of tree debris was present against the flashing.
- Soon after water testing was initiated, water began draining down the roof along the sloped roof-to-wall interface. Water was visible on the surface of the stucco below the end of the gutter and the KOD flashing.
- Upon inspection, the KOD flashing appeared to have been a field-formed section of "L"-shaped step flashing that had been bent to form a diversionary-type flashing. However, the seams created during the forming process were not watertight.

In this particular scenario, forming the KOD flashing in the field created a type of funnel, channeling water behind the stucco cladding and into the exterior wall assembly. Had flashing been constructed and installed properly (i.e., proper angle, watertight seams, or a seamless flashing component, etc.), water would have been diverted into the nearby gutter and properly drained away.

Note that although kick-out flashing was discussed with regard to stucco and stone veneer, it is considered best practice to install it with all types of various exterior claddings.

9.3.3 EXTERIOR INSULATION FINISH SYSTEMS (EIFS)

Exterior Insulation and Finish Systems, otherwise known by its acronym EIFS, combines a textured and colored finished layer with a layer of rigid exterior insulation. The most common type (polymer-based) consists of a reinforced basecoat applied to the insulation that consists of closed expanded polystyrene (EPS) and is either adhesively or mechanically attached to the exterior sheathing. The rigid insulation is then covered with a lamina composed of a modified cement basecoat with glass-fiber reinforcement after which the finish coat is applied.[55–58]

9.3.3.1 Historic Problems with EIFS

EIFS originally gained popularity in the 1980s due to its increased thermal performance and insulating qualities, ease of installation, and relatively low cost. However, early installations of EIFS attempted to employ a face-sealed approach to water management, which resulted in poor performance with significant water intrusion and rotting of wall cavity members.

Subsequently, class action lawsuits stating that the EIFS system was fundamentally flawed were filed.[56,57]

It was the opinion of many experts, validated by a review of best practices, that face-sealed EIFS claddings were "inherently defective and unfit for use as an exterior cladding system where moisture sensitive components are used without a provision for drainage, or in locations and assemblies without adequate drying."[57] The evidence supporting this opinion included the observations that[57]:

- Face-sealed is an approach to water management that essentially depends on the single, exterior-most layer to control all rain water penetration.
- A face-sealed approach relies on perfect workmanship and materials, which is contrary to historical experience.
- It is nearly impossible to prevent rain water from penetrating any one of the thousands of joints, penetrations, and/or cracks that are present at some point during the service life of EIFS.
- A face-sealed "perfect" barrier approach also relies on "perfect" sealant material installed in a "perfect" fashion onto surfaces that have been prepared "perfectly." This is improbable, bordering on impossible, to expect from even a properly trained tradesman or technician who needs to perform this "perfect" task thousands of times in a row for an EIFS-clad building.
- Even if the impossible was possible, a face-sealed barrier approach relies on the work of several different trades involved during the construction of a building in which any one problem could lead to future water-related issues. This poses a particular problem when moisture-sensitive materials are used such as wood framing.
- This approach also assumes that the window and door units are designed and manufactured to be leak-free over their service lives; experience and studies have shown that this is not true. In fact, a survey[46] of over 3,500 vinyl windows was conducted, and they were discovered to begin to leak to some degree within 2 years of the manufacturing date.
- Cracks are also an issue over the expected service life of the EIFS due to long-term weathering, aging, and inevitable building movement such as settling cracks.

Although many negative reactions are now elicited when it comes to EIFS exterior claddings, these systems can be successfully installed in most climates and exposures given new best practices that evolved after these early failures. Best practices assume that modern EIFS systems are designed, installed, and maintained properly and utilize a drainage space and an integrated WRB and flashings, just like all exterior assemblies discussed thus far.

9.3.3.2   EIFS Water Management Details

The importance of water management details, particularly drainage, cannot be emphasized too much for EIFS systems. Although EIFS does provide some protection from moisture at the basecoat level, sources of water infiltration typically occur along interfaces and around openings in the wall envelope.[57] When water inevitably infiltrates behind the EIFS cladding, whether it be through a crack from building settlement or below a leaky window, the water must be drained out of the wall. If not properly accounted for, history contains several examples of the rapid rate at which trapped moisture behind EIFS systems can rot conventional wood-frame construction.

Specific details regarding proper water management details for EIFS are provided by manufacturers such as Dryvit®,[55] which happened to be the company that introduced EIFS in the United States in the late 1960s, and from best practices organizations such as the EIFS Industry Members Association (EIMA).[58] It is beyond the scope of this book to introduce and analyze the wide variety of interface and joint details that are needed for EIFS-clad buildings. Therefore, recommendations from these best practices should be consulted in order to determine proper detailing for site-specific conditions.

However, some important points regarding modern EIFS water management details are illustrated in Figure 9.24.

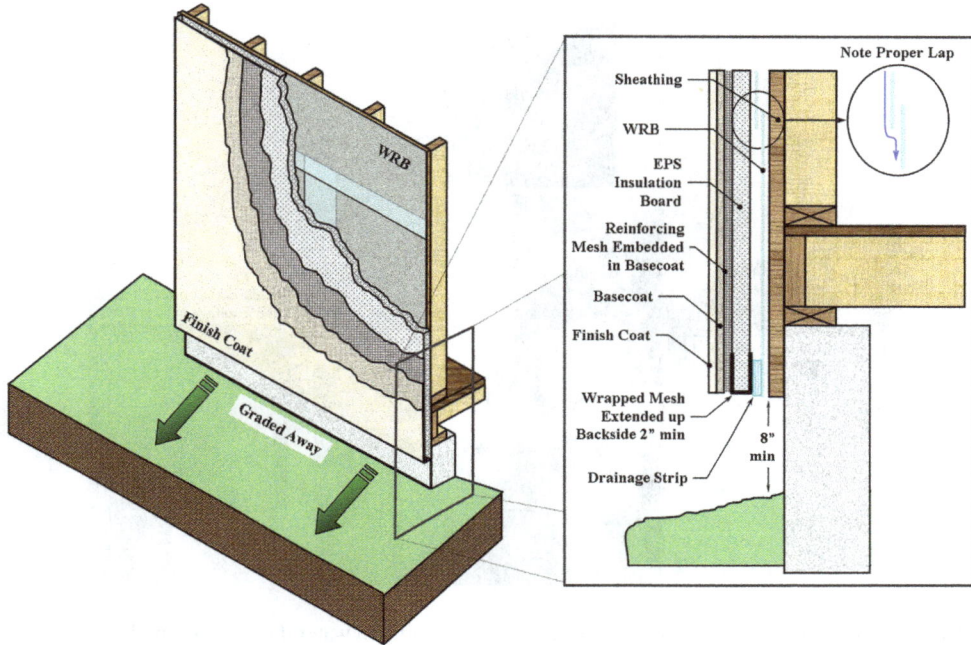

FIGURE 9.24   General construction and water management details – typical EIFS-clad wall.

These EIFS water management details are summarized as follows:

- Adequate clearance of EIFS from: (1) finished grade, (2) intersecting roof surfaces, and (3) concrete sidewalks, porches, driveways, and foundations must be maintained.
- Sufficient drainage space between the back side of the EIFS insulation board and the continuous drainage plane (i.e., WRB and flashings) should be present. According to best practices installation instructions from Dryvit®,[55] this drainage space is created by applying the adhesive to the back side of the insulation board in a vertical, notched trowel configuration, creating channels for water drainage.
- Proper flashing at locations such as: (1) roof intersections, (2) windows, doors, and other miscellaneous openings in the building envelope, (3) interfaces with other claddings, (4) at locations of decks and balconies, and (5) adjacent EIFS joint interfaces must be installed.
- Proper sealant selection, application, and surface preparation along interfaces with: (1) windows and doors, (2) service penetrations, (3) other cladding types, (4) decks and balconies, and (5) adjacent EIFS joint interfaces must occur.

It should be noted that for sealant application, the sealant should be applied to the EIFS basecoat and not the finish coat. The finish coat is somewhat porous and if sealant is applied to it, moisture can travel through and past the finish coat bypassing the sealant.[59]

Case Study 9.9

A site investigation was performed to determine possible cause(s) for water damage and cracking to the exterior EIFS of the home. It did not take long to see that many critical details pertaining to water management were not followed during the installation of the EIFS, as shown in Figures 9.25 and 9.26.

FIGURE 9.25   Lack of proper water management details leading to water damage behind EIFS cladding.

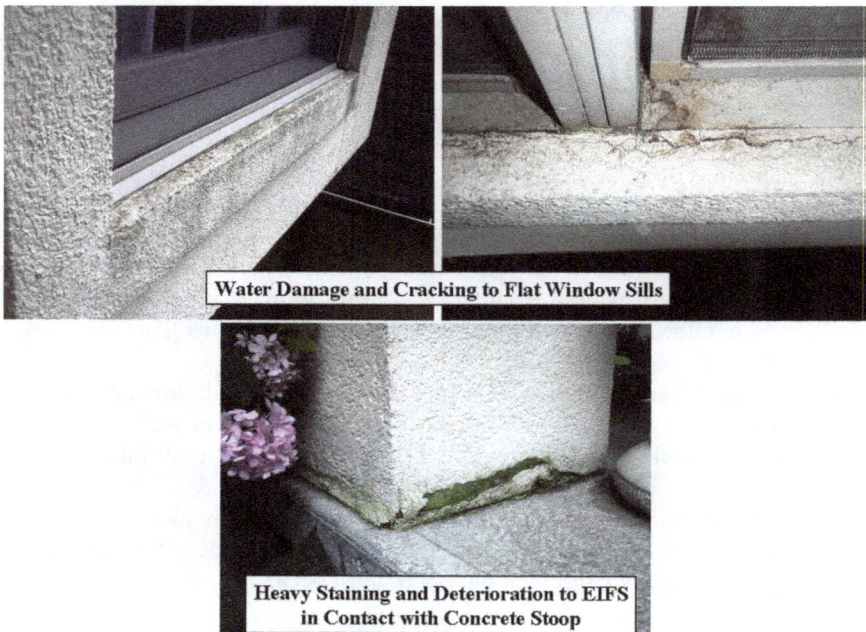

FIGURE 9.26   Water-damage staining, cracking, and deterioration to EIFS due to improper management details.

These deficiencies included the following:

- The EIFS was in contact with the asphalt shingles along the sloped roof-to-wall interfaces. Along each of these interfaces, the EIFS finish coat was discolored and eroded/water-damaged, exposing the mesh reinforcement beneath.
- No KOD flashing was present where the roof eave intersected the exterior EIFS-clad walls. Consequently the EIFS at and below these locations was heavily discolored, stained, and damaged (i.e., cracked).
- The EIFS finish coat on the windowsills was cracked and stained. The sills were flat and not sloped away from the windows for proper drainage.
- Especially in the area of the front porch stoop, the EIFS in contact with the concrete was heavily stained and deteriorated, exposing the mesh reinforcement. Best practices require a minimum gap of ¾ of an inch between EIFS and concrete to allow for drainage and to prevent water from possibly wicking up into the EIFS.

9.3.4 SIDING

Siding is available in a wide range of materials, styles, profiles, colors, and textures. A large number of modern-day residential homes are clad in some type of siding, including vinyl, aluminum, wood, or fiber cement. The most popular of these materials is vinyl siding due to its cost-effectiveness, durability, performance, ease of installation and maintenance.[60]

9.3.4.1 Siding Water Management Details

The installation of the various types of siding may differ somewhat from product to product based on the specific material, manufacturer, and style, but the water management details are generally the same.[60–63] Each siding system incorporates a WRB and integrated flashing behind the cladding in order to maintain a continuous drainage plane (Figure 9.27).

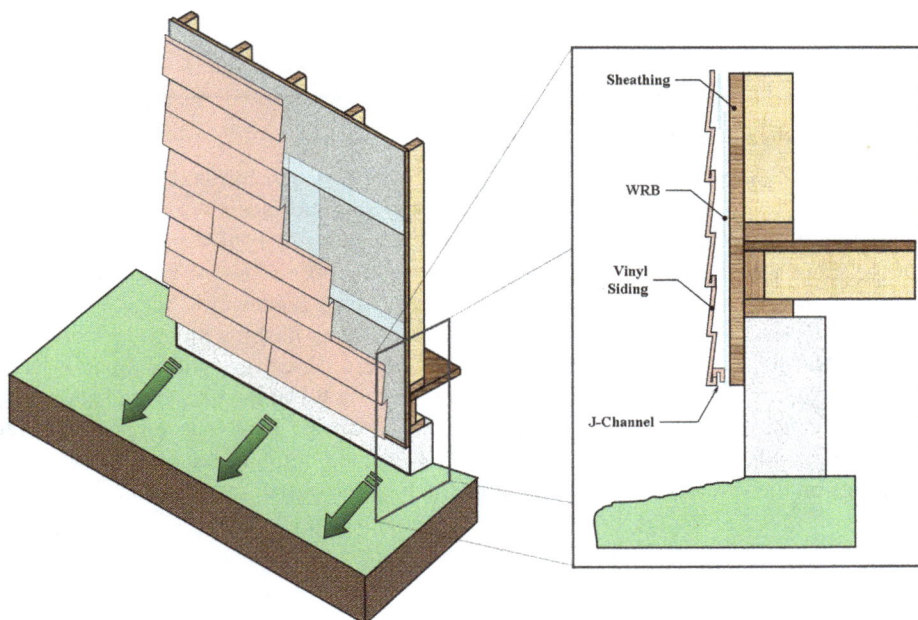

FIGURE 9.27 General construction and water management details – vinyl siding.

Wood siding differs slightly in that it utilizes a rainscreen wall construction with treated, vertically oriented wood furring strips.[46,61] Creating this air space between the back side of the siding and the WRB with the furring strips helps to ensure that the wood will dry evenly after becoming wet (Figure 9.28).[40,61]

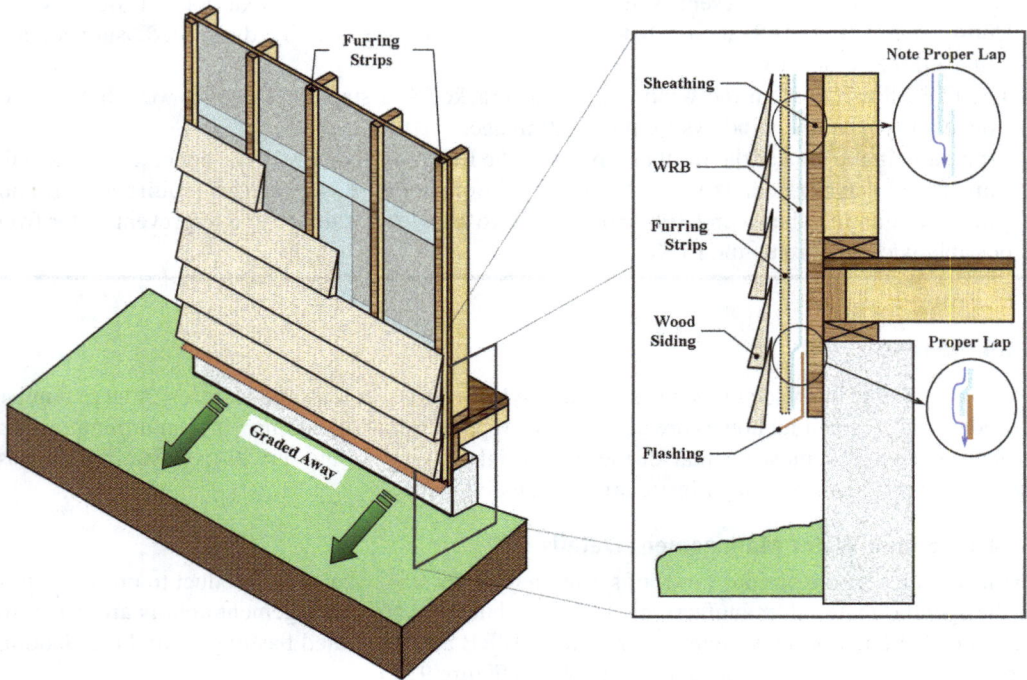

FIGURE 9.28   General construction and water management details – wood siding utilizing rainscreen wall construction.

Case Study 9.10

The condominium unit shown in Figure 9.29 was experiencing water intrusion issues at a couple of different locations. The first location was along a ceiling/wall interface above the staircase leading to the second floor of the unit. Based on inspection observations and subsequent water testing, the source of the water was determined to originate between a loose section of vinyl siding and wood fascia above the lower roofline. From there, the water traveled downward over the surface of the exterior sheathing until it encountered the interlaced metal step flashing along the sloped roof-to-wall interface. Normally this would not be cause for concern. However, in this situation, no WRB was present over the sheathing. Therefore, the water traveling down the sheathing continued down and behind the vertical leg of the step flashing since an overlapped WRB was not present to redirect it onto the flashing where it could have been properly drained out onto the roof surface. Note the chalk markings on the shingled surface indicating the location of the interior stain.

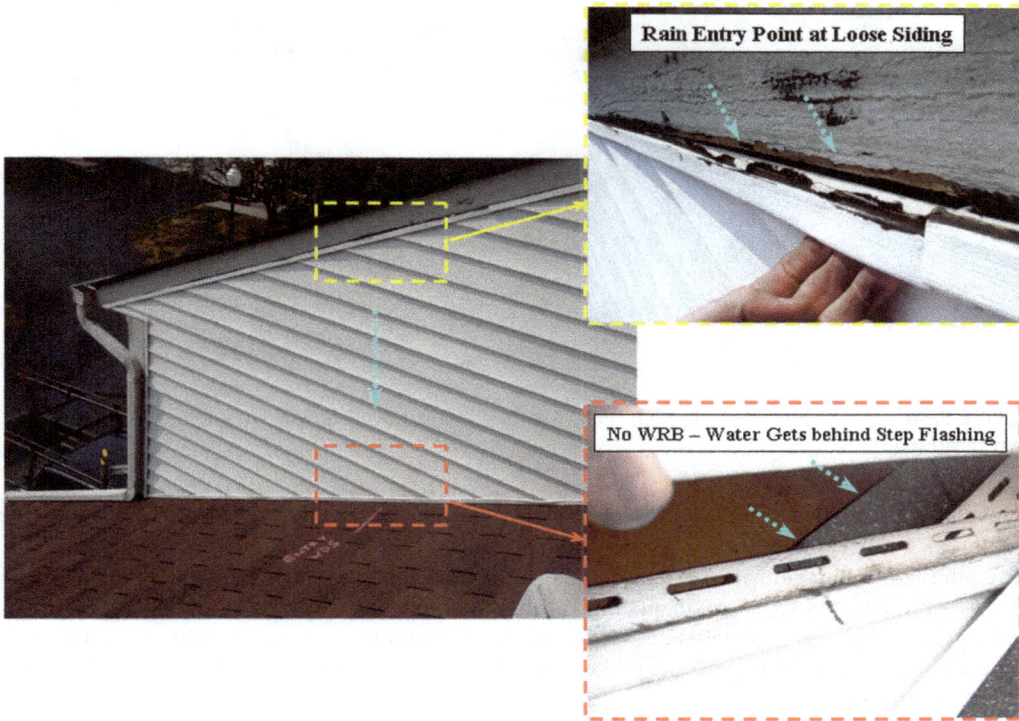

FIGURE 9.29 Rain water infiltrating through loose siding interface along fascia travels downward and enters behind roof step flashing due to no WRB.

The second area of reported water infiltration was occurring through the top of the sliding glass patio door on the side of the unit. The initial visual inspection of the exterior surfaces above the door did not reveal a significant deficiency. Water testing was conducted and directed along the top of the door to induce a leak. However, these attempts were unsuccessful. An inspection of the window above the patio door revealed historic sealant along the wooden trim interfaces. The decision was made to water test the window. After a few short minutes, droplets of water began forming along the top of the door. In order to determine exactly what was occurring, the siding immediately above the door was temporarily loosened for observations. Once loose, the problem was evident (Figure 9.30).

FIGURE 9.30 Water infiltration through second-story window travels downward and enters behind gap in foam sheathing.

Insulated foam sheathing was present over the exterior sheathing; however, there was a rather large break between adjacent foam panels above the head flashing along the top of the door. The drip-patterned water stains on the surfaces above the break and the stained condition of the sheathing further supported what was happening. Water had entered around the second floor window above and traveled over the surfaces of the foam sheathing until it encountered the gap in the foam, where it was able to travel inward and make its way behind the head flashing of the door and through to the top of the door.

9.4 WATER MANAGEMENT DETAILS FOR COMMON FENESTRATION ELEMENTS

Fenestration, or the openings within the building envelope (i.e., windows and doors), is typically the leading cause of water infiltration issues. These openings interrupt the continuous drainage plane created behind the exterior cladding. In order to maintain the principle of water management properly, the drainage plane must be modified to divert the flow of water around these susceptible areas. This is successfully accomplished with a series of lapped elements (i.e., WRB and flashings) installed in a fashion so as to prevent water from getting past them and into the opening. If done properly, this reduces the opportunity for water intrusion at these locations and the consequential rot of wood and corrosion of the fasteners, which can weaken the frame of the fenestration.

9.4.1 WINDOWS AND DOORS

Water leakage associated with windows and doors can occur between the units and their frames but the predominant leakage paths, based on experience and experimental study results, are those associated with the window and door-to-wall interfaces.[6,64-66] Infiltrating water through these framing/wall interface leakage paths can cause considerable amounts of damage to the wooden framing, which, oftentimes, is concealed until the water damage has become much more extensive. Experience indicates that the primary cause for these issues is the lack of adherence to: (1) relevant codes, (2) relevant standards, and (3) industry best practices.

9.4.1.1   Codes and Standards for Windows and Doors

Performance and construction requirements for exterior windows and doors are governed by Section R612 of the current IRC.[4] Regarding water management, the relevant code refers to the general flashing requirement (see Section 9.2.3.1.3) and requires that installation and flashing installed follow manufacturers' instructions. Experience with window and door-related water claims suggests that many times adherence to manufacturers' instructions can be sufficient. On the other hand, these instructions can be inferior to details provided by industry standards and best practices.[38]

Details for window and door installation are provided by the ASTM Standard E 2112-19c "Standard Practice for Installation of Exterior Windows, Doors & Skylights,"[44] which focuses on detailing and installation procedures intended to minimize water infiltration. It is beyond the scope of this book to analyze the comprehensive detailing for the variety of windows and doors available; the reader is referred to the ASTM E 2112-19c for the specific details that would be appropriate for the window or door and local conditions that apply.

9.4.1.2   Window and Door Water Management Details

The water management details for windows and doors follow the same general principles as those for exterior wall claddings discussed earlier. Likewise, in most instances, the window and door details for one type of exterior cladding will typically apply to most other claddings; however, slight modifications may be needed in order to deal with the specific water control needs based on the drainage behavior of the wall. As a general rule of thumb, the force of gravity is constant and acts in one direction, down. Therefore, the design, installation, and maintenance for a particular wall assembly should be performed with this consideration constantly in mind.

Water management details for windows and doors utilize proper details for sills/thresholds, WRBs, flashings, caulking, and a proper integration with the water-resistive system and continuous drainage plane of the wall.

9.4.1.2.1   Window Sill and Door Thresholds

Window sill and door thresholds that lack of a positive slope (i.e., away) are a common occurrence in the field and contribute greatly to potential water intrusion, particularly when sill pan flashing is omitted in the subsill portion of the wall below the window/door units (see later section). Windowsills and door thresholds that are nearly horizontal, or worse yet, sloped back toward the interior of the building, create relatively large ledges that can collect water and expedite deterioration of sealants, which lead to water intrusion.

Best practices recommend a pronounced slope that aids in the prompt drainage of water, thus deflecting it away from susceptible sill interfaces. The BIA[47] recommends a slope of 15° away from exterior windows and doors for brick veneer applications.

Another best practices design detail is the creation of a groove on the underside of the sill/threshold roughly an inch from its outside face, which serves as a capillary break, or drip edge, and stops the continuation of water (due to surface tension) along the underside of the sill and back to the exterior wall.[44] Figure 9.31 provides an example photograph of improper sloping of a brick windowsill where the slope was only 3°–4°, well below the BIA recommended value of 15°.

FIGURE 9.31 Brick window sill improperly sloped.

9.4.1.2.2 Water-Resistive Barriers (WRBs)

WRBs are a critical drainage element for the entire exterior wall envelope and are especially important around openings such as windows and doors, which create interruptions in the drainage plane. Special attention should be paid to the details around window and door openings during the design and installation of the WRB around these openings to ensure proper water management in these areas. Important points to remember with regard to the installation of the WRB are the proper preparation of the opening with house wrap, felt, or building paper (Figure 9.32) and its proper integration with the sill, jamb, and head flashings (see later sections and pertinent figures) in order to maintain the continuous drainage plane.

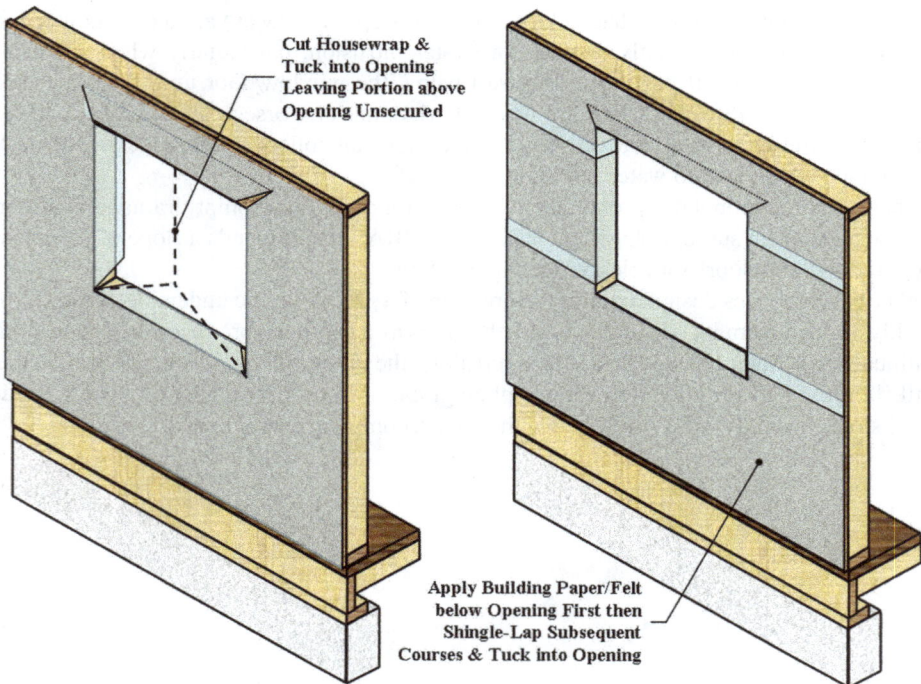

Cut Housewrap &
Tuck into Opening
Leaving Portion above
Opening Unsecured

Apply Building Paper/Felt
below Opening First then
Shingle-Lap Subsequent
Courses & Tuck into Opening

FIGURE 9.32 Window opening preparation using housewrap and building paper or felt.

9.4.1.2.3 Pan Flashing

Pan flashing serves to protect the subsill framing of the wall and interior portions of the building beneath the window or door openings that are susceptible to leakage. This susceptibility to leakage is typically at the lower corners of the rough opening in the wall, usually due to preparation of the openings where the WRB coverage is minimal at these corners. Subsill drainage (provided by pan flashings) is "the single most significant recommendation in achieving improved performance of installed windows"[64] even though it is rarely encountered in actual practice.

Best practices have recommended the use of pan flashings beneath windows and doors for years based on their propensity to leak; however, recent IRC[4] has adopted the use of pan flashings, making it a requirement under their general flashing statement in Section R703.8; however, they then resort to manufacturers' installation and flashing instructions for proper details. This change in the code tenor by the code bodies acknowledges that they recognize the issues in this area and are moving to more prescriptive code language on how buildings should be constructed near window and door openings.

Subsill drainage can be properly accomplished in a couple of different ways. The first would be the use of a preformed metal pan flashing with soldered, watertight joints (Figure 9.33).

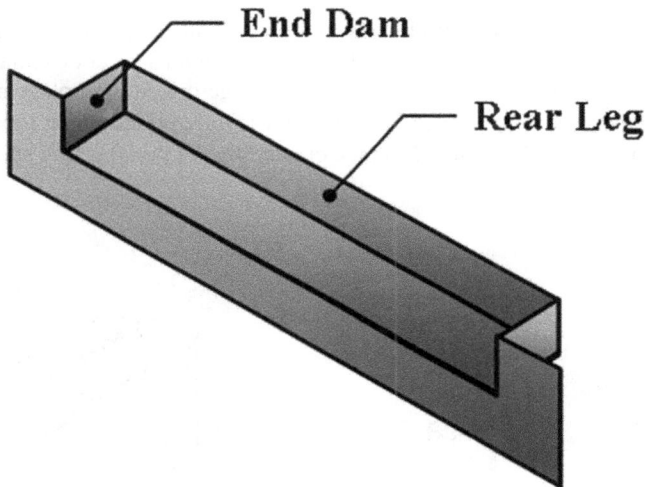

FIGURE 9.33 Typical preformed metal window/door sill pan flashing with soldered, watertight joints.

This form of flashing consists of a rear leg at the back of the sill, end dams at the jamb interfaces, and a front portion that laps over the WRB below the window/door opening to facilitate drainage. Essentially, this pan flashing serves as a sort of basin to catch much of the water that infiltrates around the window and then allows it to be directed back to the outside harmlessly. Pan flashings are installed after the WRB and before the windows are installed. A continuous bead of sealant is applied to the pan flashing to seal it to the WRB.

Standards, such as ASTM E2112,[67] recommend that the height of the end dams and rear leg of the pan flashing extend up a maximum of 2 inches. However, this Standard only specifies the use of pan flashings with non-finned windows, and it states that the use of higher end dams/rear legs is "not usually needed because of weather history indicating that high rain and wind are usually not simultaneous." This is not the case in all situations, however. Best practices (and now current IRC) state the need for pan flashings with all fenestrations regardless of fins. Additionally, for coastal climates, such as the southeastern United States, in which the likelihood for storms producing very high winds and rain is greater, more height on the pan flashings is needed and recommended to be a minimum of 3–4 inches where wind speeds are capable of reaching 110 MPH or greater.[59]

Since pan flashings, by design, extend through much of the wall's thickness and transition from cooler outdoor temperatures to more warmer, humid indoor environments, there is a potential for sheet metal pan flashings, such as those described in previous paragraphs, to act as thermal bridges.[67] If left unaccounted for by the designer or installer in cold-weather climates, where higher levels of humidity are often present, the formation of condensation within the wood-framed wall cavity could lead to future and unintended concealed wood decay problems. Fortunately, the benefits of subsill pan flashing in these climates or situations can be accomplished with the use of modern-day flexible peel-and-stick membrane flashings, such as DuPont™ and FlexWrap™.[46,68] The membrane is first applied to the central portion of the sill and then worked outward toward and up the sides of the jambs, typically a minimum of 6 inches, paying particular attention to work it into the corner interfaces between the sill and jambs (see Figure 9.34 for example).[68] This eliminates the potential for condensation formation created by the typical metal pan flashings. However, no rear leg is created with the use of peel-and-stick membrane flashings. To help combat this issue, oftentimes installers will provide a rectangular or beveled material beneath the membrane, creating a dam or sloped surface to divert water to the exterior side of the opening.[46]

9.4.1.2.4 *Jamb and Head Flashings*

Jamb and head flashings are applied following the placement of the window or door within the rough opening in the wall. It should be noted that a *discontinuous* bead of sealant is applied either to the WRB or to the backside of the window nailing flange prior to the placement of the window and the head/jamb flashings (Figure 9.34). The sealant is *not* applied along the bottom of the sill. This break in the sealant allows for any drainage created by the pan flashing should it be needed.

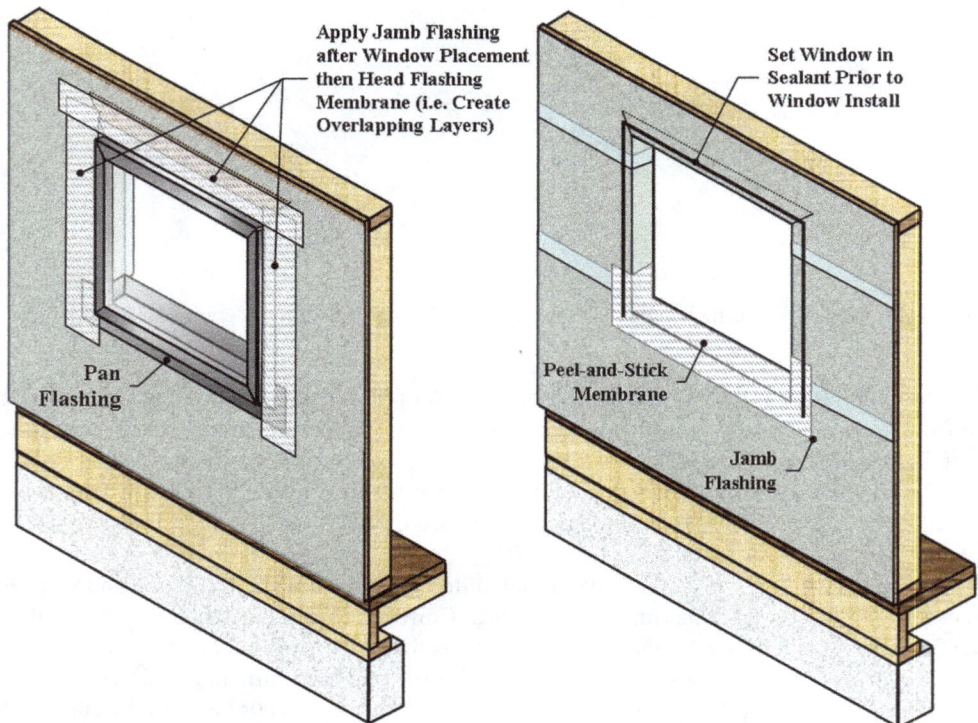

FIGURE 9.34 Jamb and head flashing membranes applied after window placement into rough opening.

Following the placement of the window (or door) unit, strips of peel-and-stick membrane flashing are applied over the nailing fins along both window jambs. This jamb flashing is recommended by best practices to be a minimum of 9 inches in width and extend a minimum of 8-1/2 inches above and below the rough opening making sure to extend past the lower edge of the sill flashing.[67] After the installation of the jamb flashings, a strip of peel-and-stick membrane flashing along the window head should be applied (Figure 9.35).

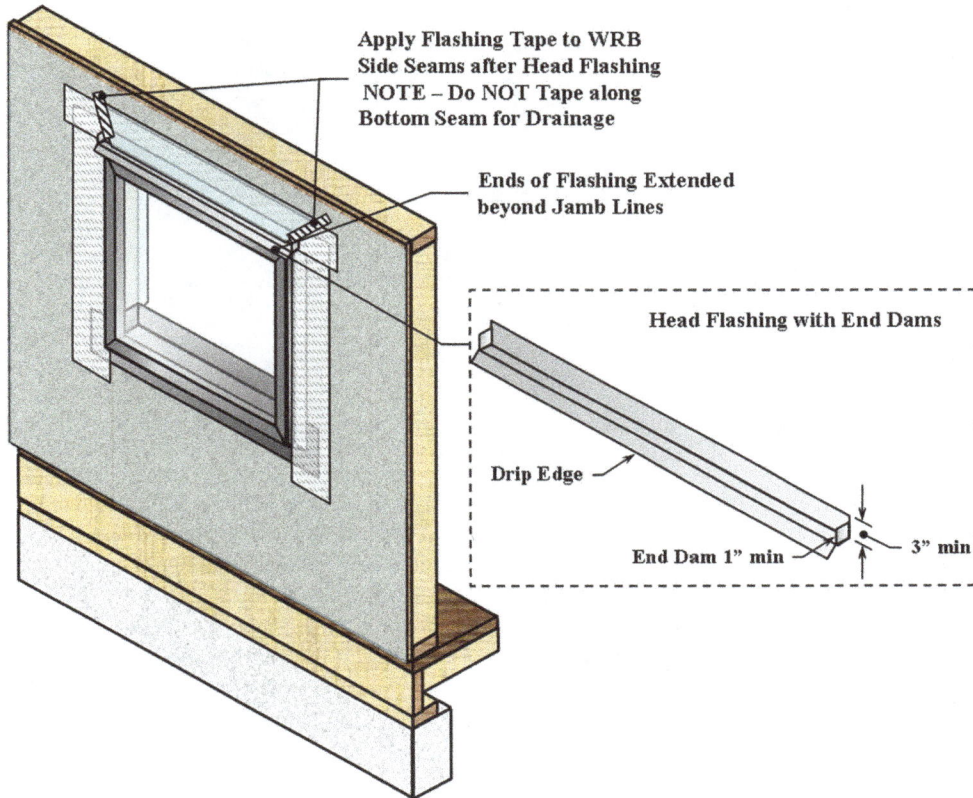

FIGURE 9.35 Installation of rigid head flashing with end dams and taping of WRB seams overlapping head flashing.

This head flashing is recommended by best practices to be a minimum of 9 inches in width also and extend a minimum of 1 inch beyond the outer edges of both jamb flashings.[67] Following the application of the peel-and-stick membrane head flashing along the top of the window or door, best practice is to install a rigid, typically metal, head flashing over the peel-and-stick flashing. The vertical leg of this rigid head flashing is then overlapped by the WRB, which was originally left unsecured above the window or door. In order to resecure the WRB, flashing tape is applied to the side seams after it has been properly lapped over the flashing. Note that the bottom seam of the WRB is not taped, thus allowing for any drainage should it be needed.

Best practices recommend that the vertical leg of the rigid metal head flashing extends up a minimum of 3 inches from the top of the window/door opening, extends beyond the surface of the exterior cladding, and is formed to provide a drip edge. Prior to its installation, the ends of the rigid head flashing are turned upward in order to create end dams, which are then soldered to create watertight seams (Figure 9.36).

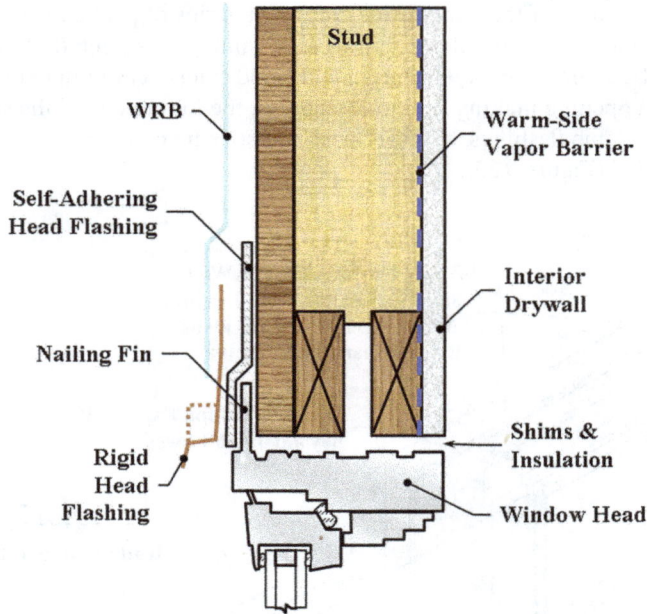

FIGURE 9.36 Side view of water management details along window head.

These end dams are particularly important to properly redirect any infiltrating water from above the window/door out to the exterior side of the wall and over the window/door instead of allowing it to travel horizontally along and over the ends of the flashing and along the window/door jambs where it could possibly find its way through the wall opening.

One last issue often encountered is where wood trimming wider than the flashing is installed around window or door openings.[69] When a gap exists between the edge of the trim and cladding (e.g., siding), above or beyond the edge of the flashing, water has a direct entry to the wall system without the ability to exit. This causes extensive decaying of materials susceptible to water damage.

9.4.1.3 Window and Door Leak Case Studies

A few examples of site investigations in which water infiltration had occurred through windows and/or doors are given in subsequent subsections to illustrate deviations from the principles just covered.

Case Study 9.11

Case study 9.11 began as an investigation to determine the cause of warped and buckled hardwood flooring along the entire west wall of a home in a particular upscale suburb known to have had homes constructed with poor brick veneer water management details (two masonry companies went bankrupt due to complaints about workmanship). Through the course of the investigation, diagnostic moisture meter testing of the water-damaged flooring indicated that the heaviest affected areas were located in close proximity to exterior windows. An inspection of the exterior revealed historic, weathered, and deteriorated caulking along the brick sill interfaces. A water test was conducted to simulate wind-driven rains onto the subject windowsill. Shortly after commencement of the water test, visual observations through an HVAC supply air register in the floor revealed water dripping down the surfaces of a floor joist directly below the window and near areas where historic drip-patterned water staining was present. Efflorescence was present on the HVAC duct and the sill plate (Figure 9.37).

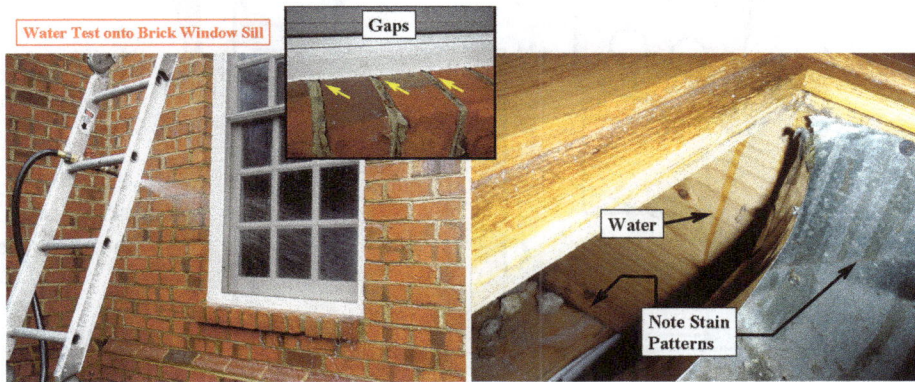

FIGURE 9.37 Water infiltration through brick window sill interface.

A combination of poor detailing concerning the water management principles of deflection and drainage, as well as poor maintenance, led to the observed water infiltration for this particular home. First, and foremost, as discussed in earlier chapters with respect to brick veneer, one must assume that water *will* infiltrate and enter the wall assembly behind the veneer. Therefore, the design and installation of windows and doors with brick veneer cladding should provide for an adequate means to control and redirect the water from affecting interior surfaces. In the case of this home, a lapse in maintenance of the sealant along the brick windowsill interfaces revealed the drainage deficiencies of the windows and the brick wall. Although destructive testing of the wall would be needed to determine exact construction details, it is a sure bet that pan flashing or through-wall flashing was not present beneath this window. Further, the brick sill was not adequately sloped away from the window (i.e., 15° according to best practices[27]), and no evidence of weep holes or through-wall flashing was present below the brick sill or above finished grade. Thus, no means was provided to discharge any water that will inevitably get into the wall. As might be expected, water infiltration around this window (west elevation) and those on the south elevation were the first to show water entry issues since they faced the predominant wind and weather patterns. The cross-sectional view of the brick windowsill in Figure 9.38 depicts two options of subsill drainage.

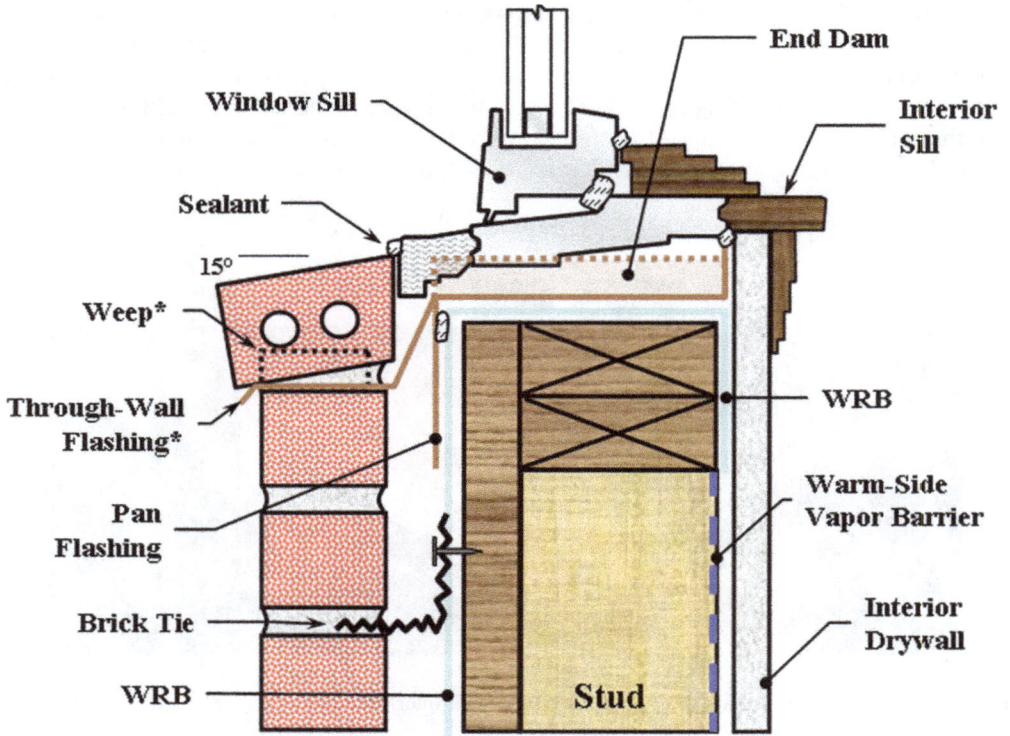

FIGURE 9.38   Cross-sectional view of brick window sill – proper water management details.

As discussed earlier, pan flashing with formed, watertight end dams that drains water onto the surface of the WRB below the window would suffice, but is dependent on whether through-wall flashing and weeps are present at the base of the exterior wall assembly. Another option would be to extend the through-wall flashing through the veneer just below the brick sill. Of course, if this detail were utilized, weeps would be needed to provide drainage. Regardless of which subsill drainage element is used, through-wall flashing and weeps are required at the base of the wall; neither was used in this case example, leading to rotting of the wood members in the wall cavities.

Case Study 9.12

Case study 9.12 for the home shown in Figure 9.39 provides a classic example of the importance for extending the rigid metal head flashing past the window jamb and the need for an end dam formed at the end of the flashing. This particular house was experiencing problems in the form of water dripping down their dining room windows for well over a year. During the original site investigation, it was determined that the source of the water was occurring at the interfaces between the manufactured stone veneer and the second story window above the dining room. Two months later, a return visit to the home was scheduled coinciding with the removal of the stone veneer. This remediation immediately gave a visual explanation where the source of the water was coming from, as the wooden framing between the first and second story windows was heavily water-damaged and rotted.

FIGURE 9.39   Lack of end dam on metal flashing – head flashing with no end dam.

Following the patterning of the water damage led to the following discoveries with respect to the second floor window:

- No evidence of subsill flashing or jamb flashing was present. The wooden sill plate for the window and studs directly below were heavily rotted from years of water infiltration.
- Staining patterns clearly indicated that water had traveled down the jamb lines for the window toward the sill.
- The rigid metal head flashing terminated at the jamb lines for the window.
- No end dams were formed at the ends of the rigid head flashing.

The lack of end dams in the rigid metal head flashing along the top of the second story window allowed for water, which inevitably infiltrated through the manufactured stone, to travel horizontally and empty out directly onto the window jambs where no jamb flashing was present. From there, the water likely traveled downward along the jamb lines for the window opening and settled on the windowsill plate where no form of subsill flashing or drainage mechanism was present. Based on the level of water damage and deterioration to the wooden framing below the window, there was a significant imbalance between the rates of wetting and drying of these wooden members and likely dated back to the original date of construction of the home 10 years earlier. The builder of this home went out of business and is no longer building homes.

Case Study 9.13

Case Study 9.13 for the home shown in Figure 9.40 emphasizes the need for pan flashings or through-wall flashings for doors. First, as is often the case, sealant appeared to have been relied upon as one of the primary means for deflecting water from the interfaces between the front door threshold, framing, and brick sill. When the sealant reached the end of its useful service life and began to deteriorate, cracks and gaps formed and revealed a complete lack of subsill drainage for this front door and sidelight unit. Gaps, measuring up to 1/4 of an inch in width, were present between the metal threshold and the jamb for this door.

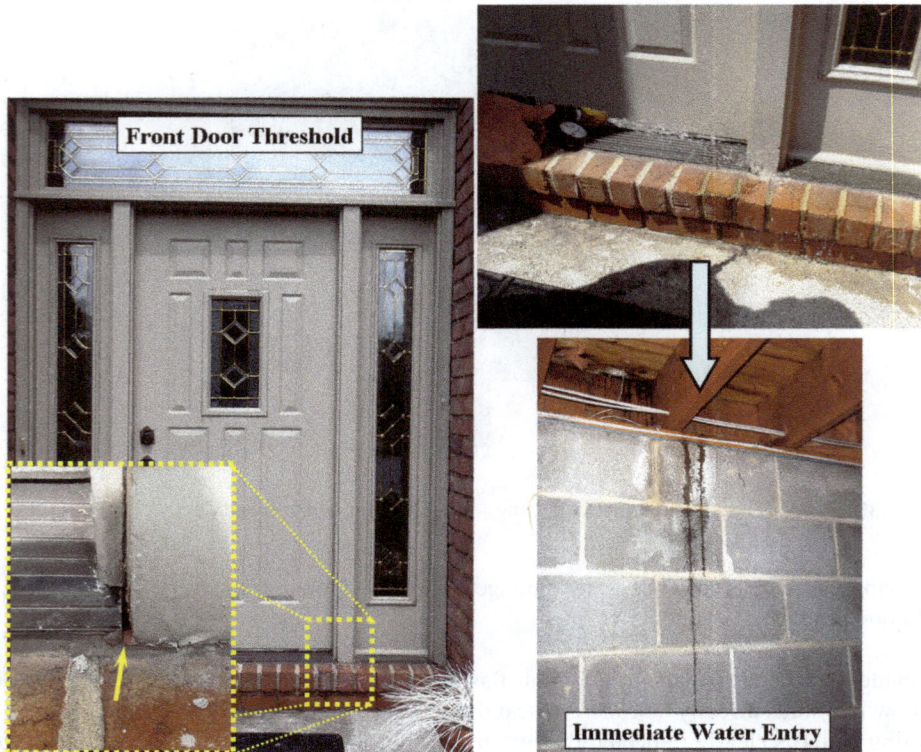

FIGURE 9.40　Gap at corner of front door brick threshold.

Water testing was conducted with the water spray directed toward the lower corner of the door where this gap was located. Immediately water began to pour into the basement, over the band board, and onto the foundation wall. The water test was terminated and the owners of this home were notified of the deficiency.

This situation could have been prevented had pan flashing or through-wall flashing and weeps been installed prior to the installation of the door/sidelight unit. Then whenever the sealant eventually gave way, any water intrusion could have been collected (managed) and then redirected harmlessly back to the exterior. Unfortunately, this deficiency is commonplace in forensic field investigations.

Case Study 9.14

Case Study 9.14 for the home shown in Figure 9.41, again, emphasizes the need for proper flashing details around windows this time with vinyl siding. The owner of this home noticed water on the garage floor below south windows whenever rains would come from that direction.

FIGURE 9.41 No sill flashing beneath window.

Approved destructive testing was performed on the drywall below one of the windows and revealed "classic" staining patterns on the OSB sheathing below the lower corners of the window. From the exterior side of the window, the loose J-channel just below the windows showed that no form of sill flashing was present. Further, the upper edge of the house wrap was loose and showed evidence of moisture and water staining. A water test directed along the J-channel/sill interface produced water infiltration on the sheathing below the lower corner of the window.

The "classic" staining patterns referred to in the previous paragraph are those that have been discovered numerous times during field investigations for water losses involving windows. These patterns are almost always located on the exterior sheathing starting at either the upper or the lower corners of the window and then spread out and downward.

Case Study 9.15

Case Study 9.15 shown in Figure 9.42, again, shows these "classic" staining patterns below windows. This particular field investigation was discussed in a previous section on KOD flashing, but issues regarding the proper water management details around the windows were also of concern.

FIGURE 9.42 "Classic" staining patterns on OSB sheathing below lower corners of window and center mullion.

During the construction of the condominiums in this particular complex, there was a lack of coordination between the trade personnel (i.e., window, window trim, and stone veneer installers), which led to the installation of the window trim around the windows prior to the application of the WRB and manufactured stone veneer. Due to this mishap in scheduling, no form of flashing or WRB was present around the windows. Then, once the stone veneer installation contractors started their work, they attached WRB to the sheathing around the windows up to the edge of the window trim and then applied the cultured stone veneer. To make matters even worse, no form of sealant was applied along the interfaces between the windows and the window trim. This allowed for water to infiltrate at the unprotected trim/stone interface. Water entering at this point had a straight path behind the trim and stone veneer near the window openings. What resulted were the "classic" staining patterns on the OSB sheathing below the lower corners and center mullions of the windows.

Along with giving a prime example of the "classic" staining patterns below mismanaged windows, this case study also showed what can happen if the different trades are not on the same page and are not properly sequenced by the general contractor. In this case, the consequence was that the components of the exterior wall envelope were not installed in proper order, which is essential for proper detailing around such critical elements such as windows, doors, and roof-to-wall interfaces, leading to rotting of the interior wall systems. Again, experience has shown that this happens more often than not as a result of construction completion time constraints.

9.4.2 COMMON WALL PENETRATIONS

A typical residential or light commercial building will contain one or more of many different types of miscellaneous wall penetrations including dryer vents, bathroom exhaust fans, exterior light fixtures, and gas lines, to name a few. Figure 9.43 gives typical flashing instructions for a general wall penetration.

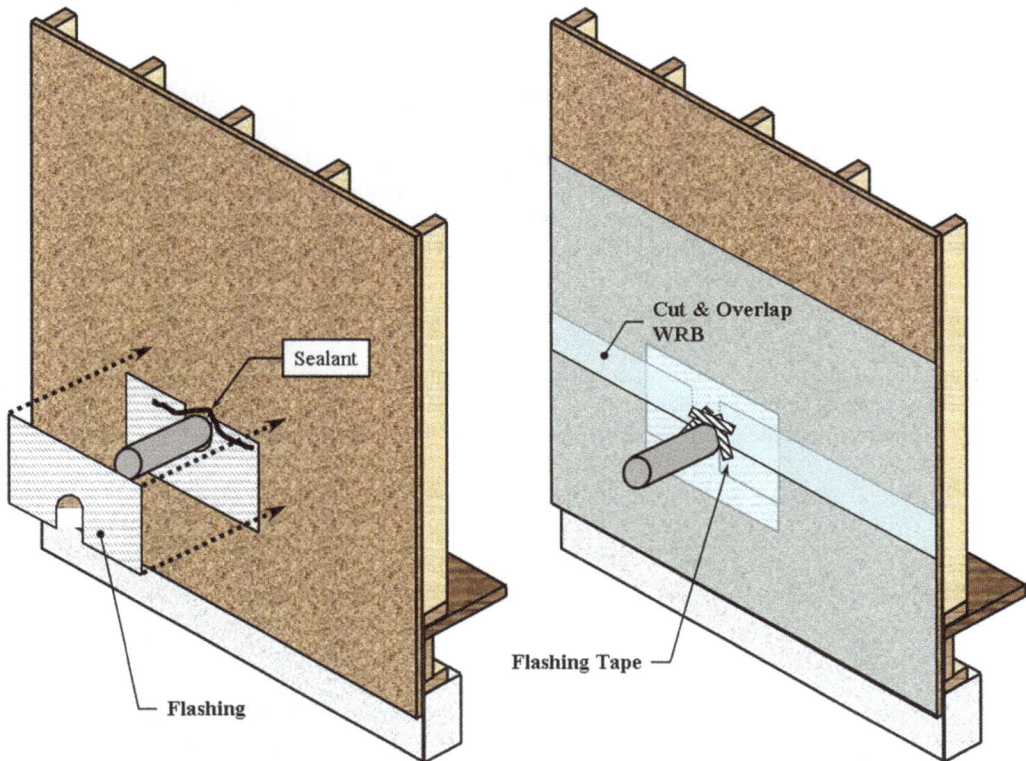

Sealant

Cut & Overlap
WRB

Flashing Tape

Flashing

FIGURE 9.43   General flashing guidelines for typical wall penetration.

As illustrated, with the use of a little common sense and the knowledge gained thus far, these penetrations can be properly flashed in order to maintain the continuity of the drainage plane.

9.5   INSPECTION METHODOLOGY FOR EXTERIOR WATER CAUSE AND ORIGIN INVESTIGATIONS

The general methodology for cause and origin investigations has been given and discussed at length in previous chapters, most notably Chapters 1 and 8. Areas of focus pertaining to the determination of the primary cause(s) of water infiltration through exterior finishes are listed below.

9.5.1 INTERVIEW WITH OWNER/POINTS OF CONTACT

Following an introductory meeting, an interview is conducted with the building owner, the owner's representative, and/or other parties who have intimate knowledge of the issue(s) at hand. Oftentimes, the information obtained during the interview portion of the site investigation can help provide invaluable information that can help focus the investigation and the determination of the likely causes for water issues.

A list of additional example questions for exterior finish water cause and origin investigations beyond those normally asked during forensic investigation interviews (Chapter 1) is given below:

- Where are the damaged areas located?
- When was the damage *first* noticed? Where there any special conditions that occurred that might explain the formation of the damage?
- Has any further damage occurred following the initial discovery?
- Is the damage noticed only during specific storm events (i.e., only during heavily wind-driven rains, rain events from a particular direction, heavy rains, extended periods of rain, snow accumulation followed by periods of warmer weather and melting, etc.)?
- Have there been any recent modifications or improvements prior to the discovery of the damage that might explain why the damage formed (i.e., any changes to the exterior cladding itself and/or a change in the exposure such as the removal of a large tree previously obstructing portions of the wall)?
- Have any contractors or third-party inspectors been hired that have investigated the issue? If so, what were their findings?
- Does the owner or point of contact have an opinion on what they think may be the issue?
- Who was the builder of the home?
- Have any of the homes in the neighborhood experienced similar problems?

The key is to determine the timeline of events that led from the discovery of the water loss up until the date of the investigation. Similar to the method of performing the actual visual investigation of the building, the interview must be methodical and systematic and attempt to provide specific information that may serve to be useful in focusing the inspection and later analysis.

9.5.2 INTERIOR INSPECTION

After a clear understanding of the issue has been gathered during the initial interview, the process of the site investigation typically begins with a visual inspection of the interior surfaces of the building pertinent to the areas of concern. Begin first by sketching out the area of concern. If more than one area of concern exists, or it extends to multiple floors, sketch out the entire floor or multiple floors. Take all pertinent dimensions and record them on the sketch. This will give an accurate representation of the layout of the home and can aid in locating site-specific observations obtained later in the investigation and in the creation of any computer-aided drawings completed following the inspection.

One should then conduct a systematic and methodical visual inspection of the structure, beginning with the lowest impacted elevation first. For each level, a space-by-space investigation should be completed, documenting the general conditions of the interior finished surfaces. Begin the floor-level space-by-space investigation by first taking an overview photograph of the area and documenting general information and conditions. Then systematically divide the area into subparts and inspect them further for information that is more detailed. For example, for an investigation beginning on the first floor of a home, begin by documenting the general construction, finish, appearance, slope, and then any specifics regarding possible signs of water presence for each space. For a given space, begin taking observations of the floor surfaces, then move to the walls and the ceiling, following the same template for a visual inspection.

During the course of the water cause and origin investigations, special attention should be paid to the patterning of the water-damaged areas. Ask yourself the following questions:

- Does the patterning appear to originate from a single point? Is it widespread?
- What are the conditions of the damaged areas? Are they lightly stained, heavily stained? Is the area deteriorated, soft, or wet? Feel the areas by hand and document if they are wet, soft, or cool to the touch.
- Are there signs of long-term and historic wetting? Typically, heavily affected areas over prolonged periods of time will begin to soften, deteriorate, and show efflorescence and/or concentric areas of staining indicative of periodic wetting and drying. Is the wood rotted? If so, to what extent?
- Measure all affected areas to help establish the severity of water damages.
- Locate any areas of concern on schematics by measuring these areas from reference points (i.e., corners, walls, windows, etc.). This can be helpful later in the investigation when lining up this area on other floors, attic, roof, or exterior surfaces opposite the interior damaged areas.
- Use diagnostic tools such as moisture meters, boroscopes, and infrared survey instruments to provide further detail of the extent and severity of the affected areas. Generally, it is also a best practice to obtain sample readings and surveys of normal, unaffected areas in order to establish background levels of values such as normal moisture meter readings in apparently unaffected areas.

The key to the interior portion of the investigation is to proceed through the structure in a methodical, systematic, and controlled fashion. Always record observations (in writing and photographically) on general conditions associated with the space and then on specific conditions associated with that same space. Oftentimes, observations regarding general conditions can provide insights into the nature of the surroundings and help to explain certain phenomena.

9.5.3 Exterior Inspection

By the time the inspector has obtained all necessary background information from the owner or point of contact and has then systematically surveyed the interior portions of the building, a good foundation of knowledge and details regarding where the key water cause and origin situations exist has likely been established. Based on the level of experience of the inspector, possible causes for water infiltration or areas common to deficiencies can be formulated and then either validated or eliminated. It is important to remember that oftentimes, issues regarding water infiltration are very experiential. The more times the process has been extensively conducted and completed, a better understanding is gained on what the important aspects are to help narrow focus and attention to detail. This usually is greatly enhanced by all pertinent background information such as that which has been discussed within previous sections.

During the investigation of the exterior surfaces of the building, be sure to identify the areas that are opposite and above the locations of reported and/or observed interior water-related damages. Begin with documentation of the general conditions for a given elevation and then systematically focus more on specific detailed information on construction, installation, design, and maintenance for that elevation. For water infiltration cause and origin investigations, pay special attention to the four principles of water management discussed earlier in this chapter and how they pertain to areas of concern. Example considerations following the principles of water management are given below:

- *Deflection*: Determine the building's exposure to precipitation events and the ability of the exterior cladding to resist any water impact and possible infiltration.
  - Note the elevation in which the exterior areas of concern are located. Does it face predominant winds and weather patterns?

- Observe and document the locations of any nearby obstructions (buildings, trees, shrubs, vehicles, etc.) that may limit the ability of the wall to be impacted by wind-driven rains.
- Note locations and impact of roof overhangs, canopies, balconies, and decks. Do they help to shield the exterior wall from rain impact?
- Check to make sure that gutters are present along the roof eaves and if they are properly cleaned and maintained, if not, this may limit water runoff from the exterior cladding.
- Check the condition of sealant along wall interfaces with rough openings in the wall envelope. Use your finger to check the condition of sealant. Does the sealant appear historic? Is it cracked, brittle, or missing? Are there gaps? If so, check the appearance of the gaps. Cracks and gaps that have been present for prolonged periods of time will typically exhibit a dull, more weathered color and evidence of debris such as dirt and cobwebs, which will eventually find their way into the crevices.
- *Drainage:* Determine the ability of the exterior wall to drain water properly out and away from the wall assembly as to not create a hazard that may lead to potential interior or structural water damage.
  - Check the sills and thresholds for adequate slope away from the building.
  - Check to see that flashings are present in required and recommended locations. Measure all components of visible flashings for further comparison and analysis to building codes and best practices.
  - Although many of the key elements to the proper management and drainage of moisture from the wall assembly are located behind the exterior cladding, the presence of WRBs and flashings can often be determined through visible gaps or displaced portions of cladding.
  - Check to see if the lower edge of the exterior cladding is terminated above finished grade. In addition, note the direction and severity of the slope for the local finished ground surfaces near the home.
  - Check the typical locations for drainage outlets (i.e., base of walls, above windows and doors, along roof interfaces and roof eave intersections, etc.) to make sure they are present, unobstructed, free of debris, and left to drain freely.
  - Check for evidence that the exterior cladding may be retaining moisture and unable to drain freely. Look for evidence of discoloration and cracks, noting their locations and patterning.
- *Drying*: The drying potential of a particular wall system is dependent on local climate conditions, including air temperature and humidity levels, but also is greatly affected by the presence of an air space between the cladding and the backing wall, such as with rain-screen wall construction.
  - Check the construction details of the wall and type of wall system utilized.
  - Check for evidence that the wall is retaining moisture and not adequately drying. Note that oftentimes, particularly for stucco, the claddings on the north and east elevations of a building may appear to retain moisture longer due to their decreased sun exposure and exposure to predominant winds, which both aid greatly in the drying potential for the wall.
- *Durability*: The long-term durability of the exterior wall construction is directly dependent on the water management details listed above and the quality of design, construction, installation, materials, and maintenance. Check for general conditions of key envelope construction components, paying special attention for any signs of deterioration.

9.6   EXTERIOR FINISHES WATER CAUSATION AND ORIGIN INSPECTION REPORT

Experience and best practices recommend the creation of a formal written report in a timely manner following the site investigation. The general outline for this report was given in Chapter 1 and has been referenced several times in previous chapters. Important aspects of the formal inspection report pertaining to exterior water infiltration follow:

- *Scope of work:* The inspection report should clearly delineate the scope of work and questions to be answered as part of this specific forensic investigation.
- *Summary of information obtained from interview(s):* Oftentimes some of the most crucial evidence needed to determine accurately the primary cause of water infiltration is a comprehensive and detailed interview with the parties most closely involved with the issue at hand. Accurately provide the pertinent information on the history of events, making sure to detail all information discussed during the site inspection. Note that the information given in this section of the report is directly dependent on the thoroughness of the interview.
- *Summary of interior observations (by floor level):* Provide all of the pertinent information and observations regarding the areas associated with the reported water loss. Make special note to describe in detail observations made by space for each level. A schematic for each level should be provided with key findings identified on the schematic. The schematic is often quite helpful in illustrating findings later should questions arise from the client and/or owner. Observations should be organized by the most important first, followed by less important observations next.
- *Summary of exterior walk-around:* Observations from the inspection pertaining to the exterior areas of concern should be described in detail in this portion of the report. Describe the conditions of sealants, windows, doors, exposure conditions (roof overhangs, gutters, nearby trees/buildings, etc.), flashings, and specifics regarding the exterior cladding, paying particular attention to the manner in which it was installed. Provide a summary of measurements and water testing results made to facilitate conclusions made later regarding the primary cause(s) of water infiltration.
- *Discussion/analysis of observations:* This is one of the most important and significant sections of the formal report. It helps to answer the question of why water infiltration occurred. Analyze observations pertinent to the area(s) of concern identified during the investigation and compare them with standards and methods consistent with best practices. Provide clear distinctions on what was supposed to have been done (i.e., particular installation detail or requirement) and compare it to what was actually done (i.e., the detail that was deficient and likely created the conditions that led for water infiltration to occur).
- *Conclusions:* Should restate what has already been determined to be the primary cause for the reported and observed water infiltration. Give supporting evidence that validates the conclusions.
- *Photographs and figures:* Provide additional photographs of pertinent areas and details, which may or may not have been included within the body of the report.
- *Evidence and/or supporting documents:* Provide any additional literature or supporting documents that were relied upon during the analysis and determination of the particular issue(s) at hand.

IMPORTANT POINTS TO REMEMBER

- Moisture within an exterior wall assembly can come from both exterior (i.e., rain, snow, air, etc.) and interior (water vapor, air, etc.) sources.
- Controlling the moisture content of the building materials is essential in controlling the presence of and the rates of decay/rot/deterioration.
- With exterior walls, oftentimes water damage and extensive deterioration are concealed within the wall assembly due to factors such as water-resistant materials and the force of gravity.
- A building must be properly designed, constructed, and maintained in order to allow for the proper management of moisture within the wall assembly. This is accomplished by implementing the four principles of water management (i.e., the four D's):
  - Deflection (exposures, overhangs, exterior finishes, etc.).
  - Drainage (WRBs, flashings, weeps, ground slope, etc.).
  - Drying (air cavity, rainscreen construction, material properties, etc.).
  - Durability (expected service lives of materials, decay-resistant materials, etc.).
- Secondary water management details such as caulks and sealants are important components to the overall water management system, but should not be relied upon as the primary means of preventing any water from incidentally infiltrating the exterior finishes.
- Best practices approaches to water management details are the assumption that water *will* infiltrate through the exterior cladding and designing a drainage mechanism accordingly will divert and redirect water from affecting moisture-sensitive building materials.
- Common causes for water infiltration through exterior finishes are primarily deficiencies in: (1) design, (2) construction/installation, and/or (3) maintenance and typically occur at locations of openings (i.e., windows, doors, etc.), cladding interfaces (i.e., differing adjacent exterior finishes), and roof intersections (i.e., roof eaves, sloped roof-to-wall, etc.).
- The site investigation to determine the primary causes of water infiltration should consist of: (1) a comprehensive interview with the owner/points of contact knowledgeable of the events, (2) a methodical and systematic visual inspection of both the interior and exterior pertinent to the areas of concern, (3) diagnostic testing, and (4) water testing in accordance with standards and best practices.
- Following the site investigation, best practices are to create a formal report that summarizes the inspection findings, including information obtained from the owner/points of contact, inspection observations (interior and exterior), a discussion on observational analysis, conclusions with supporting evidence, and other pertinent information used in the determination process.

REFERENCES

1. Insurance Information Institute. "Insurance Industries Statistics." Accessed May 20, 2020, http://www.iii.org/facts_statistics/homeowners-insurance.html.
2. Goldman, Mark E. "Investigating Water Loss Origins." *Claims Magazine*, August 22, 2005.
3. Straube, John F. "Moisture Control and Enclosure Wall Systems." A Thesis Presented to the University of Waterloo in fulfilment of the Thesis Requirements for the Degree of Doctor of Philosophy in Civil Engineering, Waterloo, Ontario, Canada, 1998.
4. International Code Council. "2018 International Residential Code (IRC)." 2017. Accessed June 9, 2020, https://codes.iccsafe.org/content/IRC2018/chapter-7-wall-covering.

5. American Forest & Paper Association, Inc. "Design of Wood Frame Structures for Permanence." 2006. Accessed June 9, 2020, https://www.awc.org/pdf/codes-standards/publications/wcd/AWC-WCD6-Permanence-ViewOnly-0604.pdf.

6. FEMA – Federal Emergency Management Agency. "FEMA P-762: Local Officials Guide for Coastal Construction – Design Considerations, Regulatory Guidance, and Best Practices for Coastal Communities." February 2009. Accessed June 9, 2009, https://www.fema.gov/media-library-data/20130726-1706-25045-9843/logcc_rev1.pdf.

7. Lstiburek, Joseph W. "Learning from mistakes." *ASHRAE Journal*, pp. 60–64, February 2010.

8. APA – The Engineered Wood Association. "Avoiding Moisture Accumulation in Walls." Form Number A530A, September 2016. Accessed June 9, 2020, https://www.apawood.org/publication-search?q=a530.

9. Day, Kevin C. "Risk Management of Moisture in Exterior Wall Systems." Morrison Hershfield Limited, Toronto, Canada. Accessed June 9, 2020, https://www.irbnet.de/daten/iconda/CIB2410.pdf.

10. APA – The Engineered Wood Association. "Controlling Decay in Wood Construction." Technical Note Number R495D, December 2009. Accessed June 9, 2020, https://www.apawood.org/publication-search?q=R495d.

11. Morris, P. "Understanding the Biodeterioration of Wood in Structure." Forintek and British Columbia Building Envelope Council, 1998. Accessed June 9, 2020, https://cwc.ca/wp-content/uploads/aboutdecay-biodeterioration.pdf.

12. Canada Mortgage & Housing Corporation. "Best Practice Guide for Wood Framed Envelopes in the Coastal Climate of British Columbia." 1998.

13. Canada Mortgage & Housing Corporation. "Survey of Building Envelope Failures in the Coastal Climate of British Columbia." Technical Series 98–102, 1999.

14. Inculet, D. and Surrey, D. "Simulation of Wind-Driven Rain and Wetting Patterns of Buildings." Prepared for Canada Mortgage & Housing Corporation, November 1994, Revised February 1995.

15. Straube, John. "Simplified Prediction of Driving Rain on Buildings: ASHRAE 160P and WUFI 4.0." Building Science Digest 148, November 15, 2010. Accessed June 9, 2020, https://www.buildingscience.com/documents/digests/bsd-148-wufi-simplified-driving-rain-prediction.

16. Lstiburek, Joseph W. "WUFI – barking up the wrong tree?" *ASHRAE Journal*, pp. 62–70, October 2015.

17. Lstiburek, Joseph W. "Double play" *ASHRAE Journal*, pp. 62–67, April 2019.

18. Ge, Hue and Ronald Krpan. "Wind-Driven Rain Study in the Coastal Climate of British Columbia - Final Report." Submitted to: Mr. Silvio Plescia, P. Eng., Sustainable Housing and Communities Policy and Research Division, Canada Mortgage and Housing Corporation, 700 Montreal Road, Ottawa, Ontario, K1A 0P7; Dr. Denisa Lonescu, Manager, Research and Education, Homeowner Protection Office, PO Box 11132, Royal Centre, Suite 2270, 1055 W. Georgia Street, Vancouver, British Columbia, V6E 3P3; and Mr. Brian Simpson, British Columbia Housing Management Commission, 800–5945 Kathleen Avenue, Burnaby, British Columbia, V5H 437, British Columbia Institute of Technology (BCIT).

19. Lstiburek, Joseph W. "Flow-through assemblies." *ASHRAE Journal*, pp. 46–51, December 2015.

20. BIA – Brick Institute of America. "Technical Note 7A – Water Penetration Resistance – Materials." December 2005.

21. BIA – Brick Institute of America. "Technical Note 7B – Water Penetration Resistance – Construction and Workmanship." December 2005.

22. Partnership for Advancing Technology in Housing (PATH). "Windows & Doors." Volume 4, U.S. Department of Housing and Urban Development, Washington, D.C. 20410-6000, May 1999.

23. Lstiburek, Joseph W. "They all laughed… at Fulton and his steamboat, Hershey and his chocolate bar." *ASHRAE Journal*, pp. 56–63, January 2017.

24. Lstiburek, Joseph W. "Evolution of the residential air barrier – forty years of air barriers." *ASHRAE Journal*, pp. 48–56, March 2015.

25. Lstiburek, Joseph W. "Interior spray foam." *ASHRAE Journal*, pp. 60–67, February 2020.

26. Lstiburek, Joseph W. "Inward drive-outward drying." *ASHRAE Journal*, pp. 76–82, June 2017.

27. Lstiburek, Joseph W. "Drain the rain, on the plane: The drainage plane." *ASHRAE Journal*, pp. 60–64, August 2019.

28. Brock, Linda. *Designing the Exterior Wall: An Architectural Guide to the Vertical Envelope*. John Wiley & Sons, Inc.: Hoboken, NJ. June 20, 2005.

29. FEMA. "Moisture Barrier Systems." FEMA 499 – Home Builder's Guide to Coastal Construction Technical Fact Sheet No. 9, June 2005.

30. E.I. du Pont de Nemours and Company. "DuPontTM Tyvek® Water-Resistive and Air Barriers Installation Guidelines." March 2020. Accessed June 12, 2020, https://www.dupont.com/content/dam/dupont/amer/us/en/performance-building-solutions/public/documents/en/K16282-Residential-WRB-Install.pdf.

31. Fortifiber Building Systems Group. "Installing Jumbo Tex®", September 11, 2015. Accessed June 12, 2020, https://www.buildsite.com/pdf/fortifiber/Fortifiber-Installing-Jumbo-Tex-Applies-to-All-Grades-Installation-Instructions-1408716.pdf.

32. SMACNA. *"Architectural Sheet Metal Guidelines"*, Sixth Edition. Sheet Metal and Air Conditioning Contractors' National Association, Inc. (SMACNA): Chantilly, VA, September 2003.

33. Salonvarra, Mikael, et al. *"Air Cavities Behind Claddings – What Have We Learned?"* ASHRAE: Atlanta, 2007.

34. BIA – Brick Institute of America. "Technical Note 28 – Brick Veneer/Wood Stud Walls." November 2012. Accessed June 11, 2020, https://www.gobrick.com/docs/default-source/read-research-documents/technicalnotes/28-brick-veneer-wood-stud-walls.pdf?sfvrsn=0.

35. BIA – Brick Institute of America. "Technical Note 27 – Brick Masonry Rain Screen Walls – Revised." August 1994. Accessed June 11, 2020, https://www.gobrick.com/docs/default-source/read-research-documents/technicalnotes/27-brick-masonry-rain-screen-walls.pdf?sfvrsn=2.

36. Lstiburek, Joseph W. "Rain control: Drained, barrier and mass." *ASHRAE Journal*, pp. 76–81, April 2020.

37. Lstiburek, Joseph W. "Built wrong from the start." *Fine Homebuilding*, pp. 52–57, April/May 2004.

38. Green Building Rating System Guides. "Wood Specification: Durability." Accessed June 12, 2020, https://www.naturallywood.com/sites/default/files/documents/resources/building-green-with-wood-toolkit-rating-system-guides-all-in.pdf.

39. U.S. Green Building Council. "LEED Program Information." Access June 12, 2020, https://www.usgbc.org/leed.Journal>.

40. Marshall, Robert. "Delivering durable building envelopes." *PCI Journal*, 53(1) January-February 2008.

41. Canadian Green Building Council. "LEED Canada Program Information." Accessed June 12, 2020, https://www.cagbc.org/CAGBC/About_Us/CAGBC/AboutUs/About_Us.aspx.

42. Canadian Standards Association: Mississauga, Canada. "CSA S478-19: Guidelines on Durability in Buildings." 1995 (Reaffirmed Year: 2007).

43. Steven Winter Associates, Inc. "The Rehab Guide Volume 2: Exterior Walls." U.S. Department of Housing and Urban Development, August 1999. Accessed June 12, 2020, https://www.huduser.gov/portal//Publications/pdf/walls.pdf.

44. Carson Dunlop & Associates Limited. "Principle of Home Inspection: Exteriors." Dearborn Real Estate Education: Chicago, IL, 2003.

45. TecnoCem USA. "Stucco Guidelines - Drawing Details Information courtesy of Northwest Wall & Ceiling Bureau."

46. Lstiburek, Joseph W. "Water-managed wall systems." *The Journal of Light Construction (JLC)*, March 2003.

47. The Brick Industry Association (BIA). "Technical Note 7: Water Penetration Resistance – Design and Detailing." September 2017. Accessed June 9, 2020, https://www.gobrick.com/docs/default-source/read-research-documents/technicalnotes/7-water-penetration-resistance-design.pdf?sfvrsn=0.

48. Hodgin, Derek A. "Problems with Code-Compliant Brick Veneer in Residential Construction." Interface Magazine, September 2008, pp. 7–13.

49. Building Stone Institute. "Recommended Best Practices." 2010. Accessed June 12, 2020, https://www.stonesofnorthamerica.com/technical/BSI_Recommended_Practices.pdf.

50. Building Stone Institute. "Natural Stone Veneer – Information Guide." March 2017. Accessed June 12, 2020, https://fieldstoneveneer.com/blog/natural-stone-veneer-installation-guide-by-bsi.

51. Lstiburek, Joseph W. "Stoned – problems with thin stone and thin manufactured stone veneer, aka, 'lick and stick stone.'" *ASHRAE Journal*, pp. 70–74, October 2019.

52. Building Stone Institute. "Adhered Natural Stone Veneer Installation Guide." February 2012. Accessed June 12, 2020, https://lompocstone.com/images_2009/brochures/2012_02_Lompoc_TV_Install_Guide.pdf.

53. Owens Corning. "Cultured Stone – Best Practices for Flashing Details." June 2007. Accessed June 12, 2020, http://www.inspectorpaul.com/sitebuildercontent/sitebuilderfiles/Owens-Corning-Cultured-Stone-Installation-Instructions.pdf.

54. Centurion Stone Products, Inc. "Centurion Stone – Manufacturer's Installation Instructions." October 2013. Accessed June 12, 2020, https://www.centurionstone.com/docs/pdf/installation.pdf.

55. Dryvit. "Outsulation Plus MD System (formerly Outsulation Plus) Installation Details." September 18, 2017. Accessed June 12, 2020, https://www.dryvit.com/media/174889/ds137_outsulation-plus-md-system-specifications.pdf.

56. Zwayer, Gary L. "Whole Building Design Guide: Building Envelope Design Guide – Exterior Insulation and Finish System (EIFS)". National Institute of Building Sciences, May 10, 2016. Accessed June 12, 2020, https://www.wbdg.org/guides-specifications/building-envelope-design-guide/wall-systems/exterior-insulation-and-finish-system-eifs.

57. Lstiburek, Joseph W. "EIFS – Problems and Solutions." Building Science Digest 146, July 11, 2007. Accessed June 12, 2020, https://www.buildingscience.com/documents/digests/bsd-146-eifs-problems-and-solutions.

58. EIFS Industry Members Association (EIMA). "Guide to Exterior Insulation & Finish System Construction." 2000. Accessed June 12, 2020, https://www.allseasonswindows.com/wp-content/uploads/2015/04/EIMA-Guidelines-EFIS-installation.pdf.

59. Home Builder's Guide to Coastal Construction. "FEMA P-499 – Technical Fact Sheet No. 6.1: Window and Door Installation." December 2010. Accessed June 12, 2020, https://www.fema.gov/media-library-data/20130726-1538-20490-2983/fema499web_2.pdf.

60. Yumpu. *Vinyl Siding Installation Manual*. Vinyl Siding Institute: Washington, DC, December 2010.

61. Western Red Cedar Lumber Association. "How to Install Western Red Cedar Siding." July 2015. Accessed June 12, 2020, https://www.realcedar.com/wp-content/uploads/2014/05/Real_Cedar-How_to_install_Western_Red_Cedar_Siding-2015_07_20.pdf.

62. *Metal Siding Installation Manual*. Gentek Building Products, Inc.: Avenel, NJ, 2016. Accessed June 12, 2020, http://www.rooftopquotes.com/Images/Pdf/MetalSidingInstall_Cae.pdf.

63. James Hardie Siding Products. "Best Practices – Installation Guide Siding and Trim Products." Version 9.1, December 2019. Accessed June 12, 2020, https://www.jameshardiepros.com/getattachment/9a1017e1-853d-4574-b3e9-7afb1a5a472d/intro-tools-hz5-us-en.pdf.

64. RDH Building Engineering Limited. "Water Penetration Resistance of Windows – Study of Manufacturing, Building Design, Installation and Maintenance Factors." Vancouver, BC, December 31, 2002.

65. Lstiburek, Joseph W. "Punched openings." *ASHRAE Journal*, pp. 80–86, March 2018.

66. Day, Kevin C. and W. C. Brown. "Challenges with using ASTM E2112 in North American climate zones." *ASHRAE Journal*, pp. 1–7, 2007.

67. American Society for Testing and Materials. "ASTM E 2112 – 19c Standard Practice for Installation of Exterior Windows, Doors & Skylights." West Conshohocken, PA, 2019.

68. E.I. du Pont de Nemours and Company. "DuPontTM Flashing Systems Installation Guidelines." 2008.

69. Hodgin, Derek A. "What about the trim." *Interface,* pp. 4–10, September 2012.

10 Water Infiltration into Basements

Ronald L. Lucy
EES Group, Inc.

CONTENTS

PURPOSE/OBJECTIVES

The purpose of this chapter is to:

- Present common causes for water intrusion into basements.
- Discuss basement code requirements and best practices for preventing water intrusion into basements.
- Provide a methodology for completing basement water causation and origin forensic inspections.

Following the completion of this chapter, you should be able to:

- Conceptually understand why water intrusion into a basement occurs.
- Understand code requirements and industry best practices designed to eliminate water intrusion into basements.
- Have a basic understanding on corrective actions needed to eliminate basement water intrusion.

- Conduct a methodical and systematic inspection to locate and identify the source and cause for water intrusion into a basement.
- Be able to complete a basement water causation and origin written report.

10.1   INTRODUCTION

Since, by definition, basements are located below grade, it is inevitable that they are susceptible to water intrusion. Statistics from home inspectors suggest that basement leakage is one of the most common problems found in homes and that over 90% of all basements will suffer from some form of moisture and/or water infiltration issues at some point in time.[1]

Basements have become much more than a space to hold mechanical equipment, laundering utilities, and general storage. Modern basements have been transformed into elaborate finished spaces used for many purposes. Basements can be converted into bedrooms or living spaces, game rooms, home theatres, playrooms, gyms, home offices, workshops, "man caves," and lounge areas with wet bars. These transformations, if not properly constructed, can conceal underlying water intrusion until the damages become extensive and visible though the presence of rot and visible mold. The underlying problem often was not addressed prior to build-out of the basement spaces and results in costly repairs not only to fix the original cause of water intrusion but also to restore the interior finished spaces.

Basement water intrusion or elevated moisture levels can result from: (1) surface drainage through the foundation walls, (2) groundwater intrusion through foundation walls, (3) improper drainage of roof water away from foundation walls, (4) improper grade of ground surfaces near foundations, (5) failed sump pumps, (6) failures of water supply or sanitary lines, (7) backups to sanitary lines, (8) improper water management details at windows and doors, porch, deck, or stoop interfaces with wall systems at/near foundation walls, and 8) condensation of accumulated interior moisture. The purpose of this chapter is to provide: (1) an understanding on the causes of water intrusion, (2) knowledge of the steps to take to identify the causes of basement water intrusion, and (3) a methodology for basement water cause and origin inspections and reports.

The following sections outline common causes of water intrusion into basements.

10.2   REVIEW OF CAUSES FOR WATER INTRUSION AND WATER DAMAGE TO BASEMENT SURFACES

Common causes for water intrusion and water damage to basement surfaces are associated with the following:

- Drainage.
- Sump Pump Failures.
- Supply and Sanitary Line Breaks.
- Sanitary Line Backups.
- Failures with Exterior Wall Flashing Details.
- Condensation.

These are discussed at length below:

10.2.1   DRAINAGE

Experience has shown that the most common cause of basement water intrusion is due to poor drainage. During precipitation events, water can accumulate against a basement wall from the ground, roof, or poorly designed/missing gutters and downspouts. This can result in surface and groundwater flowing against and through foundation walls and then into the basement of a structure. Figure 10.1 illustrates basement water intrusion pathways.

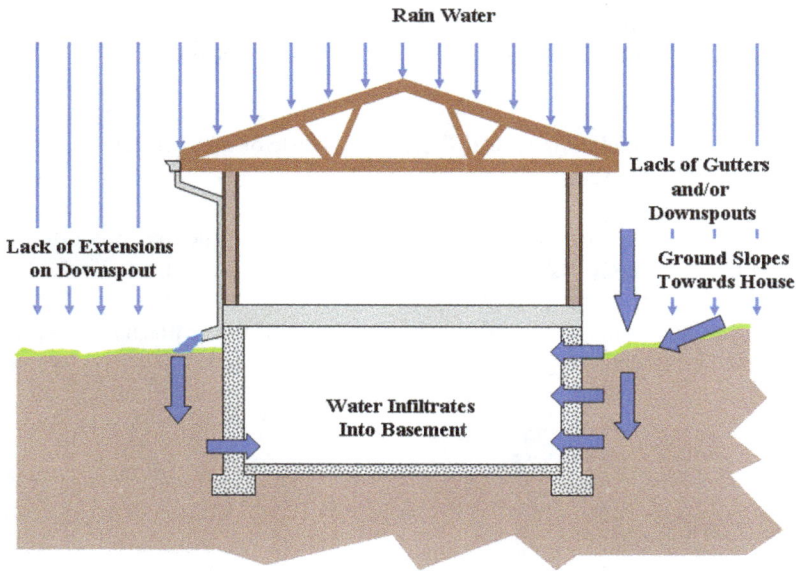

FIGURE 10.1   Basement water intrusion pathways.

To minimize drainage against and through foundation walls, surface water and groundwater must be collected and drained away from the foundation walls of a structure. The term used for this process is water management. Figure 10.2 illustrates some basic water management methods to move water away from the foundation walls of a structure.

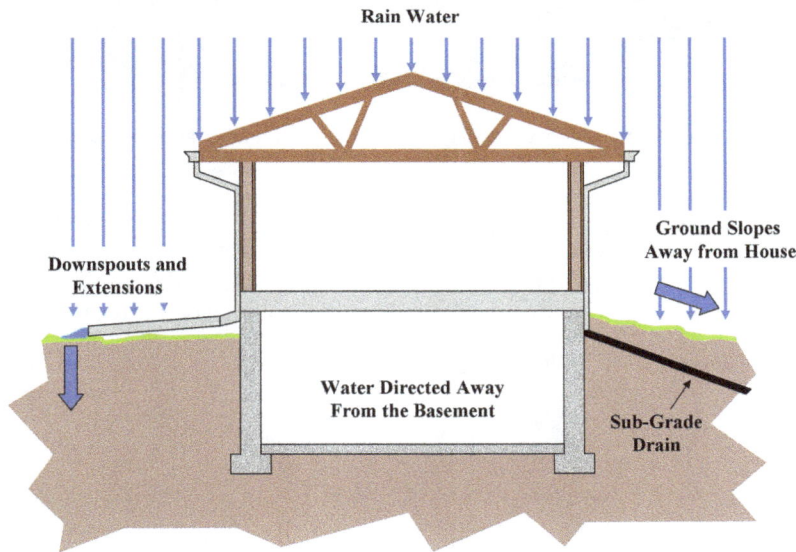

FIGURE 10.2   Drainage directed away from the home.

To prevent basement water intrusion from surface water and groundwater, proper drainage (water management) must be present for the roof system, gutter/downspout systems, wall systems, surface grading, and subsurface drainage systems. For instance, if the ground surface is not graded with a slope down and away from the structure, it is likely that some level of basement water intrusion can be expected. In the following sections, the principles, code requirements, and industry best practices for surface, roof, and foundation (subsurface) drainage will be outlined.

10.2.1.1 Surface Drainage

Surface drainage against foundation walls is the greatest cause of water intrusion into the basements of structures. This can be the result of the following situations:

- When backfill is placed around the foundation walls of a structure, it is often not sloped away from the structure or is sloped away but was not well compacted, which will settle and form low spots near the foundation walls.
- When landscaping around a home alters the designed drainage patterns of the ground slope near the foundation walls (Figure 10.3).

FIGURE 10.3   Landscaping bed against foundation trapping water and creating a grade toward the home.

- Improper surface ground slope when a driveway, sidewalk, or patio is installed directly against the foundation. When these structures (typically concrete) are not poured such that they slope down and away from the foundation wall or settle such that they slope toward the foundation wall, surface water will flow toward the foundation walls (Figure 10.4).

FIGURE 10.4   Water intrusion occurring at driveway/wall interface.

Sometimes the location of the structure, combined with the higher elevation of the neighboring property, does not allow for the surface grade to be properly sloped away from the foundation of the structure (Figure 10.5).

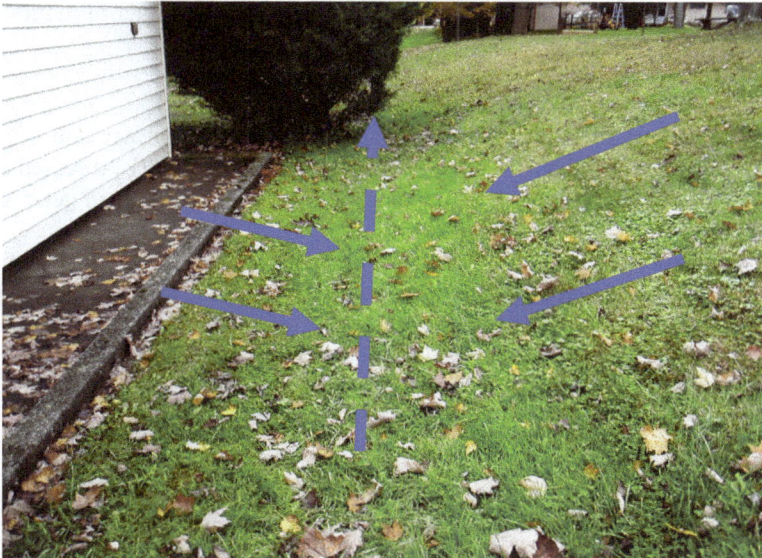

FIGURE 10.5   Neighboring property at higher elevation – drainage swale catching water.

In these cases, the installation of a swale (Figure 10.5) can be an effective method to divert water away from a foundation wall. A swale is essentially a trench that creates a low spot to collect surface water that would otherwise be directed toward the home.

Ground surface slope requirements are set forth in Section R401.3 of the 2018 International Residential Code (IRC)[2]:

> *R401.3 Drainage*: Surface drainage shall be diverted to a storm sewer conveyance or other *approved* point of collection that does not create a hazard. *Lots* shall be graded to drain surface water away from foundation walls. The *grade* shall fall not fewer than 6 inches (152 mm) of fall within 10 feet (3,048 mm), drains or swales shall be constructed to ensure drainage away from the structure. Impervious surfaces within 10 feet (2,048 mm) of the building foundation shall be sloped not <2% away from the building.
>
> *Exception*: Where *lot lines*, walls, slopes, or other physical barriers prohibit 6 inches (152 mm) of fall within 10 feet (3,048 mm), drains or swales shall be constructed to ensure drainage away from the structure. Impervious surfaces within 10 feet (3,048 mm) of the building foundation shall be sloped not <2% away from the building.

The requirements include the following:

- Surface drainage must be diverted to an area that does not create a hazard.
- Lots around foundations shall be graded 6 inches within the first 10 feet or ~2.9° down and away from the foundation.
- When lot lines and other barriers prevent 6 inches of fall per 10 feet, drains or swales are required to drain water away from the foundation.
- Impervious surfaces within 10 feet are also required to be sloped a minimum of 2% away from the building.

10.2.1.2   Roof Drainage

Improper roof drainage will lead to water flowing onto the ground surface and often against the foundation walls of a structure. Thus, water from roof surfaces must also be directed away from the foundation. One inch of rainfall on a 2,500 square foot roof surface area can generate ~208 cubic feet or ~1,556 gallons of water to be managed.

Different types of roof construction lead to different water runoff and water management systems and issues. Gable roof surfaces (two-elevation roof) concentrate the roof water typically at two eaves, whereas hip-style roof surfaces with typically four roof elevations will distribute water more evenly at four eaves. Areas near dormers will also concentrate roof water. The types or roof areas and varying roof drainage patterns/loadings at eaves are illustrated in Figure 10.6.

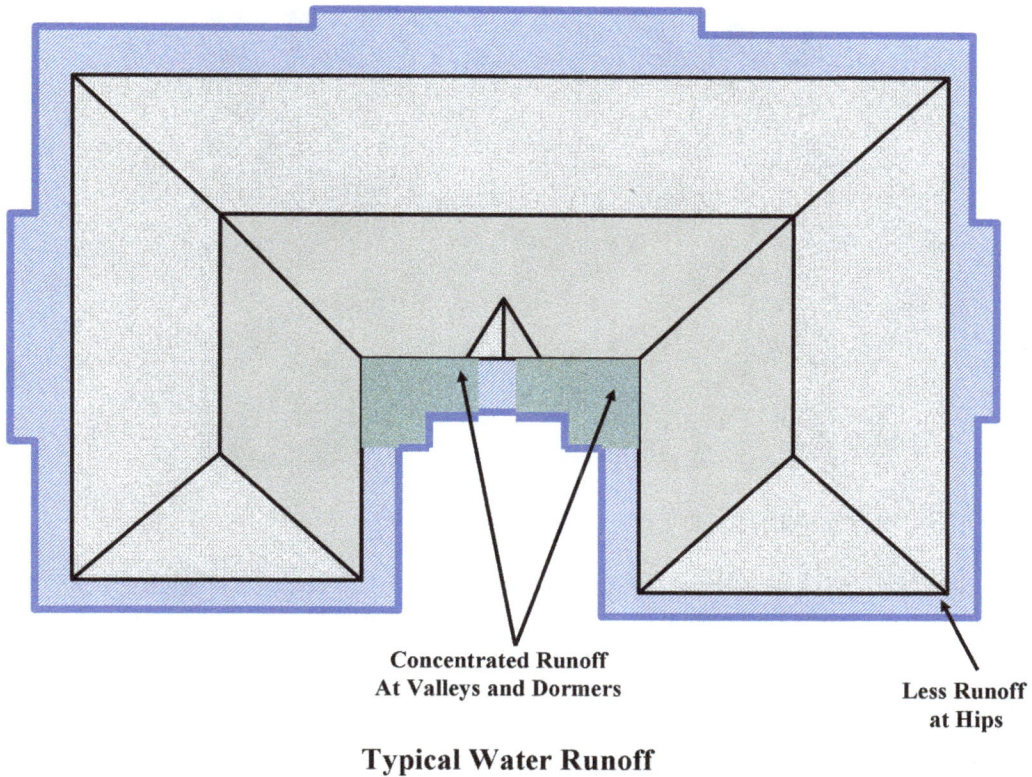

**Concentrated Runoff
At Valleys and Dormers**

**Less Runoff
at Hips**

Typical Water Runoff

FIGURE 10.6   Roof drainage patterns.

To prevent the runoff of water from accumulating near the foundation, one must ensure a proper surface design, ensure that gutters and downspouts are properly sized and maintained, and ensure that subgrade drainage systems are properly designed and maintained. As discussed in Chapter 8, gutters and downspouts must be sized in accordance with the local rainfall intensities and roof surface areas. If gutters or downspouts are undersized, water can overwhelm the gutter and/or downspout and flow directly onto the ground surface below at/near the foundation walls.

Even if properly sized, gutters and downspouts can still deposit roof water onto the ground surface due to installation and/or maintenance deficiencies. These include: (1) the gutter is installed with a slope that is too flat or away from the downspout opening, (2) the gutter slope, while originally proper, had been impacted such that it slopes away from the downspout opening by ice/snow loads, mechanical damage from impacts with items such as tree limbs, or settlement, (3) downspout discharges at/near the foundation wall rather than a subgrade drain or downspout run-out along the ground, (4) clogged, broken, or undersized subgrade drains, and/or (5) clogged gutters and downspouts due to the lack of maintenance.

Best practices recommend that gutters slope down toward the downspout opening 1 inch for every 40 feet of run (0.12°) to ensure adequate drainage.[3] This slope is also intended to prevent standing water and debris from accumulating in the gutter, which will restrict or stop roof water flows and can cause sagging and displacement of the gutter (Figure 10.7).

FIGURE 10.7   Improperly sloped gutter containing standing water and debris.

As illustrated in Figure 10.6, a large amount of roof water will be directed toward roof valleys during heavy periods of rainfall and/or snow/ice melt. The resulting flow of water in the valleys can result in water spilling over the front edge of the gutter onto the ground surface below. Figure 10.8 illustrates water pouring over the front edge of a gutter and onto the ground surface due to water draining around a large chimney cricket valley.

FIGURE 10.8   Water running over gutter at valley onto ground surface.

To avoid this situation, diverter baffles can be installed to keep the roof water from the valley flowing into the gutter. Best practices suggest that the installed diverter baffle should extend to a height of 4–6 inches above the front edge of the gutter.

If the downspout leader discharges water too close to the foundation, is not aligned with the subgrade drain, or the subgrade drain is broken or clogged, the water discharged from the roof and gutter will discharge relatively near the foundation wall (Figure 10.9).

FIGURE 10.9 Downspouts directed to drain water near the foundation wall.

Oftentimes, downspouts are located near the corners of a structure or tucked away between wall abutments. The foundation walls in these areas are prone to settlement cracking, thus increasing the risk for water intrusion if roof waters are not properly managed.

There are no known codes or standards that specify the exact distance in which downspout leaders should be placed away from a foundation wall. However, to meet the surface drainage code requirements (discussed earlier in this chapter), a leader should discharge water onto a ground surface with the necessary slope (~6 inches/10 feet down and away from the home, or ~2.9°). If not, the leader should be extended to a location where this grade requirement is met or to a swale or subgrade drain capable of handling design roof waters discharged from the downspout.

Case Study

A homeowner was experiencing water intrusion to the corner basement walls near where a downspout was discharging water to the ground near the corner foundation walls. To correct this problem, the homeowner extended the leader out to a location about 12 feet away from the corner of the home. During heavy periods of rain, water intrusion continued to occur at this same corner. Water testing and yard-level measurements performed during the subsequent forensic investigation revealed that while the water was being discharged 12 feet away, the ground surface in this area sloped back toward the foundation. Within 12 minutes of starting the water test, water was observed flowing against the corner foundation walls of the home. Thus, the cause of water entry was the lack of proper grading of the ground surface near the foundation walls, allowing surface water from the roof to flow toward the foundation walls.

The most effective way to manage roof water discharging from downspouts is to connect the discharge to a subgrade drainage system. Subgrade drainage systems are typically located near the footer of foundation walls and collect water near the footer and from downspouts. The collected water is then gravity drained to nearby streets or to storm water sewer systems. However, while normally very effective for moving water away from foundation walls, even this type of system can be subject to problems. Settling of soils where the drainage piping is buried near the foundation walls often occurs. This settlement can cause separation, breaking, and even clogging of subgrade drain piping. Figure 10.10 illustrates a subsoil drain that settled in relationship to the downspout leader, allowing roof water to partially miss the subgrade drain and pour against the foundation walls.

FIGURE 10.10 Roof water partially missing subgrade drain and flowing against foundation walls.

Significant basement water intrusion was observed on foundation walls in this corner of the basement.

In addition, it is not unusual for tree roots or vegetation to clog, break, or dislocate these subgrade drains. This condition can be difficult to determine and oftentimes requires the use of a boroscope with a camera to locate the cause of damage to the subgrade drain. Figure 10.11 shows a still photograph from footage taken during a forensic investigation where the subgrade drain was broken.

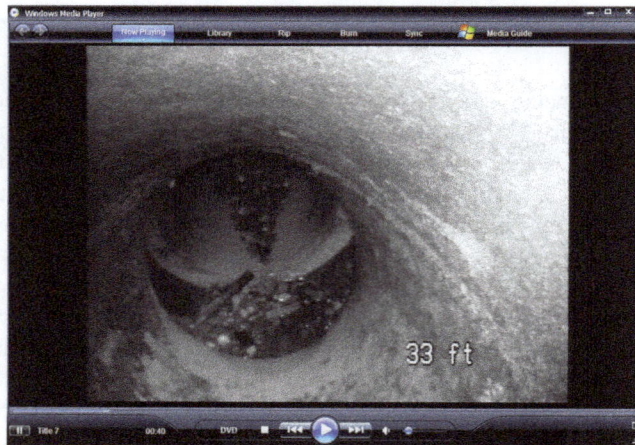

FIGURE 10.11 Linear camera photograph showing subsoil line breakage.

Another method that sometimes works to determine if a blockage is present in a subgrade drain is simply run water through a garden hose into the suspected blocked drain (Figure 10.12).

FIGURE 10.12 Water flowing from subgrade drain outlet at the street – no apparent blockage present in the line.

If water flows freely from the subgrade drain outlet at the street, then blockage of the line is unlikely. If, on the other hand, water back-flows out of the subgrade line near the downspout entrance, the subgrade drain line is likely blocked. The difficulty here is that the exact location or nature of the blockage may not be known without either using a boroscope or excavating the line. In addition, during times when grass or vegetation is not dormant, more robust and/or darker green spots in the lawn may provide an approximate location of the blockage.

Another one of the more common problems associated with roof water drainage and management is the improper maintenance of the gutters, downspouts, and subgrade drain lines. This often manifests itself in the clogging of these systems with leaf and tree debris. If leaf and tree debris is not cleaned from gutters and downspouts, the blockage will result in roof water flowing onto the ground surface below, often then flowing toward the foundation walls (Figure 10.13).

FIGURE 10.13 Water pouring over clogged gutter.

Depending on the arrangement of flashings, this roof water can also flow behind the gutter and cause damage and rot to the fascia and wall structures below. In colder climates, standing water in the clogged gutters freezes adding weight and additional blockage. This in turn can cause the gutter to sag and detach, leading to further water-related damage such as that from ice damming conditions. Long-term debris accumulation can even lead to active growth of vegetation in the gutters (Figure 10.14).

FIGURE 10.14   Vegetation growing in gutter.

Lack of maintenance can result in consequential issues such as rotting of the fascia that supports the gutters. In turn, these conditions, coupled with the added weight of debris, ice, or standing water, can ultimately detach and collapse the gutter (Figure 10.15).

FIGURE 10.15   Collapsed gutter due to weight of water and debris and fascia rot.

At a minimum, gutters should be cleaned out once per year and preferably at the end of fall. However, normally the gutters will have to be cleaned multiple times each season if nearby trees and bushes are present. Many types of gutter guards/screens are available to facilitate keeping gutters free from debris with varying degrees of success.

10.2.1.3 Foundation Drainage

Basement foundation drains are normally installed at the time of the construction and were touched on briefly, earlier in this chapter as part of the overall subgrade drainage system designed to collect and move water away from foundation walls. Foundation drains are installed along the footing at the interior and/or exterior side of the basement wall. The location for placement of the foundation drains (inside or outside the foundation walls) appears to be regional in nature. For example, in the Midwest, they are installed inside the footer, whereas in the northwest, they are installed outside of the footer.

An example of typical code language for the installation of foundation drains can be found in the 2018 IRC,[2] Chapter 4 (Foundations), which is reproduced in part below:

> *R405.1 Concrete or masonry foundations*: Drains shall be provided around all concrete or masonry foundations that retain earth and enclose habitable or usable spaces located below *grade*. Drainage tiles, gravel or crushed stone drains, perforated pipe, or other approved systems or materials shall be installed at or below the top of the footing or below the bottom of the slab and shall discharge by gravity or mechanical means into an *approved* drainage system. Gravel or crushed stone drains shall extend at least 1 foot beyond the outside of the footing and 6 inches above the top of the footing and be covered with an *approved* filter membrane material. The tops of open joints of drain tiles shall be protected with strips of building paper. Except where otherwise recommended by the drain manufacturer, perforated drains shall be surrounded with an approved filter membrane or the filter membrane shall cover the washed gravel or crushed rock covering the drain. Drainage tiles or perforated pipe shall be placed on not <2 inches (51 mm) of washed gravel or crushed rock not less than one sieve size larger than the tile joint opening or perforation and covered with not <6 inches (152 mm) of the same material.
>
> *Exception*: A drainage system is not required when the foundation is installed on well-drained ground or sand–gravel mixture soils according to the Unified Soil Classification System, Group I soils, as detailed in Table R405.1.

The drainage language is intended to keep surface water from flowing against foundation wall surfaces.

Foundation drains consist of perforated pipes that collect water flowing toward or at the foundation wall surfaces. Water collected by foundation drains is moved away from the foundation walls in one of two ways: (1) if the property is located on a hill or above a storm-water drain, the collected water can be drained away by gravity or (2) through the use of mechanical means such as a sump pump located in a basement. These two types of drainage discharge methods for foundation drains are shown in Figure 10.16.

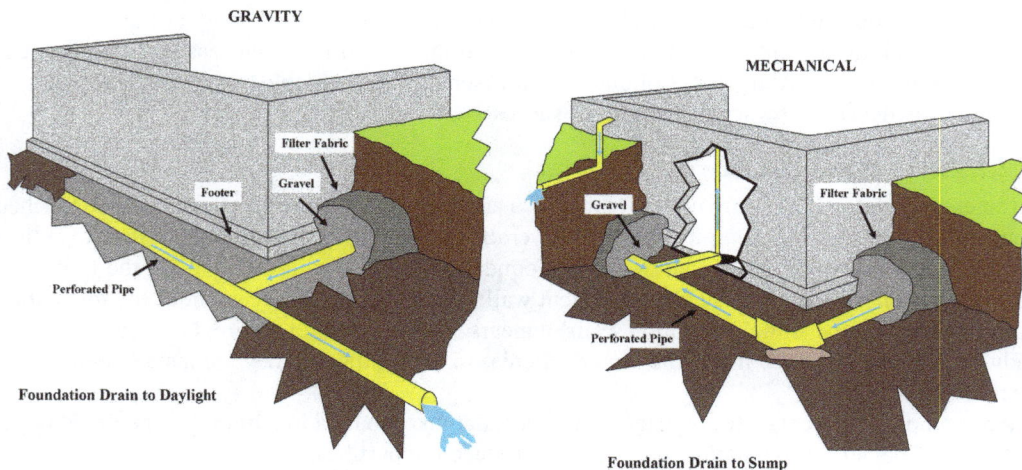

FIGURE 10.16 Foundation drainage – gravity vs mechanical methods.

When a sump pump discharge system is used, the discharge outlet is connected to the same subgrade drainage system that handles the downspout drainage. It is important to remember that this can be problematic should the subgrade drains break or become clogged or blocked (see section 10.2.2). The reader is also referenced to Lstiburek,[4] who illustrates the proper solutions to these drainage problems.

10.2.2 SUMP PUMPS

In most residential settings, a sump pump, commonly located in the basement, is utilized to remove water from around the foundation. Collected water flows into a sump pump pit where it is then pumped into the subgrade drain lines or to the ground surface where it flows away from the foundation walls by gravity. Detailed analysis of sump pump failures is described further in Chapter 23 on Equipment and Installation Failures.

10.2.3 WATER SUPPLY LINE AND SANITARY LINE BREAKS OR LEAKS

Water from water line and sanitary line breaks is a common cause of water entering a basement or adding to the total water vapor load in the air that can result in water damage from condensation. The most common locations for these leaks are failures from the plumbing line to freezer icemakers and failures of water and sanitary lines to/from toilets. These leaks can be readily observed by water-staining patterns to the floor decking and wood members forming the basement ceiling. If the leaks are active, dripping will be observed and moisture meter readings will be on the high end of the scale.

10.2.4 SANITARY LINE BACKUPS

While not a common occurrence, some basement water intrusion issues can be related to sanitary drainage line backups. This can cause sewage or gray water to flood the basement. Besides causing water damage and possible fungal growth, the backups can also result in possible bacterial contamination (see Chapter 13).

Based on experience, property owners will often report the water intrusion originated from the floor drains due to sewage line backups. However, out of hundreds of basement water intrusion investigations, only one case had been encountered where the water backup was related to a sanitary sewage line backup.

When considering whether or not the reported sanitary line backup is related to surface and/or ground water or sanitary issues, it is important to consider these key points:

- Plumbing codes require that water from all internal drains be directed to a sanitary sewer system. This includes bathrooms, kitchens, laundry, and basement floor drains. These drains are also separate from the drains handling surface or ground water.
- Sanitary sewage is almost always accompanied by a distinct "sewage" type odor and is rarely clear in appearance.
- Conversely, surface water or groundwater is clear and typically free of odors.
- The sump pump system is separate and isolated from the sanitary piping serving the floor drains.
- If properly designed, basement floor drains are located at the lowest points in the floor and contain a trap designed to prevent sewer gas from entering the home. When water enters from the perimeter walls or from a sump pump backup, the water will flow toward these drains. This would explain why these floor areas near floor drains are sometimes damp.

In summary, if the water entering the basement is clear and free of odors, then the source of the water likely was not from a sanitary line source.

10.2.5 Exterior Wall Surface Water Management Issues

Aside from the surface water and/or groundwater intrusion occurrences and from piping leak sources, water can also enter into a basement as a result of water management at porch and deck attachments, around windows, and with the exterior walls.

When water from these sources works its way into a basement, staining patterns will typically be most noticeable along the wood members forming the top plate or rim joist for the first floor construction. When the staining patterns are located above the surface of the exterior grade, these are the likely sources of the water entering the basement. Chapter 9 explores water management from window and door openings at length.

10.2.5.1 Porch or Deck Interfaces

Porch or deck interfaces have been found to be a reoccurring location for water entry into a basement. This primarily occurs as a result of improper attachment of the porch or deck to the structure and is related to the following sequence of events related to construction:

1. A porch or deck is either constructed or abutted against the exterior wall of a home.
2. The porch or deck is typically at or above the top of the foundation wall.
3. Porches or decks (especially porches) experience settlement over time. The settlement can provide a separation between the wall interfaces, or it can settle to where the water runoff is directed toward the home.
4. It is rare to find flashing at the interfaces. The primary water-proofing component is typically a sealant that is usually poorly maintained.

The IRC[2] requires flashing at locations where porches, decks, or stairs are attached to walls or floors in wood-framed construction (see Chapter 16 for detailed requirements on porch/deck installations). In lieu of flashing, sometimes contractors install a strip of sealant along the structure/wall interface. These sealants degrade and split with time, leaving openings directly into the interior.

Another problem often seen in this area is related to settlement, especially with concrete porches. The settlement results in a separation along the sealant. If the porch sinks toward the home, surface water will run toward the interface, if the porch sinks away from the home, a larger separation occurs allowing for larger volumes of water to infiltrate into the wall cavity (Figure 10.17).

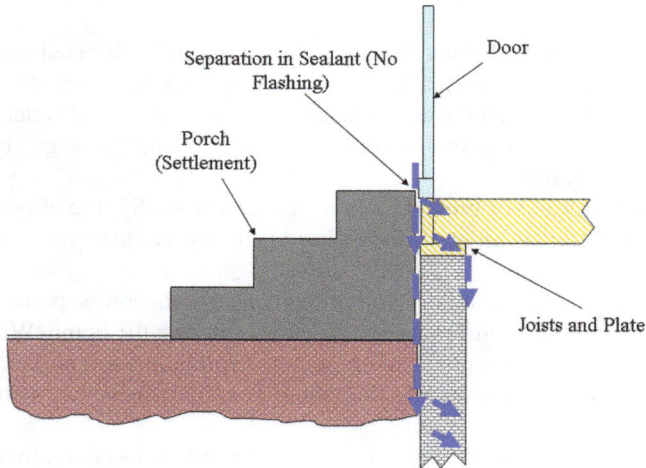

FIGURE 10.17   Water intrusion pathway at porch interfaces.

Since the elevation of porches or decks is typically higher than the beginning of the wood frame construction, the water can damage and/or rot the top plate, joists, and subflooring of the main structure (Figure 10.18).

FIGURE 10.18   Water damage to flooring and framing below stoop.

Additionally, the water that does not enter at the top of the wall will essentially be trapped behind the appurtenance, resulting in water accumulation and hydrostatic pressure against the foundation wall.

In one particular case, a wood deck was found to be attached directly to the rim joist of a vinyl sided home. Water reportedly poured into the basement during periods of rains. The homeowner's contractor believed that this was occurring due to groundwater intrusion. A water test was performed on the deck surface and water was observed flowing down the foundation wall and onto the floor of the basement (Figure 10.19).

FIGURE 10.19 Water test resulting in water intrusion at deck attachment.

This water intrusion situation could have been prevented had a section of sheet metal flashing been installed where the deck attached to the porch (see Figure 10.20).

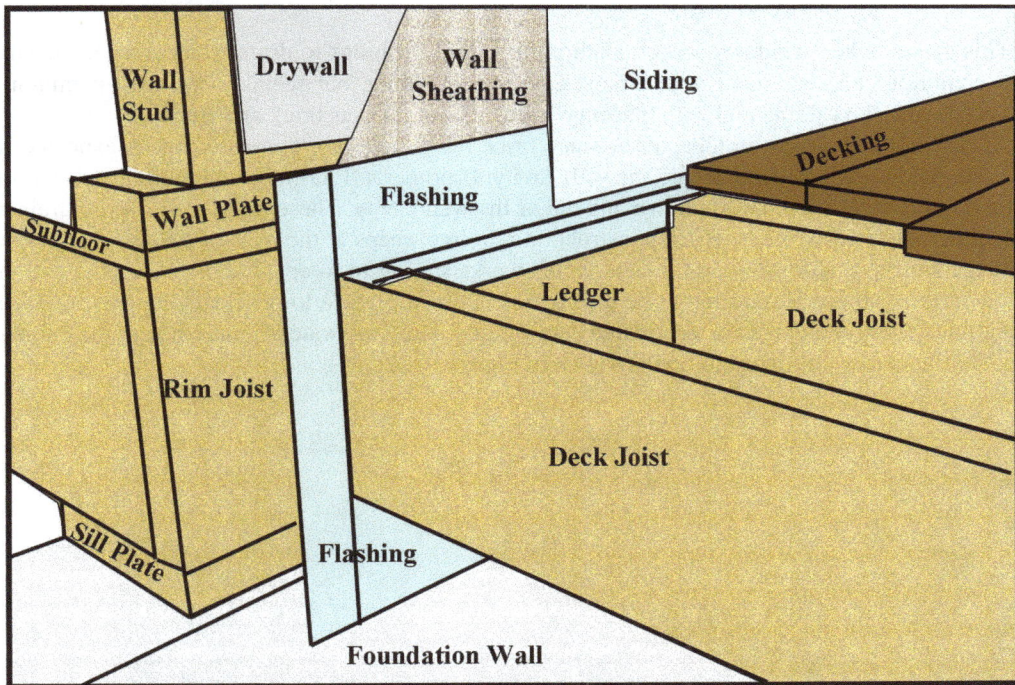

FIGURE 10.20 Flashing detail at deck attachment.

10.2.5.2 Windows and Walls

When poor water management details are present around windows and wall cavities, the water has the ability to travel down the wall cavity and into the basement. A good indicator of water entry from window and door openings is localized staining to the rim joists and top plate below these openings (Figure 10.21).

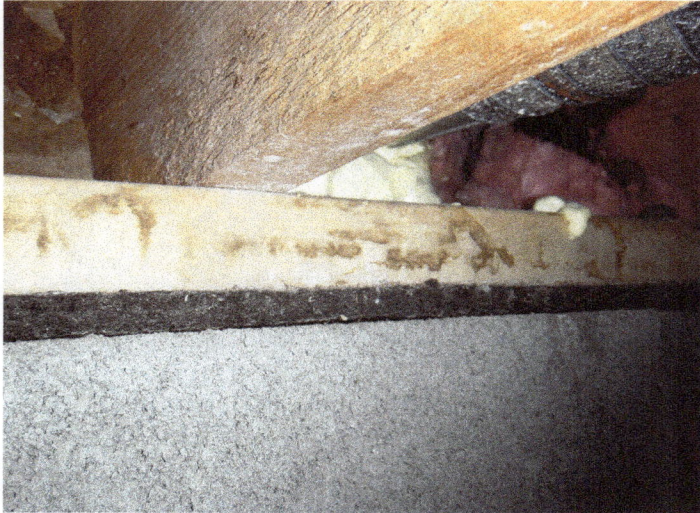

FIGURE 10.21   Water staining to top plate below window.

This occurs when windows do not contain proper sill flashings to drain water from the cavity. This condition can also occur with almost any type of siding, but seems to be most commonly observed with vinyl siding and with masonry wall surfaces such as brick and stucco.

As discussed at length in Chapter 9, since brick and stucco are porous, water is expected to infiltrate behind the veneer and into the wall cavity. If proper sill flashings and building paper are installed, the water should drain down and out of the wall cavity. These types of walls are drained with either weeps or screeds. Brick construction requires weeps at the base of the wall for proper drainage. Stucco requires a metal screed, which works similar to weeps. In both cases, the bottom of the brick or stucco is required to be terminated above the grade to provide adequate drainage. Oftentimes, these claddings are terminated below grade. Thus, the water drains out near the foundation. An illustration of this condition is shown in Figure 10.22.

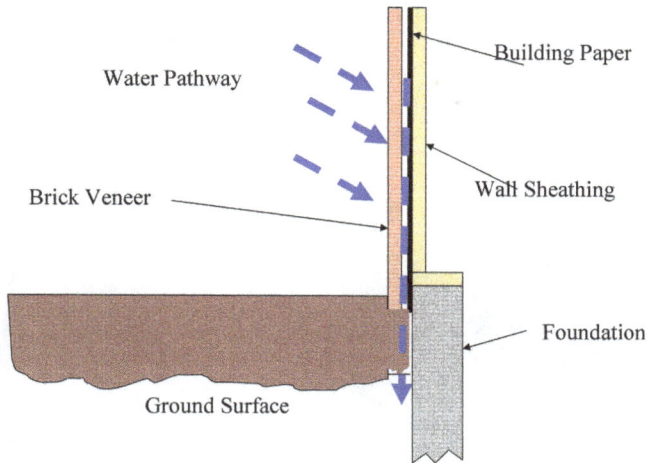

FIGURE 10.22   Water pathway when brick is terminated below grade.

If the brick veneer shown in Figure 10.22 was terminated above grade and weeps were present at the base of the wall, the water would be directed out to the ground surface rather than against the foundation wall.

10.2.6 CONDENSATION

In this chapter, we have touched on various sources of water that can cause issues in basements. As discussed in greater detail in Chapters 11–13, condensation can occur when water in its vapor form comes in contact with cooler surfaces and condenses out to its liquid form. Thus, often rot and visible mold seen in basements will not be from water intrusion per se, but from moisture that is trapped and then condenses in the basement.

Since basements are typically below grade, year-round moist conditions can be present. Homes with forced-air type HVAC systems typically dehumidify the air in the air-conditioning mode and lower the humidity by bringing in some fresh air, which aid in removing moisture from a basement. When poor ventilation is present, excess moisture can build up and eventually condense out on cooler surfaces in the basement. Indicators of excess moisture levels in basements are condensate on cold water supply lines and warped or buckled wood paneling (Figure 10.23).

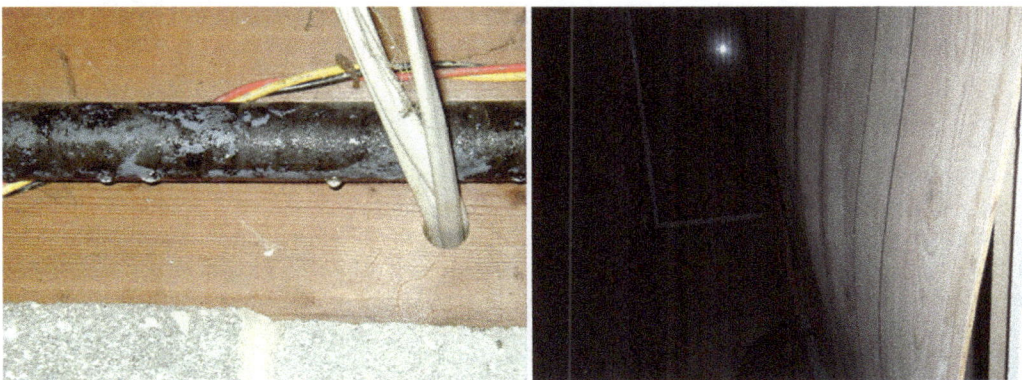

FIGURE 10.23   Indicators of moisture issues – condensate on water line and warped paneling.

These conditions can also be verified using an indoor air quality meter (see Chapter 11).

Moisture issues are typically accompanied with surficial mold growth on walls and ceiling areas. This is especially the case with walkout basements. The exposed walls on the walkout portions become an ideal condensing surface for the trapped moisture (Figure 10.24).

Exposed North Wall of Basement   **Condensate and Mold Growth to Interior**

FIGURE 10.24   Exposed wall providing a colder surface for condensation and mold growth.

Two methods to control this problem would be to add a dehumidifier or add insulation and a vapor retarder to the wall surfaces (see Chapter 9).

In cases where homes are vacant for extended periods of time and the HVAC system is shut down (e.g., snow birds), the lack of air movement will often lead to the buildup of high levels of moisture in the basement. In these situations, it is not uncommon to find widespread surficial mold growth on basement surfaces. If the growth is surficial (this can be verified by a test cut into the wall cavity), then the cause was likely interior moisture rather than leakage from surface water and/or ground-water intrusion.

10.3   METHODOLOGY FOR BASEMENT WATER CAUSATION AND ORIGIN FORENSIC INSPECTIONS

The basic steps in performing a cause and origin inspection have been discussed in Chapters 1 and 9. These same methodologies, analyses, and inspection tools can be used to investigate basement water intrusion issues. The following sections provide additional information to consider when performing the basement water causation and origin forensic inspection.

10.3.1   INTERIOR INSPECTION

All visual wall surfaces should be viewed when performing the walk-around portion of the site investigation. Any deficiencies or issues should be documented in the field notebook. It is also important to measure out the locations of the deficiencies for later correlation to possible findings on exterior finishes. Some of the more common things to look for on basement surfaces include water damage staining, efflorescence, visible mold, and evidence of moisture condensation on pipes or HVAC vents.

Water damage staining patterns and rot are often the best indicators of determining where the water is originating. If most of the staining and rot were isolated to the perimeter of the basement and below the wood framing, then the likely source of the water would be from surface water and/

or groundwater. When most of the water damage is found on lower wall surfaces, including interior wall surfaces, this is often indicative of a past flooding situation or event where moisture affected only the lower portions of the walls. It is also not uncommon to observe historic water damage to the subflooring around plumbing line penetrations especially in older homes; this suggests plumbing leaks as a cause of the water damage.

Once visual observations are made, the damaged surfaces should be tested with a moisture meter. The moisture meter can aid in determining whether the leaks are active at the time of the investigation.

In some cases with finished basements, test cuts can be an important step to evaluate what is going on in the wall cavity.

The best indicator of surface water or groundwater intrusion through masonry foundation walls is the presence of efflorescence. Efflorescence is light-colored crystalline deposits that are left behind from evaporated water or moisture (see Figure 10.25).

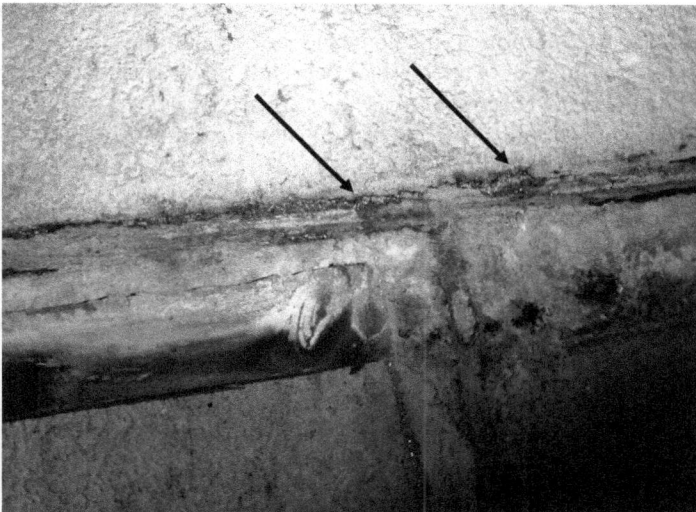

FIGURE 10.25   Heavy efflorescence to CMU wall.

Efflorescence typically forms in porous masonry construction from water movement through the wall cavity. This can be associated with moisture wicking through masonry construction, or in more serious cases, where water is found to be entering through openings. The presence of efflorescence on wall surfaces is typically indicative of long-term water or moisture intrusion events. In most cases, efflorescence is harmless and can be cleaned using mild acid solutions. In more severe cases, efflorescence can lead to spalling or damage to masonry wall surfaces.

Rust or corrosion to nails, wall ties, and reinforcements can also provide evidence that the water intrusion has been occurring for a long time (months/years vs days/weeks).

Wall repairs such as patching, sealing, or cement parging can also suggest that water intrusion has been a historic issue.

Once the inspection is complete, the following testing is recommended:

- Moisture probe the affected surfaces to determine if the leakage is active.
- Measure the indoor air quality to identify moisture and/or ventilation issues.
- Scan the wall surfaces with a FLIR™ camera (if available).
- Perform water testing on the exterior surfaces to confirm the source of water intrusion.

10.3.2 Exterior Inspection

The exterior inspection includes documenting the conditions of the ground surface, visible portions of the foundation, window flashings, deck or porch attachments, and roof drainage appurtenances. When performing the inspection, the following is recommended:

- Document the surface grade around the home. This can be done visually, with a digital level, or with a transit.
- Observe the visible portions of the foundation wall. Look for signs of staining, efflorescence, cracks, and repairs.
- Check for wall drainage (i.e., weeps or screeds) when the exterior finishes consist of brick masonry, stucco, or other masonry finishes.
- Investigate the flashing details around windows, doors, or decks.
- Look for staining patterns on the outside surfaces of gutters, soffits, and fascias. Staining would suggest past overflows.
- Measure and record dimensions of gutters, downspouts, and roof surface areas. This information can be used to determine if the home has sufficient drainage.
- Check to make sure the gutters are clear of debris. If linear washout were present in the soil, this would indicate reoccurring overflows. If necessary, run water in the gutter to make sure the downspouts are clear and water is flowing out of the subsoil drains.

10.4 METHODOLOGY FOR BASEMENT WATER CAUSATION AND ORIGIN FORENSIC INSPECTION REPORTS

The forensic report for basement water cause and origin inspections should follow the same methodology outlined in Chapters 1 and 9 with modifications to adjust for basement-specific observations and findings.

IMPORTANT POINTS TO REMEMBER

- The ground surface should be graded away from the foundation at least 6 inches within the first 10 feet. This is a code requirement.
- Gutter and downspouts need to be sized in accordance with the roof area and rainfall intensities.
- Gutters should be sloped slightly toward the downspouts to work effectively.
- Downspouts should be directed to drain at least 10 feet away from the foundation. A subsoil drainage system is effective.
- Gutters and downspouts must be cleaned, as necessary, to keep debris from affecting their performance.
- Code requires that drains are present around foundations. The collected water needs to be discharged down and away from foundation walls. This can be done with either gravity or mechanical means (i.e., sump pump).
- Sump pits should be covered and kept clear of debris.
- Sump pumps should be powered on a dedicated electrical circuit.
- Sump pumps must have check valves and relief holes to prevent air lock, which can cause cavitation.
- Flashing deficiencies at porches, decks, windows, and doors can lead to water entry.
- Condensation and surficial mold growth usually occur due to excess moisture levels and poor ventilation in the basement.
- Efflorescence on foundation wall surfaces is probably the best indicator of surface or groundwater intrusion.

REFERENCES

1. "Basement Leakage." last modified February 13, 2012.
2. International Code Council. "2018 International Residential Code for One- and Two-Family Dwellings (IRC)." Accessed June 8, 2020, https://codes.iccsafe.org/content/IRC2018/chapter-4-foundations.
3. Partnership for Advancing Technology in Housing. "The REHAB Guide ROOFS." March 1999, p. 64. Accessed June 8, 2020, https://www.huduser.gov/portal/publications/destech/roofs.html.
4. Lstiburek, Joseph W. "Keeping water out of basements." *ASHRAE Journal*, pp. 62–68, March 2019.

11 Indoor Environmental Quality

Stephen E. Petty
EES Group, Inc.

CONTENTS

PURPOSE/OBJECTIVES

The purpose of this chapter is to:

- Define the term Indoor Environmental Quality (IEQ).
- Define the term Indoor Air Quality (IAQ).
- Explain how IEQ and IAQ measurements can be used in forensic investigations.
- Example of IEQ – Heating – Delivered Air Temperature vs Air Velocity.
- Define normal IAQ values and/or range of values.
- Lessons learned from the field.
- What does "Adequate Ventilation" mean?
- What air cleaning technologies are available?

Following the completion of this chapter, you should be able to:

- Understand the terms IEQ and IAQ and the differences between these terms.
- Recognize forensic investigation scenarios where these parameters should be measured and may be helpful in the investigation.
- Be able to interpret IEQ and IAQ parameters.
- Be able to recognize where IAQ parameters may be valuable in forensic investigations.

11.1   INTRODUCTION

Broadly, the term Indoor Environmental Quality (IEQ) as defined by the Centers for Disease Control and Prevention (CDC) "refers to the quality of the air in an office or other building environments."[1] Generally, IEQ is broadly tied to building-related symptoms experienced by individuals that are associated with building characteristics, including dampness, cleanliness, and ventilation. Specifically, conditions and contaminants to be considered when evaluating a building's IEQ include the following:

- Indoor (dry-bulb) temperatures.
- Radiant temperatures.
- Air velocities.
- Ventilation (airflow) rates, including fresh outdoor air vs recycled return air.
- Dampness and humidity levels.
- Water-damaged/decaying building materials.
- Contaminants in the air and/or on surfaces:
  - Emissions from office machines.
  - Emissions from cleaning products.
  - Dusts and vapors from construction activities.
  - Emissions from carpeting and furnishings.
  - Perfumes.
  - Cigarette smoke.
  - Insect debris.
  - Animal debris.
  - Mold and mold by-products (mycotoxins).
  - Bacteria (including *Legionella*).
  - Viruses.
  - Asbestos.
  - Formaldehyde (CH_2O).
  - Carbon dioxide (CO_2).
  - Carbon monoxide (CO).
  - Volatile organic carbons (VOCs) such as benzene.
  - Pesticides.
  - Herbicides.
  - Radon.
  - Ozone.

These environmental agents, singularly or in combination, can affect human comfort and/or health. The total of these environmental agents are broader than just those factors associated with indoor air (i.e., those listed as contaminants in the air) and commonly described as Indoor Air Quality (IAQ) parameters.

Since litigation continues to become a more important issue as to whether or not analyses are performed, the standard of care (i.e., the typical quality required of an IEQ investigation) must be met by professionals in this field; Neumann[2] provides guidance in this area.

The American National Standards Institute (ANSI) and ASHRAE, the American Society of Heating, Refrigerating and Air-Conditioning Engineers, provide two standards, ASHRAE 55 and ASHRAE 62, often cited regarding best practices for IEQ and IAQ parameters. ANSI/ASHRAE Standard 55 – Thermal Environmental Conditions for Human Occupancy[3–7] defines environmental factors as temperature, thermal radiation, humidity, and air speed affecting personal comfort and separates IEQ parameters from IAQ parameters in their ASHARE 62 IAQ Standards.[8–20] The first version of ASHRAE 62 was issued in 1973 and was primarily designed to address IAQ issues through ventilation. Recommended ventilation levels and methods to determine levels have changed and become more complex with time. In 2004, ASHRAE 62 was split into two standards (62.1 – Ventilation for Acceptable IAQ and 62.2 – Ventilation and Acceptable IAQ in Low-Rise Residential Buildings) both of which continued to be updated.[8–14] The scope of ASHRAE 62.1 "Applies to all spaces intended for human occupancy except those within single-family houses, multifamily structures of three stories or fewer above grade, vehicles, and aircraft," whereas the scope of ASHRAE 62.2 "Applies to spaces intended for human occupancy within single-family houses and multifamily structures of three stories or fewer above grade, including manufactured and modular houses. This standard does not apply to transient housing such as hotels, motels, nursing homes, dormitories, or jails." ASHRAE 62 began to define acceptable IAQ levels using parameters such as carbon dioxide levels (i.e., should not exceed 1,000 ppm) in 1989, but has backed off from setting human health effects levels in subsequent versions of the standards.[21] However, in 2010 (revised 2015), ASHRAE provided guidance (ASHRAE Guideline 24) on these IAQ parameters along with acceptable levels for these parameters.[22] A review of U.S. and international IAQ parameters and their recommended levels can be found in the 2005 Canadian IAQ Guidelines and Standards – RR-204 document[23] and in other sources.[22,24,25]

11.2 IEQ

ASHRAE 55 sets boundaries for IEQ parameters (e.g., temperature, thermal radiation, humidity, and air speed affecting personal comfort based on clothing) as conditions where either 80% or 90% of the subjects tested find the environment acceptable (i.e., occupant acceptability limits). ASHRAE noted in the standard "The 80% acceptability limits are for typical applications and should be used when other information is not available. It is acceptable to use the 90% acceptability limits when a higher degree of thermal comfort is desired." A summary of IEQ parameters from ASHRAE 55-2017 along with recommended values for these parameters is provided in Table 11.1; the reader is referenced to the standard for additional detail and because values in this standard have changed with time.

While ASHRAE 55-2017 (Table 11.1) provides an effective resource for defining IEQ parameters and acceptable values for many of these parameters, it does not provide detailed information on acceptable lower limits of humidity for drying of the skin and mucus membrane tissues, nor values for indoor air contaminants.

TABLE 11.1

Overview of ASHRAE 55-2017 IEQ Parameters and Recommended Values

IEQ Parameter	Recommendations on Values or Range of Values	Discussion/Comments
Temperature (indoors) based on temperature (outdoors)	80% Acceptability: From ~50°F outdoor temp. (indoor acceptable range from ~64.4°F to ~75.4°F) to ~93°F outdoor temp. (indoor acceptable range from ~75.2°F to ~88.7°F). 90% Acceptability: From ~50°F outdoor temp. (indoor acceptable range from ~65.3°F to ~74.3°F) to ~93°F outdoor temp. (indoor acceptable range from ~78.8°F to ~87.8°F)	Figure 5.4.2 acceptable operative temperature ranges based on 80% or 90% Occupant Satisfaction. These temperatures can increase with increasing average air speed as follows: Avg. air speed allow Δ in temp – °F 118 fpm (0.6 m/s) 2.2°F (1.2°C) 177 fpm (0.9 m/s) 3.2°F (1.8°C) 236 fpm (1.2 m/s) 4.0°F (2.2°C)
Temperature (indoors) change with time	Cyclic variation: 2°F in 15 minutes	Under Temperature Variations in Time (Section 5.3.5.2).
Temperature (indoors) change with time	Drifts or ramps: 2°F in 15 minutes 3°F in 30 minutes 4°F in 60 minutes 5°F in 2 hours 6°F in 4 hours	Under Temperature Variations in Time (Section 5.3.5.3)
Vertical temperature (indoors) between head and ankles	<5.4°F for seated occupants and <7.2°F with standing occupants	Allowable difference under Local Thermal Discomfort Provisions (Section 5.3.4.1)
Radiant temperature differences	Cold ceiling: <25.2°F Warm ceiling: <9.0°F Cold wall: <18.0°F Warm wall: <41.4°F	Allowable radiant temperature differences under Local Thermal Discomfort Provisions (Section 5.3.4.2)
Floor surface temperatures - range	66.2°F to 84.2°F	Allowable floor surface temperatures with seated occupants with feet on floor under Local Thermal Discomfort Provisions (Section 5.3.4.4)

One example of IEQ is illustrated by examining the often-heard marketing slogan, "gas heat is warm heat." In homes, the temperature of the heat delivered at the register by a gas furnace is typically above the body temperature of 98.6°F, whereas the temperature of heat delivered by an electric heat pump is below the body temperature (Figure 11.1).

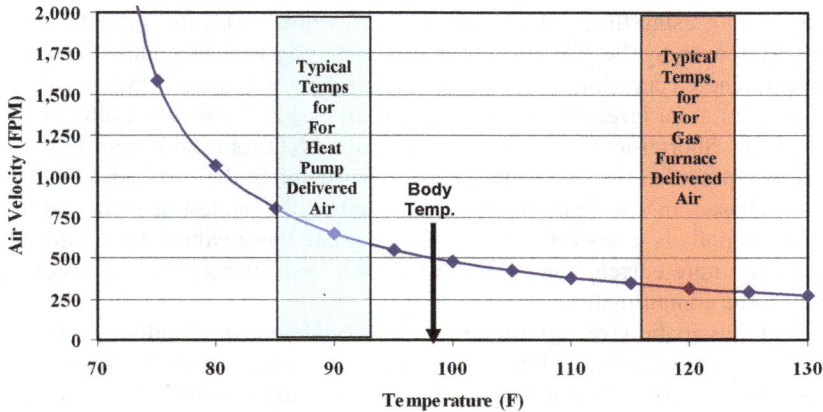

FIGURE 11.1 Residential setting – typical delivered air temperature and velocity for delivered heated air into a space.

The other important comfort factor other than delivered air temperature between an electric heat pump and a gas furnace is the delivered air velocity between the two technologies. Assuming the same delivered energy, the energy per cubic foot of air is less for a heat pump than for a gas furnace due to relative delivered air temperatures. Thus, more air must be delivered by a heat pump to provide the same total heat as a gas furnace. Given the fixed area of the registers in most homes, this implies a higher delivered air velocity (~2×) for a heat pump versus that for a gas furnace. Thus, not only is the heat pump affecting human IEQ comfort by delivering air below the body temperature, the air is being delivered at a higher velocity, making it feel even colder. While not advocating one technology over the other, this example illustrates the concept of IEQ (temperature and air velocity) and why the slogan "gas heat is warm heat" makes sense and is understood by the public.

11.3 IAQ

IAQ is a subset of IEQ and specifically evaluates the levels of contaminants in the air one breathes. These contaminants can take the form of vapors (e.g., gases such as carbon dioxide, carbon monoxide, formaldehyde, and benzene), fumes (i.e., gases that can condense – for example, gases from fires), and particulates (e.g., asbestos and silica). Biological contaminants such as mold, bacteria, and other biotoxins (e.g., anthrax), which are particulates, have also received considerable attention and can contaminate indoor air. Molds have the ability to off-gas metabolic products called mycotoxins, which can affect air quality (see Chapter 13). Oak Ridge National Laboratory (ORNL) reported that IAQ problems exist in up to one in five schools.[25]

Many organizations, both in and outside of the United States, have attempted to set acceptable levels for many of these contaminants, but the values vary among the various code and standards bodies so these recommendations have tended to decrease with time. For example, the ACGIH (American Conference of Governmental Industrial Hygienists) Threshold Limit Value

(TLV – 8-hour recommended value) for benzene dropped from 100 ppm in the 1940s to 0.5 in the 1970s, a factor of 200; using the NIOSH recommended value of 0.1, the value has dropped by a factor of 1,000. Moreover, by the 1980s, benzene was declared a probable human carcinogen by the U.S. Government. The Occupational Safety and Health Administration (OSHA) sets Permissible Exposure Limits (PELs) for levels of substances in the air (see 29 CRF 1910.1000). PELs are based on an 8-hr. workday. Shorter-term 15-minute values called STELs (Short-Term Exposure Limits) are also sometimes provided for some substances. Unfortunately the PELs and STELs are dated with many values based on 40+ year old data. The number one request of industrial hygienists in Synergist magazine polls is a request to have OSHA update these values. Even OSHA has stated one should consider more current values from California and Europe when considering airborne exposures to airborne contaminants.

The process of anticipation, recognition, evaluation, and control of indoor air hazards are the basic tenants of the field of industrial hygiene (IH). While the focus of the American Industrial Hygiene Association (AIHA – IH trade association) is on occupational (industrial) exposures, the books and manuals it has published,[26–28] including The Industrial Hygienist's Guide to Indoor Air Quality Investigations and others,[29,30] provide excellent guidance on performance of IAQ investigations.

In 1989, ASHRAE dabbled with setting IAQ levels in their ASHRAE 62 standard by setting a recommended limit for carbon dioxide (CO_2) of 1,000 ppm. However, ASHRAE later withdrew from this area citing it was not within their scope to establish health-related standards. The recommended limit for carbon dioxide (CO_2) of 1,000 ppm in the past has been replaced by an equation. This equation for the limit is 700 plus the ambient outdoor level of CO_2, which results in a value somewhat >1,000 ppm. In either case, ASHRAE has justified this value based on CO_2 being a surrogate for likely buildup of indoor contaminants.[21]

ASHRAE developed Guideline 24 "Ventilation and Indoor Air Quality in Low-Rise Residential Buildings" (as opposed to a standard; most recent revision 2015) for recommended indoor air levels of various contaminants.[22] ASHRAE noted under their scope "The guideline primarily applies to ventilation and IAQ for human occupancy in residential buildings three stories or fewer in height above grade, including manufactured and modular homes" and defined an IAQ contaminant as "a constituent of air that may reduce acceptability of that air." Further, they note "application of industrial exposure limits would not necessarily be appropriate for other indoor settings, occupancies, and exposure scenarios" likely due to the fact that industrial limits are based on 8-hour/day exposure times, whereas residential exposures could be up to 24 hours/day.

Table 11.2, based primarily on data in References,[8,22–24,31,32] provides examples of recommended levels for various IAQ contaminants; however, it should be noted that these levels are considered minimums and not necessarily safe.

TABLE 11.2
IAQ Contaminants and Recommended Limiting Values[a]

IEQ Parameter	NAAQS/EPA USA	OSHA PEL/STEL USA (2020)	NIOSH REL USA (2020)	ACGIH TLV/STEL USA (2020)[b]	MAK German (2000)	Canada (2020)	WHO/Europe (2010)	Hong Kong (2020)
Benzene		1 5 [15 minutes]	0.1 1 [15 minutes]	0.5 2.5 [15 minutes]	No safe level	Keep indoor levels of Benzene as low as possible[c]	No safe level can be recommended	
Carbon Dioxide		5,000	5,000 30,000 [15 minutes]	5,000 30,000 [15 minutes]	5,000 10,000 [15 minutes]	3,500 [L]		<800/<1,000 [8 hours][d]
Carbon Monoxide	9 [8 hours][e] 35 [1 hour][e]	50	35 200 [°C]	25	30 60 [15 minutes]	10 [24 hours] 25 [1 hour]	90 [15 minutes] 31 [30 minutes] 8.7 [1 hour] 6.1 [8 hours]	<2/<10 μg/m³ [8 hours][d]
Formaldehyde	0.4[f]	0.75 2 [15 minutes]	0.016 0.1 [15 minutes]	0.1 (TLV) 0.3 [STEL]	0.3 1.0[g]	0.1 [L] 0.01 [L][h]	0.081 (0.1 mg/m³) [30 minutes]	<0.03/<0.1 μg/m³ [8 hours][d]
Lead	0.15 μg/m³ [rolling 3 month avg.]	0.05 mg/m³	0.05 mg/m³ [10 hours]	0.05 mg/m³	0.1 μg/m³ 1 μg/m³ [30 minutes]	Minimize Exposure	0.5 μg/m³ [1 year]	
Nitrogen Dioxide	0.053 [1 year] 0.100 [1 hour]	5 [Ceil.]	1.0 [15 minutes]	0.2	0.5 1 [15 minutes]	0.011 [24 hours] 0.09 [1 hour]	0.1 [1 hour] 0.02 [1 year]	<0.04/<0.15 μg/m³ [8 hours][d]
Ozone	0.075 [8 hours – annual 4th highest value avg. over 3 years]	0.1	0.1 [Ceil.]	0.05 – HW 0.08 – MW 0.1–LW 0.2–Any work [2 hours]	Carcinogen No max. value established	0.02	0.02 (40 μg/m³) [8 hour]	<50/<120/ μg/m³ [8 hours][d]
Particles (<2.5 μm)	12 μg/m³ [1 year – Primary std.] 15 μg/m³ [1 year – secondary std.] 35 μg/m³ [24 hours – both stds.]	5 μg/m³		3 mg/m³ (respirable)	1.5 mg/m³ for <4 μm	Keep indoor levels as low as possible		0.05/0.12 mg/m³ [8 hours][d]

(Continued)

TABLE 11.2 (*Continued*)
IAQ Contaminants and Recommended Limiting Values[a]

IEQ Parameter	NAAQS/EPA USA	OSHA PEL/STEL USA (2020)	NIOSH REL USA (2020)	ACGIH TLV/STEL USA (2020)[b]	MAK German (2000)	Canada (2020)[c]	WHO/Europe (2010)	Hong Kong (2020)
Particles (<10 μm)	50 μg/m³ [1 year – primary std.] 150 μg/m³ [24 hours]			10 mg/m³ (inhalable)	4 mg/m³		12 μg/m³ [1 year] 35 μg/m³ [24 hours]	<20/<180 μg/m³ [8 hours][d]
Particles total		15 μg/m³ 5 μg/m³ respirable						
Radon	4 pCi/L [1 year]					200 Bq/m³ [1 year]	2.7 pCi/L [1 year]	<150/<200 Bq/m³ [8 hours][d]
Sulfur dioxide	0.075 [1 hour – primary std.] 0.500 [3 hours – secondary std.] 0.14 [24 hours][d]		2 5 [15 minutes]	0.25	0.5 1.0[g]	0.38 [5 minutes] 0.019	0.048 [24 hours] 0.012 [1 year]	

[a]　Unless otherwise specified, values are given in parts per million (ppm). Where no time limit given, time is 8 hours. Numbers in brackets [ ] refer to either a ceiling value or an averaging time of ≤8 hours (C, ceiling; L, long term, LW, light work; MW, moderate work; and HW, heavy work).

[b]　From 2020 ACGIH TLVs and BEIs.

[c]　From https://www.canada.ca/en/health-canada/services/air-quality/residential-indoor-air-quality-guidelines.html.

[d]　First value is for guideline value for Excellent Class and second value is guideline value for Good Class.

[e]　Not to be exceeded more than once per year.

[f]　Value from 24 CFR Part 3280 for manufactured homes.

[g]　Never to be exceeded.

[h]　Target level is 0.05 ppm because of its carcinogenic potential (Note – NTP in 2011 declared it a carcinogen).

As can be observed in reviewing the data in Table 11.2, world-recommended limits tend to be lower than U.S. recommended limits since most of the U.S. values have not been updated since the 1970s and are based on data from the 1950s and 1960s. Additional detailed information on indoor air contaminants, including levels, sources, and health effects can be found in references[32–36] (Table 11.3).

TABLE 11.3
IAQ Typical and Trigger Conditions for Nonindustrial Spaces – Burton[36]

Markers	"Typical" Conditions[a]	"Trigger" Conditions[b]
Human Markers		
Comfort issues	10% or less complaining of any one thermal condition. 20% or less complaining of more than one thermal condition	Greater percentages than these
Health issues	Infectious diseases: 100% protected asthma/allergies/MCS: 95-99% satisfied	Any less than listed
Physical Markers		
Temperature (air)	68°F–76°F	Temperature outside of this range
Δ Temperature	1–2°F	More than 2°F in a space served by a thermostat (controlled space)
Relative Humidity (RH)	30%–60%; except up to 70% in Southeast United States	<30% or >60%
Air velocity	>20 fpm but <50 fpm	Conditions different than those listed (80%–90%) satisfaction levels. In warmer climates, air velocities of 100–300 fpm ok
Outdoor Air (OA) delivered	Offices: ~15 to ~60 cfm/person; Classrooms: ~15 to ~40 cfm/person	<15–20 cfm/person
Total Supplied Air (SA) delivered	Offices: 1–2 cfm/square feet; 6–15 air changes per hour (ACH)	Lower SA than listed
Chemical Markers		
Carbon monoxide	0.25–1 ppm	>1–2 ppm
Carbon dioxide	600–1,100 ppm	700 ppm+background (typically 400) – total at/near 1,100 ppm
Formaldehyde	0.005–0.03 ppm	>0.03 ppm (40 µg/m^3). Note acceptable levels in California lower. Note these level contrast with concentrations from emissions of laminate flooring of 189 µg/m^3 (50th percentile) to 929 µg/m^3 (95th percentile)[c]
Nitrogen dioxide (NO$_2$)	0.01–0.5 ppm	>0.05 ppm
Mold	See Chapter 13; no visible mold or mold odors	Visible mold, musty, earthy, fishy, moldy odors
Odors	"None" (acclimated persons – not a visitor to the space)	Any detectible odor for over 5–15 minutes
Ozone	0.005–0.01 ppm	>0.01 ppm
Particles (<2.5 µm)	Up to 15 µg/m^3	>15 µg/m^3
Particles total	Up to 50 µg/m^3	>50 µg/m^3

(Continued)

TABLE 11.3 (*Continued*)

IAQ Typical and Trigger Conditions for Nonindustrial Spaces – Burton[36]

Markers	"Typical" Conditions[a]	"Trigger" Conditions[b]
Radon	<0.5 pCi/ℓ	>1–4 pCi/ℓ (US EPA recommended Standard is 4)
TVOC's (excludes benzene, toluene, ethylbenzene, or xylenes).	1–2 ppm (100–300 µg/m³)	>1 to >2 ppm

[a]   "Typical" levels of contaminants in problem-free spaces, or better.
[b]   "Trigger" suggests a range of actions, depending on the professional judgment of the OHS professional and ranging from "take note" to "investigate" to "correct measured level of a contaminant.
[c]   See Offermann.[37]

A detailed discussion of mold and bacteria is presented in Chapter 13.

11.4   APPLICATION OF IAQ TO FORENSIC INVESTIGATIONS AND LESSONS FROM FIELD INVESTIGATIONS

While any of the IEQ parameters discussed above may be helpful in conducting forensic investigations, the most common IAQ parameters measured, and of the most value, are the following:

- Temperature.
- Humidity (relative and specific).
- Carbon monoxide.
- Carbon dioxide.
- Dew point.

At times, particulates, formaldehyde, and VOCs will also be measured when conducting forensic investigations. The most common reason for measuring these parameters is for water cause-and-origin investigations where mold may be present. When both humidity and carbon dioxide levels in a space are elevated, the water, or moisture, responsible for the probable visible mold or water damage may be due to condensation of interior moisture. These parameters can be used to support such a conclusion.

A table summarizing typical indoor air measurements (i.e., temperature, carbon monoxide, carbon dioxide, humidity – relative humidity (RH), specific humidity (SH), and dew point (DP)) using portable IAQ meters is shown in Table 11.4.

TABLE 11.4

Typical IAQ Report Results

Time	Sampling Location	Temp (°F)	CO (ppm)[a]	CO_2 (ppm)[a]	Humidity[b] RH (%)	Humidity[b] SH (grains/#)	Dew Point (°F)
11:05 A.M.	1. Basement – living room	65.3	0.0	617	55.9	51.2	49.0
11:06 A.M.	2. Basement – furnace room	65.3	0.0	568	54.9	51.0	48.1
11:07 A.M.	3. Basement – bathroom	65.3	0.0	624	54.5	50.4	48.4
11:10 A.M.	4. 1st floor – living room	68.5	0.0	1,091	55.0	56.6	51.7
11:11 A.M.	5. 1st floor – dining room	70.4	0.0	993	51.2	56.9	51.8
11:12 A.M.	6. 1st floor – kitchen	69.4	0.0	1,137	53.9	58.4	52.3
11:24 A.M.	7. 2nd floor – master bedroom	70.9	0.0	1,052	48.9	54.7	50.8
11:24 A.M.	8. 2nd floor – NW bedroom	71.1	0.0	1,017	48.2	53.9	50.3
11:24 A.M.	9. 2nd floor – SE bedroom	70.9	0.0	1,004	47.9	53.3	50.0
11:27 A.M.	10. Outdoors	54.7	0.0	396	60.7	36.5	40.2

[a] Parts per million.
[b] Portable dehumidifier present in the laundry room, but not operating.

IAQ test results are then compared with typical values, or range of values, to determine whether or not they fall within these values and if not, what they may imply of conditions within the home. In this particular example of IAQ data (Table 11.4), the home was likely tight (lack of fresh outdoor air) based on elevated carbon dioxide and humidity levels.

Typical values or range of values used for comparison purposes follow.

11.4.1 TEMPERATURE AND HUMIDITY – RANGE OF VALUES FOR HUMAN COMFORT

Acceptable recommended limits for human comfort for temperature (and RH) have been continually revised by ASHRAE. Values from ASHRAE Standard 55-2004 and Standard 55-1992 are shown below to illustrate values provided in the Standard during different seasons.

11.4.1.1 Revised ASHRAE Standard (55-2004)

Summer (May 1–September 30):
   Air Temperature ~74°F–80°F
   RH ~64.4%–84.5%; equates to a SH of 84 grains/lb

Winter (October 1–April 30):
   Air Temperature ~67°F–80°F
   RH ~54.6%–66.6%; equates to a SH of 84 grains/lb.

11.4.1.2 ASHRAE Standard (55-1992)

Summer (May –September 30):
   Air Temperature ~72°F–81°F
   RH ~20.64%–81.84%; equates to a SH range of 23.91 – 130.92 grains/lb.

Winter (October 1–April 30):
   Air Temperature ~67°F–76°F
   RH ~24.33%–85.28%; equates to a SH range of 23.75 – 115.32 grains/lb.

Similarly, the recommended temperature and humidity human comfort ranges for commercial and institutional buildings from OSHA[35] are: (1) temperature – 68°F–78°F and (2) RH – 30%–60%. Burton[36] (Table 11.3) recommended temperatures between 68°F and 76°F and RH between 30% and 60%. It should be noted, these levels are for human comfort and not specifically for levels that would inhibit the amplification of mold growth.

11.4.2 CARBON MONOXIDE (CO)

Carbon monoxide is generally considered a product of incomplete combustion. It acts as a systemic chemical asphyxiate, replacing oxygen on red blood cells, thus reducing the amount of oxygen transported to the organs and other tissues in the body. Nonsmoking humans exhale ~0.5–2 ppm CO when breathing, whereas smokers exhale ~6–10 ppm.[36] The Threshold Limit Value (TLV) recommended by the American Conference of Governmental Industrial Hygienists (ACGIH) for an 8-hour exposure to industrial workers is 25 parts per million (ppm). However, this is for an exposed population of otherwise healthy adults between 20 and 65 years old. The National Ambient Air Quality Standard (NAAQS) for all populations published by the U.S. EPA is 9-ppm exposure averaged over an 8-hour period. A 9 ppm is also the "Limit for Acceptable Indoor Air Quality" recommended by the U.S. EPA. Recommended actions based on the levels of indoor CO were provided by Bergmann[38] (Table 11.5).

TABLE 11.5
Recommendations for Various Levels of Indoor CO

CO Level (ppm[a])	Response Description
1–9	Normal levels within the building. If there are no smokers, investigation is recommended
10–35	Advise occupants, check for symptoms and check all un-vented appliances, furnace, hot water tank, and/or boiler
36–99	Recommend fresh air, check for symptoms, ventilate the space, and recommend medical attention
100+	Evacuate the building and contact emergency medical services (911). Don't attempt to ventilate the space. Short-term exposure to these levels can cause damage

[a] Parts per million.

Sometimes, outdoor or occupational limits are used as surrogate values for residential or commercial indoor CO levels. Regarding use of occupational levels, one must recall that these are typically based on 8 hours/day exposures, whereas exposures in homes can be upward of 23 hours/day. Thus, one may want to decrease the occupational value by up to a factor of 3 (24/8 hours) for comparison purposes to account for residential vs occupational exposures to CO.

11.4.3 CARBON DIOXIDE (CO₂)

Carbon dioxide originates from products of combustion as well as from biological activity (i.e., human respiration). It acts as a simple asphyxiate by displacing oxygen in the air, thus reducing the amount of oxygen available for consumption. At relatively low levels, carbon dioxide can cause an increase in pulse rate, breathing problems, headaches, and abnormal fatigue. At higher concentration levels, the symptoms can include nausea, dizziness, and vomiting, and at extremely high levels, loss of consciousness.

Several organizations provide various recommendations for limits on CO_2 levels in the air; some are based on health effects and some (like ASHARE's) as surrogates for buildup of indoor air contaminants. The ACGIH threshold limit value (TLV) for an 8-hour occupational exposure is 5,000 ppm. However, the ASHRAE recommends a target level of 700 ppm plus ambient carbon dioxide value (~369) for odor control purposes for a total value of ~1,069. A 1,000 ppm value is the "Limit for Acceptable Indoor Air Quality" recommended by the U.S. EPA and others.[39] The acceptable range for carbon dioxide in air according to AIHA[40] is <850 ppm. Also, the Occupational and Environmental Health Directorate (a division of Armstrong Laboratory located at Brooks Air Force Base in San Antonio, Texas) reported in 1992 that CO_2 concentrations in excess of 600 ppm cause significant physiological effects, such as fatigue, drowsiness, lack of concentration, and sensations of breathing difficulty.[41] These researchers state they found that between 15% and 33% of the population will have symptoms from CO_2 exposure at 600–800 ppm, 33%–50% will have symptoms at 800–1,000 ppm, and 100% will show symptoms at ≥1,500 ppm. Further, this report claims that humans will experience an increase in breathing rate from just a slight change in CO_2 level above the normal ambient CO_2 level of 300–400 ppm. Based on these findings, this report recommends that CO_2 concentrations not exceed 600 ppm, a level that can be achieved with a minimum of 40 cfm (cubic feet per minute) per person. If the CO_2 concentration exceeds 600 ppm, complaints of drowsiness, fatigue, difficulty in concentrating, and difficulty in breathing can be expected. More recently, Allen et al.[42] evaluated the effects of levels of CO_2, VOCs, and ventilation on office workers' cognitive function scores and found that: "On average, a 400-ppm increase in CO_2 was associated with a 21% decrease in a typical participant's cognitive scores." Further, they found that: "a 20-cfm increase in outdoor air per person was associated with an 18% increase in these scores and a 500-µg/m³ increase in TVOCs [Total Volatile Organic Compounds] was associated with a 13% decrease in these scores."

Finally as detailed in papers by Burton[43] and others,[39,44] carbon dioxide measurements can be used as screening tools for preliminary IEQ investigations (surrogate for buildup of other indoor contaminants).

11.4.4 INDOOR HUMIDITY

Indoor humidity levels can be reported under various units; the most common are RH, SH, and DP. Each of these measures of humidity is discussed in the following sections.

11.4.4.1 Relative Humidity (RH)

RH levels measure the amount of moisture in the air relative to a fixed temperature and can provide an initial indication whether or not the indoor environment has either too much or too little moisture. This author believes this term for humidity is overrated. Various values of RH can only be compared if they are taken at a fixed temperature; SH and DP values are more useful terms for comparing humidity values. Nevertheless, for indoor environments where temperatures remain relatively constant, RH values are reported and used in the literature.

Acceptable ranges of RH in forensic investigations are used for two primary purposes: (1) comfort conditions and (2) levels above which mold spores will amplify in indoor environments.

For comfort conditions, ASHRAE and others tend to recommend a range between 30% and 60% RH. Indoor values below ~30% RH tend to result in drying out of mucus membrane tissues, resulting in comfort issues. On the other hand, values above ~60% RH result in human discomfort due to a feeling of increased "wetness" of the skin. Others have looked in depth on the effects of low humidity on human comfort[45] and means to control humidity for indoor pools.[46] High levels of RH can result in higher levels of molds, mildew, and bacteria. Amplification of mold in indoor environments appears to be limited to when the indoor humidity levels are maintained below RH levels of 55% (ranges of 50%–60%).[47–54]

11.4.4.2   Specific Humidity (SH)

SH levels measure the absolute level of moisture in the air and provide an initial indication whether or not the indoor environment has either too much or too little moisture. Conventional practice is to report this moisture level in terms of grains per pound of air where 7,000 grains equal one pound. For typical indoor conditions (temperatures of 68°F–74°F) and levels of RH from 55% to 60% (i.e., levels above which mold spores tend to amplify in indoor environments), the SH should be below 56–76 grains per pound. The basis for the SH values is shown in Table 11.6.

TABLE 11.6
Basis for Recommended Limits on Indoor Air Specific Humidity (SH) for Amplification of Mold Growth

Temperature (°F)	Relative Humidity (%)	Specific Humidity (grains/#)
68	55	56
	60	61
74	55	69
	60	76

11.4.4.3   Dew Point (DP)

The DP is the temperature at which if the air were cooled, the water vapor in the air would condense out or change from a gas to a liquid. Typical examples of DPs are: (1) a glass of ice water or (2) a cold winter window surface. Under these scenarios, condensation of the moisture in indoor air can often be observed. The practical application of DP in forensic applications is that outdoor wall or ceiling surfaces, especially if uninsulated, can be at temperatures below the DP temperature of the indoor air, which allows moisture in the air to condense out onto these cold surfaces. Often, visible mold will be seen on north and east exterior indoor wall or ceiling surfaces where water from condensation is present. Under these scenarios, destructive testing (i.e., removal of the drywall) and/or moisture meter readings will indicate that the water source is not from the exterior environment, but from condensation of interior moisture.

11.4.4.4   Volatile Organic Hydrocarbons (VOCs)

While some instrumentation and sample results report total VOCs, it is important to evaluate both the individual VOCs and their totals against target action levels. Often an individual contaminant such as benzene may be at a small fraction of the total VOCs and yet exceeds recommended levels. Often these contaminants arise from historic contaminated soils and/or groundwater and flow into buildings; Medina[55] outlines methods to evaluate contaminants arising from such circumstances.

11.5   USE OF THE TERM "ADEQUATE VENTILATION"

In many labels and warnings of the hazards of products, the term "Use with Adequate Ventilation" is used. However, most users of such products do not know what this term means. Francis Offermann (Budd) and Mark Nicas addressed this issue in their May 2018 ASHRAE paper entitled "Use with Adequate Ventilation?"[56] The authors noted:

> ANSI C400/Z129.1 defines "adequate ventilation" as a condition falling within either or both of the following categories: 1) *Ventilation to reduce concentrations below that which may cause personal injury or illness*, and 2) Ventilation sufficient to prevent accumulation to a concentration in excess of 25% of the level set for the lower flammable limits.

The authors go on to say:

> In other words, having "adequate ventilation" is a good thing, the air is safe to breathe and explosions will be prevented.
>
> However, the product information offered by manufacturers to consumers never contains guidance regarding the volume flow rate of ventilation (e.g., cfm [m^3/h]) that constitutes "adequate ventilation." Further confusing consumers, is that the same recommendation, "use with adequate ventilation," is provided by manufacturers for both relatively non-toxic products and highly toxic products such as paint strippers.

Finally, the authors test for the ventilation adequate for a paint stripper MSDS stating it contains between 60% and 100% methylene chloride when applied to 4.7″×5.5″ painted piece of plywood. Based on the chemical emission rates of methylene chloride measured, they concluded that the "adequate ventilation" required would be:

380 cfm/square feet of product application (6,950 m^3/hour-m^2), or
203 cfm/ounce of product applied (12 m^3/hour mℓ).

Using a box fan rated at 1,000 cfm (1,700 m^3/h), the authors concluded that only a small work area of 2.63 square feet (0.24 m^2), which would use a small amount of the stripper (4.9 oz.), could be used before the concentration would exceed health criteria.

The authors concluded for this scenario that:

> Paint strippers containing methylene chloride require impractically large ventilation rates for paint stripping activities involving more than a very small area, and thus cannot be conducted indoors with "adequate ventilation.

Regarding the term "adequate ventilation," the authors concluded that:

> Chemical emission rate testing for wet-applied products under actual or simulated use conditions can be determined by relatively simple testing and should be used by manufacturers to provide specific guidance to consumers as to how much ventilation constitutes "adequate ventilation, and,
>
> Providing consumers with the required ventilation rates and product quantity limitations for indoor applications of paints, cleaning chemicals, and adhesives should significantly reduce adverse health impacts associated with the use of these products.

Thus, the term "adequate ventilation" is not actionable without the manufacturer specifying how much product can be used for a given ventilation rate. Moreover, it appears that for volatile products, very little product can be used indoors before exceeding health criteria.

11.6 AIR CLEANING TECHNOLOGIES

With increasing requirements for energy efficiency required by building energy codes [e.g., ASHRAE Standard 90.1-2019 (other than low-rise residential buildings); ASHRAE Standard 90.2-2018 (low-rise residential buildings); ASHRAE Standard 100-2018 (existing buildings) and the 2018 International Energy Conservation Code (IECC)], the ability to use dilution as a method for lowering indoor contaminant loads is becoming less of an option and air cleaning technologies for various contaminants are increasingly being explored. For example, Arzbaecher and Hurtado evaluated the following technologies for air purification[57]:

- Media filters – particulates.
- High-efficiency particulate arrestance filters (HEPA) – particulates.
- Antimicrobial filters – mold and bacteria.
- Electrofiltration (i.e., electrostatic filters) – particulate.
- UVGI and filter systems – particulates, mold, and bacteria.

- Gas sorption and filter systems – gases and particulate contaminants.
- Bipolar ionization and filter systems – gases, particulate, mold, and bacteria.
- Photocatalytic oxidation and filter systems – gases, particulate, mold, and bacteria.
- UV/ozone catalytic oxidation and filter systems – gases, particulate, mold, and bacteria.

Filtering media are limited to removing particulates to varying degrees. Ultraviolet (UV) systems address non-particulate contaminants but generate ozone and typically require annual replacement of the UV bulb. Sorption technologies use absorbents such as activated carbon, which must be replaced at times. Early bipolar ionization systems generated ozone and carbon dioxide; newer systems have greatly reduced ozone generation and have become increasingly popular.[58]

11.7 LESSONS FROM THE FIELD

IAQ measurements can be used to determine cause and origin of water intrusion or the reasons for the presence of probable visible mold in some forensic investigations. Examples of field investigation scenarios, IAQ measurements observed, and interpretation of such results to explain scenarios based on experiences in the field are summarized in Table 11.7.

TABLE 11.7

Lessons from the Field – Examples Using/Interpreting IAQ Measurements

Scenario	IAQ Parameters Measured	IAQ Measurement Results	Reason(s) for Measurements and Interpretation of Results
Mold on exterior walls of home – especially north and east elevations	Carbon dioxide, humidity, and dew point	$CO_2 > \sim 1,000\,ppm$ $RH > 50\%$ and/or $SH > 56$ grains DP at or approaching temperatures of wall surfaces	Condensation on exterior walls and trapped indoor humidity. Combination of elevated CO_2 and humidity suggests lack of adequate ventilation/fresh outdoor air. The cold uninsulated walls are at or below the dew point temperature, thus providing a condensing surface
Presence of vent-free combustion product – gas log or fireplace venting into space Reports of lethargy, headaches and high humidity/visible mold	Carbon dioxide, carbon monoxide and humidity	$CO > 20\,ppm$ $CO_2 > \sim 2,000\,ppm$ $RH > 50\%$	Verification of impact of vent-free product on IAQ. Elevated CO and CO_2 readings indicative of either a vent-free product or failed gas furnace heat exchanger – see below. Condensation on windows and mold on walls and window frames often the result of operating a vent-free product. A typical 30,000 BTU/hour. vent free product will emit ~4 gallons of water in a space if operated 12 hours during a day.
Reports of high humidity/visible mold	Humidity	$RH > 50\%$ $SH > 56$ grains	Interior moisture sources – APA41: Source: Amount shower 0.5 pints/5 minutes shower Clothes dryer 4.7–6.2 pints/load – vented indoors Cooking Dinner 1.2–1.6 pints per family for 4 Dishwashing 0.7 pints per family of 4 House plants 0.9 pints/6 plants

(Continued)

TABLE 11.7 (Continued)
Lessons from the Field – Examples Using/Interpreting IAQ Measurements

Scenario	IAQ Parameters Measured	IAQ Measurement Results	Reason(s) for Measurements and Interpretation of Results
Visible mold on attic surfaces	Humidity and dew point	RH > 50% SH > 56 grains DP at or approaching temperatures of attic surfaces. (SH a better measurement in this scenario)	Often indicative of inadequately ventilated attic space and/or incremental sources of moisture into the attic (e.g., ventilation of bathroom/other vents into the attic and/or roof leaks). The cold surfaces are typically at or below the dew point temperature, thus providing a condensing surface
Reports of headaches and lethargy	Carbon dioxide, carbon monoxide	CO > 20 ppm CO_2 > ~2,000 ppm Values greater in the ductwork than in the open spaces	With the furnace on, measure the CO and CO_2 levels in delivered air registers. If the levels are higher than indoors, then the furnace heat exchanger should be inspected by a reputable HVAC serviceman
Visible mold on cooler wall and ceiling surfaces elevated Moisture in the home – sub-slab furnace ductwork	Humidity and dew point	Values greater in the ductwork than in the open spaces. DP at or approaching temperatures of wall surfaces	Look for evidence of water and water deposits in the sub-slab ductwork. Elevated readings and reports of "gurgling" from the ductwork during periods of heavy rain and ice/snow melts suggest water in present ductwork from ground/surface water intrusion. Surface temperatures at or below the dew point temperature would provide a condensing surface

As illustrated in Table 11.7, IAQ instrument readings can be used to rule in or rule out interior condensation, vent-free products, and water in sub-slab ductwork as sources of moisture for water damage and probable visible mold on interior surfaces. They can also be used to determine potential causes of unsafe levels of carbon monoxide such as cracked gas furnace heat exchangers and poorly operating vent-free products.

IMPORTANT POINTS TO REMEMBER

- The differences between IEQ and IAQ. IAQ are a subset of IEQ with a focus on contaminants in the air.
- The most common IAQ parameters measured by forensic professionals are indoor temperature, carbon monoxide, carbon dioxide, humidity levels, and DP temperature. These are often used to rule in or rule out sources of moisture responsible for interior water damage and/or mold growth on surfaces.
- Low and high indoor humidity values affect human comfort and high indoor humidity levels can result in amplification of mold growth on interior surfaces.
- IEQ parameters such as air velocity and delivered air temperatures can affect human comfort and often manifest themselves in health complaints.
- The term "Adequate Ventilation" is not well understood by users of products and not actionable unless use and ventilation rates are defined for a given product.

REFERENCES

1. Centers for Disease Control and Prevention. "Indoor Environmental Quality." last modified May 17, 2013. Accessed May 13, 2020, https://www.cdc.gov/niosh/topics/indoorenv/default.html.
2. Neumann, Alan J. "Standards of care for indoor environmental investigations, indoor environment," *Connections*, pp. 16–18, 2011.
3. ANSI (American National Standards Institute)/ASHARE (American Society for Heating, Refrigeration and Air Conditioning Engineers). "Standard 55-2017: Thermal Environmental Conditions for Human Occupancy." Accessed May 13, 2020, https://ashrae.iwrapper.com/ViewOnline/Standard_55-2017.
4. ANSI (American National Standards Institute)/ ASHARE (American Society for Heating, Refrigeration and Air Conditioning Engineers). "Standard 55-2010: Thermal Environmental Conditions for Human Occupancy." 2010.
5. ANSI (American National Standards Institute)/ ASHARE (American Society for Heating, Refrigeration and Air Conditioning Engineers). "Standard 55-2004: Thermal Environmental Conditions for Human Occupancy." 2004.
6. ANSI (American National Standards Institute)/ ASHARE (American Society for Heating, Refrigeration and Air Conditioning Engineers). "Standard 55-1995: Thermal Environmental Conditions for Human Occupancy." 1995.
7. ANSI (American National Standards Institute)/ ASHARE (American Society for Heating, Refrigeration and Air Conditioning Engineers). "Standard 55-1992: Thermal Environmental Conditions for Human Occupancy." 1992.
8. ANSI/ASHRAE, ANSI (American National Standards Institute)/ ASHRAE (American Society for Heating, Refrigeration and Air Conditioning Engineers). "Standard 62.1-2019: Ventilation for Acceptable Indoor Air Quality." 2019. Accessed May 13, 2020, https://ashrae.iwrapper.com/ViewOnline/Standard_62.1-2019.
9. ANSI/ASHRAE, ANSI (American National Standards Institute)/ ASHRAE (American Society for Heating, Refrigeration and Air Conditioning Engineers). "Standard 62.2-2019: Ventilation and Acceptable Indoor Air Quality in Low-Rise Residential Buildings." 2019. Accessed May 13, 2020, https://ashrae.iwrapper.com/ViewOnline/ Standard_62.2-2019.
10. ANSI/ASHRAE, ANSI (American National Standards Institute)/ ASHRAE (American Society for Heating, Refrigeration and Air Conditioning Engineers). "Standard 62.1-2010: Ventilation for Acceptable Indoor Air Quality." 2010.
11. ANSI/ASHRAE, ANSI (American National Standards Institute)/ ASHRAE (American Society for Heating, Refrigeration and Air Conditioning Engineers). "Standard 62.2-2010: Ventilation and Acceptable Indoor Air Quality in Low-Rise Residential Buildings." 2010.
12. ANSI/ASHRAE, ANSI (American National Standards Institute)/ASHRAE (American Society for Heating, Refrigeration and Air Conditioning Engineers). "Standard 62.1-2007: Ventilation for Acceptable Indoor Air Quality." 2007.
13. ANSI/ASHRAE, ANSI (American National Standards Institute)/ ASHRAE (American Society for Heating, Refrigeration and Air Conditioning Engineers). "Standard 62.2-2007: Ventilation and Acceptable Indoor Air Quality in Low-Rise Residential Buildings." 2007.
14. ANSI/ASHRAE, ANSI (American National Standards Institute)/ASHRAE (American Society for Heating, Refrigeration and Air Conditioning Engineers). "Standard 62.1-2004: Ventilation for Acceptable Indoor Air Quality." 2004.
15. ANSI/ASHRAE, ANSI (American National Standards Institute)/ ASHRAE (American Society for Heating, Refrigeration and Air Conditioning Engineers). "Standard 62.2-2004: Ventilation and Acceptable Indoor Air Quality in Low-Rise Residential Buildings." 2004.
16. ANSI/ASHRAE, ANSI (American National Standards Institute)/ ASHRAE (American Society for Heating, Refrigeration and Air Conditioning Engineers). "Standard 62-2001: Ventilation for Acceptable Indoor Air Quality." 2001.
17. ASHRAE (American Society for Heating, Refrigeration and Air Conditioning Engineers). "Standard 62-1999: Ventilation for Acceptable Indoor Air Quality." 1999.
18. ASHRAE (American Society for Heating, Refrigeration and Air Conditioning Engineers). "Standard 62-1989: Ventilation for Acceptable Indoor Air Quality." 1989.
19. ASHRAE (American Society for Heating, Refrigeration and Air Conditioning Engineers). "Standard 62-1981: Ventilation for Acceptable Indoor Air Quality." 1981.
20. ASHRAE (American Society for Heating, Refrigeration and Air Conditioning Engineers). "Standard 62-1973: Standards for Natural and Mechanical Ventilation." 1973.

21. "ASHRAE 62 – Interpretation of CO2 – Carbon Dioxide". Accessed May 13, 2020, https://www.ashrae.org/File%20Library/Technical%20Resources/Standards%20and%20Guidelines/Standards%20Intepretations/IC-62-1999-03.pdf.

22. ASHRAE (American Society for Heating, Refrigeration and Air Conditioning Engineers). "Guideline 24-2015: Ventilation and Indoor Air Quality in Low-Rise Residential Buildings." 2015.

23. Charles, Kate, Robert J. Magee, Doyun Won, and Ewa Lusztyk. "Indoor Air Quality Guidelines and Standards, RR-204." National Research Council Canada, March 2005. Accessed May 13, 2020, https://nrc-publications.canada.ca/eng/view/fulltext/?id=c597c638-536c-4ed9-b99c-20eb102a3bc.0.

24. IDPH. "Illinois Department of Public Health Guidelines for Indoor Air Quality, Environmental Health Fact Sheet," updated May 2011, Accessed May 13, 2020, http://dph.illinois.gov/topics-services/environmental-health-protection/toxicology/indoor-air-quality-healthy-homes/idph-guidelines-indoor-air-quality.

25. Bayer, Chrales, W., Sydney A. Crow, and John Fischer. "Causes of Indoor Air Quality Problems in Schools – Summary of Scientific Research." Oak Ridge National Laboratory (ORNL), ORNL/M-6633, January 1999.

26. Mulhausen, John R. and Joseph Damiano. *A Strategy for Assessing and Managing Occupational Exposures*, 2nd Edition. Virginia: American Industrial Hygiene Association Press, 1998.

27. Ignacio, Joselito S. and William H. Bullock. *A Strategy for Assessing and Managing Occupational Exposures*, 3rd Edition. Virginia: American Industrial Hygiene Association Press, 2006.

28. Jahn, Steven D., William H. Bullock, and Joselito S. Ignacio. *A Strategy for Assessing and Managing Occupational Exposures*, 4th Edition. Virginia: American Industrial Hygiene Association Press, 2015.

29. Rafferty, Patrick J. ed. *The Industrial Hygienist's Guide to Indoor Air Quality Investigations*, The American Industrial Hygiene Association Technical Committee on Indoor. Environmental Quality. Virginia: American Industrial Hygiene Association Press, 1993.

30. SAIF, "Indoor Air Quality Investigations." Accessed May 14, 2020, https://www.google.com/url?sa=t&rct=j&q=&esrc=s&source=web&cd=2&ved=2ahUKEwid9fKA3bPpAhXVXc0KHU2jCc8QFjABe gQIBBAB&url=https%3A%2F%2Fwww.saif.com%2FDocuments%2FSafetyandHealth%2FIndoorAir Quality%2FSS436_indoor_air_quality_investigation.pdf&usg=AOvVaw1mzao9LIAvodb8rKkHEsCC, September 2018.

31. Meyer, Christian, Dr. "Overview of TVOC and Indoor Air Quality", March 2020. Accessed May 14, 2020, https://www.idt.com/us/en/document/whp/overview-tvoc-and-indoor-air-quality.

32. World Health Organization (WHO). "WHO Criteria for Indoor Air Quality – Selected Pollutants", 2010. http://www.euro.who.int/_data/. assets/pdf_file/0009/128169/e94535.pdf.

33. Wadden, Richard A. and Peter A. Scheff. *Indoor Air Pollution – Characterization, Prediction and Control.* New York: John Wiley & Sons, 1983.

34. Hays, Steve M., Ronald V. Gobbell, and Nicholas R. Ganick. *Indoor Air Quality – Solutions and Strategies*, New York: McGraw-Hill, Inc., 1995.

35. Occupational Safety and Health Administration, OSHA. "Indoor Air Quality in Commercial and Institutional Buildings." 2011. http://www.osha.gov/Publications/3430indoor-air-quality-sm.pdf.

36. Burton, D. Jeff "Using triggers to avoid IAQ problems," *Synergist*, 2017.

37. Offermann, F. "Formaldehyde emissions from laminate flooring," *ASHRAE Journal*, pp. 102–105, March 2017.

38. Bergmann, James "What does that analyzer do, anyway?" *Air Conditioning, Heating & Refrigeration News*, pp. 24–26, January 23, 2006. http://www.achrnews.com/articles/what-does-that-analyzer-do-anyway.

39. Kudlinski, David and Steven Rupkey. "Office IEQ – using carbon dioxide air concentrations as a screening tool during preliminary IEQ evaluations." *Synergist*, pp. 38–40, May, 2011.

40. Rafferty, Patrick *The Industrial Hygienist's Guide to Indoor Air Quality Investigation.* Virginia: American Industrial Hygiene Association Press, 1993.

41. Bright, P. Diane, Michael J. Mader, David R. Carpenter, and Ivette Z. Herman-Cruz. "Guide for Indoor Air Quality Surveys, Occupational and Environmental Health Directorate." Brooks Air Force Base, Armstrong Laboratory, May 1992. http://www.dtic.mil/cgi-bin/GetTRDoc?Location=U2&doc=GetTR Doc.pdf&AD=ADA251638.

42. Allen, Joeseph G., Piers MacNaughton, Usha Satish, Soresh Santanam, Jose Vallarino, and John D. Spengler. "Associations of cognitive function scores with carbon dioxide, ventilation, and volatile organic compound exposures in office workers: A controlled exposure study of green and conventional office environments," *Environmental Health Perspectives*, vol. 124, no. 6, pp. 805–812, 2016.

43. Burton, D. Jeff "Six ways to approximate airflow," *Synergist*, June/July, 2018.

44. Persily, Andrew K. and Lillian De Jonge. "Carbon dioxide generation and building occupants," *ASHRAE Journal*, pp. 64–66, July 2017.

45. Derby, Melanie M. and Roger M. Pasch. "Effects of low humidity on health, comfort & IEQ," *ASHRAE Journal*, pp. 44–51, September 2017.

46. Lochner, Gary and Lynne Wasner. "Ventilation requirements for indoor pools," *ASHRAE Journal*, pp. 16–24, July 2017.

47. U.S. Environmental Protection Agency. "A Brief Guide to Mold, Moisture, and Your Building." Washington, DC, September 2012. www.epa.gov/iaq.

48. USEPA. "Mold Course Chapter 2, Lesson 3 Humidity." Accessed June 10, 2020, https://www.epa.gov/mold/mold-course-chapter-2.

49. Harriman, Lou "Damp Buildings, Human Health, and HVAC Design" Report of the ASHRAE Multidisciplinary Task Group, Damp Buildings, 2020.

50. American Lung Association, American Medical Association, U.S. Consumer Product Safety Commission, and U.S. Environmental Protection Agency. "Indoor Air Pollution: An Introduction for Health Professionals," 2014.

51. U.S. Environmental Protection Agency. "Indoor Air Quality." April 1995. Accessed June 9, 2020, https://www.epa.gov/indoor-air-quality-iaq.

52. Burton, D .Jeff *IAQ and HVAC Workbook*, 4th Edition. Virginia: American Industrial Hygiene Association Press, 2011.

53. AIHA. *Recommendations for the Management, Operation, Testing, and Maintenance of HVAC Systems: For Maintaining Acceptable Indoor Air Quality in Non-Industrial Employee Occupancies through Dilution Ventilation*. Virginia: American Industrial Hygiene Association Press, Table A7.1, January 2004.

54. APA. "Mold and Mildew - Controlling Mold and Mildew." APA (American Plywood Association) Publication A525, The Engineered Wood Association Headquarters. http://www.apawood.org/level_b.cfm?content=pub_main.

55. Medina, Enrique "Vapor intrusion: Environmental and IAQ challenge," *The Synergist*, pp. 43–46, February 2007.

56. Offerman, Francis (Budd) J. and Mark Nicas, "Use with adequate ventilation?" *ASHRAE Journal*, pp. 70–76, May 2018.

57. Arzbaecher, Cecilia, Patricia Hurtado, and Ammi Amarnath. "Indoor Air Purification Technologies that Allow Reduced Outdoor Air Intake Rates While Maintaining Acceptable Levels of Indoor Air Quality, 2008 ACEEE Summer Study on Energy Efficiency in Buildings." Accessed June 8, 2020, https://www.aceee.org/files/proceedings/2008/data/papers/3_281.pdf.

58. Agopian, Nick "Five concerns over air ionization for enhancing IAQ – especially in schools," *RenewAire*, March 9, 2017. Accessed June 8, 2020, https://www.renewaire.com/wp-content/uploads/2017/09/MAR_LIT_104_RGB_Concerns_with_Air_Ionization.pdf.

12 Attic and Crawlspace Ventilation

Stephen E. Petty
EES Group, Inc.

CONTENTS

PURPOSE/OBJECTIVES

The purpose of this chapter is to:

- Demonstrate how to determine what areas of ventilation are of concern in residential and commercial structures.
- Discuss why ventilation is important and what the consequences are of poor ventilation.
- Demonstrate how to calculate proper/adequate attic ventilation.
- Demonstrate how to calculate proper/adequate crawlspace ventilation.
- Demonstrate how to use and interpret IAQ measurements.

Following the completion of this chapter, you should be able to:

- Understand why proper attic and crawlspace ventilation is important.
- Understand the basis for determining whether or not attic and crawlspace ventilation is adequate for a given structure.

- Be able to obtain information from the field needed to calculate probable attic and/ or crawlspace ventilation.
- Be able to calculate estimated attic and crawlspace ventilation levels and compare with recommended ventilation requirements.
- Be able to use and interpret IAQ parameters relevant to ventilation.

12.1 INTRODUCTION

Ventilate comes from the Latin word *ventilo*, meaning "to fan." Proper ventilation of an attic and crawlspace is important because, if inadequate, it can lead to premature failure or reduction of the life of a roofing system, amplification of mold growth on attic and crawlspace surfaces, and degradation of structural members in attics and crawlspaces, which may impact roof and interior structural systems.[1–3] For example, when attic ventilation is inadequate, it is common to observe thermal degradation (i.e., shrinkage, cracking, and blistering) of asphalt roof shingles on the south and west-facing elevations (Figure 12.1), resulting in reduced life of the shingles.

FIGURE 12.1   Illustration of thermally degraded asphalt shingles caused by inadequate attic ventilation.

With outdoor conditions at 90°F, unvented or poorly vented attic spaces can reach 140°F with roof surfaces reaching up to 170°F; whereas, under these same outdoor conditions, well-vented attic spaces will typically be held to 115°F during warm-weather months.[1] Similarly, Stewart[2] noted that when the air temperature outdoors ranges from 95°F to 97°F, proper ventilation [air exchanged rates of 30–60 air changes per hour (ACH)] can reduce attic temperatures from 160°F to 155°F and from 106°F to 101°F. Lstiburek[3] suggests that poorly ventilated attic spaces can reduce the useful service life of asphalt shingles by 10%. Conversely, attic temperature can approach outdoor temperatures during the colder months, which helps to prevent ice dams. Additionally, it is very common to observe probable visible mold and water damage staining on attic roof wood decking and rafter members (Figure 12.2) when attic ventilation is inadequate.

FIGURE 12.2   Illustration of visible mold on attic roof decking caused by inadequate attic ventilation.

It is also common to observe condensation droplet splatter marks on attic floor surfaces (Figure 12.3) when attic ventilation is inadequate.

FIGURE 12.3   Illustration of water condensation drip staining on attic floor boards caused by inadequate attic ventilation.

Condensation conditions typically occur when the moisture in warmer humid attic air will condense on colder roof decking surfaces during the nighttime hours (spring and fall) or during the colder winter months. Inadequately vented crawlspace ventilation can also result in water damage staining, mold growth, and structural damage/failure of crawlspace wood joists and subfloor decking (Figure 12.4).

FIGURE 12.4 Illustration of visible mold and wood joist and wood subfloor degradation caused by inadequate crawlspace ventilation.

Such conditions encourage the presence of wood-destroying insects such as ants and termites that accelerate the destruction of structural support members in homes and commercial buildings.

12.2 ATTIC VENTILATION

In an article entitled "Principles of Attic Ventilation: A comprehensive guide to planning The Balanced System™ for Attic Ventilation," Air Vent, Inc. states, "During warmer months, ventilation helps keep attics cool. During colder months, ventilation reduces moisture to help keep attics dry. It also helps prevent ice dams."[1] The article then states, "Several purposes of an attic ventilation system are to provide added comfort, to help protect against damage to materials and structure, and to help reduce energy consumption–during all four seasons of the year."

Ventilation, simply defined, is the movement of air to control moisture and heat buildup in attic spaces. The two methods of ventilation are "passive ventilation" and "mechanical ventilation." Passive ventilation implies that energy-consuming mechanical components, such as pumps and fans, are not used and that the air movement is caused by natural convection from differences in air density caused by differences in air temperature and wind blowing around a building.[2,3] Mechanical ventilation includes electric or wind-driven fans to force air movement in the direction(s) desired using applied energy.[2,3] This chapter will focus on passive attic and crawlspace ventilation since most buildings use this type of ventilation.

Attics are typically passively vented using a variety of vent types/openings. These are illustrated in Figure 12.5.

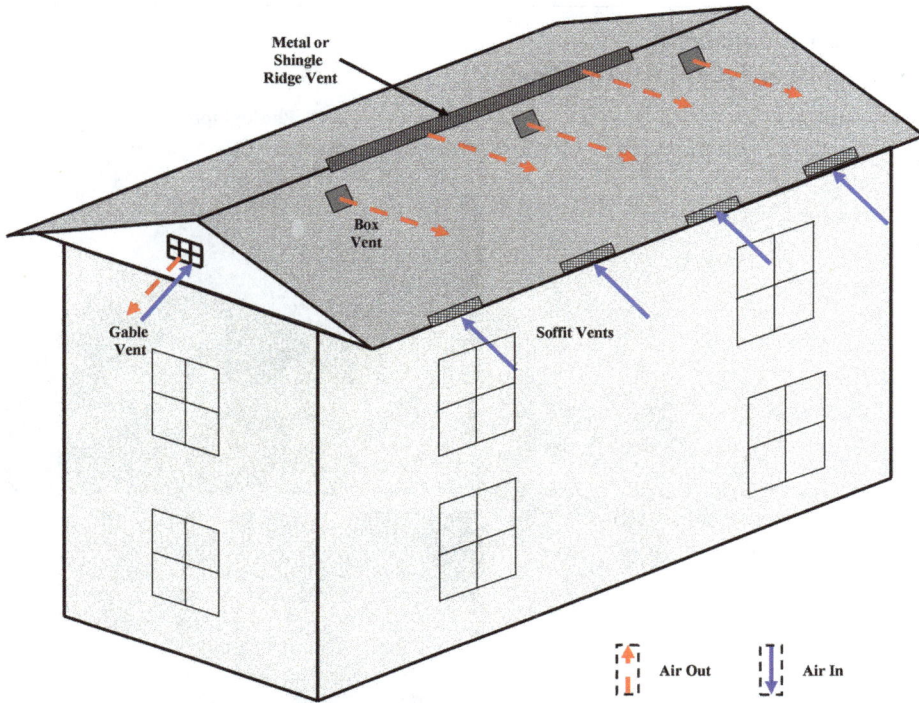

FIGURE 12.5   Illustration of passive ventilation openings for a typical attic.

The concept of passive ventilation is to introduce cooler outside air at intake positions lower in the attic and discharge warmer air near the peak of the roof to take advantage of the chimney effect associated with heated air. Typically, the intake fresh (cooler) air is introduced at the soffit (i.e., soffit vents) of a building or sometimes through gable vents. Conversely, exhaust (warm) air is typically discharged from an attic through box vents, ridge vents (metal or shingle-covered), or gable vents. Mechanical exhaust ventilation is sometimes observed; typical vents of this type include power vents and turbine vents. Examples of each of these vent types are illustrated in Tables 12.1–12.3.[1–9]

TABLE 12.1
Photographs of Typical Intake Vents

Intake Vent Types	Photograph
Lanced soffit vents – continuous panels	

(Continued)

TABLE 12.1 (*Continued*)
Photographs of Typical Intake Vents

Intake Vent Types	Photograph
Lance soffit vents – ~4″ × ~16″ panels	
Strip soffit vents (note these have been partially painted over)	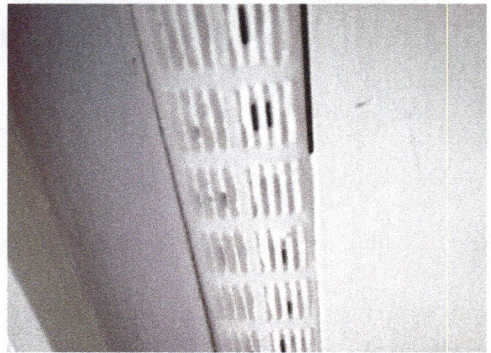
Round soffit vent plugs	
Gable vent (can be either an intake or an exhaust vent)	

TABLE 12.2
Photographs of Typical Exhaust Vents

Exhaust Vent Types	Photograph
Metal ridge vent	
Shingled ridge vent	
Box vent – square	
Box vent – slant backed	

(*Continued*)

TABLE 12.2 (*Continued*)
Photographs of Typical Exhaust Vents

Exhaust Vent Types	Photograph
Gable vent (can be either an intake or an exhaust vent)	

TABLE 12.3
Photographs of Typical Mechanical Exhaust Devices

Mechanical Exhaust Vent Types	Photograph
Power vent	
Turbine vent	

Aside from lack of adequate ventilation, the most common issues observed regarding venting are as follows:

- Clogging of the soffit vents (either by insulation from the attic side or by painting over the vents from the outdoor side).
- Adding box vents on a roof to vent an attic previously ventilated using soffit and gable vents.

This last situation results in the gable vent becoming an intake vent and effectively shutting down the soffit intake vents.[10] The net result is that the cooling air bypasses much of the attic, effectively cooling only the upper portions of the roof system and attic. In this situation, moisture condensation and mold growth can occur near the eaves or in the area that receives less air flow.

12.2.1 Attic Ventilation Requirements

Information regarding attic ventilation requirements can be found in most modern residential and commercial building codes and in best practices documents. The basis for these requirements are historical experience and simple psychometrics (moisture in air),[3] but, nevertheless, somewhat arbitrary.

Typical code language for ventilation, like that from the 2018 International Residential Code (IRC) – Chapter 8,[11] Section R806; the 2018 International Building Code (IBC) – Chapter 12 (Interior Environment),[12] Section 1203; and Manufactured Housing Ventilation Requirements – 24 CFR Subpart F – Thermal Protection Subsection 3280.504 – Condensation Control and Installation of Vapor Barriers[13] codes are similar, which are reproduced below.

The 2018 IRC[11] ventilation code states:

R806.1 Ventilation required: Enclosed attics and enclosed rafter spaces formed where ceilings are applied to the underside of roof rafters shall have cross-ventilation for each separate space by ventilating openings protected against the entrance of rain and snow. Ventilation openings shall have a least dimension of 1/16 inches (1.6 mm) minimum and 1/4 inches (6.4 mm) maximum. Ventilation openings having a least dimension larger than 1/4 inches (6.4 mm) shall be provided with corrosion-resistance wire cloth screening, hardware cloth, perforated vinyl, or similar material with openings having a least dimension of 1/16 inches (1.6 mm) minimum and 1/4 inches (6.4 mm) maximum. Openings in roof framing members shall conform to the requirements of Section R802.7. Required ventilation openings shall open directly to the outside air and shall be protected to prevent the entry of birds, rodents, snakes, and other similar creatures.

R806.2 Minimum vent area: The minimum net free ventilation area shall be 1/150 of the area of the vented space.

Exception: The minimum net free ventilation area shall be 1/300 of the vented space provided one or more of the following conditions are met:

1. In Climate Zones 6, 7, and 8, a Class I or II vapor retarder is installed on the warm-in-winter side of the ceiling.
2. Not <40% and not more than 50% of the required ventilating area is provided by ventilators located in the upper portion of the attic or rafter space. Upper ventilators shall be located not more than 3 feet (914 mm) below the ridge or highest point of space, measured vertically. The balance of the required ventilation provided shall be located in the bottom one-third of the attic space. Where the location of wall or roof framing members conflicts with the installation of upper ventilators, installation more than 3 feet (914 mm) below the ridge or highest point of the space shall be permitted.

R806.3 Vent and insulation clearance: Where eave or cornice vents are installed, blocking, bridging, and insulation shall not block the free flow of air. Not <1-inch (25 mm) space shall be provided between the insulation and the roof sheathing and at the location of the vent.

R806.4 Installation and weather protection: Ventilators shall be installed in accordance with manufacturer's installation instructions. Installation of ventilators in roof systems shall be in accordance with the requirements of Section R903. Installation of ventilators in wall systems shall be in accordance with the requirements of Section R703.1.

Under Section R806.5 "Unvented attic and unvented enclosed rafter assemblies," detailed requirements are defined to allow for these conditions.

The 2018 IBC[12] ventilation code states:

1202.2 Roof ventilation: Roof assemblies shall be ventilated in accordance with this section or shall comply with Section 1202.3.

1202.2.1 Ventilated attics and rafter spaces: Enclosed attics and enclosed rafter spaces formed where ceilings are applied directly to the underside of the roof framing members shall have cross-ventilation for each separate space by ventilating openings protected against the entrance of rain and snow. Blocking and bridging shall be arranged so as to not interfere with the movement of air. A minimum of 1 inch (25 mm) of airspace shall be provided between the insulation and the roof sheathing. *The net free ventilating area shall not be <1/150th of the area of the space ventilated*. Ventilators shall be installed in accordance with manufacturer's installation instructions.

Exceptions: The net free cross-ventilation area shall be permitted to be reduced to 1/300 provided both of the following conditions are met:
1. In Climate Zones 6, 7, and 8, a Class I or II vapor retarder is installed on the warm-in-winter side of the ceiling.
2. At least 40% and not more than 50% of the required venting area is provided by ventilators located in the upper portion of the attic or rafter space. Upper ventilators shall be located not more than 43 feet (914 mm) below the ridge or highest point of the space, measured vertically, with the balance of the *ventilation* provided by eave or cornice vents. Where the location of wall or roof framing members conflicts with the installation of upper ventilators, installation of more than 3 feet (914 mm) below the ridge or highest point of the space shall be permitted.

1202.2.2. Openings into attic: Exterior openings into the attic space of any building intended for human occupancy shall be protected to prevent the entry of birds, rodents, snakes, and other similar creatures. Openings for ventilation having a least dimension of not <1/16 inches (1.6 mm) and more than 1/4 inches (6.4 mm) shall be permitted. Openings for ventilation having a least dimension larger than 1/4 inches (6.4 mm) shall be provided with corrosion-resistance wire cloth screening, hardware cloth, perforated vinyl, or similar material with openings having a least dimension of 1/16 inches (1.6 mm) minimum and 1/4 inches (6.4 mm) maximum.

Under Section 1202.3 "Unvented attic and unvented enclosed rafter assemblies," detailed requirements are defined to allow for these conditions.

The MHRA[13] Section 3280.504(c) of the HUD code ventilation language for manufactured homes requires:

(c) Attic or roof ventilation
1. Attic and roof cavities shall be vented in accordance with one of the following:
   i. A minimum free ventilation area of not <1/300 of the attic or roof cavity floor area. At least 50% of the required free ventilation area shall be provided by ventilators located in the upper portion of the space to be ventilated. At least 40% shall be

provided by eave, soffit, or low gable vents. The location and spacing of the vent openings and ventilators shall provide cross-ventilation to the entire attic or roof cavity space. A clear air passage space having a minimum height of 1 inch shall be provided between the top of the insulation and the roof sheathing or roof covering. Baffles or other means shall be provided where needed to ensure the 1 inch height of the clear air passage space is maintained.

ii. A mechanical attic or roof ventilation system may be installed instead of providing the free ventilation area when the mechanical system provides a minimum air change rate of 0.02 cubic feet per minute (cfm) per square ft. of attic floor area. Intake and exhaust vents shall be located so as to provide air movement throughout space.

iii. Single-section manufactured homes constructed with metal roofs and having no sheathing or underlayment installed are not required to be provided with attic or roof cavity ventilation provided that the air leakage paths from the living space to the roof cavity created by electrical outlets, electrical junctions, electrical cable penetrations, plumbing penetrations, flue pipe penetrations, and exhaust vent penetrations are sealed.

2. Parallel membrane roof sections of a closed cell-type construction are not required to be ventilated.

3. The vents provided for ventilating attics and roof cavities shall be designed to resist entry of rain and insects.

Codes basically require 1 square foot of net free ventilation area for each 150 square feet of space (plan view) to be ventilated. The ratio can be reduced to 1 in 300 if the ventilation is balanced (50% of area intake and exhaust) and moisture entry is limited by using a rated vapor retarder. However, in most real-world construction scenarios, coupled with today's tighter homes, conditions that would allow for this lower ratio are not present, requiring the use of the 1:150 ratios. In addition, one must remember that codes are typically minimum requirements, so as a practical matter, the use of the 1:300 ratios should be avoided for determining whether or not the attic ventilation is adequate.

Humbarger[14] commented on the history of these ventilation ratios noting, "The 1/300 net-free ventilating area requirement was first promulgated in 1942 by the Federal Housing Administration (FHA) with very little research to back it up."

Interestingly, Rose[15–17] in a detailed review of the history of these ventilation factors of 1:150 and 1:300, concluded that:

The attic ventilation ratio "1/300" is an arbitrary number selected by the writers of FHA (1942) with no citations or references. One might speculate that it is based on Rowley's 1939 research, which showed a slight performance difference between openings with vent ratios of 1/288 and 1/576. However, other evidence indicates it was not based on Rowley.

The asphalt shingle industry began to link installation practices to recommended and code required venting practices in the mid-1980s.

Professionals in the building industry—design, codes and construction—may view the support for the current regulations, described in this paper, as being strong or weak. In the opinion of the author, the support is weak, and a strict interpretation of 1/300 compliance is not appropriate. Indeed, the building industry may wish to question whether ensuring moisture control is an appropriate duty and responsibility of the building codes, and, if it is, whether prescriptive venting regulations are the best way to provide it. Perhaps the building codes may wish to study removing some of the overly-exact provisions of the code, and instead rely on the industry to provide vapor control, as it relies on the roofing industry to provide weather protection against rain.

By the 1960s, the 1/300 and later the 1/150 requirement had been adopted into all of the building codes, but the numbers were still arbitrary." Nevertheless, as illustrated in Table 12.4, industry best practice documents such as the APA,[18–20] ARMA,[21,22] and APA[23] parallel code (i.e., ICC International Residential Code and International Building Code) requirements for ventilation.

TABLE 12.4
APA Ventilation Recommendations

Location	Construction	Natural Ventilation[a] – Net Free Area Opening as a Proportion of Attic or Floor Area
Attic and structural spaces[b]	Class I or Class II vapor retarder in ceiling	1/300
	At least 40% and not more than 50% of required vent area in upper portion of space to be ventilated at least 3 feet above eave or cornice vents.[c]	1/300
	Other ventilated attic configurations	1/150
Crawl spaces[d]	No vapor retarder and one vent opening within 3 feet of each corner of the building	1/150
	Class 1 vapor-retarder ground cover and one vent opening within 3 feet of each corner of the building	1/1500

[a] Note that where power attic vents are used, they should provide at least 0.7 cfm per square foot of attic area (15% more for dark roofs), and air intake of 1 square foot of free opening should be provided for each 300 cfm of fan capacity. Although intended to exhaust warm summer air, power vents should also operate during cold months to help prevent condensation.

[b] There are some instances where unvented conditioned attic assemblies are permitted by the code. The requirements for such an assembly are highly dependent upon the regional climatic zone. See the International Residential Code (IRC) for additional information.

[c] Certainly the code provision should not be interpreted to violate a reasonable balance between low and high vents. For natural ventilation systems, some experts recommend that 60% of net free area should be provided at eaves and 40% at the ridge or high gables. To meet the code provision for minimum 50% high vents, this would require that the free opening of high vents total 1/600 and low vents total 1/400 of attic area, for an overall ratio of 1/240.)

[d] Ventilation openings are not required for crawlspaces when a Class I vapor retarder is used in conjunction with insulated perimeter walls. In addition, one of the following shall be provided in the crawlspace:

- A continuously operated mechanical exhaust ventilation system at a rate equal to 1 cfm for each 50 square feet of crawlspace floor area including an air pathway to the common area.
- A conditioned air supply sized to deliver at a rate equal to 1 cfm for each 50 square feet of crawlspace floor area including an air pathway to the common area
- Plenum complying with the appropriate requirements for the code if the underfloor space is used as a plenum.

ARMA[21] warns that attics not meeting this "minimum" attic ventilation ratio of 1:150 may result in the following thermal and moisture-related problems:

- Premature failure of the roofing, including blistering.
- Buckling of roofing shingles due to deck movement.
- Rotting of wood members.
- Moisture accumulation in the deck and/or building insulation.
- Ice dam formation in cold weather.

Other best practices support a ratio of 1:150 or greater for ventilation area.[2,14,10]

Note that Lstiburek[3] recommends a supplemental ventilation requirement for vented cathedral ceiling assemblies have a minimum 2-inch air between the roof decking and the top of the insulation to ensure adequate ventilation above cathedral ceilings.

12.2.2 Concept of Net Free Area (NFA) or Net Free Vent Area (NFVA)

The overall effectiveness of a vent opening is accounted for in a term known as Net Free Area (NFA) or Net Free Vent Area (NFVA). Typically, the NFVA accounts for the portion of the gross vent opening that restricts passive air flow through an otherwise unrestricted opening. If an opening was completely wide open, its NFVA would be 100% of the gross vent opening space. On the other hand, if the opening were completely closed (e.g., painted over lanced soffit vents), the NFVA would be 0% of the gross vent opening. Since most vents are screened over to prevent entry of debris, insects, birds, and other animals, the NFVA of most vents is <100%. Manufacturers of vents[1] and others like the APA[18–20] provide default values for the NFVA associated with certain vent types; these are illustrated in Table 12.5.

TABLE 12.5
NFVA Guidelines for Vents and Screens

Ventilator Type	Gross Area (square inch)	Net Free Vent Area (square inch)
Ridge roof vent	N/A	$18 \times$ linear feet
Box vents		
Square metal and slant backed	Height $\times$ width	Area $\times$ 0.3470
Square plastic	Height $\times$ width	Area $\times$ 0.4236
Roof screen button cap of jacks	Vent pipe area ($\pi d^2/4$)	Area $\times$ 0.6
Gable or foundation (louvered and screened)		
Rectangular	Height $\times$ width	Area $\times$ 0.44
Triangular	$\frac{1}{2} \times$ Height $\times$ width	Area $\times$ 0.44
Soffit vents		
General	# Sections $\times$ area/section	Area $\times$ 0.3000
$16'' \times 8''$ under eave	# Sections $\times$ area/section	Area $\times$ 0.4375
$16'' \times 6''$ under eave	# Sections $\times$ area/section	Area $\times$ 0.4375
$16'' \times 4''$ under eave	# Sections $\times$ area/section	Area $\times$ 0.4375
Continuous soffit vent – 1' length	# Sections $\times$ area/section	Area $\times$ 0.0625
Vented drip edge – 1' length	# Sections $\times$ area/section	Area $\times$ 0.0625
Perforated aluminum – 1 square feet	# Sections $\times$ area/section	Area $\times$ 0.0972
Lance aluminum – 1 square feet		
Open	#Sections $\times$ area/section	Area $\times$ 0.0486
Average	# Sections $\times$ area/section	Area $\times$ 0.0382
Clogged	# Sections $\times$ area/section	Area $\times$ 0.0278
Open unscreened opening	Height $\times$ width	Area $\times$ 1.0
Screens – codes require corrosion resist. Steel		
1/16 mesh	Height $\times$ width	Area $\times$ 0.5
1/8 mesh	Height $\times$ width	Area $\times$ 0.8
1/16 mesh and louvers	Height $\times$ width	Area $\times$ 0.33
1/8 mesh and louvers	Height $\times$ width	Area $\times$ 0.44

Care must be taken to ensure that proper units are used in NFVA calculations since the calculations are often made in terms of square inches and must be converted to square feet by dividing by a conversion factor of 144 ($12/1 \times 12/1$).

12.2.3 EXAMPLE ATTIC VENTILATION NFVA CALCULATION

In order to determine whether or not an attic is properly ventilated (i.e., ventilation area ratio 1:150 NFVA met), the following information must be collected:

- Plan area of attic in square feet (basis for NFVA required).
- Intake vent types and areas (or # and area/vent).
- Exhaust vent types and areas.

An example of an NFVA calculation follows in the next section.

12.2.3.1 Example Attic Ventilation (NFVA) Calculation – Attic Area and Required NFVA

The plan (i.e., floor) dimensions of an attic space must be measured or obtained from drawings to determine the proper NFVA. This area calculation is the basis for intake and exhaust ventilation considerations. An example attic for use in this example is shown in Figure 12.6.

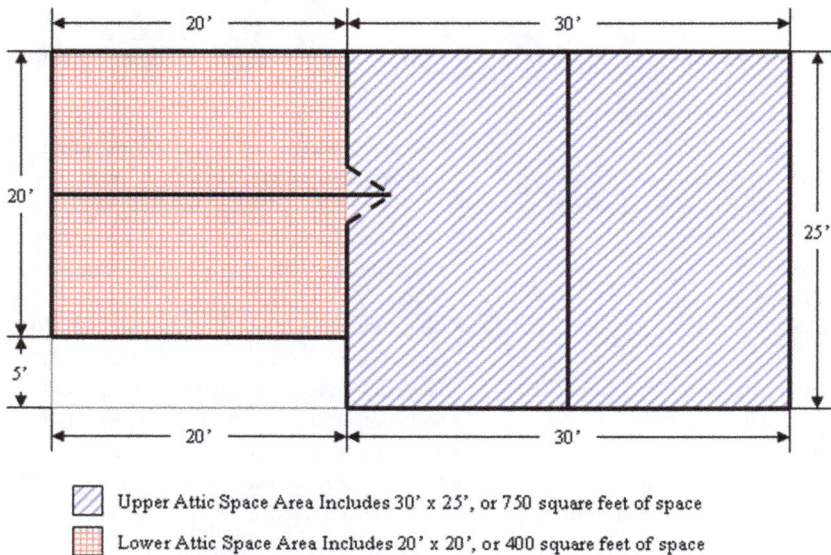

Upper Attic Space Area Includes 30' x 25', or 750 square feet of space

Lower Attic Space Area Includes 20' x 20', or 400 square feet of space

FIGURE 12.6 Example calculation attic area.

Assuming that the attic spaces are continuous and a vapor barrier is not present, the total plan area of the attic from Figure 12.6 is 1,150 square feet (i.e., 750+400). Using a ratio of 1:150 for NFVA, the total NFVA for the attic space would be ~7.67 square feet (1,150/150). If split equally between intake and exhaust areas, the intake and exhaust NFVA needed would be ~3.83 square feet (7.67/2) each.

12.2.3.2 Example Attic Ventilation (NFVA) Calculation – Actual Intake Ventilation vs Required Intake Ventilation

During the inspection, intake ventilation is observed to consist of 16 soffit vents, measuring 16″ × 4″ each, spaced uniformly around the home. The total area of these soffit vents is calculated to be 1,024 [16 × (16 × 4)] square inch or 7.11 square feet. Using an effectiveness factor of 0.4375 from From Table 12.5, the total intake NFVA is calculated to be 3.11 (0.4375 × 7.11) square ft. Note that this is less than the ~3.83 square feet needed, suggesting that the intake ventilation for this attic is inadequate. To meet the intake NFVA needed (~3.83 square feet), the areas/numbers of the three most common intake vents encountered in the field were computed below:

- Perforated Soffit Vents (effectiveness rating of 0.0972): ~40 total square feet.
- Lanced Soffit Vents (effectiveness rating of 0.0486): ~79 total square feet.
- Under Eave Vents (i.e., $16'' \times 8''$, $16'' \times 6''$, and $16'' \times 4''$) (effectiveness rating of 0.4375): would require the following:
  - $16'' \times 8''$: ~10 total vents evenly spaced around the home.
  - $16'' \times 6''$: ~13 total vents evenly spaced around the home.
  - $16'' \times 4''$: ~20 total vents evenly spaced around the home.

Intake vent installations can use multiple vent types. However, recall that when intake vents are placed higher on the roof than those placed under the soffit/eave, they have the potential to short-circuit the attic ventilation, causing the upper attic spaces to be ventilated while effectively diminishing or eliminating the lower ventilation.

12.2.3.3   Example Attic Ventilation (NFVA) Calculation – Actual Exhaust Ventilation vs Required Exhaust Ventilation

During the inspection, exhaust ventilation is observed to consist of eight metal box vents, measuring $12'' \times 12''$ each. The total area of these metal box vents is calculated to be 1,728 [12 × (12 × 12)] square inch or 12.00 square feet. Using an effectiveness factor of 0.3470 from Table 12.5, the total exhaust NFVA is calculated to be 4.16 (0.3470 × 12.00) square feet. Note that this (i.e., 4.16 square feet) is greater than the ~3.83 square feet needed, suggesting that the exhaust ventilation for this attic is adequate. To meet the exhaust NFVA needed (~3.83 square feet), the areas/numbers of the three most common exhaust vents encountered in the field were computed below:

- Metal Box Vents ($12'' \times 12''$ and an effectiveness rating of 0.3470): ~11 total box vents.
- Ridge Vents (Area = 18 in square inch × lineal feet): ~31 [3.83 × 144/18] total lineal feet of ridge venting.
- Gable Vents/Louvers (i.e. $12'' \times 18''$ or $24'' \times 30''$) would require the following:
  - $12'' \times 18''$ (0.380 effectiveness rating): ~7 total vents.
  - $24'' \times 30''$ (0.450 effectiveness rating): ~2 total vents.

12.2.3.4   Example Attic Ventilation (NFVA) Calculation – Net Results

Net results from the example calculations were as follows:

- Total NFVA required, based on a 1:150 ratio, was 7.67 square feet for the example attic; actual NFVA was 7.27 (3.11 + 4.16) square feet. Thus, the total actual NFVA was less than the desired NFVA.
- Intake NFVA required, based on a 1:150 ratio, was 3.83 square feet for the example attic; actual NFVA was 3.11 square feet. Thus, the actual intake NFVA was less than the desired NFVA.
- Exhaust NFVA required, based on a 1:150 ratio, was 3.83 square feet for the example attic; actual NFVA was 4.16 square feet. Thus, the actual exhaust NFVA exceeded the desired NFVA.

It is important to note that if the intake ventilation is inadequate and the total NFVA is adequate, the attic may not ventilate as desired since air is restricted from getting into the attic in the first place. If air cannot get in, it will not be available to be exhausted.

As noted earlier, improper ventilation of attic spaces can expedite the aging of construction components, including roof finish materials such as asphalt shingles, effectively reducing the service life.

12.2.4 EXAMPLES OF ATTIC VENTILATION ISSUES OBSERVED IN THE FIELD

Examples of issues found in the field resulting in lower than desired NFVA or mold formation and water damage in attics include the following:

- Intake Ventilation Issues:
  - Clogging of intake vents.
  - Inadequate under-eave or soffit ventilation.
  - Short-circuiting of the ventilation process.
  - False vents where soffit vents are present, but no openings are cut into the soffits.
- Exhaust Ventilation Issues:
  - Lack of exhaust ventilation.
  - Mixing exhaust vent types (short-circuiting).
  - False vents where no opening had been cut into the sheathing or the opening is covered over with roof materials such as underlayment and shingles.
- Other Attic Ventilation Issues:
  - Bathroom and/or clothes dryer vents exhausting into the attic.
  - Flue vents from furnace or water heater exhausting into the attic.

Attic insulation that is blown in, or rolled into the eaves (Figure 12.7), will strongly reduce intake ventilation flow into an attic. When adding blown-in insulation, baffles should be installed to keep the insulation back from the eaves.

FIGURE 12.7   Attic eaves clogged with insulation.

Two other factors that are often found to reduce attic intake ventilation is when: (1) soffit vents are painted or partially painted closed during maintenance activities (Figure 12.8) and (2) soffit vents are eliminated by additions, such as a closed-in porch or an added garage.

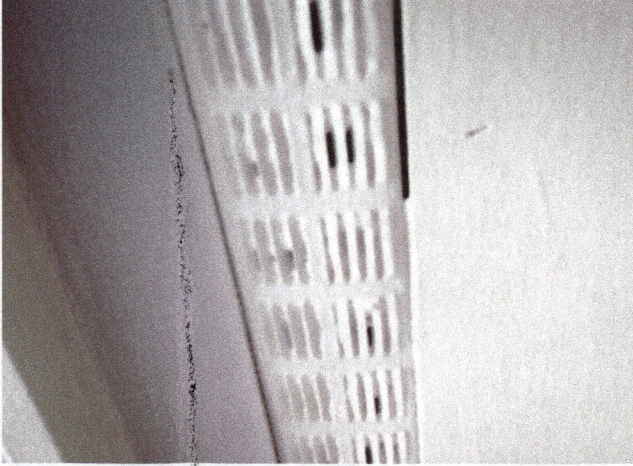

FIGURE 12.8   Soffit intake vents partially painted closed.

Ventilating plumbing (soil stack pipes – Figure 12.9), bathroom vents (Figure 12.10), clothes dryer vents, and flue vents exhausting into an attic space generally result in moisture loads that cannot be handled by typical natural convection attic ventilation and often lead to mold formation on the roof decking. These situations should be noted during the inspection and recommendations made to extend these vents to the outdoors.

FIGURE 12.9   Soil stack vented to an attic space.

FIGURE 12.10 Bathroom vent vented to an attic space.

A final issue that has been surprisingly encountered during several cases has been the installation of false exhaust and intake vents. For example, one home had continuous ridge vents installed along every ridge on the exterior of the home. However, the newly installed roof sheathing developed extensive mold growth in a 3-month timeframe. When the attic space was observed, the ridge vent opening was found to be cut into the sheathing, but the shingles and underlayment were installed over the opening (Figure 12.11). In a similar case with attic mold growth, the eaves of a home contained continuous vents around the entire perimeter of the exterior of the home that appeared to provide adequate ventilation, yet no light or insulation blockage was observed in the attic. Some of the soffit vents were removed, and it was observed that no openings had been cut into the soffit area (Figure 12.12).

FIGURE 12.11   Ridge vent opening covered with shingles.

FIGURE 12.12   No opening cut into soffit for vent.

While the previous examples focused on vented attic spaces, codes and best practices have increasingly looked at non-vented attic designs, especially in certain very cold or very moist climates,[24] see also code requirements regarding such designs (i.e., 2018 IRC R806.5 and 2018 IBC 1202.3).

12.3   CRAWLSPACE VENTILATION

Crawlspace ventilation is needed for many of the same reasons as have been covered previously with attic ventilation. Inadequate underfloor ventilation may allow moisture to accumulate, and in cooler winter months, it may create frost and/or icing conditions. As most homes are built with wood frames, wood members are especially susceptible to moisture because of the porosity of wood. Earthen floors without vapor retardation act as an additional moisture source not included for attic spaces. Ground and/or surface water intrusion into underfloor spaces is another contributor to moisture in these spaces.

Like with attic ventilation, the ventilation in a crawlspace can be accomplished passively using natural convection or can be accomplished using mechanical ventilation or insulation with conditioning of the air in the space.[4–10,13,14,21,25–27] The focus of this section is on adequate passive ventilation for crawlspaces.

Crawlspaces are typically passively vented using screened rectangular openings in the upper foundation walls. This type of ventilation is illustrated in Figure 12.13.

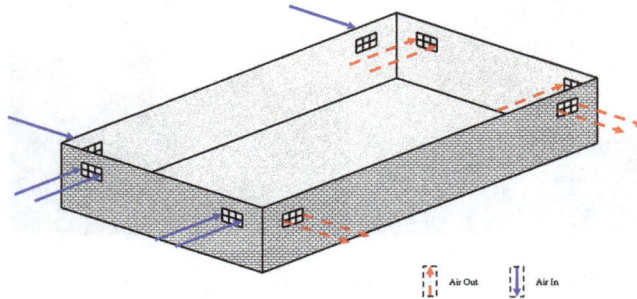

FIGURE 12.13 Illustration of passive ventilation openings for a typical crawlspace.

Note that in this case, those vents that serve as intake vents and those that serve as exhaust vents typically depend on the pressurization of the home above and the wind direction or speed.

12.3.1 CRAWLSPACE VENTILATION REQUIREMENTS

Information regarding crawlspace ventilation requirements can be found in most modern residential and commercial building codes and in best practices documents. Typical code language for crawlspace ventilation from the 2018 International Residential Code – Chapter 4[11] and the 2018 International Building Code (IBC),[12] Portions of Chapter 4 (Under-Floor Space), Section R408, from the 2018 IRC Code pertinent to crawlspaces is reproduced below:

R408.1 Ventilation: The underfloor space between the bottom of the floor joists and the earth under any building (except space occupied by a *basement*) shall have ventilation openings through the foundation walls or exterior walls. The minimum net area of ventilation openings shall be not <1 square foot ($0.0929 \, m^2$) for each 150 square feet ($14 \, m^2$) of underfloor space area, unless the ground surface is covered by a Class I vapor retarder material. Where a Class 1 vapor retarder material is used, the minimum net area of ventilation openings shall not be <1 square foot ($0.0929 \, m^2$) for each 1,500 square feet ($140 \, m^2$) of underfloor space area. One such ventilation opening shall be within 3 feet (914 mm) of each corner of the building.

R408.2 Openings for underfloor ventilation: The minimum net area of ventilation openings shall not be <1 square foot ($0.0929 \, m^2$) for each 150 square feet ($14 \, m^2$) of underfloor area. One ventilation opening shall be within 3 feet (915 mm) of each corner of the building. Ventilation openings shall be covered for their height and width with any of the following material provided that the least dimension of the covering shall not exceed 1/4 inches (6.4 mm):

1. Perforated sheet metal plates not <0.070 inches (1.8 mm) thick.
2. Expanded sheet metal plates not <0.047 inches (1.2 mm) thick.
3. Cast-iron grill or grating.

4. Extended load-bearing brick vents.

5. Hardware cloth of 0.035 inches (0.89 mm) wire or heavier.

6. Corrosion-resistant wire mesh, with the least dimension being 1/8 inches (3.2 mm) thick.

Exception: The total area of ventilation openings shall be permitted to be reduced to 1/1,500 of the underfloor area where the ground surface is covered with an *approved* Class I vapor retarder material and the required openings are placed to provide cross-ventilation of the space. The installation of operable louvers shall not be prohibited.

R408.3 Unvented crawlspace: Ventilation openings in underfloor spaces specified in Sections R408.1 and R408.2 shall not be required where the following items are provided:

1. Exposed earth is covered with a continuous Class I vapor retarder. Joints of the vapor retarder shall overlap by 6 inches (152 mm) and shall be sealed or taped. The edges of the vapor retarder shall extend not <6 inches (152 mm) up the stem wall and shall be attached and sealed to the stem wall or insulation.

2. One of the following is provided for the underfloor space:

   2.1 Continuously operated mechanical exhaust ventilation at a rate equal to 1 cubic foot/minute (0.47 L/s) for each 50 square feet (4.7 m^2) of crawlspace floor area, including an air pathway to the common area (such as a duct or transfer grill), and perimeter walls insulated in accordance with Section N1102.2.11 of this code.

   2.2 Conditioned air supply sized to deliver at a rate equal to 1 cubic foot/minute (0.47 L/s) for each 50 square feet (4.7 m^2) of underfloor area, including a return air pathway to the common area (such as a duct or transfer grille), and perimeter walls insulated in accordance with Section N1102.2.11 of this code.

   2.3 Plenum in existing structures complying with Section M1601.5, if underfloor space is used as a plenum.

   2.4 Dehumidification sized to provide 70 pints (33 liters) of moisture removal per day for every 1,000 square feet (93 m^2) of crawlspace area.

Portions of Chapter 12 (Interior Environment), Section 1203, from the 2018 IBC Code pertinent to crawlspaces are reproduced below:

1202.4 Underfloor ventilation: The space between the bottom of the floor joists and the earth under any building, except spaces occupied by basements or cellars, shall be provided with ventilation in accordance with Section 1202.4.1, 1202.4.2, and 1202.4.3.

1202.4.1 Ventilation openings: Ventilation openings through foundation walls shall be provided. The openings shall be placed so as to provide cross-ventilation of the underfloor space. The net area of ventilation openings shall be in accordance with Section 1202.4.1.1 or 1202.4.1.2. Ventilation openings shall be covered for their height and width with any of the following materials, provided that the least dimension of the covering shall be not >1/4 inches (6 mm):

1. Perforated sheet metal plates not <0.070 inches (1.8 mm) thick.

2. Expanded sheet metal plates not <0.047 inches (1.2 mm) thick.

3. Cast-iron grilles or gratings.

4. Extruded load-bearing vents.

5. Hardware cloth of 0.035 inches (0.89 mm) wire or heavier.

6. Corrosion-resistant wire mesh, with the least dimension not >1/8 inches (3.2 mm).

7. Operable louvers, where ventilation is provided in accordance with Section 1202.4.1.2.

1202.4.1.1 Ventilation area for crawlspaces with open earth floors: The net area of ventilation openings shall not be <1 square foot for each 150 square feet (0.67 m^2 for each 100 m^2) of crawlspace area.

1202.4.1.2 Ventilation area for crawlspaces with covered floors: The net area of ventilation openings for crawlspaces with ground surface covered with a Class I vapor retarder shall be not <1 square foot for each 1,500 square feet ($0.67\,m^2$ for each $1,000\,m^2$) of crawlspace area.

1202.4.2 Ventilation in cold climates: In extremely cold climates, where a ventilation opening will cause a detrimental loss of energy, ventilation openings to the interior of the structure shall be provided.

1202.4.3 Mechanical ventilation: Mechanical ventilation shall be provided to crawlspaces where the ground surface is covered with a Class I vapor retarder. Ventilation shall be in accordance with Section 1202.4.3.1 or 1202.4.3.2.

1202.4.3.1 Continuous mechanical ventilation: Continuously operated mechanical ventilation shall be provided at a rate of 1.0 cubic fool per minute (cfm) for each 50 square feet ($1.02\,L/s$ for each $10\,m^2$) of crawlspace ground surface area and the ground surface shall be covered with a Class I vapor retarder.

1202.4.3.2 Conditioned space. The crawlspace shall be conditioned in accordance with the International Mechanical Code and the walls or the crawlspace shall be insulated in accordance with the International Energy Conservation Code.

1202.4.4 Flood hazard and high-wind areas: For buildings in flood hazard areas as established in Section 1612.3, the openings for underfloor ventilation shall be deemed as meeting the flood control requirements of ASCE 24 provided that the ventilation openings are designed and installed in accordance with ASCE 24. Like with attics, the ratio of ventilated open area to floor plan area recommended is 1:150 unless the space is mechanically ventilated or in a flood zone (see for example, 2018 IRC Sections R408.2/3, 2018 IBC Sections 1202.3/4 and Lstiburek[28]). This value is also recommended by industry when no vapor retarder is present (Table 12.4).[18-20] NAHB[29] provides guidance on venting keeping water out of vent systems in high -wind regions.

12.3.2 Example Crawlspace Ventilation Calculation

In order to determine whether or not a crawlspace is properly ventilated (i.e., ventilation area ratio 1:150 met), the following information must be collected:

- Plan area of crawlspace in square feet.
- Vent area (# and area/vent).
- Screen type over vent openings.

Crawlspace ventilation requirements are simpler to calculate than attic ventilation requirements because the vents are typically of a single type and the concern is the total vent area rather than intake, exhaust, and total area. An example of an NFVA calculation follows.

12.3.2.1 Example Crawlspace Ventilation Calculation – Crawlspace Area and Required Vent Area

The plan (i.e., floor) dimensions of a crawlspace must be measured or obtained from drawings to determine the ventilation area. An example of a simple crawlspace used in this example is shown in Figure 12.14.

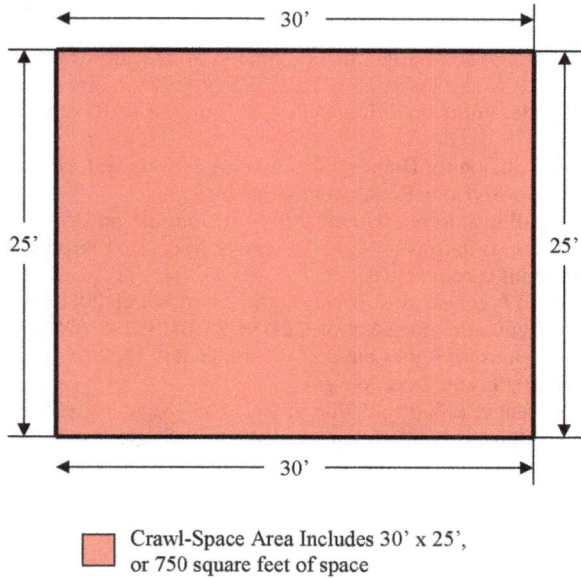

Crawl-Space Area Includes 30' x 25', or 750 square feet of space

FIGURE 12.14 Example calculation crawlspace area.

Assuming that the crawlspace area is continuous and a vapor barrier is not present, the total plan area of the crawlspace from Figure 12.14 is 750 square feet. Using a ratio of 1:150, the total ventilation opening needed for this crawlspace would be ~5.0 square feet (750/150). Assuming that the screen type was approximately 1/16″ mesh, the effectiveness factor of the opening would be 0.5 (Table 12.5), resulting in an effective vent opening area needed of 2.5 (5 × 0.5) square feet. A variety of combinations of number of openings and area per opening could be used to satisfy this total ventilation requirement. Regardless, one would want to have at least one opening on each elevation (and preferably two) spaced as evenly as possible.

IMPORTANT POINTS TO REMEMBER

- The most common attic and crawlspace ventilation encountered in the field is passive ventilation.
- The consequences of inadequate attic or crawlspace ventilation are water damage, formation of mold, and possible loss of structural integrity to structural wood members.
- The effect of poor attic ventilation can shorten roof shingle life and encourage the presence of insects such as ant and termites, which accelerate damage to a structure.
- NFVA is a term used to represent the effective area of a given vent type.
- Commonly recommended minimum effective vent area to attic and crawlspace floor (plan) areas is a ratio of 1:150 although the history for selecting this value is hazy.
- Attic intake vents are commonly restricted by insulation or painting activities.

REFERENCES

1. Air Vent, Inc. "Principles of Attic Ventilation: A Comprehensive Guide to Planning the Balanced System™ for Attic Ventilation." 2018. Accessed May 12, 2020, http://www.airvent.com/index.php/ventilation-resources/literature-sales-tools/downloads/21-principles-of-attic-ventilation-technical-booklet/file.

2. Stewart, B.R. "Attic Ventilation for Homes." Texas A&M University. Last updated November 13, 2011. http://www.factsfacts.com/MyHomeRepair/ventilation.htm.

3. Lstiburek, J. "Understanding attic ventilation." *ASHRAE Journal*, pp. 36–45, April 2006.

4. Air Vent, Inc. "Specifications for Intake Vents." Accessed May 2020, http://www.airvent.com/products/intake-vents/the-edge-vent#specifications.

5. Improvenet. "Ventilation: A House Must Breathe." Accessed March 2011, http://www.improvenet.com/projecttools/estimators/roofing/re_facts4.html?CEID=17730713278646764447.

6. Air Vent, Inc. "Specifications for Roof Louvers." Accessed May 12, 2020, http://airvent.com/index.php/products/exhaust-vents/static-vents/roof-louvers.

7. Air Vent, Inc. "VenturiVent Plus™ Installation Instructions." September 2013. Accessed May 12, 2020, http://airvent.com/index.php/ventilation-resources/literature-sales-tools/installation/76-venturivent-plus-installation-instructions/file.

8. Air Vent, Inc. "Filtervent® Installation Instructions." 2003. Accessed May 12, 2020, http://airvent.com/index.php/ventilation-resources/literature-sales-tools/installation/55-filtervent-installation-instructions/file.

9. Air Vent, Inc. "Attic Ventilation Inspection Form." December 2016. Accessed May 12, 2020, http://www.airvent.com/index.php/ventilation-resources/literature-sales-tools/downloads/1-attic-ventilation-inspection-form/file.

10. Air Vent, Inc. "Attic Ventilation: Tips and Answers from the Experts." November 2018. Accessed May 12, 2020, http://www.airvent.com/index.php/ventilation-resources/literature-sales-tools/downloads/30-tips-and-answers-booklet-from-ask-the-expert-seminar/file.

11. International Code Council. "2018 International Residential Code for One- and Two-Family Dwellings (IRC)." Accessed May 11, 2020, https://codes.iccsafe.org/content/IRC2018P3/chapter-8-roof-ceiling-construction and https://codes.iccsafe.org/content/IRC2018/chapter-4-foundations.

12. International Code Council. "2018 International Building Code (IBC)." Accessed May 11, 2020. https://codes.iccsafe.org/content/IBC2018/chapter-12-interior-environment.

13. Authenticated U.S. Government Information GPO. "Title 24- Housing and Urban Development, Part 3280: Manufactured Home Construction and Safety Management, Subpart F: Thermal Protection, Section 3280.504 Condensation Control and Installation of Vapor Barriers." Accessed May 12, 2020, https://www.govinfo.gov/content/pkg/CFR-2010-title24-vol5/pdf/CFR-2010-title24-vol5-part3280.pdf.

14. Humbarger, R.W. "Attic and Cathedral Ceiling Ventilation and Ice Dam Protection." *Interface Magazine*, RCI Institute, January 2009, pp. 32–35.

15. Rose, W.B. "Early History of Attic Ventilation." Research Architect, Building Research Council-School of Architecture, University of Illinois at Urbana-Champaign. Accessed May 11, 2020, http://docserver.nrca.net/technical/7877.pdf.

16. Rose, W.B. "The History of Attic Ventilation Regulation & Research." Research Architect, Building Research Council-School of Architecture, University of Illinois at Urbana-Champaign, Thermal Envelopes VII Moisture and Air Leakage Control II-Practices. Accessed May 11, 2020, https://web.ornl.gov/sci/buildings/confarchive/1995%20B6%20papers/016_Rose.pdf.

17. TenWolde, A. and W.B. Rose. "Issues Related to Venting of Attics and Cathedral Ceilings, ASHRAE Research Project CH-99-11=4." 1999. Accessed May 11, 2020, https://www.fpl.fs.fed.us/documnts/pdf1999/tenwo99a.pdf.

18. APA (American Plywood Association). "Condensation Causes and Control." APA (American Plywood Association) Publication X485H. The Engineered Wood Association Headquarters, January 1991.

19. APA (American Plywood Association). "Condensation Causes and Control." APA (American Plywood Association) Publication X485P. The Engineered Wood Association Headquarters, January 2007.

20. APA (American Plywood Association). "Condensation Causes and Control." APA (American Plywood Association) Publication X485Q. The Engineered Wood Association Headquarters, October 2017.

21. Asphalt Roofing Manufacturers Association (ARMA). "Ventilation and Moisture Control for Residential Roofing." Technical Bulletin, Form No. 209 RR-66, August 2019. Accessed May 12, 2020, https://www.asphaltroofing.org/media/2019/09/Ventilation-and-Moisture-Control-for-Residential-Roofing-FINAL.pdf.

22. Asphalt Roofing Manufacturers Association (ARMA). "Attic Ventilation Best Practices for Steep Slope Asphalt Shingle Roof Systems," Technical Bulletin, May 2017. Accessed May 12, 2020, https://www.asphaltroofing.org/media/2017/11/Ventilation-Best-Practices.pdf.

23. American Plywood Association. "How to Minimize Buckling of Asphalt Shingles," APA Publication K310S, Tacoma, WA, July, 2005.

24. Lstiburek, J. "Understanding Attic Ventilation," Building Science Digest 102, December 2013.

25. Graham, F. "Crawl Space Ventilation." Mississippi State University Extension Service, Information Sheet 1488, 2009. http://msucares.com/pubs/infosheets/is1488.htm.

26. Warren, B. "Crawl Space Research Project" Presentation, Advanced Energy, June 2003. http://www.advancedenergy.org/buildings/knowledge_library/healthy_homes_research/.

27. Baechler, M.C., Z.T. Taylor, R. Barlett, T. Gilbride, M. Hefty, and H. Stewart. "Building America Best Practices Series: Volume 4. Builders and Buyers Handbook for Improving New Home Efficiency, Comfort and Durability in the Mixed-Humid Climate." National Renewable Energy Laboratory, Report 09-2005, Project # NREL/TP-550-38448, 2005.

28. Lstiburek, J.W. "Crawlspaces: Either in or out." *ASHRAE Journal*, pp. 64–69, January 2020.

29. Home Builder's Guide to Coastal Construction. Federal Emergency Management Agency (FEMA) "Minimizing Water Intrusion through Roof Vents in High-Wind Regions." Technical Fact Sheet No. 7.5. Accessed July 27, 2020, https://www.fema.gov/media-library-data/20130726-1537-20490-9128/fema499_7_5.pdf.

13 Mold and Bacteria

Stephen E. Petty
EES Group, Inc.

Herbert D. Layman
U.S. Micro Solutions

CONTENTS

PURPOSE/OBJECTIVES

The purpose of this chapter is to:

- Demonstrate methodologies for the evaluation of mold and bacteria inspections.
- Provide an overview on the mold and bacteria sampling processes.
- Describe how one interprets mold results.
- Describe how one interprets bacteria results.
- Address other biological contaminants that one may encounter while performing forensic inspections.

Following the completion of this chapter, you should be able to:

- Understand when and how to sample for mold and bacteria.
- Understand how to interpret mold results.
- Understand how to interpret bacteria results.
- Understand when a formal mold remediation specification is needed.

13.1   INTRODUCTION

Significant press coverage of biological contamination of homes and business has occurred over the past 25–30 years, especially coverage of reported problems with mold (fungi) contamination of air and surfaces, even though reports of mold and bacteria health effects have been reported back into antiquity. Mold and bacteria, in their aerosol forms, can cause health effects to individuals and are from a broader class of materials known as bioaerosols.[1] Bioaerosols include the following:

- Amoebae.
- Pollen.
- Algae.
- Arthropods and Arthropod Antigens (e.g., mites).
- Bacteria (e.g., *E. coli*, *Legionella*, *Mycobacterium tuberculosis*, *Staphylococcus*, and *Streptococcus*).
- Fungi/Mold (e.g., *Aspergillus* spp., *Penicillium* spp., and *Stachybotrys* spp.).
- Mammals and Mammalian Antigens (e.g., cat and dog allergens).
- Viruses (e.g., Influenza, HIV, COVID-19).

The focus of this chapter will be limited to mold (fungi) and bacteria even though other bioaerosols may have to be considered when responding to reports of health effects.

Bacteria and mold are essentially ubiquitous; in other words, they are present nearly everywhere.[2] Niemeier et al.[3] report that up to 40% of the homes in the United States and that between 20% and 40% of the homes in Canada and Europe also have mold problems. Contamination of the indoor environment is demonstrated from work by Sahay et al.,[4] who reported the following specific findings regarding bacteria and mold found indoors based on 10 years of IAQ sampling:

Bacteria: 11,463 air samples, 3,946 different sites belonging to 623 buildings: % indoor and outdoor air containing the following culturable bacterial taxa:

- 340 bacterial taxa.
- *Micrococcus luteus* (13.25%).
- Gram-negative bacilli (12.8%).
- *Bacillus* species (11.8%).
- *Staphylococcus* species (11.5%).
- *Kytococcus sedentarius* (11.1%).

Mold (Fungi): 6,119 air samples, 3,898 different sites belonging to 616 buildings: % indoor and outdoor air containing the following culturable fungal taxa:

- *Cladosporium cladosporioides* (29.6%).
- *Cladosporium* species (23.6%).
- *Penicillium* species (17.9%).
- *Mycelia sterilia* or unidentified fungi (13.5%).
- *Penicillium brevicompactum* (13.0%).

Of course, the extent to which given taxa are found for a given location is highly dependent on local environmental conditions. Examples of surfaces contaminated by mold and bacteria are illustrated in Figures 13.1 and 13.2.

FIGURE 13.1 Probable visible mold on dining room wall surface – Hole into wall cavity.

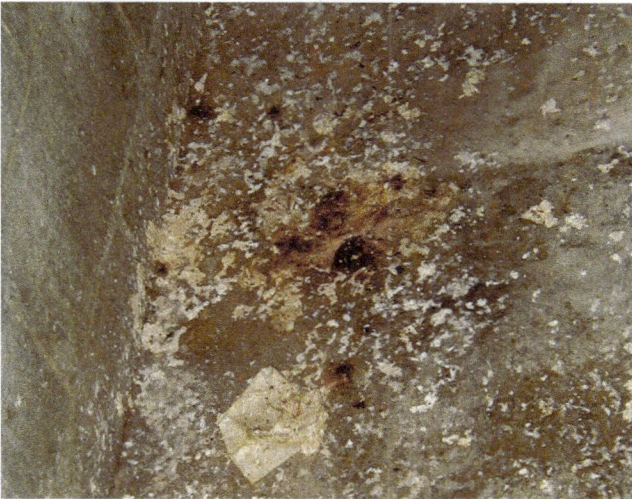

FIGURE 13.2 Sewage spill on basement floor below failed sanitary line.

Both mold and bacteria exposures have been reported to result in significant health effects, often associated with the respiratory tract [i.e., rhinitis (nasal congestion), pharyngitis (cough), dyspnea (shortness of breath)] along with other symptoms (conjunctival irritation, headache, or dizziness, lethargy, fatigue, malaise, nausea, vomiting, anorexia, rashes, fever, and chills).[3,5] The health effects are associated with responses not only to the bacteria or mold, but in the case of mold, but also to their metabolites (Table 13.1).

TABLE 13.1
Metabolites by Type of Mold (Fungi)

Mold/Fungi	Metabolite(s)
Aspergillus	Aflatoxin
Aspergillus parasiticus	Sterigmatocystin
Aspergillus flavus	Patulin, Citrinin
Aspergillus versicolor	
Aspergillus terreus	
Fusarium	Zearalenone
Fusarium moniliforme	Trichothecenes
Fusarium spp.	
Penicillium	Ochratoxin
Penicillium viridicatum	Citrinin, patulin
Penicillium spp.	
Stachybotrys	Trichothecenes
Stachybotrys chartarum (atra)	

It should be noted that with the emphasis in the public discourse on health hazards associated with exposure(s) to mold, the likely greater health hazard to occupants and contractors is exposure to bacteria (e.g., sewage spills). Many bacteria can cause severe health effects, including death, whereas mold exposures tend to result in increased respiratory health effects but rarely death.

The science of anticipation, recognition, evaluation, and control of biological contamination of the air and surfaces[1,4,5] requires specific expertise, often that of a Certified Industrial Hygienist (CIH), an experienced industrial hygienist, or an experienced certified indoor environmental consultant. Methods for assessing and controlling bioaerosols, in general, can be found in textbooks by Macher[1] and others.[6]

In addition, the methods for assessing and controlling mold have been written about extensively and can be found in ASTM Standards[7,8] (e.g., ASTM D7338-10, 2014, Standard Guide for Assessment of Fungal Growth in Buildings and ASTM E2418-06 Standard Guide for Readily Observable Mold and Conditions Conductive to Mold in Commercial Buildings: Baseline Survey Process), IICRC,[9–13] and other publications.[14–30]

A full understanding of these areas is outside the scope of this book. Nevertheless, it is helpful to be aware of how to recognize biological growth, have an understanding of reported health symptoms, why it occurs, how it is quantified, how to interpret laboratory results, and how biological growth is controlled and remediated. This chapter is intended to provide an overview of these issues and topics.

13.2 MOLD

13.2.1 INTRODUCTION TO MOLD

Molds, mushrooms, mildews, and yeasts are all classified as fungi, a kingdom of organisms different from plants and animals.[31] Molds (fungi) are present almost everywhere[31–36] with hundreds of types of different molds found indoors. The Kingdom Fungi has been estimated at around 1.5 million species, with about 5% of these having been formally classified. For the purposes of this book, the less informal, but more common term "mold" will be used to describe fungi even though fungi is the more proper term scientifically. Of interest, the word "mold" is derived from the word "fuzzy" in the obsolete Old Norse language, which is how mold appears on surfaces when it amplifies to levels where it is readily visible. The mold life cycle and common mold terminology are illustrated in Figure 13.3.

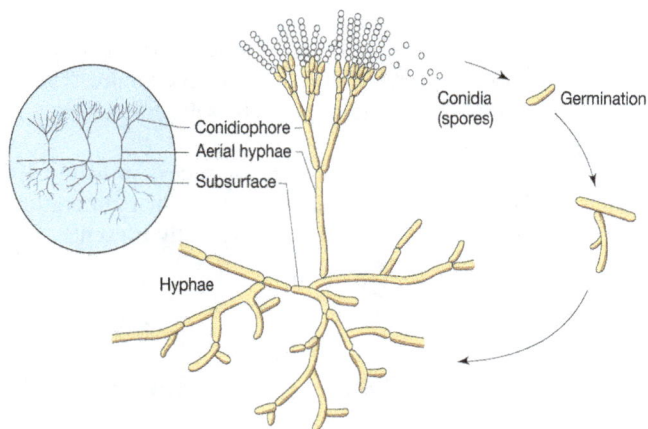

FIGURE 13.3   Mold life cycle.

The structure of mold contains subsurface elements called hyphae, surface elements called aerial hyphae, stem-like elements called conidiophores, and seed-like materials for reproduction known as conidia or more commonly spores. The release of spores allows for procreation or amplification of mold when environmental conditions are conducive for growth. Thus, the term amplification is used to describe mold growth when conditions such as excess moisture are present and visible mold forms on surfaces. Note that the growth is geometric rather than arithmetic. Common molds, and where they are often found, are summarized in Table 13.2.

TABLE 13.2
Common Molds and Locations Found

Mold Genus	Common Locations Found
Alternaria	Common outdoors, moist windowsills, walls
Ascospores	Outdoors, if found indoors in spore trap may be due to improperly filtered outdoor air since they won't grow on most indoor materials
Aspergillus	First-level colonizer – carpet and chronically damp locations. Damp wood, potting soil, and wallpaper glue
Aureobasidium plullulans	Bathroom and kitchen cellulose walls
Basidiospores	Outdoors, if found indoors in spore trap may be due to improperly filtered outdoor air since they won't grow on most indoor materials
Chaetomium	Third-level colonizer – wall joists and building timber
Cladosporium	Second-level colonizer – common outdoors; cellulose-containing products
Curvularia	Common outdoors
Epicoccum	Common outdoors
Fusarium	Ventilation systems
Mucor	Dust-rich carpets
Penicillium	First-level colonizer – carpet and chronically damp locations. Damp wallpaper and behind paint
Rhizopus	Dust-rich carpets
Stachybotrys	Third-level colonizer – ceiling tiles, wet carpet, and sheetrock
Trichoderma	Wall joists and building timber

Background levels of mold on interior wall surfaces tend to be on the order of <10,000 colony forming units per square inch (CFU/square inch), whereas levels on surfaces, when visible mold is present, tend to be on the order of 1,000,000 CFU/square inch or more. Thus, the term amplification is used to describe mold growth when conditions such as excess moisture are present and much greater/visible mold forms on surfaces.

As illustrated in Figure 13.4, the presence of water or excess moisture levels is critical[31–35,37] to the amplification of mold since all other elements (i.e., right temperature range, food, mold spores, and areas with limited air velocity and light) are typically present somewhere in the indoor environments.

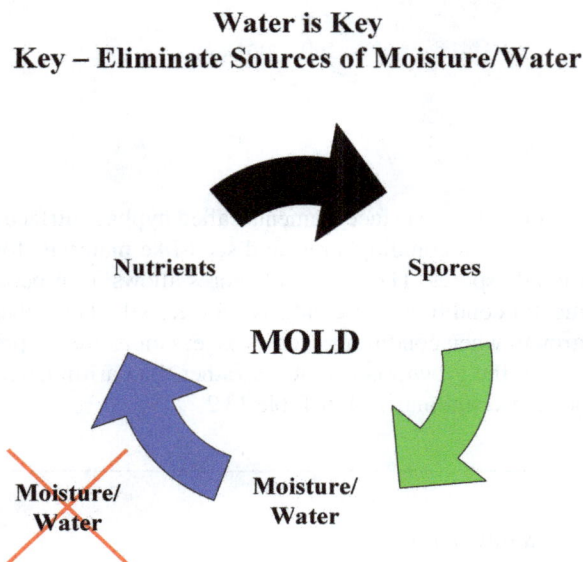

Water is Key
Key – Eliminate Sources of Moisture/Water

Nutrients   Spores

MOLD

Moisture/   Moisture/
Water   Water

FIGURE 13.4   Environmental factors affecting the amplification of mold.

Elevated moisture levels in buildings are typically associated with one of four situations: (1) rain and groundwater entry, (2) elevated relative humidity levels, causing condensation on building surfaces from both interior and exterior air, (3) construction moisture present in either the building material or as a result of exposure to the weather prior to close-in, and (4) interior water due to breakage in a plumbing fixture.[38] The types of molds found typically depend on the amount and progression of moisture present:

- Primary colonizers (e.g., *Penicillium* spp. and *Aspergillus* spp.).
- Secondary colonizers (e.g., *Cladosporium* spp.).
- Tertiary colonizers (e.g., *Stachybotrys chartarum* or *Chaetomium* spp.) on wet materials.

The presence of *Stachybotrys* or *Chaetomium*, for example, implies the presence of high moisture levels for extended periods of time. Provided one reduces or eliminates the source(s) of water/moisture indoors, the amplification of mold is greatly reduced or stopped. Methods to minimize the formation of mold include[37,39] the following:

- Dry any wet or damp materials or areas within 24–48 hours after a leak or spill occurs.
- Clean and repair roof gutters regularly.

- Ensure that the ground slopes away from the building foundation, so that water does not enter or collect around the foundation.
- Keep air-conditioning drip pans clean and the drain lines unobstructed and flowing properly.
- Reduce potential for interior condensation by:
  - Reducing the relative humidity (indoor levels to 35%–50%).
  - Increasing the ventilation or air movement by opening doors and/or windows, when practical. Use fans as needed.
  - Covering cold surfaces, such as cold water pipes, with insulation.
  - Increasing the air temperature.

13.2.2 Mold and Health Effects

Much debate has occurred regarding the extent to which exposure to mold results in health effects to humans. The technical literature does associate *some*, but not all, of the health effects listed above, with increased indoor moisture and elevated indoor mold levels.[28,31,35,40–45]

The National Academy of Sciences[41] found *sufficient evidence of an association* between exposure to damp indoor environments and some respiratory health outcomes: upper respiratory tract (nasal and throat) symptoms, cough, wheeze, and asthma symptoms in sensitized asthmatic persons. Epidemiologic studies also indicate that there is sufficient evidence to conclude that the presence of mold (otherwise unspecified) indoors is associated with upper respiratory symptoms, cough, wheeze, asthma symptoms in sensitized asthmatic persons, and hypersensitivity pneumonitis (a relatively rare immune-mediated condition) in susceptible persons.

Limited or suggestive evidence was found for an association between exposure to damp indoor environments and dyspnea (the medical term for shortness of breath), respiratory illness in otherwise healthy children, and the development of asthma in susceptible persons.

Inadequate or insufficient information was identified to determine whether damp indoor environments or the agents associated with them are related to a variety of other health outcomes. Included among these is acute idiopathic pulmonary hemorrhage in infants (AIPHI). The committee concluded that the available case-report information constitutes inadequate or insufficient information to determine whether an association exists between AIPHI and the presence of *Stachybotrys chartarum* or exposure to damp indoor environments in general.[41]

The complexity of determining associations between mold exposure and disease lies with the large numbers of molds present, the lack of understanding of individual molds and their mycotoxins on disease, and other cofounders present in a given environment. The New York State Toxic Mold Task Force[35] summarized this situation as follows:

> Although the presence of indoor mold is associated in many studies with respiratory effects such as cough or asthma exacerbation, a single causative agent is not clearly identified from these studies. Many studies suggest fungal allergens are probably important exposure agents. However, mold growth in damp buildings almost always involves multiple fungal species, many of which may produce multiple allergen proteins that vary in their ability to cause allergic reactions. Furthermore, the allergenicity of many fungal species that can occur in wet buildings has not been studied, so that focusing on one or a few well-characterized fungal allergens will not adequately describe exposure (and therefore health risks) in many cases.

Health effects have also been observed in workers performing mold remediation activities.[22]

In addition, much discussion has taken place in public forums regarding the terms "toxic mold," "black mold," and "Stachybotrys," suggesting that some mold species are more toxic than other mold species. Available data do not support a distinction between "toxic mold species" health effects and "non-toxic mold species" health effects,[35] suggesting that all molds should be treated equally with respect to health effects and damage to structures.

Mold claims and litigation associated with mold claims have often occurred over the past 20–25 years (Table 13.3).

TABLE 13.3
Examples of Mold Litigation – Cases, Locations, and Results

Case	Location	Status/Award
Melinda Ballard v. Farmers Insurance Group	State Court, Austin, Texas	Melinda Ballard and her family were awarded $32.1 million in June 2001. The $32.1 million award represents $6.2 million for replacement of the home and contents, $5 million for mental anguish, $12 million in punitive damages, and $8.9 million for legal fees. On appeal, this was reduced to $4 million plus legal fees
Thomas Anderson v. Allstate Insurance Company	California	California jury ordered Allstate Insurance to pay a policyholder $18.5 million in a coverage dispute over mold in the plaintiff's home in Placerville, California. The award included a compensatory $500,000 in damages and $18 million in punitive damages. The trial judge reduced the award to $3 million. Decided on September 3, 2002, the district court reduced the punitive damages award to five times the compensatory damages. Allstate appealed both awards and Anderson cross-appealed the district court's reduction of punitive damages. The compensatory award was confirmed and a reverse of the punitive damages was awarded
Charles Blum et al. v. Chubb Custom Insurance Co., Chubb Group of Insurance Companies, and Texas Windstorm Insurance Association	Texas Dist., Nueces Co.	Claimed that the insurer(s) denied, delayed, or failed to pay or properly investigate claims stemming from accidental plumbing leaks and roof damage. The case went to trial; and after 2-1/2 weeks, the case was settled for $1.5 million on December 18, 2000.
Martin County, Florida v Centex-Rooney Construction Co., Inc., et al.	Florida	Martin County sued its construction manager for dampness that promoted mold growth and excessive humidity in a courthouse. Fifteen employees in the building alleged injuries caused by exposure to the mold. The source of the water problem was the exterior insulation finish system (EIFS). On an appeal, the appeals court affirmed the $14 million verdict against the construction manager. Martin County also secured out-of-court settlements worth $3 million from other defendants.
Elizabeth Stroot v. New Haverford Partnership, et al. and New Haverford Partnership, et al. v. Elizabeth Stroot, et al.	Delaware Superior Court	In the original lawsuit and its appeal, Elizabeth Stroot and three other plaintiffs were awarded damages for medical expenses, permanent impairment, and pain and suffering associated to exposure to various mycotoxins, bacteria, fungi, and other toxins while living in an apartment complex owned by New Haverford Partnership. In May 1999, a jury awarded $1 million in damages to Stroot and $40,000 to Joletta Watson. In addition, the jury awarded damages for expenditures made for substitute housing: $5,000 to Stroot, $1,500 to Angela McCarthy, and $3,700 to Lois Schlindler. In May 2001, the Delaware Supreme Court upheld the award to the residents.
Crocker v. Jeffcoat Builders and Gordon Plastering Co.	South Carolina	November 7, 2005 – Contractors settle construction South Carolina defect/mold lawsuit for $870,000 The house was allegedly contaminated by mold to such an extent that it had to be destroyed. The lawsuit also alleges that the wife has become sensitized to mold exposure curtailing her lifestyle. The settlement will pay for living expenses and mortgage payments while the couple was living elsewhere, medical expenses, and $300,000 of personal property that had to be destroyed. The contractors' insurers were Harleysville Mutual Ins. Co. and Zurich American Ins. Co.

(Continued)

TABLE 13.3 (*Continued*)

Examples of Mold Litigation – Cases, Locations, and Results

Case	Location	Status/Award
O'Hara v. Jeff Stangland, Contractor Harvey and Son, and designer Michael Cockram	Oregon	Family sued for $3.5 million, alleging that faulty construction led to the growth of mold in their home and subsequent adverse health effects. Shortly before the start of the trial, Harvey & Son reached an undisclosed settlement with O'Hara. Shortly after the beginning of the trial, the claim was dropped and a cash settlement was decided. The attorney for O'Hara said that the settlement would be paid by Stangland's Insurance Company
Homeowners v. Trinity Homes and Beazer Homes	Indiana	Trinity Homes and Beazer Homes, its parent, agreed to fix about 2,000 houses. On October 21, 2004, a Hamilton County judge approved a $24 million settlement between an Indianapolis home builder and the owners of more than 2,000 houses potentially affected by moisture and mold. (The Indianapolis Star, October 21).
Doe Homeowners v. Roe Seller	California	New owners of a house in California sued the sellers in 1997, alleging that toxic mold caused bodily injuries and property damage. The case was settled for $1,353,000.
Willard v. Wren	North Carolina	November 7, 2005 – HVAC contractor pays $120,000 to settle North Carolina mold contamination lawsuit. An HVAC contractor agreed to pay $120,000 to settle a mold contamination suit filed by a North Carolina homeowner. The lawsuit alleged that during a remodeling project, the contractor negligently installed an HVAC zoning system leading to excessive sweating and mold contamination. The contractor settled a week before trial was to begin. The contractor's insurer was Interstate Fire & Casualty.
Crane v. Bank of America	Ohio	An Ohio hotel manager sued the hotel owners alleging that he experienced adverse health effects subsequent to participating in remediation of toxic mold in the hotel.
Andrejevic et al. v. Board of Education of Wheaton-Warrenville School District No. 200.	DuPage County, Illinois	This class-action suit was filed by ~1,700 students, parents, and teachers. The plaintiffs are seeking $67 million for injuries caused by exposure to toxic mold and other indoor pollutants following a flood at the elementary school. The lawsuit claims that the school district did not properly remediate flood damage, resulting in growth of the mold.
Reber v. ServiceMaster and Reber v. Allstate	Indiana	The Reber's filed a lawsuit alleging that ServiceMaster did a poor job removing moisture from their 4,600-square-foot home, causing mold to grow throughout the house. According to the Indianapolis Star report, attempts to clear the mold have already cost $43,000 and current estimates predict the cost of removal to be $100,000. Allstate allegedly would only partially cover the cost of remediation efforts.
Marina Eddy et al. v. C.B. Richard Ellis Inc., Henry Knott, AMG Realty Partners LP, Kronos Property Holdings N.V. and Maritime Reality Corp.	Maryland	Three plaintiffs filed a suit claiming personal injuries stemming from exposure to mold and fungi in their workplace, an office building in Maryland. Injuries claimed include asthma and reactive airways disorder. The suit alleges that mold and fungi "were allowed to flourish" in the building's heating, ventilation, and air-conditioning system.
Robert E. Coiro et al. v. Dormitory Authority of the State of New York	New York	The plaintiffs sought punitive and exemplary damages in the sum of $50 million as well as an additional $5 million for services lost. Coiro alleged that he suffered from personal injuries and pain and suffering as a result of employment with LaGuardia Community College at the premises owned by Dormitory Authority of the State of New York. The complaint, filed in Queens County Supreme Court, maintained that toxic mold and fungus, water leaks, unsafe and unsanitary conditions, improper ventilation, and other dangerous conditions in the building created "an unsafe, contaminated and dangerous environment, all to the plaintiff's detriment and loss."

While much of the litigation has focused on damages from health effects, these damages can be difficult to prove given current knowledge on the association between mold exposures and health effects; the more easily proven case is the presence of mold and damage to structures.[46]

Some reduction in attention to mold has occurred most recently due to increased public awareness of the hazards associated with mold and a decrease in coverage by insurance companies with standard exclusions or limitations on mold coverage.

13.2.3 OVERVIEW OF THE MOLD INSPECTION AND REMEDIATION PROCESSES

Excellent sources of information regarding mold, taken from the New York State Toxic Mold Task Force,[35] are summarized in Table 13.4.

TABLE 13.4

Sources of Information on Moisture and Mold

Designation	Title
American Industrial Hygiene Association http://www.aiha.org	
AIHA Mold Guideline	Assessment, Remediation and Post-remediation Verification of Mold in Buildings
AIHA HVAC Workbook	Indoor Air Quality and HVAC Workbook
AIHA Air Quality	The Industrial Hygienist's Guide to Indoor Air Quality Investigations
ASTM International http://www.astm.org	
ASTM C 1338	Standard Test Method for Determining Fungi Resistance of Insulation Materials and Facings
ASTM D 2020	Standard Test Methods for Mildew (Fungus) Resistance of Paper and Paperboard
ASTM D 3273	Standard Test Method for Resistance to Growth of Mold on the Surface of Interior Coatings in an Environmental Chamber
ASTM D 4300	Standard Test Methods for Ability of Adhesive Films to Support or Resist the Growth of Fungi
ASTM D 4445	Standard Test Method for Fungicides for Controlling Sapstain and Mold on Unseasoned Lumber (Laboratory Method)
ASTM D 1151	Standard Practice for Effect of Moisture and Temperature on Adhesive Bonds
ASTM D 1860	Moisture and Creosote – Type Preservative in Wood
ASTM D 2065	Standard Test Method for Determination of Edge Performance of Composite Wood Products under Surfactant Accelerated Moisture Stress
ASTM D 2118	Assigning a Standard Commercial Moisture Content
ASTM D 2247	Standard Practice for Testing Water Resistance of Coatings In 100% Relative Humidity
ASTM D 2987	Standard Test Method for Moisture Content of Asbestos Fiber
ASTM D 4442	Standard Test Method for Direct Moisture Content Measurement of Wood and Wood-Base Materials
ASTM D 4502	Test Method of Heat and Moisture Resistance of Wood-Adhesive Joint
ASTM D 4610	Standard Guide for Determining the Presence of and Removing Microbial (Fungal or Algal) Growth on Paint and Related Coatings
ASTM D 4933	Standard Guide for Moisture Conditioning of Wood and Wood-Based Materials
ASTM D 6403	Test Method for Determining Moisture in Raw and Spent Materials
ASTM MNL 18	Moisture Control in Buildings
ASTM MNL 40	Moisture Analysis and Condensation Control in Building Envelopes
ASTM E 2267	Standard Guide for Specifying and Evaluating Performance of Single Family Attached and Detached Dwellings – Indoor Air Quality (IAQ)
ASTM D 5157	Standard Guide for Statistical Evaluation of Indoor Air Quality Models
ASTM D 5791	Standard Guide for Using Probability Sampling Methods in Studies of Indoor Air Quality in Buildings

(Continued)

TABLE 13.4 (Continued)

Sources of Information on Moisture and Mold

Designation	Title
ASTM D 6245	Standard Guide for Using Indoor Carbon Dioxide Concentrations to Evaluate Indoor Air Quality and Ventilation
ASTM D 7391	Standard Test Method for Categorization and Quantification of Airborne Fungal Structures in an Inertial Impaction Sample by Optical Microscopy
ASTM STP 1205	Modeling Of Indoor Air Quality and Exposure
ASTM WK3792	Guide for Assessment of Fungal Growth in Buildings (work item in progress as of August, 2009)
American Conference of Governmental and Industrial Hygienists http://www.acgih.org	
ACGIH Indoor Air Quality	Indoor Air Quality 2nd Edition
ACGIH Bioaerosols	Bioaerosols: Assessment and Control
American Society of Heating, Refrigerating and Air-Conditioning Engineers http://www.ashrae.org	
ASHRAE STD 55	Thermal Environmental Conditions for Human Occupancy
ASHRAE STD 62	Ventilation for Acceptable Indoor Air Quality
American Association of Textile Chemists and Colorists http://www.aatcc.org	
AATCC 100	Assessment of Antibacterial Finishes on Textile Materials
AATCC 30	Antifungal Activities Assessment on Textile Materials: Mildew and Rot Resistance of Textile Materials
Sheet Metal and Air Conditioning Contractors' National Association http://www.smacna.org	
SMACNA 1637	Indoor Air Quality – A System Approach 3rd Edition
Technical Association of the Pulp and Paper Industry http://www.tappi.org	
TAPPI T 487	Fungus Resistance of Paper & Paperboard
ANSI/Greenguard Environmental Institute http://www.greenguard.org/Default.aspx?tabid=115	
ANSI/GEI – MMS1001	ANSI/GREENGUARD Environmental Institute. Mold And Moisture Management Standard For New Construction

A visual assessment by qualified personnel (e.g., CIH or equivalent) is the most important step in identifying the extent of mold contamination and for setting the baseline for information needed to provide guidance on remedial activities, assuming that such activities are needed. As noted in the New York City Mold Guidelines[34]: "Environmental sampling is not usually necessary to proceed with remediation of visually identified mold growth or water-damaged materials. Decisions about appropriate remediation strategies can generally be made on the basis of a thorough visual inspection,"[34,35] and mold sampling is not needed[30,31] since the scope of remedial activities is typically set by the areas observed to be contaminated by probable visible mold. Initial sampling should be limited to situations where either: (1) concerns exist for individuals living in the environment with immune-compromised systems (e.g., HIV or transplant patients), (2) some reason exists for verification of the probable visible mold observed, or (3) the perception of mold is present but cannot be observed. Other than these situations, testing of remediated surfaces for residual mold present on them is the scenario where mold testing should be completed.

Assessments of mold-contaminated structures can be performed using the guideline document ANSI/IICRC S520-2015: Standard for Professional Mold Remediation, 3rd Edition.[9] This Standard document describes the procedures and precautions to be followed when performing mold remediation in residential, commercial, and institutional buildings and of the systems and personal property contents of those structures. The S520-2015 document can be used to develop a mold assessment plan, which can be performed by an indoor environmental professional (IEP) or other qualified individuals (e.g. CIH or Council-certified Indoor Environmental Consultant (CIEC)) that would include the following:

1. Evaluation of data collected from a building history.
2. Formulation of one or more hypotheses about the origin, description, location, and extent of a *Condition 1, 2, or 3 as defined for indoor environments relative to mold.
3. Construction of a sampling plan.
4. Determination of appropriate equipment and monitoring tools for sample collection and evaluation of indoor environmental parameters, e.g., temperature, humidity, moisture.
5. Interpretation of data.
6. Development of a remediation plan.

*Condition: for the purpose of this Standard, Conditions 1, 2, and 3 are defined as follows:

- Condition 1 (normal fungal ecology): an indoor environment that may have settled spores, fungal fragments, or traces of actual growth whose identity, location, and quantity are reflective of a normal fungal ecology for a similar indoor environment.
- Condition 2 (settled spores or fungal fragments): an indoor environment that is primarily contaminated with settled spores or fungal fragments, which were dispersed directly or indirectly from a Condition 3 area and which may have traces of actual growth.
- Condition 3 (actual growth): an indoor environment contaminated with the presence of actual mold growth, associated spores, and fungal fragments. Actual growth includes growth that is active or dormant, visible, or hidden.

The AIHA[28–31] and others[9–19,22,23,29–36,37–39] provided excellent reviews of the proper methods and procedures to be used to remediate mold on surfaces and in the air. The US EPA and the latest NYC Guidelines[32] suggest that mold remediation of <10 square feet can be completed by a home-owner; whereas, larger areas should be completed by a professional contractor properly trained and equipped to perform mold remediation activities.[37,39] The New York City guidelines,[32-35] first issued in 1993, initially recommended that professional contractors be utilized for areas covered with >30 square feet of visible mold and continued with this recommendation in 2000 and 2002, but have increased this limit to >100 square feet in 2008.[34] Based on current trends, formal mold remediation specifications performed, utilizing experienced mold contractors, probably are not needed until areas of probable visible mold >100 square feet are encountered.

Respiratory protection (e.g., N-95 disposable respirator), worn in accordance with the OSHA respiratory protection standard (29 CFR 1910.134), is still recommended for remediation completed on areas <100 square feet of visible mold.

Typically, the indoor mold remediation would dictate negative pressure in the room or areas where the remediation is scheduled to take place. However, if the mold is present on the outside of a structure, positive (not negative as sometimes observed) pressure should be applied to the inside spaces rather than negative pressure so that the mold is not drawn indoors into living spaces.

13.2.4 SAMPLING FOR MOLD

Regarding mold sampling, the New York City Mold Guidelines[35] provide excellent insights on mold sampling when they state: "If environmental samples will be collected, a sampling plan should be developed that includes a clear purpose, sampling strategy, and addresses the interpretation of results. Many types of sampling can be performed (e.g. air, surface, dust, and bulk materials) on a variety of fungal components and metabolites, using diverse sampling methodologies. Sampling methods for fungi are not well standardized, however, and may yield highly variable results that can be difficult to interpret. Currently, there are no standards or clear and widely accepted guidelines with which to compare results for health or environmental assessments."[34] They further noted that such sampling must be conducted by individuals trained in the sampling methods and who is aware of the limitation of such methods. Recent work by New York State confirms a lack of established criteria for interpretation of mold sampling results.[35] AIHA[30] and others suggest that mold sampling, in general, is not needed during the initial inspection.[31]

The types, pros and cons, and an explanation of mold sampling techniques are presented and discussed at length by Prezant et al.[28] and are summarized in Table 13.5.

TABLE 13.5
Mold Sampling Methods

Medium Sampled	Sampling Method	Analytical Method	Description of Method
Bulk material	Collection	Cultured/ counts	Measures viable mold in units of CFU/gram of bulk material
Surface	Tape lift	Microscopic/ counts	Qualitative or total mold in spores/square inch
	Swab – wetted	Cultured/ counts	Measures viable mold in units of CFU/square inch
	MycoMeter™	Digested/ colorimetric	Measures egosterol levels as associates with total mold levels
	Carpet Vac – dust	Cultured/ counts	Measures viable mold in units of CFU/ft3.
Bulk air	Pump and filter (e.g.,≈Air-O-Cell™)	Microscopic/ counts	Measures total (viable and non-viable) mold in units of spores/m^3; cannot distinguish between *Aspergillus* and *Penicillium*
	Pump and filter (Wall-Chek™)	Microscopic/ counts	Measures viable mold in units of spores/m^3; cannot distinguish between *Aspergillus* and *Penicillium*. Method not recommended
	Pump and auger medium (e.g., Anderson™ Impactor)	Cultured/ counts	Measures viable mold only in units of CFU/m^3

Methods used to measure mold levels depend on whether a bulk sample, the contaminated surface, or the air is sampled for mold. Niemeier et al.[3] and others recommend that multiple methods be used to identify the types and concentrations of species present in a given environment. Bulk samples are simply samples of the contaminated material packaged in a plastic bag and sent to a lab for analysis of the mold content per mass of material submitted. Surface samples are generally taken using a wetted swab with results typically reported in terms of colony forming units (CFU) per square inch of area sampled. Surface methods such as MycoMeter™ have the advantage of being a real-time method but do not provide speciation information. However, this method is particularly effective for post-remediation real-time clearance sampling.

Mold levels in carpets can be determined using a Carpet Vac method wherein the mold in a fixed area (either ~6 inch × 6 inch or 12 inch × 12 inch) is vacuumed using a pump drawing the air across a 0.4–0.8 μ sterile filter cassette; results are reported in terms of (CFU)/square feet of carpet area.

Air samples are generally taken using a pump and collection device; results are reported in terms of either spores/per cubic meter of air (non-culturable) or CFU per cubic meter of air (culturable). The collection device is either a cassette containing a sticky glass coverslip to collect particulates or an agar plate. An agar plate is a sterile Petri dish that contains agar plus nutrients (media), used to culture (grow) molds. Many types of agars are available and to various degrees, are specific for certain mold species; the most commonly used for collecting molds are inhibitory mold agar (IMA), potato dextrose agar (PDA), malt extract agar (MEA), and Sabouraud dextrose agar (SDA).

Wall cavity air can be monitored using a wall check air sampling method but is not recommended except as a qualitative tool or when a supplemental method such as visual wall cavity inspection is also completed in conjunction with this type of sampling. This method was ruled unreliable and inadmissible[44] by a Texas court due to the fact that the source of the air is not known (i.e., indoor air, wall air, or outdoor air) *and* no secondary validation of the wall cavity conditions was completed. Caution regarding this method is also mentioned by the AIHA.[28]

Recommendations regarding sampling based on experience follow:

- Calibrate sampling pumps before and after sampling events. Use mid-point flow values for chain of custody forms when reporting air flow rate unless the values vary widely. If the results vary widely, then one should resample since this situation calls into question the reliability of the pump flow during sampling.
- Ensure that the media used for sampling has not expired. This is a common problem since media such as agar plates have a shelf life as little as 30 days. Use of expired media can lead to false results and almost certainly, the results would be thrown out if the data were ultimately involved in litigation.
- Conduct mold sampling using multiple methods. Typically, collection of a combination of surface and air samples (both spore trap and agar methods) should be taken if a mold sampling event is scheduled.
- For surface samples, samples should not only be taken on surfaces with visible mold, but occasionally (1 in 5 to 1 in 10) on apparently clear surfaces with 12–18 inches of surfaces sampled containing visible mold.
- Experience suggests that taking air samples using both a spore trap (total – viable and nonviable) and using agar plates (viable) provides complimentary information regarding airborne mold levels. When using either of these methods, two additional samples must always be taken; an outdoor sample for background mold levels and a blank to ensure the media were not contaminated. Thus, if five indoor spore trap air samples were taken, a total of seven samples should be taken (five plus one outdoor and one blank).
- Provisions should be made for portable power (vehicle inverter or generator) to power sampling pumps for some scenarios where power may not be available (e.g., home subject to fire and water damage).
- Carefully label all media and transfer all field information to a field chain of custody form to ensure that samples are properly labeled and accounted for. Clean chain of custody forms can be completed as part of the process of sending samples to a lab, but the biggest mistakes seem to be made by taking chain-of-custody information in the field.
- All samples, once properly packaged, should be sent along with a properly completed chain-of-custody form, to an American Industrial Hygiene Association (AIHA) Environmental Microbiology Laboratory Accreditation Program (EMLAP) accredited laboratory. Other laboratories may be perfectly capable of providing the necessary analysis, but should the results ultimately be used in litigation, one would be at a disadvantage under court proceedings explaining why he/she did not use an EMLAP-certified laboratory.

One should be aware of the fact that air-sample results can vary widely during a 24-hour period for a given space based on changes in indoor humidity associated with changes in outdoor conditions (i.e., rainy or sunny days, windy or snow-covered days). Connell[47] argues that: "the interpretation of airborne fungal results is one of the most misconstrued, possibly even the most abused and misunderstood aspect of fungal exposures in buildings and the outdoors" and states: (1) there is no correlation between indoor and outdoor concentrations for closed building conditions and (2) it is impossible to determine the indoor concentrations of airborne microorganisms based on one or two or even three short duration samples. An example of data produced illustrating this variation is shown in Table 13.6.[47]

TABLE 13.6
Variation in Airborne Mold Sampling Results over 24 Hours

Time of Sample	Spore Count
08:00	213
09:30	1,195
11:00	393
12:30	567
14:00	900
15:30	3,257

As can be seen, the variation in reported airborne spore counts in a typical Colorado home is over an order of magnitude for an approximate 8-hour period.

Further, aside from the actual variation and/or errors in sampling airborne mold levels, an added level of error occurs at the laboratory, even in EMLAP laboratories, in accurately and consistently analyzing the samples. Again, Connell[47] reported that one study from 2011 revealed that only 75% of the accredited laboratories could consistently identify *Cladosporium*, the most common mold in the environment. Furthermore, *Aspergillus/Penicillium*-like spores, the most common mold category related to water intrusion, were identified by only 50% of the accredited laboratories. The authors concluded, "This research reveals that precision of spore trap analyses, even among laboratories involved with analytical proficiency testing, lack precision and should be interpreted with caution."

Thus, laboratories have difficulty in identifying and quantifying even the most commonly found molds adding further to the inaccuracy of mold results.

As initially stated in this chapter, sampling for mold should be conducted only under limited circumstances, and even when sampled, one should understand that mold sampling remains an imprecise science. Nevertheless, it is the state of the science at this time and provides the best basis for identifying levels and types of mold present in an environment for situations where needed (e.g., immune-compromised individuals or other situations where health effects are of concern).

13.2.5 Interpreting Mold Results

Despite all the limitations associated with mold sampling and the results from mold sampling, one may be faced with the need to interpret, or understand an interpretation, of mold results. This too is an imprecise science, yet the client will want to know whether or not they have a mold issue once samples are taken, regardless of whether or not firm interpretation levels exist or not. The New York State Toxic Mold Task Force Report[33] provided a detailed table on guideline values for mold and bacteria levels provided in various studies, but then stated: "As the relations between dampness, microbial exposure and health effects cannot be quantified precisely, *no quantitative health-based guideline values or thresholds can be recommended for acceptable levels of contamination with microorganisms*. Instead, it is recommended that dampness and mould-related problems be prevented. When they occur, they should be remediated because they increase the risk of hazardous exposure to microbes and chemicals." Thus, with the understanding that no correlation of mold levels with health effects exists, one is faced with either telling a client that the results cannot be interpreted or with providing them some guidance based on available (albeit imprecise) literature in order to make some actionable decision. This section provides some of this guidance.

13.2.5.1   Interpreting Mold Results – General

Any interpretation of molds should be prefaced with a warning such as the following: "There are currently *no regulatory numeric standards* for airborne or surface microbes indoors. Interpretation of mold results is based on a comparison of indoor/outdoor concentration ratios, compliant vs non-compliant areas, and predominant fungal genera. In addition, the data should be interpreted with caution and used as a *screen* for performance evaluation and not for health criteria since different individuals react to various allergens in different concentrations. These guidelines are intended to be a 'reactionary threshold' to incite further investigation." Assuming that the limitation of interpretation of mold results is known, the following general approach for analyzing surface and air mold sampling results should be followed.[48–52]

Surface Samples Results:

1. Compare total mold levels with screening levels for total mold.
2. Compare ratio of results for surface with and without visible mold present.
3. Identify species of interest (i.e., those not typically found indoors, indicating high moisture levels in the building).

For MycoMeter™ results, compare the results with Category 1, 2, or 3 levels.

Air Sample Results:

1. Compare total mold counts with screening levels for total mold counts.
2. Compare indoor/outdoor concentration ratios for individual mold species.
3. Compare compliant vs noncompliant areas or affected vs nonaffected areas.
4. Identify species of interest (i.e., those not typically found indoors, indicating high moisture levels in the building).
5. Consider air exchange rates, activity levels, and moisture and humidity measurements in a building structure.
6. Consider current weather trends and season of the year.

One should determine if "indicator" molds are present indoors. Some of the predominant indicator molds include *Stachybotrys*, *Chaetomium*, *Aspergillus*, *Penicillium*, and *Trichoderma*. The isolation of fungi such as *Chaetomium*, *Mucor*, *Rhizopus*, *Stachybotrys*, *Trichoderma*, and yeast-like fungi, such as *Rhodotorula* species, often suggests excess moisture and/or increased relative humidity within the areas sampled. One should look for a possible indoor source of moisture if these species are found, but keep in mind that these species may also come from the outdoors.

Alternaria, *Aspergillus*, *Cladosporium*, and *Penicillium* species are widely distributed in nature and can be isolated in low concentrations from the indoor environment. These fungi are known to colonize the ventilation systems of homes, schools, and office buildings. However, these fungi can grow indoors on water-damaged drywall, ceiling tiles, wallpaper, and fiberglass insulation duct liners.

It has also been reported in several studies that elevated concentrations of certain molds (e.g., *Aspergillus versicolor*, *Penicillium*, and *Stachybotrys*) may produce metabolic products such as mycotoxins and volatile organic compounds (VOCs).

13.2.5.2   Interpreting Mold Results – Example of Typical Surface and Air Sampling Results

13.2.5.2.1   Example Mold Data

Example mold data from an actual mold-sampling event are summarized in Tables 13.7–13.10 for illustrating how to interpret mold-sampling results:

TABLE 13.7
Example Air (Agar) Sampling Mold Results

Sample ID	Sample Description	Total Fungal Count (CFU/m³ Air)	Major Species (%)
Agar-1	First floor: hallway	2,200	*Alternaria* spp. *Aspergillus versicolor* *Cladosporium* spp. *Penicillium* spp. – (~50%) *Scopulariopsis* spp.
Agar-2	First floor: kitchen	294	*Non-sporulating hyaline fungi* – 32% *Penicillium* spp. – 24% *Cladosporium* spp. – 20% *Paecilomyces* spp. – 8% *Ustilago* spp. – 4% *Acrodontium* spp. – 4% *Unidentified Coelomycete* – 4% *Non-sporulating dematiaceous fungus* – 4%
Agar-3	Basement: sitting room	200	*Cladosporium* spp. – 41% *Non-sporulating hyaline fungi* – 35% *Penicillium* spp. – 18% *Aspergillus versicolor* – 6%
Agar-4	Outdoors	576	*Cladosporium* spp. – 45% *Epicoccum* spp. – 10% *Penicillium* spp. – 8% *Alternaria* spp. – 8% *Aspergillus fumigatus* – 4% *Ustilago* spp. – 4%
Agar-5	Blank	0	No growth of fungi

Detection Limit for Agar Results: 12 CFU/M³ of air.

TABLE 13.8
Example Air (Spore Trap) Sampling Mold Results

Sample ID	Sample Description	Total Fungal Count (CFU/m³ Air)	Major Species (%)
AOC-1	First floor hallway	45,851	*Alternaria* – 0% *Ascospores* – 0% *Aspergillus/Penicillium*-like – 2% *Basidiospores* – 36% *Cladosporium* – 2% *Smuts/Myxomycetes* – 60%
AOC-2	First floor: kitchen	1,222	*Ascospores* – 7% *Aspergillus/Penicillium*-like – 7% *Basidiospores* – 64% *Cladosporium* – 11% *Rusts* – 1% *Smuts/Myxomycetes* – 9% *Unidentified dematiaceous conidia* – 1%

(Continued)

TABLE 13.8 (*Continued*)

Example Air (Spore Trap) Sampling Mold Results

Sample ID	Sample Description	Total Fungal Count (CFU/m³ Air)	Major Species (%)
AOC-3	Basement: sitting room	1,079	*Ascospores* – 7%
			Aspergillus/Penicillium-like – 16%
			Basidiospores – 57%
			Cladosporium – 11%
			Epicoccum – 1%
			Pithomyces/Ulocladium – 1%
			Rusts – 7%
AOC-4	Outdoors	7,345	*Alternaria* – 1%
			Ascospores – 8%
			Aspergillus/Penicillium-like – 4%
			Basidiospores – 76%
			Cladosporium – 8%
			Epicoccum – 1%
			Nigrospora – 0%
			Peronospora – 1%
			Pithomyces/Ulocladium – 1%
			Rusts – 0%
			Smuts/Myxomycetes – 1%
AOC-6	Blank	0	No growth of fungi

Detection Limit for AOC Results: 13 spores/m³ of air.

Note: Percentages may not equal 100% due to rounding.

TABLE 13.9

Example Swab Surface Sampling Mold Results

Sample ID	Sample Description	Total Mold Count (CFU/square inch surface)	Major Species
Swab-1	Living room – ceiling streak	100	*Mucor* spp. – 100%
Swab-2	Living room – clean area	<100	No growth
Swab-3	First floor hallway ceiling	50,000	*Ulocladium* – 60% *Penicillium* – 40%
Swab-4	First floor hallway ceiling – clean area	<100	No growth
Swab-5	Blank	<10	No growth

Detection limit for Swab sample results: 25 CFU/square inch surface.

Note: Percentages may not equal 100% due to rounding.

TABLE 13.10

Example MycoMeter™ Surface Sampling Mold Results

Sample #	Location	Mycometer™ Value
M-1	Basement – bottom of main furnace supply duct surface	100
M-2	Basement – side of main furnace return duct surface	188
M-3	First floor – north kitchen register duct surface	<25
M-4	Crawlspace – stained ceiling OSB near master bedroom bathroom toilet sanitary line	245

13.2.5.2.2   Mold Screening Levels

Mold screening levels, by sample type and recommended precautionary language, are summarized below.[47–51]

Air (Agar) Screening Levels: Screening levels for air (agar) sampling mold results are summarized in Table 13.11; precautionary and interpretation language follows the table.

TABLE 13.11
Air (Agar) Sampling Mold Results Screening Levels

No Growth	Low (Normal Growth)	Borderline/Moderate	Active Growth	Very Active Growth
	<1,000 CFU/m^3 of air[a]		>1,000 CFU/m^3 of air[b]	
	<250 CFU/m^3	250–1,000 CFU/m^3	>1,000 CFU/m^3	>5,000 CFU/m^3

[a] Low contamination.
[b] High contamination; depending on outdoor levels.

Precautionary and Interpretation language:

For the ease of identifying areas of concern, sample results falling above either the "active" or "very active" level (i.e., 1,000 CFU/m^3) are used as total airborne (agar) screening level. Bioaerosol levels above 2,000 CFU/m^3 may suggest an indoor air quality issue but not necessarily a health issue. Recall, it should be noted that no regulatory standards on acceptable levels have yet to be established. These levels should not be used for "safe" or "unsafe" level determinations but should be used in context with the other parameters discussed earlier.

It has also been reported that elevated concentrations of certain fungi (e.g., *Aspergillus versicolor*, *Penicillium*, and *Stachybotrys*) may produce metabolic products such as mycotoxins and VOCs.

Sahay et al.[4] recommend a total airborne culturable guideline for indoor air mold of <350 CFU/m^3 indoors assuming that no pathogenic species are present.

Air (Spore Trap) Screening Levels: Screening levels for air (spore trap) sampling mold results are summarized in Table 13.12; precautionary and interpretation language follow the table.

TABLE 13.12
Air (Spore Trap) Sampling Mold Results Screening Levels

No Growth	Low (Normal Growth)	Borderline/Moderate	Active Growth	Very Active Growth
	<2,000 spores/m^3 of air[a]		>2,000 spores/m^3 of air[b]	
<100 CFU/g	<25,000 CFU/g	25,000–200,000 CFU/g	200,000–1,000,000 CFU/g	>1,000,000 CFU/g

[a] Low contamination.
[b] High contamination; depending on outdoor levels.

For the ease of identifying areas of concern, sample results falling above a level of 2,000 spores/m^3 are used as a total airborne (spore trap) screening level. Recall, it should be noted that no regulatory standards on acceptable levels have yet to be established. These levels should not be used for "safe" or "unsafe" level determinations but should be used in context with the other parameters discussed earlier.

Surface (Swab) Screening Levels: Screening levels for surface (swab) sampling mold results are summarized in Table 13.13; precautionary and interpretation language follow the table.

TABLE 13.13
Swab Surface Sampling Mold Results Screening Levels

No Growth	Low (Normal Growth)	Borderline/ Moderate	Active Growth	Very Active Growth
<1,000 CFU/square inch of surface[a]	1,000–10,000 CFU/square inch of surface[b]	>10,000 CFU/square inch of surface[c]		
<100 CFU/square inch	<10,000 CFU/square inch	10,000–100,000 CFU/square inch	100,000–1,000,000 CFU/square inch	>1,000,000 CFU/square inch

[a] Low concentration.
[b] Medium concentration.
[c] High concentration.

For the ease of identifying areas of concern, sample results falling below a level of 10,000 CFU/square inch are used as a screening level; results below this level are considered to be at normal background levels. Recall, it should be noted that no regulatory standards on acceptable levels have yet to be established. These levels should not be used for "safe" or "unsafe" level determinations but should be used in context with the other parameters discussed earlier.

In addition, it is common practice to compare surfaces containing visible mold (noncompliant) with adjacent areas with no visible mold (compliant surface). If the ratio of the total noncompliant mold level to the compliant mold level exceeds 10:1, the noncompliant area is generally viewed as containing elevated mold levels.

Surface (MycoMeter™ Swab) Screening Levels: Screening levels for surface (MycoMeter™) sampling mold results are summarized in Table 13.14; precautionary and interpretation language follow the table.

TABLE 13.14
MycoMeter™ Surface Sampling Mold Results Screening Levels

Category	Value	Comments
A	MV < 25	The level of mold is not above normal background level
B	25 < MV < 450	The level of mold is above normal background level. This is typically due to high concentrations of spores in dust deposits but may, in some cases, indicate the presence of old mold damage (mold growth)
C	MV > 450	The level of mold is high above normal background level due to growth of molds

For the ease of identifying areas of concern, sample results falling at/below a MycoMeter™ value (MV) of 25 suggest that the level of surface mold is not above normal background levels. Recall, it should be noted that no regulatory standards on acceptable levels have yet to be established. These levels should not be used for "safe" or "unsafe" level determinations but should be used in context with the other parameters discussed earlier.

13.2.5.2.3 Interpretation of Example Mold Sample Results

With the exception of MycoMeter™ example mold sampling results, example mold levels observed above the three screening criteria (1,000 CFU/m^3 of air for viable mold (agar), 2,000 spores/m^3 of air for viable and nonviable mold (spore trap), and 10,000 CFU/square inch for surface swab samples) are summarized in Table 13.15.

TABLE 13.15

Example Mold Results – Sample Results above Screening Levels

Sample Type	Sample #	Location	Results (CFU/m³ – air) and (CFU/square inch – swabs)	Dominant Genera
Air - Agar	Agar-1	First Floor Hallway	2,200	*Alternaria* spp.[a]
				Aspergillus versicolor[a]
				Cladosporium spp.[a]
				Penicillium spp. – (~50%)
				Scopulariopsis spp.[a]
Air – Spore Trap	AOC-1	First Floor Hallway	45,851	*Alternaria* – 0%
				Ascospores – 0%
				Aspergillus/Penicillium-like – 2%
				Basidiospores – 36%
				Cladosporium – 2%
				Smuts/Myxomycetes – 60%
Swab	Swab-3	First Floor Hallway Ceiling	50,000	*Ulocladium* – 60%
				Penicillium – 40%

[a] Unable to calculate percentages due to overgrowth of competing fungal genera.

Note: Percentages may not equal 100% due to rounding.

In this example, airborne samples in the hallway and a swab sample on the hallway ceiling were above screening levels. In addition, the total airborne mold results were found to be much higher than the viable mold results taken at the same location, suggesting that much of the mold was older, nonviable mold.

In addition, one should compare absolute indoor mold levels to outdoor levels and identify mold genera consistent with elevated indoor moisture levels. For the example data, this analysis is summarized in Table 13.16.

TABLE 13.16

Example Mold Results – Indoor/Outdoor Ratios and Genera of Interest

Sample Type	Sample #	Location	Ratio: Indoor to Outdoor Concentrations	Dominant Genera
Agar	Agar-1	First-floor hallway	N/A	*Alternaria* spp.
			N/A	*Aspergillus versicolor* – 2%
			Not found outdoors	*Cladosporium* spp.
			21.7 [(0.5 * 2,000)/ (0.08 * 576)]	*Penicillium* spp. – (~50%)
			Not found outdoors	*Scopulariopsis* spp. – 3%
Agar	Agar-2	First floor: kitchen	Not found outdoors	*Acrodontium* spp. – 4%
			0.88	*Cladosporium* spp. – 20%
			Not found outdoors	*Paecilomyces* spp. – 8%
			1.53	*Penicillium* spp. – 24%
			Not found outdoors	*Unidentified Coelomycete* – 4%
			Not found outdoors	*Ustilago* spp. – 4%

(Continued)

TABLE 13.16 (*Continued*)
Example Mold Results – Indoor/Outdoor Ratios and Genera of Interest

Sample Type	Sample #	Location	Ratio: Indoor to Outdoor Concentrations	Dominant Genera
Agar	Agar-3	Basement: sitting room	Not found outdoors	*Aspergillus versicolor* – 6%
			0.32	*Cladosporium* spp. – 41%
			0.78	*Penicillium* spp. – 18%
			Not found outdoors	Non-sporulating hyaline fungi – 35%
Spore Trap	AOC-1	First-floor hallway	N/A	*Alternaria* – 0%
			N/A	*Ascospores* – 0%
			3.12	*Aspergillus/Penicillium*-like – 2%
			2.96	*Basidiospores* – 36%
			1.56	*Cladosporium* – 2%
			374.6	*Smuts/Myxomycetes* – 60%
Spore Trap	AOC-2	First floor: kitchen	0.15	*Ascospores* – 7%
			0.29	*Aspergillus/Penicillium*-like – 7%
			0.14	*Basidiospores* – 64%
			0.23	*Cladosporium* – 11%
			Not found outdoors	*Rusts* – 1%
			1.49	*Smuts/Myxomycetes* – 9%
Spore Trap	AOC-3	Basement: sitting room	0.13	*Ascospores* – 7%
			0.59	*Aspergillus/Penicillium-ike* – 16%
			0.11	*Basidiospores* – 57%
			0.20	*Cladosporium* – 11%
			0.15	*Epicoccum* – 1%
			0.15	*Pithomyces/Ulocladium* – 1%
			Not found outdoors	*Rusts* – 7%
Swab	Swab-1	Living room – ceiling streak	N/A	*Mucor* spp. – 100%
Swab	Swab-3	First-floor hallway ceiling	N/A	*Ulocladium* – 60%
				Penicillium – 40%

Whenever the ratio of indoor mold levels for a given mold exceeds outdoor levels, the ratio will be greater than 1. Ratios above 1 are of interest except when the absolute mold levels are low or for species such as basidiospores, conidia, and rusts. Genera of interest are underlined in Table 13.16; these genera, if elevated, typically are associated with elevated indoor moisture levels.

For MycoMeter™ surface sample results, any values above 25 imply that the surface has mold levels above normal background levels. For example, results in Table 13.10 imply that samples M-1, M-2, and M-4 have elevated surface mold levels while sample M-3 is likely at/below normal background levels. If elevated MycoMeter™ levels are encountered as part of mold remediation activities, this would imply that additional remedial activities are needed on these sampled surfaces.

13.2.5.3 Interpreting Mold Sampling Results – Case Study

The following is a case study adapted from an actual mold assessment performed by CIEC. A mold investigator was contacted by a homeowner regarding a potential indoor air quality issue in her home. As related to the investigator, the homeowner had experienced upper respiratory symptoms (including headaches, throat and eye irritation, and general malaise) over the past month especially when spending time in the basement and a first-floor computer room. The homeowner wanted the investigator to perform a visual assessment of the home and determine whether air and/or surface samples for mold should be collected for analysis. The home was a one-story ranch with a basement built in 2000 with a forced gas HVAC system. No musty or moldy odors were noted upon entry into the home. The homeowner relayed to the investigator that the dehumidifier in the basement had to be emptied daily. A crawlspace was located in the southeast corner of the basement, and the investigator noted efflorescence on the concrete block wall in the crawlspace and on the adjacent basement wall. Efflorescence is a crystalline or powdery deposit of salts resulting from the movement of water through a structure. A moisture meter was used to probe the affected basement wall; several areas tested >20% wood moisture equivalence (%WME). Total retained moisture content is measured in units of %WME. Generally, surfaces testing below 17% WME are considered "dry." Materials measuring 17%–19% WME are considered at "risk" for developing mold. Measurements >20% WME are considered "wet." Four spore trap air samples were collected, one in the southeast corner of the basement, one near the crawlspace, one in the computer room directly above the crawlspace on the first floor, and one outdoors near the front entrance. A blank was also included for quality assurance purposes to ensure that no cross-contamination occurred during the handling or analytical processes.

The spore trap air samples revealed high levels of *Aspergillus/Penicillium*-like conidia (spores) in the basement area of the home (Table 13.17). The level of mold spores in the basement represented a Condition 3 or "abnormal fungal ecology" as stated in the *Institute of Inspection, Cleaning and Restoration Certification's ANSI/IICRC S-520-2015 Standard for Professional Mold Remediation*, 3rd Edition.[9] Condition 3 is defined as "an indoor environment with the presence of actual growth, associated spores, and fungal fragments. Actual growth includes growth that is active or dormant, visible or hidden." The predominance of *Aspergillus/Penicillium*-like conidia (spores) noted in the Computer Room (Table 13.7) (located above the crawl space) may be explained by the migration of the mold spores via the forced gas HVAC system and/or by the stairway from the basement to the first floor of the home. Based upon the spore trap results and visual assessment, a remediation plan was formulated.

TABLE 13.17

Spore Trap Results – Case Study

Area/Room	Sample #	Volume of air (L)	Allergenco-D Spore Trap (spores/m³ of air)	Rank Order Assessment Predominant Mold Genera
Computer room – first floor	A1	75	1,638	90% *Aspergillus/Penicillium*-like conidia 6% *Cladosporium* spp. 2% Basidiospores 2% *Ascospores*
Basement – SE corner	A2	75	36,084	100% *Aspergillus/Penicillium*-like conidia <1% Basidiospores <1% *Cladosporium* spp.
Outdoor – front entrance	A3	30	330	70% Basidiospores 30% Ascospores
Blank control	A4	-	0	No visible trace, no particulates noted

13.3 BACTERIA

13.3.1 Introduction to Bacteria

AIHA[36] and others[6] provide detailed information on the process of biological monitoring in the field. Sampling for tuberculous and nontuberculous mycobacteria (NTM) is a specialty activity and will not be covered here. The most common situations likely to be encountered during a forensic investigation are sewage spills and potable water systems contaminated with *Legionella*.

Fecal coliforms (including *E. coli*) and fecal *Enterococcus* spp. are bacteria that may be present with sewage spills. Coliforms are common environmental bacteria that may be found in soil, on hands, on equipment surfaces, in water, and in other environments. These organisms can be detected from water, swab, and air samples by a variety of methods utilizing substrate fermentation and enzymatic activity.

Specific information on *Legionella* follows in Section 13.3.4.

13.3.2 Sampling for Bacteria

With the exception of bulk water samples, the process for sampling bacteria on surfaces and in the air is similar to that for mold, using swabs for surfaces and bioaerosol pumps/samplers (e.g., SAS or Anderson impactors) with agar plates for air sampling. Typical air volumes collected are ~100–500 liters, depending on the sampling device. As with mold, many types of agars are available for sampling bacteria, including trypticase soy agar (TSA), MacConkey agar (MAC), R2A agar, and sheep blood agar (SBA).[2] Specific agars are typically selected by the industrial hygienist in collaboration with the laboratory for sampling selected bacteria of interest. If water samples are collected from a chlorinated system, they must be treated with sodium thiosulfate to neutralize the chlorine. Swab samples for bacterial culture must use a transport media to preserve the sampled bacteria. In all cases of bacteria, including *Legionella*, sample preservation methods recommended by the laboratory should be followed.

13.3.3 Interpreting Bacterial Sampling Results

The following are bioaerosol and surface sample results (Tables 13.18–13.20) from a sampling event in an office building, which illustrate typical organisms that may be present and the difficulty in assigning potential significance to their presence.

TABLE 13.18
Bioaerosol (Air) Sample Results – Office Building

Sample Description	Total Bacterial Count (CFU/m³ air)	Organisms (Rank Order)
Lunchroom	325	*Micrococcus* spp.
		Coagulase-negative *Staphylococcus* spp.
		Bacillus spp.
Office 3	247	*Micrococcus* spp.
		Bacillus spp.
		Coagulase-negative *Staphylococcus* spp.
Waiting room	510	*Micrococcus* spp.
		Coagulase-negative *Staphylococcus* spp.
		Bacillus spp.
Outdoor front entrance	64	*Micrococcus* spp.
		Coryneform bacillus

TABLE 13.19
Surface Swab Sample Results – Office Building

Sample Description	Total Bacterial Coliforms (CFU/square inch surface)	Organisms (Rank Order)
Lunchroom - A/C vent	360,000	*Bacillus* spp.
		Micrococcus spp.
		Coagulase-negative *Staphylococcus* spp.
Office 3 - A/C vent	280,000	Gram-negative rods
Waiting room - A/C vent	>1,000,000	Coagulase-negative *Staphylococcus* spp.
		Bacillus spp.
Waiting room – under carpet	>1,000,000	Coagulase-negative *Staphylococcus* spp.
		Coryneform bacillus

TABLE 13.20
Example Bacteria Results – Sample Results Containing Fecal Coliforms/Fecal *Enterococcus* spp.

Sample Type	Sample #	Location	Results (CFU/m³ – air) and (CFU/square inch – swabs)	Dominant Genera
AGAR	BACT-2*	BM: Main Room	871	Fecal *Enterococcus* spp.
				Bacillus spp. 1[a]
				Coagulase-negative *Staphylococcus* spp.
				Bacillus spp. 2[a]
				Bacillus spp. 3[a]
SWAB	SWAB-2	BM: SW Room: SE Floor	11,000	Fecal coliforms not *E. coli* isolated – 11,000
				Fecal *Enterococcus* spp. isolated – 76,000
SWAB	SWAB-3	BM: Main Room: Floor	0	No fecal coliforms isolated
				No *E. coli* isolated
				Fecal *Enterococcus* spp. isolated – 140
SWAB	SWAB-4	BM: Main Room: Return Air Duct	0	No fecal coliforms isolated
				No *E. coli* isolated
				Fecal Enterococcus spp. isolated – 240
SWAB	SWAB-5	1st Floor: Kitchen: Pantry Closet Floor	114,000	Fecal coliforms not *E. coli* isolated – 101,000
				E. coli isolated – 13,000
				Fecal *Enterococcus* spp. isolated – 1,200,000

[a] Unable to calculate percentages due to overgrowth of competing fungal genera.

The bioaerosol samples from selected areas of the office building revealed no elevated levels of bacteria. Air concentrations of bacteria depend upon the following factors:

- Number of persons present in the occupied space.
- Activity levels of persons in the occupied space.
- Types of clothing persons are wearing.
- Ventilation rate to an occupied space.
- Presence of animals, especially in a residential setting.
- Presence of water after a water intrusion event where standing water becomes an issue.

According to Abeysekera,[2] in the absence of standards, a mixture of Gram-positive bacteria (cocci and rods) is considered typical for normal office and residential environments. Gram-positive bacteria such as *Bacillus* spp., *Micrococcus* spp., *Staphylococcus* spp. (coagulase-negative), and coryneform bacilli (including *Corynebacterium* spp.) typically dominate indoor air in occupied buildings and are abundant in dust and on surfaces. *Bacillus* is typically of environmental origin (e.g., dust, soil), whereas the others are found commonly on the skin of humans and animals.

Gram-negative bacteria were noted in a surface sample of the A/C vent in Office 3. Gram-negative rods are ubiquitous in the environment and can be free-living or originate from human and animal sources. They thrive in damp conditions such as high relative humidity, standing water, or a heavy water or sewage intrusion event. Detection of Gram-negative rod bacteria in a surface swab sample of the A/C vent in Office 3 may signify moisture within the A/C ducts. Elevated counts of bacteria (>1 million CFU/swab) were detected in two swab samples collected from the waiting room area. These types of organisms are expected in indoor environments, and large numbers could be due to the factors listed above in addition to inadequate cleaning. The elevated levels should prompt routine housekeeping of the space.

In the sampling event above, it is difficult to attribute any significance to the majority of bacterial types found in the surface samples. However, in other scenarios, there are certain organisms that carry potential significance. Bacteria indicative of possible sewage contamination include *E. coli*, coliforms, *Enterococcus* spp., and *Bacteroides* spp. Specific standards do not exist for levels of *E. coli* and fecal *Enterococcus* spp. in the air or on surfaces. Where standards do exist (i.e., drinking water), the acceptable contamination level for *E. coli* is zero. Most coliforms are not harmful (pathogenic), but if a test detects their presence, it is considered to be an indication of unsanitary conditions.

13.3.4 *Legionella* Bacteria

Legionellosis (Legionnaires' disease and Pontiac fever) is caused by Gram-negative, rod-shaped bacteria of the genus *Legionella*. *Legionella* was recognized as respiratory pathogen at the initial outbreak during the American Legion convention in 1976 at a Philadelphia, PA, hotel. Pontiac fever is a self-limiting, influenza-like syndrome, whereas Legionnaires' disease is a more severe and potentially fatal illness, with pneumonia as the predominant clinical finding.

The Centers for Disease Control and Prevention (CDC) reported 10,000 cases of Legionnaires' disease in the United States in 2018; but many more cases remain undiagnosed. The rate of infection from 2000 to 2017 in the United States increased by a factor of 5.5 (~0.4 to 2.4 cases/100,000 population).[53,54] Currently, there are ~58 *Legionella* species and over 70 serogroups. Of these species, 25 are known to cause human disease. Most infections are caused by *Legionella* pneumophila serogroup 1. Lindahl et al.[55] note that "The bacteria are not transmitted person-to-person, or from normal ingestion of water. Transmission of *Legionella* to humans is by inhalation and aspiration of contaminated water."

Outbreaks of Legionnaires' disease have been associated, but not limited to, a variety of aerosol-producing devices such as cooling towers, whirlpool spas, decorative fountains, mist machines, air scrubbers, as well as faucets and showerheads. Risks factors for Legionellosis include heavy cigarette smoking, chronic lung disease, diabetes, and immunosuppression (especially caused by corticosteroid therapy and organ transplantation).

Recent evidence has shown that as many as 22% of *Legionella* cases in outbreaks (particularly community outbreaks) have no apparent risk factors for the disease.

13.3.4.1 Sampling, Culturing, and Identification of *Legionella* Bacteria

Swab samples of aerosol-producing devices (e.g., faucets and showerheads) and water samples (250 ml or more) are generally collected for *Legionella* culture. Other samples such as ice from machines and soil from potted plants have been collected for *Legionella*. The gold standard for detection and quantification of *Legionella* bacteria is culture. Potable water is concentrated by filtration before plating the sample to agar. Non-potable water rarely requires concentration and can be plated directly to agar. Swab and other bulk samples are plated to a special medium called buffered charcoal yeast extract (BCYE) agar, both with and without selective agents. Samples can be heat treated and/or treated with a low pH buffer and then plated to agar to remove high numbers of contaminating microorganisms.

Environmental cultures for *Legionella* are normally incubated at 35°C for seven days before reporting as negative for *Legionella*.

Positive cultures for *Legionella* may be apparent as early as 3–5 days of incubation. Suspect colonies are plated to BCYE agar and an SBA plate, incubated for 1–2 days, and then examined for growth. If growth does not occur on SBA, then the suspect colonies on BCYE agar are definitively identified by direct fluorescent antibody (DFA), latex agglutination, and/or MALDI-TOF mass spectrometry methods.

There are no federal action levels for *Legionella*; however, some states, agencies, and organizations have published interpretive guidelines for *Legionella* quantitative culture results.[56-58] The tables below were modified from *Legionella* 2019: "A Position Statement and Guidance Document" published by the Association of Water Technologies (Tables 13.21 and 13.22).

TABLE 13.21
Legionella Control Programs: Potable Water[55,57]

Regulatory Authority	Test Frequency	Legionella Conc. (CFU/mℓ)	Remediation
NY city	No data		No data
NY state[a]	Every 90 days first year, annually thereafter	<30% positive results of sites tested	Maintain Water Management Plan (WMP)
		≥30% positive results of sites tested	Immediate short-term control levels. Retest. Persistent ≥30% results – institute long-term control levels.
AIHA	2×/year	1–9	Continue monitoring and review WMP
		10–100	Identify infection source and online disinfection
		>100	Identify infection source and offline disinfection
OSHA	Not stated	10–99	Online disinfection
		>100	Online disinfection
Australia and New Zealand	Not data	No data	No data
France – MSS	No data	>0.025	Action required in healthcare facilities
		>0.1	Action required in healthcare facilities with severely immunocompromised patients
		>1.0	Action required in non-healthcare facilities
Germany – DVGW	No data	1.0	No data
Holland – VROM	No data	0.1	No data
United Kingdom – HSE	Start monthly and adjust as per test results	>0.1, <1.0, <50% positively	Review WMP
		>0.1, <1.0, >50% positively	Review WMP and consider disinfection
		≥1.0	Review WMP and consider disinfection

[a] Regulation addresses covered facilities (general hospitals and residential healthcare facilities).

TABLE 13.22

Legionella Control Programs: Cooling Towers and Evaporative Condensers[55–57]

Regulatory Authority	Test Frequency	Legionella Conc. (CFU/mℓ)	Remediation
NY city[a]	At system startup and every 90 days thereafter	<10	Maintain water chemistry and biocide levels
		≥10 to <100 ≥100 to <1000	Online disinfection within 24 hours. Retest
		≥1,000	Online disinfection within 24 hours. Offline disinfection within 48 hours. Retest. Notify DOH within 24 hours of results.
NY state[b]	At system startup and every 90 days thereafter	<20	Maintain water chemistry and biocide levels
		≥20 to <1,000	Online disinfection immediately. Retest. Review WMP
		≥1,000	Online disinfection immediately. Retest. Review WMP. Any retest >1000 offline disinfection immediately
AIHA[c]	Monthly	10–99	Review WMP and retest until <10 CFU/mℓ
		100–1,000	Review WMP and conduct an online disinfection until consistently <10 CFU/mℓ
		>1,000	Review WMP and conduct an offline disinfection until consistently <10 CFU/mℓ
OSHA[d]	Not stated	100–999	Online disinfection
		>1,000	Online disinfection
Australia and New Zealand – AS/NZS 366.3[e]	Quarterly	10–99	On-line disinfection and add biodispersant. Retest 3–7 days
		100–1,000	On-line disinfection with oxidizing biocide. Retest 3–7 days
		>1,000	Off-line disinfection with oxidizing biocide. Retest 3–7 days
Canada – PW and GSC[f]	Every 2 months	1,000–10,000 non-Lp	Online disinfection within 48 hours
		1–100 Lp sg 2–15	Online disinfection within 48 hours
		1–10 Lp sg 1	Online disinfection within 48 hours
		>1,000 non-Lp	Online disinfection immediately
		>100 Lp sg 2–15	Online disinfection immediately
		>10 Lp sg 1	Online disinfection immediately
France	No data	No data	No data
Germany – DVGW	No data	No data	No data

(Continued)

TABLE 13.22 (*Continued*)

Legionella Control Programs: Cooling Towers and Evaporative Condensers[55–57]

Regulatory Authority	Test Frequency	Legionella Conc. (CFU/mℓ)	Remediation
Holland – VROM	No data	No data	No data
United Kingdom – HSE[g]	Quarterly	0.1–1.0	Review WMP and resample.
		≥1.0	Review WMP, resample, and disinfect

[a] NY city: Chapter 8 (Cooling Towers) of Title 24 of the Rules of the City of New York (2015).

[b] NY state: Protection Against Legionella (2016).

[c] AIHA: American Industrial Hygiene Association. Recognition and Control of Legionella in Building Water Systems (2015).

[d] OSHA: Occupational Safety and Health Administration. Technical Manual, Legionnaires' disease, Section III, Chapter 7.

[e] AS/NZS 366.3 Australia and New Zealand: AS/NZS 366.3 Code of Practice, Prevention and Control of Legionnaires' Disease (2010).

[f] PW and GSC Canada: Public Works and Government Services Canada. Control of Legionella in Mechanical Systems, Mechanical Design 15161 (2013).

[g] HSE UK: Health and Safety Executive, UK. Legionnaires' disease. The control of Legionella bacteria in water systems, Approved Code of Practice and Guidance (2009).

It must be noted that ultimately it is the responsibility of a qualified professional to select actions that are appropriate for the situation being evaluated.

13.3.4.2 *Legionella* and ASHRAE

ASHRAE, the American Society of Heating, Refrigeration and Air Conditioning Engineers, has become increasingly proactive in the area of *Legionella* over the past 20 years. Beginning with papers and a guidance document on the topic (ASHRAE Guideline 12-2000 – Minimizing the Risk of Legionellosis Associated with Building Water Systems), ASHRAE has developed a Standard (Standard 188) for *Legionella* that was effective from July of 2012. Since then, the ASHRAE *Legionella* Guidance document was updated in 2020 and the Standard in 2018. As stated by ASHRAE, the scope of the standard is to present practices for the prevention of Legionellosis with building water systems. It is a risk management Standard that establishes minimum requirements for prevention of Legionellosis associated with building water systems. Except for single-family residential buildings, it covers the following human occupied buildings:

1. Commercial.
2. Multiunit Residential.
3. Industrial.

Within these buildings, the types of water systems covered are as follows:

1. Potable water systems.
2. Open and closed-circuit cooling towers or evaporative condensers that provide cooling, refrigeration, or both.
3. Whirlpools or spas in the building or on site.
4. Ornamental fountains, misters, atomizers, air washes, humidifiers, or other non-potable water systems either in the building or on site.

The reader is referenced to the ASHRAE Guidance D Document, Standard 188-2018 and the Lindahl et al.[54] paper for more information from ASHRAE on this topic.

IMPORTANT POINTS TO REMEMBER

- Mold and bacteria are present nearly everywhere in the environment, including the indoor environment.
- Amplification of mold (fungi) occurs indoors with the presence of elevated moisture or water levels. The term amplification of mold is used because when conditions exist for mold growth, it grows geometrically rather than arithmetically.
- The causal relationship between mold exposures and health effects has yet to be proven although an association between moisture and mold with some health effects has been documented. Issues limiting a determination of this causal effect include the large number of mold species and their metabolic products along with cofounder causes that often exist when mold is present.
- No consensus standards exist on threshold levels for mold exposure; however, methods can be used to interpret mold results based on literature available today. Inevitably, one will be asked to interpret mold-sampling results and asked whether the environment is acceptable or not.
- While less discussed in the public discourse, the hazards of bacteria (for example, from a sewage spill) often exceed those from mold. *Legionella* continues to be an increasing public health concern. Forensic investigators should protect themselves with appropriate personal protective equipment (PPE) when exposed to such environments.

REFERENCES

1. Macher, Janet, Ed. *Bioaerosols: Assessment and Control.* Washington, DC: American Conference of Governmental Industrial Hygienists, 1999.
2. Abeysekera, Tharanga "Bacteria: A general overview on sampling." *The Environmental Reporter.* Vol. 5, Issue 1, 2007. http://www.emlab.com/s/sampling/env-report-01-2007.html.
3. Niemeier, R. Todd, Satheesh K. Sivasubramani, Tiina Reponen, and Sergey A. Grinshpum. "Assessment of fungal contamination in moldy homes: Comparison of different methods." *Journal of Occupational and Environmental Hygiene*, 2006, Vol. 3, pp. 262–273.
4. Sahay, Ph.D., Rajiv, R., and Alan L. Wozniak. "Air quality guidelines established for microbiological assessment of residential and commercial buildings." Mold, July 5 2005.
5. The United States Environmental Protection Agency. "Indoor Air Pollution: An Introduction for Health Professionals." last updated November 4, 2010. http://www.epa.gov/iaq/pubs/hpguide.html.
6. Department of Health and Human Services, Centers for Disease Control and Prevention, National Institute for Occupational Safety and Health (NIOSH). "Guidance for Filtration and Air-Clearing Systems to Protect Building Environments from Airborne, Chemical, Biological, or Radiological Attacks." DHHS Publication No. 2003-136, April 2003.
7. ASTM. "D7338-14: Standard Guide for Assessment of Fungal Growth in Buildings." 2014.
8. ASTM. "E2418-06: Standard Guide for Readily Observable Mold and Conditions Conductive to Mold in Commercial Buildings: Baseline Survey Process." 2006.
9. Institute for Inspection, Cleaning and Restoration Certification (IICRC). "ANSI/IICRC Standard S520: Standard for Professional Mold Remediation", 3rd Edition, 2015.
10. Institute for Inspection, Cleaning and Restoration Certification (IICRC). IICRC Standard S520: Reference Guide for Professional Mold Remediation," 3rd Edition, 2015.
11. Institute for Inspection, Cleaning and Restoration Certification (IICRC). "ANSI/IICRC Standard S100: Standard and Reference Guide for Professional Carpet Cleaning," 6th Edition, 2015.

12. Institute for Inspection, Cleaning and Restoration Certification (IICRC). "IRCRC Standard S300: Standard and Reference Guide for Professional Upholstery Cleaning," 2000.

13. Institute for Inspection, Cleaning and Restoration Certification (IICRC). "ANSI/IICRC Standard S500: Standard and Reference Guide for Professional Water Damage Restoration," 4th Edition, 2015.

14. U.S. Environmental Protection Agency. "Learn about Mold," updated January 16, 2020. Accessed May15, 2020, https://www.epa.gov/mold.

15. Occupational Safety & Health Administration (OSHA). "Building Assessment, Restoration, and Demolition – Mold Remediation." Accessed May 15, 2020, https://www.osha.gov/SLTC/etools/hurricane/mold.html.

16. National Institute of Environmental Health Sciences. "Mold Clean-up and Treatment: Health and Safety Essentials for Workers, Volunteers, and Homeowners." Accessed May 15, 2020, https://www.aft.org/sites/default/files/niehs_moldbooklet_2013.pdf, September 2013.

17. National Institute of Environmental Health Sciences."NIEHS Disaster Recovery – Mold Remediation Guidance." Accessed May 15, 2020, http://elcosh.org/record/document/3671/d001213.pdf, May 2013.

18. U.S. Army Public Health Center (APHC). "Technical Guide 278 – Industrial Hygiene Public Health – Mold Assessment Guide", October 2018 Accessed May 15, 2020, https://phc.amedd.army.mil/PHC%20Resource%20Library/TG278.pdf.

19. National Toxicology Program (NTP). "Mold," June 2013. Accessed May 16, 2020, https://www.niehs.nih.gov/health/materials/mold_508.pdf.

20. Cole, Eugene C. Pamela Dulaney, Keith Leese, Karin Foarde, Deborah Franke, and Michael Berry, "Bio-pollutant sampling and analysis of indoor surface dusts: Characterization of potential sources and sinks." *Methods for Characterizing Indoor Sources and Sinks*, In: B.A. Tichenor, ed., Characterizing Sources of Indoor Air Pollution and Related Sink Effects, ASTM STP 1287, American Society for Testing and Materials, 165. 1996, pp. 153–166.

21. EPA. "Building Air Quality: A Guide for Building Owners and Facility Managers. EPA/400/1-91/033 and DHHS (NIOSH) Publication No. 91-114, December 1991.

22. WETP. "Guidelines for the Protection and Training of Workers Engaged in Maintenance and Remediation Work Associated with Mold." Washington DC: The National Clearinghouse for Worker Safety and Health Training Operated by MDB, Inc., May 20, 2005. http://www.wetp.org.

23. Enterprise Community Partners, Inc. and the National Center for Healthy Housing. Creating a Healthy Home: A Field Guide for Clean-up of Flooded Homes, 2006. http://www.enterprisecommunity.org.

24. Jahn, Steven D., William H. Bullock, and Joselito S. Ignacio. *A Strategy for Assessing and Managing Occupational Exposures*, 4th Edition. Virginia: American Industrial Hygiene Association Press, 2015.

25. Ignacio, Joselito S. and William H. Bullock. *A Strategy for Assessing and Managing Occupational Exposures*, 3rd Edition. Virginia: American Industrial Hygiene Association (AIHA) Press, 2006.

26. Mulhausen, John R. and Joseph Damiano. *A Strategy for Assessing and Managing Occupational Exposures*, 2nd Edition. Virginia: American Industrial Hygiene Association (AIHA) Press, 1998.

27. Rafferty, Patrick J., Ed. *The Industrial Hygienist's Guide to Indoor Air Quality Investigations*. Virginia: American Industrial Hygiene Association (AIHA) Press, 1993.

28. Prezant, Bradley, Donald Weekes, and J. David Miller, Eds. *Recognition, Evaluation, and Control of Indoor Mold*. Virginia: American Industrial Hygiene Association (AIHA) Press, 2008.

29. *AIHA Guideline 1-Biological Monitoring: A Practical Field Manual*. Virginia: American Industrial Hygiene Association (AIHA) Press, 2004.

30. *AIHA Guideline 3-Assessment, Remediation, and Post-Remediation Verification of Mold in Buildings*. Virginia: American Industrial Hygiene Association (AIHA) Press, 2004.

31. The CDC Mold Work Group. "Mold Prevention Strategies and Possible Health Effects in the Aftermath of Hurricanes Katrina and Rita." National Center for Environmental Health National Center for Infectious Diseases. National Institute for Occupational Safety and Health, Centers for Disease Control and Prevention, October 2005.

32. New York City Department of Health & Mental Hygiene. "How to Prevent and Get Rid of Mold in the Home", November 2019. Accessed May 16, 2020, https://www1.nyc.gov/site/doh/health/health-topics/mold.page.

33. New York State Department of Public Health. "New York State Toxic Mold Task Force – Final Report to the Governor and Legislature." December 2010.

34. New York City Department of Health, Bureau of Environmental and Occupational Disease Epidemiology. "Guidelines on Assessment and Remediation of Fungi in Indoor Environments." 2008.

35. New York City Department of Health, Bureau of Environmental and Occupational Disease Epidemiology. "Guidelines on Assessment and Remediation of Fungi in Indoor Environments." April 2002.

36. AIHA. *"Report of the Microbial Growth Task Force."* Virginia: American Industrial Hygiene Association (AIHA) Press, May 2001.

37. EPA."Mold Remediation in Schools and Commercial Buildings." United States Environmental Protection Agency, Office of Air and Radiation, Indoor Environments Division, EPA 402-K-01-001, March 2001. www.epa.gov/iaq/molds/graphics/moldremediation.pdf.

38. Powell, Kevin "Mold & Moisture Intrusion Case Study Report." NAHB Research Center, Inc., January 2004.

39. EPA. "A Brief Guide to Mold, Moisture and Your Home." EPA 402-K-02–003, U.S. Environmental Protection Agency Office of Air and Radiation Indoor Environments. www.epa.gov/iaq.

40. Cox-Ganser, Jean M., Sandra K. White, Rebecca Jones, Ken Hilsbos, Eileen Storey, Paul L. Enright, Carol Y. Rao, and Kathleen Kreiss. "Respiratory morbidity in office workers in a water-damaged building." *Environmental Health Perspectives*, Vol. 113, Issue 4, 2005, pp. 485–490.

41. Damp Indoor Spaces and Health. "Committee on Damp Indoor Spaces and Health – Board on Health Promotion and Disease Prevention." Washington, DC: The National Academies Press, 2004. http://www.nap.edu/openbook/0309091934/html/R1.html.

42. Mendell, Mark J., Myrna Cozen, Quanhon Lei-Gomez, Howard S. Brightman, Christine A. Erdmann, John R. Girman and Susan E. Womble, "Contamination of U.S. office buildings as risk factors for respiratory and mucous membrane symptoms: Analysis of the EPA base data." *Journal of Occupational and Environmental Hygiene*, 2006, Vol. 3, pp. 225–233.

43. Dillon, H. Kenneth, Patricia A. Heinsohn, and J. David Miller, eds. *Field Guide for the Determination of Biological Contaminants in Environmental Samples.* Virginia: American Industrial Hygiene Association (AIHA) Press, 1996.

44. Order on Proposed Expert Testimony, Olga Salinas and Martin Villarreal vs Allstate Texas Lloyd's Company, United State District Court for the South District of Texas – McAllen Division, Civil Action No.: M-02-272, July 24, 2003.

45. WHO (World Health Organization). "Guidelines for Indoor Air Quality: Dampness and Mould." World Health Organization (WHO) Regional Office for Europe, 2009.

46. Petty, Stephen E. "Proving Damages Caused by Mold Infestation in Ohio – Make a Mold Claim and Litigate the Case." *Paper and Presentation to the National Business Institute,* Cleveland, OH, April 21, 2006.

47. Connell, Caoimhin P. "Is Testing for Moulds Necessary?" 2012. http://forensic-applications.com/moulds/sampling.html.

48. "U.S. Micro Solutions Report Guidelines Webpage." Accessed March 2011 and May 16, 2020, http://www.usmicro-solutions.com/reportguidelines.html.

49. Indoor Air Quality Association, Inc. (IAQA). "Recommended Guidelines for Indoor Environments." January 2000.

50. Rao, Carol, Harriet Burge, and John C. S. Chang, "Review of quantitative standards and guidelines for fungi in indoor air." *Journal of the Air and Waste Management Association*, 46, 1996, 899–908.

51. Tiffany, John A., and Howard A. Bader, *"Industrial Hygiene and Clearance Considerations for a Microbial Remediation Project."* Titusville, NJ: Tiffany-Bader Environmental, Inc., 2002.

52. Clark, G. "Assessment and sampling approaches for indoor microbiological assessments." *The Synergist*, November 2001.

53. Center for Disease Control and Prevention (CDC). "Legionella Fast Facts." Accessed May 16, 2020, https://www.cdc.gov/legionella/fastfacts.html.

54. Cline, Daryn and Sarah Ferrari. "Evidence shows need to address pathogenic bacteria in U.S. drinking water systems." *ASHRAE Journal*, pp. 62–66, April 2020.

55. Lindahl, Paul, Bill Pearson, and R. Lee Miller. "Risk management for Legionellosis." *ASHRAE Journal*, pp. 14–21, October 2015.

56. U.S. Micro Solutions, Inc. "Interpretation of Legionella Culture Results", June 4, 2020. Accessed October 19, 2020, https://www.usmslab.com/wp-content/uploads/2020/09/QLT-02-Form-5-Legionella-non-NY-Report-Guidelines-v1.pdf.

57. U.S. Micro Solutions, Inc. "Interpretation of Legionella Culture Results – New York", June 4, 2020. Accessed October 19, 2020, https://www.usmslab.com/wp-content/uploads/2020/09/QLT-02-Form-3-Legionella-NYS-Report-Guidelines-v1.pdf.

58. American Water Technologies (AWT). "Legionella 2019: A Position Statement and Guidance Document", February 2018. Accessed May 16, 2020, https://www.awt.org/pub/?id=035C2942-03BE-3BFF-08C3-4C686FB7395C.

14 Forensic Inspection Assessments of Residential Wood Framing Systems

Bryan E. Knepper and Stephen E. Petty
EES Group, Inc.

CONTENTS

PURPOSE/OBJECTIVES

The purpose of this chapter is to:

- Describe roof, wall, and floor framing systems typically utilized in residential construction.
- Discuss current residential building code requirements for roof and floor framing systems.
- Discuss cosmetic versus structural allowable wall tilt or "out-of-plumb."
- Describe commonly found damage to wood framing members/systems.
- Understand the differences between structural damage and natural defects.
- Document a methodology for assessing structural damage claims to residential framing systems.
- Discuss commercially available structural analysis software.

Following the completion of this chapter, you should be able to:

- Understand residential wood roof, wall, and floor framing systems and terminologies.
- Understand current residential building code requirements for roof and floor framing systems.
- Know and understand allowable wall tilts or "out-of-plumb" compliance values and basis for values.
- Be able to recognize and differentiate structural damage from common natural defects in wood framing members.
- Be able to perform a thorough visual inspection of damages to roof, wall, and floor framing systems of residential structures using a systematic methodology outlined in this chapter.
- Be aware of commercially available structural analysis software.

14.1 INTRODUCTION

Wood framing is the method predominantly utilized for building residential structures in the United States and is increasingly being used in light commercial construction. Wood-framed structures are economical to build and maintain. Further, wood construction is adaptable to many building styles. There are various types of framing systems that can be used to build conventional wood-framed structures. These framing systems must transfer the gravity/vertical and lateral loads to the foundations. Working in conjunction, the framing systems and the foundations provide strength and stability for the structure. The most common type of wood-framed construction uses roof framing, exterior and interior load-bearing walls, beams, girders, posts, and floor framing to resist the gravity/vertical loads. This type of wood-framed construction also employs a system of horizontal diaphragms (roof and floors) and shear walls (vertical exterior sheathed walls) to resist the lateral loads. Modern building codes classify these types of structures as bearing wall systems. The overall integrity of the structure depends upon not only the strength of the components in the framing systems, but also the adequacy of the connections between them. Critical connections are present throughout the structure, but the most critical connections are typically where the roof system is connected to the bearing walls, where the bearing walls are connected to the floor framing, and where the walls/floor framing are connected to the foundations.[1–3]

Considerable structural damage to residential roof, wall, and floor framing members/systems can be caused by a variety of circumstances. National design standards and state building codes require residential roof framing systems to be designed to withstand forces generated from specific vertical and lateral loads without incurring damage. Unfortunately, events occur that result in roof framing members becoming damaged structurally. These events may include but are not limited to impact from fallen trees and tree limbs; vehicular impact; water infiltration, which weakens members and makes them susceptible to mold, fungus, and/or insect infestation; fire; blasts/explosions; improper modifications to the framing members by unqualified individuals; improper design and/or construction; and an exceedance of design loads (e.g., excessive wind, water, snow, or ice loads). Damaged framing members jeopardize the integrity of the structure and potentially create conditions where consequential damages (i.e., water damage) can occur if there is significant movement in the structural framing system.

Assessing damage to conventional wood framing members/systems typically requires a determination and evaluation of the following: load history on the structure, species and grades of the affected wood members, and physical properties and conditions of the affected wood members and connections. As previously indicated, the physical conditions of wood framing members and connections may be influenced by many factors. There are several methods that can be utilized to assess damage to wood framing members such as coring, drilling, laboratory testing, load testing, moisture meter surveying, probing, radiographic study, sounding, stress wave propagating, and visual inspection. Further descriptions, applications, and limitations of these assessment methods are outlined in the American Society of Civil Engineers (ASCE) Standard 11 "Guideline for Structural Condition Assessment of Existing Buildings".[4]

This chapter will focus only on the visual inspection method for assessing damage to wood framing members/systems and include the following topics:

- Common Conventional Wood Framing Members/Systems and Terminologies.
- Building Code Requirements and Specifications.
- Acceptable Wall Tilt (Out-of-Plumb).
- Common Causes of Structural Damage and Types of Damage to Structural Wood Framing Members.
- Inspection Methodologies for Completing a Forensic Inspection of Residential Wood Framing, Including Analysis of Framing Members/Systems with Commercially Available Structural Analysis Software.

14.2   COMMON ROOF FRAMING SYSTEMS

14.2.1   Pre-Engineered Press-Plated Wood Trusses

This framing system utilizes pre-engineered press-plated wood trusses uniformly spaced across the structure. These trusses are typically constructed with nominal 2× dimensional lumber for the top and bottom members called "chords" and the diagonal members called "webs." The members are typically held together with metal gusset plates at each joint. The plates are secured to the wood members by rolling the entire assembly through a hydraulic press. The trusses with this type of framing system typically span from bearing wall to bearing wall with no intermediate support. An example of a pre-engineered press-plated wood truss roof framing system is shown in Figure 14.1.

FIGURE 14.1   Roof framing system – pre-engineered press-plated wood trusses.

14.2.2 Ridge Beam and Rafters

This framing system utilizes a load-bearing ridge beam, spanning bearing wall to bearing wall and/or intermediate support posts, typically above the center of the room/structure, with rafters perpendicular to and uniformly spaced along each side. The rafters on either side of the ridge beam are not required to be aligned or similar in length. The rafters are typically nominal 2× dimensional lumber members that span from the bearing wall to the ridge beam. The ridge beam is a main structural member capable of carrying load. Ceiling joists may or may not be present; however, if joists are present, they are typically not sufficiently attached to the rafters to resist any thrust (horizontal force) created at the bottom of the rafters (see discussion in Section 14.2.3). This type of construction is generally referred to as "stick framing." An example of a roof framing system consisting of a ridge beam and rafters is shown in Figure 14.2.

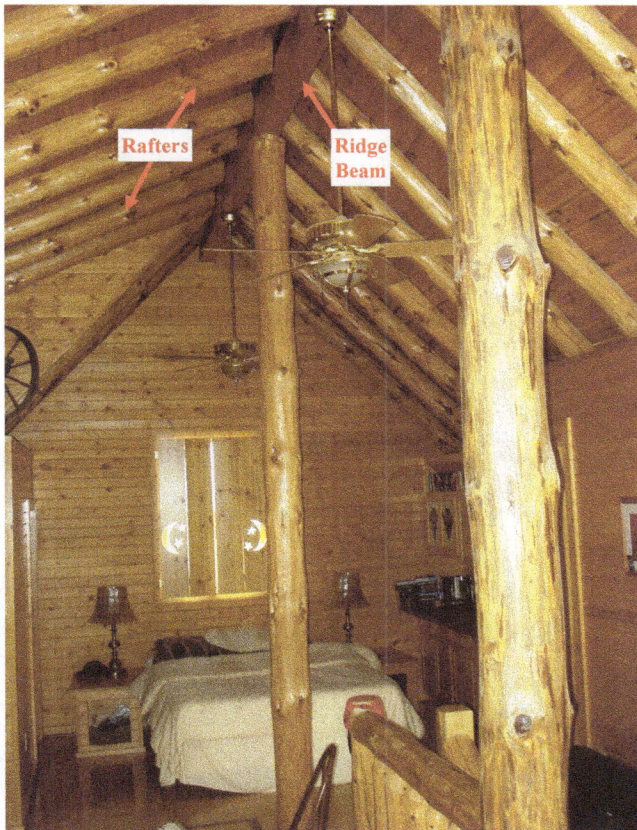

FIGURE 14.2 Roof framing system – ridge beam and rafters.

14.2.3 Rafters with Ceiling Joists or Rafter Ties

This framing system utilizes rafters uniformly spaced along the structure, tied in pairs to one another by ceiling joists or rafter ties. The rafters are typically nominal 2× dimensional lumber of relatively the same length on either side of the ridge and aligned with one another. The ceiling joist or rafter tie is typically a nominal 2× dimensional lumber member spanning horizontally between the two rafters, sufficiently attached to each rafter, in essence, creating a "truss." Once created, the "truss" spans from bearing wall to bearing wall with no intermediate support. A ridge board may be present, but it has no load-carrying capacity; it is merely there to provide a surface for the rafters to bear against and to create a straight ridgeline. This type of construction is also generally referred to as "stick framing."

A common misconception is to refer to the rafter ties as collar ties. Collar ties and rafter ties are both horizontal framing members typically found in roof stick framed construction; however, each has a different purpose, location, and attachment requirement. Rafter ties are designed to tie the bottoms of opposing rafters together.[5] When a sloped roof is framed with rafters butting against a ridge board and subjected to loads, vertical and horizontal forces are created at the bottom of the rafters. The vertical forces are resisted by the exterior bearing walls. However, the exterior bearing walls cannot resist the horizontal forces (thrust) exerted on them. The thrust must be resisted through adequately connected continuous ceiling joists or ties across the building. Rafter ties form the bottom chord of a simple triangular roof truss. They should be placed as low as possible in the roof framing, ideally in the lower third of the roof rafter. Collar ties are designed to tie the tops of opposing rafters together. This helps brace the roof framing against uplift caused by wind. Collar ties must be placed in the upper third of the roof rafters.[5] An example of a roof framing system consisting of rafters and ceiling joists/rafter ties along with collar ties is shown in Figure 14.3.

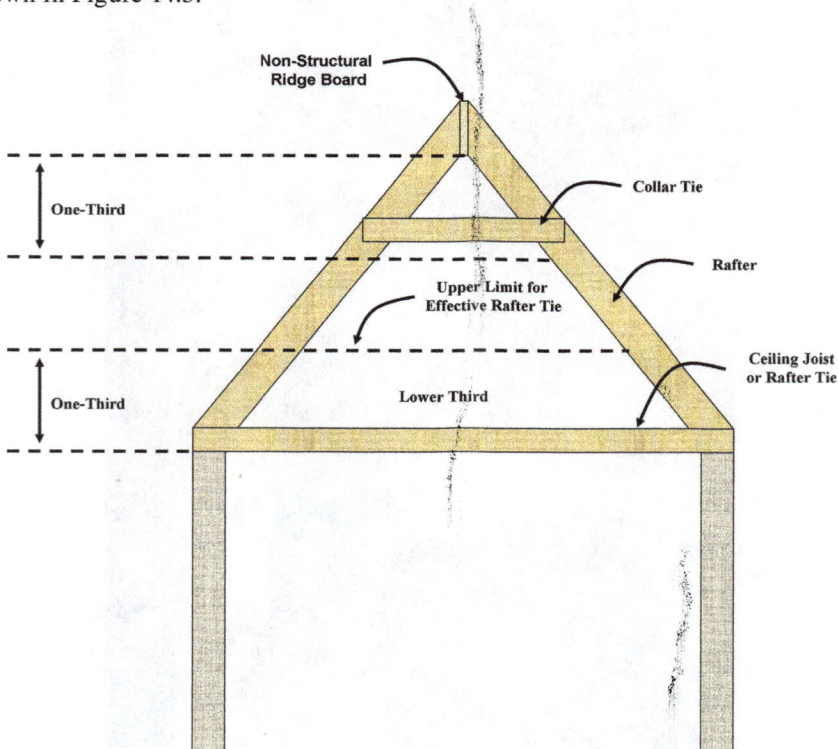

FIGURE 14.3 Roof framing system – rafters with ceiling joists/rafter ties and collar ties.

14.3 COMMON WALL/FLOOR FRAMING SYSTEMS

14.3.1 BALLOON WALLS

The studs in balloon walls are continuous from the sill plate supported by the foundation wall to the top plates of the second floor walls supporting the roof-framing members. The second floor joists are typically supported by members called "ribbon strips" that are "let-in" or inserted into notches cut in the inside edges of the studs.[1] This type of construction results in continuous cavities between the wall studs from the first floor sill plate to the second floor ceiling unless completely sealed fire blocking is installed between the studs at the second floor joists. Figure 14.4 shows a detail of balloon wall framing.

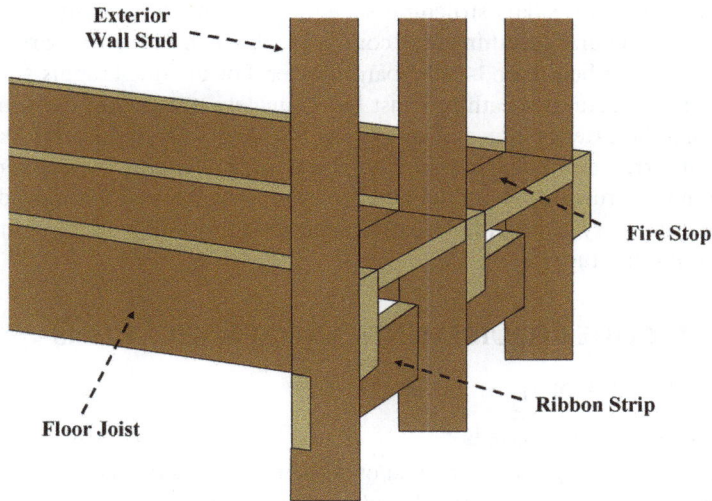

Typical Floor Balloon Framing Detail

FIGURE 14.4 Balloon wall framing.

14.3.2 PLATFORM FRAMING

In platform framing systems, the studs are only one story in height with the floor joists for each story either resting on the top plates of the story below or on the sill plate supported by the foundation wall. The floor joists are completely covered with subflooring (a rough floor that serves as the base for the finish floor), which forms a "platform." The bearing walls and partitions of each successive story rest upon the subfloor or "platform" of the story below. Platform framing is the most common type of framing system utilized in current residential construction.[1]

14.4 COMMON FLOOR JOIST MEMBERS

Common wood floor framing members utilized in residential and some light commercial construction are summarized below:

- Nominal 2× dimensional lumber.
- Wood I-Joists: I-Joists are fabricated framing members that consist of a top and bottom flange composed of dimensional or structurally composite lumber and a continuous web of plywood or oriented strand board (OSB) separating the flanges.
- Pre-Engineered Press-Plated Trusses: Floor trusses are similar to roof trusses with one major exception: the chord and web members are typically configured with the wide faces in a horizontal position as opposed to roof trusses where the wide faces of the chord and web members are configured vertically.

14.5   DIAPHRAGMS AND SHEAR WALLS

The majority of wood-framed residential and light commercial structures utilize horizontal diaphragms (roof and floor systems covered with structural sheathing) and vertical shear walls (exterior walls covered with structural sheathing commonly referred to as braced wall lines or braced wall panels in modern building codes and standards) provide stability and resist the lateral loads imposed on the structure. The structural sheathing commonly consists of plywood or OSB panels. However, the structural sheathing may consist of other panels such as particle board, wafer-board, structural insulated board, or 1-inch board lumber. For the diaphragms and shear walls to perform properly, the structural sheathing must be adequately attached to the framing members, most critically along the exterior edges of the panels. Further, when the lateral loads imposed on a structure are transferred to the diaphragms and shear walls, compression and tension forces are induced throughout the structure. These induced forces must be resisted through adequate connections between the framing systems. A failure of a connection or member of a framing system could result in failure of the structure.[1–3]

14.6   BUILDING CODE REQUIREMENTS AND SPECIFICATIONS

14.6.1   ROOF SYSTEMS LOADING

14.6.1.1   Gravity and Vertical Loads

When designing roof framing members and/or systems for residential and light commercial structures, the following gravity and vertical loads must be considered:

- *Dead load*: The weight of all materials that are permanently attached to the roof framing system including the weight of the framing members.
- *Roof live load*: Model building codes specify minimum design roof live loads based on roof slope and a member's tributary loaded area. Roof live loads are specified to take into account miscellaneous loads that may occur on a roof such as reroofing operations. Minimum design roof live loads can be found in Table R301.6 of the 2018 International Residential Code (IRC).[6]
- *Snow load*: Model building codes typically specify that roof framing members are to be designed for roof snow loads in accordance with Chapter 7 of the ASCE Standard 7 "Minimum Design Loads for Buildings and Other Structures".[7] Additionally, the effects of potential drifting and/or sliding snow, as well as unbalanced loading conditions on a hip or gabled roof, must be considered when designing for roof snow loads. Further, the effects of potential partial loading conditions on continuous beams, as may be the case in a ridge beam and rafter roof framing system, must also be investigated. Finally, rain-on-snow surcharge loading must be considered when applicable. Ground snow loads, which are the basis for roof snow loads, vary in magnitude and are dependent on the geographic region in which the subject roof system is located. Ground snow loads are shown geographically in Figure R301.2(6) of the 2018 IRC[6] and affect primarily the northern half of the United States. For example, the ground snow load ranges from 20 to 25 pounds per square foot for the majority of the states of Ohio, Indiana, and Illinois. However, local building officials should be contacted on a job-by-job basis to verify if local jurisdiction roof snow load requirements are more stringent than those specified by the IRC and/or state codes.
- Wind Load: Model building codes specify that roof-framing members be designed for the vertical component of pressures (wind loads) associated with a basic wind speed for the subject roof pitch. Basic wind speeds vary in magnitude and are dependent on the

geographic region for a given roof system. For example, these basic wind speeds are shown geographically in Figure R301.2(5)A of the 2018 IRC[6] and Figure 6.1 in the ASCE Standard 7 "Minimum Design Loads for Buildings and Other Structures".[7] The basic wind speeds represent the 3-second gust wind speeds in miles per hour (MPH) at 33 feet (10 m) above ground level for Exposure C category. For example, in much of the Midwest, a basic wind speed of 90 MPH should be used when determining the design wind loads on a structural framing system, whereas higher basic wind speeds are specified for the coastal areas. Note that the wind speed figures in the 2016 edition of the ASCE Standard 7[7] indicate the ultimate design wind speeds as opposed to the basic wind speeds presented in pre-2010 editions. When utilizing allowable stress design (ASD), these ultimate design wind speeds may be reduced by the square root of 0.6 (Vasd = Vult × √0.6), which results in the basic wind speeds indicated in pre-2010 editions of the ASCE Standard 7.[7] To date, not all model building codes have adopted the changes.

Tabulated design wind loads (pressures) for varying zones, effective wind areas, and basic wind speeds are also typically provided in model building codes. One such table is Table R301.2(2) of the 2018 IRC.[6] The tabulated loads are based on a structure with a mean roof height of 30 feet located in an area designated with an Exposure Category of "B" and act normal or perpendicular to the effected surface. Further explanation of the tabulated design wind load table follows:

- The zone column represents the area of the structure in which the framing member of interest is located. The surfaces of a structure are divided into five zones: a roof interior zone, a roof end zone, a roof corner zone, a wall interior zone, and a wall corner zone. These zones are designated to account for the increased wind forces that occur at areas of discontinuity such as corners, eaves, rakes, and ridges. The zones for a typical residential structure are shown in Figure R301.2(8) of the 2018 IRC.[6]
- The effective wind area column represents the area (in square feet) used to determine the wind load on a framing member. This effective area is the larger of the following:
  - The tributary area of the framing member
  - The length of the framing member squared divided by 3 ($L^2/3$)
- Exposure categories were developed to adequately reflect the characteristics of ground surface irregularities such as variations in surface elevation, vegetation, and/or structures. Exposures are divided into four categories. Highlights from Section R301.2.1.4 of the 2018 IRC[6] are as follows:
  - *Exposure category A*: Large city centers with at least 50% of the buildings having a height in excess of 70 feet (21,336 mm).[6]
  - *Exposure category B:* Urban and suburban areas, wooded areas, or other terrain with numerous closely spaced obstructions having the size of single-family dwellings or larger. Exposure category B shall be assumed unless the site meets the definition of another type of exposure.[6]
  - *Exposure category C*: Open terrain with scattered obstructions, including surface undulations or other irregularities, having heights generally <30 feet (9,144 mm) extending more than 1,500 feet (457 m) from the building site in any quadrant. This category includes flat, open country and grasslands.[6]
  - *Exposure category D*: Flat, unobstructed areas exposed to wind flowing over open water, smooth mud flats, salt flats, and unbroken ice for a distance of not <5,000 feet (1,524 m). This exposure shall apply only to those buildings and other structures exposed to the wind coming from over the unobstructed area. Exposure D extends downwind from the edge of the unobstructed area a distance of 600 feet (183 m) or 20 times the height of the building or structure, whichever is greater.[6]

The tabulated design wind loads given in Table R301.2(2) of the 2018 IRC[6] must be multiplied by an adjustment coefficient when the mean roof height is >30 feet above the ground surface and/or the structure is located in an area that would be classified other than Exposure Category "B." Appropriate height and exposure adjustment coefficients for these situations are given in Table R301.2(3) of the 2018 IRC.[6]

14.6.1.2 Lateral Loads

The following lateral loads must also be considered in the design of roof framing members and/or systems:

- *Wind load*: Model residential building codes specify that roof-framing members be designed for the horizontal component of the pressures for the subject roof pitch associated with a basic wind speed (see earlier discussion for information on determining design wind loads).
- *Earthquake (seismic) load*: Lateral loads developed during an earthquake. The Seismic Design Categories for Site Class D are shown geographically in Figure R301.2(2) of the 2018 IRC.[6] Modern building codes specify that seismic design requirements apply only to buildings constructed in Seismic Design Categories C, D_1, D_2, or E.[6]

14.6.2 FLOOR SYSTEMS LOADING

The following gravity and vertical loads must be considered when designing floor framing members and/or systems for residential structures:

- *Dead load*: The weight of all materials that are permanently attached to the floor framing system, including the weight of the framing members.
- *Live load*: Model building codes specify minimum design floor live loads based on intended use, i.e., sleeping room vs room other than sleeping room. Minimum design floor live loads can be found in Table R301.5 of the 2018 IRC.[6]

14.6.3 LOAD COMBINATIONS

Model building codes specify that framing members be designed for different combinations of the above-specified loads. The load combinations have varying multipliers to account for the reduced probability that certain design loads will act in combination with other full design loads. Load combinations to consider while designing roof and floor framing members and/or systems (nominal loads using allowable stress design) per ASCE Standard 7[7] are as follows:

- D
- $D + L$
- $D + (Lr \text{ or } S)$
- $D + 0.75(L) + 0.75(Lr)$
- $D + (W \text{ or } 0.7E)$
- $D + 0.75(W \text{ or } 0.7E) + 0.75(L) + 0.75(Lr \text{ or } S)$
- $0.6D + (W \text{ or } 0.7E)$

Where

- D = Dead Load
- L = Floor Live Load
- Lr = Roof Live Load

- S = Roof Snow Load
- W = Wind Load
- E = Earthquake Load

14.6.4 Deflection Criteria

Model building codes establish deflection limitations for framing members that are not to be exceeded under certain loads. These deflection limits are intended to ensure user comfort and to prevent excessive cracking of finish materials. Both the span and functionality of a framing member are considered when determining its allowable deflection limit. Allowable deflection limits for roof, wall, and floor framing members of residential structures are set forth in Table 1 R301.7 of the 2018 IRC.[6]

14.6.5 Wood Member and Fastener Allowable Loads and Adjustment Factors

Model residential building codes indicate that the National Design Specification for Wood Construction (NDS) should be used as the governing code for the design of wood framing members. The NDS[8] tabulates design values for wood structural members and fasteners typically utilized in conventional wood-framed construction. These tabulated design values must be multiplied by all applicable adjustment factors to determine the allowable design values. Adjustment factors are a result of material properties that are unique to wood as a structural member. Applicable adjustment factors such as Load Duration Factor, Beam Stability Factor, Repetitive Member Factor, Size Factor, and so forth for wood members can be found in Chapter 2 of the NDS[8] and Chapter 4 of the NDS Supplement.[9] Additionally, adjustment factors such as Wet Service Factor and Toe-Nail Factor for fasteners can be found in Chapter 7 of the NDS.[8]

14.6.6 Allowable Wood Floor Joist and Rafter Spans

14.6.6.1 Floor Joists

Tables with allowable horizontal spans for dimensional lumber floor joists based on loading, spacing, member size and species, and deflection criteria are provided in modern residential building codes (Chapter 5 Tables R502.3.1(1) and R502.3.1(2) of the 2018 IRC[6]).

14.6.6.2 Roof Rafters

Tables with allowable horizontal spans for dimensional lumber rafters based on loading, spacing, member size and species, and deflection criteria are provided in modern residential building codes (Chapter 8 Tables R802.4.1(1) through R802.5.1(8) of the 2018 IRC[6]).

14.6.7 Rafter Tie and Connection Requirements

As noted in Section 14.2.3, when a sloped roof is framed with rafters butting against a ridge board and subjected to loads, vertical and horizontal forces are created at the bottom of the rafters. The vertical forces can be resisted by the exterior bearing walls, but the exterior bearing walls cannot resist the horizontal forces (thrust) exerted on them. However, this thrust can be resisted by using continuous ceiling joists or ties across the building that are adequately attached to the rafters. If no continuous ceiling joists or ties are present, a ridge beam must be provided to support the ends of the rafters at the ridge, eliminating the thrust at the bottom of the rafters. If adequately attached continuous ceiling joists or ties across the building or a load-bearing ridge beam is not provided, the rafters tend to push out the exterior bearing walls and the non-load-bearing ridge board becomes a load-bearing ridge beam. The ridge board is usually not sufficient or intended to carry any load and therefore deflects (sags) excessively and/or fails as a result (Figure 14.5).

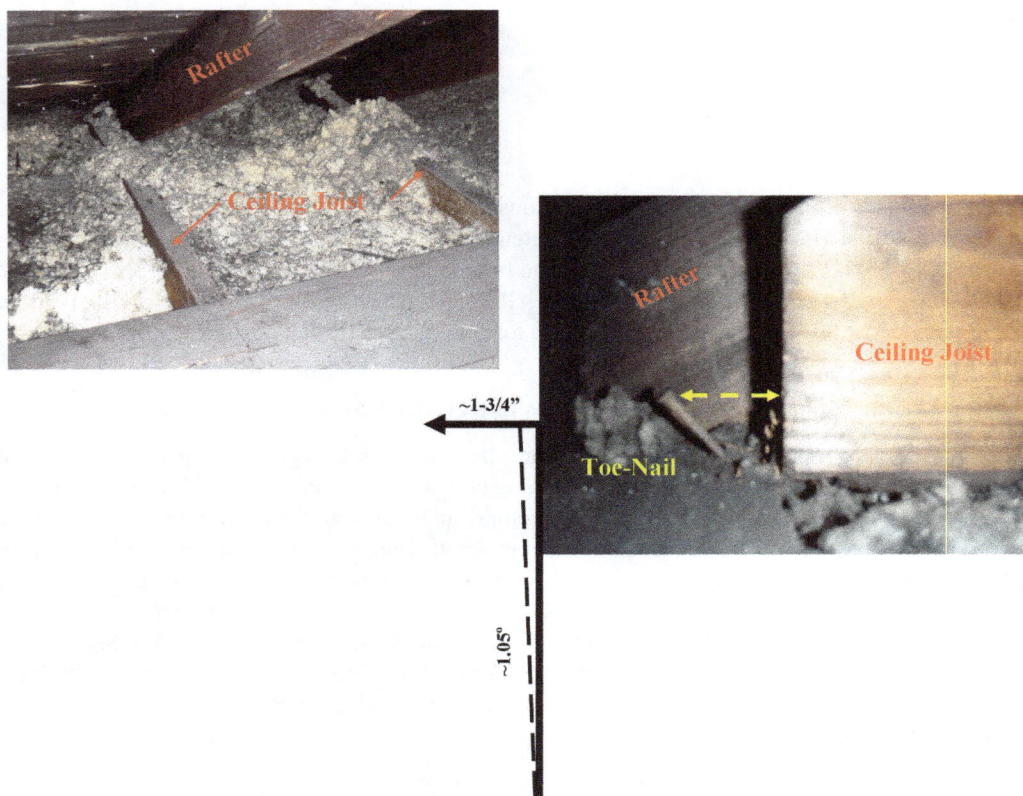

FIGURE 14.5   Failure of ceiling joist connection and resulting displacement of exterior bearing wall.

Modern residential building codes have certain requirements for roof and ceiling construction. These requirements were developed based upon past success and failure of different framing configurations. Chapter 8 of the IRC[6] governs roof and ceiling construction; highlights follow:

R802.4.2 Framing details: Rafters shall be framed not more than 1-1/2 inches (38 mm) offset from each other to a ridge board or directly opposite from each other with a collar tie, gusset plate, or ridge strap in accordance with Table R602.3(1). Rafters shall be nailed to the top wall plates in accordance with Table R602.3(1) unless the roof assembly is required to comply with the uplift requirements of Section R802.11.

R802.5.2 Ceiling joist and rafter connections: Where ceiling joists run parallel to rafters, they shall be connected to rafters at the top wall plate in accordance with Table R802.5.2.

Where ceiling joists are not connected to the rafters at the top wall plate, they shall be installed in the bottom third of the rafter height in accordance with Figure R802.4.5 and Table R802.5.2. Where the ceiling joists are installed above the bottom third of the rafter height, the ridge shall be designed as a beam. Where ceiling joists do not run parallel to rafters, the ceiling joists shall be connected to top plates in accordance with Table R602.3(1). Each rafter shall be tied across the structure with a rafter tie or a 2-inch by 4-inch (51 mm × 102 mm) kicker connected to the ceiling diaphragm with nails equivalent in capacity to Table R802.5.2.[6]

14.6.8 Modifications to Wood Structural Framing Members

14.6.8.1 Floor Framing Members

Modern residential building codes have certain limitations regarding modifications that can be made to floor framing members. Highlights from Chapter 5 of the 2018 IRC[6] are as follows:

> *R502.8 Cutting, drilling, and notching*: Structural floor members shall not be cut, bored, or notched in excess of the limitations specified in this section. See Figure R502.8.
>
> *R502.8.1 Sawn lumber.* Notches in solid lumber joists, rafters, and beams shall not exceed one-sixth of the depth of the member, shall not be longer than one-third of the depth of the member, and shall not be located in the middle one-third of the span. Notches at the ends of the members shall not exceed one-fourth the depth of the member. The tension side of members 4 inches (102 mm) or greater in nominal thickness shall not be notched except at the ends of the members. The diameter of the holes bored or cut into members shall not exceed one-third the depth of the member. Holes shall not be closer than 2 inches (51 mm) to the top or bottom of the member or to any other hole located in the member. Where the member is also notched, the hole shall not be closer than 2 inches (51 mm) to the notch.
>
> *R502.8.2 Engineered wood products.* Cuts, notches, and holes bored in trusses, structurally composite lumber, glue-laminated members, or I-joists are prohibited except where permitted by the manufacture's recommendations or where the effects of such alterations are specifically considered in the design of the member by a registered design professional.
>
> *R502.11.3 Alterations to trusses.* Truss members and components shall not be cut, notched, spliced, or otherwise altered in any way without the approval of a *registered design professional. Alterations* resulting in the addition of load (e.g., HVAC equipment, water heater) that exceeds the design load for the truss shall not be permitted without verification that the truss is capable of supporting the additional loading.[6]

14.6.8.2 Wall Framing Members

Modern residential building codes have certain limitations regarding modifications that can be made to wall framing members. Highlights from Chapter 6 of the 2018 IRC[6] are as follows:

> *R602.6 Drilling and notching of studs*: Drilling and notching of studs shall be in accordance with the following:
>
> 1. *Notching*: Any stud in an exterior wall or bearing partition may be cut or notched to a depth not exceeding 25% of its width. Studs in nonbearing partitions may be notched to a depth not to exceed 40% of a single stud width.
> 2. *Drilling*: Any stud may be bored or drilled, provided that the diameter of the resulting hole is no more than 60% of the stud width, the edge of the hole is no more than 5/8 inch (16 mm) to the edge of the stud, and the hole is not located in the same section as a cut or notch. Studs located in exterior walls or bearing partitions drilled over 40% and up to 60% shall also be doubled with no more than two successive double studs bored. See Figures R602.6(1) and R602.6(2).
>
> *Exception*: Use of *approved* stud shoes is permitted when they are installed in accordance with the manufacturer's recommendations.
>
> *R602.6.1 Drilling and notching of top plate*: When piping or ductwork is placed in or partly in an exterior wall or interior load-bearing wall, necessitating cutting, drilling, or notching of the top plate by more than 50 percent of its width, a galvanized metal tie not <0.054 inch thick (1.37 mm) (16 ga) and 1-1/2 inches (38 mm) wide shall be fastened across and to the plate at each side of the opening with not less than eight 10*d* (0.148 inch diameter) having a minimum length of 1-1/2 inches (38 mm) at each side or equivalent. The metal tie must extend a minimum of 6 inches past the opening. See Figure R602.6.1.

Exception: When the entire side of the wall with the notch or cut is covered by wood structural panel sheathing.[6]

14.6.8.3   Roof Framing Members

Modern residential building codes have certain limitations regarding the modifications that can be made to roof framing members. Highlights from Chapter 8 of the 2018 IRC[6] are as follows:

> *R802.7 Cutting, drilling, and notching*: Structural roof members shall not be cut, bored, or notched in excess of the limitations specified in this section.
>
> *R802.7.1 Sawn lumber.* Cuts, notches, and holes in solid lumber joists, rafters, blocking, and beams shall comply with the provisions of R502.8.1 except that cantilevered portions of rafters shall be permitted in accordance with Section R802.7.1.1.
>
> *R802.7.1.1 Cantilevered portions of rafters*: Notches on cantilevered portions of rafters are permitted provided the dimensions of the remaining portion of the rafter is not <3-1/2 inches (89 mm) and the length of the cantilever does not exceed 24 inches (610 mm) in accordance with Figure R802.7.1.1.
>
> *R802.7.1.2 Ceiling joist taper cut*: Taper cuts at the ends of the ceiling joists shall not exceed one-fourth the depth of the member in accordance with Figure R802.7.1.2.
>
> *R802.7.2 Engineered wood products.* Cuts, notches, and holes bored in trusses, structural composite lumber, structural glue-laminated members, or I-joists are prohibited except where permitted by the manufacture's recommendations or where the effects of such alterations are specifically considered in the design of the member by a registered design professional.
>
> *R802.10.4 Alterations to trusses.* Truss members shall not be cut, notched, drilled, spliced, or otherwise altered in any way without the approval of a registered *design professional.* Alterations resulting in the addition of load such as HVAC equipment water heater that exceeds the design load for the truss shall not be permitted without verification that the truss is capable of supporting such additional loading.[6]

An effective resource on wood wall bracing based on the 2018 IRC is published by the International Code Council and APA, the Engineered Wood Association, titled "A Guide to the 2018 IRC Wood Wall Bracing Provisions."

14.6.9   Wood Structural Panel Diaphragms and Shear Walls

14.6.9.1   Wood Structural Panel Sheathing in Roof and Floor Diaphragms

Tables with maximum allowable spans and fastener schedules for wood structural panel roof and floor sheathing are provided in modern residential building codes (Chapter 5 Table R503.2.1.1(1) and Chapter 6 Table R602.3(1), respectively, of the 2018 IRC[6]).

14.6.9.2   Wood Structural Panel Sheathing in Shear Walls

Requirements for wall bracing including lengths and locations of braced wall lines and specifications for types of braced wall panels are outlined in Chapter 6 Section R602.10 of the 2018 IRC.[6]

Tables with minimum thickness and fastener schedules for wood structural panel sheathing in shear walls are provided in modern residential building codes (Chapter 6 Tables R602.10.4 and Table R602.3(3) of the 2018 IRC[6]). Further, the wood sill plates of braced wall lines should be anchored to the concrete or masonry foundations as prescribed in Chapter 4 Section R403.1.6 and Chapter 6 Section R602.11.1 of the 2018 IRC.[6]

14.7 ALLOWABLE WALL TILT (OUT-OF-PLUMB)

The question arises when completing forensic investigations as to when the tilt or out-of-plumb of a wall is within tolerance, could be visibly observed, or may be unstable (Figure 14.6).

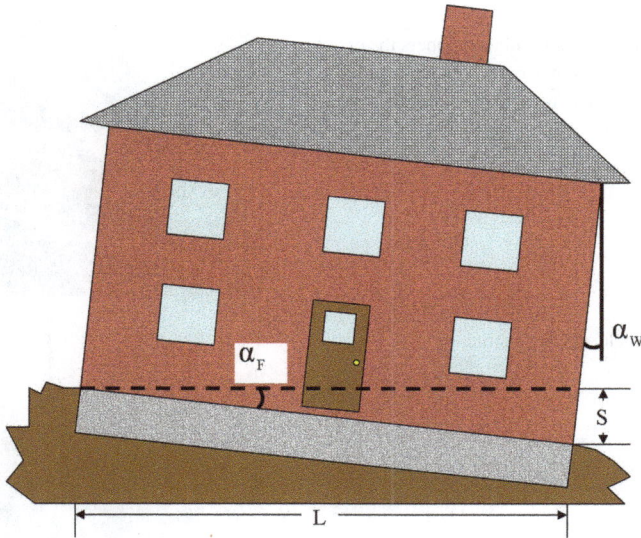

FIGURE 14.6 Wall tilt schematic.

In other words, at what point does the tilt of a wall transition from being merely a cosmetic issue that is readily visible yet does not greatly impact the structural stability of the wall and become a structural/stability issue that is out of tolerance. Dickinson et al. stated that an "out-of-plumb" of a wall (i.e., tilted wall) will not be noticeable (cosmetically) until the angle of the tilt compared with a vertical line (expressed as a ratio of $L/\#$) is in the order of $L/200$–$L/250$ (~0.23° to ~0.29°).[10] On the other hand, many sources (see Chapter 15 for detailed information on this topic) cite wall slope tolerances of up to 0.59° or ~0.6°. Therefore, from a cosmetic or tolerance standpoint, walls out-of-plumb or tilted less than or equal to 0.6° are acceptable.

14.8 COMMON CAUSES OF STRUCTURAL DAMAGE

When completing forensic investigations, it is important to understand basic framing terminology (e.g., studs, plates, headers, joist, trusses, rafters, and beams) and to be aware of typical issues found during inspections so that this information is properly described, identified, and recorded as part of the investigation. To facilitate this process, examples and photographs of structural damage to framing members/systems are illustrated in Table 14.1 and discussed in the paragraphs that follow.

TABLE 14.1
Examples of Structural Damage to Framing Systems

Structural Damage	Photograph
Impact damage due to fallen tree (note hole in decking and broken upper rafter members) Always check all members, including ridge members the entire length of the attic space, not just where the impact occurred – Subtle breaks of main beams can be determined using a change in slope between rafters using an electronic level	
Impact damage due to vehicle striking a structure	
Water infiltration and rot damage to wood members	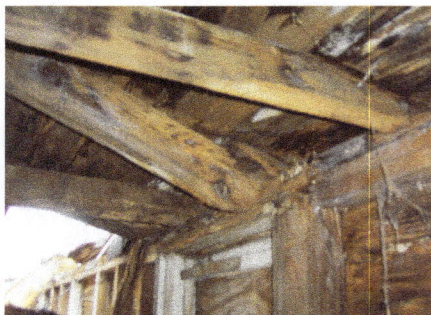
Fire damage to roof framing system	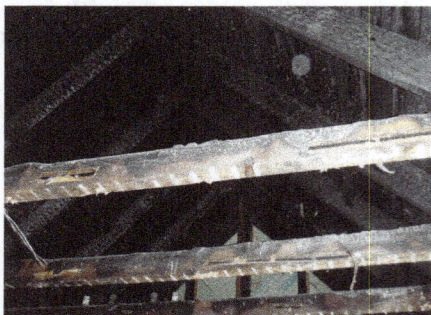

(Continued)

TABLE 14.1 (*Continued*)
Examples of Structural Damage to Framing Systems

Structural Damage	Photograph
Damage to floor framing system due to propane gas explosion within residence	
Localized failure of floor framing system (improper repairs/modifications)	
Tilting wall (inadequate lateral support/bracing)	
Failed roof rafter system due to improper construction (inadequate splice in hip beam member)	

(*Continued*)

TABLE 14.1 (*Continued*)
Examples of Structural Damage to Framing Systems

Structural Damage	Photograph
Failed under-designed roof truss framing system	
Sagging floor system (improper support piers)	

14.8.1 Structural Damage Caused by Impact from Fallen Objects or Vehicles

Structural framing members can become damaged from the impact of a fallen object. Based on experience, this typically involves damage from a fallen tree. When an object such as a tree or tree limb falls from some height onto a roof, the load imposed on the framing members can be exponentially greater than just the weight of the fallen object. Therefore, when a falling object lands on the framing members, the actual stresses in the members may increase such that the members' capacities are exceeded, causing them to twist, deform, and/or break and fail. Model building codes such as the IRC 6 do not have provisions requiring residential roof framing members and/or systems to be designed for impact loads.

The investigator is cautioned not to just inspect/document the damage at/near the impact zone, but throughout the entire framing system as well. Due to the interconnectedness of most framing systems, damage can occur at distances well beyond the impact zone. For example, in one case where a roof system was damaged due to impact from a fallen tree, the main structural composite wood ridge beam was visibly damaged at the impact location. However, during further investigation, the ridge beam was also found to be cracked ~25′ away from the direct impact zone. The roofing contractor, owner, and insurance agent had not observed this crack. The consensus prior to the engineering inspection had been to reinforce the main beam at the impact location and remove/replace damaged rafters and roof decking in the impact zone. Had this been done, a roof system with a cracked ridge beam would have been left in place. The key was to inspect the entire attic for any damaged members; in this case, the ridge beam contained damage far away from the impact zone.

Damage due to the impact of a vehicle is discussed in further detail in Chapter 18 – Impact Damage.

14.8.2 Structural Damage Caused by Water Infiltration/Moisture

Structural framing members can become damaged through decay due to water infiltration. Water weakens wood framing members by providing the needed moisture for colonies of bacteria and fungi to grow, which digest the organic material (e.g., lignin) within the wood. This is known as wood rot. Additionally, wet and moist wood framing members are more susceptible to insect infestation, where the insects attracted to wood in these conditions. The insects bore into the members and feed on the organic materials within the wood.[11]

Wood framing members with a moisture content of 20% or less typically will not decay/rot (Table 14.2).[12]

TABLE 14.2
Wood Decay Based on Moisture Content[12]

Moisture Content (Mass %)	State of Wood Decay
>35%	Rapid rotting occurs
>30%	Rotting occurs
20< × <30	Variable
25–27	Rotting can commence
20–22	Rotting terminates
<20%	No rotting occurs

However, decay/rot will occur in wood members when a moisture content of 22%–24% or greater is reached. Once decay/rot has begun in a wood framing member, the process continues until the moisture content in that member decreases to around 15%. However, the fungi causing the rot may not necessarily be killed as they can survive up to 9 years in wood at moisture contents around 12%. If the wood should become wet again, the decay process can restart. Wood framing members that are in atmospheric conditions normally found inside of structures and protected from water intrusion and/or excessive moisture typically will contain moisture contents that do not exceed 15% and therefore decay/rot should not occur.[13,14]

Another result of water infiltration is that it causes the wood members to swell and soften. If a member is carrying a load when it absorbs excess water, it can permanently deform in response to the load. Secondary moments associated with eccentric loading may now form in the deformed member. These secondary moments are created when a member is misshaped and the applied loads are no longer centered or symmetric, causing the member to twist or flex. Furthermore, wood members that have absorbed excess moisture and soften may allow fasteners to loosen and/or pull out, especially if the wood goes through wet and dry cycles over time.[11]

Model residential building codes have specific requirements regarding the protection of residential framing systems from water infiltration. See Chapters 8–10 for further discussion on this topic.

The water vapor content of the ambient air within an attic space is another source of moisture that can potentially cause damage to roof framing members. Elevated levels of water vapor can accumulate within an attic space due to roof leaks or other means of infiltration, particularly when the attic space is inadequately ventilated (see Chapter 12 – Ventilation). Condensation will occur on the underside of the roof decking when the temperature of the roof decking drops below the dew point temperature of the ambient air in the attic space. This condition occurs frequently in northern climates in the winter months and at nights during the shorter months of fall and spring.

Roof and/or attic ventilation should be designed so that outside air is taken in at the intake or soffit vents, flows on the underside of the roof decking, and is removed at or near the peak by exhaust vents. This circulation of air removes excessive heat and/or moisture that would otherwise accumulate in the attic space and potentially cause damage to the roof decking. See Chapter 12 for further discussion on this topic.

14.8.3 STRUCTURAL DAMAGE CAUSED BY FIRE

When structural framing members are damaged due to fire, the exposed outer layers of the members are oxidized away by the fire[15]. In cases where a structural member is only partially consumed by fire, an insulating char layer would have formed, which protected the core of the member. The relative extent of the char layer determines whether or not a significant portion of the original strength of the member has been lost due to fire. Heavy timbers (members 5″ x 5″ or greater) with larger cross sections are more likely to be structurally adequate after a fire than a member with a smaller initial cross section.[14] The fire-damaged member will have a smaller than original cross section (i.e., remaining core) and therefore less load-carrying capacity. Depending on the original design, this capacity may be sufficient to carry the imposed loads or may not be sufficient, resulting in a member deflecting excessively and/or failing. Model building codes do not have any provisions requiring residential framing systems to be sustainable after exposure to fire. Therefore, the rule of thumb in structural forensic investigations of residential structures is that any charred wood members should be removed and replaced (see Chapter 17 for further discussion on this topic).

14.8.4 STRUCTURAL DAMAGE CAUSED BY BLASTS/EXPLOSIONS

Structural framing members/systems can become damaged from blasts and explosions. This type of damage primarily occurs from one of three events: (1) blasts from near-by quarrying (surface) operations, (2) blast from nearby mining (surface and subsurface) operations, and (3) propane and natural gas explosions. Damages to nearby residential structures from blasts associated with these three events can be minimal to catastrophic (see Chapter 20 for further discussion on this topic).

14.8.5 STRUCTURAL DAMAGE CAUSED BY MODIFICATIONS AND/OR IMPROPER CONSTRUCTION

Structural framing members can become damaged due to modifications to the framing members themselves (i.e., removal of a truss web), modifications to adjacent framing members (i.e., shifting or removing rafters/joists/studs), and/or improper construction. When members such as rafters or joists in a framing system are modified or improperly installed (see Section 14.8.2 above), other members of that system receive additional loads and/or possibly eccentric loads for which they were not originally designed. A particular member may twist, warp, deflect excessively, or in the worst case, break and fail as a result of overstressing of that member within the system. Model residential building codes have specific requirements in regard to proper construction and allowable modification of framing members. See Sections 14.6.6 through 14.6.9 for specifications and limitations per the 2018 IRC.[6]

14.8.6 STRUCTURAL DAMAGE CAUSED BY ACTUAL LOADS EXCEEDING DESIGN LOADS OR UNDER-DESIGNED FRAMING MEMBERS

Structural framing members can become damaged as a result of the actual loads imposed on the framing system exceeding the design loads for the system. This excessive loading may be a result of the system originally being designed for loads less than the minimums specified in the model building codes (see Sections 14.6.1 and 14.6.2 above) and/or an unforeseen and/or altered building condition. For example, an addition to a structure may inadvertently create roof elevations of differing heights and/or valleys that were not previously present. The new roof configuration may result in localized snowdrift conditions where increased loads are imposed on the original roof system. Again, the structural framing members may twist, warp, deflect excessively, and/or in worst cases, break and fail as a result of the increased loading.

14.8.7 ACCUMULATION OF FIBER DAMAGE

The wood fibers of a structural framing member can accumulate damage due to previous overloading. Fiber damage occurs at an even greater degree in cross-grain tension, typically induced at connections. All or a combination of the aforementioned events can cause framing members and/or their connections to become overstressed. It is in this high or overstressed range that fiber damage may accumulate and reduce the capacity of wood framing members and/or connections to resist loads. When a wood member has accumulated fiber damage such as along the outer edges, that portion of the member becomes ineffective and its effective section is reduced. The reduced section has less ability to resist the imposed loads and the stresses in the member increase. If the stresses reach significant levels as the member is continuously loaded and unloaded over time, the accumulation of fiber damage continues to reduce the effective section until failure occurs. Hence, wood framing members and ultimately the framing systems can fail under lesser loads than previously carried with the accumulation of fiber damage/reduction of effective section in the members. It is theoretical that a wood member could fail sometime after the snow/live load was removed or under just the self-weight of construction materials present (dead load).[16,17]

14.9 NATURAL DEFECTS IN WOOD FRAMING MEMBERS

Natural defects, generally resulting from growth or drying, may exist in wood framing members. Natural defects include knots (developed from branch growth), shakes (separations in the wood grain that develop between growth cycles), and checks (discussed in further detail below). Natural defects such as knots and shakes are typically easily detectable during structural condition assessments. Knots and shakes can create localized stress concentrations, which may lead to damage and/or failure when loads are imposed on the structural framing members. Another defect that adversely affects the structural integrity of wood framing members is deformation (bowing, twisting, crooking, and cupping).[16] Secondary moments (see Section 14.8.2 above) likely will be created in the deformed member. The forces associated with these secondary moments may result in damage and/or failure of the member/framing system.

Checking, the separation of continuous wood fibers along the grain,[18] is a naturally occurring consequence of the seasoning process of wood. The outer fibers lose moisture to the surrounding atmosphere and attempt to shrink, but the inner portion of the timber member loses moisture at a much slower rate. The different rates of shrinkage can cause the wood to check or split. Rapid drying increases the differential moisture content between the inner and outer fibers and thus increases the propensity for checking in the timber. The checking (and shrinkage) process will stabilize as the moisture content of the member reaches equilibrium with the surrounding environmental conditions.[19]

For a column or post, the only time a check becomes a structural concern is if it develops into a full-length split (the entire height of the column or post, on both sides of the post). In this very unusual case, the length-to-depth (or L/d) ratio used in the design of columns will change and the resulting structural capacity of the column should be confirmed by a qualified design professional. A partial check is not a structural concern.[19]

Shear is the tendency of two equal and parallel forces acting in opposite directions to cause the adjoining surface layers of a member to slide one on the other. In a top-loaded wood framing member, this internal stress tends to tear the beam in half longitudinally. This force is known as horizontal shear stress. The horizontal shear strength in a wood framing member is dependent on the species of wood and the extent of the check(s) present in the wide face of the member. In a member with multiple checks, or that is split full length, a major portion of the shear stress is redistributed and carried by the upper and lower halves of the member. Research has established that this "two-beam" shear action allows a beam containing longitudinal splits/checks to carry loads for which it would appear to be inadequate in terms of horizontal shear. Tabulated allowable values listed in model building codes are established conservatively as if the members were split in half longitudinally. Hence, the values in these tables may be conservative for bending members with no splits or checks.[20] Therefore, checks in the vertically oriented face of rafters, joists, and beams have a negligible effect on the structural performance of the bending member and pose no structural concern.

14.10 METHODOLOGY FOR A STRUCTURAL FRAMING DAMAGE ASSESSMENT INSPECTION

Based upon hundreds of inspections, the following section outlines a methodology for completing structural wood framing damage assessments. When a request to complete such an investigation is received, the overall protocol outlined in Chapter 1 should be followed with specific attention paid to damage and/or displacement to finished surfaces and structural framing members. This section provides details and examples of specific activities to be recognized and followed during structural framing damage assessment inspections.

The inspection should begin with an interview of the point of contact(s) and then proceed with an inspection detailing observations of the structure floor by floor, as well as the roof framing system if required. The inspection should conclude with detailed observations of the exterior elevations (and the roof surface when required). The optimal approach is to begin the inspection indoors working up from the lower levels to the attic. Then complete the investigation by moving outdoors to inspect damage to exterior elevations and finally the roof surfaces. This process avoids soiling the indoors from outdoor dirt and debris and also allows the owner to leave should they have other business. During this process, one should be recording (in writing and with measurements and photographs) any evidence pertaining to the cause(s) and extent of the structural damage.

The specific activities for a structural framing damage investigation within the context of the overall inspection process outlined in Chapter 1 follow.

14.10.1 INTERVIEW WITH THE POINT(S) OF CONTACT

The first step in the process of determining structural damage is the interview. The point of contact, typically the owner or owner's representative, can be quite helpful in describing the history of the building and the history of the structural issue being investigated. The interview provides information about the subject structure and possibly prompts the point(s) of contact to provide critical details concerning the root cause(s) of the structural damage. Many times the information gathered can be helpful in terms of focusing on certain items or areas during the inspection. Structural failures are frequently associated with modifications (cuts, notches, holes, etc.) to framing members; knowledge of when and why these modifications were done helps to answer causation questions regarding the failure.

14.10.1.1 Building Information
- When was the structure built and when did the current owner purchase the property?
- What is the approximate square footage of the structure?
- Have there been any recent modifications to the structure (i.e., room additions, remodeling, etc.)?

14.10.1.2 Roof Covering Information
- What type of roof covering (i.e., asphalt shingles, wood shakes, metal roof panels, etc.) is installed?
- What is the approximate age of the roof coverings?
- Is there any indication of water leaks (i.e., water damage staining) to the interior surfaces of the structure? If so, where are they located, when were they first discovered, and if possible, when do the leaks seem to occur?

14.10.1.3 Damage History Information

- What known damage has occurred to the framing system and where is the damage located?
- What known damage has occurred to the finished surfaces of the structure and where is the damage located?
- When was the visible damage first discovered?
- What known events are believed to have caused the damage to the framing systems and/ or finished surfaces?

14.10.1.4 Storm History Information (when applicable)

- Was the property owner at the structure at the time of the weather event(s)?
- When did the weather event(s) occur (i.e., date and time)?
- From which direction did the weather event(s) arrive?

14.10.2 Interior Inspection of the Structure

After the interview has been completed, the structural damage assessment starts by performing an inspection of the interior spaces first. This interior inspection begins by making a sketch of the area being inspected (a floor plan). A floor plan is a diagram of the relationships between rooms, spaces, and other physical features at one level of a structure as viewed from directly overhead. The floor plan is drawn to scale as closely as possible, and approximate measurements are taken and recorded. This sketch supplies the approximate dimensions of the structure and a space to record the types and locations of damages to the interior finished surfaces observed during the inspection. An example of a computer-aided schematic for an interior floor plan, along with identifiers indicating key observations applicable to the subject structure, is shown in Figure 14.7.

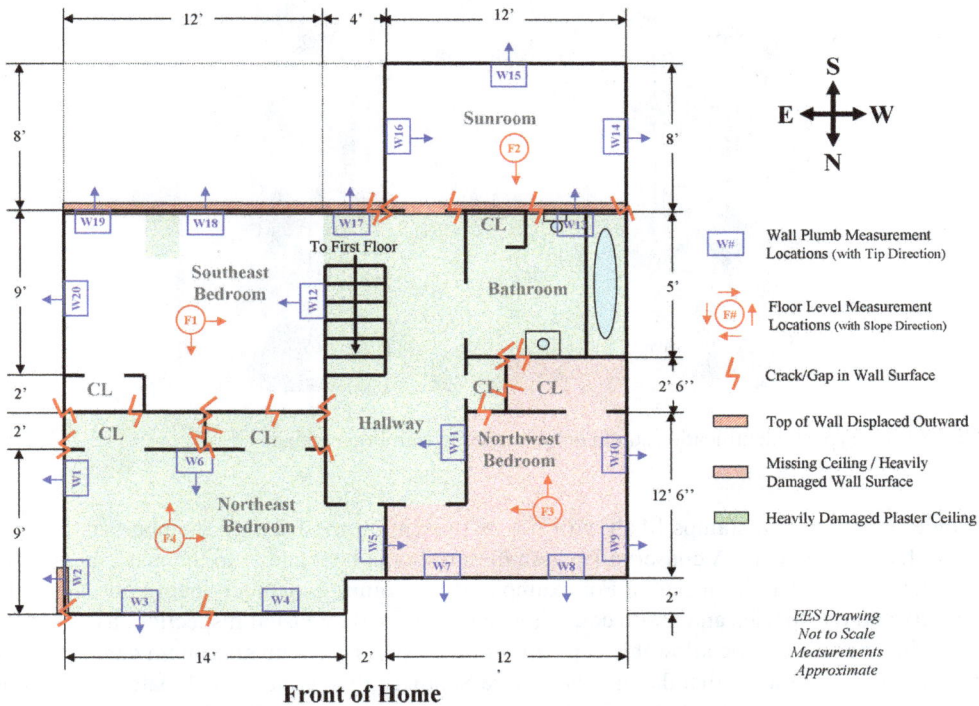

FIGURE 14.7 Computer-aided schematic – floor plan with key observations (developed from on-site measurements).

If multiple levels are to be inspected, begin with the lowest level first (i.e., basement or crawl-space) and then move upward to the top floor and attic (if accessible). Experience has shown this process to be the most effective in ultimately determining causes for structural damage and that most problems are illuminated in the crawlspace/basement and attic spaces.

Once the sketch of the floor plan is finished, a room-by-room inspection of each floor is completed. Inspection observations are documented in writing in the field book and visually with digital photography. Included in the inspection are general observations of the room and the damage(s) observed to the readily observable visible finished surfaces and/or framing members when exposed.

14.10.2.1 General Observations

For each interior elevation, the following general observations should be taken and recorded.

- Finishes: Generally describe the finished surfaces of floor, walls, and ceiling.
- Framing Members: When the walls and/or ceiling are unfinished and the framing members are exposed, generally describe the framing, including member sizes, spacing, orientation, and supports. In addition, attempt to determine/verify the species and grade of the wood framing members.

In newer construction, the species, grade, seasoning condition at the time of manufacturing, producing mill number, and grading rules writing agency will be stamped periodically onto the wide face of the wood members.[1] An example of this stamp is shown in Figure 14.8.

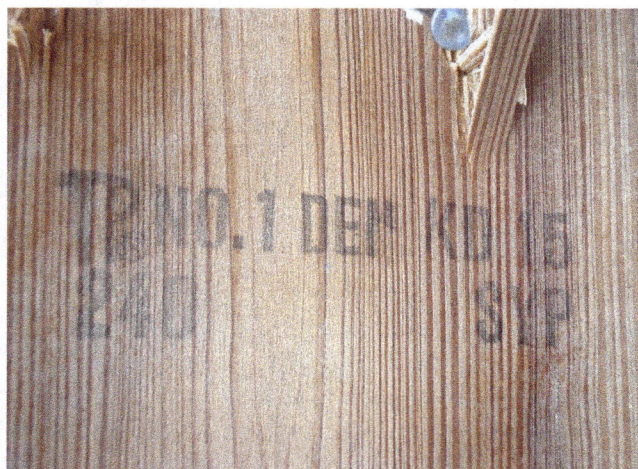

FIGURE 14.8 Typical identification stamp on newer wood framing member.

In older construction, stamps likely will not be present on the framing members and assumptions will have to be made. A conservative assumption should be made as to the material and grade of the framing members in question. For example, if the framing members being investigated are roof rafters and the species and grade could not be determined by visual inspection, a conservative approach in determining the allowable capacities of the rafters (i.e., the maximum capacities of the rafters) would be to assume that the members were Southern Pine Grade No. 2 material as opposed to Spruce-Pine-Fir Grade No. 2 material. However, if more accurate allowable material properties of the wood framing members are required, the method outlined in pages 131 through 151 of ASTM STP 901[21] may be used.

- *Damage area(s)*: Provide an overview of where the structural damage is present in the structure.

14.10.2.2 Damage(s) Observed to Visible Finished Surfaces

For each interior surface (i.e., floor, wall, and ceiling), the following specific observations should be taken and recorded.

- *Water damage staining:* Document the size (i.e., length by width or diameter) and location of any water damage staining on the floor, wall, and/or ceiling surfaces.
- *Cracks*: Document any cracks in the floor, wall, and/or ceiling surfaces. The documentation should include width of the crack, orientation (i.e., horizontal, vertical, diagonal, or stair-stepped), location, relative movement of the wall panel portions comparatively on either side of the crack (vertical, diagonal, and stair-stepped), and relative age. Good indicators in determining the age of a crack are the color and sharpness of the crack surfaces and the condition of the crack. If the crack surfaces/edges are sharp and lighter in color than the wall surface surrounding the crack and/or debris or paint is not present in the crack, the crack is likely to be relatively fresh. If the crack surfaces/edges are rounded and duller/darker in color and/or debris or paint is present in portions of the crack, the crack is likely older. Fresh cracks usually occur within a period of days or weeks while older cracks would likely have been present for months or years (see Chapter 15 for a detailed examination of cracks).
- *Gaps*: Document any gaps between finish materials (i.e., baseboard molding and wall surfaces). The documentation should include width of the gap, location, and relative age. The same principles described above can be utilized to determine the relative age of the gaps.
- *Holes*: Document any holes in the wall and/or ceiling surfaces. The documentation should include the size of the hole (i.e., length by width or diameter), location, and relative age. The same principles described above can be utilized to determine the relative age of the hole.
- *Char*: The solid carbonaceous substance that remains on a solid material after it has been partially burnt. Document the area (i.e., length by width or diameter) and location of any charring to the finished surfaces. Sometimes old fire damage is present, but has only partially been restored and is covered up by new finishes.
- *Soot*: The residue, usually black in color, produced during incomplete combustion. Document the area (i.e., length × width or diameter) and location of the finished surfaces coated with soot.
- *Irregularities*: Document any other irregularities observed in the wall and/or ceiling surfaces.
- *Wall plumb and floor levels*: Often, the movement of floors and/or walls can be quantified by taking level/plumb measurements. These should be completed using an electronic level or a conventional level and a tape measure. Results, including the location where the measurement was taken and the degree and direction of tilt/slope, should be recorded in tabular format in the field notebook. Wall plumb measurements should be taken at a minimum of two points on the wall (i.e., high and low) at each tested location to determine if the wall is plumb, bowed, or tilted monotonically. If the results of the plumb measurements indicate that the wall might be snaked or contain a compound tilt, a third plumb measurement should be taken at mid-height of the wall to verify the condition.

14.10.2.3 Damage(s) Observed to Exposed Framing Members

Detailed inspection observations and recordings (in writing in the field notebook and photographically) of damage to the exposed framing members should be included in the inspections for the following conditions (examples of damage to structural framing members are illustrated in Table 14.3).

TABLE 14.3
Examples of Damage to Structural Framing Members

Type of Damage	Photograph
Water damage staining and heavy rotting conditions to floor framing member	
Crack in attic rafter – note honeying on rafter member (indication of poor attic ventilation)	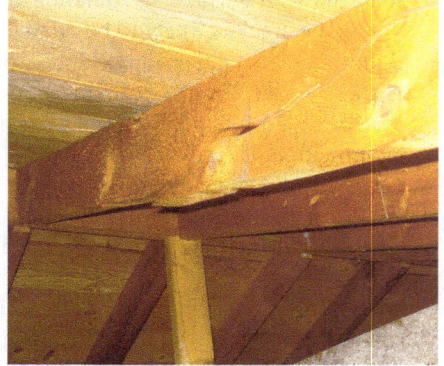
Cracked and bowed roof decking	
Cracked basement joist	

(Continued)

TABLE 14.3 (*Continued*)
Examples of Damage to Structural Framing Members

Type of Damage	Photograph
Older in appearance cracked in ceiling joist	
Fresh in appearance split in framing member	
Check in support post	

(*Continued*)

TABLE 14.3 (*Continued*)
Examples of Damage to Structural Framing Members

Type of Damage	Photograph
Gap between support beam and support post	
Modification to framing member (notch cut out of box beam)	

- *Water damage staining*: Document the size (i.e., length by width or diameter) and location of any water damage staining, in particular staining to the decking.
- *Rotting conditions*: Document the area(s) of any rotting conditions to the framing member(s) and the location(s) of the member(s).
- *Cracks*: Document any cracks in the structural framing members. The documentation should include width of the crack, orientation (i.e., horizontal, vertical, or diagonal), location, and relative age. If possible, measure the width of the crack to the nearest 1/32 of an inch. When the width varies throughout the length of the crack, record the crack width as being up to the maximum width. The same principles described in Section 14.10.2.2 can be utilized to determine the relative age of the crack.
- *Splits*: Document any splits (separation of the member along the grain lines caused by tearing apart of the wood cells) in the structural framing members. The documentation should

include width of the split (measure to nearest 1/32″ as described above), location, and relative age. The same principles described in Section 14.10.2.2 can be utilized to determine the relative age of the split.

- *Checks*: Document any checks in the structural framing members. The documentation should include width of the check (measure to nearest 1/32″ as described above), location, and relative age. The same principles described in Section 14.10.2.2 can be utilized to determine the relative age of the check.
- *Gaps*: Document any gaps between framing members (i.e., rafter ends and face of ridge board). The documentation should include width of the gap, location, and relative age. The same principles described in Section 14.10.2.2 can be utilized to determine the relative age of the gap.
- *Modifications and/or Irregularities*: Document any other modifications made to and/or irregularities observed in the structural framing members.
- *Char*: Document the area (i.e., length by width) and location of any charred members.
- *Soot*: Document the area (i.e., length by width) and location of any members coated with soot.
- *Framing member plumbs and levels*: The use of a level on framing members looking for changes in slope or plumb can identify damaged framing members and/or systems. Measurements should be taken and recorded similar to that discussed in Section 14.10.2.2.

This process of sketching the floor plan, thoroughly inspecting each room/area, and recording the observations (including plumb and level measurements) is repeated on each floor of the subject structure as required, again moving from the lowest level to the upper levels.

14.10.3 Roof Framing System Inspection

Following completion of the interior inspection, an inspection of the roof framing system should be performed when applicable, beginning with the attic and then moving outdoors to inspect the roof system from above.

The inspection of the roof framing system is performed in order to determine the extent of the damage and overall condition of the roof framing members and/or point to the likely cause(s) of the structural damage(s). Additionally, the inspection can uncover any defects and/or deficiencies present in the roof framing system, which should be communicated to the property owner. The owner can then address the issues and possibly prevent future damages.

The inspection of the roof framing begins with sketching the attic space, including lines to represent the structural framing members. The attic space is drawn to scale as closely as possible and approximate measurements are taken and recorded. This sketch provides a template to record the locations of key observations found during the inspection. Examples of computer-aided schematics for roof framing plans, along with identifiers indicating key observations applicable to the subject structure, are shown in Figures 14.9 and 14.10.

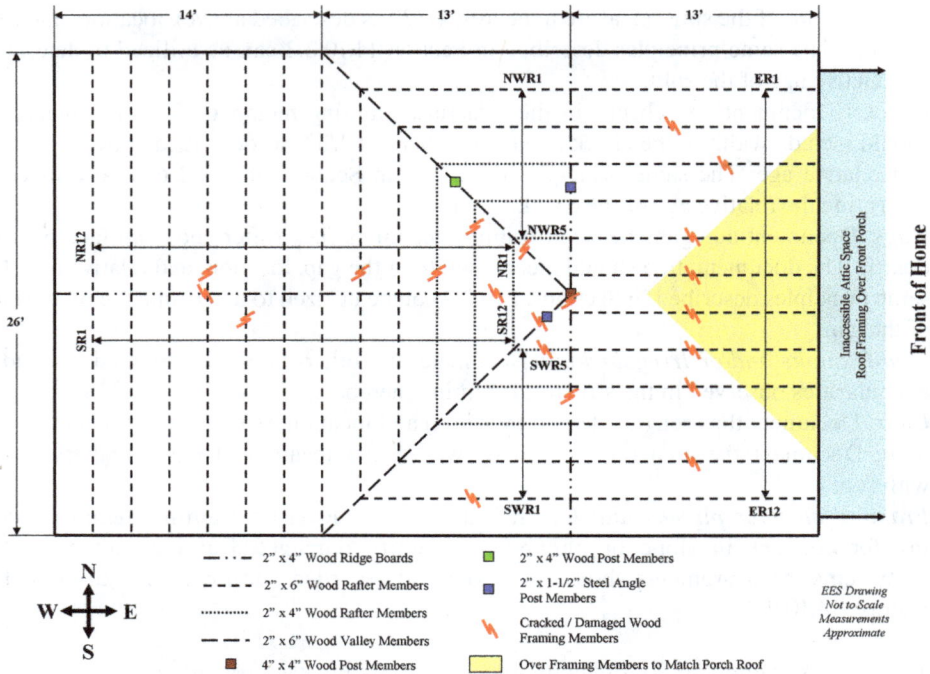

FIGURE 14.9 Computer-aided schematic – roof framing plan (rafters with ceiling joists system) with key observations (developed from on-site measurements).

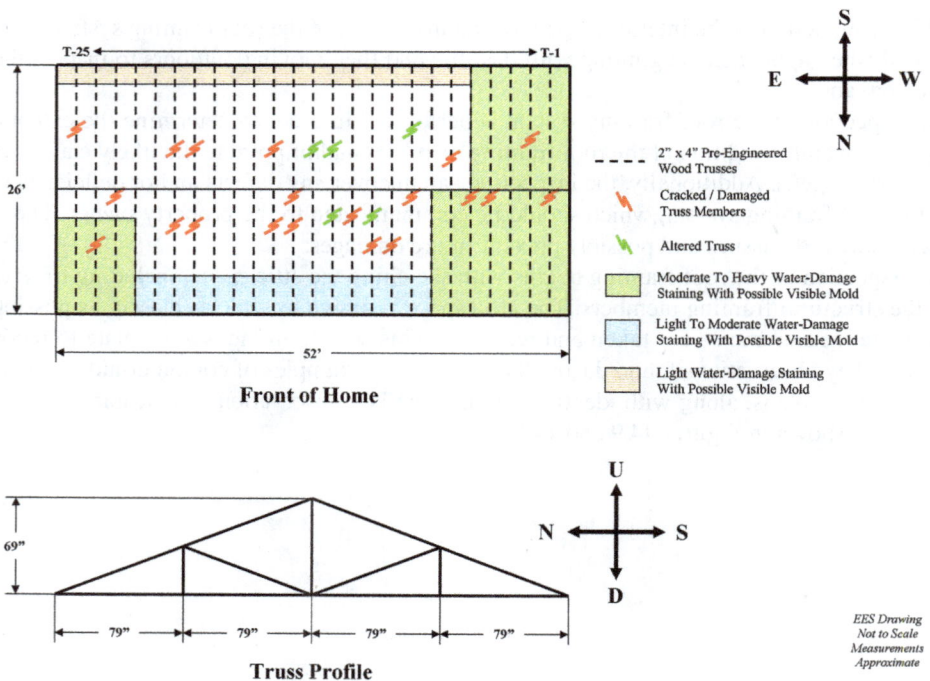

FIGURE 14.10 Computer-aided schematic – roof framing plan (roof truss system) with key observations (developed from on-site measurements).

Once the sketch of the attic space and structural framing members is completed, a thorough inspection is performed with the observations documented in writing in the field book and visually with digital photographs. Included in the inspection are observations of the attic construction, damage(s) to the framing members, and in certain cases, the method of providing attic ventilation.

14.10.3.1 General Attic Observations

General observations of the roof/attic construction should be recorded and should include the following:

- *Roof and attic construction*: Documentation of the roof and attic construction should include, but is not limited to, the framing member sizes, lengths, orientations, spacing, and support conditions. If the framing system is comprised of pre-engineered trusses, the truss configuration and member sizes should be documented. Again, attempt to determine/verify the species and grade of the wood framing members/components (see Section 14.10.2.1). The floor construction, when applicable, and the roof decking material (including the presence or absence of edge clips) should also be noted.
- *Attic insulation*: Document the type and thickness of the insulation covering the attic floor when applicable.

14.10.3.2 Damage(s) Observed to Framing Members

Detailed inspection observations of damage to the framing members similar to that indicated in Section 14.10.2.3 should be taken and recorded. Examples of damage to roof structural framing members observed from attic spaces are illustrated in Table 14.4.

TABLE 14.4
Examples of Damage to Roof Structural Framing Members

Type of Damage	Photograph
Water damage staining to roof decking	
Heavy rotting conditions to roof framing member	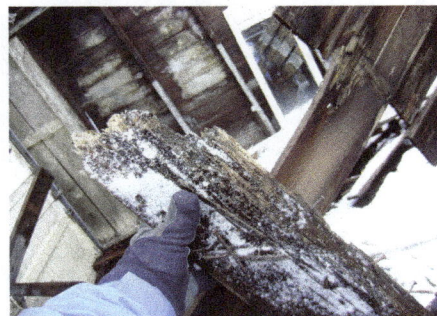

(Continued)

TABLE 14.4 (*Continued*)

Examples of Damage to Roof Structural Framing Members

Type of Damage	Photograph
Crack in roof rafter	
Separation of roof truss member at gusset plate	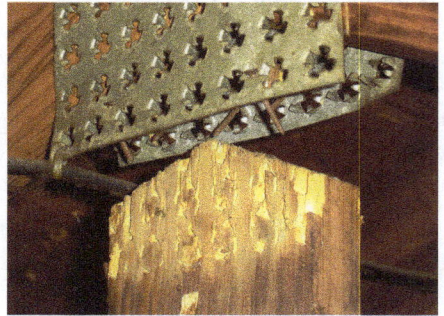
Check in roof framing member	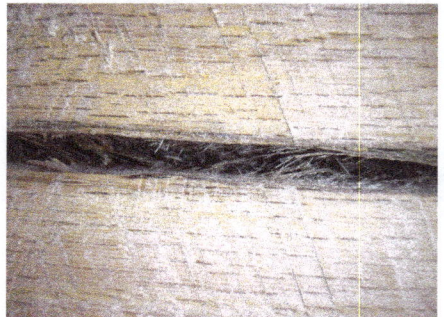
Gap between rafter and ridge beam of roof framing system	

(*Continued*)

TABLE 14.4 (*Continued*)

Examples of Damage to Roof Structural Framing Members

Type of Damage	Photograph
Modification to roof framing member (pre-engineer press-plated truss web cut and removed)	
Fire damage to roof decking and framing members	
Roof framing members coated with soot deposits	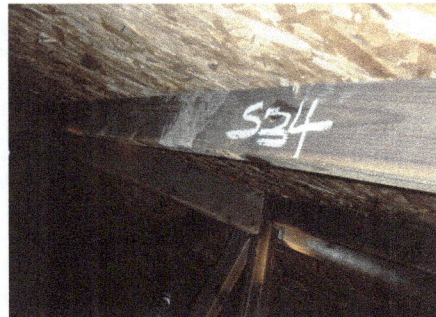

14.10.3.3 Attic Ventilation (When Applicable)

Detailed inspection observations and recordings of attic ventilation should include the following:

- *Baffles*: Document the presence or absence of baffles between the roof framing members along the eaves when soffit vents are present.
- *Gable end vents*: Document the size (i.e., length by width or diameter) and locations (i.e., the elevations in which the vents are located) of gable end vents if present.
- *Box, ridge, and ventilator vents*: Document, the number, location, size, and condition of the vents. In addition, inspect for signs of water entry and damage to wood members near the vents since the roof members may have been cut or partially cut at the time of vent installation.

14.10.4   EXTERIOR WALK-AROUND

Although the actual purpose of the on-site inspection is to determine the cause(s) and extent of the structural damage that has occurred to the framing system(s), a general inspection of the exterior surfaces and surroundings of the structure should be performed for completeness. A more in-depth inspection of the exterior surfaces of the structure is completed in cases where more information is required. Included in the exterior walk-around inspection are general exterior observations, documentation of any observed damage(s) to the visible finished surfaces, and in certain cases, determination of the method used for roof ventilation.

14.10.4.1   General Exterior Observations

General inspection observations of the exterior of the structure should include the following elements:

- *Exterior finish of structure*: Generally describe the finished surfaces of each elevation of the structure.
- *Surroundings*: Generally describe the conditions of the area around the structure as they pertain to the structural damage claim.

14.10.4.2   Damage(s) Observed to Visible Finished Surfaces

The inspection should document the cracks, gaps, holes, and/or other observed damage(s) to the exterior finished surfaces of the structure including the width or size, location, and relative age of the damage. The same principles described in Sections 14.10.2.2 can be utilized to determine the relative age of the damage(s).

14.10.4.3   Roof Ventilation Inspection

Inadequate roof attic ventilation can be critical to the premature aging and/or damage to a roof system. Ventilation elements visible during the inspection of the exterior surfaces should be documented when ventilation appears to be a contributing factor to the failure of roof framing members. Vents to be documented should include the following:

- *Soffit vents*: Document the type (i.e., continuous, under-eave, perforated, or lanced), size (i.e., length by width or diameter), and quantity of the vents present in the soffits.
- *Gable vents*: Document the number, location, and size (i.e., length by width or diameter) of the vents present.

14.10.5   ROOF INSPECTION

Following the interior inspections of the structure, including the inspection of the roof framing from the attic, an inspection of the exterior roof surfaces should be completed. The inspection of the roof begins by first sketching a plan view of the roof area. A plan view is a view of the roof surfaces from directly overhead; all hips and ridges are drawn as solid lines and all valleys are drawn as dashed lines. Along with the actual roof surfaces, all miscellaneous appurtenances located on the roof, such as box vents, furnace vents, ridge vents, soil stacks, and/or chimneys, are included in the sketch. The structure is drawn to scale as closely as possible and approximate exterior measurements are performed. Typically, the required exterior measurements include all ridge lengths, eave lengths, valley lengths, elevation slopes, and elevation pitches (measured in inches of rise per linear foot). An example of a computer-aided schematic of a roof plan, along with identifiers indicating key observations applicable to the subject residence, is shown in Figure 14.11.

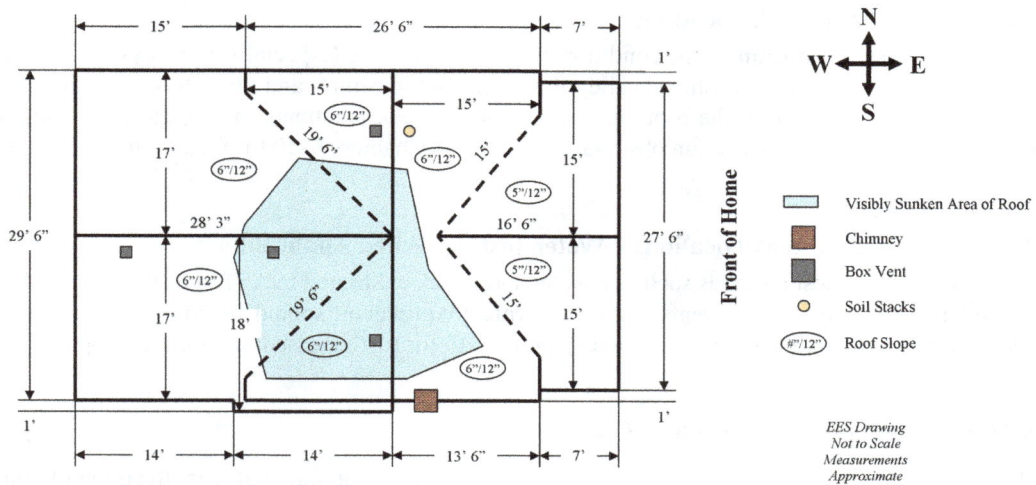

FIGURE 14.11 Computer-aided schematic – roof plan with key observations (developed from on-site measurements).

An inspection of the roof is performed following the completion of a roof plan sketch. Included in the inspection are general observations of the roof assembly, damage(s) observed to the roof surfaces, and when applicable, observations of probable leak location(s).

14.10.5.1 General Roof Observations

Inspection observations of the general roof construction and condition should include the following elements:

- *Roof assembly*: Documentation of the roof assembly should include but is not limited to whether or not drip edge molding is installed along the eave, whether or not felt underlayment is installed over the roof decking, the type of roofing material(s) present, and the number of layers of roofing material installed on the structure. Common roofing materials used on residential and light commercial structures include three-tab or dimensional/laminated composition (asphalt) shingles (which are the most common), wood shakes or shingles, slate tiles, clay or concrete tiles, and metal panels.
- *Valley construction*: Documentation of the valley construction (open, closed-cut, or woven).
- *Roof condition*: Documentation of the condition of the roof should include but is not limited to observations regarding the general condition/appearance of the roof coverings, noting of the feel of the decking (firm or soft), and citing of any visible depressions in the roof surfaces. The relative age of cited depressions should also be determined. The condition of the roof coverings is a good indicator in determining the age of a depression in a roof. The depression in the roof is likely older if the roof covering appears to be installed such that it generally follows the contours of the sunken roof surface. However, the depression in the roof is likely relatively fresh if the roof coverings are buckled, raised, or otherwise displaced and debris is not present in the gaps created by the displaced roof coverings.

14.10.5.2 Damage(s) Observed to Roof Surfaces

Documentation of the observed damage(s) to the roof surfaces should include the type of damage(s), the dimensions or extent of the damage(s), and the location of the damage(s).

14.10.5.3   Probable Leak Locations

The inspection should document the condition of the roof covering (especially at valleys), the installation of the soil stack boot flashing, valley flashing, brick flashing, and other flashings, and any other deficiencies present on the roof surface at probable leak location(s) based upon the interior and roof framing system inspection observations. Refer to Chapters 8–10 for further discussions on this topic.

14.10.5.4   Probable Leak Locations – Water Testing (When Applicable)

Water testing, using test methods such as a water spray rack calibrated to ASTM E1105 Standards[22] to simulate wind-driven rain events and/or equivalent equipment, should be completed to verify leaks at suspected leak locations. Refer to Chapters 8–10 for further discussions on this topic.

14.10.6   Analysis of Information Collected

Information collected during the inspection, including information obtained from the point of contact, should be analyzed to determine the answers to the following questions:

- What was the extent of the structural damage (e.g., what structural framing members were damaged and where are they located)?
- What was the cause of the structural damage?
- What damages to the framing members/systems were preexisting?
- What framing members should be removed/replaced or reinforced versus those that require no remedial action?

   The rule of thumb in structural forensic investigations of residential structures is that framing members that contain cracks, splits, and/or other visible damages beyond those of typical checking (discussed in Section 14.9) should be either removed and replaced or reinforced as required. A typical reinforcement technique commonly known as "sistering" is to add a new structural member onto the side of an existing member left in place. The new member should be adequate to support/resist all of the imposed loads/forces as well as be sufficiently attached to transfer all of the loads/forces from the original member to the "sistered" member.[16]

   Further, be aware when assessing damage to framing members/system due to impact from fallen trees, tree limbs, and/or vehicles, that damage may have occurred to members adjacent to the heavily damage area(s) even though visible signs of damage (cracks, splits, and/or gaps) are not present. This is particularly true in roof systems with pre-engineered wood trusses. Trusses are relatively stiff and therefore can absorb a large amount of impact energy. In situations where framing members/portions of the framing system have been heavily damaged due to impact, it is better to make a conservative judgment and remove/replace the apparently undamaged framing members on either side of the obviously damaged members as well.[16]

- What was the extent of cosmetic damage to the finished surfaces (e.g., which floor, wall, and/or ceiling surfaces contained damage and where was the damage located)?
- What damage to the finished surfaces was preexisting?

14.10.6.1   Determination of Cause of Damage

It is important to consider all possible scenarios when determining the likely cause(s) of damage to the structural framing members/systems being investigated. Most often, the unlikely cause(s) of damage can be eliminated by reviewing the information gathered during the inspection, weather data, building codes, industry literature, and/or result of an analysis, and a convergence can be made as to the most likely cause(s) of damage. Ultimately, the observations made during the inspection must support the likely cause(s) of damage as reported by the inspector. The key to many structural damage investigations is to compare the actual loads at the time the damage occurred to the in-place capacities of the structural framing members/systems.[16]

The cause of damage is relatively straightforward when it is due to recent impact from fallen trees and tree limbs, vehicular impact, water infiltration/moisture, or fire/blasts/explosions. Damage indicators to wood structural framing that can result from events discussed above follow:

- Damage due to recent impact will consist of cracks, splits, gaps, and/or other visible damage that is fresher in appearance as discussed in Section 14.10.2.3.
- Damage due to water infiltration/moisture will be indicated by wood members that likely contain evidence of historic moisture intrusion and long-term degradation (rot).
- For further discussion on damage to structural framing members as the result of a fire or blasts/explosions, refer to Chapter 17 or 20, respectively.

When the primary cause of damage is reported to be resultant from exceeding the design load for the structure, such as an exceptional amount of snow accumulation on a roof system, it is critical to research weather data and verify the actual loading conditions (snow accumulation) that likely existed on the structure at the time the damage occurred. See Section 1.5.2.1 for websites and sources where local weather data can be readily obtained. This weather data can be used to determine the maximum amount of snow accumulation and hence the maximum load likely present on a roof system at the time the damage occurred. This can then be compared with the requirements of the building code/local jurisdiction to verify if the minimum design loads were exceeded at the time of an unusual weather event.

In addition to design guidance documents for residential, commercial, and mobile homes,[23–27] software is available to analyze wood structural framing members and/or systems when the cause of damage is suspected to be the result of modifications, inadequate design/construction, and/or excessive loading conditions. Software such as Forte™ Software by Weyerhaeuser NR Company, ENERCALC software by ENERCALC, Inc., and StruCalc™ software by StruCalc, Inc. can be utilized to analyze single framing members such as a joist, rafter, column, or beam. Analysis and design programs such as RISA-3D by RISA Technologies, LLC, STAAD.Pro by Bentley Systems, Inc., and Multiframe 3D by FormSys (Formation Design Systems Pty Ltd.) can be utilized to analyze complex framing members such as trusses or entire framing systems. Examples of analyses where Forte™ Software and the RISA-3D program were utilized in determining the cause of damage/failure are provided below:

Example 1: An example of an analysis where Forte™ Software was utilized to determine if the cause of damage was due to inadequate design/construction follows:

An inspection was conducted to determine the cause(s) of the sagging floor surface and cracks in the finished wall and ceiling surfaces throughout the south portion of a home. During the interior inspection, performed as outlined in Section 14.10, the patterning of the cracks in the wall surfaces and the floor-level measurements indicated that the floor was sagging/settling toward the center of the home.

The crawlspace beneath the home was inspected following the interior inspection. The home was measured to be ~24 feet in width. The first floor framing system consisted of nominal 2 × 8 wood joists spaced on ~16-foot centers and configured in east/west orientations. The floor joists were supported by the east and west concrete masonry unit (CMU) walls and a central two-ply nominal 2 × 8 flush wood beam. The flush wood beam was supported by the north and south CMU walls and intermediately by CMU piers that measured ~8 feet by 16 feet and were spaced typically on ~8-foot centers with the exception of the south pier, which was located ~11 feet north of the south CMU wall.

Based upon the observations, it was suspected that the central flush wood beam was undersized and ultimately the cause of the damage observed to the first floor of the home.

Following the inspection, the south end of the central two-ply nominal 2 × 8 flush wood beam was analyzed using Forte™ Software to verify the likely cause of damage observed to the first floor of the home. The software analysis verified that the central flush wood beam was indeed undersized. The assumptions and results of the analysis were as follows:

Assumptions:

- Dead load equaled 10 pounds per square foot (total weight of the construction materials).
- Live load equaled 40 pounds per square foot (residential building code minimum).
- The nominal 2 × 8 wood beam plies were Southern Pine grade No. 2 members.

Results:

- The calculation indicated the two-ply nominal 2 × 8 wood flush main support beam at the south end of the central beam line was well undersized to carry the current code minimum design loads based on the observed configuration of the framing and support piers.
- The maximum total load deflection in the ~11′ span portion of the member (~0.675″) was up to ~3.3 times the maximum total load deflection in the ~8′ span portion of the member (~0.204″).

Output from the Forte™ Software version v3.5 for this example is shown in Figure 14.12.

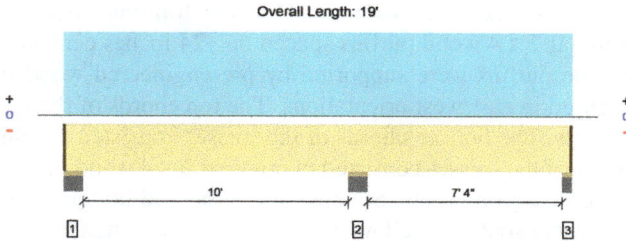

FORTE MEMBER REPORT *Level, Central Beam - South End* **FAILED**

2 piece(s) 2 x 8 Southern Pine No. 2

Overall Length: 19'

All Dimensions Are Horizontal; Drawing is Conceptual

Design Results	Actual @ Location	Allowed	Result	LDF	Load: Combination (Pattern)
Member Reaction (lbs)	7027 @ 11'	10200	Passed (69%)	--	1.0 D + 1.0 L (All Spans)
Shear (lbs)	3240 @ 10' 3/4"	2538	Failed (128%)	1.00	1.0 D + 1.0 L (All Spans)
Moment (Ft-lbs)	-6706 @ 11'	2629	Failed (255%)	1.00	1.0 D + 1.0 L (All Spans)
Live Load Defl. (in)	0.559 @ 5' 5 1/8"	0.349	Failed (L/224)	--	1.0 D + 1.0 L (Alt Spans)
Total Load Defl. (in)	0.675 @ 5' 4 5/8"	0.523	Failed (L/186)	--	1.0 D + 1.0 L (Alt Spans)

System : Floor
Member Type : Flush Beam
Building Use : Residential
Building Code : IBC
Design Methodology : ASD

* Deflection criteria: LL (L/360) and TL (L/240).
* Bracing (Lu): All compression edges (top and bottom) must be braced at 6" o/c unless detailed otherwise. Proper attachment and positioning of lateral bracing is required to achieve member stability.
* Applicable calculations are based on NDS 2005 methodology.

Supports	Bearing Length			Loads to Supports (lbs)			
	Total	Available	Required	Dead	Floor Live	Total	Accessories
1 - Plate on concrete - SPF	8.00"	6.75"	2.30"	591	2410/-99	3001/-99	1 1/4" Rim Board
2 - Plate on concrete - SPF	8.00"	8.00"	5.51"	1457	5571	7028	None
3 - Plate on concrete - SPF	4.00"	2.75"	1.60"	336	1770/-433	2106/-433	1 1/4" Rim Board

* Rim Board is assumed to carry all loads applied directly above it, bypassing the member being designed.

Loads	Location	Tributary Width	Dead (0.90)	Floor Live (1.00)	Comments
1 - Uniform(PSF)	0 to 19'	12'	10.0	40.0	Residential - Living Areas

Weyerhaeuser Notes

SUSTAINABLE FORESTRY INITIATIVE

Weyerhaeuser warrants that the sizing of its products will be in accordance with Weyerhaeuser product design criteria and published design values. Weyerhaeuser expressly disclaims any other warranties related to the software. Refer to current Weyerhaeuser literature for installation details. (www.woodbywy.com) Accessories (Rim Board, Blocking Panels and Squash Blocks) are not designed by this software. Use of this software is not intended to circumvent the need for a design professional as determined by the authority having jurisdiction. The designer of record, builder or framer is responsible to assure that this calculation is compatible with the overall project. Products manufactured at Weyerhaeuser facilities are third-party certified to sustainable forestry standards.

The product application, input design loads, dimensions and support information have been provided by Forte Software Operator

Forte Software Operator	Job Notes
Patrick Slattery EES Group, Inc. (614) 798-4123 pslattery@eesinc.cc	

3/2/2012 11:52:37 AM
Forte v3.5, Design Engine: V5.5.3.2
Beam.4te

Page 1 of 1

FIGURE 14.12 Example output from Forte™ Software. (Courtesy of Weyerhaeuser.)

Based upon the observations, it was suspected that the visible sag in the kitchen floor surface at the subject home was likely the result of excessive deflection of the undersized flush main support member at the south end of the central beam line.

Example 2: An example of an analysis where the RISA-3D program was utilized to determine if the cause of a collapse was due to inadequate design/construction follows:

An inspection was conducted to determine the cause(s) of the partial collapse of the roof system at a light framed commercial building that was reportedly due to excessive roof snow load.

An addition was constructed on the north end of the building, which likely created snow drift conditions on the portion of the roof that collapsed. The depth of the snow accumulation on the remaining portion of the roof surface was recorded. The roof framing system consisted of OSB decking supported by nominal 2×4 wood purlins spaced on ~24 inches centers and configured in north/south orientations. The purlins were supported by pre-engineered wood trusses spaced on ~8-foot centers and configured in east/west orientations. The top chords of the trusses consisted of nominal 2×8 wood members. The bottom chords of the trusses consisted of nominal 2×6 wood members. The web members of the trusses consisted of nominal 2×4 wood members. Metal gusset plates were present at the connection points. Stamps on the top and bottom chords of the trusses indicated that the members were southern yellow pine grade number 1 material. The truss configuration was obtained through measurements. Based upon the observations, it was suspected that the partial collapse of the roof system was likely due to undersized/inadequate roof trusses.

Following the inspection, the local weather records were researched to obtain snow accumulation depths for the months prior to the collapse. The weight of the maximum snow accumulation based on the weather records, as well as the observed depth of the snow on the remaining roof surfaces recorded during the inspection, was calculated to determine if the actual loads on the roof system at the time of the collapse exceeded the minimum code required design loads.

Based on the weather data from the local weather stations, the weight of the total depth of snow accumulation reported in the area up to the date of loss was less than the design roof snow load required by the current governing building code. Based on the depth of the snow accumulation observed on the roof of the subject building at the time of the inspection, the roof snow load was less than the design roof snow load required by the current governing building code.

Next, a typical roof truss was analyzed using the RISA-3D program to verify that the likely cause of the collapse was failure of the undersized/inadequate roof trusses. The basis, assumptions, and results of the analysis were as follows:

Basis and Assumptions:

- The truss configuration and member sizes were based on field observations and measurements.
- The design dead load of the ceiling (total weight of the construction materials and ceiling finish) was 5 pounds per square foot.
- The design dead load of the roof (total weight of the construction materials and roof coverings) was 10 pounds per square foot.
- The uniform design roof snow load was 25 pounds per square foot (current code minimum design roof snow load).
- Unbalance snow load conditions were ignored in the analysis.
- Snowdrift loads were ignored in the analysis.
- Top chord member properties were based on nominal 2×8 Southern-Yellow-Pine (SYP.) grade No. 1 sections.
- Bottom chord member properties were based on nominal 2×6 SYP grade No. 1 sections.
- Web member properties were based on nominal 2×4 SYP grade No. 2 sections.
- The load duration adjustment factor (CD) of 1.15 was used for load combinations including roof snow load.
- The allowable member stresses were as specified per the National Design Specification for Wood Construction (NDS).[8]

Results:

- The maximum combined axial and bending stress in the bottom chord was ~2.26% or 126% over the allowable stress.

- The maximum combined axial and bending stress in the top chord was ~2.96% or 196% over the allowable stress.
- The maximum axial stress in the inner diagonal web members was ~1.72% or 72% over the allowable stress.
- The maximum vertical total load deflection of the truss was ~1.41″ or ~L/340 for a ~40′ span.
- The maximum vertical snow load deflection of the truss was ~0.87″ or ~L/551 for a ~40′ span.

Output from the RISA-3D program version v9.0.1 for this example is shown in Figure 14.13.

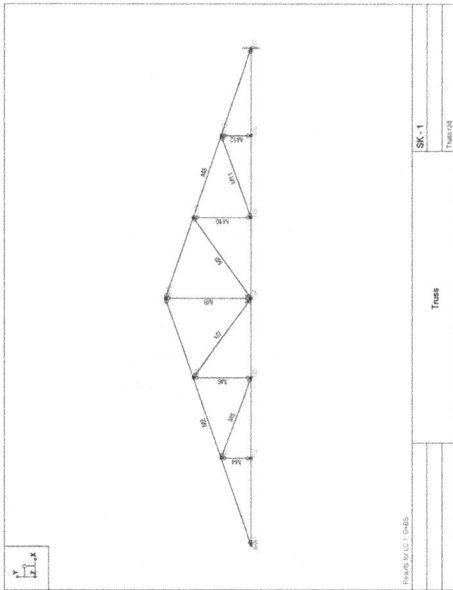

FIGURE 14.13 Example RISA-3D program output. (Courtesy of RISA Technologies.)

The analysis indicated that the pre-engineered roof trusses above the original portion of the subject building were inadequate to support the current code minimum design loads based on the observed truss spans and configurations. Further, this analysis did not include the additional weight of snowdrifts that likely formed on the roof above the north end of the original portion as a result of the addition constructed onto the north end of the subject building. The pre-engineered roof trusses above the north end of the original portion of the subject building were most likely grossly over-stressed as a result of the framing configuration in combination with the snow load. Although the base snow that had accumulated on the north end of the original portion of the subject building roof did not likely exceed the code minimum design roof snow load, the amount of snow at that location may have been exacerbated by drifting conditions. As a result, the pre-engineered roof trusses above the north end of the original portion of the subject building failed.

14.10.7 INSPECTION REPORT

A report of the findings should be prepared as outlined in Chapter 1 based on the interview and site inspection information, review of pertinent building code requirements and technical literature, knowledge of industry best practices, analysis, and professional experience.

The report should include conclusions that address the extent of damage (structural and cosmetic) and the likely cause(s) of the damage. Further, recommendations are commonly provided regarding the removal/replacement or reinforcement of damaged framing members and the necessity for shoring and/or other temporary supports prior to the completion of such repairs. Recommendations should emphasis that all remediation work should be completed by a qualified professional. Finally, any situations where imminent danger to life and/or property exists should be clearly emphasized in the report and also be forwarded to the responsible parties at/near the time of the inspection.

IMPORTANT POINTS TO REMEMBER

- Typical wall/floor framing approaches include balloon wall framing and platform framing. Typical roof framing approaches include pre-engineered trusses, ridge beam and rafter system, and rafters with ceiling joists or rafter ties.
- Building codes provide minimum loads that a structure must support, minimum requirements for framing members and systems, and limitations regarding the modification of structural members.
- Loads imposed on members/systems are defined as gravity/vertical loads and lateral loads. Gravity/vertical loads may include "dead load," "live load," "snow load," and/or a component of the "wind load." Lateral loads may include "wind load" and/or "earthquake (seismic) load."
- Structural failures most often occur due to improper modification to structural members (i.e., holes, cuts, removal of members), improper design/construction, lack of maintenance (i.e., rot from water leaks), and inadequate ventilation of crawlspace and attic spaces (i.e., premature aging/decay from excess temperature and rotting from excess moisture).
- Damage to structural framing members includes rot, cracks, splits, checks, gaps, and char. Detailed observation and documentation of the framing members and the damages therein can provide insight into the cause and origin of structural damage to wood framing systems.
- Commercial software can be used to determine whether or not the failure of a wood framing member/system occurred due to excessive loading and/or improper design/construction.

REFERENCES

1. American Forest & Paper Association. "Details for Conventional Wood Frame Construction," 2001. Accessed June 5, 2020, https://www.awc.org/pdf/codes-standards/publications/wcd/AWC-WCD1-ConventionalWoodFrame-ViewOnly-0107.pdf.
2. FEMA – Federal Emergency Management Agency. "FEMA P-762: Local Officials Guide for Coastal Construction – Design Considerations, Regulatory Guidance, and Best Practices for Coastal Communities," February 2009. Accessed June 5, 2020, https://www.fema.gov/media-library-data/20130726-1706-25045-9843/logcc_rev1.pdf.
3. Breyer, D.E., K. Cobeen, and Z. Martin. *Design of Wood Structures*, Eight Edition. New York: McGarw-Hill, Inc., 2020.
4. American Society of Civil Engineers. "Guideline for Structural Condition Assessment of Existing Buildings," SEI/ASCE 11-99, 2000.
5. Gromicko, N., R. London, and K. Shepard. "Collar Ties vs Rafter Ties." International Association of Certified Home Inspectors. Accessed June 5, 2020, http://www.nachi.org/collar-rafter-ties.htm.
6. International Code Council. "2018 International Residential Code (IRC)." Accessed June 5, 2020, https://codes.iccsafe.org/content/IRC2018/chapter-3-building-planning.
7. American Society of Civil Engineers. "Minimum Design Loads for Buildings and Other Structures." ANSI/SEI 7-16, 2010.
8. American Forest & Paper Association. "National Design Specification® (NDS®) for Wood Construction." ANSI/AWC NDS-2015, 2015 Edition, Approved September 30, 2014. Accessed June 5, 2020, https://www.awc.org/pdf/codes-standards/publications/nds/AWC-NDS2015-ViewOnly-1603.pdf.
9. American Forest & Paper Association. "Design Values for Wood Construction." NDS® Supplement, 2015 Edition, Approved November 2014. Accessed June 5, 2020, https://www.awc.org/pdf/codes-standards/publications/nds/AWC-NDS2015-Supplement-ViewOnly-1411.pdf.
10. Dickinson, P.R. and N. Thorton. *Cracking and Building Movement*. Coventry, UK: RICS Business Services Limited, 2006.
11. Noon, R.K. *Forensic Engineering Investigation*. New York: CRC Press, 2001.
12. Day, K.C. *Risk Management of Moisture in Exterior Wall Systems*. Toronto, Canada: Morrison Hershfield Limited.
13. American Forest & Paper Association, Inc. "Design of Wood Frame Structures for Performance," 2006. Accessed June 5, 2020, https://www.awc.org/pdf/codes-standards/publications/wcd/AWC-WCD6-Permanence-ViewOnly-0604.pdf.
14. Morris, P.I. "Understanding Biodeterioration of Wood in Structures." Fortintek Canada Corp., 1998. Accessed June 5, 2020, https://cwc.ca/wp-content/uploads/aboutdecay-biodeterioration.pdf.
15. American Forest & Paper Association. "Technical Report 10 – Calculating the Fire Resistance of Exposed Wood Members," May 2020. Accessed June 5, 2020, https://awc.org/pdf/codes-standards/publications/tr/AWC_TR10_20200520_AWCWebsite.pdf.
16. Ratay, R.T. *Forensic Structural Engineering Handbook*, Second Edition. New York: The McGraw-Hill Companies, Inc., 2010.
17. Nielsen, L.F. "Crack Propagation in Linear-Viscoelastic Materials." Bygningsstatiske Meddelelser, 1978.
18. American Institute of Timber Construction. "Checking in Glue Laminated Timber – AITC," 1987, Technical Note 11, Accessed June 27, 2020, https://qbcorp.com/wp-content/uploads/2018/04/aitc_tn_11.pdf.
19. APA – The Engineered Wood Association. Form No. EWS F450: Owner's Guide to Understanding Checks in Glued Laminated Timber, March 2006.
20. Hall & Foreman, Inc. "Horizontal Cracking in Heavy Timber." Accessed January 24, 2012, http://www.hfinc.com/markets/investigative-services-pace/horizontal-cracking-in-heavy-timber.html.
21. Jedrzejewski, M.J. "A Method for Determining the Allowable Strength of In-Place Wood Structural Members." Building Performance: Function, Preservation, and Rehabilitation, ASTM STP 901, G. Davis, Ed., American Society for Testing and Materials, Philadelphia, PA, pp. 136–151.
22. ASTM International. "ASTM Standard E1105-15: Field Determination of Water Penetration of Installed Exterior Windows, Skylights, Doors, and Curtain Walls by Uniform or Cyclic Static Air Pressure Difference," 2015.
23. American Plywood Association. "Brace Walls with Wood," APA Publication G440, Tacoma, WA, 2008.
24. American Plywood Association. "Design/Construction Guide – Non-Residential Roof Systems," APA Publication A310, Tacoma, WA, May 1996.

25. American Plywood Association. "Panelized Roofs", APA Publication G630, Tacoma, WA, December 2006.
26. American Plywood Association. "Roof Diaphragms for Manufactured Homes," APA Publication N435, Tacoma, WA, September 1993.
27. American Plywood Association. "APA Structural-Use Panels for Mobile Home Roofs and Floors," APA Publication N435, Tacoma, WA, August 1997.

15 Forensic Inspection Assessments of Foundation Walls

Stephen E. Petty
EES Group, Inc.

CONTENTS

PURPOSE/OBJECTIVES

The purpose of this chapter is to:

- Define terminology and construction methods for rubble stone, concrete masonry unit (CMU), and concrete foundation walls.
- Understand residential code language for the design and construction of foundation walls.
- Describe typical foundation wall failure mechanisms and causes.
- Formulate a standard for acceptable wall distortion and wall crack widths.
- Discuss the correlation between cracks and building movement.
- Define the difference between cosmetic damage and structural damage.
- Demonstrate methodologies for inspecting foundation walls.
- Illustrate reporting approaches for foundation wall inspections.

Following the completion of this chapter, you should be able to:

- Understand foundation terminology and code language for foundation walls.
- Understand typical foundation wall failure mechanisms.
- Recognize acceptable wall distortion levels for cosmetic and structural stability.
- Understand the correlation between cracks and building movement.
- Be able to inspect and document foundation wall failure cause(s) and origin(s).

15.1   INTRODUCTION

There are various types of materials that can be used to build foundation walls. The three most common types of foundation construction include rubble stone, concrete masonry unit (CMU), and poured concrete.

Rubble stone foundations are the most common type of foundations in homes built before 1915.[1] The construction of a rubble stone wall begins by placing larger stones on top of one another, then filling the interstices with spalls or chips of stones, and finally adding mortar to fill in the crevices. A vertical bed filled with gravel or sand should be placed along the exterior of the wall to allow a drainage path for groundwater.[2]

CMU or concrete block foundation walls are commonly found in the United States due to their strength, durability, and fire resistance. The CMU foundation wall is constructed by stacking masonry units, typically in a staggered pattern with one-half of the unit above overlapping one-half of the unit below, commonly referred to as "running bond," to form continuous horizontal layers of masonry units known as courses. During the stacking process, the units are bonded together with a bed of mortar placed between each course, commonly referred to as bed joints, and between the masonry units of a course, commonly referred to as head joints. If the hollow cores of the CMU wall are partially or fully grouted and reinforced with steel, the wall will have a tendency to act more like concrete. If the hollow cores of the wall are not grouted or properly reinforced, the wall will act more like a brick or stone masonry wall.[3]

A continuous poured concrete foundation/footer is constructed by first setting forms spaced to achieve wall widths required by local codes. Reinforcing steel bars are then added within the forms, which serve to strengthen the wall. Finally, concrete is poured into the forms and anchor bolts or sill plate strap anchors are inserted into the concrete before it sets, which will anchor the building to the foundation.[4]

The foundation of the structure transfers the load of the building to the footer. The footer then transfers the load to the soil/gravel/rock below. The soil-bearing capacity is the capability of the soil to support the loads applied to the ground. A structure built on a solid foundation is critical to the integrity of the whole structure.

The type of soil around the foundation should be noted before the foundation is constructed to determine if the soil properties can withstand the forces imposed by the structure. According to the Military Soils Engineering FM 5-410,[5] soils are broken down into two main divisions, course-grained and fine-grained soils. Each soil division is broken down and given a soil classification symbol and description. The soil is rated on the respective drainage characteristics, frost action, and volume change potential (i.e., compressibility and expansion). Different soil types (e.g., rock/granite, gravel, sand, silt, and clay) also apply different lateral pressures to a foundation wall and may require a structural engineer to design footings and foundation walls based on local soil types and conditions.[6] Footers, unless very specifically designed, should not be installed into saturated soils or soils consisting of fill, since both conditions either reduce the allowable bearing strength of the soils or represent soils of unknown/uncertain bearing strengths. Buildings fabricated on footers and foundations set in saturated soils or improperly compacted soils will most likely have settlement-related problems. Further, in areas where winter frost levels are set by local/state codes, footers must be set below the frost line to avoid freeze/thaw-related movement and failures of the foundation systems.

Another consideration in foundation design is that the structure is made of a porous material (e.g., rubble stone or CMU) or material that may crack (e.g., poured concrete) and will likely be in contact with groundwater. The following are a few important principles in foundation design[7]:

- Water runs downhill and will flow toward lower foundation wall elevation(s).
- Dammed-up water (improper draining systems or ground slope around foundation walls) causes hydrostatic pressure against the foundation wall.
- Concrete, mortar, and masonry blocks are inherently porous materials.

In addition to the construction materials, a properly designed foundation system relies on limiting the amount of surface/groundwater flowing toward the foundations walls. Proven methods to limit the amount of surface/ground water that flows toward the foundation walls include providing adequate ground slope near foundation walls such that the ground surface slopes down and away 6 inches over 10 feet, the use of swales to direct surface water away from a structure, or installation of subsoil drainage systems. The subsoil drainage system around the foundation of a structure provides a means of removing water that accumulates at/near the footing/foundation wall. Typically, a perforated drainpipe surrounded by a coarse gravel bed is installed around the perimeter of the footing structure. Removing water away from the structure reduces forces from hydrostatic pressure and/or freeze/thaw cycles.

The performance of the foundation wall is directly related to the amount of water/moisture penetration and the presence of water/moisture around the foundation and/or footer. Typical failures associated with improper water management manifest themselves in cracked, leaning, and/or bowed walls. Total wall collapses have occurred in the most severe cases of improper water management.

This chapter will address the following facets of foundation walls:

- Building code requirements for stone rubble, CMU, and concrete foundation walls.
- Correlation of cracks and building movement.
- Causes of foundation wall distortion.
- Differences between cosmetic damage and structural damage.
- Inspection methodologies.
- Common failure modes and causes for these systems observed in the field.

15.2 FOUNDATION CODE DESIGN REQUIREMENTS

Information regarding foundation requirements can be found in most modern residential and commercial building codes and in best practice documents. The basis for these requirements is long-term historical experience by those practicing in their specific fields of expertise. Codes typically provide more general (and minimal) guidance, whereas best practice documents will tend to be more prescriptive by providing guidance that is more detailed. When codes are prescriptive, this often reflects conditions where problems have occurred in the past and the codes officials want to ensure such problems are reduced or eliminated. For example, more recent code language is quite specific regarding requirements for fill materials, whereas past versions of the code were silent in this area.

An example of typical code language for foundations can be found in the 2018 International Residential Code (IRC),[8] Chapter 4 (Foundations), which is reproduced in part below.

R401.3 Drainage: Surface drainage shall be diverted to a storm sewer conveyance or other *approved* point of collection that does not create a hazard. Lots shall be graded to drain surface water away from foundation walls. The *grade* shall fall a minimum of 6 inches (1,152 mm) within the first 10 feet (3,048 mm).

Exception: Where *lot lines*, walls, slopes, or other physical barriers prohibit 6 inches (152 mm) of fall within 10 feet (3,048 mm), drains or swales shall be constructed to ensure drainage away from the structure. Impervious surfaces within 10 feet (3,048 mm) of the building foundation shall be sloped a minimum of 2% away from the building.

R405.1 Concrete or masonry foundations: Drains shall be provided around all concrete or masonry foundations that retain earth and enclose habitable or usable spaces located below grade. Drainage tiles, gravel or crushed stone drains, perforated pipe, or other *approved* systems or materials shall be installed at or below the top of the footing or below the bottom of the slab and shall discharge by gravity or mechanical means into an *approved* drainage system. Gravel or crushed stone drains shall extend at least 1 foot (305 mm) beyond the outside of the footing and 6 inches (152 mm) above the top of the footing and be covered with an *approved* filter membrane material. Except where otherwise recommended by the drain manufacturer, perforated drains shall be surrounded with an *approved* filter membrane or the filter membrane shall cover the washed gravel or crushed rock covering the drain. Drainage tiles or perforated pipe shall be placed on not <2 inches (51 mm) of washed gravel or crushed rock not less than one sieve size larger than the tile joint opening or perforation and covered with not <6 inches (152 mm) of the same material.

Exception: A drainage system is not required when the foundation is installed on well-drained ground or sand-gravel mixture soils according to the United Soil Classification System, Group 1 soils as detailed in Table R405.1.

The drainage language is intended to keep surface water from flowing against foundation wall systems. Water flowing against foundation walls increases the hydrostatic loads on the wall(s) and decreases the strength of the soils below the footer. Both factors can adversely affect the integrity of foundation walls.

R404.1.8 Rubble stone masonry: Rubble stone masonry foundation walls shall have a minimum thickness of 16 inches (406 mm), shall not support an unbalanced backfill exceeding 8 feet (2,438 mm) in height, shall not support a soil pressure >30 pounds per square foot per foot (4.71 kPa/m), and shall not be constructed in Seismic Design Categories D_0, D_1, D_2, or townhouses in Seismic Design Category C, as established in Figure R301.2(2).

Note that the IRC requires a minimum thickness of 16 inches (406 mm) for a rubble stone masonry foundation and that the unbalanced backfill is less than or equal to 8 feet. As stated in the introduction, many rubble stone foundations were constructed prior to 1915 or prior to the development and enforcement of modern building codes.

Most foundation wall failures observed occur with rubble stone or CMU walls that did not contain reinforcement and/or had saturated soils exerting pressure. As discussed below, code language for these two scenarios is not typically conservative; but reinforcement of rubble stone walls is difficult from a practical standpoint, and many structures built with CMU foundation walls were constructed before the advent of modern state building codes in the 1990s.

Sections R404.1.1 (design of concrete and masonry foundation walls), R404.1.2 (design of masonry foundation walls), and R404.1.3 (concrete foundation walls) in the 2018 IRC describe the appropriate design provisions for the respective foundation walls. In accordance with R404.1.2, design masonry foundation walls shall also be in compliance with provisions of the TMS 402 and concrete walls with provisions in ACI 318/ACI 332/ or PCA 100. In accordance with R404.1.3.1, masonry foundation walls shall be constructed in compliance with Table R404.1.1(1), R404.1.1(2), R404.1.1(3), R404.1.1(4) and also shall comply with Sections R606. An example of typical code language contained in Table R404.1.1(1) for an 8-foot-high plain masonry foundation wall is provided in Table 15.1.[8]

TABLE 15.1

Portion of Table R404.1.1(1) for an 8-Foot Plain Masonry Foundation Wall

	Eight-Foot Plain Masonry Foundation Wall		
	Plain Masonry[b] Minimum Nominal Wall Thickness (inches)		
	Soil Classes[c]		
Maximum Unbalanced Backfill Height[a] (feet)	GW, GP, SW and SP	GM, GC, SM, SM-SC and ML	SC, MH, ML-CL and Inorganic CL
4	6 Solid[d] or 8	6 Solid[d] or 8	8
5	6 Solid[d] or 8	10	12
6	10	12	12 Solid[d]
7	12	12 Solid[d]	See Footnote[e]
8	10 Solid[d]	12 Solid[d]	See Footnote[e]

Source: Courtesy of the International Code Council.[8]

For SI: 1 inch = 25.4 mm, 1 foot = 304.8 mm, 1 pound per square inch = 6.895 Pa.

[a] Unbalanced backfill height is the difference in height between the exterior finish ground level and the lower of the top of the concrete footing that supports the foundation wall of the interior finish ground level. Where an interior concrete slab-on-grade is provided and is in contact with the interior surface of the foundation wall, measurement of the unbalanced backfill height from the exterior finish ground level to the top of the interior concrete slab is permitted.

[b] Mortar shall be Type M or S and masonry shall be laid in running bond. Ungrouted hollow masonry units are permitted except where otherwise indicated.

[c] Soil classes are in accordance with the Unified Soil Classification System. Refer to Table R405.1.

[d] Solid grouted hollow units or solid masonry units.

[e] Wall construction shall be in accordance with Table 404.1.1(2), Table 404.1.1(3), Table 404.1.1(4), or a design shall be provided.

[f] The use of this table shall be prohibited for soil classifications not shown.

Table 15.1 tabulates the minimum nominal wall thickness required for an 8-foot-high plain (unreinforced) masonry foundation wall based on maximum unbalanced backfill heights and classification of the soils retained. Note that the maximum unbalanced backfill guidelines set by the IRC are typically exceeded in real-world situations encountered. Typical plain masonry foundation walls inspected are ~8 feet tall and constructed with 8-inch nominal CMU. According to Table 15.1, the maximum unbalanced backfill height for a wall as previously described (8 feet tall with 8-inch nominal CMU) surrounded by a clay-like soil is 4 feet. In practice, the backfill is not limited to halfway up the foundation wall (~4′), but commonly extends up to one or two courses below the top of the wall (up to ~7′ of unbalanced backfill). Thus, plain (unreinforced) masonry foundation walls, typically encountered, inherently exceed the code requirements. Many times, these walls fail due to the imposed loads from saturated soils (hydrostatic pressure).

Examples of typical code language contained in Tables R404.1.1(2–4) for an 8-foot-high reinforced masonry foundation wall are provided in Table 15.2.[8]

TABLE 15.2
Portions of Tables R404.1.1(2–4) for an 8-Foot Reinforced Masonry Foundation Wall

Eight-Foot Reinforced Masonry Foundation Wall

Wall Thickness (inches)	Distance of Rebar from Exterior of Wall (*d*)[a,b] (inches)	Height of Unbalanced Backfill[c]	Minimum Vertical Reinforcement and Spacing (inches)[b,d]		
			Soil Classes and Lateral Soil Load[e] (psf per foot below grade)		
			GW, GP, SW and SP	GM, GC, SM, SM-SC and ML	SC, MH, ML-CL and inorganic CL
8	>5	4 feet (or less)	#4 at 48	#4 at 48	#4 at 48
		5 feet	#4 at 48	#4 at 48	#4 at 48
		6 feet	#4 at 48	#5 at 48	#5 at 48
		7 feet	#5 at 48	#6 at 48	#6 at 40
		8 feet	#5 at 48	#6 at 48	#6 at 32
10	>6.75	4 feet (or less)	#4 at 56	#4 at 56	#4 at 56
		5 feet	#4 at 56	#4 at 56	#4 at 56
		6 feet	#4 at 56	#4 at 56	#5 at 56
		7 feet	#4 at 56	#5 at 56	#6 at 56
		8 feet	#5 at 56	#6 at 56	#6 at 48
12	>8.75	4 feet (or less)	#4 at 72	#4 at 72	#4 at 72
		5 feet	#4 at 72	#4 at 72	#4 at 72
		6 feet	#4 at 72	#4 at 72	#5 at 72
		7 feet	#4 at 72	#5 at 72	#6 at 72
		8 feet	#5 at 72	#6 at 72	#6 at 64

Source: Courtesy of the International Code Council.[8]

For SI: 1 inch=25.4 mm, 1 foot=304.8 mm, 1 pound per square foot=0.157 kPa/mm.

a   Mortar shall be Type M or S and masonry shall be laid in running bond.

b   Vertical reinforcement shall be Grade 60. The distance, *d*, from the face of the soil side of the wall to the center of vertical reinforcement.

c   Unbalanced backfill height is the difference in height between the exterior finish ground level and the lower of the top of the concrete footing that supports the foundation wall of the interior finish ground level. Where an interior concrete slab-on-grade is provided and is in contact with the interior surface of the foundation wall, measurement of the unbalanced backfill height from the exterior finish ground level to the top of the interior concrete slab is permitted.

d   Alternative reinforcing bar sizes and spacing having an equivalent cross-sectional area of reinforcement per lineal foot of wall shall be permitted provided the spacing of the reinforcement does not exceed 72 inches.

e   Soil Classes are in accordance with the Unified Soil Classification System and design lateral soil loads are for moist conditions without hydrostatic pressure. Refer to Table R405.1.

f   The use of this table shall be prohibited for soil classifications not shown.

Table 15.2 tabulates the minimum size and spacing of vertical reinforcement required for an 8-foot-high masonry foundation wall based on the maximum unbalanced backfill height, wall thickness, and classification of the soils retained. Note that the designed lateral soil loads and subsequent reinforcing requirements are for moist conditions without hydrostatic pressure. Foundation walls undergo increased stress due to hydrostatic pressure caused by wet/saturated soils; thus the reason for code language for proper ground slope and the need to direct surface water and groundwater away from foundation walls.

Table 15.3 tabulates the minimum size and spacing of vertical reinforcement for an 8-foot-high concrete foundation wall based on the maximum unbalanced backfill height, wall thickness, and classification of the soils retained. Note per Section R403.1.3.5.2, the center of the vertical reinforcement in basement walls as specified in Tables R404.1.2(2–7) shall be located at the center-line of the wall.[8] Also note that the designed lateral soil loads and subsequent reinforcing require-ments are for moist conditions without hydrostatic pressure.

TABLE 15.3
Portions of Tables R404.1.2(2–4) for an 8-Foot Flat Reinforced
Concrete Foundation Wall

Eight-Foot Flat Reinforced Concrete Foundation Wall

| | | Minimum Vertical Reinforcement – Bar Size and Spacing (inches)[b-f,g,h] | | |
| | | Soil Classes and Lateral Soil Load[i,j] (psf per foot below grade) | | |
Wall Thickness (inches)	Height of Unbalanced Backfill (feet)[a]	GW, GP, SW and SP	GM, GC, SM, SM-SC and ML	SC, MH, ML-CL and Inorganic CL
6	4	Not required	Not required	Not required
	5	Not required	6 @ 39	6 @ 48
	6	5 @ 39	6 @ 48	6 @ 35
	7	6 @ 48	6 @ 34	6 @ 25
	8	6 @ 39	6 @ 25	6 @ 18
8	4	Not required	Not required	Not required
	5	Not required	Not required	Not required
	6	Not required	Not required	6 @ 37
	7	Not required	6 @ 36	6 @ 35
	8	6 @ 41	6 @ 35	6 @ 26
10	4	Not required	Not required	Not required
	5	Not required	Not required	Not required
	6	Not required	Not required	Not required
	7	Not required	Not required	Not required
	8	6 @ 48	6 @ 35	6 @ 28

Source: Courtesy of the International Code Council.[8]

For SI: 1 inch = 25.4 mm, 1 foot = 304.8 mm, 1 pound per square foot = 0.1571 kPa²/m, 1 pound per square inch = 6.895 kPa.

[a] Where walls will retain 4 feet or more of unbalanced backfill, they shall be laterally supported at the top and bottom before backfilling.

[b] Table values are based on reinforcing bars with a minimum yield strength of 60,000 psi concrete with a minimum specified compressive strength of 2,500 psi and vertical reinforcement being located at the centerline of the wall. See Section 404.1.2.3.7.2.

[c] Vertical reinforcement with a yield strength of <60,000 psi and bars of a different size than specified in the table are permit-ted in accordance with Section R404.1.2.3.7.6 and Table 404.1.2(9).

[d] NR indicates no vertical reinforcement is required.

[e] Deflection criterion is $L/240$, where L is the height of the basement wall in inches.

[f] Interpolation is not permitted.

[g] See Section R404.1.2.2 for minimum reinforcement required for basement walls supporting above-grade concrete walls.

[h] See Table R611.3 for tolerance from nominal thickness for flat walls.

[i] Soil Classes are in accordance with the Unified Soil Classification System. Refer to Table R405.1.

[j] The use of this table shall be prohibited for soil classifications not shown.

Again, foundation walls undergo increased stress due to hydrostatic pressure caused by wet/satu-rated soils; thus the reason for code language of proper ground slope and the need to direct surface water and groundwater away from foundation walls.

15.3   EXTERIOR FORCES ON FOUNDATION WALLS

The foundation of a structure accomplishes two main objectives: it must evenly transfer the floor and roof loads to the footing below, and it must resist the lateral loads due to soil pushing on the basement walls.[9]

The vertical loads on a basement wall and the lateral loads from the soil on the outside of the wall vary due to the nature of the structure and soil properties. A well-designed and constructed foundation wall will be capable of sustaining and resisting these loads.[9,10]

Typical forces exerted on a foundation wall include the following:

- Lateral Soil Pressure.
- Surcharge Pressure.
- Hydrostatic Pressure.
- Tree Roots.
- Differential Settlement (Chapter 16 Section 16.1).
- Frost Heave (Chapter 16 Section 16.1).

Lateral soil pressure is the horizontal load generated by the weight of the soil retained by the wall. The lateral soil pressure increases linearly with depth, producing a triangular horizontal pressure gradient pattern as illustrated in Figure 15.1.

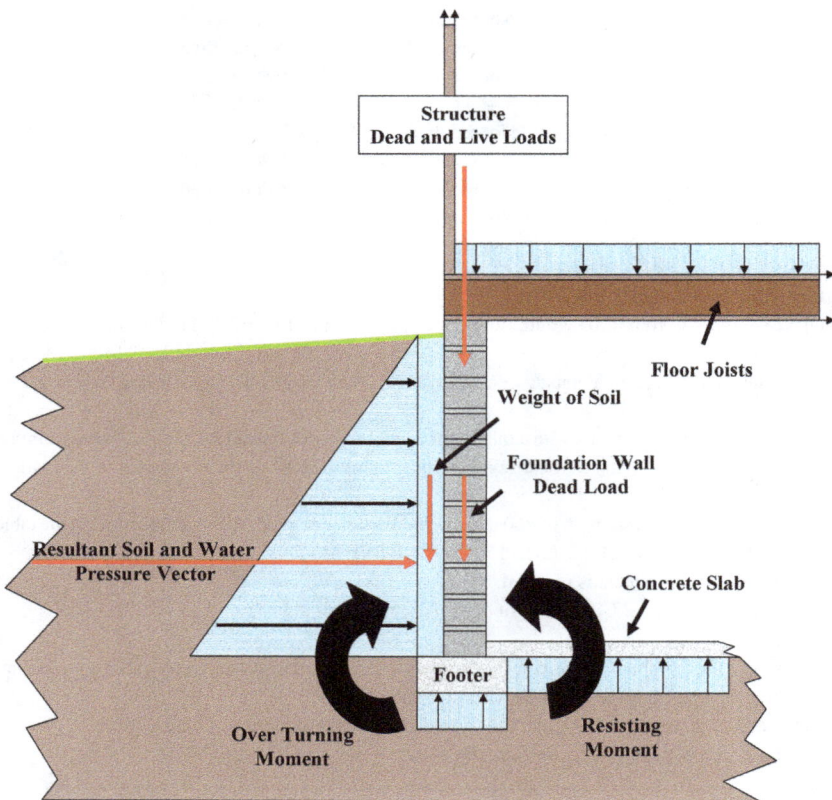

Typical Foundation Wall

FIGURE 15.1   Lateral soil pressure on foundation wall.

Surcharge pressures result from loads that are applied to the ground surface adjacent to the foundation wall. Due to soil characteristics, a vertical load applied to the surface of retained soils, such as that to the outside of a basement foundation wall, will result in a lateral pressure applied to that wall. Therefore, any increase in load on the ground surface (e.g., vehicles parked on driveways near the foundation) causes the total stress in the retained soils to increase, resulting in a surcharge pressure imposed on the foundation walls. However, unlike the lateral soil pressure, the surcharge pressure remains constant over the entire height of the wall (i.e., does not increase linearly with depth).

Hydrostatic pressure is the force generated by the accumulation of undrained groundwater in the soils outside of the wall. This situation can occur if no perimeter drainage system is present, or if the perimeter drainage system is clogged, or becomes inundated simply due to the volume of surface water and/or groundwater, often from improper grading or from overflowing gutters (see Chapter 8).

Typically, foundation walls must resist all, or a combination of, soil surcharge and hydrostatic pressures imposed on the walls. These pressures typically range from 30 to 62.4 pounds per square foot per foot of depth (psf/feet).[11] Similar to lateral soil pressure, hydrostatic pressure increases linearly with depth, producing a triangular horizontal pressure gradient pattern. For example, the pressure at the bottom of a 10′ deep foundation wall correlates to a hydrostatic pressure of 62.4 psf/feet × 10 feet, or 624 psf, in all directions.[11] Also, hydrostatic pressure will cause an upward (buoyancy) force on the basement concrete slab when the groundwater level is higher than the elevation of the concrete floor slab.

Additionally, trees and some bushes can grow roots that also exert pressure on foundation walls. Tree roots have been known to grow as far away from the tree as the tree is tall. Though, direct root action on a foundation wall is not as common, settlement or lateral pressure caused by shrinking/swelling action of the soil due to the presence of trees near the structure is a common occurrence. The presence of trees near a structure influences the amount of water in the soil around the foundation walls. The roots of trees resemble water pumps. The roots pull water from the ground to provide hydration and nutrients to the tree above. Hence, the hydrostatic pressure imposed on the foundation walls can fluctuate with seasonal changes or with the removal of trees. Additionally, if the root system is deep enough, the soil beneath a foundation/footer can shrink due to the water loss and cause settlement. For these reasons, guidelines such as the "Kew 'Tree To House Safe Distances' Table"[12,13] and tables produced by The Institution of Structural Engineers[12] and the National House Building Council (NHBC)/others[13] have been developed. Each tree species has a typical root spread based on its water demand. Subsequently, recommendations have been developed regarding the placement of common species of trees relative to the foundation of a structure. For example, an appropriate planning distance for an oak or willow tree from a foundation wall typically ranges from ~43 to ~60 feet (13–18 m).[12]

15.4   WALL DISTORTION

Foundation wall distortion occurs when the overturning moment of the soil and/or water pressure is greater than the resisting moment of the wall. The resisting moment must be greater than the overturning moment to avoid foundation wall failures. The resisting moment typically depends on the physical properties (e.g., width, reinforcement), dead load of the wall, and the interaction with the footing at the bottom of the wall and floor framing at the top of the wall. The overturning moment is any combination of lateral, surcharge, and/or hydrostatic pressures along with any other form of stress on the soil around the foundation wall. Figure 15.1 depicts the respective resisting and overturning forces on a typical foundation wall.

As illustrated in Figure 15.2, the overturning forces exceeding the resistive forces can cause a wall to lean/tilt, bow, and/or shear/displace; in a worst-case scenario, these forces will cause the foundation wall to collapse.

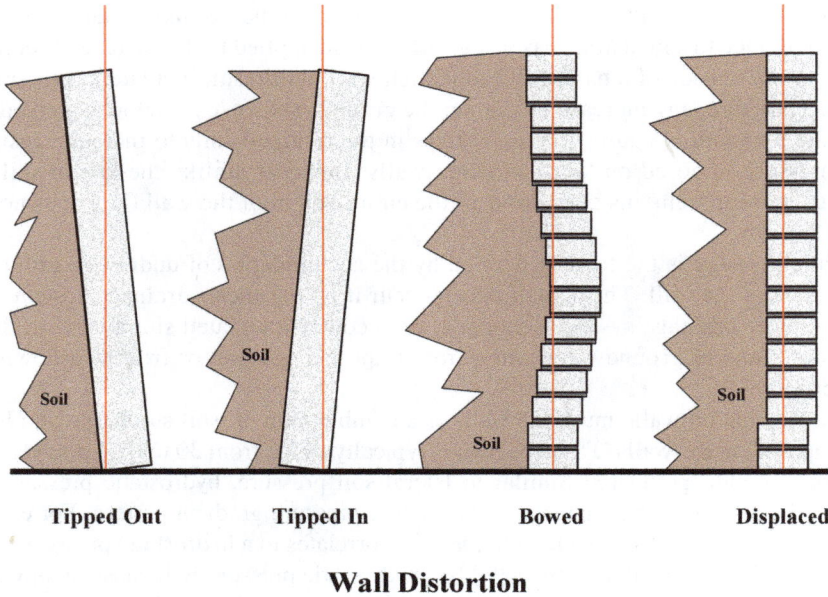

Wall Distortion

FIGURE 15.2   Four typical types of wall distortion.

Bowing and displaced walls are typically of more concern than tilted walls unless the wall tilt is approaching the recommended stability limit.

A properly designed foundation is expected to remain reasonably plumb and provide structural integrity. The structural integrity of the foundation wall is the ability of the foundation to support/resist the designed loads, including lateral pressures. Several indications of reduced structural integrity are commonly defined by excessive deflection, cracking, partial collapse, loss of a section, material deterioration, or demonstrated by calculations.[14]

A leaning/tilting foundation wall can compromise the integrity of a single portion of a structure or the entire building. A foundation wall is designed to be able to endure some tilting without overstress.[14] The stability criteria of a wall, defined in the *1997 Uniform Code for the Abatement of Dangerous Building*, published by the International Conference of Building Officials, is described as stable if the masonry's center of gravity falls inside of the middle one-third of the base of the bearing area.[15] This means a wall is deemed dangerous when a plumb line is attached to the top of a wall and the displacement of the plumb bob from the wall is >1/6 of the wall's bearing width. Figure 15.3 depicts a wall that is deemed dangerous due to the center of gravity falling outside of the middle third of the bearing area.

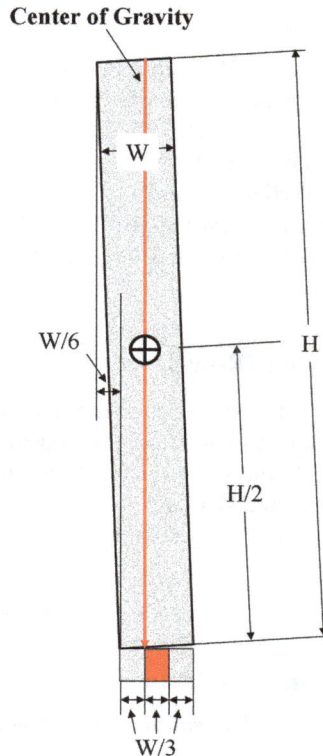

Unstable Wall

FIGURE 15.3 Visual depiction of the one-third rule for wall tilt.

A bowed and/or sheared foundation wall can also affect the structural integrity of a portion of the structure or the entire building. A bowed foundation wall commonly occurs due to increased lateral pressures, typically increased hydrostatic pressure, causing all or a portion of the soil to exert a larger load on the foundation wall. A sheared/displaced foundation wall occurs less frequently than a tilted and/or bowed wall; however, the conditions leading to the displacement are the same. The sheared/displaced wall typically occurs along a mortar joint in a CMU wall. The bowed and/or sheared foundation wall is deemed dangerous in the same manner as the tilted wall, when the center of gravity falls outside the middle third of the bearing area.

15.5 DIFFERENTIATING BETWEEN COSMETIC TILTS AND DAMAGE VS STRUCTURAL TILTS AND DAMAGE

In order to differentiate between cosmetic and structural damage to a foundation wall, typical construction performance guidelines were consulted to formulate a basis of standards. In a perfect world, walls are constructed plumb or 0° from vertical. To differentiate between cosmetic and structural damage, three classifications are assigned to express the degree of tilt/damage: (1) cosmetic, (2) acceptable (repair), and (3) significant (unstable).

A discussion for each of these classifications follows.

15.5.1 Cosmetic Tilt and Damage

The Residential Construction Academy (RCA) in association with the National Association of Home Builders defined residential carpentry standards. The RCA established national standards for the residential construction industry that reflected the industries skill requirements. The minimal acceptable measures for plumb in framing and concrete forms standards are ±1/4 inch in 8 linear feet.[16] This equates to ~0.15° from vertical. According to the RCA standards, the allowable new construction wall tilt of an 8-foot wall is ±1/4 inch or 0.15°. Cracking was reported to begin in a single wythe of brick masonry when the flexural deflection of about $L/2000$ is reached,[17–19] which is ~0.03°.

15.5.2 Acceptable Tilt and Damage (Repair)

Performance guidelines are established by agencies as a basis for coverage under the insured warranty programs. The guidelines offer contractors and their customers a benchmark that deals with the performance of the provided goods and services. The Construction Performance Guidelines for the Ontario Home Building Industry,[20] the National Association of Home Builders,[21] and the BRE Digest 475 Building Research Establishment[22] were researched and tabulated in Table 15.4.

TABLE 15.4
Classification of Acceptable Wall Tilt Tolerances

Classification of Acceptable Wall Tilt Tolerances				
		Structural Columns		
References	**Wall**	**Wood**	**Concrete or Masonry**	**Steel**
Construction Performance Guidelines For the Ontario Home Building Industry Ontario New Home Warranty Program[20]	Tolerance 25/2,400 mm Degrees 0.597	Tolerance 25/2,400 mm Degrees 0.597	Tolerance 25/2,400 mm Degrees 0.597	Tolerance 25/2,400 mm Degrees 0.597
Residential Construction Performance Guidelines Consumer Reference Third Edition National Association of Home Builders[20]	Tolerance 1/96 inch Degrees 0.597	Tolerance 0.75/96 inch Degrees 0.448	Tolerance 1/96 inch Degrees 0.597	Tolerance 0.375/96 inch Degrees 0.224
BRE Digest 475[21] Building Research Establishment[21]	Tolerance 1/100 Degrees 0.573	Tolerance 1/100 Degrees 0.573	Tolerance 1/100 Degrees 0.573	Tolerance 1/100 Degrees 0.573

The three independent agencies suggest that some method of repair (i.e., restoration, alteration, or partial or full replacement of materials) is necessary once the wall/column is out of plumb more than ~0.6°

The Building Research Establishment[21] classifies some remedial action when tilt to a structure is 1 unit in 100 units or ~0.573°.

As the American Society for Testing and Materials (ASTM) STP 992 articulated, in a 1988 review of the literature on cracks in masonry,[17,22,23] no repairs were recommended on a structural basis for walls not displaced by more than 1 inch for a normal story height or one-half inch for a bulge over a normal story height. For an 8-foot-high wall, this would equate to a slope of 0.597°.

15.5.3 Significant Tilt and Damage (Unstable)

The "one-third rule" stated in the 1997 Uniform Code for the Abatement of Dangerous Buildings provides an example of what is considered to be an unstable wall tilt. A wall is deemed stable (Section 15.4) if the center of gravity falls inside of the middle one-third of the base of the bearing area, or conversely dangerous if "the exterior walls or other vertical structural members list, lean, or buckle to such an extent that a plumb line passing through the center of gravity does not fall inside the middle one third of the base."[15] Thus, a stable wall is able to tilt up to 1/6 of the bearing width. Note that the one-third (or one-sixth) criterion is independent of wall height. In order to determine a basis for an unstable wall tilt using the "one-third rule," an analysis of three different wall thicknesses (i.e., 8, 10, and 12 inches) for an 8-foot wall was completed. The results of this analysis are summarized in Table 15.5.

TABLE 15.5
Analysis of "One-Third Rule" on an 8-Foot-High Wall

	"One-Third Rule" Analysis of an 8-Foot-High Wall					
	Wall Thickness					
	8 inches		**10 inches**		**12 inches**	
Stability	**Degrees from Plumb**	**Horizontal Displacement (inches)**	**Degrees from Plumb**	**Horizontal Displacement (inches)**	**Degrees from Plumb**	**Horizontal Displacement (inches)**
Unstable criteria	Actual	Actual	Actual	Actual	Actual	Actual
	0.796	1.33″	0.995	1.67″	1.19	2″
	Rounded	Rounded	Rounded	Rounded	Rounded	Rounded
	0.8	1-3/8″	1.0	1-5/8″	1.2	2″

As shown in this table, the "one-third rule" predicts that an 8-inch thick wall remains stable until the wall is tilted 1.33″ (~1.3/8″) or ~0.8°. For 10″ and 12″ thick walls, the allowable tilt increases to 1.0° and 1.2°, respectively. Note that the lateral displacement of the wall is an absolute value based on the bearing width of the foundation wall. The slope of the wall tilt is based on the height of the wall and changes with height.

Research also reported[18] that clay masonry wall instability occurs when the wall tilt reached 80% of the wall thickness. For an 8-inch thick wall (~8″×~16″ CMU wall ~8 feet in height) this would imply a wall tilting at 3.8°.

15.5.4 Recommended Wall Tilts/Slopes by Damage Classification

Based on information reviewed in this section and experience, recommendations regarding wall tilts by damage classification are presented in Table 15.6.

TABLE 15.6
Recommended Wall Tilt/Slopes by Damage Classification

Wall Tilt Classification	Range of Wall Tilt
Minor to acceptable	0.0°–<0.6°
Moderate (repair) damage	0.6°–1.0°
Severe (unstable) damage	>1.0°

Acceptable recommended ranges for cosmetic/acceptable wall tilt/slopes are from 0.0° to <0.6°. Recommended wall tilts/slopes for repairs range from 0.6° to 1.0°, and for instability are tilts/slopes >1°. Similar tilts/slope classifications are recommended for horizontal surfaces such as ceilings, floors, and slabs with the exception of slabs where water drainage is needed or required. In those cases, the minimum slope for the slab should be a 1/8 inch drop per foot away from the structure.

15.6 DISCUSSION OF CRACKS

Crack analysis is often coupled with wall/tilt analysis and ultimately is critical for helping in the determination of where, when, and why the damage occurred and the magnitude of the damage (e.g., minor, modest, or significant/major). Cracks form when materials move differently to the provisions of their design. Masonry and concrete used in foundation walls are strong in compression but are weak in tension.[9,10] A crack in a wall might form due to differential settlement, lateral displacement, water infiltration, and/or frost heave. Typical crack analyses include the following elements:

- Directionality[10]:
  - Vertical cracks.
  - Stair-stepped cracks.
  - Horizontal cracks.
- Age (older cracks vs newer cracks).
- Location.
- Movement:
  - Displacement (width).
  - Shearing.
  - Compression.
  - Relative movement of the wall on either side of the crack.
  - Movement in relation to the exterior ground slope.

By definition, the presence of a crack is a symptom of differential movement, not the cause. The directionality, age, and movement of a crack provide the inspector/engineer with valuable information about the history of the wall movement. Thus, careful inspection and documentation of the cracks are an integral part of wall assessments.

The types of cracks encountered and their probable causes are summarized below:

- Vertical Crack: A vertical crack typically passes through masonry units instead of following the mortar joints and proceeds in a mostly vertical direction. A mostly vertical direction may result in a stair-stepped crack and may not be perfectly vertical. This type of crack is typically the result of shrinkage and/or thermal movement of the wall.[10]
- Horizontal Crack: Especially for foundation walls, a horizontal crack causes the most concern for inspectors/engineers. It suggests that the lateral soil loads exceed the strength of the foundation wall, causing a bowing and/or displacement action.[10] Horizontal cracks

typically are of no more concern than the other two types of cracks for non-foundation walls.

- Diagonal Crack: A crack commonly found on drywall, plaster, and CMU wall surfaces that typically are wider at one end compared with the opposite end of the crack. This type of crack is typically the result of uneven or differential settlement on either side of the crack.[12]

- Stair-Step Crack: A stair-stepped crack involves a diagonal zigzag pattern that typically follows the mortar joints. Again, this type of crack is typically the result of uneven or differential settlement on either side of the crack.[10]

A final category of cracks would be a combination of two or more of these four types of individual cracks, a condition often encountered during forensic inspections. For example, a crack may be horizontal near the center of a wall, then transition into a stair-stepped crack near the end of the wall.

Another aspect of a crack inspection is the relative age of the crack. Often in forensic investigations, the client will want to know whether the damage was recent, associated with a claimed incident, or preexisting. Such information may be needed in situations that include damage associated with foundation failures, blasts and explosions, fire, tornado or other weather-related events, and general damage to walls and ceilings.

The age of the crack is determined by close inspection. A magnifying lens can be used to increase the intricacies and allow for a better visual inspection of the crack. Terminology for the age of a crack is divided into two categories: newer in appearance and older in appearance (i.e., days/weeks vs months/years). A newer-in-appearance crack will be clean, will have sharp crisp edges, and is almost always brighter in color than adjacent surfaces. An older crack will have dirt/debris/cobwebs and/or paint in the crack and have softer eroded edges that appear duller in color.[12] The age of the crack is important in differentiating which cracks are the result of recent movement/events and which cracks are historically static. Table 15.7 depicts both old and new types of, vertical, stair-stepped, and horizontal cracks.

TABLE 15.7
Examples of Crack Types and Newer/Fresh vs Older Cracks

Examples of Crack Types and Newer/Fresh vs Older Cracks		
Crack Description	Fresh/New Crack	Older Crack
Vertical crack		

(Continued)

TABLE 15.7 (*Continued*)
Examples of Crack Types and Newer/Fresh vs Older Cracks

Examples of Crack Types and Newer/Fresh vs Older Cracks

Crack Description	Fresh/New Crack	Older Crack
Horizontal crack		
Diagonal crack		
Stair-stepped crack		

Note that a shear crack, such as a vertical or stair-stepped crack, will have a relative displacement along the crack with similar features on either side of the crack, whereas a compressive crack will show spalling and flaking of the masonry surface at the edges of the crack.[12] Hence, another aspect of crack analysis is to determine the relative movement of wall sections comparatively on either side of the crack. This provides information regarding settlement or movement of the wall. For example, it is common for the downhill (relative to outside ground slope) wall section to be settling or lower than the uphill wall section when comparing wall sections on either side of a crack. The ideal location for obtaining this information is where the crack changes direction (e.g., the change from vertical to horizontal direction in a stair-step crack). This is illustrated in Figure 15.4.

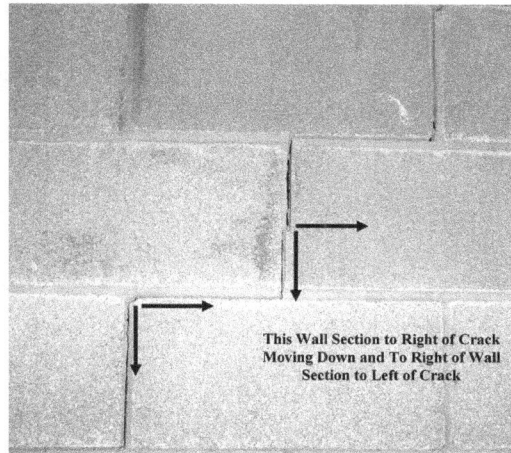

FIGURE 15.4 Relative movement of wall sections on either side of crack.

The final and perhaps most important aspect of cracks is the width of the crack. Terms used to describe crack widths include minor, moderate, and severe. A review of information on classification of crack widths is summarized in Table 15.8.

TABLE 15.8
Literature on Classification of Crack Widths

References	Classification of Visible Damage		
	Minor	**Moderate**	**Severe**
British Royal Institute of Chartered Surveyors (RICS) Degree of Damage per W.H. Ransom (RICS)[24]	Single isolated cracks – Fine cracks <5 mm (0.3″) wide, slightly sticking doors and/or windows	5–15 mm (0.2″–0.6″) wide, point up brick, some local replacement, doors/windows stick, pipes may break, not weather tight	>15 mm (0.6″) to >25 mm (0.98″), walls likely to lean, bulge, may require shoring; beams may lose bearing; windows distort, glass may break, pipes probably break. External repairs needed, partial or complete rebuild
Building Research Establishment[25]	Fine cracks that can be treated easily using normal decoration. Damage generally restricted to internal wall finishes; cracks rarely visible in external brickwork. Typical crack widths up to 1 mm (0.04″)	Cracks which require some opening up and can be patched by a mason. Re-pointing of external brickwork and possibly a small amount of brickwork to be replaced. Doors and windows sticking. Service pipes may fracture. Weather-tightness often impaired. Typical crack widths are 5–15 mm (0.20″–0.59″), or several of, say, 3 mm (0.12″)	Structural damage which requires a major repair job, involving partial or complete rebuilding. Beams lose bearing, walls lean badly and require shoring. Windows broken with distortion. Danger of instability. Typical crack widths are greater than 25 mm (0.98″), but depends on number of cracks
Movement control in the fabric of buildings[17,22,26]	Very slight: <1 mm (0.04″) Slight: 1–5 mm (0.04″–0.20″)	Moderate: 5–15 mm (0.20″–0.59″)	Severe – >15 mm (0.059″)

As illustrated in Table 15.8, crack widths are typically defined as minor, moderate, and severe. Minor cracks were typically <0.3 inch (~5/16″) wide. Moderate cracks were typically 0.3 inch (~5/16″) to 0.6 inch (~5/8″) in width. Severe cracks were typically >0.6 inch (~5/8″) wide.

Based on information reviewed in this section and experience, recommendations regarding crack widths by classification (minor, moderate, and severe) are presented in Table 15.9.

TABLE 15.9
Recommended Crack Widths by Crack Classification

Crack Width Classification	Range of Widths
Minor (repair)	0.0–0.125 inch
	0–1/8 inch
	0–3.18 mm
Moderate (repair)	0.125–0.500 inch
	0–1/2 inch
	3.18–12.70 mm
Severe (replace)	>0.500 inch
	>1/2 inch
	>12.70 mm

The recommended definition of a minor crack ranges from 0.0 to 0.125 inch (1/8″) in width. The recommended definition of a moderate crack ranges from 0.125 inch (1/8″) to 0.50 inch (1/2″) in width. The recommended definition of severe cracks is >0.50 inch (1/2″) in width.

15.7   FOUNDATION DAMAGE ASSESSMENT METHODOLOGY

Most foundation damage assessments encountered will be associated with either failed or failing foundation walls as shown in Table 15.10.

TABLE 15.10
Examples of Structural Damage to Foundation Walls

Cause of Damage	Photograph
Collapsed CMU foundation Caused by a combination of poorly drained silty-clay soil and a lack of steel in the wall	

(Continued)

TABLE 15.10 (*Continued*)
Examples of Structural Damage to Foundation Walls

Cause of Damage	Photograph
Impending failure of CMU foundation Caused by water intrusion due to undersized gutters, local ground slopes toward the structure, and lack of steel in the wall system	
Collapsed stone rubble foundation Caused by the improper construction of the foundation wall	
Severe buckling/bowing crack (~1″ wide) to CMU foundation wall	

(*Continued*)

TABLE 15.10 (*Continued*)
Examples of Structural Damage to Foundation Walls

Cause of Damage	Photograph
Bulged/tilted CMU foundation wall Caused by lateral pressure exerted on the exterior side of the walls exceeding the resistive strength of the walls which bowed/displaced them inward	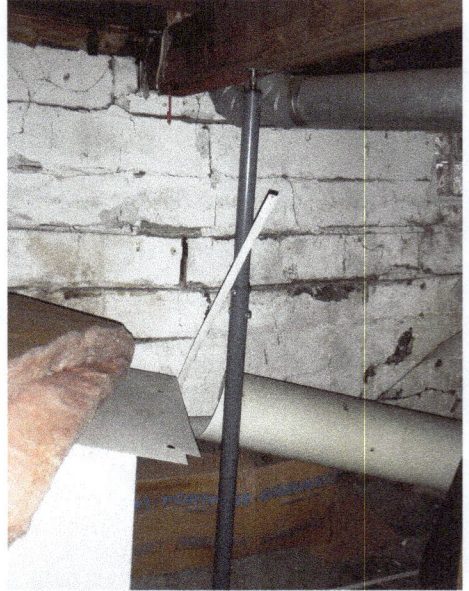
Vertical crack Caused by local ground subsidence	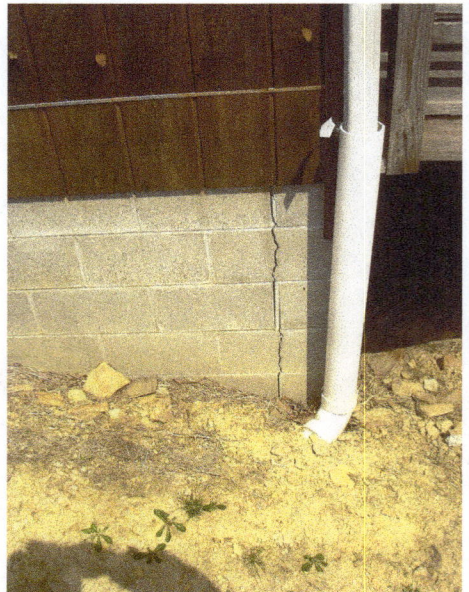

The inspection methodology for foundation damage assessment follows that outlined in Chapter 1 with the following modifications:

1. The inspection will likely be limited to the basement and/or crawlspace and exterior observations.
2. The focus will include measurements on foundation wall construction details, wall tilt/slope measurements, and crack analysis.

The methodology for completing foundation wall damage assessments includes the following, sequential elements:

- Interview of the property owner or their representative.
- Site inspection (both interior and exterior).
- Analysis of information collected.
- Written report, summarizing findings.

Details regarding each of these elements follow:

15.7.1 Foundation Damage Assessment Methodology – Interview of the Property Owner or Owner's Representative

Upon arriving at the site, the inspection process should follow the overall approach outlined in Chapter 1. Specific interview questions for foundation damage should include the following:

- When did the foundation damage occur? Record recollections regarding date and time if possible.
- What event(s) led to/caused discovery of the known damage to the foundation wall?
- What is the construction of the foundation wall(s)? Are any drawings regarding the foundation wall/footers available? If so, take pictures of key foundation/footer design features and obtain a copy of the drawings, if possible.
- Have there been any modifications to the foundation wall? If so, what, when, and by whom?
- Have there been any modifications to the exterior grade around the foundation wall? If so, what, when, and by whom?
- Have there been any recent heavy rain events coinciding with the foundation damage? If so, what was the date of the rainstorm?
- What are the damages (existing and recent), if known, to the foundation wall? Record information regarding displacement, spalling, cracks, staining, or other damage.
- Does the owner or owner's representative have opinions regarding the cause of the damage? If so, record those opinions.
- Have there been any moisture intrusion or flooding events into the basement or crawlspace? If so, when?

The sum of this information will assist in areas to focus on during the inspection and may help provide information forming the basis for opinions on the cause and origin of the failed or impending failure of a foundation system.

15.7.2 Foundation Damage Assessment Methodology – Site Inspection

The site inspection should follow the overall approach outlined in Chapter 1. All observations should be recorded in writing in a field notebook and key observations documented with photographs.

Particular attention should be spent documenting wall tilts/slopes and cracks to both interior and exterior readily accessible visible surfaces. Similar details should be recorded for finished floors where applicable.

During the inside portion of the inspection, the typical inspection methodology outlined in Chapter 1 should be followed; however, it will be necessary to record information not only by space, but also by walls, floors, and ceilings within each of the spaces. For each of these space surfaces, crack and distortion (i.e., tilted, bowed, displaced) observations should be documented.

Cracks on each surface should be documented by location, directionality, width, and crack characteristics (e.g., brightness of color and sharpness of crack surfaces, whether or not debris is present in the crack, and/or whether or not paint or other coatings are present on the edges of the crack). All observed and recorded damages should be photo-documented. The methodologies for recording and analyzing cracks discussed earlier in this chapter should be followed.

For illustration purposes, two example drawings of crack observations are shown in Figures 15.5 (floor crack observations) and 15.6 (wall crack observations):

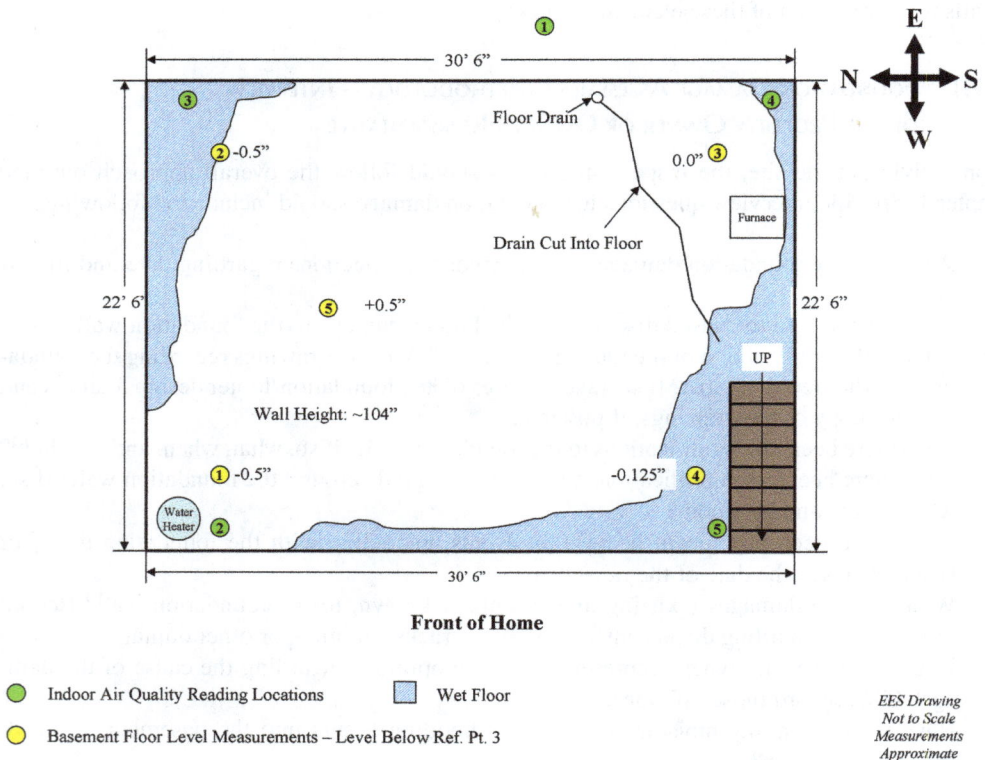

FIGURE 15.5 Illustration of documenting floor crack observations.

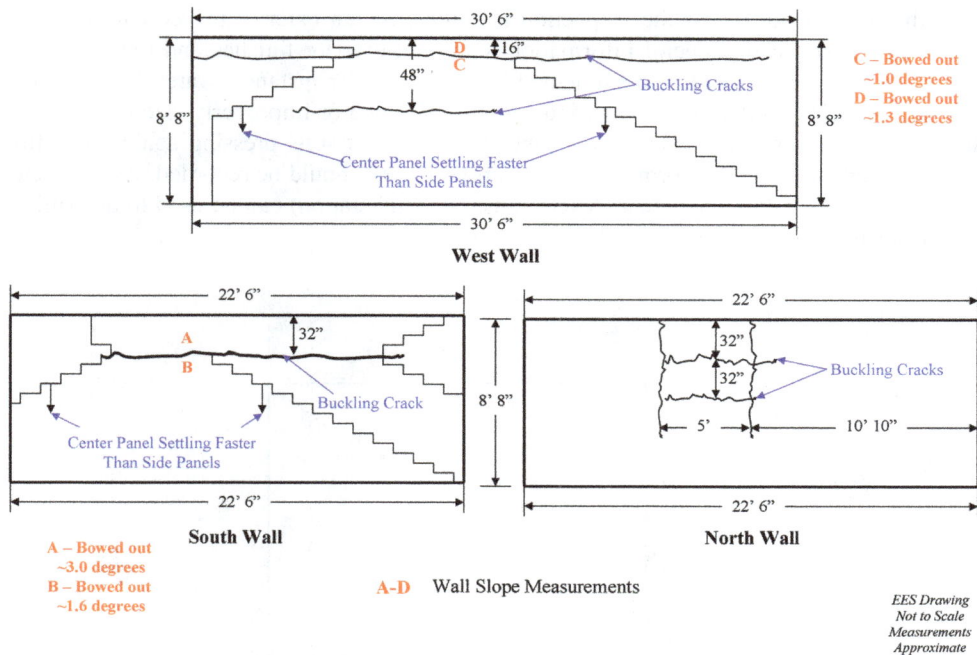

FIGURE 15.6 Illustration of documenting wall crack observations.

During a forensic inspection, the location of the crack in relation to a point of reference (i.e., floor, ceiling, and/or wall) and the starting/ending points of the crack are documented to show the directionality of the crack. If possible, measure the crack width to the nearest 1/32 of an inch. Cracks that are <1/32 of an inch are described as "hair-line" cracks. The width of the crack may vary throughout the length of the crack. If this is the case, record the crack width as being up to the maximum width or record the estimated average width of the crack noting which ends of the crack are larger/smaller.[12] At ideal crack locations, as discussed earlier in this chapter, determine and record relative movement of wall sections comparatively on either side of the crack.

In addition to documenting cracks, any anomalies should also be documented, such as:

- Fresh cracks in wood framing members such as joists.
- Gaps associated with movement between structural members.
- Gaps or wracking of windows and doors.
- Warping or bulging of floors.

The interior inspection should conclude with the recording (in tabular format) of wall plumb and floor level measurements to determine if the walls are plumb, bowed, or tilted/sloped and/or the floor surfaces are sloped. Floor-level measurements should be taken in two directions (e.g., north/south and east/west. Wall plumb or tilt/slope measurements should be performed following the methodology discussed in Chapter 14. Walls with tilt/slopes <0.6° are within the range of slopes commonly encountered in residential and light commercial construction. Hence, these walls likely were not damaged due to the claimed incident when the finished surface (e.g., drywall or plaster) is not freshly and/or significantly cracked.

Once the interior portion of the inspection is completed, an exterior inspection of the structure should be performed. General information regarding exterior finishes and the conditions of downspouts and gutters should be documented by elevation. For instance, water drip lines on the ground surface below gutters may be an indication of clogged or improperly sized gutters and/or downspouts. These conditions may be contributing to the wet soils pressing against a failing or failed foundation wall. Next, information on the ground slope should be recorded. As illustrated in Figure 15.7, surveying equipment (e.g., level, transit, or total station) can be used to determine the change in elevation around the structure.

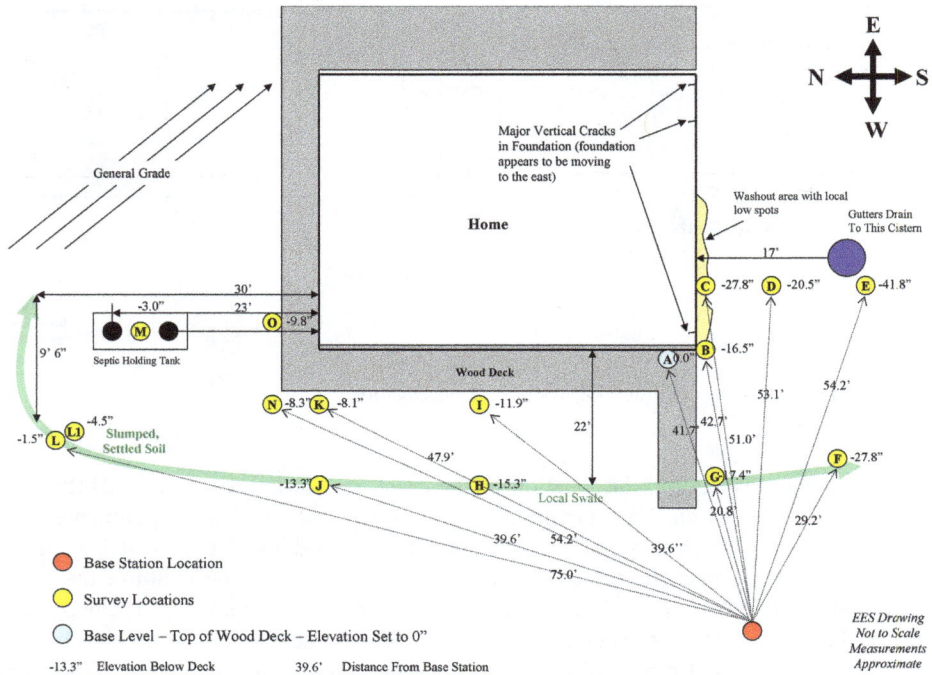

FIGURE 15.7 Illustration of exterior ground elevation and slope observations.

Once exterior ground slope and elevation observations have been taken, attention should be directed to recording cracks to the visible portion of the stone, concrete, brick masonry, and/or CMU foundation walls. These should be recorded following the methodology discussed earlier in the chapter.

15.7.3 FOUNDATION DAMAGE ASSESSMENT METHODOLOGY – ANALYSIS OF INFORMATION COLLECTED

The crack information along with the wall plumb and floor-level measurements should be analyzed and compared with the recommendations discussed earlier in this chapter to determine if the damage present in the foundation walls would be classified as minor, moderate, or severe. The analysis of additional information collected during the inspection should follow methodologies outlined in earlier chapters such as Chapters 1 and 14.

An example of floor-level measurement recordings for analysis is illustrated in Figure 15.8 and Table 15.11.

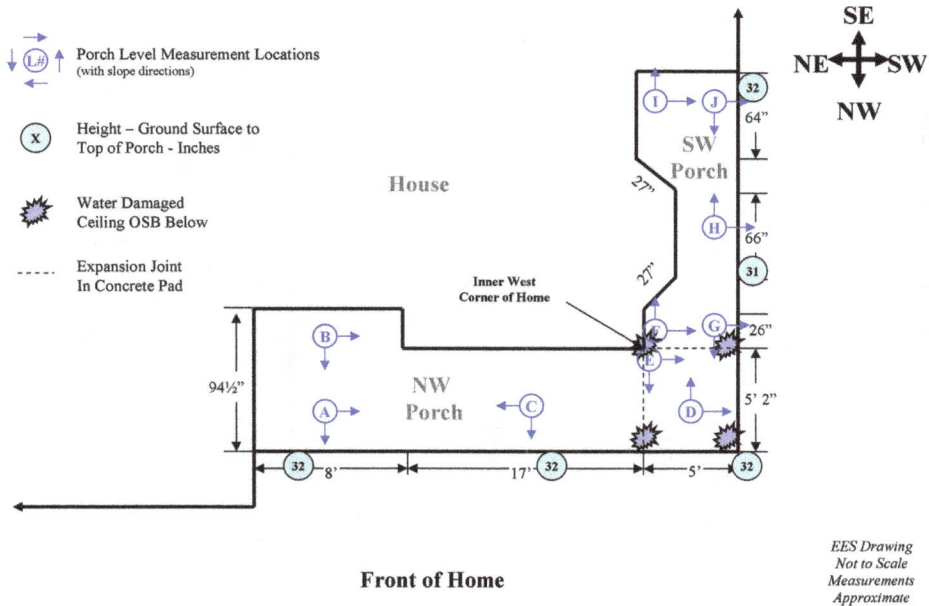

FIGURE 15.8   Illustration of floor-level measurements.

TABLE 15.11
Illustration of Reported Floor-Level Measurements

ID	Location	SE/NW Direction Direction Tipping Down	Slope	SW/NE Direction Direction Tipping Down	Slope
A	NW portion – NE end – NW side	Northwest	0.6°	Southwest	0.4°
B	NW portion – NE end – SE side	Northwest	0.5°	Southwest	0.1°
C	NW portion – central	Northwest	0.4°	Northeast	0.2°
D	Outer west corner	Southeast	0.3°	Southwest	0.1°
E	Inner west corner – NW side	Northwest	0.6°	Southwest	0.5°
F	Inner west corner – SE side	Southeast	0.1°	Southwest	0.5°
G	SW portion – NW end – SW side	Northwest	0.2°	Southwest	0.4°
H	SW portion – central	Southeast	0.2°	Southwest	0.7°
I	SW portion – SE end – NE side	Southeast	0.1°	Southwest	0.4°
J	SW portion – SE end – SW side	Northwest	0.3°	Southwest	0.9°

In this example, the following conclusions were drawn regarding the floor-level measurements:

- In general, the porch sloped down and away from the home in the northwest and southwest directions; however, the slope was very shallow.
- The slope of the porch concrete pad near the inner west corner was essentially level in the northwest/southeast plane.

An example of wall plumb measurement recordings for analysis is illustrated in Figure 15.9 and Table 15.12.

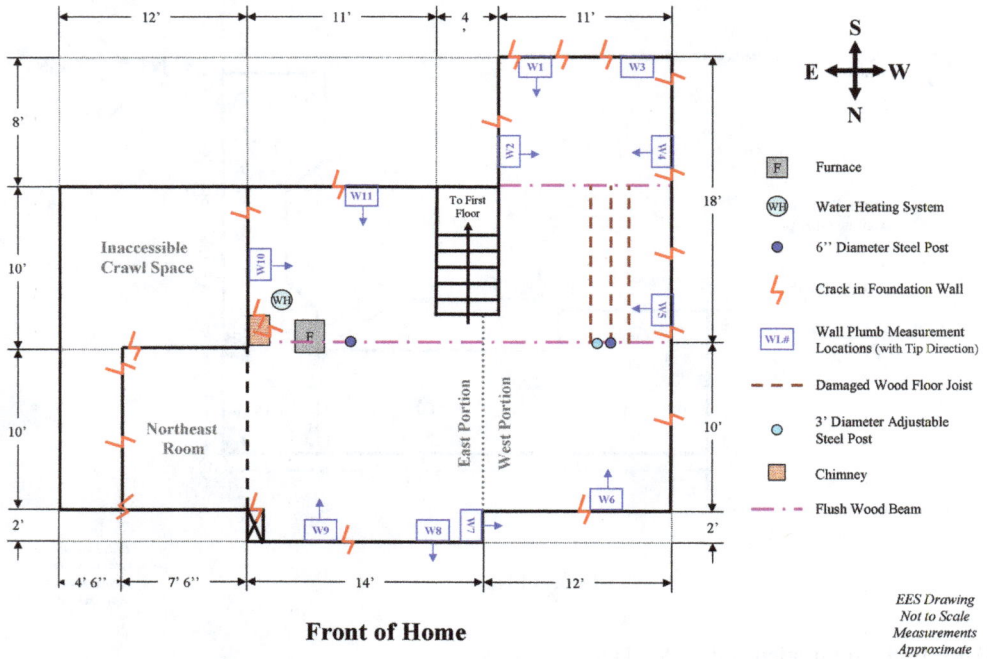

FIGURE 15.9　Illustration of wall plumb measurements.

TABLE 15.12
Illustration of Reported Wall Plumb Measurements

ID	Slope	Direction Top Tips
WL1	~0.40°	North
WL2	~1.00°	West
WL3	~0.00°	Plumb
WL4	~0.40°	East
WL5	~1.10°	East
WL6	~0.20°	South
WL7	~0.50°	West
WL8	~0.30°	North
WL9	~0.20°	South
WL10	~1.50°	West
WL11	~1.00°	North

In this example, where questions were raised as to whether or not a tree impact had caused damage to the foundation walls, the following conclusions were drawn based on a combination of wall plumb measurements and crack analysis:

The measurements indicated that the foundation walls were predominately rotated/leaning into the home. Observations of the cracks present in the foundation walls suggested that this rotation/leaning most likely occurred over a long period of time as a result of the lateral earth and hydrostatic pressure pushing on the exterior surfaces of the foundation walls rather than attributable to the impact of the fallen tree striking the home.

15.7.4 FOUNDATION DAMAGE ASSESSMENT METHODOLOGY – INSPECTION REPORT

Based on interview information, site inspection information, analysis, technical literature, and experience, a report of findings can then be prepared as outlined in Chapter 1.

The report should provide conclusions as to the likely cause(s) and severity of the damage to the subject foundation walls. Further, a competent contractor with proper experience should be recommended with regard to the removal/replacement of failed or severely damaged foundation walls or repairs for a moderately damaged foundation wall. When making recommendations, note that small hairline cracks (up to ~1/16″) can be repaired using epoxy injection and painting. Larger cracks (>1/16″) to CMU, concrete, and rubble stone foundation walls can be repaired with tuck-pointing using Portland cement mortar. Temporary shoring or supports designed/specified by a structural engineer should also be recommended as necessary. Any situations where imminent danger to life and property exists should be clearly emphasized (this information should be forwarded to responsible parties at/near the time of the inspection).

IMPORTANT POINTS TO REMEMBER

- Foundation wall systems are typically constructed from rubble stone, CMU, and poured concrete, or a combination of these materials.
- Residential and commercial codes provide guidance on proper construction of foundation walls and footers. However, rubble stone and unreinforced CMU foundation walls are often not built to these modern code requirements. Upon excessive soil and hydrostatic loads, these wall types will fail (the most common failures seen in forensic investigations).
- Surface water or groundwater saturating soils near foundation walls, and/or flowing against foundation walls, normally occurs due to improper grading of the ground surface near the home and/or faulty water management systems (i.e., gutters, downspouts, or subgrade drains).
- Wall tilts/slopes can cover the range from plumb, to cosmetic, to those being structurally unstable. Recommended values for ranges of wall tilts/slopes have been provided in this chapter.
- Crack analysis often provides a basis for helping to determine the direction, severity, and movement of walls.

REFERENCES

1. Kibble, Bill. "Historic Buildings – Stone Foundations," 2001–2009. Accessed June 4, 2020, http://historicbldgs.com/stonefoundations.htm.
2. The Colliery Engineer Co. "A Treatise on Architecture and Building Construction Vol. 2: Masonry, Carpentry, Joinery," 1899. Accessed June 4, 2020, http://chestofbooks.com/architecture/Building-Construction-V2/index.html.
3. Chylinski, Richard and Timothy P. McCormick. "5 Foundations." *Seismic Retrofit Training*, Chapter 5 - Foundations, pp. 65–70.
4. IRC and Home Building Answers. "House Foundations," 1990–2008. Accessed January 30, 2012, http://www.home-building-answers.com/house-foundations.html.
5. FM 5-410. "Military Soils Engineering." Department of the Army, updated June 4, 1997. Accessed June 4, 2020, http://www.bits.de/NRANEU/others/amd-us-archive/fm5_410(97).pdf.
6. Adams, David K. *"The Structural Engineer's Professional Training Manual."* New York: McGraw Hill, 2008.
7. The Engineered Wood Association. "APA (American Plywood Association) Publication A520 – Foundations", September 2017.

8. International Code Council. "2018 International Residential Code for One- and Two-Family Dwellings (IRC)." Accessed June 4, 2020, https://codes.iccsafe.org/content/IRC2018/chapter-4-foundations.

9. Das, Braja M. *"Principles of Foundation Engineering,"* 6th Edition. Toronto, ON: Thomson Canada Ltd., 2007.

10. Mattson Macdonald Young. "Basement Wall Cracking" Basset Creek Business Center: Minneapolis, 2015. Accessed June 4, 2020, http://www.mattsonmacdonald.com/wp-content/uploads/2015/12/Basement-Wall-Cracking.pdf.

11. Henshell, Justin. *"The Manual of Below-Grade Waterproofing Systems,"* 2nd Edition. New York: John Wiley & Sons, Inc., 2016.

12. Dickinson, Peter R., and Nigel Thornton. *"Cracking and Building Movement,"* 2nd Edition. United Kingdom: The Royal Institute of Chartered Surveyors, 2016.

13. Mercer, Giles, Alan Reeves, and Dealga O'Callaghan. "The Relationship between Trees, Distance to Buildings and Subsidence Events on Shrinkable Clay Soil." *Arboricultural Journal*, Vol. 33, pp. 229–245, 2011.

14. Texas Section of American Society of Civil Engineers. "Guidelines for the Evaluation and Repair of Residential Foundations," Version 2, adapted May 1, 2009. Accessed June 4, 2020, https://www.texasce.org/wp-content/uploads/2018/11/Guidelines-for-the-Evaluation-and-Repair-of-Residential-Foundations.pdf.

15. The International Conference of Building Code Officials. "1997 Abatement of Dangerous Buildings." March 1997. Accessed June 4, 2020, https://www.crystalriverfl.org/sites/default/files/fileattachments/planning_amp_development/page/4241/1997.pdf.

16. National Association of Home Builders. "Residential Carpentry Standards." Residential Construction Academy, National Association of Home Builders, Home Builders Institute Based on the Residential Construction Performance Guidelines for Professional Builders and Remodelers. 2nd Edition.

17. Masonry Indusiries. "Assessing the Condition and Repair Alternatives of Fire-Exposed Concrete and Masonry Members." Fire Protection Planning Report. Building Construction Information from the Concrete and Masonry Industries, National Codes and Standards Council of the Concrete and Masonry Industries: Skokie, Illinois, August 1994. Accessed June 4, 2020, https://pdfs.semanticscholar.org/19ce/ac79f6baf7481d1aa908a373db069c853937.pdf.

18. Grim, C. T. Masonry: Materials, Design, Construction and Maintenance. "Masonry Cracks: A Review of the Literature." ASTM STP 992. American Society for Testing and Materials. Harry A. Harris: Philadelphia, PA, 1988, pp. 257–280.

19. Structural Clay Products Research Foundation Research. "Compressive, Transverse, and Racking Strength Tests for Four-Inch Brick Walls. Report No. 9, Reston, VA: Brick Institute of America, August 1965, p. 17.

20. ONHWP (Ontario New Home Warranty Program). *"Construction Performance Guidelines for the Ontario Home Building Industry,"* Third Edition, January 1, 2013.

21. National Association of Home Builders. *"Residential Construction Performance Guidelines – Consumer Reference,"* Fourth Edition. Washington, DC: BuilderBooks, 2011.

22. Building Research Establishment. "BRE (Building Research Establishment) Digest 475 – Tilt of Low-Rise Buildings with Particular Reference to Progressive Foundation Movement," 2003.

23. American Society of Civil Engineers. "Guideline for Condition Assessment of the Building Envelop." SEI/ASCE 30-00, 2000.

24. Friedman, Daniel. "The Foundation Crack Bible – Foundation Cracks, Leans, Bulges, Settlement: Inspecting Foundations for Structural Defects – Detection, Diagnosis, Cause, Repair." Accessed June 6, 2019, https://inspectapedia.com/structure/Foundation_Crack_Dictionary.php.

25. Building Research Establishment. "BRE (Building Research Establishment) Digest 251 – Assessment of Damage in Low-Rise Buildings with Particular Reference to Progressive Foundation Movement," 1995.

26. Rainger, Phillip. *Mitchell's Movement Control in the Fabric of Buildings.* New York: Nichols Publishing Co., 1983, p. 67.

16 Forensic Inspection Assessments of Brick Masonry Chimneys, Veneer Walls, and Porches/Decks

Stephen E. Petty
EES Group, Inc.

CONTENTS

PURPOSE/OBJECTIVES

The purpose of this chapter is to:

- Define terminology and construction methods for brick masonry chimneys, veneer walls, porches, and decks.
- Understand the differences between residential and commercial code language and industry best practices for the design and construction of brick masonry chimneys, veneer walls, porches, and decks.
- Describe typical mechanisms and causes associated with the failure of brick masonry chimneys, veneer walls, porches, and decks.
- Demonstrate inspection methodologies for inspecting brick masonry chimneys, veneer walls, porches, and decks.
- Illustrate reporting approaches for brick masonry chimneys, veneer walls, porches, and decks.

Following the completion of this chapter, you should be able to:

- Understand terminology, code requirements, and best practices for construction of brick masonry chimneys, brick veneers, porches, and decks.
- Understand typical forensic failure mechanisms commonly found with brick masonry water management and related design/workmanship issues.
- Be able to determine and document the cause and origin of failure associated with brick masonry chimneys, brick veneers, porches, and decks.

16.1 INTRODUCTION

Masonry construction is widely used to build various structures, including houses, pathways, steps, decks, porches, and chimneys. In general terms, masonry structures are constructed of individual units laid in and bound together by mortar.[1] Familiar materials of masonry construction include brick, concrete block, and stone rubble.

All masonry construction, regardless of the material, utilizes mortar. Mortar is a malleable paste used to bind masonry units together and is typically composed of sand, water, and a binder (e.g., Portland cement or lime).[2] The paste hardens as it sets, forming a rigid aggregate structure with the masonry units. Due to the natural makeup of masonry construction, water is able to absorb into and through masonry units, which can cause damage if deviations are present in the product or if improper installation practices occur.

Aside from the quality of the masonry units, the strength and performance of masonry systems are dependent on two main components: water management details and design/workmanship. Variations in either of these two areas can potentially allow excessive moisture or water entry into the masonry system that could lead to degradation. Exterior masonry chimneys or veneer walls are constantly exposed to water/moisture. In some cases, freezing and thawing of this water/moisture occur.

The performance of a masonry chimney or veneer wall is directly related to both limiting the amount of water/moisture penetration and controlling or managing the water/moisture that does enter through the brick masonry. Typical failures in proper water management manifest themselves in spalling, efflorescence, cracking of the masonry, and/or rust oxidation, which results in "oxide jacking" of the embedded steel members. These failures are often caused by a lack of adequate weep holes and improper flashing details. Other design/workmanship issues regarding masonry include under-designed lintels and improperly designed and/or constructed footings, brick ties, and/or expansion joints.

In colder climates, mortar performance can be negatively affected by compositions mixed outside of the normal ranges and the insertion of excessive antifreeze materials.

This chapter will address:

- Brick Masonry Chimneys/Veneer Walls and Porches/Decks:
  - General Construction.
  - Building Code and Best Practice Requirements.
  - Inspection Methodologies.
  - Common Failure Modes and Causes for these Systems Observed in the Field.
- How to write a report consistent with the approaches outlined in Chapters 1, 14, and 15 once the inspection is complete.

16.2 BRICK MASONRY CHIMNEYS

The word "chimney" comes from the Latin word *caminus*, which means "furnace".[3] Man's desire to stay warm and remove combustion by-products caused the chimney to continually evolve. Chimneys are designed and constructed to provide fire protection and safely convey combustion by-products to the exterior of the structure at a rate that does not unfavorably influence the combustion process.[4]

Several typical chimney elements are listed and depicted in Figure 16.1.

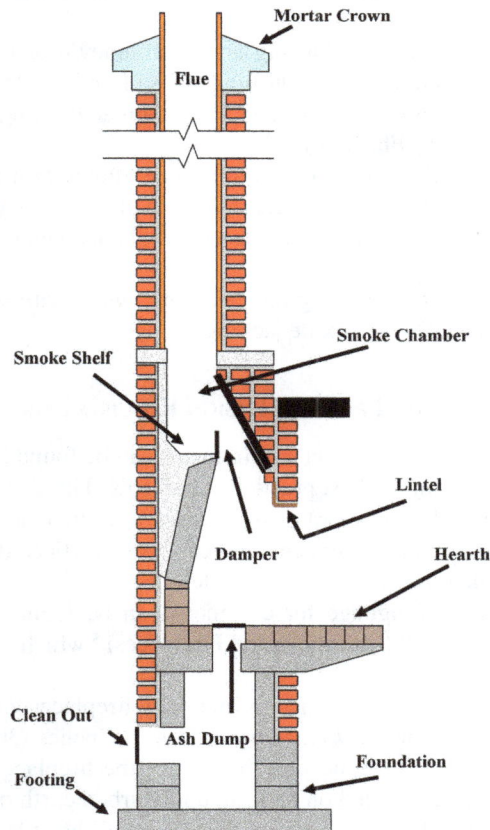

Typical Brick Masonry Chimney

FIGURE 16.1 Illustration of typical brick masonry chimney.

The most common issue encountered in forensic field investigations for the failure of a chimney is associated with the footings where the chimney appears to have rotated away from the structure. Often, when an insurance claim is made on a tilted chimney, the homeowner or the insurance adjuster attributes the leaning chimney to a windstorm event. However, two common reasons for the rotation at the footing are settlement of the ground below the footing due to inadequate footings or poor load-bearing soils and frost heave caused by the location of the footer above the frost line. Differential settlement of a foundation occurs when the soil below one portion of the footing settles more than another portion due to differences in the soil-bearing capacity. To prevent differential settlement, the footing should be founded upon natural, undisturbed earth or engineered fill below the frost depth. Frost heaving is the process by which the freezing of water-saturated soil causes the deformation and upward thrust of the ground surface. Frost heaving can cause foundations to crack and may lead to differential settlement of the foundation once the frozen ground has thawed. To prevent frost heaving, the bottom of a footing should be placed below the frost line. Each state and county typically will specify frost line depths in climates where this is an issue.

Aside from rotation at the footing, other common issues regarding brick masonry chimneys originate from:

- Spalling of the exterior brick due to water infiltrating into brick, then freezing/thawing.
- Efflorescence of the brick masonry due to interior or exterior moisture absorbing/wicking and transporting minerals and salts (i.e., calcium carbonate, sodium sulfate, and potassium sulfate) to the surface.
- Improper installation of chimney flashing at the chimney/roof interface and/or lack of proper repairs to damaged or degraded flashing. This often includes the flashing not being properly tucked into the mortar joints or lack of counter flashing (see Chapter 9 Water Infiltration – Brick Masonry Flashing).
- The lack of a cricket at the roof/chimney interface allows rain and meltwater to flow directly against the masonry and/or the accumulation of ice/snow at this interface.
- Gaps and cracks in the chimney cap coupled with an inadequate means of allowing water infiltration to escape.
- The lintel above the fireplace opening has been in contact with water/moisture and has subsequently oxidized leading to "oxide jacking."

16.2.1 Chimney Code and Best Practices Design Requirements

Information regarding chimney construction requirements can be found in most modern residential and commercial building codes and in best practice documents. The basis for these requirements is long-term historical experience by those practicing in their specific fields of expertise. Codes typically provide general (and minimal) guidance, whereas best practices documents will tend to be more prescriptive and provide guidance that is more detailed.

An example of typical code language for chimneys can be found in the 2018 International Residential Code (IRC), Chapter 10 (Chimneys and Fireplaces),[5] which is reproduced below:

R1001.2 Footings and foundation: Footings for masonry fireplaces and their chimneys shall be constructed of concrete or *solid masonry* at least 12 inches (305 mm) thick and shall extend at least 6 inches (153 mm) beyond the face of the fireplace or foundation wall on all sides. Footings shall be founded on natural, undisturbed earth or engineered fill below frost depth. In areas not subject to freezing, footings shall be at least 12 inches (305 mm) below finished *grade*.

R1001.7 Lintel and throat: Masonry over a fireplace opening shall be supported by a lintel of noncombustible material. The minimum required bearing length on each end of the fireplace opening shall be 4 inches (102 mm). The fireplace throat or damper shall be located a minimum of 8 inches (203 mm) above the lintel.

R1003.9.1 Chimney caps: Masonry chimneys shall have a concrete, metal, or stone cap, sloped to shed water, a drip edge, and a caulked bond break around any flue liners in accordance with ASTM C 1283. The concrete, metal, or stone cap shall be sloped to shed water.

R1003.9.3 Rain caps: Where a masonry or metal rain cap is installed on a masonry chimney, the net free area under the cap shall be not less than four times the net free area of the outlet of the chimney flue it serves.

The 2018 International Building Code (IBC), Chapter 21 (Masonry), Section 2111[6] contains similar code language.

These codes basically require that the foundation be at least 12 inches (305 mm) thick and extend a minimum of 6 inches (153 mm) beyond each face of the masonry bearing on the footing. The bottom of the footing should also be set below the frost line to reduce the possibility of the ground freezing below the footing, causing the foundation to heave.

The noncombustible lintel above the fireplace opening should bear on the brick masonry at least 4 inches (102 mm) at each end. The chimney cap should be sloped to drain water, with drip edge and a caulked bond break around the flue liner. Notably, codes are typically minimum requirements. Therefore, as a practical matter, industry best practice documents should be reviewed for a more detailed construction description.

Industry best practice documents, such as those developed by the Brick Industry Association (BIA), parallel code requirements (i.e., ICC International Residential Code and IBC) for chimney and fireplace design and construction,[4,7] but provide more detailed information on construction details based on industry experience. BIA[4,7] provides additional information about chimneys and fireplaces that goes above the "minimum" code requirements and may result in a reduced probability of common chimney and fireplace issues. The BIA provides the following guidance for brick masonry chimneys and fireplaces:

- When placing the lintel above the fireplace opening, a compressible, noncombustible material, such as insulation of a fibrous nature, should be placed at the end of the lintel where the lintel is embedded in the masonry. This precaution is a means of dealing with the dissimilar expansion characteristics of masonry and steel, which tend to induce stresses in the masonry, causing cracking.[7]
- When choosing a chimney cap, prefabricated caps generally provide superior performance as compared with the cast-in-place type. Adequate reinforcement should be placed in the cap to help control cracking due to shrinkage and thermal movements. Additional reinforcement may be necessary in the portion of the cap that overhangs the face of the chimney.[4]

BIA best practices documents also provide supplemental information regarding how to reduce lintel cracking and recommend a reinforced, prefabricated chimney cap to reduce several common issues that occur in chimneys and fireplaces. In many cases, the documents provide excellent schematics of construction details.

16.2.2 Methodology for Chimney Inspections

Chimney damage inspections on residential homes are conducted using a methodology similar to the one defined and illustrated in Chapter 1. When performing brick masonry inspections, special equipment is needed such as a ladder, a tape measure, a shovel, a pitch gauge, and an electronic level.

16.2.2.1 Interview – Obtain Pertinent Information from the Property Owner, Owner's Representative, or Occupant

During the interview, pertinent information should be obtained from the property owner, owner's representative, or occupant regarding the damage to the chimney and the local storm history. The following questions can assist in collecting background information for the investigation:

- Have there been any recent modifications to the brick masonry or chimney? If so, when and by whom?
- When did the wind event(s) occur (i.e., date and time)?
- From which direction did the windstorm arrive?
- What are the damages, if known, to the chimney?
  - Damages on the exterior surfaces of the chimney (i.e., displacement from the home, cracking, spalling, etc.).
- When were the damages first observed?
- What event(s) led to the caused/known damage to the chimney?

16.2.2.2 Create or Obtain a Basic Plan-View Sketch of the Chimney

After interview-related information regarding the structure and the storm history has been gathered, the chimney assessment process begins by sketching out the dimensions of the chimney in an inspection field notebook. Cracks, gaps, and other issues should be drawn in and noted in the field notebook, and any key observations should be documented with photographs.

16.2.2.3 Complete Chimney Inspection – Document Observations in Writing with Measurements and Photographs

The interior and exterior surfaces of the chimney should be inspected in order to perform a complete and comprehensive assessment of damage from wind or other forces. The interior of the chimney should be inspected before the exterior to avoid bringing debris into the home.

The interior of the chimney should be measured and inspected for cracks and deformations. The type, location, and measurements of any damaged components or defects should be documented in the field notebook. The crack type (horizontal, vertical, and/or stair-step), size (width), location, and which direction the crack is oriented should be noted. The condition of the crack and crack surfaces should also be recorded in writing and documented with photographs (see Chapter 15 for how to record and interpret crack observations). After the interior chimney inspection is complete, the exterior of the chimney should be inspected in the same manner.

Once the visual inspections of the interior and exterior surfaces of the chimney are completed, plumb measurements on the exterior surfaces of the chimney should be taken to determine the direction of rotation or buckling. A digital 2-foot level gives the user immediate knowledge of how many degrees the chimney is rotated. Plumb measurement typically should be taken on 3–4 heights of all accessible elevations of the chimney to determine the slope of each elevation at various heights. These measurements will verify whether or not the chimney is level, sloped uniformly in one direction, or is bowed. An example of uniformity of slope from a given chimney is presented in Table 16.1.

TABLE 16.1

Tabulating the Slope of Each Chimney Elevation

| Location | Degrees Tipped Out by Elevation | | |
	East[a]	South	North
1	1.8	0.0	0.0
2	1.8	0.0	0.0
3	1.8	0.0	0.0

[a] Note that the chimney was monotonically tipped or rotated to the east.

Plumb measurements were taken on the east, south, and north chimney faces, at elevations of ~3′, ~10′, and ~16′ above the footer. These locations were defined as "1," "2," and "3," respectively, and are illustrated in Figure 16.2.

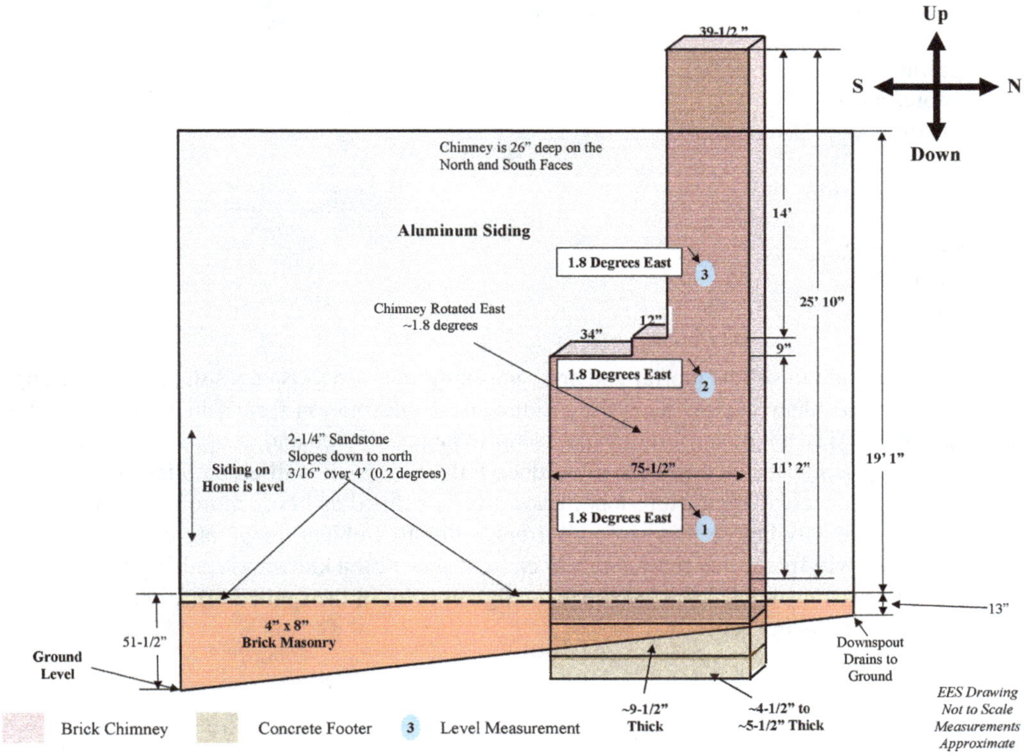

FIGURE 16.2 Typical chimney schematic (developed from on-site measurements).

Note that the chimney was monotonically tipped or rotated to the east. The uniformity of the slope (1.8°) along the east elevation of the chimney informs the inspector that the focal point of rotation is at the footer beneath the chimney.

After the exterior plumb measurements have been taken, the next step is to determine the depth and width of the footing by digging a hole in the ground next to the chimney. Special care should be considered when digging the hole. Check that there are no buried utilities or foundation drains near the chimney before digging. The goal of digging the hole is to determine the following:

- Depth of the bottom of the footer below the ground surface (serves as basis for whether or not the bottom of the footer is below the local frost line).
- Thickness of the footer.
- Width of footer outside the vertical plane of the chimney.
- Soil type (is it top soil or a soil of likely bearing strength?).

Once the footing of the chimney is uncovered, the parameters should be recorded in a field notebook and the key findings documented with photographs. Afterward, backfill the excavation; compact the soils as the excavation is filled; and attempt to leave the site as it was entered.

16.2.3 EXAMPLES OF COMMON CHIMNEY ISSUES OBSERVED IN THE FIELD

Water Management:

- Spalling.
- Efflorescence.
- Flashing.

Design/Workmanship:

- Chimney Caps.
- Lintels.
- Footings.

Spalling and degradation of the mortar and brick masonry units are likely a result of historic trapped moisture and freeze/thaw cycles. According to technical information from Old Virginia Brick on weatherproofing, "The single most important factor to be understood and designed for in the severe northern climate is moisture control. Seldom does failure of brick wall assemblies occur without the presence of an excessive moisture load. Excessive is defined as being more moisture than can dry out of the assembly before that assembly freezes due to ambient temperature conditions. The trapped moisture will freeze and thaw and may cause deterioration known as spalling, in which part of the brick surface may break away."[8] Notice the lighter color of the spalled brick in comparison to the surrounding bricks in Figure 16.3.

FIGURE 16.3   Spalled brick and mortar.

Efflorescence is another factor often found on field inspections of chimneys. Efflorescence is a light-colored crystalline deposit left behind from evaporated water or moisture and typically forms in porous masonry/concrete construction from water movement through the wall cavity. Four elements are required during the formation of efflorescence: free salt, a path to travel, moisture, and evaporation. The removal of any one of these four elements will typically prevent its formation.

Efflorescence is evidence of moisture wicking through masonry construction, or in more serious cases, water entry through openings. The presence of efflorescence on a wall is typically indicative of long-term water or moisture intrusion. In most cases, efflorescence is harmless and can be cleaned using mild acid solutions. In more severe cases, efflorescence can lead to spalling or damage to masonry wall cavities.[9] Figure 16.4 depicts an example of light-colored efflorescence deposits on the exterior of a spalled brick:

FIGURE 16.4   Efflorescence (white deposits) and spalling on brick masonry unit.

Another factor often found on field inspections of chimneys is improperly installed flashing. According to the Brick Industry Association (BIA)[10] and The Engineered Wood Association,[11] to prevent moisture infiltration into roof structures, the roof/chimney flashing should extend at least 4 inches up the sides of the chimney and under the roof, with counter flashing extending down at least 3 inches and embedded into the mortar joint at least 1-1/2″.[10,11] The counter flashing is then sealed with Portland cement mortar.[11] Chapter 9 illustrates these proper flashing methods. Figure 16.5 depicts an example of improperly installed chimney flashing. Notice the grooved mortar joints above the installed flashing. According to the APA, the counter flashing should be embedded into the mortar joint.

FIGURE 16.5 Improperly installed chimney flashing.

Cracks around the chimney cap and flue liner are other factors often found on field inspections of chimneys. The chimney cap (crown) is sloped to keep rain from seeping down into the masonry and deteriorating the chimney. Most mortar crowns crack almost immediately after installation because of shrinkage and become vulnerable to freeze/thaw actions. Figure 16.6 depicts an example of a cracked mortar chimney cap.

FIGURE 16.6 Crack in mortar chimney cap.

Another factor often found during field inspections of chimneys is rusting of the chimney metal. A cracked mortar chimney cap can result in water traveling down the chimney liner and into the fireplace. When a structural element comprised of steel or ferroalloy is exposed to water/moisture, the iron atoms within the structural element undergo an oxidation process and are converted into iron oxides (oxide scale). Oxidation (rusting) of the lintel over the fireplace opening leads to a process that culminates in "oxide jacking." This oxide scale can grow up to ten times the thickness of the original structural element. In other words, a steel lintel embedded in a masonry wall and supporting the brick above an opening can grow in height from 0.375 (3/8) inch to nearly 3.75 inches. As the oxide scale grows or expands, it places very large forces on surrounding materials. The upward or outward growth of the ferrous scale can easily move entire veneer sections, or if restricted, crush the masonry in contact with it. This movement of the masonry by the structural elements is called "Oxide Jacking." Once oxide jacking has begun, the only way to stop it is to remove the masonry and replace the steel or ferroalloy structural element. Oxide jacking is a long-term process that can be spotted well in advance of catastrophic failure of the masonry system. Unfortunately, since this process takes a long time to occur, it is often ignored or missed. Figure 16.7 depicts an example of oxide jacking to a metal lintel above a fireplace opening.

FIGURE 16.7   Chimney – oxide jacking to metal lintel.

16.3   BRICK MASONRY VENEER WALLS

Brick masonry veneer construction consists of a nominal 3″ or 4″ thick brick wythe anchored to a backing system with metal ties. The metal ties provide a clear air space between the veneer and the backing system. The backing system is typically constructed of wood framing, steel framing, concrete, or masonry.

A veneer wall is constructed of masonry units or other weather-resisting, noncombustible materials, securely attached to the backing, but not so bonded as to intentionally exert common action under load to the backing. Brick veneer is designed to carry the loads due to its own weight and is not designed to resist other loads.[12]

Several typical brick masonry veneer wall elements are listed and depicted in Figure 16.8.

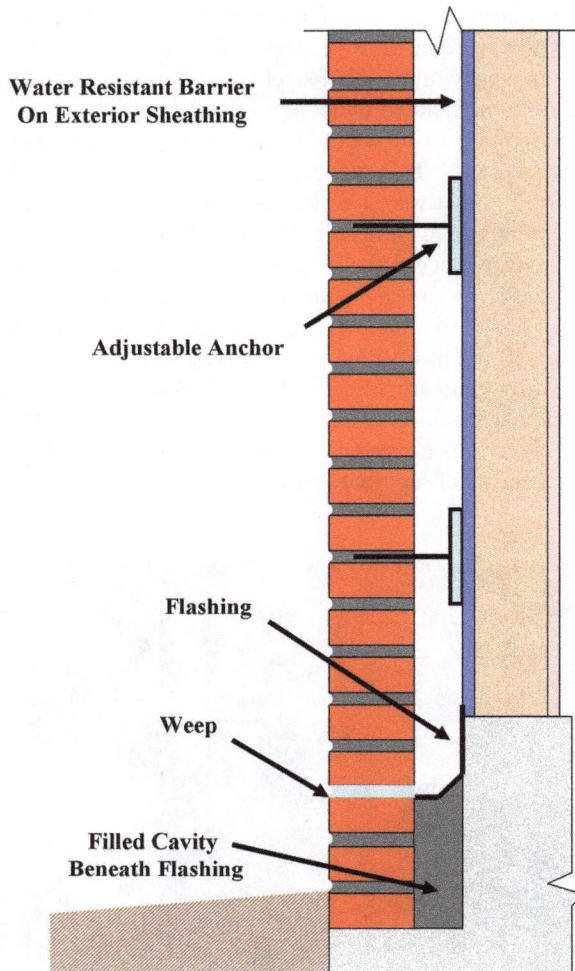

Typical Brick Masonry Wall Construction

FIGURE 16.8 Illustration of typical brick masonry veneer wall cross-section.

The most common issues encountered in the field regarding brick masonry veneer walls include problems with lateral support (i.e., masonry ties) or water management (i.e., through-wall flashing and weep holes). Lateral support of a brick masonry veneer wall is controlled by metal anchors, which are fastened between a solid structural member and the mortar of the brick masonry veneer. Ties must be of proper materials, strength, and spacing, and the method of attachment is set typically by codes and industry best practice documents. Water management of a brick masonry veneer wall is controlled through the installation of through-wall flashing and weep holes at points of support and along the bottom of the veneer above finished ground level. It is recognized by the building sciences that water/moisture will at times enter building wall systems. This issue is addressed by attempting to set limits on water/moisture infiltration and manage the water that does enter the wall system.

Aside from lateral support and water management-related brick masonry wall system failures, the most common failures observed (similar to Chimney Section 16.2.3) originate from the following:

- Spalling of the exterior brick due to water infiltrating into brick, then freezing/thawing.
- The occurrence of efflorescence on the brick masonry due to interior or exterior moisture wicking and transporting minerals and salts (i.e., calcium carbonate, sodium sulfate, and potassium sulfate) to the surface.
- Improper sill slopes around windows, allowing water to travel down into the wall cavity.
- The lintel above an opening in the wall is in contact with water and oxidizes causing "oxide jacking".
- Lack of expansion joints between the brick masonry units.

16.3.1 Brick Masonry Veneer Wall Code and Best Practices Design Requirements

Information regarding brick masonry veneer requirements can be found in most modern residential and commercial building codes and in best practice documents. The basis for these requirements is long-term historical experience by those practicing in their specific fields of expertise. Codes typically provide more general (and minimal) guidance, whereas best practices documents will tend to be more prescriptive, providing guidance that is more detailed. An example of typical code language for brick masonry veneer walls can found in the 2018 International Residential Code (IRC),[5] IRC Chapter 7 (Wall Covering) is reproduced below:

> *R703.8.4 Anchorage*: Masonry veneer shall be anchored to the supporting wall studs with corrosion-resistant metal ties embedded in mortar or grout and extended into the veneer a minimum of 1-1/2 inches (38 mm), with not <5/8 inch (15.9 mm) mortar or grout cover to outside face. Masonry veneer shall conform to Table R703.8.4(1). For masonry veneer tie attachment through insulating sheathing not >2 inches (51 mm) in thickness to not <7/16 performance category wood structural panel, see Table R703.8.4(2) – see Tables 16.2 and 16.3.

TABLE 16.2
2018 IRC Table R703.8.4(1) Tie Attachment and Air Space Requirements

Backing and Tie	Minimum Tie	Minimum Tie Fastener[a]	Air Space[b]
Wood stud backing with corrugated sheet metal	22 U.S. gage (0.0299 inch) × 7/8 inch wide	8d common nail[c] (2-1/2 inches × 0.131 inch)	Nominal 1 inch between sheathing and veneer
Wood stud backing with metal strand wire	W1.7 (No. 9 U.S. gage; 0.148 inch) with hook embedded in mortar joint	8d common nail[c] (2-1/2 inches × 0.131 inch)	Minimum nominal 1 inch between sheathing and veneer.
Cold-formed steel stud backing with adjustable metal strand wire		No. 10 screw extending through the steel framing a minimum of three exposed threads	Maximum 4-1/2 inches between backing and veneer

Source: Courtesy of International Code Council.

For SI: 1 inches = 25.4 mm.

[a] In seismic design category D_0, D_1, or D_2, the minimum tie fastener shall be an 8d ring-shank nail (2-1/2 inches × 0.131 inches) or a No. 10 screw extending through the steel framing a minimum of three exposed threads.

[b] An airspace that provides drainage shall be permitted to contain mortar from construction.

[c] All fasteners shall have rust-inhibitive coating suitable for the installation in which they are being used, or be manufactured from material not susceptible to corrosion.

TABLE 16.3

2018 IRC Table R703.8.4(2) Required Brick Spacing for Direct Application to Wood Structural Panel Sheathing[a-c]

Required Brick-Tie Spacing (Vertical-Tie Spacing/Horizontal Tie Spacing) (inches/inches)

Fastner Type[d]	Size (Dia or Screw #)	110 mph V Ultimate			120 mph V Ultimate			130 mph V Ultimate			140 mph V Ultimate		
		Zone 5 Expose. B	Zone 5 Expose. C	Zone 5 Expose. D	Zone 5 Expose. B	Zone 5 Expose. C	Zone 5 Expose. D	Zone 5 Expose. B	Zone 5 Expose. C	Zone 5 Expose. D	Zone 5 Expose. B	Zone 5 Expose. C	Zone 5 Expose. D
Ring shank nails	0.091	16/16, 16/12, 12/16, 12/12	16/12, 12/16, 12/12	12/12	16/16, 16/12, 12/16, 12/12	16/12, 12/16, 12/12	12/12	16/12, 12/16, 12/12	12/12	–	12/12	–	–
	0.148	24/16, 16/24, 16/16, 16/12, 12/16, 12/12	16/16, 16/12, 12/16, 12/12	16/16, 16/12, 12/16, 12/12	24/16, 16/24, 16/16, 16/12, 12/16, 12/12	16/16, 16/12, 12/16, 12/12	16/16, 16/12, 12/16, 12/12	16/16, 16/12, 12/16, 12/12	16/12, 12/16, 12/12	16/12, 12/16, 12/12	16/16, 16/12, 12/16, 12/12	16/12, 12/16, 12/12	12/12
Screws	#6	24/16, 16/24, 16/16, 16/12, 12/16, 12/12	16/16, 16/12, 12/16, 12/12	16/16, 16/12, 12/16, 12/12	24/16, 16/24, 16/16, 16/12, 12/16, 12/12	16/16, 16/12, 12/16, 12/12	16/16, 16/12, 12/16, 12/12	16/16, 16/12, 12/16, 12/12	16/12, 12/16, 12/12	16/12, 12/16, 12/12	16/16, 16/12, 12/16, 12/12	16/12, 12/16, 12/12	12/12
	#8	24/16, 16/24, 16/16, 16/12, 12/16, 12/12	24/16, 16/24, 16/16, 16/12, 12/16, 12/12	16/16, 16/12, 12/16, 12/12	24/16, 16/24, 16/16, 16/12, 12/16, 12/12	16/16, 16/12, 12/16, 12/12	16/16, 16/12, 12/16, 12/12	24/16, 16/24, 16/16, 16/12, 12/16, 12/12	16/16, 16/12, 12/16, 12/12	16/12, 12/16, 12/12	16/16, 16/12, 12/16, 12/12	16/12, 12/16, 12/12	16/12, 12/16, 12/12

(Continued)

TABLE 16.3 (Continued)
2018 IRC Table R703.8.4(2) Required Brick Spacing for Direct Application to Wood Structural Panel Sheathing[a-c]

Required Brick-Tie Spacing (Vertical-Tie Spacing/Horizontal Tie Spacing) (inches/inches)

Fastener Type[d]	Size (Dia or Screw #)	110 mph V Ultimate			120 mph V Ultimate			130 mph V Ultimate			140 mph V Ultimate		
		Zone 5 Expose. B	Zone 5 Expose. C	Zone 5 Expose. D	Zone 5 Expose. B	Zone 5 Expose. C	Zone 5 Expose. D	Zone 5 Expose. B	Zone 5 Expose. C	Zone 5 Expose. D	Zone 5 Expose. B	Zone 5 Expose. C	Zone 5 Expose. D
#10		24/16	24/16	24/16	24/16	24/16	16/16	24/16	16/16	16/16	24/16	16/16	16/12
		16/24	16/24	16/24	16/24	16/24	16/12	16/24	16/12	16/12	16/24	16/12	12/16
		16/16	16/16	16/16	16/16	16/16	12/16	16/16	12/16	12/16	16/16	12/16	12/12
		16/12	16/12	16/12	16/12	16/12	12/12	16/12	12/12	12/12	16/12	12/12	
		12/16	12/16	12/16	12/16	12/16		12/16			12/16		
		12/12	12/12	12/12	12/12	12/12		12/12			12/12		
#14		24/16	24/16	24/16	24/16	24/16	24/16	24/16	24/16	16/16	24/16	16/16	16/16
		16/24	16/24	16/24	16/24	16/24	16/24	16/24	16/24	16/12	16/24	16/12	16/12
		16/16	16/16	16/16	16/16	16/16	16/16	16/16	16/16	12/16	16/16	12/16	12/16
		16/12	16/12	16/12	16/12	16/12	16/12	16/12	16/12	12/12	16/12	12/12	12/12
		12/16	12/16	12/16	12/16	12/16	12/16	12/16	12/16		12/16		
		12/12	12/12	12/12	12/12	12/12	12/12	12/12	12/12		12/12		

Source: Courtesy of International Code Council.

For SI: 1 inches = 25.4 mm, 1 mph = 0.447 m/s.

a. This table is based on attachment of brick ties directly to wood structural panel sheathing only. Additional attachment of the brick tie to lumber framing is not required. The brick ties shall be permitted to be place over any insulating sheathing, not to exceed 2 inches in thickness. Wood structural panel sheathing shall be a minimum 7/16 performance category. The table is based on a building height of 30 feet or less.

b. Wood structural panels shall have a specific gravity of 0.42 or greater in accordance with NDS.

c. Foam sheathing shall have a minimum compressive strength of 15 psi in accordance with ASTM C578 or ASTM C1289.

d. Fasteners shall be sized such that the tip of the fastener passes completely through the wood structural panel sheathing by not <¼ inch.

R703.8.4.1 Size and spacing: Veneer ties, if strand wire, shall not be less in thickness than No. 9 U.S. gage [(0.148 inch) (4 mm)] wire and shall have a hook embedded in the mortar joint, or if sheet metal, shall be not less than NO. 22 U.S. gage by [(0.0299 inch) (0.76 mm)] 7/8 inch (22 mm) corrugated. Each tie shall support not more than 2.67 square feet (0.25 m^2) of the wall area and shall be spaced not more than 32 inches (813 mm) on center horizontally and 24 inches (635 mm) on center vertically.

Exception: In Seismic Design Category D$_0$, D$_1$, or D$_2$ or townhouses in Seismic Design Category C or in wind areas of more than 30 pounds per square foot pressure (1.44 kPa), each tie shall support not more than 2 square feet (0.2 m^2) of wall area.

R703.8.4.1.1 Veneer ties around wall openings: Additional metal ties shall be provided around wall openings >16 inches (406 mm) in either dimension. Metal ties around the perimeter of openings shall be spaced not more than 3 feet (9,144 mm) on center and placed within 12 inches (305 mm) of the wall opening.

R703.8.5 Flashing: Flashing shall be located beneath the first course of masonry above finish ground level above the foundation wall or slab and at other points of support, including structural floors, shelf angles, and lintels when masonry veneers are designed in accordance with Section R703.8. See Section R703.4 for additional requirements.

R703.8.6 Weep holes: Weep holes shall be provided in the outside wythe of masonry walls at a maximum spacing of 33 inches (838 mm) on center. Weep holes shall not be <3/16 inches (5 mm) in diameter. Weep holes shall be located immediately above the flashing.

The 2018 IBC, Chapter 14 (Exterior Walls), Section 14016, and the Building Code Requirements and Specifications for Masonry Structures (BCRMS)[13] contain similar code language. Further, the BCRMS provides supplementary information on brick masonry anchors and how the metal anchors should be attached to different backings.

Industry best practice documents such as those from the Brick Industry Association Technical Notes (BIA) parallel code requirements (i.e., ICC International Residential Code and International Building Code) for brick masonry veneer wall design and construction. The BIA[14] best practices generally provide more details than those found in residential and commercial codes and are more explicit on how to construct masonry wall systems better. The residential and commercial codes should be considered minimum standards. BIA provides tabulated data and recommendations on the tie spacing requirements per wall type (see Table 16.4).

TABLE 16.4

2003 BIA Table 1 Tie Spacing Requirements[a,b]

Wall Type	Tie System and Material	Maximum Cavity Width[c], inches (mm)	Maximum Area Per Tie, square feet (m²)	Maximum Vertical Spacing, inches (mm)	Maximum Horizontal Spacint, inches (mm)
Cavity (both wythes designed to resist out of plane stresses)	Unit tie W1.7 W2.8	4-1/2 (114)	2.67 (0.25) 4.50 (0.42)	24 (610)	36 (914)
	Standard joint reinforcement W1.7 W2.8	4-1/2 (114)	2.67 (0.25) 4.50 (0.42)	24 (610)	16 (406)
	Unit adj. double eye & pintle	4-1/2 (114)	1.77 (0.16)	16 (406)	16 (406)
	Unit adj. joint reinforcement	4-1/2 (114)	1.77 (0.16)	16 (406)	16 (406)
Brick veneer/ wood stud	Corrugated	1 (25)	2.67 (0.25)	18 (457)	32 (813)
	Other than corrugated adj. 2 piece W1.7	4-1/2 (114)	2.67 (0.25) 3.50 (0.33)	18 (457)	32 (813)
Brick veneer/ steel stud	Adj. unit veneer ties	4-1/2 (114) (2 inches (50 mm) recommended	2.67 (0.25) (2.0 square feet 0.18 m²) recommended	18 (457)	32 (813) (24 inches recommended)
Brick veneer/ concrete or CMU backing	Adj. unit and W1.7 Sheet metal and W2.8	4-1/2 (114)	2.67 (0.25) 3.50 (0.33)	18 (457)	32 (813)
Multi-wythe masonry composite	Unit ties W1.7 W2.8	No cavity	2.67 (0.25) 4.50 (0.42)	24 (610)	36 (914)
	Joint reinforcement W1.7 W2.8		2.67 (0.25) 4.50 (0.42)	24 (610)	36 (914)

Source: Reprinted with permission© the Brick Industry Association, Table 1, Technical Notes 44B-Wall Ties for Brick Masonry.

[a] Masonry laid in running bond. Consult application building code for special bond patters such as stack bond.

[b] Based on the requirements in the 2002 MSJC code.

[c] Maximum allowable distance between inside face of veneer and framing material, per MSJC code, unless noted otherwise.

The BIA[14] provides additional information about brick masonry veneer that tabulates the wall tie spacing requirements for different backing wall types. The BIA suggested requirements for anchored brick masonry veneer walls to wood-framed construction, which follow:

- There should be one tie for every 2-2/3 square feet (0.25 m²) of wall area with a maximum spacing of 24 inches (600 mm) o.c. in either direction. The nail attaching a corrugated tie must be located within 1/2 inch (13 mm) of the bend in the tie. The best location of the nail is at the bend in the corrugated tie, and the bend should be 90°.[14]

- Wire ties must be embedded at least 1-1/2 inches (38 mm) into the bed joint from the air space and must have at least 5/8 inch (16 mm) cover of mortar to the exposed face. Corrugated ties must penetrate to at least half the veneer thickness or 1½ inches (38 mm) and have at least 5/8 inch (16 mm) cover. Ties should be placed so that the portion within the bed joint is completely surrounded by the mortar.[14]
- Weep holes must be located in the head joints immediately above all flashing. Clear, open weep holes should be spaced no more than 24 inches (600 mm) o.c. Weep holes formed with wick materials or with tubes should be spaced at a maximum of 16 inches (400 mm) o.c.
- If the veneer continues below the flashing at the base of the wall, the space between the veneer and the backing should be grouted to the height of the flashing. Flashing should be securely fastened to the backing system and extend through the face of the brick veneer. The flashing should be turned up at least 8 inches (200 mm). Flashing should be carefully installed to prevent punctures or tears. Where several pieces of flashing are required to flash a section of the veneer, the ends of the flashing should be lapped a minimum of 6 inches (150 mm) and the joints properly sealed. Where the flashing is not continuous, such as over and under openings in the wall, the ends of the flashing should be turned up into the head joint at least 2 inches (50 mm) to form a dam.[14]

Note that BIA[14] recommends supplemental additions to the brick masonry tie spacing beyond "minimum" code requirements to reduce several common issues that occur to brick veneer walls.

16.3.2 Inspection Methodology for Brick Veneer Inspections

Brick masonry veneer wall inspections follow a methodology similar to that of chimney inspections detailed in Section 16.2.2. The methodology for brick veneer inspections follows.

16.3.2.1 Interview – Obtain Pertinent Information from the Property Owner, Owner's Representative, or Occupant

As part of an interview with the owner, owner's representative, or occupant, obtain background information regarding the apparent damage to the brick masonry veneer wall.

- Have there been any recent modifications to the brick masonry veneer (i.e., tuck pointing, brick or mortar repairs)? If so, when and by who?
- What were the damages, if known, to the brick masonry veneer wall?
  - Damages on the exterior surfaces of the brick masonry veneer wall (i.e., cracking, spalling, displacement, etc.).
- When were the damages first observed?
- What event(s) led to/caused the known damage to the brick masonry veneer wall?
- Did you notice any moisture intrusion into the structure prior to the damaged brick masonry veneer?

16.3.2.2 Create or Obtain a Basic Plan-View Sketch of the Building

After interview-related information regarding the structure has been gathered, the brick masonry veneer wall assessment process begins by sketching out a plan view of the building in an inspection field book. The dimensions of the building, cracks, bulges, missing bricks, and other issues should be indicated on the sketch and noted in the field notebook. Key observations should also be documented with photographs.

An example of a typical brick masonry veneer wall layout, with key details included, is provided in Figure 16.9.

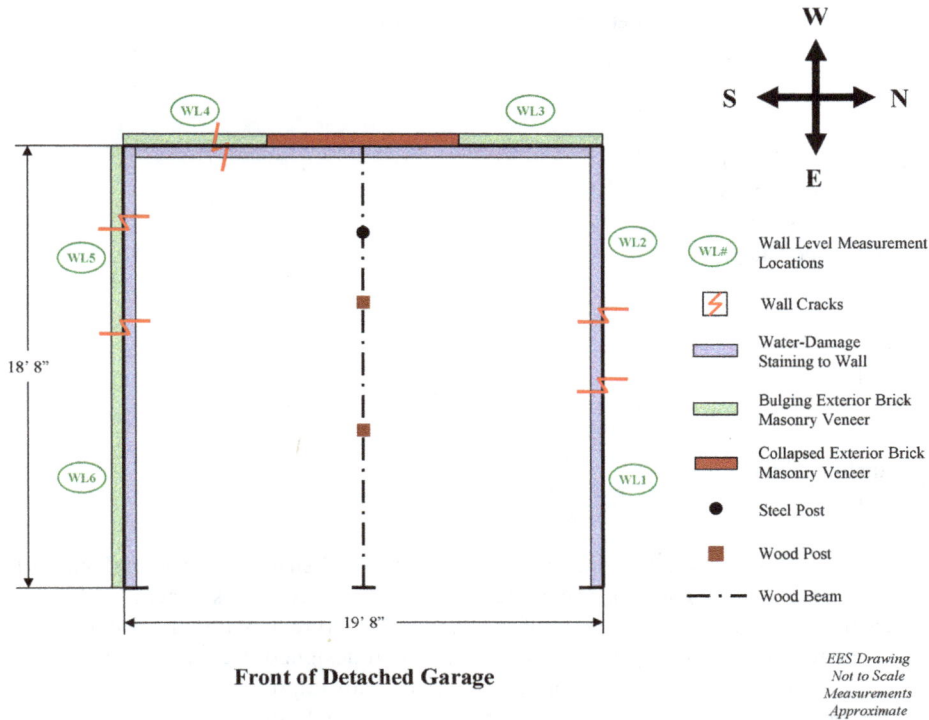

FIGURE 16.9   Typical brick masonry veneer wall schematic (developed from on-site measurements).

16.3.2.3   Complete Brick Veneer Wall Inspection – Document Observations in Writing with Measurements and Photographs

The exterior faces of the veneer wall should be thoroughly inspected. All observations should be documented, including information regarding defects and their locations along with proper measurements of damaged components. Inspect for and document the presence of weep holes and brick ties along the wall. Oftentimes, a boroscope with an attached digital camera is helpful for accessing and documenting the features and conditions within the air space behind the veneer. Weep hole and tie spacing, locations, and their visual characteristics (i.e., clogged, missing, displaced, corroded) should be documented. Next, lintels and sills above and below the wall openings should be inspected. Document the span of the lintel, the extent to which the lintel stretches beyond the window or door opening, the slope of the sill, and the condition of the lintel or sill. The exterior veneer wall should also be inspected for cracks and measured (i.e., take dimensions) for deformations. The crack type (horizontal, vertical and/or stair-step), size (width), location, and which direction the crack is oriented should be noted. In addition, the condition of the crack and crack surfaces should be recorded in writing in a field notebook and documented with photographs (see Chapter 15 for how to record and interpret crack observations).

Exterior wall plumb measurements (i.e., uniformity or nonuniformity of slope) should be taken and analyzed following the methodologies discussed in Chapters 14 and 15 to determine if the wall is bowing, tilting uniformly, or plumb.

If movement from settling or possible freeze/thaw action is suspected, excavate the ground near the wall following the approach as outlined in Section 16.2.2.3 above to rule in, or out, such possibilities for wall movement or damage.

16.3.3   Examples of Brick Veneer Wall Issues Observed in the Field

Examples of issues found in the field:
  Water Management:

- Spalling (see Section 16.2.3).
- Efflorescence (see Section 16.2.3).
- Weep holes (see Chapter 9).
- Flashing (see Chapter 9).

Design/Workmanship:

- Brick Ties.
- Sills.
- Lintels.
- Expansion Joints.
- Footings.

Lack of brick ties and/or sporadic spacing causes an uneven distribution of lateral support to the brick masonry veneer wall. The proper construction of an anchored brick veneer wall ensures that the wall functions as a complete system with the backing structure. Commonly, water management issues cause the metal brick ties to corrode, weakening their designed strength. Then, when a lateral load (e.g., wind load, seismic load) is applied to a wall with improperly spaced and/or corroded brick ties, the brick masonry has an increased chance of failure. Figure 16.10 depicts an example of a brick masonry veneer wall system that failed due to improperly spaced brick ties, ~32 inches on centers both vertically and horizontally or ~7.11 square feet of brick veneer wall area.

FIGURE 16.10   Veneer wall failure due to improperly spaced brick ties.

Inadequately sloped sills are another factor often found on field inspections of brick masonry veneer walls. According to the BIA publication on increasing water penetration resistance, through-wall flashing should be installed before the window is installed.[15,16] The brick masonry sill should be sloped down and away from the structure a minimum of 15° to drain water away from the window opening. Figure 16.11 depicts an example of a masonry sill that is sloped <15°. Note that the slope is ~2° away from the structure.

FIGURE 16.11   Low sloped sill (<15°).

Another factor often found on field inspections of brick masonry veneer walls is oxidation (rusting) of the metal lintel over openings causing "oxide jacking" (see Section 16.3.3). Figure 16.12 depicts an example of oxide jacking to a metal lintel above an opening in the brick masonry wall. Note the stair-stepped cracks at the top left and right corners of the lintel.

FIGURE 16.12   Oxide jacking to metal lintel.

Expansion and contraction of the brick masonry units are another factor that causes issues with brick masonry veneer walls. Bricks absorb moisture due to the natural permeability of brick masonry. The expansion process commences once the brick masonry is exposed to atmospheric moisture. The magnitude of the expansion depends on the nature of the raw materials, the method of manufacture, and the temperature and duration of the firing.[17] The vertical cracks caused by expansion/contraction are likely near the midpoint of the brick veneer wall or near the corner of the structure. Figure 16.13 depicts an example of cracking near the corner of a brick masonry wall due to a lack of expansion joints.

FIGURE 16.13   Cracks at corner probably due to lack of expansion joints.

16.4   PORCHES AND DECKS

The word "porch" originally derives from the Latin word *porticus* or the Greek word *portico*, both of which signify the columned entry to a classical temple.[18] The porch of a home is a common place to find people sitting and enjoying the outdoors under the comforts of a roof. By definition, porches are exterior structures at building entrances, typically composed of a roof, steps, and a guardrail. Posts/columns, the house wall, and/or foundations are typically used to support the porch and roof.

Decks, another exterior structure attached to a building at an entrance point, are slightly different by design. Decks typically are not covered by a roof and are usually raised slightly above grade level. The flooring system is open to allow water to pass though the floor and onto the ground surface. The flooring system of a deck is commonly supported by posts, a house wall, and/or foundations.[19]

Several typical porch/deck elements are listed and depicted in Figure 16.14.

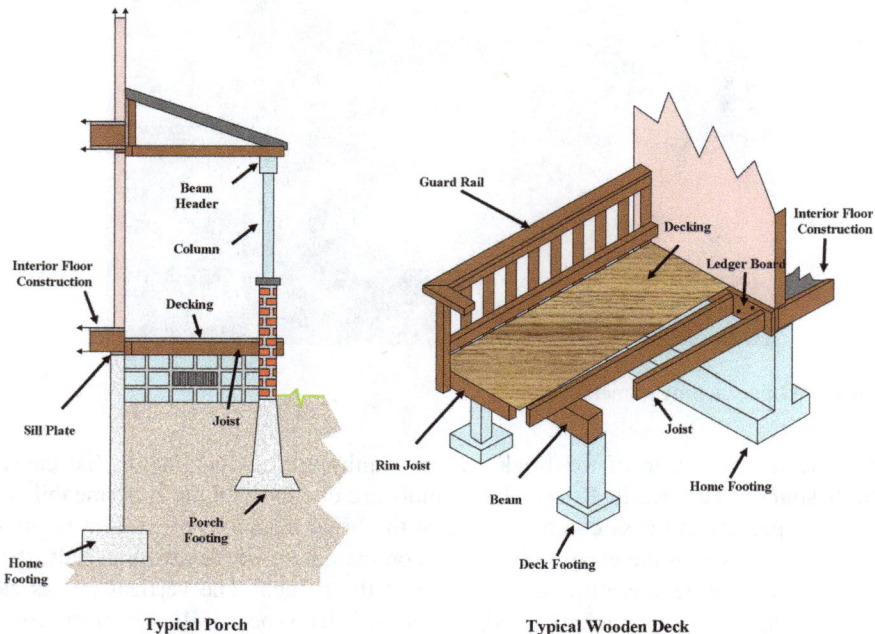

FIGURE 16.14   Illustration of typical porch/deck.

Three of the most common issues encountered in forensic investigations regarding reported porch/deck failures are due to settling or heaving, moisture damage, and/or improper installation.

Settlement of the ground below the footing or frost heave caused by inadequate depth of the footing below the frost line is manifested in movement or rotation of the structure at or near the footer/foundation supporting the porch or deck. Differential settlement of the footer/foundation occurs when the soil below one portion of the footing settles more than another portion due to differences in the soil-bearing capacity. This loss of bearing capacity is associated with: (1) improper soils (e.g., top soil vs bearing soils or engineered fills), (2) saturated soils and/or (3) placement of the footer above the frost line, subjecting it to freeze/thaw cycles. To prevent differential settlement, the footing should be founded upon adequate soil-bearing undisturbed earth or engineered fill that is kept dry and is located below frost depth (if applicable). Frost heaving is the process by which the freezing of water-saturated soil causes the deformation and upward thrust of the ground surface. Frost heaving can cause foundations to crack and cause differential settlement of the foundation walls once the frozen ground has thawed.

In addition to settlement and heaving-related movements of the footing/foundation, other common causes for porch/deck failures observed in forensic investigations are as follows:

- Rotted wood members from contact with moisture and/or the soil.
- Poor end support due to building shifting, settling columns, rot, and/or sagging floor beams.
- Poorly secured ledger boards (i.e., have insufficient fasteners and/or attached to a rotted rim joist). This is often caused by improper flashing and attachment to the building rim joist.
- Porches and decks not properly sloped away from the home, allowing water intrusion.

16.4.1   Porch and Deck Code and Best Practices Design Requirements

Information regarding porch/deck construction requirements can be found in most modern residential and commercial building codes and in best practice documents. Codes typically give general (and minimal) guidance, whereas best practices documents will tend to be more prescriptive, providing guidance that is more detailed. An example of typical code language for porches/decks can be found in the 2018 IRC[5]; note that this Section of the Code has been essentially rewritten from the 2012 version of the code. This suggests area codes officials encountered problems in the past and modified the code to be more prescriptive in this area. Portions of Section R507 (Exterior Decks) contained in Chapter 5 is reproduced below:

R507.1 Decks: Wood-framed decks shall be in accordance with this section. For decks using materials and conditions not prescribed in this section, refer to Section R301.

R507.2.4 Flashing: Flashing shall be corrosion-resistant metal of nominal thickness not <0.019 inch (0.48 mm) or *approved* nonmetallic material that is compatible with the substrate of the structure and the decking materials.

R507.3 Footings: Decks shall be supported on concrete footings or other approved structural systems designed to accommodate all loads in accordance with Section R301. Deck footings shall be sized to carry the imposed loads from the deck structure to the ground as shown in Figure R507.3. The footing depth shall be in accordance with Section R403.1.4.

Exceptions: Free-standing decks consisting of joists directly supported on grade over their entire length.

R507.3.1 Minimum size: The minimum size of concrete footings shall be in accordance with Table R507.3.1, based on the tributary area and allowable soil-bearing pressure in accordance with Table R401.4.1.

R507.8 Vertical and lateral supports: Where supported by attachment to an exterior wall, decks shall be positively anchored to the primary structure and designed for both vertical and lateral loads. Such attachment shall not be accomplished by the use of toenails or nails subject to withdrawal. For decks with cantilevered framing members, connection to exterior walls or other framing members shall be designed and constructed to resist uplift resulting from the full live load specified in Table R305.1 acting on the cantilevered portion of the deck. Where positive connection to the primary building structure cannot be verified during inspection, decks shall be self-supporting.

R507.9 Vertical and lateral supports at band joist: Vertical and lateral supports for decks shall comply with this section.

R507.9.1 Vertical supports: Vertical loads shall be transferred to band joists with ledgers in accordance with this section.

R507.9.1.1 Ledger details: Deck ledgers shall be a minimum 2-inch by 8-inch (51 mm by 203 mm) nominal, pressure-preservative-treated Southern pine, incised pressure-preservative-treated hem-fir, or approved, naturally durable, No. 2 grade or better lumber. Deck ledgers shall not support concentrated loads from beams or girders. Deck ledgers shall not be supported on stone or masonry veneer.

R507.9.1.2 Band joist details: Band joists supporting a ledger shall be a minimum 2-inch-nominal (51 mm), solid-sawn, spruce-pine-fir, or better lumber or a minimum 1-inch by 9-1/2-inch (25 mm × 241 mm) dimensional, Douglas fir or better, laminated veneer lumber. Band joist shall bear fully on the primary structure capable of supporting all the required loads.

R507.9.1.3 Ledger and band joist details: Fasteners used in deck ledger connections in accordance with Table R507.9.3(1) shall be hot-dipped galvanized or stainless steel and shall be installed in accordance with Table R507.9.1.3(2) and Figures R507.9.1.3(1) and R507.9.1.3(2) – see Figure 16.15.

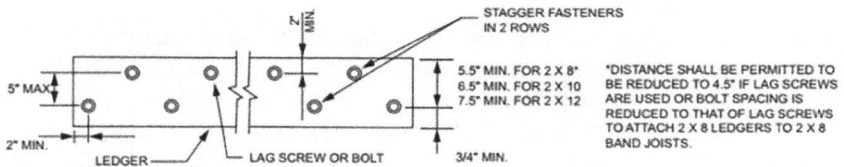

For SI: 1 inch = 25.4 mm.

FIGURE R507.9.1.3(1)

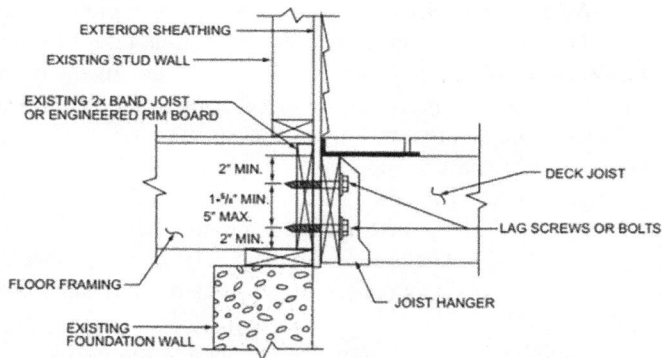

For SI: 1 inch = 25.4 mm.

FIGURE R507.9.1.3(2)

FIGURE 16.15   IRC Figures R507.9.1.3(1) and R507.9.1.3(2).

R507.9.1.4 Alternative ledger details: Alternative framing configurations supporting a ledger constructed to meet the load requirements of Section R301.5 shall be permitted.

R507.9.2 Lateral connection: Lateral loads shall be transferred to the ground or to a structure capable of transmitting them to the ground. Where the lateral load connection is provided in accordance with Figure R507.9.2(1), hold-down tension devices shall be installed in not less than two locations per deck, within 24 inches (610 mm) of each end of the deck. Each device shall have an allowable stress design capacity of not <1,500 pounds (6,672 N). Where the lateral load connections are provided in accordance with Figure R507.9.2(2), the hold-down tension devices shall be installed in not less than four locations per deck, and each device shall have an allowable stress design capacity of not <750 pounds (3,336 N).

The 2018 IBC,[6] Section 1604.8.3 (Decks) in Chapter 16, contains similar code language.

In general, where porches or decks are attached to the home, they must be installed with footers/foundations. In some cases where the porch or patio is not attached to the home, a footer/foundation may not be required. Decks must be decay, termite, and fire-resistant.

Industry best practice documents, such as those published by the American Wood Council's Design of Code Acceptance (AWC),[20] parallel code requirements (i.e., ICC International Residential Code and International Building Code) for deck design and construction. Industry best practices documents usually provide detailed illustrations and are more explicit than the "minimum" requirements set by code. The AWC[20] provides valuable design and construction information based on the IRC.[5] Specifically the AWC[20] provides the following guidance for wood deck construction:

- Minimum requirements for single-level residential wood decks.
- Data tables and pictorials that extrapolate from the IRC[5].
- Details and cross sections of specific wood deck elements.
- Commentary and explanations of code, design, and construction requirements.

Note that the *Design of Code Acceptance*[20] by the AWC provides supplemental information regarding the design and construction of single-level residential wood decks based on the IRC.[5]

16.4.2 INSPECTION METHODOLOGY FOR PORCH AND DECK INSPECTIONS

The inspection methodology for porch and deck inspections is very similar to those illustrated in Section 16.2.2 for chimney inspections and Section 16.3.2 for brick masonry veneer inspections.

16.4.2.1 Interview – Obtain Pertinent Information from the Property Owner, Owner's Representative, or Occupant

As part of an interview with the owner, owner's representative, or occupant, obtain background information regarding the apparent damage to the porch/deck:

- Have there been any recent modifications to the porch/deck (i.e., new additions)? If so, when?
- Have there been any changes to the ground surrounding the porch/deck (i.e., landscape bed)? If so, when?
- What were the damages, if known, to the porch/deck?
  - Damages on the exterior surfaces of the porch/deck (i.e., cracking, settlement, displacement, etc.)
- When were the damages first observed?
- What event(s) led to/caused the known damage to the porch/deck?

16.4.2.2 Create or Obtain a Basic Plan-View Sketch of the Porch/Deck

After interview-related information regarding the structure has been gathered, the porch/deck assessment process begins by creating a plan-view sketch of the porch/deck in an inspection field notebook (Figure 16.16).

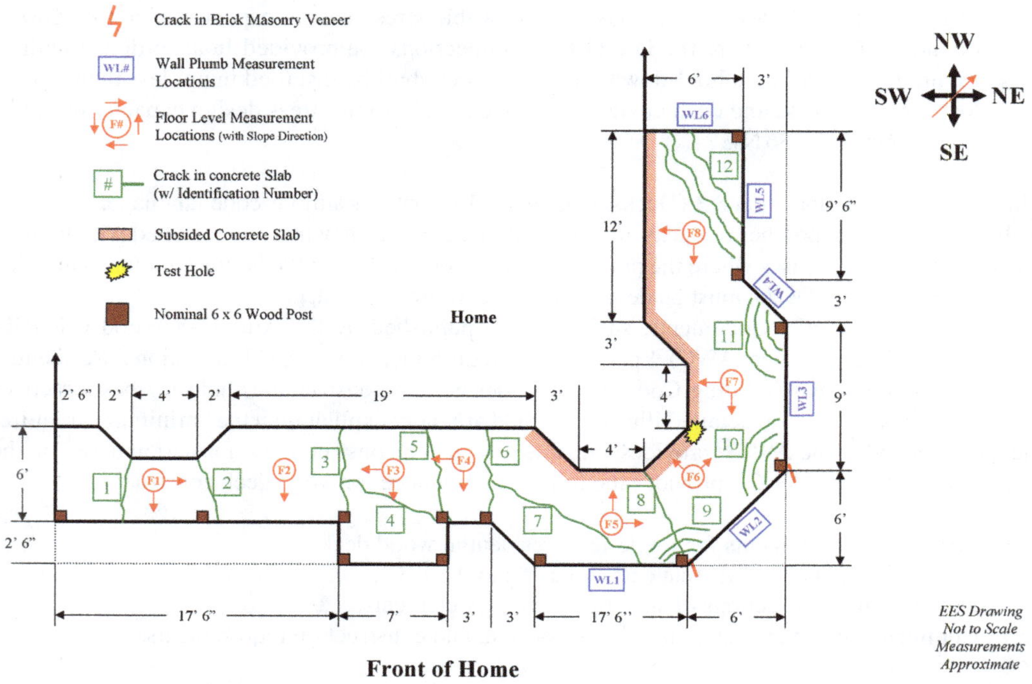

FIGURE 16.16 Typical porch/deck schematic (developed from on-site measurements).

The sketch should include porch/deck dimensions and any key features. Cracks, subsided concrete pads or support elements, tilted columns, areas of decay, method(s) of attachment to the building, ground-slope direction(s), and other issues should be drawn and documented in a field notebook.

16.4.2.3 Complete Porch/Deck Inspection – Document Observations in Writing with Measurements and Photographs

After the porch/deck schematic is completed, the inspection process should involve a systematic recording of all defects to the porch/deck by elevation, including the deck itself. These defects should be documented by location, type, and conditions. All observations should be documented in the field notebook; key observations should also be documented with photographs. Key inspection tasks to be completed during this phase of the inspection should include, if possible, the following:

- Inspecting the conditions under the porch/deck.
- Determining the method of porch/deck attachment to the building.
- Obtaining the dimensions of the ledger board and the floor joists spacing.
- Documenting the type(s) and number of fasteners attaching the porch/deck to the foundation.
- Recording the level measurements showing the displacement and the direction of settlement between the porch/deck and the structure.

In addition, the porch/deck surfaces, framing members, and foundation walls should be inspected for cracks and measured for deformation(s). The crack type (horizontal, vertical, and/or stair-step), size (width), location, and which direction the crack is oriented should be noted. Also, the condition of the crack and crack surfaces should be recorded in writing in the field notebook and documented with photographs (see Chapter 15 for how to record and interpret crack observations).

Plumb and level measurements of the porch/deck surfaces, framing members, and/or foundation walls when applicable should be taken following the methodologies discussed in Chapter 14 and 15 to determine the direction of slope/tilt, if any, of the porch/deck surfaces, framing members, and/or foundation elements. A digital 2-foot level is an excellent tool that provides instant measurements of the extent to which these elements may be sloped/tilted. Level measurements should be taken at various locations in a grid-like pattern throughout the floor. For example, Figure 16.16 illustrates the locations of floor-level measurements that were taken during an inspection. A summary of these measurements is shown in Table 16.5.

TABLE 16.5
Floor-Level Measurements

| | Perpendicular to Home Direction | | Parallel to Home Direction | |
| | Direction | | Direction | |
ID	Sloping Down	Slope	Sloping Down	Slope
F1	Southeast	~0.90°	Northeast	~0.30°
F2	Southeast	~0.90°	Level	~0.00°
F3	Southeast	~0.50°	Southwest	~0.10°
F4	Southeast	~0.90°	Southwest	~0.50°
F5	Northwest	~2.30°	Northeast	~1.00°
F6	West	~2.30°	North	~0.70°
F7	Southwest	~3.40°	Southeast	~1.00°
F8	Southwest	~2.00°	Southeast	~1.60°

In the example above, the measurements indicated that the floor surface at the southwest portion of the front porch sloped down to the southeast or away from the home while the floor surface at the northeast portion of the front porch sloped down heavily to the northwest and southwest or toward the home (see Figure 16.16 above for visual).

After these steps have been completed, and assuming a footer/foundation is present, the next step is to determine the depth and width of the footing by excavating the ground near the piers/foundation wall following the approach as outlined in Section 16.2.2.3.

- Depth of the bottom of the footer below the ground surface (serves as basis for whether or not the bottom of the footer is below the local frost line).

Once the bottom of the footing is uncovered, the footing details should be recorded in the field notebook and the key findings should also be documented with photographs. Afterward, backfill the excavation; compact the soils as the excavation is filled in; and attempt to the leave the site as it was entered.

16.4.3 Examples of Common Porch/Deck Issues/Findings Observed in the Field

Settling or Heaving:

- Shifted columns.
- Cracked flooring/steps.

Moisture Damage:

- Rotted decking and floor joists.

Improper Installation:

- Ledger board improperly installed to structure.
- Inadequate ventilation below the porch.

The lack of a solid footing, coupled with insufficient depth for frost protection, causes differential movement due to freeze/thaw. The proper construction of a footing below frost depth reduces the probability that a porch column or deck post rotates/tilts out of plane. Figure 16.17 depicts an example of a porch column that rotated/tilted out of plane due to the lack of a footing with insufficient depth. Note that the shovel is at the base of the rotated/tilted column with essentially no footer/foundation.

FIGURE 16.17   Tilted porch column (a) and lack of footing coupled with insufficient depth (b).

Another factor often found on field inspections of porches/decks is differential settlement due to improperly compacted soils. The fill material for a concrete slab on grade needs to be free of vegetation and compacted to ensure uniform support of the slab. Figure 16.18 depicts an example of a porch slab that cracked due to an improperly compacted gravel base. Note that the base of the concrete floor slab is textured brown with soil residual, suggesting that the concrete slab had originally been poured on grade.

FIGURE 16.18 Cracked porch slab (a) and settled gravel below slab (b).

Another factor often found during field inspections of porches/decks is rotted wood members due to a lack of ventilation for the enclosed space beneath the porch (see Section 12.3 "Crawlspace Ventilation" in Chapter 12). In most cases, these spaces will have cool and humid environments since they are constructed partially or entirely below grade and prone to hydrostatic pressure from surface water and/or groundwater. This pressure is created by the weight of the water in the soil surrounding the foundation walls. When the height of the water in the soil surrounding the foundation walls is higher than that in the soil of the enclosed underfloor space, the water level in the soil surrounding the foundation walls tends to fall due to gravity, which causes the water level in the soil of the enclosed underfloor space to rise until the water levels in the soils reach equilibrium. If there are no effective barriers, large quantities of water/moisture can be forced up into the underfloor space. The conditions typically found in enclosed underfloor spaces (a combination of high moisture levels, lack of air movement, and lack of light) provide an environment conducive to the amplification of mold growth. The long-term result of mold growth on wood-based materials is a significant reduction in the strength and ultimately complete degradation of these materials. Underfloor space moisture issues can normally be controlled by reducing moisture and/or providing adequate ventilation. Figure 16.19 depicts an example of subsidence to a porch due to a lack of ventilation for the enclosed space beneath the porch that caused degradation/collapse of wood framing members supporting the concrete slab.

FIGURE 16.19 Subsided porch due to a lack of ventilation that caused degradation to wood framing members.

IMPORTANT POINTS TO REMEMBER

- The performance of a masonry chimney or veneer wall is directly related to limiting the amount of water/moisture penetration into the system and then managing any remaining water/moisture that enters through the brick masonry.
- Residential and commercial codes provide guidance on proper construction of masonry chimneys and walls. However, it should be remembered that codes provide for minimum requirements. Industry best practices documents often provide above-minimum requirements based on industry experience and provide more construction details than those found in codes.
- The most common failure modes for masonry systems tend to be associated with a combination of water entry into the systems, a lack of adequate ties, and/or a lack of proper footer/foundations.
- The most common issues encountered in the field concerning porch/deck failure occur due to settling or heaving, moisture damage, and/or improper installation.

REFERENCES

1. The Brick Industry Association (BIA). "Technical Notes on Brick Construction, Technical Note 2: Glossary of Terms Relating to Brick Masonry," March 1999. Accessed June 3, 2020, https://www.gobrick.com/docs/default-source/read-research-documents/technicalnotes/2-glossary-of-terms-relating-to-brick-masonry.pdf?sfvrsn=0.
2. Masonry Institute of Washington. "Masonry Glossary", 2020. Accessed June 3, 2020, https://www.masoncontractors.org/glossary/.
3. The Chimney Balloon. "History of the Chimney", 2007–2012. http://www.chimneyballoon.ie/content/history-chimney.
4. The Brick Industry Association (BIA). "Technical Note 19B – Residential Chimney - Design and Construction," April 1998. Accessed June 3, 2020, https://www.gobrick.com/docs/default-source/read-research-documents/technicalnotes/19b-residential-chimneys---design-and-construction.pdf?sfvrsn=0.
5. International Code Council. "2018 International Residential Code for One- and Two-Family Dwellings (IRC)." Accessed June 3, 2020, https://codes.iccsafe.org/content/IRC2018/chapter-10-chimneys-and-fireplaces.
6. International Code Council. "2018 International Building Code (IBC)." Accessed June 3, 2020, https://codes.iccsafe.org/content/IBC2018/chapter-21-masonry.
7. The Brick Industry Association (BIA). "Technical Note 19A – Residential Fireplaces, Details and Construction," August 2000. Accessed June 3, 2020, https://www.gobrick.com/docs/default-source/read-research-documents/technicalnotes/19a-residential-fireplaces-details-and-construction.pdf?sfvrsn=0.
8. Old Virginia Brick, Inc. "Weatherproofing," 2003. http://www.oldvirginiabrick.com/technical/weatherproofing.html.
9. Crissinger, Joseph. "Efflorescence: The white whiskers of winter." *Interface – The Journal of RCI*, vol. XXIX, 2011, pp. 6–16.
10. Brick Institute of America (BIA). "Brick Brief: Flashing Chimneys," July 2005. Accessed June 4, 2020, https://www.gobrick.com/docs/default-source/read-research-documents/brick-briefs/flashing-chimneys.pdf?sfvrsn=0.
11. The Engineered Wood Association. "APA (American Plywood Association) Publication A535B – Designing Roofs to Prevent Moisture Infiltration," September 2017.
12. The Brick Industry Association (BIA). "Technical Note 28 – Brick Veneer/Wood Stud Walls," November 2012. Accessed June 3, 2020, https://www.gobrick.com/docs/default-source/read-research-documents/technicalnotes/28-brick-veneer-wood-stud-walls.pdf?sfvrsn=0.
13. Masonry Standards Joint Committee. "Building Code Requirements and Specifications for Masonry Structures," 2008.
14. The Brick Industry Association (BIA). "Technical Note 44B – Wall Ties for Brick Masonry," May 2003. Accessed June 3, 2020, https://www.gobrick.com/docs/default-source/read-research-documents/technicalnotes/44b-wall-ties-for-brick-masonry.pdf?sfvrsn=0.

15. The Brick Industry Association (BIA). "Technical Note 7a – Water Penetration Resistance – Design and Detailing," November 2017. Accessed June 3, 2020, https://www.gobrick.com/docs/default-source/read-research-documents/technicalnotes/7-water-penetration-resistance-design.pdf?sfvrsn=0.
16. The Brick Industry Association (BIA). "Technical Note 7b – Water Penetration Resistance – Construction and Workmanship", November 2017. Accessed June 3, 2020, https://www.gobrick.com/docs/default-source/read-research-documents/technicalnotes/7b-water-penetration-resistance-construction-and-workmanship.pdf?sfvrsn=0.
17. Taylor Lauder Bersten Engineers. "TLB Engineers Technical Note: The Expansion of Clay Brickwork", March 2009.
18. Kahn, Renee and Ellen Meagher. *"Preserving Porches."* New York: Henry Holt and Company, 1990. Accessed June 4, 2020, http://xroads.virginia.edu/~CLASS/am483_97/projects/cook/roots.htm.
19. Carson Dunlop & Associates Limited. *"Principals of Home Inspections – Exteriors."* Chicago, IL: Dearborn, 2003.
20. American Wood Council (AWC). "Design for Code Acceptance, Prescriptive Residential Wood Deck Construction Guide", 2015. Accessed June 3, 2020, https://www.awc.org/pdf/codes-standards/publications/dca/AWC-DCA62015-DeckGuide-1804.pdf.

17 Fire Damage Structural Property Assessments

Stephen E. Petty
EES Group, Inc.

CONTENTS

PURPOSE/OBJECTIVES

The purpose of this chapter is to:

- Provide information on how to complete a postfire structural forensic inspection.
- Describe how to recognize structural fire damage to concrete and concrete masonry unit (CMU) walls, rubble stone, and brick masonry walls.
- Describe how to recognize structural fire damage to steal framing members and metal decking.
- Describe how to recognize structural fire damage to wood framing members.
- Provide a methodology to determine and document structural vs cosmetic fire damage to a building.

Following the completion of this chapter, you should be able to:

- Understand the extent of fire-damage claims in terms of dollar volume.
- Be able to collect necessary information needed to determine fire structural damage to a residential or light commercial building, using the methodology specified in this chapter.
- Be able to prepare a written report with analysis and documentation regarding damage (both structural and cosmetic) to a residential or light commercial building suffered during a fire.

17.1 INTRODUCTION

The National Fire Protection Association (NFPA), a clearing house for information, statistics, and standards related to fires, provides extensive resources regarding the cost to property and persons from fires.[1] In October 2019, the NFPA reported that in 2018, fire departments responded to 1,318,500 fires that resulted in 3,655 fatalities, 15,200 injuries, and $26.5 billion in direct property losses.[2] Internationally, a review of data from 16 countries (13 from Europe, Canada, Japan, and the United States) found that the total cost of fire damage averaged 0.2%–0.3% of gross domestic product (GDP) with death rates averaging 1–2 per 100,000 inhabitants.[3]

NFPA and the U.S. Fire Administration also reported some interesting facts regarding fires at the household level, some of which are listed below[4,5]:

- Odds of a household fire in a lifetime: 1 in 15 years or 5 for an average lifetime.
- Odds of someone in household being injured by a fire over an average lifetime: 1 in 4.
- Odds of having a working carbon monoxide detector: 1 in 3.
- Odds of dying in a fire: 11.2 in 1,000,000 (2017).

The good news is that the risk of dying in a fire has dropped dramatically with time, from over 100 in 1 million persons in 1917 to around 11 per 1 million persons today (Figure 17.1).

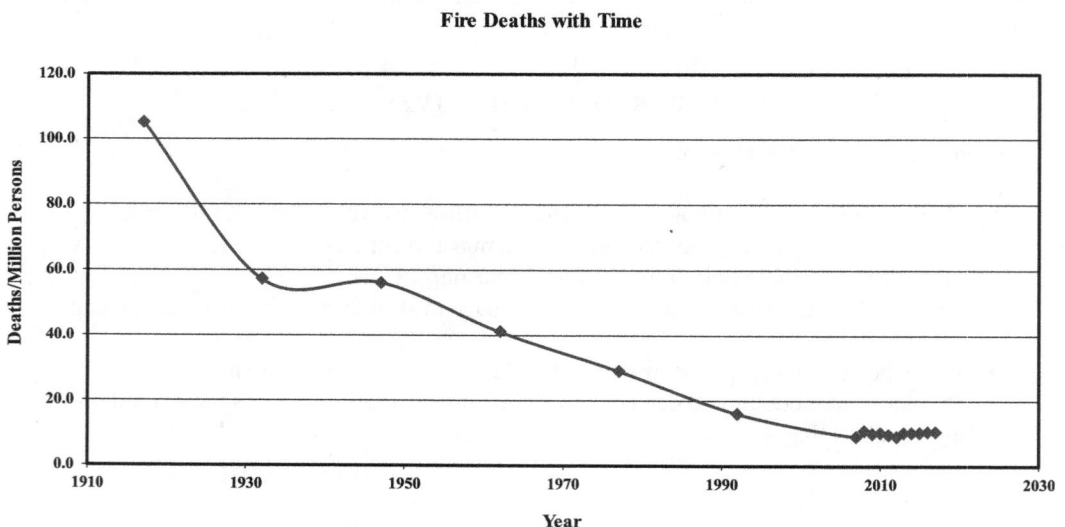

FIGURE 17.1 Risk of dying in a fire (#/million persons) with time.

Of total fire deaths, 2,695 or 77.6% were associated with individuals living in residential structures.[6] As with death rates from fires, the rates of reported fire incidents have dropped with time (Figure 17.2).[1,6]

Fire Incidents with Time

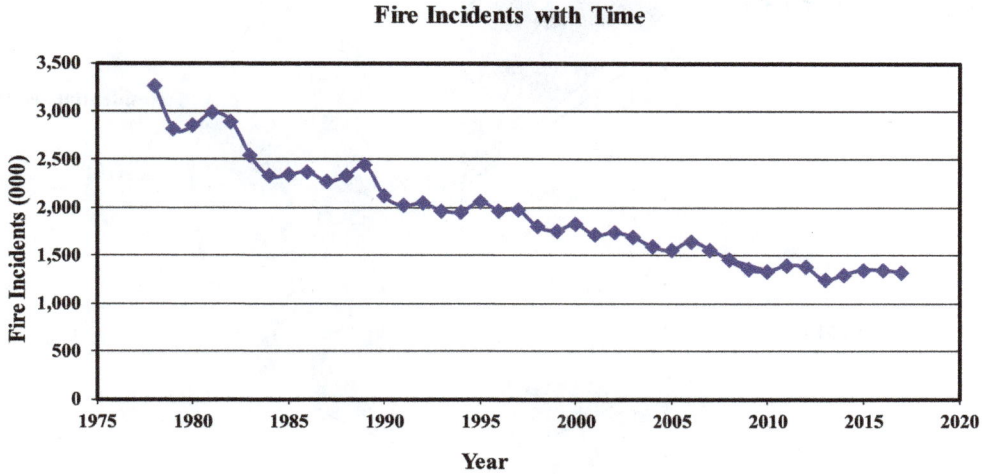

FIGURE 17.2   Fire incidents (in thousands) with time.

The number of fire incidents has dropped by ~60% over the past 40 years.
As illustrated in Figure 17.3, residential fires accounted for ~28.9% of all fires; 381,336 in 2017.

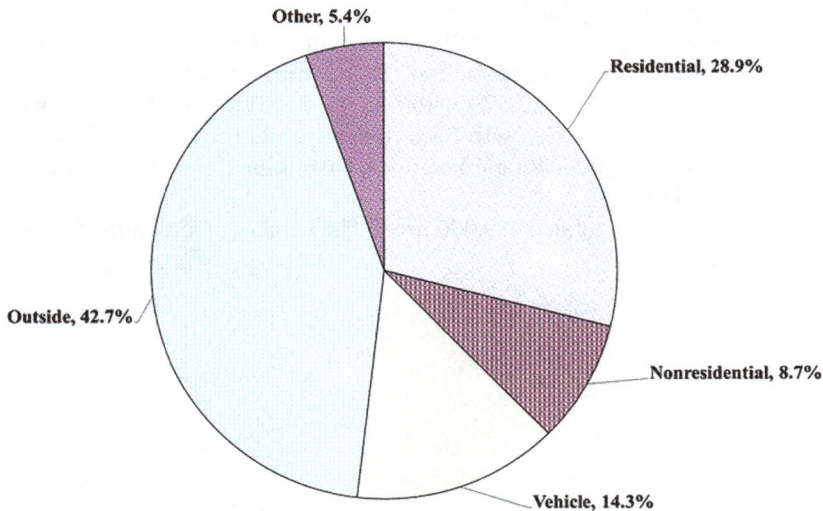

FIGURE 17.3   2017 fires – percentage of incidents by class.

Of the total fires reported in 2017, 28.9% (371,500) were associated with residential structures and 8.7% (111,000) with nonresidential structures. Residential property damage for 2017 totaled $7.8 billion and nonresidential property damage for 2017 $2.8 billion of fire-related losses. As illustrated, the vast majority of the direct property losses were associated with residential (49.7% of dollar losses) and nonresidential (31.6%) of total losses (Figure 17.4).

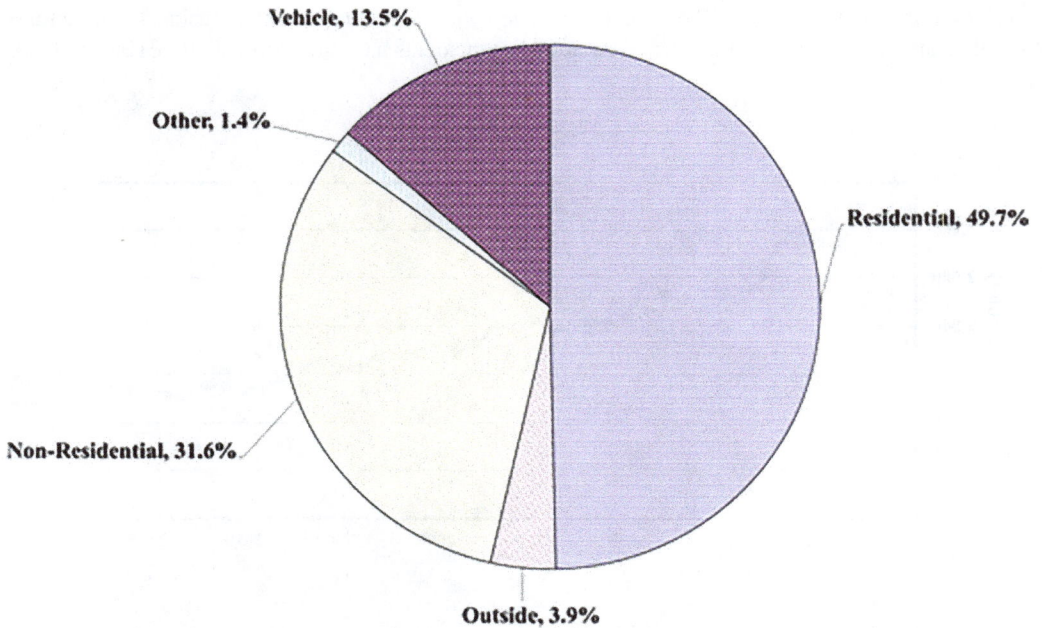

FIGURE 17.4   2017 fire losses – percentage of dollars by class.

Residential fire loss damage in 2018 in terms of dollars was dominated by one and two fam-ily homes at $6.11 billion (74.6% of total residential losses of $8.19 billion) with multifamily residential structures accounting for $1.68 billion (20.5% of total) and other residential build-ings accounting for the balance of $0.40 billion (for another 4.8% of the total. Of the nonresi-dential loss fires in 2018 totaling $2.66 billion, $638.1 million (24.0%) were associated with Stores and Offices, $536.5 million (22.5%) with Storage Facilities, $435.8 million (16.4%) with Manufacturing, $258.0 million (9.7%) with Basic Industry, and the balance (27.4%) with Eating and Drinking Establishments, Educational Facilities, Institutional Facilities, Detached Garages, Specialty, and other buildings.[6]

Total structural fires in 2018 totaled 499,000 fires.[7] The number of structural fires by property is summarized in Figure 17.5.

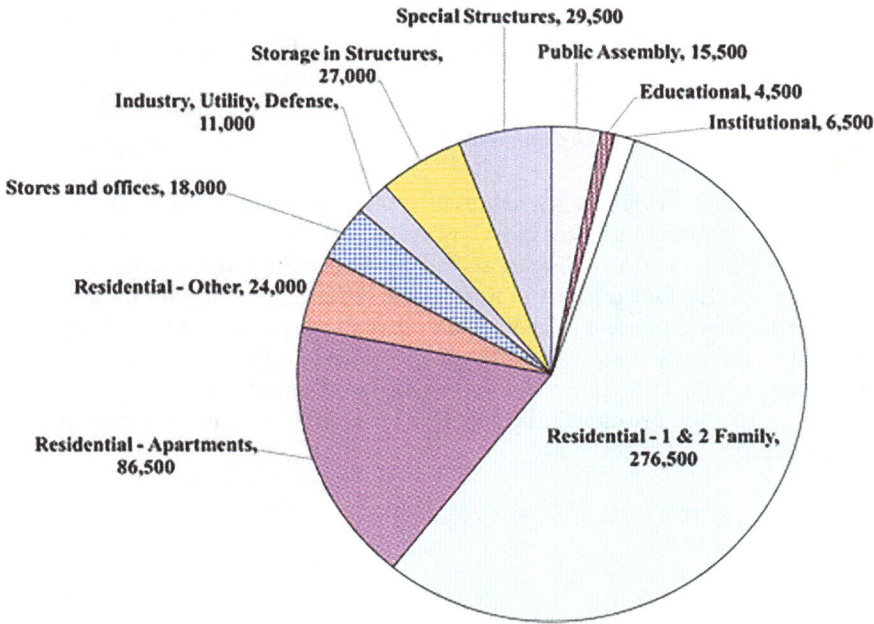

FIGURE 17.5   2018 structural fire incidents by property use.

Of the total structural fires reported in 2018, 77.6% (387,000) were associated with residential structures and 22.4% (112,000) with nonresidential structures. Within the residential fire category, one- and two-bedroom homes accounted for 266,500 of the 387,000 fires (68.9%). Structural property damage for 2018 totaled $11.066 billion. As illustrated in Figure 17.6, the vast majority of the structural property losses ($8.286 billion or 74.9%) were associated with residential properties.

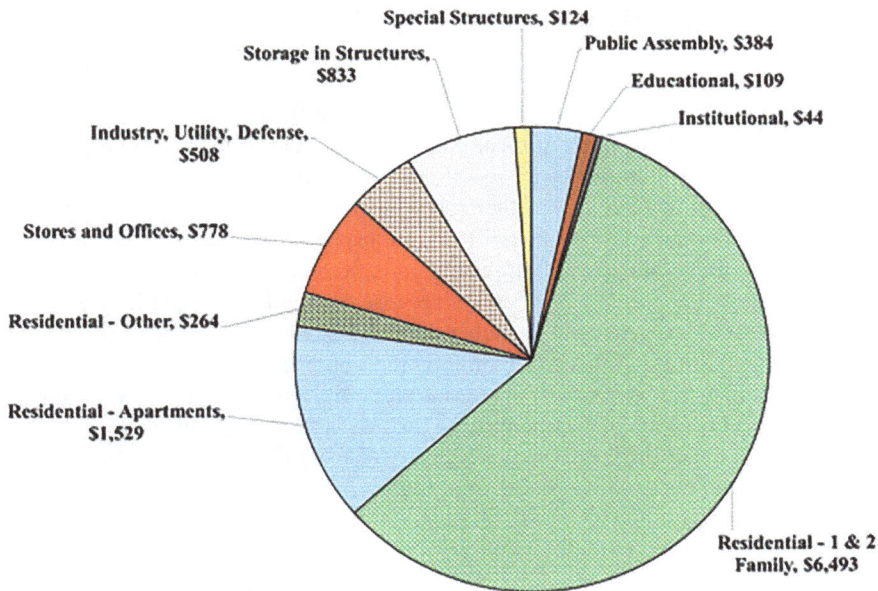

FIGURE 17.6   2018 structural fire losses ($ millions) by property use.

As illustrated, fire losses constitute thousands of incidents and billions of dollars annually. Forensic experts are utilized to determine the cause of these fires and to determine the extent to which structures can be salvaged or if they are a total loss. The topic of fire causation is a field into itself, is well covered by the NPFA in their Standard "NPFA 921 Guide for Fire and Explosion Investigations" and its companion user manual[8,9] as well as other publications[10], and therefore is not covered in this book.

Determination of what items must be removed and replaced versus those that can be repaired and/or cleaned is an area where forensics services are needed, and methods for making these determinations are not well documented. The Association of Specialists in Cleaning and Restoration's (ASCAR) and National Institute of Disaster Restoration (NIDR) provided guidance on some aspects of the inspection process and excellent advice on repairs to items from fire and smoke damage,[11] but does not fully address the topic of fire damage structural inspections. After a brief discussion of the temperature effects of fire on materials, the balance of this chapter addresses a methodology with examples for structural fire damage inspections. Structural elements to be inspected are broken down into three topics:

- Concrete and Concrete Masonry Unit (CMU) Walls.
- Steel Framing Members.
- Wood Framing Members.

17.2   FIRE TEMPERATURE EFFECTS ON MATERIALS

The survivability of various materials after exposure to fire is directly related to the characteristics of the material (i.e., softening, melting, slumping, or oxidation points) and the temperature of the fire impacting the material; temperatures of typical fires range from 1,112°F (600°C) to 2,192°F (1,200°C).[12] These temperature characteristics are used in fire investigations to determine the origin of a fire and to estimate the intensity of a fire. The intensity of a fire is dependent on four factors:

- Fuel (reducing agent).
- Oxidizing agent.
- Uninhibited chemical chain reactions.
- Heat.

These four elements are often referred to as the "fire tetrahedron" wherein fires can be prevented or suppressed by removal of one or more of these elements. The fuel is typically composed of organic materials (composed of carbon and hydrogen atoms) and can consist of wood or contents, fuels (typically defined as gases or liquids), or other solids. Typically, the oxidizer (material contributing oxygen to the combustion process) is oxygen from atmospheric air, but in some cases can consist of pure oxygen or chemicals such as ammonium nitrate, potassium nitrate, and hydrogen peroxide, which will contribute oxygen in the combustion process. It should be noted that combustion only occurs when the fuel/air (oxygen) ratio is within certain bounds. The lower limit is referred to as the lower flammability limit and the upper limit as the upper flammability limit. If too little fuel is available (lower limit), the combustion process is starved for material to be oxidized, whereas if too little air (oxidizer) is available, the combustion process is starved for oxygen to oxidize the fuel. Heat is important because if sufficient energy is given off from the combustion process, it provides the energy to sustain the combustion process for nearby fuels. Provided the proper type of fuel is present along with sufficient heat, the combustion process creates an environment where an uninhibited chain reaction takes place. The combustion process continues or accelerates until one or more of fuel, oxidizer, or heat becomes limiting.

NFPA 901 defines fires by fuel type as follows[13,14]:

Class A: Fire involving ordinary combustible materials such as paper, wood, cloth, and some rubber and plastic materials.

Class B: Fire involving flammable or combustible liquids, flammable gases, greases and similar materials, and some rubber and plastic materials.

Class C: Fire involving energized electrical equipment where safety of the employee requires the use of electrically nonconductive extinguishing media.

Class D: Fire involving combustible metals such as magnesium, titanium, zirconium, sodium, lithium, and potassium.

Class K: Fires involving commercial cooking appliances with vegetable oils, animal oils, or fats at high temperatures. A wet potassium acetate, low-pH-based agent is used for this class of fire.

Residential (Figure 17.7) and light commercial fires are typically associated with Class A and Class B fires.

FIGURE 17.7 Residential fire.

The intensity of the fire at various locations will result in localized peak temperatures, which will affect various materials, including structural materials. The estimate of the maximum temperature of the fire and/or the temperature of the fire to impact structural members of interest is critical in the determination of the potential for structural damage to steel framing members and metal decking, concrete and CMU walls, and brick masonry. Examples of materials and temperatures at which they are impacted are illustrated in Table 17.1.[8,9,12,15–20]

TABLE 17.1
Impact of Temperature on Materials

Material	Temperature Condition/Situation	°F	°C
		Temperature	

Material	Temperature Condition/Situation	°F	°C
Acetone	Ignition	869	465
Acetylene	Ignition	581	305
Aluminum – dust	Ignition	1,130	610
Aluminum	*Melting*	*1,200–1,220*	*649–660*
Aluminum	*Softens*	*752*	*400*
Aluminum alloys	Melting	900–1,200	482–650
Aluminum alloys – small machine parts, brackets, toilet fixtures, cooking utensils	Drops form	1,200	650
Benzene	Ignition	928	498
Benzene	Burning	1,690	920
Brass – yellow	*Melting*	*1,710*	*932*
Brass – red	*Melting*	*1,825*	*996*
Brass – door and furniture knobs, locks, lamp fixtures, buckles	*Drops form; sharp edges rounded*	*1,650–1,850*	*900–1,000*
Bronze – aluminum	Melting	1,800	982
Bronze – window frames and art objects	Drops form; sharp edges rounded	1,850	1,000
Cast iron – gray	*Melting*	*2,460–2,550*	*1,350–1,400*
Cast iron – pipes, radiators, pedestals, and housings	*Drops form*	*2,000–2,200*	*1,100–1,200*
Cast iron – white	*Melting*	*1,920–2,010*	*1,050–1,100*
Chromium	Melting	3,350	1,845
Coal – dust	Ignition	1,346	730
Concrete	*Spalling begins*	*482–788*	*250–420*
Concrete	*Minor (<10%) loss of strength*	*<572*	*<300*
Concrete	*Loss of strength begins*	*572*	*300*
Concrete	*Color change – pink to deep red*	*550–1,100*	*288–593*
Concrete	*Spalling begins – quartz aggregate*	*1,100*	*593*
Concrete	*Loss of load bearing capacity*	*932–1,112*	*500–600*
Concrete	*Carbonate release CO_2 –contraction and severe micro-cracking*	*1,112–1,472*	*600–800*
Concrete	*Complete disintegration of calcareous constituents – whitish-gray color and severe micro-cracking*	*1,472–2,192*	*800–1,200*
Concrete	*Melting begins*	*2,192*	*1,200*
Concrete	*Melted*	*2,372–2,552*	*1,300–1,400*
Copper	*Melting*	*1,981*	*1,082*
Copper – wire and coins	*Drops form; sharp edges rounded*	*2,000*	*1,100*
Ethanol	Ignition	685	363
Fire brick – insulating	Melting	2,980–3,000	1,638–1,650
Gasoline – 100 octane	Ignition	853	456
Gasoline	Burning	1,879	1,026
Glass	*Melting*	*1,100–2,600*	*593–1,427*

(Continued)

TABLE 17.1 (*Continued*)
Impact of Temperature on Materials

Material	Temperature Condition/Situation	Temperature °F	Temperature °C
Glass	*Softens*	*1,100–1,350*	*593–732*
Glass – window, plate, or reinforced	*Rounded*	*1,450*	*800*
	Thoroughly flawed	*1,560*	*850*
Glass block, jars, tumblers, and solid ornaments	*Softened*	*1,300–1,400*	*700–750*
	Rounded	*1,400*	*750*
	Thoroughly flawed	*1,560*	*800*
Gold	*Melting*	*1,945*	*1,063*
Kerosene	Ignition	410	210
Kerosene	Burning	1,814	990
Iron	*Melting*	*2,802*	*1,540*
JP-4	Burning	1,700	927
Lead	Melting	621	327
Lead – plumbing lead, flashing, storage batteries	Drops form; sharp edges rounded	550–650	300–350
Methanol	Burning	2,190	1,200
Natural gas	Ignition	900–1,170	482–632
Paraffin	*Melting*	*129*	*54*
ABS – plastic	*Melting*	*190–257*	*88–125*
Acrylic – plastic	*Melting*	*194–221*	*90–105*
Nylon – plastic	*Melting*	*349–509*	*176–265*
Plastics – most	*Melting or burning*	*180–300*	*82–149*
Polyethylene – plastic – bags and films	*Shrivels*	*248*	*120*
Polyethylene – plastic – bottles and buckets	*Softens and melting*	*302*	*150*
Polyethylene – plastic	*Melting*	*230–275*	*110–135*
Polyethylene – plastic	*Ignition*	*910*	*488*
Polystyrene – plastic – thin-wall food containers	*Shrivels*	*248*	*120*
Polystyrene – plastic – foam, light shades, handles, curtain hooks, radio casings	*Softens*	*248–284*	*120–140*
Polystyrene – plastic	*Melting*	*248–320*	*120–160*
Polystyrene – plastic	*Ignition*	*1,063*	*573*
PVC – plastic	*Melting*	*167–221*	*75–105*
PVC – plastic	*Ignition*	*945*	*507*
Porcelain	Melting	2,820	1,550
Pot metal	Melting	562–752	300–400
Propane	Ignition	842	450
Quartz (SiO$_2$)	Melting	3,060–3,090	1,682–1,700
Silver – jewelry and spoons	*Melting*	*1,742–1,760*	*950–960*
Silver – jewelry and coins	*Drops form; sharp edges rounded*	*1,750*	*900*
Solder (tin)	*Melting*	*275–350*	*135–177*

(Continued)

TABLE 17.1 (*Continued*)
Impact of Temperature on Materials

		Temperature	
Material	Temperature Condition/Situation	°F	°C
Steel (stainless)	*Melting*	*2,600*	*1,427*
Steel	*Blistered, discolored and flaked off coatings and paints*	*>600*	*>316*
Steel	*Mill scale tightly adhered*	*<1,200*	*<649*
Steel	*Course and eroded surface*	*>1,200*	*>649*
Steel (carbon)	*Melting*	*2,760*	*1,516*
Tin	*Melting*	*449*	*232*
Wood – softwood	Ignition	608–660	320–350
Wood – hardwood	Ignition	595–740	313–393
Wood, paper, and cotton	*Darkens*	*392–572*	*200–300*
Wood and paper	*Ignition*	*450*	*232*
Wood	*Burning*	*1,880*	*1,027*
Wax (paraffin)	*Melting*	*120–167*	*49–75*
Zinc	*Melting*	*707–790*	*375–421*
Zinc – plumbing fixtures, flashing and galvanized surfaces	*Drops form*	*750*	*400*

Materials and their affected temperatures often found in structural fire investigations are italicized in Table 17.1. For instance, it is not uncommon to find zones where the copper or aluminum wiring is melted, where plastics are melted, and where wood is burned or charred. These zones all provide an estimate of the peak temperature that existed within those zones.

The extent of damage to building materials can be classified into the following five categories identified in Table 17.2.[12]

TABLE 17.2
Fire Damage Categories

Damage Level	Description	Characteristics
1	Cosmetic	Presence of soot deposits, discoloration, and odors. Usually cleanable
2	Surface	Damage to finished surfaces. Minor spalling to concrete. Corrosion to metal
3	Minor structural damage	Cracked and/or spalled concrete. Minor deformation of metal. Moderate corrosion to metal. Lightly charred wood members
4	Major structural damage	Major cracks and spalling to concrete. Major deformation of metal; no longer fits original location. Moderately charred wood members
5	Severe structural damage	Extensive spalling of concrete, reinforcement steel in concrete exposed. Severe deformation of steel members. Severe charring of wood members

In another approach, Wang et al.[21] classified fire damage into one of three categories: no damage, slight damage, and extensive damage, while others[17,19,21] use four or more categories. Regardless of the specific classification used for fire damage assessments, these general categories allow one to define areas where cleaning/repairs versus replacement of structural members will be required.

17.3 OVERVIEW OF METHODOLOGY FOR EVALUATION OF STRUCTURAL FIRE DAMAGE TO BUILDINGS

The forensic investigator is one of the first people to visit a fire site after the fire personnel and fire causation investigators have left the premises. The completion of a preliminary fire damage inspection is the most important factor associated with the evaluation of the potential for rehabilitation of a building.[17] Moreover, a negative connotation often exists that structural members exposed to fires are always damaged and must be replaced; this is not always the case.[16]

It should be remembered that the location is inherently dangerous due to damaged structural components, glass debris, nails and wires, wet surfaces, lack of light and power, and heavy fire residue in the air. Sufficient lighting should be brought to the sight and appropriate personal protective equipment (PPE) should be worn such as a respirator, clothing (e.g. Tyvek), and shoes (e.g. steel-toe) to complete the inspection. Unstable structures should never be ventured onto or underneath without carefully ensuring that the environment is safe for completing the inspection; a cell phone should always be carried if completing the inspection alone in case of an accident.

Once a general overview of the site is obtained and unsafe areas are determined, a visual structural fire damage inspection can begin. A visual inspection is one of the most common nondestructive methods that can be utilized.[15–19] Another nondestructive method includes the use of a small hammer to conduct a tapping survey of surfaces that will detect hollow sounding delaminated material.[19] Also, destructive testing methods can be used to obtain samples to be tested in a laboratory, but these are not often used in residential and light commercial inspections. Typically, the questions to be answered based on inspection results, testing, and experience follow:

- Are structural and foundation walls fabricated from concrete, CMU, brick masonry, or stone fire damaged? Can they be repaired or must they be replaced?
- What steel framing members, if any, should be removed and replaced (versus those that can be cleaned and sealed or those that are unaffected)?
- What wood framing members, if any, should be removed and replaced (versus those that can be cleaned and sealed or those that are unaffected)?

The process used can be more extensive, as pointed out by others,[17,19,21] but the essence of a structural fire damage assessment boils down to answering these questions. In most cases, the structural inspector should leave decisions on HVAC, plumbing, and electrical systems to experts in those fields.

In this, and the following three sections that follow, the reader should be reminded that structural fire damage assessments for residential and light commercial properties typically cannot justify extensive laboratory analysis of materials, regarding whether or not they have been damaged by fire and temperatures associated with the given fire. Rather, the investigator is tasked with making relatively quick determinations of: (1) what structural materials should be removed and replaced, (2) what structural materials can be cleaned or repaired, and/or (3) what structural materials were not impacted by the fire. Thus, given limitations on resources and a difficult work environment (lack of power, light, and access), one must rely on a visual inspection to make replace, repair/clean, and/or no action required decisions based on the science of temperature effects on materials within a fire-damaged structure.

In context with the three categories of questions to be answered, the basic methods used to determine replacement for each of these classes of materials are summarized and illustrated in the following sections of this chapter.

17.3.1 SIGNS OF VISUAL STRUCTURAL DAMAGE TO CONCRETE/CMU/CLAY MASONRY/STONE FOUNDATION OR STRUCTURAL WALLS – GENERAL

The visual determination of whether or not concrete, CMU, clay masonry, and/or stone masonry will require removal and replacement due to structural damage from a fire is typically based on one or more of the following observations:

- Color change.
- Cracking (fresh cracks greater than minor in size).
- Spalling, chalkiness, or erosion of the surface.

Specific observations, and their interpretation, for concrete, CMU, clay masonry, and stone are discussed in the following sections.

17.3.1.1 Signs of Visual Structural Damage to Concrete

The visual determination of whether or not structural fire damage to concrete has occurred to the extent that removal and replacement are required is typically based on the following observations:

- Color change.
- Cracking (fresh cracks larger than minor in size).
- Spalling, chalkiness, or erosion of the surface.
- Steel reinforcement bars exposed.

Regarding color change, the color of concrete surfaces will change as the temperature to which it was exposed increases (Figure 17.8).[15–20]

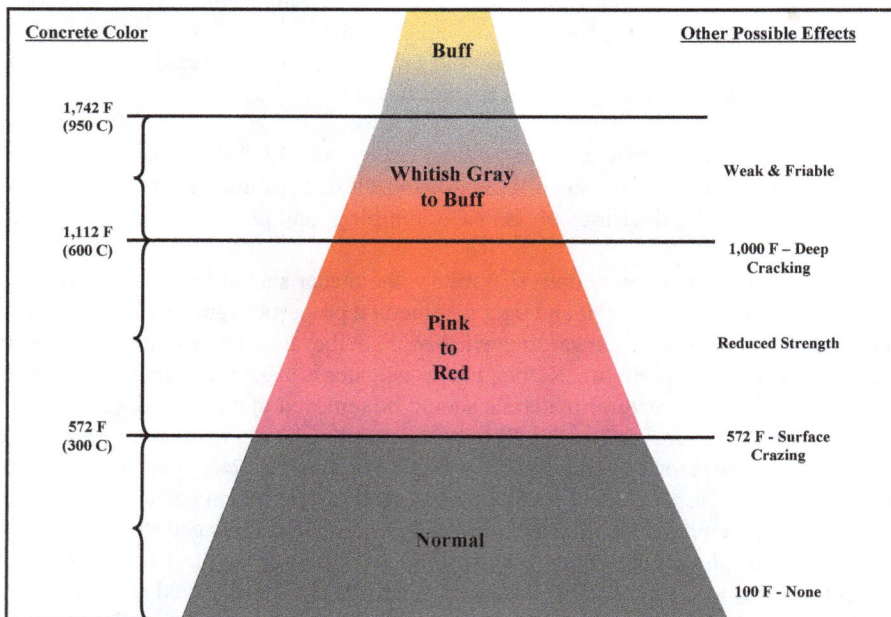

Concrete Color		Other Possible Effects
	Buff	
1,742 F (950 C)		
	Whitish Gray to Buff	Weak & Friable
1,112 F (600 C)		1,000 F – Deep Cracking
	Pink to Red	Reduced Strength
572 F (300 C)		572 F - Surface Crazing
	Normal	
		100 F - None

FIGURE 17.8   Visual color change of concrete with temperature of fire.

In general, concrete retains at least 90% of its strength and is rarely affected, until temperatures reach 550°F (300°C) (see Table 17.1). Once the temperature reaches 550°F (300°C) and until about 1,100°F (600°C), the surface of the concrete will change from a pink to a deep red color. In addition, surface craze cracking begins to occur near 550°F (300°C), with deep cracking present when temperatures reach 1,100°F (600°C). Discussions on the significance of cracks can be found in Chapter 15. Spalling of the concrete surface also occurs between temperature of ~572°F (~300°C) and ~788°F (~420°C); quartz does not spall until ~1,100°F (~593°F). Spalling can be either explosive or nonexplosive, depending on the thickness of the member (cracking), moisture, and the compressive stress level in the concrete. The moisture level (lower is better) and stress levels (lower is better) are generally believed to be the main factors associated with spalling of concrete in a fire. Concrete wall thicknesses above ~5 inches (125 mm) often does not exhibit explosive spalling. Aside from slightly reducing the thickness of the concrete, spalling can expose the reinforcement steel directly to the fire and its temperature. However, the mild steel, while weakened at the time of the fire, generally recovers most of its strength on cooling. The pink to deep red discoloration, when present, is caused by the oxidation of ferric salts in the aggregate. It is not always present and is seldom found in calcareous aggregate. In this temperature range, the structural strength of the concrete is significantly reduced. As the temperature increases from 1,100°F (600°C) to about 1,740°F (950°C), the color changes to a whitish gray color. Severe micro-cracking occurs at temperatures from ~1,112°F (~600°C) to ~1,472°F (~800°C). Chalking and powdering of the concrete occur at temperatures from ~1,472°F (~800°C) to ~2,192°F (~1,000°C). The whitish gray color change is due to calcinations (removal of oxygen) for the calcareous elements of the cement matrix. The concrete at these temperatures is weak structurally and friable (can be pulverized easily). Above this temperature (1,740°F), the color of the concrete will change to a buff color. At ~2,192°F (~1,200°C), concrete will actually begin to melt.

Another nondestructive field method for testing the integrity of concrete in structural fire damage assessments besides the visual inspection is to strike the concrete with a small hammer. When concrete is struck with the hammer, good material will sound hard or solid, whereas damaged concrete will sound hollow or muffled. Use of a hammer and chisel, while invasive, will provide information on the softness of the concrete. This further demonstrates whether or not the integrity of the concrete has been maintained and if not, the depth to which it was impacted.

In terms of forensic structural investigations, if any of the characteristics associated with temperatures above ~572°F (~300°C) are observed on a portion of the concrete wall or foundation, replacement is recommended.

17.3.1.2 Signs of Visual Structural Damage to Concrete Masonry Unit (CMU) Block

The visual determination of whether or not structural fire damage to CMU construction has occurred to the extent that removal and replacement are required, like for concrete, is typically based on the following visual observations[16]:

- Color change.
- Cracking (fresh cracks greater than minor in size).
- Spalling, chalkiness, or erosion of the surface.

The significance of these factors on CMU exposed to fire is essentially the same as for concrete walls except that until ~1990 no reinforcement will be found in many of these walls. CMU cracking may be more significant at lower temperatures than with a solid concrete wall due to the thinner thickness of the block face shells, but from a forensic standpoint, the implications for removal/replacement versus cleaning remain the same as for concrete.

17.3.1.3 Signs of Visual Structural Damage to Clay Brick Masonry

The visual determination of whether or not structural fire damage to clay brick masonry has occurred to the extent that removal and replacement are required, like for concrete, is typically based on the following visual observations:

- Color change.
- Cracking (fresh cracks greater than minor in size).
- Spalling of the surface.

However, the temperature levels at which these effects occur and the implication of these effects vary somewhat from those of concrete. Characteristic impacts of fire (temperature) on clay brick masonry are summarized as follows[17,19]:

- *Color change*: at temperatures ranging from ~482°F (~250°C) to ~572°F (~300°C), iron-bearing masonry and mortar will redden. This color change does not affect the structural stability of the masonry but is irreversible and considered cosmetic damage.
- *Mortar damage*: at temperatures ranging from ~572°F (~300°C) to ~752°F (~400°C), the mortar strength begins to be compromised. However, the depth of the damage rarely exceeds 1/2 inch (12 mm) to 3/4 inch (19 mm). In addition, fire-fighting water can thermally shock the masonry, causing it to crack at these temperatures.
- *Fusing of clay bricks*: brick masonry exposed to temperatures of ~1,822°F (~1,000°C) and above can remain structurally sound as long as they are not cracked and the mortar is not damaged; however, the brick masonry may fuse under conditions of prolonged heating.

In addition, spalling of brick masonry, especially the perforated type because of the internal pathways, can occur. Buckling of masonry walls can occur as they weaken, but experience has shown this to be a rare occurrence.

Thus, from a visual forensic structural investigation perspective, the conditions observed as they pertain to the removal/replacement versus cleaning decision remain the same as for concrete (with the exception of cosmetic lower-temperature color change). The most likely structural effects to be observed will be cracking and degradation of the mortar should the fire reach temperatures that will negatively impact the integrity of brick masonry systems.

17.3.1.4 Signs of Visual Structural Damage to Stone Masonry

The visual determination of whether or not structural fire damage to stone masonry has occurred to the extent that removal and replacement are required, like for concrete, is typically based on the following visual observations:

- Color change.
- Cracking (fresh cracks greater than minor in size).
- Spalling of the surface.

However, the temperature levels at which these effects occur and the implication of these effects vary, depending on the type of stone involved in the construction of the wall or foundation.[19] Characteristic impacts of fire (temperature) on various types of stone are summarized as follows:

- *Granite*: Cracks develop at ~1,063°F (573°C) due to quartz expansion. The thermal expansion of this stone may not be reversible if the granite is heated too rapidly below ~1,063°F (573°C). Granite begins to melt at a temperature of ~1,832°F (1,000°C).
- *Limestone*: The color of limestone changes to a pink or reddish brown at temperatures ranging from ~482°F (~250°C) to ~572°F (~300°C). The color turns more reddish at ~752°F

(~400°C). Limestone begins to chalk (calcination) at temperatures >1,112°F (>600°C) and powders to a gray-white color at temperatures ranging from ~1,472°F (~800°C) to ~1,832°F (~1,000°C). Limestone begins to melt at a temperature of ~1,832°F (1,000°C).

- *Marble*: The thermal expansion of this stone is not reversible for temperatures ranging from ~392°F (~200°C) to ~1,112°F (~600°C). At temperatures above ~1,112°F (600°C), marble degrades and becomes friable. Marble begins to melt at a temperature of ~1,832°F (1,000°C).
- *Sandstone*: The color of sandstone changes to a reddish tone at temperatures ranging from ~482°F (~250°C) to ~572°F (~300°C), but may not be visible until a temperature of ~752°F (~400°C) is reached. The degradation of the quartz grains in sandstone begins at temperatures >1,063°F (>573°C), causing it to become friable. Sandstone retains it reddish color at temperature up to ~1,832°F (~1,000°C), but begins to chalk (calcination) at temperatures ranging from ~1,472°F (~800°C) to ~1,832°F (~1,000°C). Sandstone begins to melt at a temperature of ~1,832°F (1,000°C).

Thus, from a visual forensic structural investigation perspective, the conditions observed as they pertain to the removal/replacement versus cleaning decision remain the same as for brick masonry. The most likely structural effects to be observed will be cracking and degradation of the mortar should temperatures reach the ranges that will negatively impact the integrity of stone masonry systems.

17.3.2 Signs of Visual Structural Damage to Steel

The visual determination of whether or not structural fire damage to steel framing members, including steel I-beams, steel joists/trusses, and metal floor/roof decking has occurred is generally associated with[16,20,22] the following:

- Warping, twisting, or other distortions.
- Corrosion of surfaces and loss of coatings.
- Fire residue deposits.

In residential settings, the steel damage assessment is typically associated with a basement steel I-beam and supporting steel posts, whereas in a commercial setting, steel I-beams, joists, trusses, and decking can be encountered. Fire damage to steel members has been categorized into one of three categories[16]:

Category 1: Straight members that appear to be unaffected by a fire.
Category 2: Deformed members that could be heat-treated if economical.
Category 3: Severely deformed members unlikely to be repaired.

Like with other construction materials, the extent of damage to steel framing members is related to the length of time the members are exposed to elevated temperatures. Temperature effects on steel are summarized in Table 17.1, some of which are listed below:

- Blistering, discoloration, and flaking off the surface coatings and paints occur at temperatures above 600°F (>316°C).
- Mill-scale on the surface occurs at temperatures below 1,200°F (<649°C).
- Presence of corroded or eroded surfaces suggests that temperatures exceeded 1,200°F (>649°C).
- Melting of carbon steel begins at a temperature of 2,760°F (1,516°C).

Thus, at temperatures below 600°F (<316°C), little or no degradation of steel, including coatings, would be expected. Further, if a fire has not caused distortions to hot rolled steel members, it is unlikely to have been impacted by temperatures of ~1,112°F (~600°C) or greater.[22]

In terms of forensic investigations regarding steel framing members or decking, the following should be performed:

1. Identify the area(s) where steel members and decking have been distorted.
2. Identify the area(s) where steel members and decking have any oxidation or degradation of their surfaces.
3. Define areas where steel surfaces have been coated with soot and smoke fume deposits.

In terms of actual steel members that may have been structurally compromised by fire, the rule of thumb is, "If it is still straight after exposure to fire – the steel is okay."[16,20,22] However, due to risk of future liability, the forensic structural engineer will usually designate any distorted members as well as the surfaces of any members containing oxidation/degradation for replacement even though, technically, areas with oxidized or degraded surfaces are probably not structurally damaged. Moreover, it is good practice to remove decking to the next straight member or to a member where the replacement decking will meet manufacture multi-span requirements.

Another area often overlooked in steel structural inspections is the delineation of soot and smoke fume deposits on steel members. These deposits often contain chlorides and sulfates that are corrosive and must be removed to eliminate future longer-term degradation of steel members/surfaces. Portable test kits are available (e.g., Chlor-Rid Field Test Kit) to perform this testing and should be utilized during the investigation to delineate areas with these deposits. An illustration of this test on a beam and the results and interpretation values are shown in Figure 17.9 and Table 17.3.

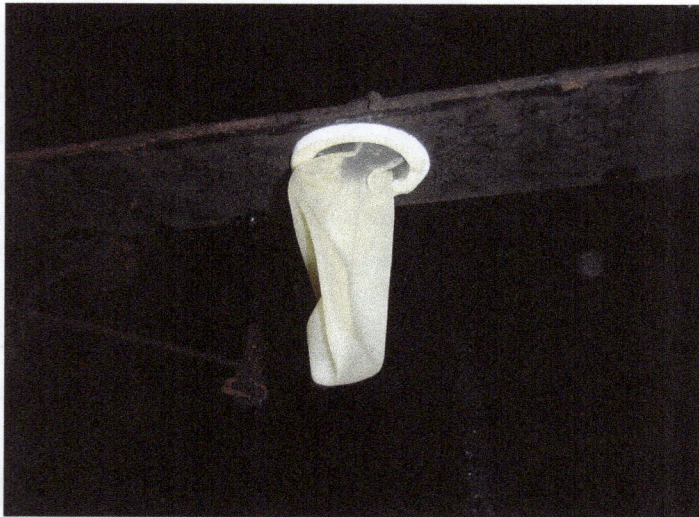

FIGURE 17.9   Chloride/sulfate test extractor on steel beam.

TABLE 17.3

Illustration of Chlorides and Sulfates Surface Tests Results on Steel Joists

ID	Location	Chlorides (μg/cm^2)	Sulfates (μg/cm^2)	Photograph
A	East roof – E2 – south end	**1–2**	8	
	East roof – E2 – north end	**5–6**	14	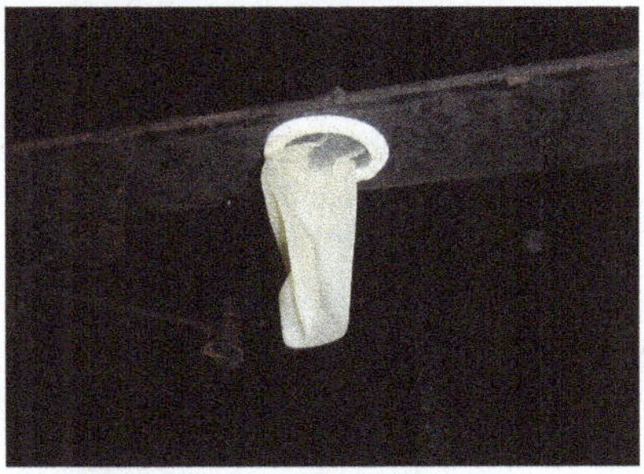
C	East roof – E8 – north end	**1–2**	2	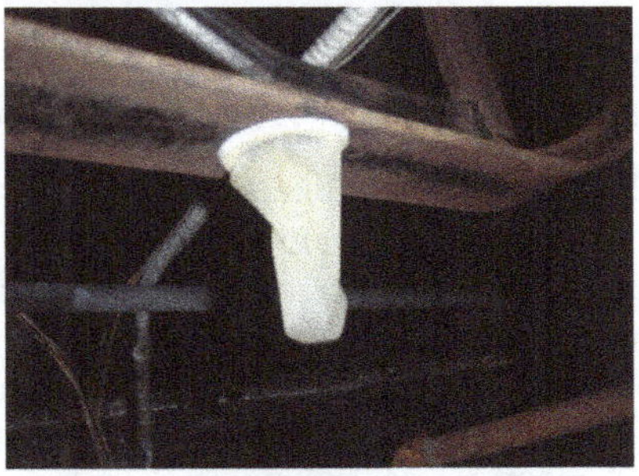
D	West roof – north end	**11–12**	85	

(Continued)

TABLE 17.3 (*Continued*)

Illustration of Chlorides and Sulfates Surface Tests Results on Steel Joists

ID	Location	Chlorides ($\mu g/cm^2$)	Sulfates ($\mu g/cm^2$)	Photograph
E	West roof – south end	4	26	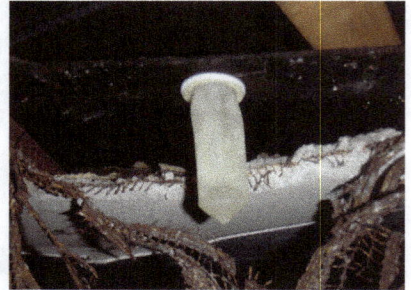

In this example (Table 17.3), the chloride sample results for the steel member D were above (7 $\mu g/cm^2$) the minimum standards set forth in the Society for Protective Coatings standard SSPC-SP 12/NACE.[22] Two of the sulfate sample results (D and E) were also at or above the minimum standards (17–50 $\mu g/cm^2$) for sulfates. As normally occurs, the sample results with high values were on the opposite end of the building from where the fire originated. This is because the smoke cools as it gets further from the fire at which point portions of the smoke fumes begin to condense out on surfaces. Thus, more smoke deposits will normally be found further away from the origin of the fire.

17.3.3 Signs of Visual Structural Damage to Wood Members

The visual determination of whether or not structural fire damage to wood members has occurred to the extent that removal and replacement are required is generally associated with the presence of charring to such members.[20] Of course, members entirely burned away would also be declared structurally damaged.

As illustrated in Table 17.1, wood begins to darken at fire temperatures of ~392°F–572°F (~200°C–300°C); ignites at a temperature of ~450°F (~232°C); and burns at temperatures of up to ~1,880°F (~1,027°C). The key in forensic structural investigations is to identify locations where the wood has burned or charred.

Char depths in wood members can also be used to determine the duration of the fire and whether or not residual strength remains within a given member. However, from a forensic structural evaluation standpoint, any charring to the member will result in a recommendation for replacement of these members. The most difficult part of the determination is where charring ends and heavy soot/smoke residual surfaces begin as well as when smoke-covered residual surfaces end.

For liability reasons, any charred wood members plus one further member are typically suggested for removal and replacement. For example, if joists set on ~16″ centers were charred, all joists with charring plus the next joist would be recommended for removal and replacement.

17.4 METHODOLOGY FOR STRUCTURAL FIRE DAMAGE ASSESSMENT OF BUILDINGS

The methodology for completing structural fire damage assessments includes the following sequential elements:

1. Interview of property owner or their representative.
2. Site inspection.
3. Acquisition of fire causation documents (if available),
4. Analysis of information collected,
5. Prepare a written report based on interview information, visual inspection observations, inspection field measurements, review of the pertinent literature, and experience.

Details regarding these elements follow.

17.4.1 FIRE DAMAGE STRUCTURAL ASSESSMENT METHODOLOGY – INTERVIEW OF PROPERTY OWNER OR OWNER'S REPRESENTATIVE

Upon arriving at the site, the inspection process should follow the overall approach outlined in Chapter 1. The intent of the inspection is to collect necessary information needed to help make the following determinations:

- Areas of the structure that may not be safe.
- Extent of structural items damaged by the fire that must be removed and replaced.
- Extent of structural items damaged by the fire that can be repaired or cleaned.
- Extent of structural items not impacted by the fire.
- Extent of nonstructural items impacted by the fire (by soot and odors). While not a structural item per se, most fire damage structural inspections will require the inspector to make judgment decisions regarding the extent of soot damage to finished surfaces and walls, ceiling cavities, and plenums. For certain types of fires, such as protein fires where odors can be quite persistent, this can be a difficult task.

Specific interview questions pertaining to the fire and resulting damage should include the following:

- Is a map or schematic of the property available? For multifamily residential (e.g., condominiums and apartment buildings) and commercial structures, a fire escape map is often useful. One should obtain three copies of these maps if possible (one for field notes, one to scan later for the report, and one as a spare).
- What date and time did the fire occur?
- Where did the fire originate, if known?
- How long did the fire burn?
- What caused the fire, if known?
- Is a fire department report available? If so, where can it be obtained?

- What areas were burned?
- Is the owner or owner's representative aware of any areas deemed unsafe? If so, where?
- Is the owner or owner's representative aware of any charred wood members? If so, what members and where are they located?
- Is the owner or owner's representative aware of any distorted steel members? If so, what members and where are they located?
- Is the owner or owner's representative aware of damaged concrete, CMU, brick masonry, or stone walls? If so, what materials and where are they located?
- What areas were smoke-damaged, but not burned?
- What areas were seemingly unaffected by the fire, if any?
- Are any pictures of the structure during or after the fire available? If so, obtain copies of the photographs. Often today, due to digital photography, these can be transferred via e-mail. If not, take pictures of the pictures using a digital camera.

The sum of this information by the owner or their representative should provide a basis for a preliminary understanding of unsafe areas, when and where the fire originated, and an overview of damages caused by the fire.

17.4.2 Fire Damage Structural Assessment Methodology – Site Inspection

The site inspection should follow the overall approach outlined in Chapter 1. However, before the detailed inspection begins, it is recommended that a half-faced respirator equipped with an N-95/VOC cartridge be worn. Note that medical clearance and training are required (29 CFR 1910.134) for the use of a respirator in occupational settings. A brief precursory inspection should be performed of the interior to determine areas that may be unsafe to walk on or walk under. Given the lack of light, presence of heavy debris, and slippery areas due to water from fire-fighting activities, it is easy to slip, fall, step on glass or nails, and at worst, fall through partially burned-through floors.

Once the inspection begins, particular attention should be spent delineating and documenting damage to structural members as outlined in Section 17.3, and when possible, attempting to delineate maximum fire temperatures by area based on criteria summarized in Table 17.1. All observations should be recorded in writing in a field notebook and photographed.

During the inside portion of the visual inspection for structural fire damage, soot deposits, fresh cracks in finished surfaces, distortion of steel members and/or support posts, and charring of wood framing members should be documented for the floor, walls, and ceiling of each room, beginning in the basement or crawlspace (if present) and moving up until the attic inspection is completed. Steel surfaces should be tested for the presence of chlorides and sulfates as illustrated in Section 17.3. Once the indoor visual inspection is completed, a visual inspection of the exterior surfaces of the building should be completed, by elevation, followed by a visual inspection of the roof.

An illustration of the types of documentation taken and reported during these types of inspections is shown in Figure 17.10 and Table 17.4.

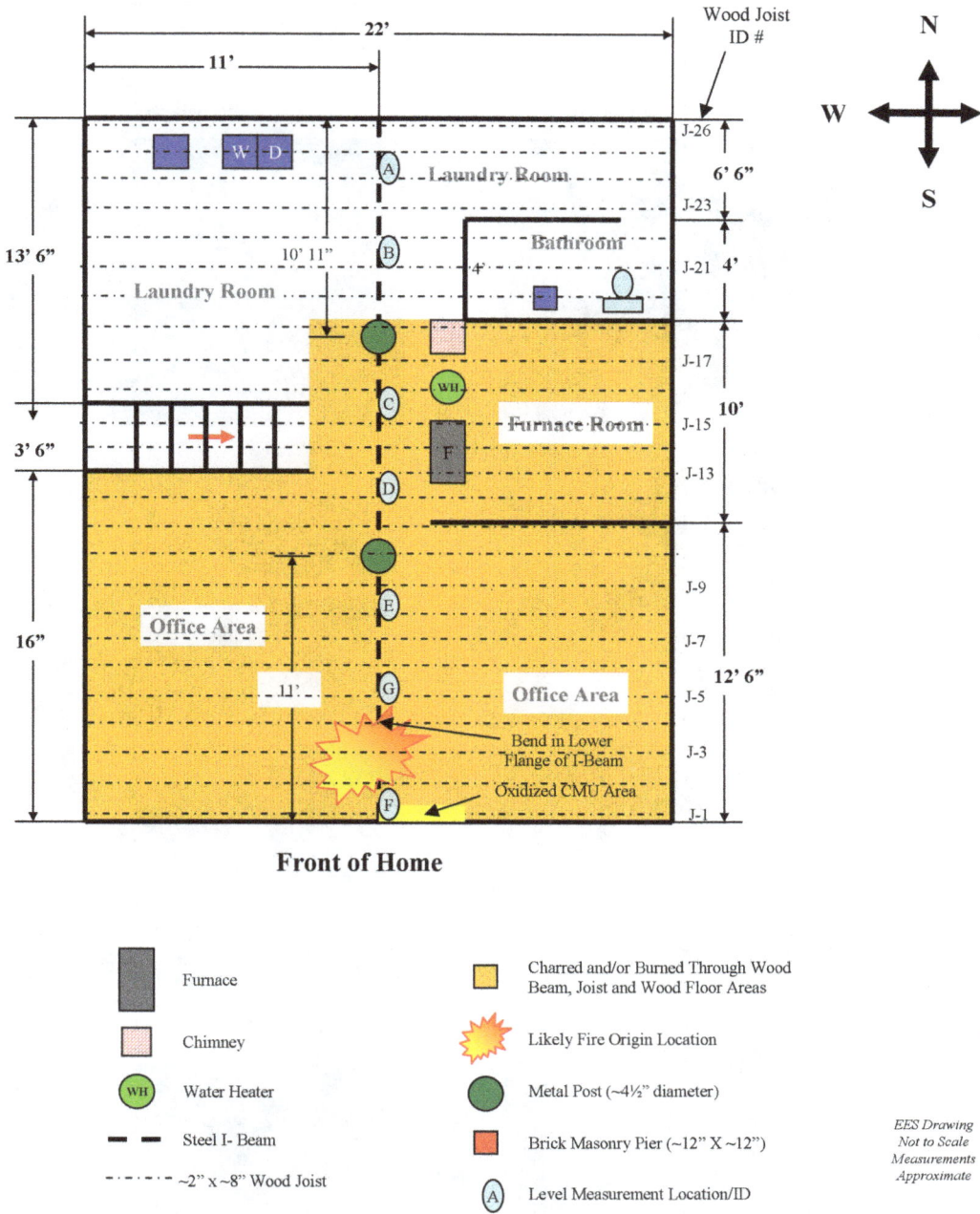

FIGURE 17.10 Fire damage assessment basement layout schematic.

TABLE 17.4
Fire Damage Materials and Structural Items

Item	Photograph	Photograph
Melted plastic on HVAC duct and melted PVC plumbing		
Melted aluminum wheel and melted solder on copper plumbing		
Spalled CMU and spalled/eroded concrete floor		
Cracked and displaced CMU		

(*Continued*)

TABLE 17.4 (*Continued*)
Fire Damage Materials and Structural Items

Item	Photograph	Photograph
Distorted steel I-beam		
Distorted steel truss		
Oxidized and melted metal decking		

(*Continued*)

TABLE 17.4 (*Continued*)
Fire Damage Materials and Structural Items

Item	Photograph	Photograph
Charred wood members		
Melted and fire damaged EPDM/ BUR room and damaged clay masonry parapet wall tile		

Information from the example basement fire schematic in Figure 17.10 illustrates the location of the origin of the fire, the area of fire charred wood members, and areas of fire-damaged CMU. Note that a designation for joists is provided so damages can be identified by joist number. The photographs in Table 17.4 illustrate damage to items such as plastics and aluminum, which helps to identify the temperature of the fire. In addition, fire damage to concrete, CMU, steel structural members and metal decking, char damage to wood framing members, and fire damage to roofing materials are illustrated in Table 17.4.

Wall and floor plumb and level measurements and crack measurements/observations should be taken for all wall, floor, and ceiling surfaces following the methodology as outlined in Chapters 14 and 15.

Although not considered structural in nature, the inspector often will be asked to delineate the extent of smoke damage. The extent of smoke damage deposits on surfaces should be defined as none apparent, light, moderate, heavy, or a combination of these categories.

The determination of soot or fume deposit levels on surfaces can be made by a white glove or cloth method and chemical analysis for soot deposits.[11]

In all structures, the wall and ceiling cavities and plenum surfaces, at distances away from areas of apparent fire damage, should be inspected and tested for evidence of smoke (soot) deposits. Destructive testing of wall finishes will be necessary to delineate the extent of fire and smoke damage. This normally is not an issue given existing damage caused by the fire and the fire-fighting activities. Sometimes the building will contain an inner wall cavity and an outer (e.g., brick masonry) wall cavity, both of which must be spot-checked for evidence of smoke deposits. If these are not removed and/or encapsulated, odor complaints will likely follow from occupants. Methods to determine the presence of soot deposits on surfaces resulting in odors include a white glove or cloth method and/or a tape lift soot analysis method.

The white glove/cloth method consists of the following steps:

1. Obtain white cloths (typically flour sack towels) and cut into ~6″ × ~6″ sections.
2. Label the upper corner of the cloth with a sample identification number.
3. Wipe a surface to be sampled. Examine the cloth and record visible findings in writing in a field notebook. Findings should include the color(s) and intensity of deposits found on the cloth.
4. Photograph the cloth sample near the surface that was sampled.
5. Place the cloth sample in a labeled plastic bag and seal the bag.
6. Repeat the sampling process for other test locations.
7. Later, when off site, open the plastic bag and smell the cloth sample for evidence of smoke odor.

The tape lift method collects surface samples for submission to a laboratory for soot analysis. The method consists of the following steps:

1. Label a white index card with a sample identification number.
2. Tear off an ~4-inch length of tape from a roll of cellophane tape.
3. Apply the tape to the surface to be sampled to capture the soot deposits. Remove the tape from the surface and attach the sample to the white index card. Examine the tape and record visible findings in writing in a field notebook. Findings should include the color(s) and intensity of deposits found on the tape.
4. Photograph tape sample near the surface that was sampled.
5. Place the tape sample and card in a labeled plastic bag and seal the bag.
6. Repeat the sampling process for other test locations.
7. When off site, prepare a chain of custody form and send card samples plus the chain of custody to a laboratory for fire residue and soot analysis. Samples should be packaged/preserved as required by the laboratory.

Experience has shown white glove/cloth sampling for fire residue deposits to be accurate, cost-effective, and timely when compared with laboratory analytical methods.

Examples of documentation for sampling locations, and white cloth and chemical sample analysis, are illustrated in Figure 17.11 and Tables 17.5 and 17.6.

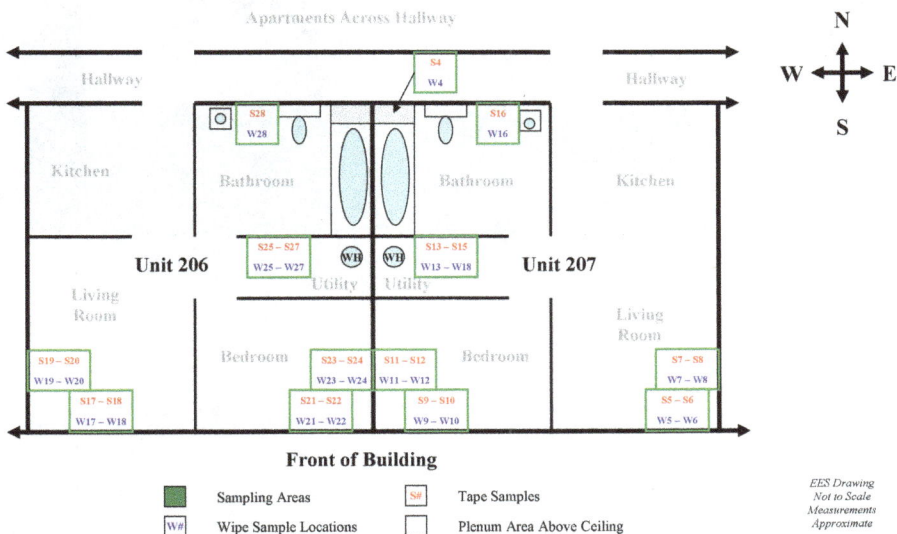

FIGURE 17.11   Smoke damage test location layout schematic.

TABLE 17.5

Illustration of Smoke Damage White Cloth Sample Results

White Cloth (W) Sample ID #s	Description of Sample Location	Deposit – None, Light, Medium, Heavy	Deposit Colors	Smoke Odor	Photograph
		White Cloth Sample Results			
12	Apartment 207 – bedroom – W. wall cavity – low	L to M	Yellow/tan dust	No	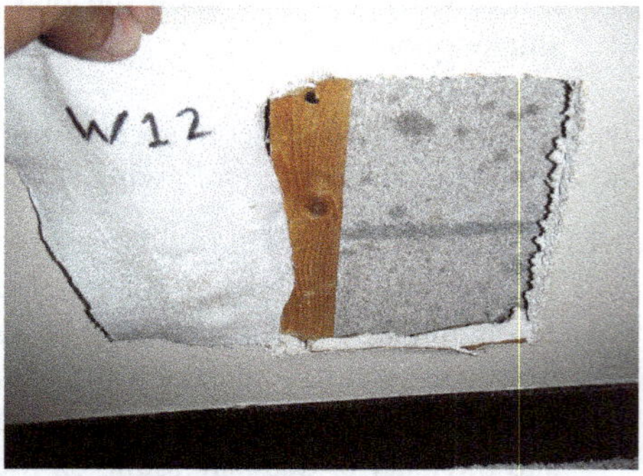
13	Apartment 207 – utility room – N. wall cavity – high	L	Brown dirt/ dust	Yes	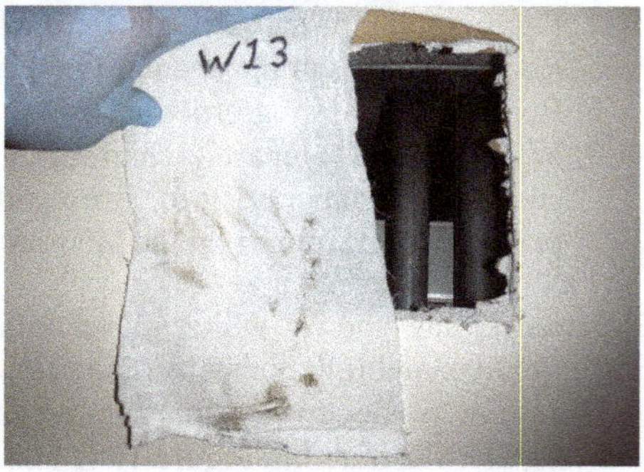
14	Apartment 207 – utility room – N. wall – ceiling cavity to N	M	Brown/Black dust; possible soot	Yes	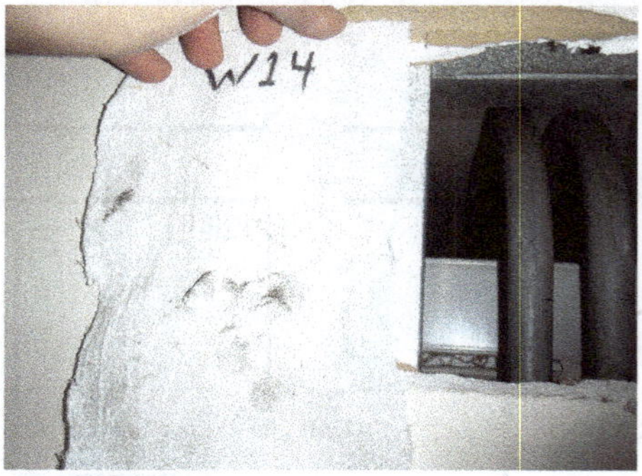

(*Continued*)

TABLE 17.5 (*Continued*)

Illustration of Smoke Damage White Cloth Sample Results

White Cloth (W) Sample ID #s	Description of Sample Location	Deposit – None, Light, Medium, Heavy	Deposit Colors	Smoke Odor	Photograph
		White Cloth Sample Results			
15	Apartment 207 – utility room – N. wall cavity – low	M to H	Brown/black dust; possible soot	Yes	
28	Apartment 206 – bathroom – N. wall cavity – central	L	Brown/tan dust	No	
29	Apartment 201 – living room – contents near east wall	H	Black soot	Yes	

TABLE 17.6
Illustration of Smoke Damage Tape Lift Laboratory Soot Sample Results

Tape Lift (TL) Sample ID #s	Description of Sample Location	Laboratory Results	Photograph
13	Apartment 207 – utility room – N. wall cavity – high	7% Carbonaceous particles. 32% Cellulose fibers 11% Epithelial (skin) cells 1% Fungal spores 7% Glass fiber 5% Gypsum dust 1% Hyphal fragments 25% Mineral particles 5% Miscellaneous 1% Plant matter 1% Pollen 1% Starch particles 3% Wood chips	
14	Apartment 207 – utility room – N. wall – ceiling cavity to N	8% Carbonaceous particles. 28% Cellulose fibers 12% Epithelial (skin) cells 3% Fungal spores 7% Glass fiber 2% Gypsum dust 20% Mineral particles 5% Miscellaneous 2% Plant matter 3% Pollen **5% Soot-like particles** 2% Starch particles 3% Wood chips	
15	Apartment 207 – utility room – N. wall cavity – low	5% Carbonaceous particles 40% Cellulose fibers 20% Epithelial (skin) cells 1% Fungal spores 2% Glass fiber 1% Gypsum dust 1% Hyphal fragments 16% Mineral particles 5% Miscellaneous 2% Plant matter 2% Pollen **3% Soot-like particles** 1% Spider web 1% Wood chips	

(Continued)

TABLE 17.6 (*Continued*)
Illustration of Smoke Damage Tape Lift Laboratory Soot Sample Results

Tape Lift (TL) Sample ID #s	Description of Sample Location	Laboratory Results	Photograph
17	Apartment 206 – living room – S. wall cavity – high	7% Carbonaceous particles 33% Cellulose fibers 17% Epithelial (skin) cells 7% Glass fiber 2% Gypsum dust 2% Insect parts 23% Mineral particles 5% Miscellaneous 1% Plant matter 2% Pollen 1% Wood chips	

As illustrated in Table 17.5, the white cloth results are reported by location, color of deposits, odor of the cloth, and a photograph of the cloth sample (both before and after the sampling) near the location that was sampled. Typically, fire residual deposits will be noticeably black (dirt and construction debris deposits are typically tan, brown, and/or yellow in color) and a smoke odor will be present when the sample in the bag is smelled. In this example, a residual smoke odor was identified in samples 13–15 and 29.

Similarly, as illustrated in Table 17.6, laboratory analysis of tape samples will, or will not, report evidence of fire residual soot-like particles depending on the location sampled. In this example, smoke residual (soot-like particles) was identified in samples 14 and 15.

17.4.3 FIRE DAMAGE STRUCTURAL ASSESSMENT METHODOLOGY – ANALYSIS OF INFORMATION COLLECTED

Information collected during the inspection, from the fire department (if available), and from the laboratory (if samples were submitted) should be analyzed to determine answers to questions initially touched upon in Section 17.3 of this chapter:

- What structural and foundation walls, constructed from concrete, CMU, brick masonry, or rubble stone, were damaged by the fire (if any)? Where is/are the damage(s) located? What was the extent of the damage? What damages to these walls were preexisting? Can these walls be repaired or must they be replaced?
- What steel members/surfaces, if any, should be removed and replaced (versus those that can be cleaned and sealed or are unaffected) as a result of fire damage? Where is/are the damage(s) located? What was the extent of the damage? What damages to these items were preexisting?
- What wood members/surfaces, if any, should be removed and replaced (versus those that can be cleaned and sealed or are unaffected) from fire damage? Where is/are the damage(s) located? What was the extent of the damage? What damages to these members were preexisting?

- What cosmetic damages associated with the fire were observed to other surfaces (e.g., plaster or drywall finished wall and ceiling surfaces)? Where is/are the damage(s) located? What was the extent of the damage? What damages to these members were preexisting?
- To what extent has fire residual been deposited on surfaces, wall cavities, and plenum spaces of the structure examined?

Site inspection observation information for damaged mineral (e.g. concrete) construction materials and steel and wood framing members is compared with damage criteria discussed in Section 17.3 to make remove/replace, repair, or clean/seal conclusions and recommendations. An example of how this information can be analyzed and summarized is shown in Table 17.7.

TABLE 17.7

Illustration of Remove/Replace vs Repair vs Clean/Seal for Basement Office Room Shown in Figure 17.8

Component (S)	Photograph	Recommendation
Foundation wall surfaces – except central south wall		Remove residual stud walls, clean and seal
South central upper foundation wall		Remove and replace top three rows of CMU – 3 blocks wide – just east of steel I-beam
Charred and burned-through wood ceiling joists, beams, and subflooring		Remove and replace; shore to support areas above West half – joists J1 to J12 East half – joists J1 to J10 Ceiling decking - all

(Continued)

TABLE 17.7 (*Continued*)

Illustration of Remove/Replace vs Repair vs Clean/Seal for Basement Office Room Shown in Figure 17.8

Component (S)	Photograph	Recommendation
South ~11′ of steel I-beam		Cut, remove, replace – tie in to remaining beam above south support metal post with plates – bolt or weld together to match the original beam integrity Shore as necessary
West window		Remove and replace
South gas meter and piping		Remove and replace
Ductwork		Clean and seal or remove and replace – select most cost-effective option

(Continued)

TABLE 17.7 (*Continued*)
Illustration of Remove/Replace vs Repair vs Clean/Seal for Basement Office Room Shown in Figure 17.8

Component (S)	Photograph	Recommendation
Natural gas black piping/copper and PVC plumbing		Plumbing surfaces could be cleaned, but it may be more efficient to remove and replace, given the ceiling joists will be removed and replaced
Soot-covered floor surfaces		Clean and seal

For reference, these example analysis conclusions and recommendations are associated with the basement office room shown in Figure 17.10. In this example, the steel I-beam, or portions of the I-beam, was recommended for removal and replacement. Other items were either recommended for replacement, repair, or cleaning and sealing.

Best practices dictate that removal/replacement or repair activities should be conducted with the oversight of a structural engineer to ensure proper shoring and/or other temporary supports are provided when necessary and that the repair or replacement of structural members is completed with properly designed adequate materials.

17.4.4   FIRE DAMAGE STRUCTURAL ASSESSMENT METHODOLOGY – WRITTEN REPORT

The written report should follow the overall approach outlined in Chapter 1. Particular attention should be spent documenting the following observations and results from the inspection in the body of the report:

- The date, time, and origin of the fire.
- The maximum temperature of the fire, by location, based on observations associated with known heat effects on materials (Table 17.1).
- Visible structural and smoke damage observations for all impacted spaces. The inspection should be completed indoors for all rooms/spaces for each impacted floor and outdoors for all elevations and the roof. The report should include a delineation of damaged areas,

including extent of damage to wall, floor, and ceiling surfaces, including structural support materials such as those formed from concrete, CMU, stone, steel, and wood. Results should be summarized by space as illustrated in Table 17.7.
- Smoke residue delineation and basis for delineation. This basis should include white glove/ fabric and/or analytical results on fire residue; results can be shown on figures and tabulated as suggested in Figure 17.11 and Tables 17.5 and 17.6.
- Plumb and level measurements on floors, walls, ceiling, and/or framing members should be reported in tabular format and any distorted surfaces noted. Results should be interpreted as outlined in Chapters 14 and 15.
- Cracks should be reported in terms of size, freshness, and location. Interpretation of crack results should follow guidelines outlined in Chapter 15.

The report should also provide guidance on fire and smoke damage repair. An excellent source of guidance in this area is the ASCR International NIDR Guidelines for Fire and Smoke Damage Repair.[11] Also, guidance on the cleaning of fire residue from HVAC systems can be found in IESO/ RIA 6001-2012.[23]

The report should close with any limitations noted. For example, often the structural fire damage inspection does not address, and is not intended to address, damage to the HVAC system, potable water systems, pluming, and electrical systems. Experts in these areas should be involved to address fire damage to these systems, as they are often outside the expertise of a structural engineer.

IMPORTANT POINTS TO REMEMBER

- Fire damage is related to the temperature and duration of the fire along with the ability of specific materials to maintain their integrity at given temperatures and exposure times.
- Most materials soften, deform, and/or melt at specific temperatures. The condition of these materials encountered during a fire damage structural inspection provides guidance on the maximum temperature of the fire at that location. This temperature can be used to determine if other materials (e.g., concrete, CMU, steel, or wood) were compromised by the fire.
- Most structural fire damage assessments conducted for residential and light commercial buildings are mainly visual inspections. However, limited destructive testing to sample wall cavities and ceiling plenums for soot/smoke residues is sometimes performed.
- Visual structural fire damage to concrete, CMU, and stone is determined by color change, cracking, and spalling of these materials.
- Visual structural fire damage to concrete, CMU, and stone is determined by color change, cracking, and spalling of these materials.
- Visual structural fire damage to steel framing members and metal decking is determined by deformation of these materials as well as oxidation/degradation of surface coatings.
- Visual structural damage to wood framing members is determined by whether or not they are charred.
- Delineation of the extent of fire residue to surfaces is usually accomplished using a white glove/cloth method.

REFERENCES

1. National Fire Prevention Association. "Overall Fire Statistics," accessed May 6, 2020. https://www. usfa.fema.gov/data/statistics/order_download_data.html#download (see FEMA 2017 and Back us_fire_ loss_data_sets_2008–2017.xlsx spreadsheet).

2. Evarts, B. and National Fire Prevention Association. "Fire Loss in the United States During 2018", October 2019, accessed May 6, 2020. https://www.nfpa.org//-/media/Files/News-and-Research/Fire-statistics-and-reports/US-Fire-Problem/osFireLoss.pdf.

3. The Concrete Centre. "Concrete and Fire: Using Concrete to Achieve Safe, Efficient Buildings and Structures." Ref. TCC/05/01, 2004. ISBN 1-904818-11-0.

4. National Fire Protection Association. "A Few Facts at the Household Level," Fire Analysis and Research Division, July 2009, accessed February 7, 2012. http://www.nfpa.org/assets/files/PDF/OS.Household.pdf.

5. U.S. Fire Administration. "U.S. Fire Deaths, Fire Death Rates, and Risk of Dying in a Fire," accessed May 6, 2020. https://www.usfa.fema.gov/data/statistics/fire_death_rates.html.

6. National Fire Prevention Association. "Overall Fire Statistics," accessed May 6, 2020. https://www.usfa.fema.gov/data/statistics/order_download_data.html#download (see FEMA 2018 Residential_nonresidential_fire_loss_estimates.xlsx spreadsheet).

7. National Fire Prevention Association. "Overall Fire Statistics," accessed May 6, 2020. https://www.usfa.fema.gov/data/statistics/order_download_data.html#download (see FEMA 2018 Structure Fires By Type of Use, 2018 (1).xls spreadsheet).

8. National Fire Protection Association (NFPA) 921. "Guide for Fire and Explosion Investigations," 2004.

9. National Fire Protection Association (NFPA) 921. "Guide for Fire and Explosion Investigations," 2005.

10. Noon, Randall K. *Forensic Engineering Investigation*. New York: CRC Press, 2001.

11. ASCR International. *NIDR Guidelines for Fire and Smoke Repair Damage*, National Institute of Disaster Restoration, 2nd Edition, Millersville, MD, 2002.

12. "Fire Damage: A Need for Guidance," 1995, accessed online July 22, 2007.

13. National Fire Protection Association (NFPA) 901. "Standard Classifications for Incident Reporting and Fire Protection Data", 2011.

14. "Fire Protection: OSHA Part 1910 Subpart L," accessed February 12, 2012. www.oshainfo.gatech.edu/ppt/FirePro-Gen.pptSimilar.

15. Erlin, Bernard, William G. Hime, and William H. Kuenning. "Evaluating Fire Damage to Concrete Structures," Concrete Construction, May 1972, accessed February 12, 2012. http://www.concreteconstruction.net/images/Evaluating%20Fire%20Damage%20to%20Concrete%20Structures_tcm45-343618.pdf.

16. Tide, Raymond H. R. "Integrity of structural steel after exposure to fire." *Engineering Journal*, First Quarter, 2003, 26–38, American Institute of Steel Construction.

17. Building Construction Information from the Concrete and Masonry Industries. National Codes and Standards Council of the Concrete and Masonry Industries. "Assessing the Condition and Repair Alternatives of Fire-Exposed Concrete and Masonry Members." Fire Protection Planning Report, August 1994.

18. Whai, Fan Foo "Presentation on Structural Appraisal & Repair of Fire-Damaged Buildings," accessed April 16, 2007.

19. Ingham, Jeremy "Forensic engineering of fire-damaged structures." *Proceedings of the Institution of Civil Engineers*, May 2009, 12–17, accessed February 12, 2012. http://www.halcrow.com/Documents/fire_safety/forensic_engineering.pdf.

20. Amin, Rizgars, Salih, Ph.D. "Structural Appraisals for Fire Damaged Buildings," accessed February 12, 2012. http://keu92.org/uploads/Search%20engineering/STRUCTURAL%20APPRAISAL%20FOR%20FIRE.pdf.

21. Wang, Yong Chang, F. Wald J. Vacha, and Monika Hajpal. "Chapter 4.1: Fire Damaged Structures." Project No. 1Mo579, accessed April 3, 2021. http://fire.fsv.cvut.cz/COST_C26_Prague/pdf/4-1_Fire%20damaged%20structures_sm.pdf.

22. TATA Steel. "Re-Use of Fire Damaged Steel," accessed online February 12, 2012. http://www.tatasteelconstruction.com/en/design_guidance/structural_design/fire/fire_damage_assesment/re_use_of_fire_damage_buildings/.

23. Indoor Environmental Standards Organization (IESO). "Standard for the Evaluation of Heating, Ventilation and Air Conditioning (HVAC) Interior Surfaces to Determine the Presence of Fire-Related Particulate as a Result of Fire in a Structure." An ANSI Standard, IESO/RIA 6001-2012, Rockville, MD, 2012.

18 Vehicle Impact Structural Property Assessments

Stephen E. Petty
EES Group, Inc.

CONTENTS

PURPOSE/OBJECTIVES

The purpose of this chapter is to:

- Provide an understanding of the forces imparted on structures when impacted by vehicles.
- Demonstrate inspection methodologies for completing a vehicle impact structural forensic inspection.

Following the completion of this chapter, you should be able to:

- Understand potential forces acting on the structural components of a building during vehicular impacts.
- Be able to collect necessary information needed to determine vehicle impact structural and nonstructural (cosmetic) damage to a residential or light commercial building using the methodologies specified in this chapter.
- Be able to prepare a written report with analysis and documentation regarding vehicle impact damage (both structural and cosmetic) to a residential or light commercial building.

18.1 INTRODUCTION

In 2018, vehicle collisions with fixed objects and animals totaled 7,422 incidents, resulting in 29,138 deaths; a portion of these occur when vehicles strike buildings.[1] In order to understand the forces involved in a vehicular impact, it is helpful to keep in mind Newton's Laws of Motion:

1. A body at rest tends to stay at rest and a body in motion tends to stay in motion unless acted upon by an external force. This is frequently referred to as the law of inertia.

2. When a force is applied to a free body, the rate at which the momentum changes is proportional to the amount of force applied. The direction in the change of momentum caused by the force is that of the line of action of the force.
3. For every action by a force, there is an equal but opposite reaction.

The third law is the one we will deal with the most. This means that when an object, in our case a moving vehicle, strikes another object, which in this case is a fixed object such as a residential or light commercial building, tremendous kinetic energy, or the energy the vehicle possesses due to its motion, is transferred from the moving car to the structure. This transfer of energy can result in significant structural damage to a building. An illustration of damage to a car and home from a vehicle crashing into the concrete porch of the home is shown in Figure 18.1.

FIGURE 18.1   Vehicle – house impact and resulting damage.

The energy associated with a vehicle impact is dependent on the weight (mass) of the car, velocity (speed) of the car, and the distance over which the deceleration occurs. The kinetic energy of the vehicle before it decelerates is given approximately by Eq. (18.1):

$$KE_{initial} = \frac{1}{2}mv^2 \tag{18.1}$$

where:
   $KE_{initial}$ = initial kinetic energy of vehicle
   m = mass of vehicle
   v = velocity of vehicle

The average force associated with deceleration of an impacting car transfers some of this kinetic energy to the object being struck with an average force equal to the kinetic energy divided by the stopping distance. Assuming all of this energy is imparted on a structure, the forces on the structure can be determined by Eq. (18.2):

$$\text{Force}_{avg} = \text{KE}_{initial}/d = \tfrac{1}{2}mv^2/d \tag{18.2}$$

where:
  Force_{avg} = average energy transferred to impacted object
  d = stopping distance

Examples of these resulting forces, in tons of force, are illustrated in Table 18.1.

TABLE 18.1

Examples of Average Vehicle Impact Forces with Speed[a]

Vehicle Speed (MPH)	Impact Energy (tons of force) by Weight of Car – Deceleration Distance ~1 feet		
	2,000 LBS	**3,000 LBS**	**4,000 LBS**
20	13.4 tons force	20.1 tons force	26.8 tons force
30	30.1	45.2	60.2
40	53.5	80.3	107.0
50	83.6	125.4	167.3
60	120.4	180.6	240.9
70	163.9	245.9	327.8

[a] Assumed deceleration distance of 1 foot.

Impact energies shown in Table 18.1, assuming that the deceleration takes place over a 1-foot distance, range from 13.4 tons of force to over 327 tons of force when speeds for 2,000–4,000 pound vehicles decelerate from speeds of 20–70 miles/hour. Forces increase proportionately by decreasing the deceleration distance and decrease proportionately by increasing the deceleration distance. For example, if the deceleration distance is decreased from 1 to 1/2 foot, or increased from 1 to 2 feet, the forces for a 3,000-pound vehicle traveling at 50 MPH change from 125.4 tons to 250.9 tons (1 to 1/2 feet stopping distances) and 125.4 to 62.7 tons (1 to 2 feet stopping distances) respectively.

While much of the vehicle impact energy in vehicle collisions is actually dissipated within the vehicle itself,[2] the calculation results shown in Table 18.1 illustrate the tremendous forces that can be imparted on structures when impacted by vehicles.

18.2   DAMAGE ASSESSMENT METHODOLOGY FOR VEHICULAR IMPACT TO A BUILDING

The methodology for completing structural damage assessments due to vehicular impact incorporates elements discussed at length in Chapters 14–17. The most significant elements of vehicular impact damage assessment inspections are the need to determine preexisting damage to the structure versus damage resulting from the vehicle impact. This need and methodology relies heavily on crack analysis, wall and/or floor slope/tilt analysis, and the ability to recognize recent (days/weeks) versus long-term (months/years) damage or defects to building systems. Examples of recent or "fresh" damage are shown in (Table 18.2).

TABLE 18.2

Examples of Fresh or Recent Damage to Structural Members

Item	Characteristics	Photograph
Split or crack in wood member	Brightness or lighter color on crack or cracked surfaces	
Cracked and displacement to concrete	Brightness or lighter color on crack or cracked surfaces	
Cracks and displacement to CMU	Brightness or lighter color on crack surfaces or cracked CMU block surfaces (note also fresh cracks to wood window frame)	
Displacement to CMU	Brightness or lighter color on crack surfaces or cracked CMU block surfaces. No debris (dirt, spider webs, algae) in cracks	

(Continued)

TABLE 18.2 (*Continued*)
Examples of Fresh or Recent Damage to Structural Members

Item	Characteristics	Photograph
Displacement to CMU	Brightness or lighter color on crack surfaces or cracked CMU block surfaces. No debris (dirt, spider webs, algae) in cracks	
Displacement and missing brick masonry	Brightness or lighter color on gap/crack surfaces. No debris (dirt, spider webs, algae) in cracks	
Cracks and displacement to brick masonry – veneer displace from window frame	Gaps or cracks at brick/mortar interface with mortar; brightness or lighter color on crack surfaces. No debris (dirt, spider webs, algae) in cracks	
Cracks and displacement to brick masonry	Gaps or cracks at brick/mortar interface with mortar; brightness or lighter color on crack surfaces. No debris (dirt, spider webs, algae) in cracks	

As illustrated in Table 18.2, the best criteria for determination of whether or not the damage is recent or "fresh" are as follows:

- A brighter or lighter color of the crack or split surfaces and/or
- Lack of debris within the crack. Debris can include dirt, paint, cobwebs, insects and insect debris, mold, algae, or lichen.

The investigator is cautioned to inspect and document damages in all rooms and spaces of the building, even those reported as not containing damage. In general, a complete assessment of all spaces within the structure, including basements/crawlspaces and attic spaces, is warranted to limit future disputes that might arise had not all spaces been assessed. If these areas are subsequently inspected, much of the "freshness" of the damage can be lost, making it much more difficult to distinguish between preexisting and the specific vehicular impact damage associated with the claim.

Like with other structural inspections, the inspection process includes the following sequential elements:

1. Interview of property owner and/or their representative.
2. Site inspection.
3. Acquisition of vehicle accident documents from local police department.
4. Analysis of information collected.
5. Prepare a written report based on interview information, inspection visual observations, inspection field measurements, a review of the pertinent literature, and experience.

Details regarding these elements follow.

18.2.1 Vehicle Impact with Building Damage Assessment Methodology – Interview of Property Owner or Owner's Representative

Upon arriving at the site, the inspection process should follow the overall approach outlined in Chapter 1. The intent of the inspection is to collect necessary information needed to help make the following determinations:

- Areas of the structure that may not be safe.
- Extent of structural damage by the vehicular impact that must be removed and replaced.
- Extent of structural damage by the vehicular impact that can be repaired (e.g., minor cracks to plaster of drywall).
- Extent of preexisting structural damage not caused by the vehicular impact with the building.

While not a structural item per se, most vehicular damage structural inspections will require the inspector to make judgment decisions regarding the extent of cosmetic damage to finished surfaces.

Specific interview questions pertaining to the impact and resulting damage should include the following:

- Is a map or schematic of the property available? For multifamily residential (e.g., condominiums and apartment buildings) and commercial structures, a fire escape map is often useful. One should obtain three copies of these maps if possible (one for field notes, one to scan later, and one as a spare).
- What date and time did the vehicle incident occur?
- What was the vehicle type and age? This provides information on the mass of the vehicle.
- Where did the vehicle impact the building?

- What was the estimated speed of the vehicle when it impacted the building? This establishes maximum forces and energies that may have been imposed on the structure.
- What were the circumstances associated with the impact? This should include questions regarding the condition of the driver and condition of the vehicle.
- What building areas and items were impacted?
- Is the owner or owner's representative aware of any areas deemed unsafe? If so, where?
- Is the owner or owner's representative aware of any damaged or distorted wood and/or steel framing members? If so, what members and where are they located?
- Is the owner or owner's representative aware of damaged concrete, CMU, brick masonry, or stone walls? If so, which walls and where are they located?
- What areas had preexisting damage? If so, what was the damage?
- What areas were seemingly unaffected by the vehicular impact, if any?
- Are any pictures of the vehicle and impact-related damage for times just after the accident occurred? If so, obtain copies of the photographs. Often today, due to digital photography, these can be transferred via e-mail. If not, take pictures of the pictures using a digital camera.

The sum of this information by the owner or their representative should provide a basis for a preliminary understanding of unsafe areas, when the accident occurred, and what and where recent damage caused by the vehicular impact is present.

18.2.2 Vehicle Impact with Building Damage Assessment Methodology – Site Inspection

The site inspection should follow the overall approach outlined in Chapter 1. The inspection should begin indoors, starting at the lowest level first and then moving upward. This normally would entail beginning the inspection in the basement or crawlspace and then moving upward, up to, and including the attic spaces. In each space, all floor, walls, ceiling, and support members should be examined for evidence of fresh displacement(s) and/or crack(s)/gap(s). Such items found should be documented in the field notebook and photographed. For reasons of inspection discipline and completeness, it is recommended that the inspection begin in spaces far away from the impact space or spaces, then work toward the impact space(s) and finally finishing with an inspection of the impact space(s).

Once the interior inspection is completed, inspections of the exterior elevations and roof surfaces should be completed. Again, the examination should be completed looking for both fresh and old damage to structural and finished surfaces and any displacement to mineral (e.g., concrete), steel, or wood structural members.

Perhaps the greatest difficulty with vehicular impact damage to buildings is hidden damage behind finished surfaces such as exterior siding, masonry or stucco, and/or interior finishes such as drywall or plaster. The areas where wall system components have been displaced, cracked, or broken at the site of impact are usually readily apparent; however, the extent to which these components are damaged beyond the obvious impact zone behind finished surfaces is often not so readily apparent. Destructive testing by removing portions of the finished surfaces would reveal these further damaged areas. If destructive testing is not allowed for a given inspection, limitations on the extent of the investigation to "readily available visible surfaces" should be clearly indicated in the report to avoid future liability for the inspector.

18.2.3 Vehicle Impact with Building Damage Assessment Methodology – Analysis and Reporting of Collected Information

Information collected during the inspection and from the police department, if available, should be analyzed to determine answers to the following questions:

- What structural and foundation walls fabricated from concrete, CMU, brick masonry, or stone were damaged by the vehicular impact (if any)? Where is/are the damage(s) located? What was the extent of the damage? What damages to these walls were preexisting? Can these walls be repaired or must they be replaced?
- What steel members, if any, should be removed and replaced due to vehicle impact damage? Where is/are the damage(s) located? What was the extent of the damage? What damages to these members were preexisting?
- What wood framing members, if any, should be removed and replaced due to vehicle impact damage? Where is/are the damage(s) located? What was the extent of the damage? What damages to these members were preexisting?
- What cosmetic damages associated with the vehicle impact were observed to other surfaces (e.g., plaster or drywall finished wall and ceiling surfaces)? Where is/are the damage(s) located? What was the extent of the damage? What damages to these surfaces were preexisting?

The written report should contain a summary of the information obtained and analyzed during the investigation including recommendations regarding damaged items that will require removal/ replacement or repairs. It should follow the overall approach outlined in Chapter 1. Particular attention should be spent documenting the following observations and analysis from the inspection in the body of the report:

- The date and time of the vehicle impacted the structure.
- The estimated speed with which the vehicle struck the building.
- The make and model of the vehicle that struck the building along with any other information regarding circumstances associated with the incident.
- Any photographs and police reports associated with the incident.
- The location where the building was struck (e.g., Figure 18.2).

FIGURE 18.2 Example of vehicle impact location schematic.

- Visible structural and cosmetic damage observations for all spaces. The inspection should be completed indoors for all spaces and outdoors for all elevations and the roof. The report should include a delineation of damaged areas, including extent of damage to wall, floor, and ceiling surfaces as well as structural framing members/materials such as those formed from concrete, CMU, stone, steel, and wood. Results should be summarized by space as illustrated in Figure 18.3 and Table 18.3.

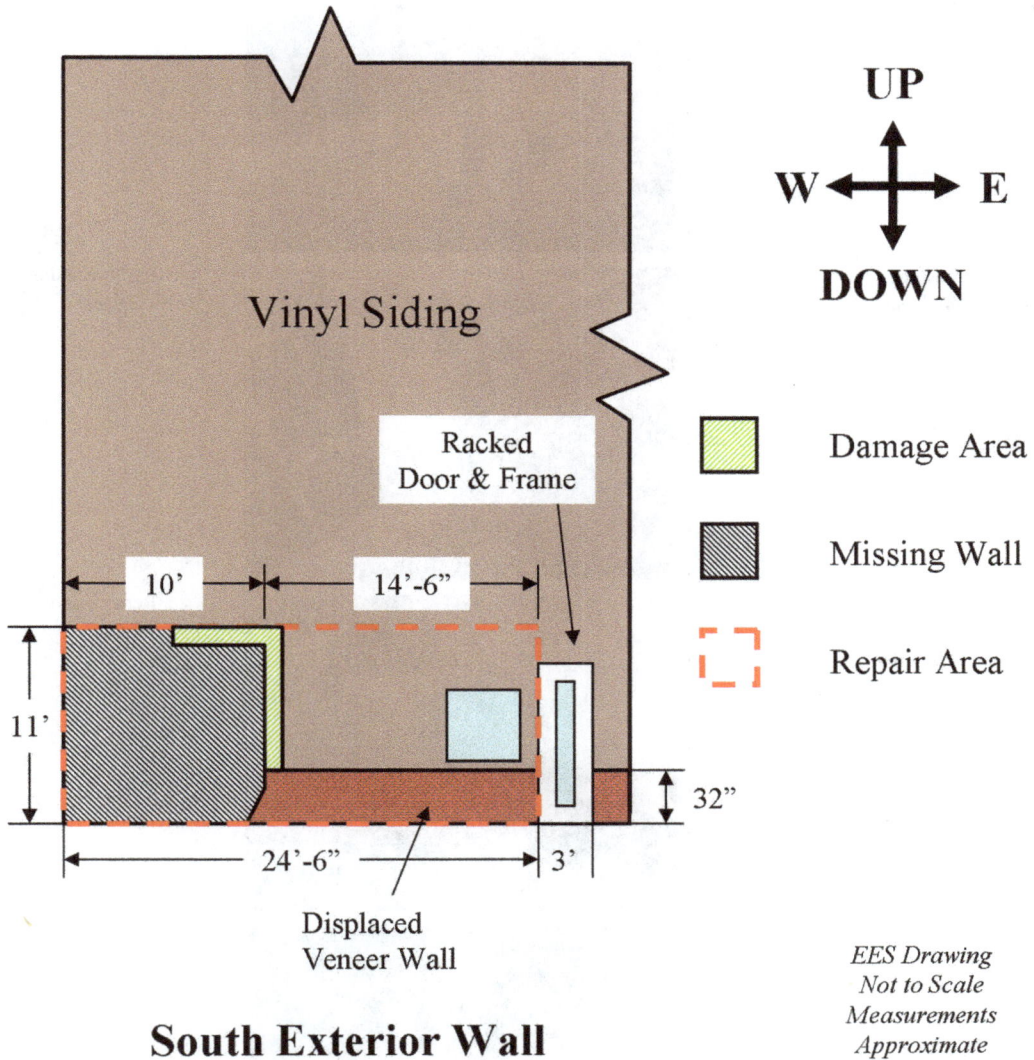

South Exterior Wall

FIGURE 18.3 Example of elevation layout schematic.

TABLE 18.3
Examples of Vehicle Damage Findings and Recommendations for a Given Elevation (see Figure 18.3)

Component(s)	Photograph	Recommendation
Red brick veneer (~32″ high×~24½′ long) from southwest corner east		Remove and/or replace
Metal cap/flashing (~8″ wide×~24½′ long)		Remove and/or replace
Vinyl siding (~24½′ wide×~7′ high) – lower southwest corner – replace more than damaged since end of panels damaged		Remove and/or replace
South bedroom window and framing		Replace

(Continued)

TABLE 18.3 (*Continued*)
Examples of Vehicle Damage Findings and Recommendations for a Given Elevation (see Figure 18.3)

Component(s)	Photograph	Recommendation
South entry door and framing		Replace
Vertical vinyl trim		Remove and/or replace

Nonstructural cosmetic damage caused by the vehicle impact should also be summarized, including locations and areas of damage.

- Cracks should be reported in terms of size, freshness, and location. Interpretation of crack results should follow guidelines outlined in Chapter 15.
- Plumb and level measurements on floors, walls, ceiling, and/or framing members should be reported in tabular format and any distorted surfaces noted. Results should be interpreted as outlined in Chapters 14 and 15. Examples of wall plumb and floor-level measurements are shown in Figure 18.4 and Tables 18.4 and 18.5.

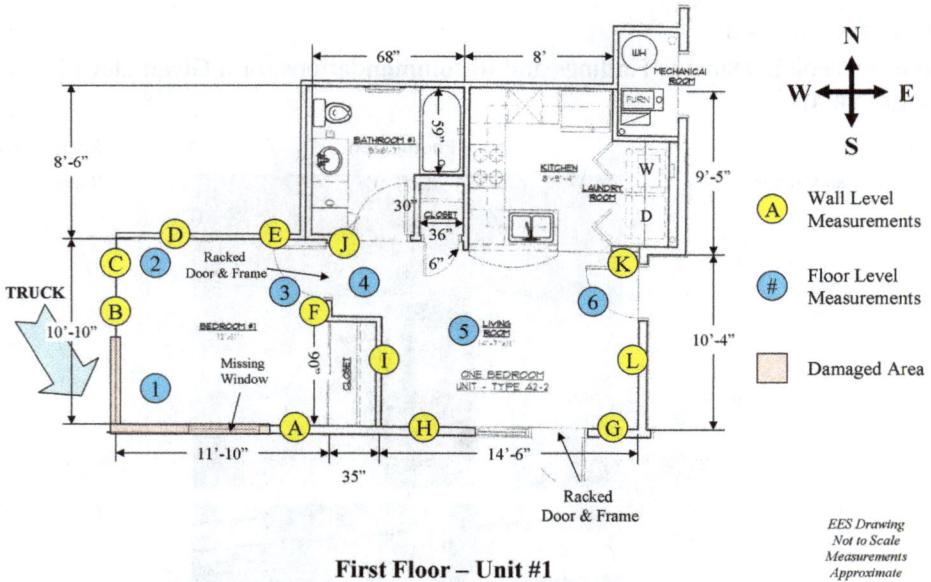

First Floor – Unit #1

FIGURE 18.4 Example of elevation plumb/level measurement schematic.

TABLE 18.4

Examples of Wall Plumb Measurement Results (see Figure 18.4)

ID	Description	High on Wall Degrees	High on Wall Top Tips	Middle of Wall Degrees	Middle of Wall Top Tips	Low on Wall Degrees	Low on Wall Top Tips
		Bedroom					
A	South wall – SE corner	~0.1	North	~0.3	North	~0.3	North
B	West wall – center	~0.0	Plumb	~0.0	Plumb	~0.6	East
C	West wall – NE corner	~0.1	East	~0.0	Plumb	~0.3	East
D	North wall – NE corner	~0.0	Plumb	~0.0	Plumb	~0.0	Plumb
E	North wall – SE corner	~0.1	North	~0.2	North	~0.0	Plumb
F	East wall – center	~0.1	West	~0.0	Plumb	~0.3	East
		Living Room					
G	South wall – SE corner	~0.4	North	~0.2	North	~0.2	North
H	South wall – SW corner	~0.2	North	~0.3	North	~0.4	North
I	West wall – center	~0.1	East	~0.0	Plumb	~0.2	West
J	North wall – NE corner	~0.3	North	~0.2	North	~0.1	North
K	North wall – SE corner	~0.3	South	~0.1	South	~0.1	North
L	East wall – center	~0.3	East	~0.1	East	~0.3	West

TABLE 18.5

Examples of Floor Level Measurement Results (see Figure 18.4)

ID	Location	North/South Direction		East/West Direction	
		Direction Sloping Down	Slope (Degrees)	Direction Sloping Down	Slope (Degrees)
		Bedroom			
1	Southwest	North	~0.6	East	~0.3
2	Northwest	Plumb	~0.0	East	~0.6
3	Northeast	North	~0.1	East	~0.3
		Living Room			
4	Northwest	North	~0.5	West	~0.2
5	Central	North	~0.2	West	~0.3
6	Southeast	North	~0.3	East	~0.4

Conclusions reached regarding these example measurements were, "The measurements indicated that all walls and floors were level within normal specification tolerance of 0.0 to 0.6 degrees."

The report should close with any limitations to the report noted. For example, if the report is based on a visual inspection only, the report should state that the observations, conclusions, and recommendations are based on an inspection of "readily available visible surfaces" at the date and time of the inspection unless specified destructive testing was undertaken to explore wall cavity and ceiling/floor plenum spaces further.

IMPORTANT POINTS TO REMEMBER

- Impact damage to building structures from vehicle impacts can be significant due to the large forces created when a portion of the kinetic energy from the vehicle is imposed on the structure. These forces can exceed 100 tons of force.
- Most damage to structures associated with vehicle impacts will be associated with distortions to structural mineral, steel, and wood framing members/systems. Cosmetic damage will also occur to interior and exterior finished surfaces.
- The entire structure should be examined for fresh damage to building components and finished surfaces associated with the vehicle impact to rule out future claims on spaces/structures/surfaces not examined since much of the visual inspection is based on differentiating preexisting damage from recent damage identified by brighter or lighter colors of the gap/crack surfaces in damaged items. These colors darken and age with time making distinctions between vehicle impact damage and other damage more difficult the further removed from the incident.
- Most vehicular damage assessments conducted for residential and light commercial structures are completed as visual inspections. Visual inspections are considered a form of nondestructive testing. Rarely is destructive testing and sample analysis done for these types of inspections for cost and timing reasons. The report should reflect these limitations.

REFERENCES

1. IIHS HLDI. "Fatality Facts 2018: Collisions with Fixed Objects and Animals." Accessed June 2, 2020, https://www.iihs.org/topics/fatality-statistics/detail/collisions-with-fixed-objects-and-animals.
2. Noon, Randall K. *Forensic Engineering Investigation*. New York: CRC Press, 2001.

19 Tornado-Related Structural Property Damage Assessments

Stephen E. Petty
EES Group, Inc.

CONTENTS

PURPOSE/OBJECTIVES

The purpose of this chapter is to:

- Be able to determine the strength and path of a tornado that impacted a structure to be evaluated.
- Be able to complete a post-tornado structural forensic inspection.
- Provide a methodology to determine and document tornado damage versus preexisting damage to a building.

Following the completion of this chapter, you should be able to:

- Understand the Fujita tornado scales.
- Be able to collect necessary information needed to determine structural and cosmetic damage to residential or commercial buildings from a tornado, using the methodologies specified in this chapter.
- Be able to prepare a written report with the analysis and documentation obtained during the inspection regarding tornado damage (both structural and cosmetic) to a residential or light commercial building.

19.1 INTRODUCTION

Tornados, by definition, are a violent destructive whirling wind accompanied by a funnel-shaped cloud that progresses in a narrow path over the land. Wind speeds within tornados can range from ~73 to ~300 mph and are among the most violent storms on the planet (Figure 19.1).[1]

FIGURE 19.1   Tornado aftermath – West Liberty, KY – March 2, 2012 – EF-3 Tornado.

The variability of tornados in terms of duration, wind speed, and toll in terms of loss of life and property makes them one of the most feared weather events. Weak tornados (wind speeds <110 mph) account for upward of 69% of all tornados, result in <5% of all tornado deaths, and last between ~1 and ~10 minutes. Strong tornados (wind speeds between 110 and 205 mph) account for upward of 29% of all tornados, result in ~30% of all tornado deaths, and last up to 20 minutes or longer. Violent tornados (wind speeds >205 mph) account for ~2% of all tornados, result in ~70% of all tornado deaths, and last up to 60 minutes or longer.

Tornado activity (>EF1) in the United States occurs most frequently in the middle part of the country (Figure 19.2).[2]

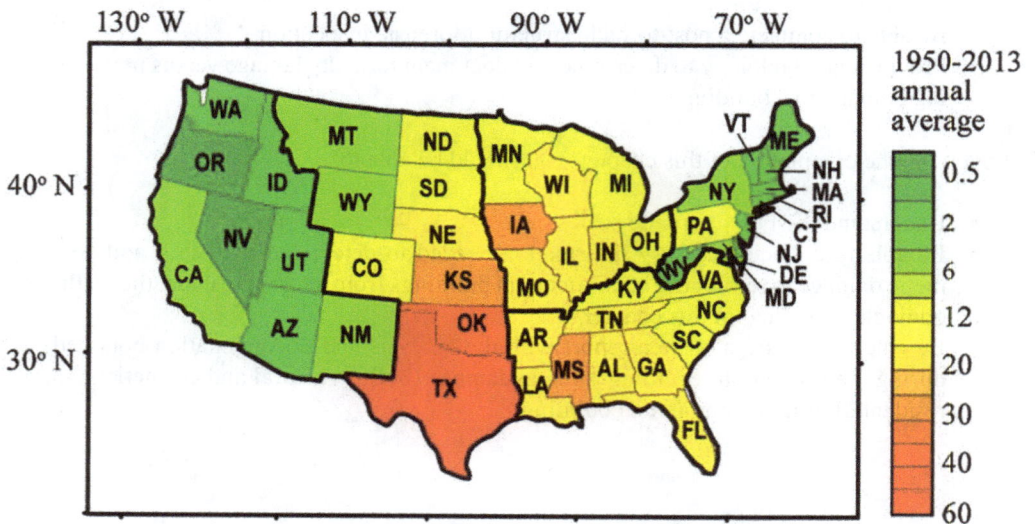

FIGURE 19.2   Annual average tornado activity in the United States by state: 1950–2013. (From Guo, Wang, and Bluestein[2] – Figure 1b.)

Areas of maximum annual tornado activity from 1950 to 2013 were in Texas, Oklahoma, Kansas, Mississippi, and Iowa.

The destructive potential to lives and properties from tornados is large and variable. The Storm Prediction Center (SPC) housed with the National Oceanic and Atmospheric Administration (NOAA), predicts, tracks, and records severe weather events and activity within the United States. As part of this work, the SPC is the repository for historical severe weather data, including that for tornados.[3] Reported tornado deaths in the United States, by strength of storm (discussed later in this chapter), from 2000 to 2011, are summarized in Table 19.1.

TABLE 19.1

Tornado Deaths by Time and Storm Strength

	Deaths by Strength of Storm (EF Scale)						
Year	EF0	EF1	EF2	EF3	EF4	EF5	Total
2000	0	2	5	21	13	0	41
2001	1	4	14	18	3	0	40
2002	0	6	17	29	3	0	55
2003	1	3	5	25	20	0	54
2004	0	8	7	52	0	0	67
2005	0	3	5	31	0	0	39
2006	0	8	7	52	0	0	67
2007	0	4	20	35	11	11	81
2008	0	4	18	53	42	9	126
2009	0	3	2	6	10	0	21
2010	0	4	3	16	22	0	45
2011	1	4	29	75	160	281	550
Totals	**3**	**53**	**132**	**413**	**284**	**301**	**1,186**
%	**0.3%**	**4.5%**	**11.1%**	**34.8%**	**23.9%**	**25.4%**	**100.0%**

Over this timeframe (2000–2011), 1,186 people lost their lives as a result of tornadoes. One might think forecasting and warnings associated with tornados would be better with time, and they may; however, in 2011, a major tornado outbreak year, 550 people died as a result of tornados. Of this total, 161 died when an EF5 tornado struck the town of Joplin, MO, in May of 2011.[4] Death rates are somewhat independent of storm strength, especially once the tornados reach EF3 strength.

In terms of locations where the tornado deaths occurred (Table 19.2), most of the deaths occurred in mobile (37.3%) or permanent (36.4%) homes.

TABLE 19.2
Tornado Deaths with Time by Location

			Deaths by Location				
Year	Mobile Home	Permanent Home	Vehicle	Business	Outside/Open	Other/Unknown	Total
2000	28	7	4	0	2	0	**41**
2001	17	15	3	3	2	0	**40**
2002	32	15	4	1	3	0	**55**
2003	25	24	0	1	3	1	**54**
2004	27	31	7	2	0	0	**67**
2005	34	3	1	1	0	0	**39**
2006	27	31	7	2	0	0	**67**
2007	52	16	2	10	1	0	**81**
2008	56	43	14	10	3	0	**126**
2009	12	7	1	1	0	0	**21**
2010	20	11	7	1	6	0	**45**
2011	112	229	34	92	7	76	**550**
Totals	**442**	**432**	**84**	**124**	**27**	**77**	**1,186**
%	**37.3%**	**36.4%**	**7.1%**	**10.5%**	**2.3%**	**6.5%**	**100.0%**

Annual property damage from tornados can approach $250 billion.[5]

19.2   OVERVIEW OF TORNADO DAMAGE SCALES AND TORNADO DAMAGE

The relative destructiveness and wind speeds of tornados today are rated by a term known as the Enhanced Fujita (EF)[6] scale. This scale ranges from EF-0 to EF-5. The EF scale was developed as an enhancement to the original Fujita scale for tornado damage and wind speeds. Both scales were designed to rate or rank the severity of tornados in terms of destructiveness and wind speeds. In order to assess damage to structures from a tornado, it is constructive to review the history and meaning of the original and enhanced Fujita scales.

In 1971, Dr. Tetsuya "Ted" Fujita of the University of Chicago developed a method to rate the intensity of tornadoes, which became known as the Fujita Scale. His research goals were to:

• Categorize each tornado by its intensity and its area.
• Estimate a wind speed associated with the damage caused by the tornado.

Based on this research, Mr. Fujita[6] developed a scale with seven categories of tornados (Table 19.3) beginning with F0 (42–72 mph) and ending with F6 (319–379 mph).

TABLE 19.3
Fuji Tornado Scale

Scale Number	Intensity Phrase	Wind Speed (mph)	Type of Damage Done
F0	Gale tornado	40–72	Some damage to chimneys; breaks branches off trees; pushes over shallow-rooted trees; damages sign boards
F1	Moderate tornado	73–112	The lower limit is the beginning of hurricane wind speed; peels surface off roofs; mobile homes pushed off foundations or overturned; moving autos pushed off the roads; attached garages may be destroyed
F2	Significant tornado	113–157	Considerable damage. Roofs torn off frame houses; mobile homes demolished; boxcars pushed over; large trees snapped or uprooted; light object missiles generated
F3	Severe tornado	158–206	Roof and some walls torn off well-constructed houses; trains overturned; most trees in forest uprooted
F4	Devastating tornado	207–260	Well-constructed houses leveled; structures with weak foundations blown off some distance; cars thrown and large missiles generated
F5	Incredible tornado	261–318	Strong frame houses lifted off foundations and carried considerable distances to disintegrate; automobile sized missiles fly through the air in excess of 100 m; trees debarked; steel reinforced concrete structures badly damaged
F6	Inconceivable tornado	319–379	These winds are very unlikely. The small area of damage they might produce would probably not be recognizable along with the mess produced by F4 and F5 wind that would surround the F6 winds. Missiles, such as cars and refrigerators would do serious secondary damage that could not be directly identified as F6 damage. If this level is ever achieved, evidence for it might only be found in some manner of ground swirl pattern, for it may never be identifiable through engineering studies

For each category, an intensity range of wind speeds and description of likely damage to structures and other items were provided. The scale, while used for 33 years, was recognized to have limitations,[7] including the following:

- Lack of damage indicators (DIs).
- No accounting for construction quality and variability.
- No definitive correlation between damage and wind speed.

This led to inconsistent ratings of tornadoes and in some cases an overestimation of tornado wind speeds. To address these inconsistencies, the Texas Tech University (TTU) Wind Science and Engineering Center (WISE) recommended a Steering Committee be assembled to attempt to get a consensus from the meteorological and engineering community to address concerns regarding the existing Fujita Scale. This Steering Committee consisted of the following six individuals:

- Jim McDonald (Professor) – TTU.
- Kishor Mehta (Director) – Wind Science & Engineering Center.
- Don Burgess (Assistant Director) – National Severe Storms Lab.
- Joe Schaefer (Director) – NOAA SPC.
- Michael Riley (Engineer) – National Institute of Standards and Technology.
- Brian Smith (Meteorologist) – National Weather Service.

The agreed-upon goals of the Steering Committee were to: (1) bring together a representative group of Fujita Scale users, (2) identify key issues, (3) make recommendations for a new or modified Fujita Scale, and (4) develop strategies for reaching a consensus from a broad cross section of users. To meet the first objective, a meeting was convened of leading tornado experts in Grapevine, TX on March 7–8, 2001, to develop a more accurate or so-called modified Fujita scale (i.e., EF scale). A total of 26 users of the Fujita Scale were invited to the conference, 20 (including those from TTU) attended. Others attending are listed below:

- Chuck Doswell – University of Oklahoma.
- Gregory Forbes – The Weather Channel.
- Joe Golden – Forecast Systems Laboratory.
- Tom Grazulis – Tornado Project.
- Rose Grant – State Farm Insurance.
- Quazi Hossain – Lawrence Livermore National Lab.
- Jeffrey Kimball – U.S. Department of Energy.
- Tim Marshall – Haag Engineering.
- Daniel McCarthy – NOAA SPC.
- Brian Peters – National Weather Service.
- Erik Rasmussen – CIMMS, Bolder, CO.
- Tim Reinhold – Clemson University.
- Thomas Schmidlin – Kent State University.
- Lawrence Twisdale – Applied Research Associates.
- Larry Vennozzi – National Weather Service.
- Roger Wakimoto – UCLA.
- Josh Wurman – University of Oklahoma.

To facilitate the meeting, and based on a request from the National Weather Service personnel, the TTU team proposed a list of 28 DIs. The basis of this modified scale was a review and detailing of these 28 DIs (buildings, structures, and trees) for which degree of damage (DOD) estimates were developed and then correlated with wind speed. The 28 DIs, by number, used by this group are summarized in Table 19.4.[8]

TABLE 19.4
Enhanced Fuji Tornado Scale Damage Indicators (DIs)

Number	Damage Indicator (DI)	Abbreviation
1	Small barns, farm outbuildings	SBO
2	One- or two-family residences	FR12
3	Single-wide mobile home (MHSW)	MHSW
4	Double-wide mobile home	MHDW
5	Apt, condo, townhouse (3 stories or less)	ACT
6	Motel	M
7	Masonry apt. or motel	MAM
8	Small retail bldg. (fast food)	SRB
9	Small professional (doctor office, branch bank)	SPB
10	Strip mall	SM
11	Large shopping mall	LSM
12	Large, isolated ("big box") retail bldg	LIRB
13	Automobile showroom	ASR
14	Automotive service building	ASB
15	School – one-story elementary (interior or exterior halls)	ES
16	School – jr. or sr. high school	JHSH
17	Low-rise (1–4 story) bldg.	LRB
18	Mid-rise (5–20 story) bldg.	MRB
19	High-rise (over 20 stories)	HRB
20	Institutional bldg. (hospital, govt. or university)	IB
21	Metal building system	MBS
22	Service station canopy	SSC
23	Warehouse (tilt-up walls or heavy timber)	WHB
24	Transmission line tower	TLT
25	Free-standing tower	FST
26	Free-standing pole (light, flag, luminary)	FSP
27	Tree – hardwood	TH
28	Tree – softwood	TS

For each of these 28 DI categories, DOD and wind speeds associated with each DOD were determined during this 2001 conference.

To illustrate the specific DODs and respective wind speeds under a specific DI, the DODs and wind speeds determined as a result of this conference for DI category #2, "One- or Two-Family Residences (FR12)," are provided in Figures 19.3 and 19.4.

DEGREE OF DAMAGE (DOD) ILLUSTRATION

[Example: DOD #2: One- and Two-Family Residences (FR12)]

Typical Construction Assumed:

• Asphalt shingles, tile, slate or metal roof covering

• Flat, gable, hip, mansard or mono-sloped roof or combinations thereof

• Plywood/OSB or wood plank roof deck

• Prefabricated wood trusses or wood joist and rafter construction

• Brick veneer, wood panels, stucco, EIFS, vinyl or metal siding

• Wood or metal stud walls, concrete blocks or insulating-concrete panels

• Attached single or double garage

DOD	DAMAGE DESCRIPTION	WIND SPEED (MPH)		
		LOWER BOUND	EXPECTED	UPPER BOUND
1	Threshold of visible damage	53	65	80
2	Loss of roof covering material (<20%), gutters and/or awning; loss of vinyl or metal siding.	63	79	97
3	Broken glass in doors and windows.	79	96	114
4	Uplift of roof deck and loss of significant roof covering material (>20%); collapse of chimney; garage door collapse inward or outward; failure or porch or carport.	81	97	116
5	Entire house shifts off foundation.	103	121	141
6	Large sections of roof structure removed; most walls remain standing.	104	122	142
7	Top floor exterior walls collapse.	113	132	153
8	Most interior walls of top story collapse	128	148	173
9	Most walls collapse in bottom floor, except small interior rooms.	127	152	178
10	Total destruction of entire building	142	170	198

FIGURE 19.3 Tornado Degree of Damage (DOD) tabled estimates for Damage Indicator (DI) #2 – One- and Two-Family Residences (FR12).

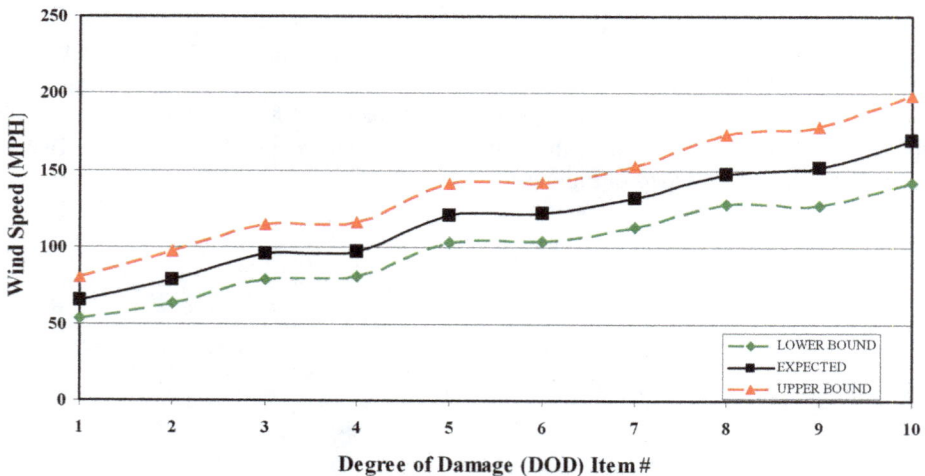

FIGURE 19.4 Plot of Tornado Degree of Damage (DOD) estimates for Damage Indicator (DI) #2 – One- and Two-Family Residences (FR12).

The variability in wind speeds for a given DOD is primarily associated with varying construction practices. The DODs provide expected (EXP), lower bound (LB), and upper bound (UB) wind speeds by degree of damage. Thus, as illustrated in Figures 19.3 and 19.4, for DI Category #2 – One and Two Family Residences (FR12), total destruction of a structure (DOD #10) was determined to occur with wind speeds from 142 to 198 mph (or expected wind speed of 170 mph). At the other end of the spectrum for this DI Category #2, the threshold of visible damage (DOD #1) was determined to be associated wind speeds from 53 to 80 mph (or expected wind speed of 65 mph). Similar tables and figures with DODs and associative wind speeds for the other 27 categories were also developed during this conference.

To develop a modified or enhanced Fujita Scale, a second expert group was selected, based on their experience with tornado damage and the existing Fujita scale:

- Bill Bunting – National Weather Service Forecast Office – Fort Worth, TX.
- Brian Peters – National Weather Service Forecast Office – Calera, AL.
- John Ogren – National Weather Service Forecast Office – Indianapolis, IN.
- Dennis Hull – National Weather Service Forecast Office – Pendleton, OR.
- Tom Matheson – National Weather Service Forecast Office – Wilmington, NC.
- Brian Smith – National Weather Service Forecast Office – Valley, NB.

While these individuals never formally met as a group, they were polled via mail to assign an old Fujita Scale value necessary to cause the damage for each of the DODs in each of the 28 DIs. The mean wind speeds (3-second gust in mph) for this second group of assigned Fujita Scale values based on the original range of Fujita Scale wind values was compared with the mean (expected) wind speed values developed by the initial group developing the DODs. These two sets of wind data were linearly correlated to compare past Fujita Scale mean wind values with Enhanced Fujita (EF) scale mean values. The correlation had a slope of 0.6246 and an intercept of 36.393 with an $R^2 = 0.9118$ (i.e., a good correlation). The slope suggested that the new damage wind speed in the EF scale would be, on average, 62.46% (or 37.54% less) of the original Fujita Scale wind speeds. Aside from lowering the wind speeds to cause known damage associated with the original Fujita Scale, this correlation allowed experts to retain the ability to correlate old Fujita Scale and new EF Scale results. The categories/wind speeds of the original Fujita Scale, the derived modified EF Scale, and the Operational (i.e., recommended for use) EF Scale are summarized in Table 19.5.

TABLE 19.5
Summary of Original and Modified Fujita Scales

	Fujita Scale		Derived EF Scale		Operational EF Scale	
F Number	Fastest 1/4-mile (mph)	Three seconds Gust (mph)	EF Number	Three seconds Gust (mph)	EF Number	Three seconds Gust (mph)
0	40–72	45–78	0	65–85	0	65–85
1	73–112	79–117	1	86–109	1	86–110
2	113–157	118–161	2	110–137	2	111–135
3	158–207	162–209	3	138–167	3	136–165
4	208–260	210–261	4	168–199	4	166–200
5	261–318	262–317	5	200–234	5	Over 200
6	319–379	>318	-	-	-	-

Note that the Operational EF scale used today has lower wind speeds by category than the original Fujita Scale and does not include a Scale 6 category.

The WISE of TTU published their work on an EF Scale in June 2004[7] and most weather experts, including NOAA, now use the Operational EF Scale as shown in Table 19.5 to describe the strength of a tornado to strike an area that caused specific damage to specific buildings, structures, trees, and other DIs. It should be noted that the overall EF rating is based on the highest estimated wind speed to have occurred corresponding to DOD observations for given DIs. Often multiple EFs are assigned to a tornado along its path.[9] An example of a tornado with multiple EF ratings along its path is shown in Figure 19.5.

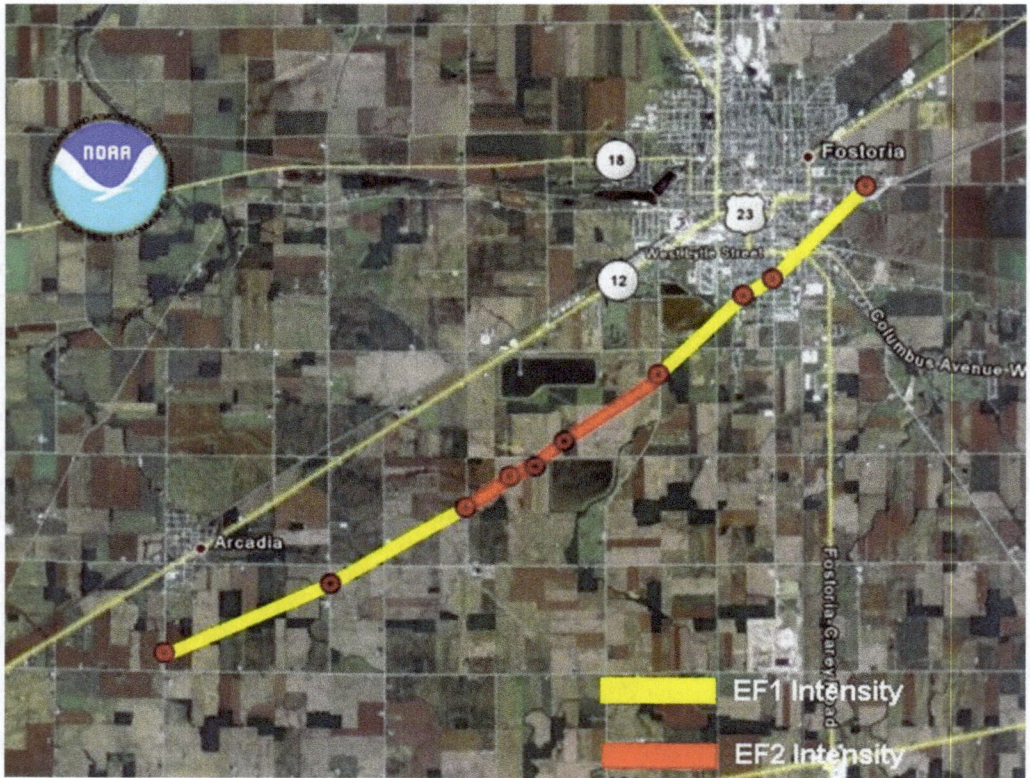

FIGURE 19.5   Path of tornado with multiple EF ratings. (Adapted from NOAA.[11])

In this example from NOAA (Figure 19.5), the initial and final rating of a tornado was EF1. However, during the middle portion of time the tornado was on the ground, it was rated as an EF2. Lower Operational EFs would apply to destruction not immediately within the path of the tornado.

The frequency of occurrence of EF tornados in the United States from 1970 to 2003 is summarized in Table 19.6.

TABLE 19.6
Occurrence of U.S. Tornados by Fujita Scale – 1970–2003[10]

Fujita Scale	Occurrence Percentage
F0	49.32
F1	32.56
F2	12.22
F3	3.43
F4	0.81
F5	0.08
Unknown	1.58

As illustrated in Table 19.6, nearly 82% of tornados were rated F0 or F1; only 4.32% were rated very strong (F3, F4, or F5).

19.3 METHODOLOGY FOR TORNADO DAMAGE ASSESSMENT OF BUILDINGS

Tornado situations encountered for structural and forensic inspections will typically fall into one of two categories:

1. Structure is a total loss.
2. Structure is a partial loss.

Under the first scenario, the inspection, if requested, is typically limited to an evaluation of whether or not the foundation walls or concrete slab is salvageable. Under these circumstances, evaluation of cracks and wall slopes (significance of size of cracks and freshness of cracks) is the basis for the analysis, and the reader is referred to Chapter 15.

Under the second scenario, the inspection follows much the same approach of fire damage inspections where a determination is made as to what components of the structure should be removed and replaced and what components can be repaired. The added complexity is the need to determine what damaged components were present before the tornado struck versus damage that occurred as a result of the storm. In almost all these cases, a portion of the structure/residence will be damaged more severely than the balance of the structure/residence. Examples of this situation are illustrated in Figure 19.6a and b.

(a)   (b)

FIGURE 19.6 Example of tornado damage – garage severely damaged with limited damage to home (left); severe tree damage to left of side of garage (right).

In Figure 19.6a, the vinyl siding had been ripped from the garage surfaces, whereas the home was relatively untouched. In this case, the tornado passed from the lower left of the photo to the left side of the garage.

The apparent pulling out of the wall system illustrated in Figure 19.7 is typical of damage from a tornado where the structure will appear to explode outward. This characteristic is because the air pressure in a tornado is low relative to the air pressure in a building, causing the structure to be displaced outward.

FIGURE 19.7 Example of tornado damage – failure of CMU wall on one side of commercial building.

The other complexity following a tornado strike to a structure is disputes arising from the loss of contents from a structure, especially farm or commercial buildings, where the contents are reported to be missing. The question arises regarding whether or not all the contents reported as missing were ever actually present.

The methodology for completing tornado damage assessments includes the following, sequential elements:

1. Interview of property owner or their representative.
2. Site inspection.
3. Acquisition of weather records.
4. Analysis of information collected, including weather data.
5. Prepare a written report based on interview information, weather records, inspection visual observations, inspection field measurements, and a review of the pertinent literature and experience.

Details regarding these elements follow.

19.3.1 Tornado Damage Assessment Methodology – Interview of Property Owner or Owner's Representative

Upon arriving at the site, the inspection process should follow the overall approach outlined in Chapter 1. The intent of the inspection is to collect necessary information needed to help make the following determinations:

- Areas of the structure that may not be safe.
- Extent of structural items damaged by the tornado that must be removed and replaced.
- Extent of structural items damaged by the tornado that can be repaired or cleaned.
- Extent of structural items not impacted by the tornado.
- Extent of missing/damaged contents.

While not a structural item per se, most tornado damage structural inspections will require the inspector to make judgment decisions regarding whether certain items were preexisting versus those caused by the tornado (e.g., foundation cracks).

Specific interview questions pertaining to the tornado and resulting damage should include the following:

- Is a map or schematic of the property available? For multifamily residential (e.g., condominiums and apartment buildings) and commercial structures, a fire escape map is often useful. One should obtain three copies of these maps if possible (one for field notes, one to scan later, and one as a spare).
- Was the owner or owner's representative home when the tornado struck?
- What date and time did the tornado occur, if known?
- What direction did the tornado arrive from, if known?
- How far away from the structure/residence was the tornado?
- Is the owner or owner's representative aware of the reported strength of the tornado? If so, what was it?
- Is the owner or owner's representative aware of any areas deemed unsafe? If so, where?
- Is the owner or owner's representative aware of any damaged wood members? If so, what members and where are they located?
- Is the owner or owner's representative aware of any distorted steel members? If so, what members and where are they located?
- Is the owner or owner's representative aware of damaged concrete, CMU, brick masonry, or stone walls? If so, where are they located?
- What areas were damaged, but were cosmetic in nature (e.g., cracks to drywall)?
- What areas were seemingly unaffected by the tornado, if any?
- Are any pictures of the tornado, or the times just after the tornado, available? If so, obtain copies of the photographs. Often today, due to digital photography, these can be transferred via e-mail. If not, take pictures of the pictures using a digital camera.
- The sum of this information by the owner or their representative should provide a basis for a preliminary understanding of unsafe areas, when and where the tornado struck and its proximity to the structure/residence, and an overview of damage caused by the tornado.

19.3.2 TORNADO DAMAGE ASSESSMENT METHODOLOGY – SITE INSPECTION

The site inspection should follow the overall approach outlined in Chapter 1; however, before the detailed inspection begins, a brief precursory inspection should be performed of the interior to determine areas that may be unsafe to walk on or walk under. Given the lack of light, presence of heavy debris and slippery areas due to water from rains that might have entered interior spaces, it is easy to slip, fall, step on glass or nails, and worse, fall through structurally damaged floors.

Once the inspection begins, particular attention should be spent delineating and documenting damage to structural members as outlined in Chapters 14, 15, and 17 [e.g., fresh cracks in wood and mineral members (e.g., concrete, CMU, brick, and stone) and distortions to steel members]. When possible, attempt to delineate structurally damaged areas. All observations should be recorded in writing in a field notebook; key observations should be photographed.

The inspection should start in the crawlspace or basement, if present, documenting any damages present in the floor, foundation walls, support beams, piers, posts, and/or floor framing members. At times, fresh movement of these items will be observed, signifying movement of the structure as a result of the tornado. The information collected should include plumb and level measurements as well as crack information as outlined in Chapters 14 and 15. The crack information (fresh versus old) will be critical to a determination of past versus recent damage to structural and nonstructural items. Next, the interior of the home/structure should be inspected. During the inside portion of the visual inspection for tornado damage, structural and cosmetic damage (older and relatively fresh in appearance) to floor, wall, and ceiling finished surfaces, as well as damage, distortion, and/or displacement to exposed structural framing members, should be documented for each room, beginning with the lowest level and moving up until the attic inspection is completed. The investigator is cautioned to inspect and document damages in all of the spaces, even those reported as not containing damage. In general, a complete assessment of all spaces within the structure, including basements/crawlspaces and attic spaces, is warranted to limit future disputes that might arise had not all spaces been assessed. If contents are reported as missing, a detailed record of contents observed by space should be completed.

Once the indoor visual inspection is completed, a visual inspection of the exterior, by elevation, of the structure should be completed followed by a visual roof inspection using the same approach as described in Chapters 6, 7, and 14. For commercial roof systems, particular attention should be paid to documenting the method of attachment and whether or not it met the code and/or manufacturer's recommended installation instructions.

In one case, the metal panel roofing of a commercial building lifted in an EF-0 tornado event. The wind speed for an EF-0 tornado is below code minimum design thresholds. It was found that the roof panels were not fully seamed as recommended by best practices documents. Thus, in this particular situation, the forces resulting from the tornado did in fact damage the roof system, however; the damage likely would not have occurred had the roofing system been properly installed according to best practices and manufacturer's recommended installation instructions (see also Ref.[10] for methods to improve structural integrity of building systems to help resist wind forces from tornados).

An illustration of the types of documentation taken and reported during these types of inspections is provided in Figure 19.8 and Table 19.7.

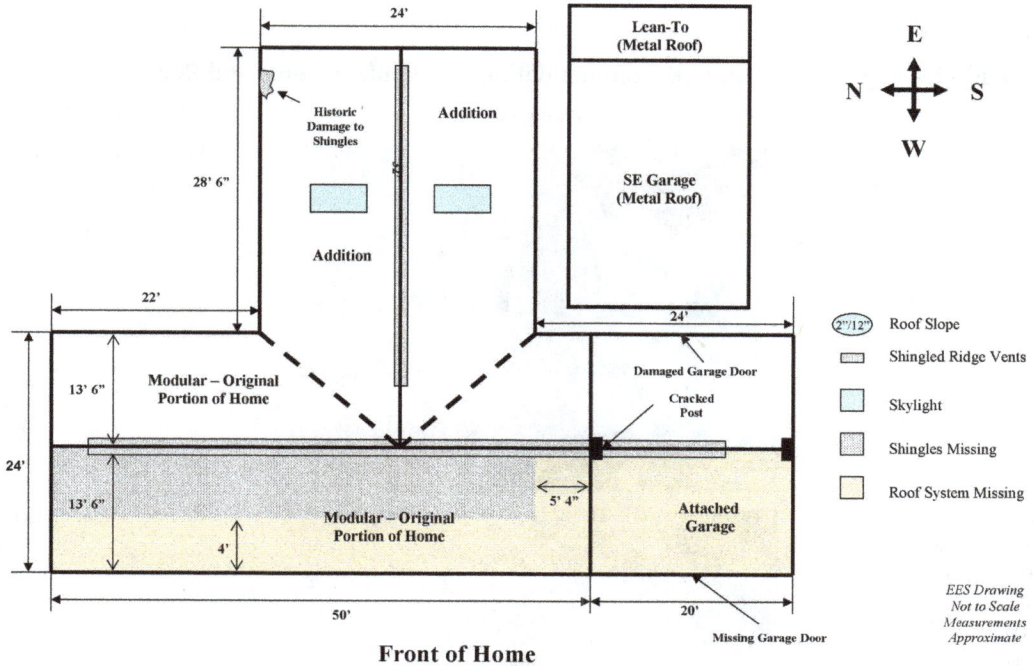

FIGURE 19.8 Tornado damage assessment roof layout schematic.

TABLE 19.7
Tornado Damage – Example of Recommendations for Replacements and Repairs

Component(s)	Photograph	Recommendation
Front and back attached garage doors Front attached garage door rails and door opener		Remove and replace

(Continued)

TABLE 19.7 (*Continued*)

Tornado Damage – Example of Recommendations for Replacements and Repairs

Component(s)	Photograph	Recommendation
Attached garage – north central post		Remove and replace
West roof elevation including rafters (see Figure 21) west roof insulation West house ceilings ridge vent		Remove and/or replace
north roof (addition) – impact damage at eave to shingles, felt, drip edge and gutter		Remove and repair/ replace
Missing siding – west side of garage and south side of home (two locations) Holes in west house and south garage vinyl siding (three locations)		Replace/Repair

(*Continued*)

TABLE 19.7 (*Continued*)

Tornado Damage – Example of Recommendations for Replacements and Repairs

Component(s)	Photograph	Recommendation
Damage to SE garage south roof – east rake trim and north gutter		Repair
Damaged/missing soffit, fascia, gutters and downspouts (see observations sections for details)		Repair or remove/replace or replace as necessary
West central bedroom floor and living room carpeting and floor damaged by water		Remove and replace
Wall and ceiling cracks/damage to interior spaces – see interior observations above for details on locations. Living room damage to floor, paneling and front door Wall/ceiling interface trim		Repair Remove/replace and repair Repair

In this illustration, major structural damage occurred to the front half of the garage and home where the entire roof systems were torn away by the tornado.

19.3.3 Tornado Damage Assessment Methodology – Analysis of Information Collected

Aside from an analysis of field information collected during the inspection, a detailed analysis of weather data should be completed. Information available from NOAA is often quite detailed regarding the strength, location, and path of a tornado along with damaged buildings and structures. Often, specific data for the structure in question is actually contained in this information and can support the extent of damage likely to be seen and reported. An illustration of the data available from NOAA's SPC for the tornado responsible for the damage to the home shown in Figure 19.8 is shown in Figure 19.9.[11]

FIGURE 19.9   NOAA February 28, 2011, tornado path and damage photographs. (Adapted from NOAA.[11])

The lowest left photograph shown in Figure 19.9 was of the home actually inspected in the example given in Figure 19.8 and is shown as the second item in the storm path.

Data from NOAA regarding the actual storm is reproduced in Figure 19.10.

NATIONAL WEATHER SERVICE REPORT ON SUBJECT TORNADO

Public Information Statement: National Weather Service – Wilmington, OH (Report dated 3/1/2011 – 1:20 PM)

- Tornado: Tornado confirmed 3 miles SSE of Millersport in Fairfield County, OH
- Date: February 28, 2011; Estimated time: 5:56 AM to 6:01 AM
- Maximum EF-Scale Rating: EF1; the wind speeds at these damage locations ranged from 95 to 105 MPH, the high end of an EF 1 tornado.
- Estimated Maximum Wind Speed: 105 MPH
- Maximum Path Width: 75 Yards
- Path Length: 5.0 miles beginning at Lat/Long: 39.854N / 82.580W & ending at Lat/Long: 39.866N / 82.492W
- Fatalities: 0
- Injuries: 0

AT THESE LOCATIONS...A MODULAR HOME HAD ITS ROOF COMPLETELY REMOVED AND TOSSED 30 YARDS. TREES WERE UPROOTED. AT A NEARBY RESIDENCE...THE TORNADO COMPLETELY UPROOTED NUMEROUS TREES...LIFTING ONE...AND SETTING IT DOWN ADJACENT TO WHERE IT WAS LIFTED FROM THE GROUND. A BARN WAS COMPLETELY DESTROYED...WITH ITS DEBRIS STREWN 500+ YARDS THROUGH NEIGHBORING FIELDS AND TREE LINES. AT THIS SECOND RESIDENCE...ONE OF THE EXTERIOR WALLS WAS BUCKLED FROM THE WIND BLOWING THE DOOR OF THE HOME OPEN...AND FORCING THE WALL TO BUCKLE. ACORN DEBRIS FROM A NEARBY FIELD WAS DRIVEN INTO THE GROUND. A STICK WAS FOUND DRIVEN INTO THE SIDING ON THE FRONT OF THE HOME. MUD SPATTER WAS FOUND ON ALL EXTERIOR WALLS OF THE HOME.

FIGURE 19.10 NOAA February 28, 2011, tornado information. (Courtesy of NOAA.[11])

The information lists the date and time of the tornado, the location, length and width of the tornado path, the tornado strength and wind speeds, and injuries and information on actual damage to homes and objects, including the home that was illustrated in this example.

If possible, information on the proximity of the structure to the tornado should be determined. An example depicting the location of a structure relative to the path of a tornado is shown in Figure 19.11.

FIGURE 19.11 Example of proximity of structure to the tornado path. (Adapted from NOAA.[11])

In this example case, the structure, a barn, was located just south of the tornado path.

Information collected during the inspection and from a review of weather records should be utilized to determine answers to the questions initially touched on in Section 19.3 of this chapter:

- What foundation walls, fabricated from concrete, CMU, brick masonry, or stone, were damaged by the tornado (if any)? Where is/are the damage(s) located? What was the extent of the damage? What damages to these walls were preexisting? Can these walls be repaired or must they be replaced?
- What steel framing members, if any, should be removed and replaced as a result of tornado damage? Where is/are the damage(s) located? What was the extent of the damage? What damages to these items were preexisting?
- What wood framing members/systems, if any, should be removed and replaced as a result of the tornado damage? Where is/are the damage(s) located? What was the extent of the damage? What damages to these members/systems were preexisting?
- What structural and/or cosmetic damages to the exterior wall and roof systems were caused by the tornado? Where is/are the damage(s) located? What was the extent of the damage? What damages to these items were preexisting? What damages will require removal and replacement of materials versus those that can be repaired?
- What cosmetic damages associated with the tornado were observed to other surfaces (e.g., plaster or drywall finished wall and ceiling surfaces)? Where is/are the damage(s) located? What was the extent of the damage? What damages to these members were preexisting? What damages will require removal and replacement of materials versus those that can be repaired?

Site inspection observation information for damaged mineral (e.g. concrete) construction materials, steel and/or wood framing members/systems should be compared with damage criteria discussed in Chapters 14, 15, and 17 to make remove/replace versus repair conclusions and recommendations. Best practices dictate that removal/replacement or repair activities should be conducted with the oversight of a structural engineer to ensure proper shoring takes place and replacement or repairs with proper materials adequately designed (as necessary) occurs.

Another issue that can occur in tornado-related property damage assessments is a dispute on contents missing from a structure as a result of the tornado. In a simple sense, the force of the wind from a tornado needed to move a specific content must overcome the force of gravity holding it in place (Figure 19.12).

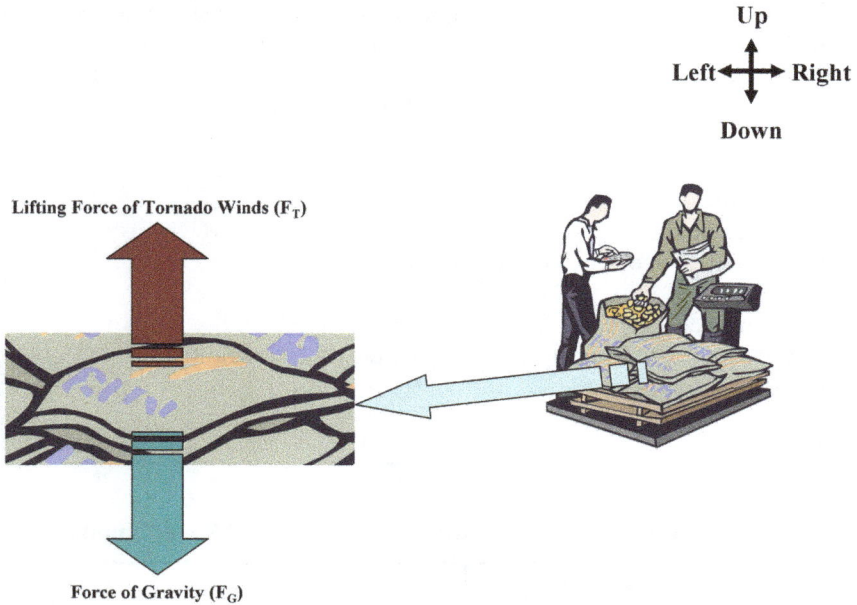

If $(F_T) > (F_G)$ or $(F_T) - (F_G) > 0$, the Content will be Lifted or Moved

EES Drawing

FIGURE 19.12 Schematic of vertical forces acting on an object during a tornado.

If the force of the wind from the tornado is greater than that of gravity, the content will move. If the wind force is substantially greater than that of gravity, the object will be lifted and moved. The distance the object moves is dependent on the wind speed of the tornado acting on the content and the mass and surface area of that content.

Equations for the lifting forces are generally associated with either movement of objects through the air (e.g., aeronautical engineering) or wind uplift associated with roof systems. Roof system uplift equations are based on equations found in American Society of Civil Engineers (ASCE) Standard ASCE-7 "Minimum Design Loads for Buildings and Other Structures" (see Chapters 26–30 of ASCE 7–10).

The uplift pressure caused by wind forces on a roof system can be used in a simplified form to estimate the lifting force (Eq. 19.1)[12,13]:

$$P_T = 0.00256v^2 \tag{19.1}$$

where:
  P_T=pressure in pounds per square foot (psf)
  v=wind speed in miles per hour.

Pressure by definition is Force per Unit Area (Eq. 19.2):

$$F_T = P_T * A_C \tag{19.2}$$

where:
  F_T=Uplift force caused by a tornado wind (Pounds force)
  P_T=pressure in pounds per square foot (psf)
  A_C=Area of contact of a content item (ft2)

Combining Eqs (19.1) and (19.2), one can derive a simplified relationship between the lifting force associated with wind from a tornado and gravitational forces (Eq. 19.3):

$$F_T = 0.00256v^2 * A_C \qquad (19.3)$$

where:

F_T=Uplift force caused by a tornado wind (Pounds force)
v=wind speed in miles per hour
A_C=Area of contact of a content item (square feet)

This is the force acting upward as illustrated in Figure 9.12.

An illustration of the range of uplift pressures by tornado classification, by wind speed, is summarized in Table 19.8.

TABLE 19.8
Wind Uplift Pressure (psf) versus Wind Speed (mph)

MPH	P (psf)	EF Scale	MPH	P (psf)	EF Scale
50	6.4	N/A	136	47.3	EF3
55	7.7	N/A	140	50.2	EF3
60	9.2	N/A	145	53.8	EF3
64	10.5	N/A	150	57.6	EF3
65	10.8	EF0	155	61.5	EF3
70	12.5	EF0	160	65.5	EF3
75	14.4	EF0	165	69.7	EF3
80	16.4	EF0	167	71.4	EF3
85	18.5	EF0	168	72.3	EF4
86	18.9	EF1	170	74.0	EF4
90	20.7	EF1	175	78.4	EF4
95	23.1	EF1	180	82.9	EF4
100	25.6	EF1	185	87.6	EF4
105	28.2	EF1	190	92.4	EF4
110	31.0	EF1	195	97.3	EF4
111	31.5	EF2	199	101.4	EF4
115	33.9	EF2	200	102.4	EF5
120	36.9	EF2	205	107.6	EF5
125	40.0	EF2	206	108.6	EF5
130	43.3	EF2	210	112.9	EF5
135	46.7	EF2			

The range of uplift pressures (PT) by EF scale and wind speed from Table 19.7 are summarized in Table 19.9.

TABLE 19.9
Wind Uplift Pressure versus Wind Speed and Tornado Strength

EF Scale	Wind Speeds (mph)	PT (psf – pounds force per square foot)
EF0	65–85	10.8–18.5
EF1	86–110	18.9–31.0
EF2	111–135	31.5–46.7
EF3	136–167	47.3–71.4
EF4	168–199	72.3–101.4
EF5	>200	>102.4

Note that for an EF2 storm, the maximum wind speed is 135 mph, resulting in a maximum lifting force of ~46.7 PSF.

As shown in Figure 19.12, the force holding an object in place is the gravitational force (FG); this force is defined in Eq. (19.4):

$$F_G = m_C * g_c = 32.174 * m_C \tag{19.4}$$

where:

m_C=mass of content item (pounds)
g_c=gravitational constant – 32.174 feet/seconds^2

Note that confusion exists between the terms pounds force (weight) and pound mass (slugs). Thus, if using weight directly, Eq. (19.4) becomes:

$$F_G = W_C \tag{19.4}$$

where:

W_C=weight of content item (pounds force)

In order for a content to be lifted, the gravitational force must be exceeded by the lifting force of the wind (i.e., $F_T > F_G$) or when $F_T - FG > 0$ Eq. (19.5):

$$F_T - F_G > 0 \text{ or } 0.00256v^2 * A_C - W_C > 0 \tag{19.5}$$

For illustration purposes, Eq. (19.6) provides the conditions for an object's area and mass to be lifted under a scenario where the wind speed during an EF2 tornado reached the worst-case speed of 135 mph:

$$46.7\left(\text{psf}\right) * A_C - W_C > 0 \tag{19.6}$$

where:

A_C=Area of contact of a content item (square feet)
W_C=Weight of content item (pounds force)

If the results of Eq. (19.6) are less than or equal to zero, a content should remain in place during an EF2 tornado event. Two examples using this equation for lifting of a tractor and a roll of insulation are summarized below:

Tractor: A model 3405 John Deere tractor weighs ~1,450 pounds and has a plan area of ~35.6″ × ~57.1″ (14.1 square feet). Placing these values in Eq. (19.6) results in a value of −791.5 pounds; thus since the value is less than zero, the tractor would not be moved by an EF2 tornado with maximum winds of 135 mph.

Roll of Insulation: A typical roll of R-13 insulation with a surface area of 1.74 square feet weighs ~19 pounds. Applying Eq. (19.6) results in a value of 62.3 pounds, suggesting that wind speeds at a maximum of 135 mph would readily lift the roll of insulation. Even at low-end EF2 wind speeds of 111 mph, the roll of insulation would have a positive uplift value of 35.8 pounds and would have lifted.

19.3.4 TORNADO DAMAGE ASSESSMENT METHODOLOGY – WRITTEN REPORT

The written report should follow the overall approach outlined in Chapter 1. Particular attention should be spent documenting the following observations and results from the inspection in the body of the report:

- A summary of information obtained from the interview.
- A summary, and analysis, of any specific tornado weather information. For example, the date, time, path, strength, proximity of the tornado to the property of interest, and reported wind speed of the tornado should be provided.
- Visible structural and cosmetic damage observations for all impacted spaces. The inspection should be completed indoors for all spaces on each impacted floor and outdoors for all elevations, including the roof, if climbable. The report should include a delineation of damaged areas, including extent of damage, to wall, floor, and ceiling surfaces, including structural support materials such as those formed from concrete, CMU, stone, steel, and wood.
- Damage (e.g., cracks, breaks, splits, and distorted members) should be reported in terms of size, freshness, and location. Interpretation of crack results should follow the guidelines outlined in Chapter 15.
- Level measurements on support beams, trusses, joists, and floor surfaces should be reported in tabular format and any distorted members noted.
- Plumb measurements on walls and support posts should be reported in tabular format and interpreted as outlined in Chapter 15.
- A summary of tornado damage and repair recommendations should be reported in tabular format and include both structural and nonstructural damages with photographs illustrating key damaged items. Results should be summarized by space as illustrated in Table 19.7 above.
- In situations where the loss of contents is disputed, an analysis of lifting forces and the types of contents likely to be carried away, as well as those that should not have been lifted, should be summarized.

As with fire damage assessments, the report should close with any limitations. For example, often the inspection addresses tornado-related damage to the structural framing members/systems and interior and exterior finishes. Experts in other areas (e.g., electrical, plumbing, and HVAC) should be involved to address tornado-related damage to these systems, which are often outside the expertise of a structural engineer.

IMPORTANT POINTS TO REMEMBER

- Tornados are some of the most powerful weather events to take place and occur mainly in the central United States.
- Tornado deaths historically occur to occupants of residences and mobile homes.
- The power of tornados is ranked by the EF Scale. Five levels of tornados are rated under this scale with wind speeds varying from 65 to >200 mph.
- For those structures not entirely destroyed by a tornado, the key issues are: (1) determination of damaged members/finished surfaces to be removed and replaced versus those that can be repaired and (2) determination of preexisting damage versus damage caused by the tornado.

REFERENCES

1. Tornado. "Hazard Mitigation Plan Update: Protecting the Region Against All Hazards, September 2010. Accessed February 15, 2012, http://www.gbra.org/documents/hazardmitigation/update/Section09-Tornado.pdf.
2. Guo, Li, Kaicun Wang, and Howard B. Bluestein. "Variability of tornado occurrence over the continental United States since 1950," *Journal of Geophysical Research: Atmospheres*, pp. 69043-6953, 2016. doi: 10.1002/2015JD024465.
3. NOAA. "Annual U.S. Killer Tornado Statistics." Accessed February 15, 2012, http://www.spc.noaa.gov/climo/torn/fataltorn.html.
4. National Society of Professional Engineers. "Tornado-Damaged Communities Face Safety Balancing Act." *PE Magazine*, January-February 2012, p. 13.
5. Brooks, Harold E. "Tornado and Severe Thunderstorm Damage." NOAA/National Severe Storms Laboratory. Accessed February 15, 2012, http://cstpr.colorado.edu/sparc/research/projects/extreme_events/munich_workshop/brooks.pdf.
6. The Tornado Project. "Fujita Scale," Accessed February 15, 2012. http://www.tornadoproject.com/fscale/fscale.htm.
7. Wind Science and Engineering Center. "A Recommendation for an Enhanced Fujita Scale (EF-Scale)," Texas Tech University, June 2004. Accessed February 15, 2012, http://www.spc.noaa.gov/efscale/ef-ttu.pdf.
8. Tornado Tim. "Enhanced Fujita Scale." Accessed February 15, 2012, http://www.tornadochaser.net/fujita.html.
9. NOAA. "Tornado Path in Fostoria, OH." Accessed June 12, 2020, https://www.weather.gov/cle/event_2008_notable.
10. American Plywood Association. "APA Publication SPE-1118, Midwest Tornadoes: Structural Performance of Wood-Frame Buildings in the Tornados of Southwest Missouri," Tacoma, WA, February 2005.
11. NOAA. "National Weather Service Forecast Office", Wilmington, OH, February 28, 2011 Tornado Event. Accessed June 12, 2020, https://www.weather.gov/iln/20110228_Photos.
12. Teitsma, Gerald. "Getting the Edge on Roof Wind Design." *Interface Magazine*, April 2003, pp. 23–29.
13. Adams, David K. *The Structural Engineer's Professional Training Manual.* McGraw-Hill Companies: New York, 2008.

20 Blast and Explosion Damage Property Assessments

Stephen E. Petty
EES Group, Inc.

CONTENTS

PURPOSE/OBJECTIVES

The purpose of this chapter is to:

- To be able to recognize the differences in blast/explosion damage inspections versus other types of forensic inspections.
- To be able to recognize the levels of explosion damage to property and persons by the strength of the explosion overpressure.
- Acquire a historical perspective on the development of "safe" parameters regarding blast overpressure and ground movement from blasts.
- Provide a methodology to determine and document whether or not blast/explosion damage has or has not occurred to a building.

Following the completion of this chapter, you should be able to:

- Understand the historical development of information used to relate actual structural blast/explosion damage to forces emanating from the blast or explosion.
- Be able to collect necessary information needed to determine whether or not blast/explosion damage has occurred to a residential or light commercial building.
- Be able to collect necessary site-specific information needed to determine the extent to which a structure sustained blast/explosion damage.
- Be able to prepare a report with analysis and documentation regarding whether or not a building suffered damage from a nearby blast or explosion.

20.1   INTRODUCTION

Claims resulting from blast damage primarily occur from one of three events: (1) blasts from nearby quarrying (surface) operations, (2) blasts from nearby mining (surface and subsurface) operations, and (3) propane and natural gas explosions. Damages to nearby structures from blasts associated with these three events can range from minimal to catastrophic. Photographic examples of the debris field associated with a destroyed home, and nearby damage to an adjacent home from a propane leak explosion, are illustrated in Figures 20.1 and 20.2.

FIGURE 20.1   Propane explosion debris field.

FIGURE 20.2 Propane explosion – damage to nearby home from explosion location.

Aside from answering the question of what caused an explosion (not part of this chapter – see Guide for Fire and Explosion Investigations, for example),[1–8] the forensic engineer is asked to document specific damage associated with explosions versus preexisting damages. Work has been done by others to define evidence factors that can be used to determine whether or not the energy from blasts was sufficient to cause specific blast damage to residential and light commercial structures. A review of the history of this work is constructive when making blast damage assessments.

An excellent overview of blast damage and the various types of shock waves (i.e., "S", "P", and body waves) caused by blasts and explosions is explained, at length, by Noon.[9] He also points out that pre- and post-blast surveys should be used to assess damage to a structure. While this approach is ideal, experience has shown that this information is typically available only in mining-related cases and not available in natural gas or LPG/propane explosions, since they are accidental by nature. When performed, pre-blast inspections and resulting seismology data are typically conducted by those responsible for the blast to ensure that false claims are not filed by owners/occupants of nearby residences or businesses.

The most common complaints associated with post-blast claims are new cracks in foundation walls, brick/stone masonry veneers, chimneys, interior drywall or plaster wall and ceiling surfaces, windows, concrete/asphalt driveways/sidewalks, and concrete/brick/stone retaining walls. The types of damage typically associated with and caused by blasting operations or other explosions follow[9]:

- Flyrock debris impact.
- Air concussion damage.
- Air shock wave (concussion) damage.
- Ground displacement due to blasting vibrations.

Debris impacts are tied to projectiles propelled by the energy of the blast affecting nearby structures. The movement of these objects typically moves in a parabolic arc where the distance traveled is associated with the amount and direction with which the energy is imparted on an object, the mass of the object, and gravitational forces driving the object back to the ground (Figure 20.3).

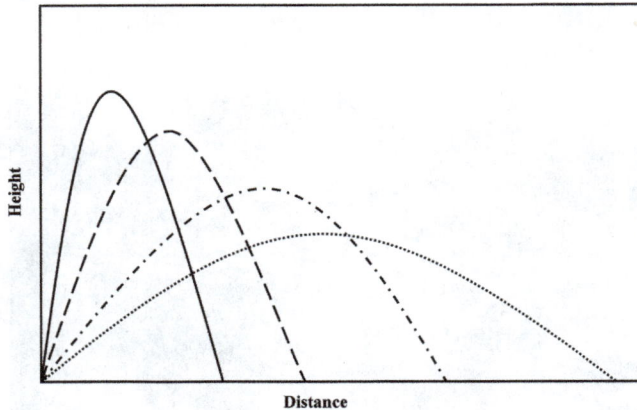

FIGURE 20.3 Debris trajectories for several initial flight directions.

In fact, the distance traveled and the mass of the objects in the debris field are often used to determine the magnitude of the blast (i.e., energy) in forensic investigations.[1-9]

Air concussion damage is associated with the pressure wave caused by the energy released in a blast/explosion, wherein the air is compressed and then expands outward. Interestingly, the pressure wave encountered by a blast/explosion can be both positive and negative (Figure 20.4).

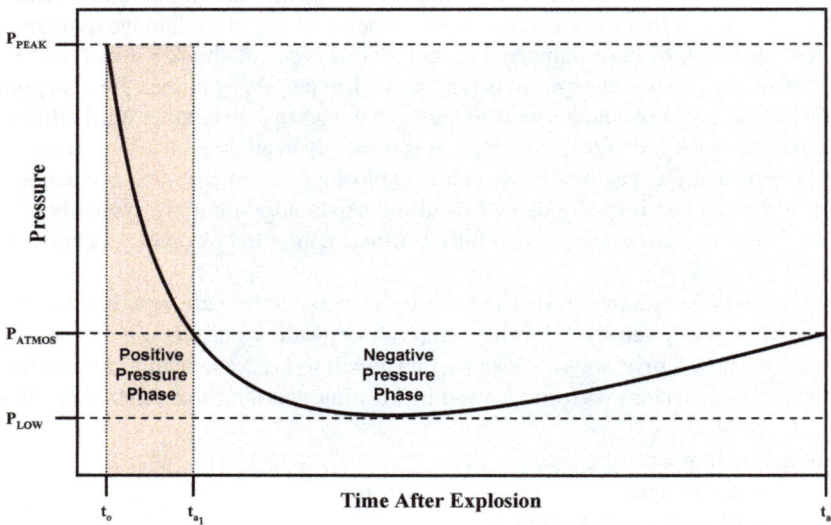

FIGURE 20.4 Time pressure history after explosion.

Initially, the blast wave will be of positive pressure and then after the initial pressure wave passes, a negative pressure wave will occur as air rushes into the vacuum created by the passing of the positive wave. Thus, one can observe damages where elements of the structure move away from the blast/explosion or move toward the blast/explosion. The initial pressure from an explosion is dependent on the type and amount of material involved in the explosion. Property damage characteristics associated with degree of overpressure in pounds per square inch (psi) are summarized in Table 20.1.

TABLE 20.1

Relationship between Peak Explosion Overpressure (psi) and Property Damage[1]

Overpressure (psi)	Property Damage	Source
0.03	Occasional breaking of large glass windows already under strain	a
0.04	Loud noise (143 dB). Sonic boom glass failure	a
0.10	Breakage of small windows under strain	a
0.15	Typical pressure for glass failure	a
0.30	"Safe distance" (probability 0.95 no serious damage beyond this value) Missile limit Some damage to house ceilings 10% window glass broken	a
0.4	Minor structural damage	a, c
0.5–1.0	Shattering of glass windows, occasional damage to window frames. One source reported glass failure at 0.147 psi (1 kPa)	a–e
0.7	Minor damage to house structures	a
1.0	Partial demolition of houses, made uninhabitable	a
1.0–2.0	Shattering of corrugated aluminum-steel paneling. Failure of wood siding panels (standard housing construction)	a–e
1.3	Steel frame of clad building slightly distorted	a
2.0	Partial collapse of walls and roofs of houses	a
2.0–3.0	Shattering of non-reinforced concrete or cinder block wall panels [(10.3 kPa (1.5 psi) according to another source]	a–d
2.3	Lower limit of serious structural damage	a
2.5	50% destruction of brickwork of house	a
3.0	Steel frame building distorted and pulled away from foundations	a
3.0–4.10	Collapse of self-framing steel panel buildings. Rupture of oil storage tanks. Snapping failure – wooden utility tanks.	a–c
4.0	Cladding of light industrial buildings ruptured	a
4.8	Failure of reinforced concrete structures	c
5.0	Snapping failure – wooden utility poles	a,b
5.0–7.0	Nearly complete destruction of houses	a
7.0	Loaded train wagons overturned	a
7.0–8.0	Shearing/flexing failure of brick wall panels [20.3–30.5 cm (8–12 inches) thick not reinforced]. Sides of steel frame buildings blown in. Overturning of rail cars.	a, b, c, d d b, c
9.0	Loaded train box cars completely demolished	a
10.0	Probable total destruction of buildings	a
30.0	Steel towers blown down	b, c
88.0	Crater damage	e

a F. Lees, Loss Prevention in the Process Industries (1996).[2]

b Brasie and Simpson (1968).[3]

c U.S. Department of Transportation (1988).[4]

d U.S. Air Force (1983).[5]

e McRae et al. (1984).[6]

Typically, one can expect failure of most light structural assemblies (e.g., non-reinforced wood siding, corrugated steel panels, and masonry block walls) at overpressures of 1–2 pound per square inch (psi). As a rule of thumb, and as illustrated by the data in Table 20.1, unless windows were broken by the force of the explosion, it is unlikely that the explosion would have caused structural damage to this same building.

The peak pressure wave dissipates with time and distance from the source of the explosion. As a rule of thumb, the pressure wave decreases by the inverse of the distance cubed. Computer models such as High Explosive and Vapor Cloud Damage Assessment Models – HEXDAM and Vapor-Cloud Explosion Damage Assessment Models (e.g., Breeze – High Explosive Damage Assessment Model (HExDAM) and Vapor Cloud Explosion Damage Assessment Model (VExDAM) and Engineering Analysis, Inc. – HEXDAM-III 5.2 and VEXDAM 5.2) are available to calculate blast/explosion peak pressures with distance more rigorously.

Like with property damage and peak overpressures, the NPFA1 has compiled data on injuries to humans from flying glass and direct overpressure effects (Table 20.2).

TABLE 20.2
Relationship between Flying Glass and Peak Explosion Overpressure (psi) to Human Injuries[1]

Overpressure (psi)	Human Injury(s) [Comments]	Source
0.6	Threshold for injury from flying glass [based on studies using sheep and dogs]	a
1.0–2.0	Threshold for skin laceration from flying glass [based on U.S. Army data]	b
1.5	Threshold for multiple skin penetrations from flying glass (bare skin) [based on studies using sheep and dogs]	a
2.0–3.0	Threshold for serious wounds from flying glass [based on U.S. Army data]	b
2.4	Threshold for eardrum rupture [conflicting data on eardrum rupture]	b
2.8	10% probability of eardrum rupture [conflicting data on eardrum rupture]	b
3.0	Overpressure will hurl a person to the ground [one source suggested an overpressure of 1.0 psi for this effect]	c
3.4	1% eardrum rupture [not a serious lesion]	d
4.0–5.0	Serious wounds from flying glass – near 50% probability [based on U.S. Army data]	b
5.8	Threshold for body-wall penetration from flying glass – bare skin [based on studies of sheep and dogs]	a
6.3	50% probability of eardrum rupture [conflicting data on eardrum rupture]	b
7.0–8.0	Serious wounds from flying glass - near 100% probability [based on U.S. Army data]	b
10.0	Threshold for lung hemorrhage [not a serious lesion – applies to a blast of long duration (over 50 m/seconds); 20.30 psi required for 3 m/seconds duration waves]	d
14.5	Fatality threshold for direct blast [primarily from lung hemorrhage]	b
16.0	50% eardrum rupture [some of the ear injuries would be severe]	d
17.5	10% probability of fatality from direct blast effects [conflicting data on mortality]	b
20.5	50% probability of fatality from direct blast effects [conflicting data on mortality]	b
25.5	90% probability of fatality from direct blast effects [conflicting data on mortality]	b
27.0	1% mortality [a high incidence of severe lung injuries – applies to a blast of long duration (over 50 m/seconds); 60–70 psi required for 3 m/seconds duration waves]	d
29.0	99% probability of fatality from direct blast effects [conflicting data on mortality]	b

Sources:
a   Fletcher, Richmond and Yelverron (1980).[7]
b   F. Lees, Loss Prevention in the Process Industries (1996).[2]
c   Brasie and Simpson (1968).[3]
d   U.S. Department of Transportation (1988).[4]

Finally, NPFA1 provides information on explosion characteristics by three fuel types likely to be seen in residential and light commercial buildings; lighter-than-air gases (e.g., natural gas), heavier-than-air gasses (e.g., propane or LPG), and liquid vapors (e.g., gasoline). These characteristics are summarized in Table 20.3.

TABLE 20.3

Characteristics of Typical Explosions by Fuel Type1

Typical Explosion Characteristic	Probability of Characteristic by Fuel Type		
	Lighter-than-Air	Heavier-than-Air	Liquid Vapors
Low-order damage (e.g., bulging of walls and laying down of structure)	3	4	4
High-order damage (e.g., shattering of structure; debris thrown great distances)	2	1	1
Secondary explosion	3	3	2
Gas/vapor pocketing	3	2	2
Deflagration	4	4	4
Detonation	1	1	1
Underground migration	2	2	2
BLEVE – boiling liquid expanding vapor explosion	2	3	5
Post-explosion fires	3	3	4
Pre-explosion fires	2	2	2
Minimum ignition energy (mJ)	0.17–0.25	0.17–0.25	0.25

Key: 0 – Never; 1 – Seldom; 2 – Sometimes; 3 – Often; 4 – Nearly Always; 5 – Always.

Fuels such as natural gas are lighter than air and will tend to move upward in a building, whereas fuels such as LPG are heavier than air and will tend to settle down into basements of structures. The settling characteristics of propane and LPG into basements can be problematic since they accumulate in areas commonly containing ignition sources such as pilots for water heaters and gas furnaces.

Of the four damage characteristics resulting from blasts and other explosions, the movement of the ground tied to potential building cracks is the area most commonly encountered in forensic investigations. For quarry and mining explosions, nearby buildings associated with claims are often located in hilly rural areas predisposed to some preexisting cracks and other damage to interior and exterior finishes. The job of the forensics engineer in these situations is to determine whether or not the energy associated with the blast had the potential to cause the reported damage and determine, document, and differentiate preexisting damage.

A review of historical literature was performed to assist the process of assessing potential damage from blasts and other explosions to nearby buildings. This background information follows.

20.2 OVERVIEW OF CRITERIA FOR EVALUATION OF BLAST DAMAGE TO BUILDINGS[1,8–13]

The development of evaluation criteria of blast damage from quarry and mining operations has primarily been led by the United States Department of Interior (DOI) Bureau of Mines (BOM) and the Office of Surface Mining Reclamation and Enforcement (OSMRE). The blast damage criteria associated with natural gas and LPG/propane explosions have been led by the National Fire Prevention Association (NFPA) and their members. These efforts, particularly by the DOI BOM and OSMRE, have in turn led to the promulgation of state rules and requirements for those conducting blasting operations. An overview of the development of blast damage criteria by the DOI and others follows.

20.2.1 OVERVIEW OF UNITED STATES DEPARTMENT OF INTERIOR HISTORY FOR EVALUATION OF BLAST DAMAGE TO BUILDINGS – BUREAU OF MINES[9–15]

The U.S. Department of Interior's Bureau of Mines (BOM), created by Act of Congress on July 1, 1910,[10] was originally formed to develop safer explosives and to eliminated black powder use in underground mines, but began to study the impacts of air and ground vibrations from quarry blasts in 1927 on residential structures and other buildings.[11] Thoenen and Windes' BOM Bulletin 442, published in 1942, was one of the early studies to summarize information on damage to residential and other structures based on ground acceleration.[14] A "no damage observed" threshold was reported when accelerations were <0.1 g. Damage would occur to a residential structure when the acceleration exceeded 1.0 g (see Table 20.4).

TABLE 20.4

Characteristics of Blasts by Peak Particle (Ground) Velocity and Extent of Damage

Referenced Study/ Parameters	Date of Study	Damage Characteristics and Ground Velocity		
		No Damage	Minor Damage	Major Damage
Rockwell quarry blasting	1927	No damage to residences 200–300 square feet away		
BOM – Bulletin 442 (quarry blasting – damage criteria based on ground acceleration rates)	1942	No damage to residential structures if ground acceleration levels <0.1 g	Minor damage (fine plaster cracks, opening of old cracks). Caution for ground acceleration levels between 0.1 and 1.0 g	Major damage (fall of plaster, serious cracking). Damage when ground acceleration levels >1.0 g
Langefors, Kilhstrom, and Westerberg (Damage criteria based on velocity of ground movement)	1957	No noticeable damage (2.8 inches/second)	Fine cracks and falling of plaster (4.3 inches/ second); Cracking of plaster (6.3 inches/second);	Serious cracking to plaster (9.1 inches/ second)
Edwards and Northwood (criteria from Canada – St. Lawrence seaway project – charges 15–30 feet below the ground)	1959	Safe (<2 inches/ second) Defined threshold damage as: opening of old cracks and formation of new plaster cracks	Minor damage – superficial, not affecting strength of the structure. Caution for levels between 2 and 4 inches/second	Major damage – serious weakening of the structure. Damage (>4 inches/ second)
BOM report 5968 by Duvall and Fogelson (BOM review of 40 papers)	1962	Low prob. of any damage – peak velocity <2.0 inches/ second	Minor damage – avg. peak velocity of 5.4 inches/second	Major damage – avg. peak velocity of 7.6 inches/second

---

In 1943, Windes[15] published the BOM investigations report #3708, *Damage from Air Blasts*, which established the first overpressure criteria for blast damage to windows. He concluded that no damage to window glass would occur when blast pressures were <0.7 lbf/square inch, while damage would occur when blast pressures were >1.5 lbf/square inch.

In 1962, Duvall and Fogelson[11] published the BOM Report of Investigations #5968 titled: *Review of Criteria for Estimating Damage to Residences from Blasting Vibrations*, which summarized blast damage criteria to residential and other structures from 1927 forward. Table 20.4 summarizes this work from their study.

Over 40 papers were reviewed by the authors, three of which they found to be the most useful. They divided damage levels into the following three categories:

- Major Damage (fall of plaster, serious cracking).
- Minor Damage (fine plaster cracks, opening of old cracks).
- No Damage (to plaster).

Damage to residential structures erected on various soil and rock conditions was classified by ground velocity. A no damage threshold was set at peak ground velocities <2.0 inches/second. Average peak velocities of 5.4 and 7.6 inches/second were given as criteria for minor and major damage respectively.

A 10-year update and expansion of the 1962 BOM work by Duvall and Fogelson[11] was completed by Nicholls, Johnson, and Duvall in 1971 and culminated in BOM Bulletin 656, *Blasting Vibrations and Their Effects on Structures*.[12] The authors reviewed data from 171 blasts from 26 sites, covering a variety of simple and complex geologies, peak ground velocities of 0.000808–20.9 inches/second, and scaled distances ranging from 3.39 to 369 square feet/lb½. The authors, in an overall conclusion, stated, "While values of 2.0 in/sec particle velocity and 0.5 psi air blast overpressure are recommended as safe blasting limits not to be exceeded to preclude damage to residential structures, lower limits are suggested to limit complaints."[12] The detailed conclusions included the following:

- Damage to residential structures from ground-borne vibrations correlates best with peak ground velocity rather than peak ground displacement or peak ground acceleration.
- A safe blasting limit of 2.0 inches/second (peak velocity) is recommended. The probability of damage to a structure at this limit is <5%. In the absence of instrumentation, a safe blasting limit of 50 square feet/lb½ may be used. Human response levels to ground vibrations, overpressure, and noise are considerably lower than levels necessary to cause damage to residential structures and will result in complaints. To avoid complaints, a lower blasting limit of 0.4 inches/second can be imposed. This should reduce complaints by at least 92%.
- A safe blasting limit of 0.5 psi air blast overpressure is recommended. If the safe blasting limit of 2.0 inches/second is met, the overpressure limit will automatically be met (this implies that windows probably would be broken by the blast overpressure before ground peak pressures above the no damage threshold would be exceeded).
- Millisecond-delay blasting will reduce vibration levels. This occurs because the maximum (peak) vibration amplitude is related to the maximum charge weight per delay interval rather than the total charge.

Regarding safe zones for peak velocity and overpressure, a detailed review of work up through 1971 was provided. The authors retained the safe zone established by Duvall and Fogelson[11] of 2.0 inches/second (see Table 20.4) while simply noting a damage zone above a peak particle velocity of 2.0 inches/second. Regarding overpressure, they again concluded that windows should not break if the peak particle velocity is in the safe zone. The safe overpressure (for window damage) was 0.5 psi, with some failure of glass at 0.75 psi and all glass expected to fail at 2.0 psi. It should be noted that unconfined open blasts produce higher pressures than partially or fully contained blasts and that Siskind et al.[13] recommended much lower threshold levels (i.e., 0.014 psi with an improbable damage to glass estimate at 0.030 psi (140 dB).

Human complaints, as a function of blast frequency (6–40 cps – predominant blast frequencies), were found to be at/below the safe zone of peak particle velocities (Table 20.5).

TABLE 20.5

Human Responses to Blast Vibrations[12]

Peak Velocity (inch/second)	Human Response
0.3–0.4	Perceptible
0.15–0.8	Unpleasant
0.6–2	Intolerable

The relationship between peak amplitude (A) and distance from blast (d) and size of charge (W) is generally given by Eq. 20.1:

$$A = k * Wb * d^n \tag{20.1}$$

where:

   A = amplitude
   d = distance
   W = charge
   k = constant
   n = constant –typically ranges from −1 to −2
   b = constant – typically ranges from 0.4 to 1.0.

Rearranging this equation and combining constants, the safe blasting distance is given by Eq. 20.2:

$$d_s = K * W^{\frac{1}{2}} \tag{20.2}$$

where:

   d_s = safe blasting distance in feet to a structure
   W = charge weight per delay (in equivalent dynamite)
   K = constant in units of ft/lb½

The authors recommended $K = 50$ ft/lb½ to ensure that the peak velocity of 2.0 inches/second would not be exceeded. Plugging this constant into Eq. 20.2 results in Eq. 20.3:

$$d_s = 50 * W^{\frac{1}{2}} \tag{20.3}$$

Thus, given a charge, the minimum safe distance could be calculated.

In addition, the peak velocity dissipates with the inverse of the square of the distance (Eq. 20.4):

$$I = k/d^2 \tag{20.4}$$

where:

   I = peak velocity
   d = distance from blast
   k = constant

Thus, the peak intensity at twice the distance decreases by a factor of 4 ($I_2/I_1 = 1/(d_2/d_1)_2 = (1/2)^2 = 4$

In 1980, Siskind et al. published the Report of Investigations #8485 titled, *Structure Response and Damage Produced by Airblast from Surface Mining*, which further addressed air blast safe levels.[13] These authors recommended a safe overpressures level of 0.014 psi for glass breakage (probability of glass breakage <1/1,000) with damage probabilities being very small, below 0.030 psi, which were much more conservative than earlier recommended values.

20.2.2 Current Blasting Formula and Criteria

As summarized by Noon[9] and used in modern state codes, it has been recognized that the closer to the blast site, higher ground frequencies dominate ground movement, whereas further out from the blast site, lower frequencies dominate as the higher frequencies are attenuated. For this reason, current recommended levels for "safe" peak velocities were expanded and adjusted from those in BOM Bulletin 6566 for blast vibration frequency. Allowable peak velocities versus frequency are summarized in Figure 20.5.

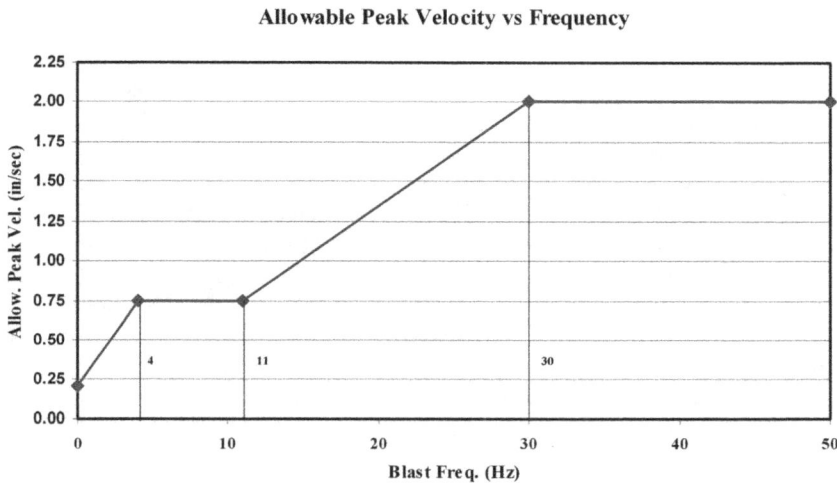

Allowable Peak Velocity vs Frequency

FIGURE 20.5 Allowable peak velocity versus blast vibration frequency.

Current "safe" blasting criteria include provisions for these lower frequencies by distance (Table 20.6). Note that one can still use an original safe limit of 2.0 inches/second if the verified measurable frequency is 30 Hz or more.

TABLE 20.6
Modern Modified "Safe" Blasting Criteria[9]

Distance from Blast (square feet)	Peak Velocity (inch/second)	"K" Scaled Distance Factor (feet/lb½)
0–300	1.25	50
301–5,000	1.00	55
>5,000	0.75	65

As illustrated in the following paragraphs, modern state codes such as the Ohio Administrative Code (1501:13-9-06 Use of Explosives) have adopted these current "safe" parameters. Generally, this code language applies to "All blasting operations, including surface blasting operations incident to underground mining, on all coal mining and reclamation operations, and on coal exploration operations"[16] and applies to blasting distances of 1,000 feet of any dwelling, public or commercial building, school, church, or community or institutional building or a distance of 500 feet of an active or abandoned underground mine.

Distance criteria from this example Ohio code parallel those cited in Table 20.6.

(7) Except as provided in paragraph (F)(8) of this rule, the maximum ground vibration shall not exceed the following limits at any dwelling, public or commercial building, school, church, or community or institutional building outside the permit area:

Distance (D), from the blasting maximum allowable site, in feet (rounded to the peak particle velocity nearest whole foot) in inches/second:

0–300	1.25
301–5,000	1.00
5,001 and beyond	0.75

(a) Ground vibration shall be measured as the particle velocity and recorded in three mutually perpendicular directions. The maximum allowable peak particle velocity shall apply to each of the three measurements.

(b) A seismographic record shall be provided for each blast. Whenever possible, the seismograph shall be located at the nearest building to be protected. Otherwise, the seismograph may be placed at some point between the blast site and the nearest building to be protected.

(8) In lieu of the requirements of paragraph (F)(7) of this rule, the ground vibration limits in the chart below may be used to determine the maximum allowable particle velocity at the nearest dwelling, public or commercial building, school, church, or community or institutional building outside the permit area. A seismographic record including both particle velocities and an electronic analysis of vibration frequencies shall be provided for each blast.

(9) In lieu of the seismographic record required by paragraph (F)(7)(b) of this rule, the scaled-distance equation, $W=(D/D_s)^2$, may be used to determine the maximum allowable charge weight of explosives that can be detonated within any 8-millisecond period, where $W=$ the maximum weight of explosives, in pounds; $D=$ the distance, in feet, from the blast site to the nearest dwelling, public or commercial building, school, church, community, or institutional building outside the permit area; and $D_s=$ the scaled-distance factor applied from the following table:

Distance (D), from the blasting scaled-distance factor site, in feet (rounded to the (D_s) to be applied nearest whole foot) without seismic monitoring:

0–300	50
301–5,000	55
5,001 and beyond	65

The use of a modified scaled-distance factor in the scaled-distance equation may be approved by the chief on receipt of a written request by the permittee, supported by seismographic records of blasts at the mine site. The modified scaled-distance factor shall be determined such that the particle velocity of the predicted ground vibration will not exceed the maximum allowable peak particle velocities prescribed in paragraph (F)(7) of this rule, at a 95% confidence level.

Example air blast guidelines are also presented from this same code below:

(5) Air blast shall not exceed the maximum limits listed below at any dwelling, public or commercial building, school, church, or community or institutional building outside the permit area, except as authorized under paragraph (F)(12) of this rule. All air blast measuring systems shall have an upper-end flat-frequency response of at least 200 Hz.

Lower frequency limit of measuring maximum level, in system, in hertz (±3 dB) decibels.

0.1 Hz or lower	Flat response 134 peak
2 Hz or lower	Flat response 133 peak
6 Hz or lower	Flat response 129 peak

As is evident, state codes tend to follow from national guidelines.

More recently, as a result of terrorist incidents, considerable work has been undertaken to determine design equations to allow buildings to withstand damage better from explosives intentionally detonated near them. Thomas Telford[17] and Ngo, Mendis, Gupta, and Ramsay[18] provide detailed equations and analytical methods for this and blast damage to buildings in general.

20.3 BLAST OR EXPLOSION DAMAGE ASSESSMENTS

The recommended methodology for completion of blast or explosion damage assessments is outlined in Sections 20.3.1 and 20.3.2.

20.3.1 BLAST DAMAGE ASSESSMENT METHODOLOGY

Most blast damage assessments encountered will be associated with nearby quarry or mining activities. The methodology for completing these types of blast damage assessments includes the following, sequential elements:

- Interview of property owner or their representative.
- Site inspection.
- Interview of blast company owner/representative.
- Acquisition of pre-blast assessment and blast-specific seismic documents.
- Analysis of information collected.
- Written report, summarizing findings.

Details regarding each of these elements follow:

20.3.1.1 Blast Damage Assessment Methodology – Interview of Property Owner or Owner's Representative

Upon arriving at the site, the inspection process should follow the overall approach outlined in Chapter 1. The intent of the inspection is to collect necessary information needed to make the following determinations:

- Distinguish between pre- and post-blast damage to the building.
- Determine post-blast damaged items that will need to be replaced.
- Determine post-blast damaged items that will need to be repaired.

For example, items damaged such that their structural integrity has been compromised (e.g., broken wood members, twisted steel I-beams or oxidized steel members, significant foundation cracks) will need to be documented for replacement. On the other hand, minor cracks to drywall can be repaired.

Specific interview questions regarding blast damage should include the following:

- Was a pre-blast survey of the property completed? If so, collect a copy of the survey (see pages of an actual pre-blast survey in Figures 20.6–20.8). In some cases, this may be available from the company or its representatives performing the blasting.

PREBLAST INSPECTION SURVEY

ABC, Incorporated

Owner or Current Occupant Name: Address of Structure:	Mr. & Mrs. Homeowner
	123 First Street Second City, OH 41234
Date of Inspection: Inspector contact number:	01/01/2000
Name of Inspector (printed): Signature of Inspector:	Mr. Inspector
Firm Represented:	ABC, Incorporated
Permitee: Permit Number:	123456

FIGURE 20.6   Example of pre-blast survey – title page.

PRE - EXISTING EXTERIOR DAMAGES
Sketches, Descriptions, and Photos

SOUTH ELEVATION

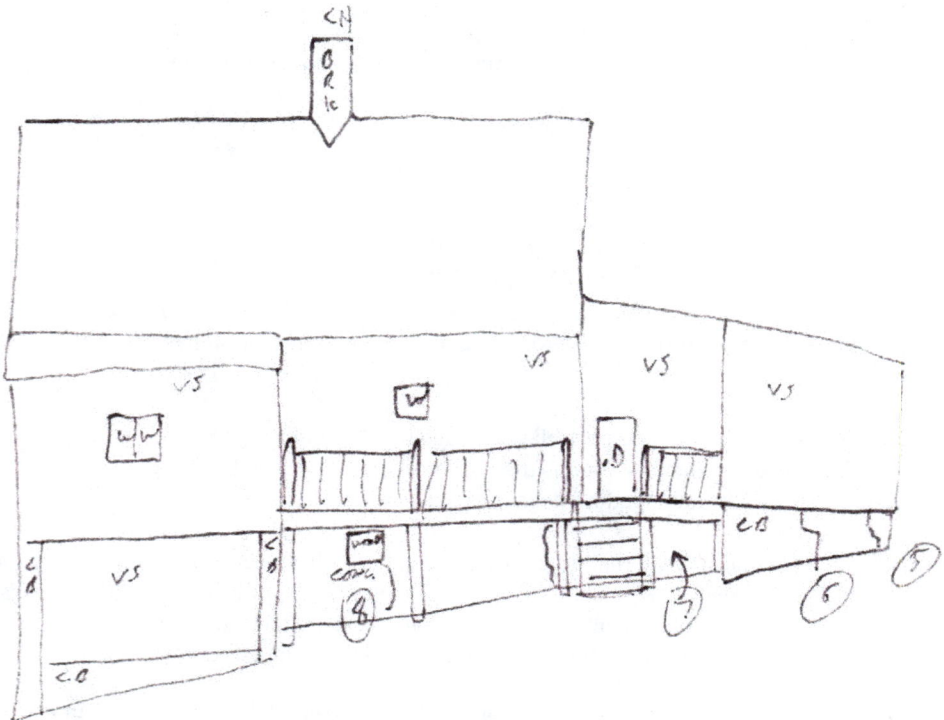

FIGURE 20.7   Example of pre-blast survey – south exterior.

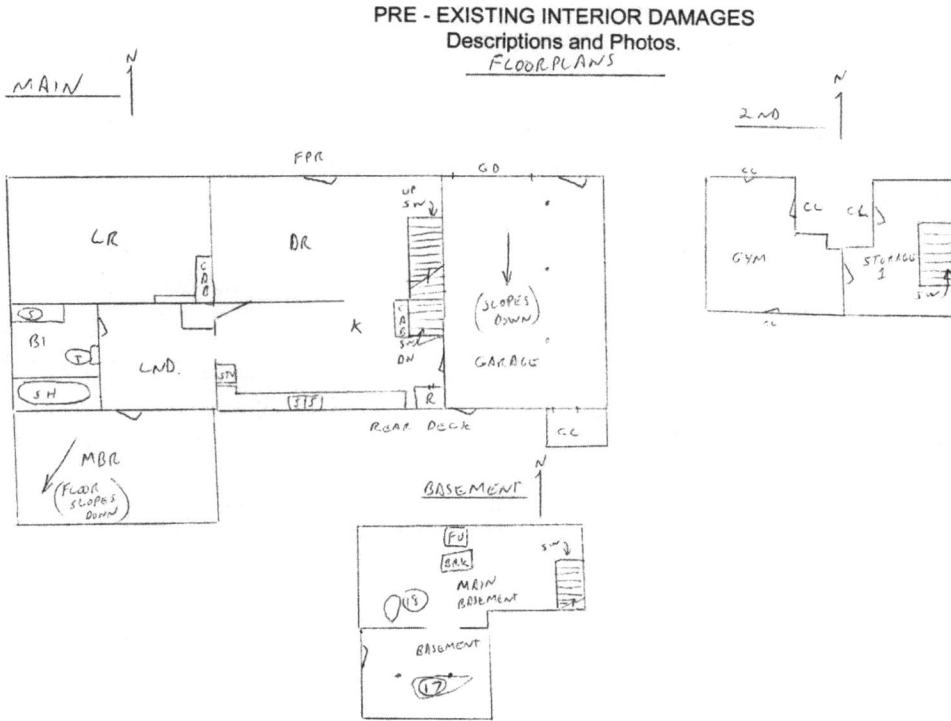

FIGURE 20.8   Example of pre-blast survey – inside – plan view.

Pre-blast surveys allow for the comparison of cracks observed during the inspection with pre-blast crack results to determine probable new cracks.

- Who is conducting the blasting (company, address, and contact)?
- What dates and times did the homeowner or representative recall the blasting occurring? Which direction and how far away was the blasting, if known?
- Were blasts on a certain date and time stronger than other blasts and/or associated with specific damage? Were any windows damaged by the blast(s)?
- What damages were believed to have occurred as a result of the blasts? Document what, when, and where. When performing this portion of the interview, remember to inquire about interior and exterior damages. Also, obtain any available photographs. Finally, inquire about preexisting damages.
- Was any post-blast survey conducted? If so, by whom? Is a copy of this survey available? If so, collect a copy of the post-blast survey; these are sometimes done by outside contractors or an insurance agent for the blasting company. Also, obtain any available photographs.
- Has the owner or representative contacted the blasting company or its representative? If so, when did this occur and what actions and activities have occurred as a result of these action(s)?

The sum of this information by the owner or their representative should provide a basis for who was conducting the blasting, when it occurred, and the damages believed to be caused by the blasting activities.

20.3.1.2 Blast Damage Assessment Methodology – Site Inspection

The site inspection should follow the overall approach outlined in Chapter 1. Particular attention should be spent documenting cracks to the interior and exterior finished surfaces. All observations should be recorded in writing in a field notebook; key observations should be photographed.

During the inside portion of the inspection, cracks and potential blast damage should be documented for the floor, wall, and ceiling finished surfaces of each room. The investigation should begin in the basement or crawlspace (if present) and continue upward, until the attic inspection is completed. Exterior inspection observations should be completed by elevation followed by the roof inspection. Particular attention should be paid to cracks in stone, poured concrete, concrete masonry units (CMU), brick masonry walls, veneers, patios, walkways, driveways, and stucco finishes if present. While performing the exterior inspection, the slope of the ground surface should also be noted for each elevation.

The need for a complete and detailed inspection lies with the need to be able to rule in, or out, cracks that were caused by the blast throughout the building thoroughly. Cracks should be documented by location, directionality, width, and crack characteristics (e.g., brightness of color and sharpness of crack surfaces, whether or not debris is present in the crack, and/or whether or not paint or other coatings are present on the edges of the crack) as discussed in Chapter 15. Additionally, wall plumb and floor-level measurements should be performed and analyzed following the methodologies discussed in Chapters 14 and 15. All of these observations should be recorded in the field notebook and photo-documented.

While on site, GPS coordinates for the location of the building should be taken and noted in the field book. This will be needed to establish the line-of-site distance between the blast location and the building for future analysis.

20.3.1.3 Blast Damage Assessment Methodology – Interview of Blast Company Owner/Representative

Following the site inspection, attempts should be made to visit the site where the blasting occurred and speak with an employee or representative from the blast company. When the blast location is identified, GPS coordinates should be taken and noted in the field book. If a company employee or representative is available, the following information should be obtained:

- Identify the person being interviewed and their role with the company or its representative. Obtain a business card if possible.
- Was a pre-blast survey of the property of interest completed? If so, collect a copy of this survey or take digital photographs of the survey.
- What days and times were blasts completed at/near the reported date(s) of loss?
- Were seismic records of these specific blasts completed? If so, collect a copy or take digital photographs of the record(s). Record key seismic facts in a field notebook such as date and time of blast, charge(s), GPS coordinates, and local ground velocities and pressure levels if available. An example of a seismic record from an investigation is shown in Figures 20.9 and 20.10.

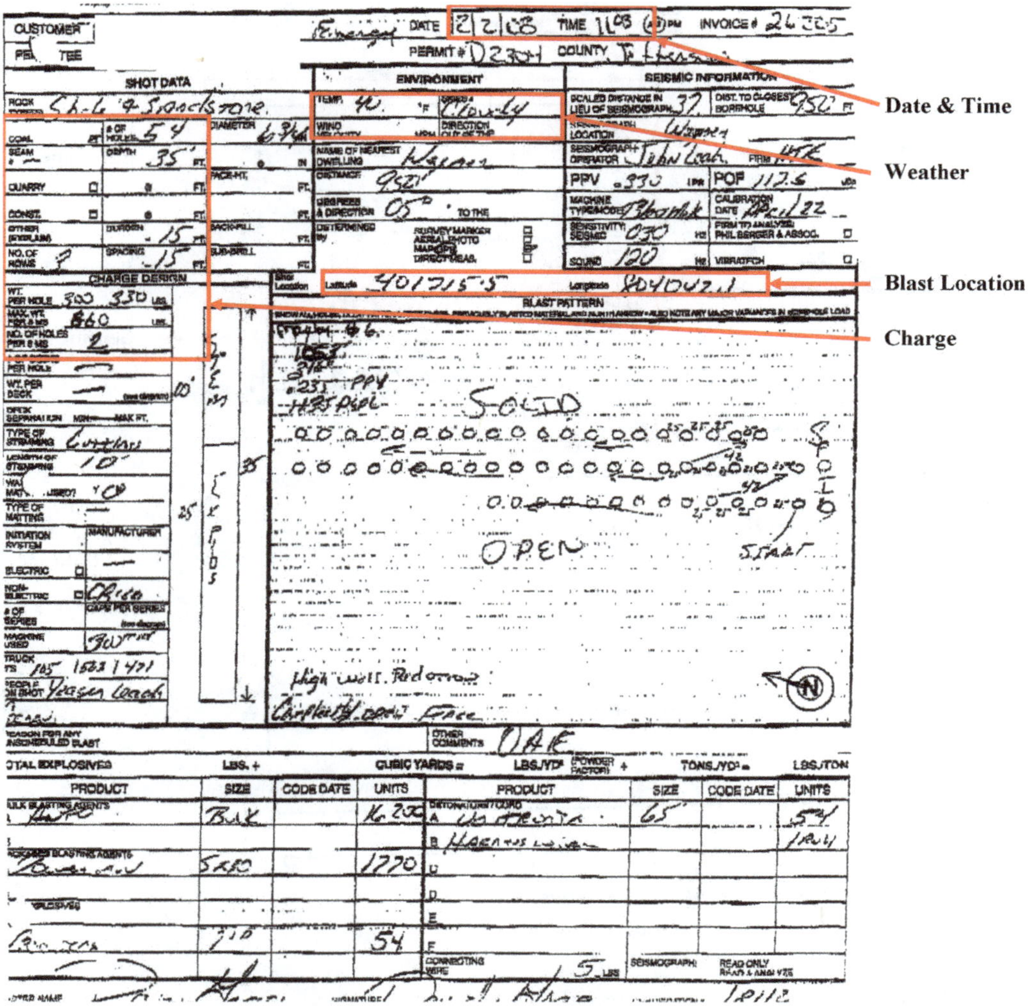

FIGURE 20.9 Example of blast seismic record – cover page.

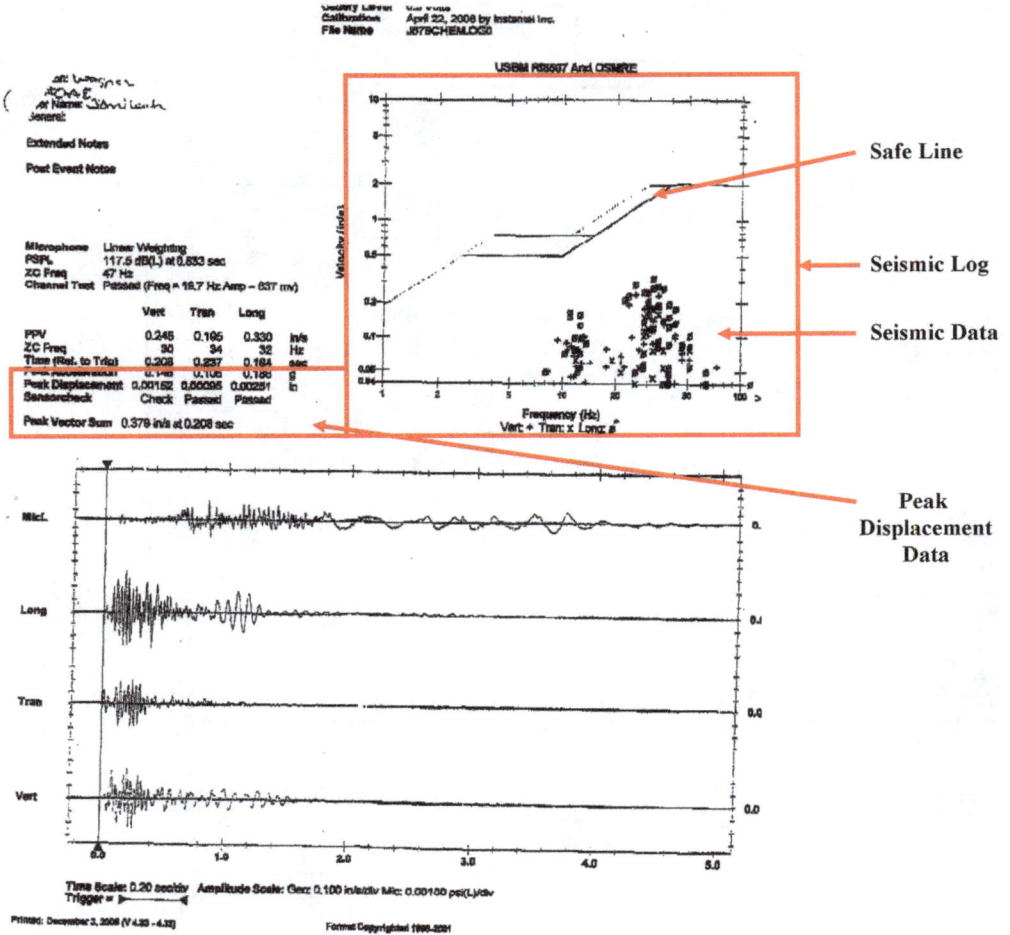

FIGURE 20.10 Example of blast seismic record – seismic page.

Figure 20.9 provides data on the date and time of the blast, the location of the blast, information on the total charge used, and weather information. The GPS coordinates listed on the record should always be checked; occasionally these forms are prefilled out and have coordinates for another location.

Figure 20.10 provides seismic data for a specific blast at a specific location and compares it with "safe" levels versus frequency. Note in the example that most of the blast data reported a frequency above 10 Hz.

- Was a post-blast survey of the property of interest completed? If so, collect a copy of this survey or take digital photographs of the survey.
- Were any complaints received from nearby owners? If so, in what direction. What were the names of the owners or owner representatives who complained? What are the addresses of those who complained?
- Are any photographs of the blast/site available? If so, attempt to obtain copies.
- Can the blast location be visited? If so, visit, photograph, and take GPS coordinates.

20.3.1.4 Blast Damage Assessment Methodology – Analysis of Collected Information

The information collected as outlined in Sections 20.3.1.1–20.3.1.3, including the coordinates of the damaged property and blasting location, along with the modern requirements for "safe" blasting operations (Section 20.2.2) should be utilized to answer the following questions:

- Are the GPS blast location coordinates consistent with the reported location of the blast(s)?
- Was the charge excessive given the distance from the home to the blast site?
- Was the peak ground velocity excessive based on seismic records?
- Were any windows broken?
- Was the blast sufficient to cause damage, be it minor or major, to the subject property?
- Are any fresh cracks and/or damage present at the subject property and/or not present on the pre-blast survey likely caused by the reported blast?

20.3.1.5 Blast Damage Assessment – Written Report

The written report should follow the overall approach outlined in Chapter 1. Particular attention should be spent documenting pre-blast versus post-blast damage and determining whether or not the damage requires the item to be removed/replaced or repaired following the methodologies outlined in Chapters 14 and 15. In addition, answers to questions posed in Section 20.3.1.4 should be included in the report.

Example seismic data collected from three blasts is shown in Table 20.7.

TABLE 20.7
Example Sample Data Taken from Seismic Reports

Date	Time	Temp. (°F)	Weather	Nearest Dwelling/ Distance (feet)	Holes/Total Explosive (#)/# Explosive/Hole	Air Overpressure in Decibels (dB)	Ground Vibrations (inch/second)
12-2-2008	11:03 AM	40	Cloudy	Wagner (952)	54/18,024/333.8	117.5	0.379
12-12-2008	4:02 PM	37	Cloudy/rain	Wagner (1,058)	74/23,374/315.9	–	–
12-18-2008	12:44 PM	30	Cloudy/rain	Wade (852)	27/6,227/230.6	116.4	0.122

Note that for the three separate blasts, data regarding the date, time, weather conditions, explosives, air blast pressure, and ground vibration is provided. The dates of the blasting should always be checked against the estimated dates and times provided in the interview. If they do not coincide, then the claim of damage from the blast should be questioned. Air blast levels and ground velocities can be checked against "safe" levels and against actual observed damage to windows and fresh cracks.

Allowable charging to the holes can also be checked by using the distance to the subject property, the actual charge, and the appropriate "K" or charge factor. The "W" allowed charge equals $(D/55)^2$. Results of this calculation for the three example blast logs to the nearest reported dwelling are shown in Table 20.8.

TABLE 20.8
Example of Calculated Allowed versus Actual Charge per Hole

Date	Time	Nearest Dwelling/ Distance (square feet)	Holes/Total Explosive (#)/# Explosive/Hole	Calculated Weight	Actual Weight
12-2-2008	11:03 AM	Wagner (952)	54/18,024/333.8	299.6	333.8
12-12-2008	4:02 PM	Wagner (1,058)	74/23,374/315.9	370.0	315.9
12-18-2008	12:44 PM	Wade (852)	27/6,227/230.6	240.0	230.6

Note that in the first instance, the actual average charge per hole (i.e., 333.8 #) exceeded the allowed charge per hole (i.e., 299.6 #). In the other two instances, the actual average charge was less than that allowed.

The type and format of typical conclusions reached in a blast inspection report are illustrated below:

Conclusions: Based on the input received from the interviews, information gained during the inspection, and our professional experience, we have arrived at the following conclusions:

- The first floor master bedroom walls and ceiling and the basement addition walls under the master bedroom contained fresh cracks and gaps, which appeared to postdate the pre-blast inspection report.
- Significant past settling of the foundation walls (i.e., older cracks) and the front sidewalk were evident during the inspection and appeared on the pre-blast inspection report.
- While damage appears to be due to a combination of blast vibrations and foundation design/installation, no broken windows were reported.
- Blasting logs provided for December 2008, and cited as the basis for the denial letter to the homeowner, do not appear to be relevant to the date of loss in January 2009. This complicates the analysis since no apparent relevant blasting logs were available for this latter timeframe. Further, the blast damage denial letter received by the homeowner on May 14, 2009 was apparently based on December 2008 blast logs and not based on relevant blasting logs from January 2009, which were not provided.
- The only December blasting log with location coordinates (even though this was before the reported date of loss in January), coupled with the GPS coordinates of the home suggested that this blast occurred 0.37 miles from the home. Coordinates for the other two December blast logs were not provided; thus, the distance between the home and the blasts could not be determined. Finally, it was reported that the date of loss for the blast was in January 2009 or more recently; however, no blast records for January 2009 or later were provided.
- State code initially states that seismic data must be taken, but then contains a provision that allows one, with written permission, to utilize a calculation method to determine whether or not the blast will be safe. Whether or not written permission was requested and/or granted remains an open question.

  On the other hand, data in the blast log could be used to determine if the blast would be safe under the state code given their assumed distance and charges detonated. Two of the three cases (December 12th and December 18th) met this criteria while one (December 2nd) did not appear to meet this criteria.
- The conclusions reached by the insurance agent in the denial letter to the homeowner stated that the highest ground displacement level at the home was 0.405 inches/second. The basis for this value is unknown since the insurance agent's report provided no information such as blasting logs, dates of blasts, and distances used in the calculation. Further, the value of 0.405 inches/second does not correlate to any values that could be found on the blasting logs provided. Given the lack of basis for the value cited and the fact that it appeared that the date of loss was in January 2009 not December 2008 (dates of blasting logs provided), the basis for the reported effects levels and denial letter is questionable.

Experience has shown that seismic logs are often not completed properly or are not available on the reported date(s) of loss. Seismic logs (data) with GPS coordinates >20 miles from the site were provided for the subject blast. These were obviously from the wrong site; no seismic data from the actual site or the date of loss was available.

Finally, in blasting damage assessments, an analysis of cracks (fresh or new vs old) in the finished building surfaces and/or the presence of broken windows provides the best approach for determining damage from nearby blasting operations. The presence of a pre-blast survey, if available, provides validation of new versus old cracks.

20.3.2 Explosion Damage Assessment Methodology

Most explosion damage assessments encountered will be associated with accidents from fuel leaks rather than from nearby quarry or mining activities as discussed above. Consequently, no pre-blast survey information or seismic records are typically available. Instead, the inspection methodology must rely on post explosion site information and literature on damage associated with specific explosion overpressures (see Tables 20.1 and 20.2 for example).

In one case, a home that was only located ~150 feet away from an explosion site was oddly found to be damage-free. Debris from the demolished home where the explosion occurred was scattered over a one-quarter mile arc. In addition, somewhat oddly, the wall surfaces nearest the explosion were pulled toward the demolished home. The explanation was that the explosion in the demolished home occurred in the corner of the basement nearest the damaged home to be inspected. Most of the overpressure from the explosion was directed away from the adjacent home and the negative pressure inflow of air (see Figure 20.4) likely caused movement of the damaged home toward the demolished home where the explosion occurred. The lesson to be learned is that the area around the building to be assessed should be canvassed by the inspector in order to become familiar with the overall site-specific conditions associated with a given explosion before focusing on the specific building in question.

The methodology for completing explosion damage assessments includes the following, sequential elements:

- Interview of the property owner or their representative.
- Site inspection.
- Analysis of information collected.
- Written report, summarizing findings.

Details regarding each of these elements follow.

20.3.2.1 Explosion Damage Assessment Methodology – Interview of Property Owner or Owner's Representative

Upon arriving at the site, the inspection process should follow the overall approach outlined in Chapter 1. In many of these cases, like the one shown in Figures 20.1 and 20.2, the damage assessment will be performed on adjacent properties since the site of the actual explosion may be a total loss. Specific questions for the interview should include the following:

- Has the cause of the explosion been determined by the local Fire Department or their representative? If available, obtain a copy of the report.
- Was the interviewee present at the time of the explosion? If so, what time did the explosion occur? What were the weather conditions? What physical effects did they, or other occupants, suffer from?

- What damages were believed to have occurred as a result of the explosion? Document what, when, and where. When performing this portion of the interview, remember to inquire about interior and exterior damages. Also, obtain any available photographs taken right after the time of the explosion if they are available. Finally, inquire about preexisting damages.

The sum of this information should provide a starting point for determining what damage was caused by the explosion and possibly the magnitude of the explosion.

20.3.2.2   Explosion Damage Assessment Methodology – Site Inspection

The site inspection should follow the overall approach outlined in Chapter 1. As mentioned in Section 20.3.1.1 on blast damage, the intent of the explosion inspection is to collect necessary information needed to make the following determinations:

- Distinguish between pre- and post-explosion damage to the building.
- Determine post-explosion damaged items that will need to be replaced.
- Determine post-explosion damaged items that will need to be repaired.

All observations should be recorded in writing in a field notebook; key observations should be photographed.

Particular attention should be spent documenting cracks in the interior and exterior finished surfaces. During the inside portion of the inspection, cracks and potential explosion damage should be documented for the floor, walls, and ceiling of each room, beginning in the basement or crawlspace (if present) and moving up until the attic inspection is completed. Exterior inspection observations should be completed by elevation followed by the roof inspection. Again, particular attention should be paid to cracks in stone, poured concrete, CMU, and brick masonry walls, veneers, patios, walkways, driveways, and stucco finishes, if present. While performing the exterior inspection, the slope of the ground surface should be noted for each elevation.

The need for a complete and detailed inspection lies with the need to be able to rule in, or out, cracks that were caused by the explosion throughout the building thoroughly. Cracks should be documented by location, directionality, width, and crack characteristics (e.g., brightness of color and sharpness of crack surfaces, whether or not debris is present in the crack, and/or whether or not paint or other coatings are present on the edges of the crack). All of the observed damages should be recorded in the field notebook and photo-documented. As part of this inspection process, any anomalies should be noted such as:

- Fresh cracks to wood members such as joists (Figure 20.11), rafter members, or studs.
- Gaps associated with movement between structural members (often change in color or where area not painted exposed), see Figure 20.11.
- Dislocated truss members from gusset plate connections (Figure 20.12).
- Disconnected plastic plumbing or wiring (often seen in attics and crawlspaces – Figure 20.12).
- Gaps or wracking of windows and doors.
- Warping or bulging of floors.
- Gaps or disconnections in heating, ventilation, and air-conditioning (HVAC) ductwork.

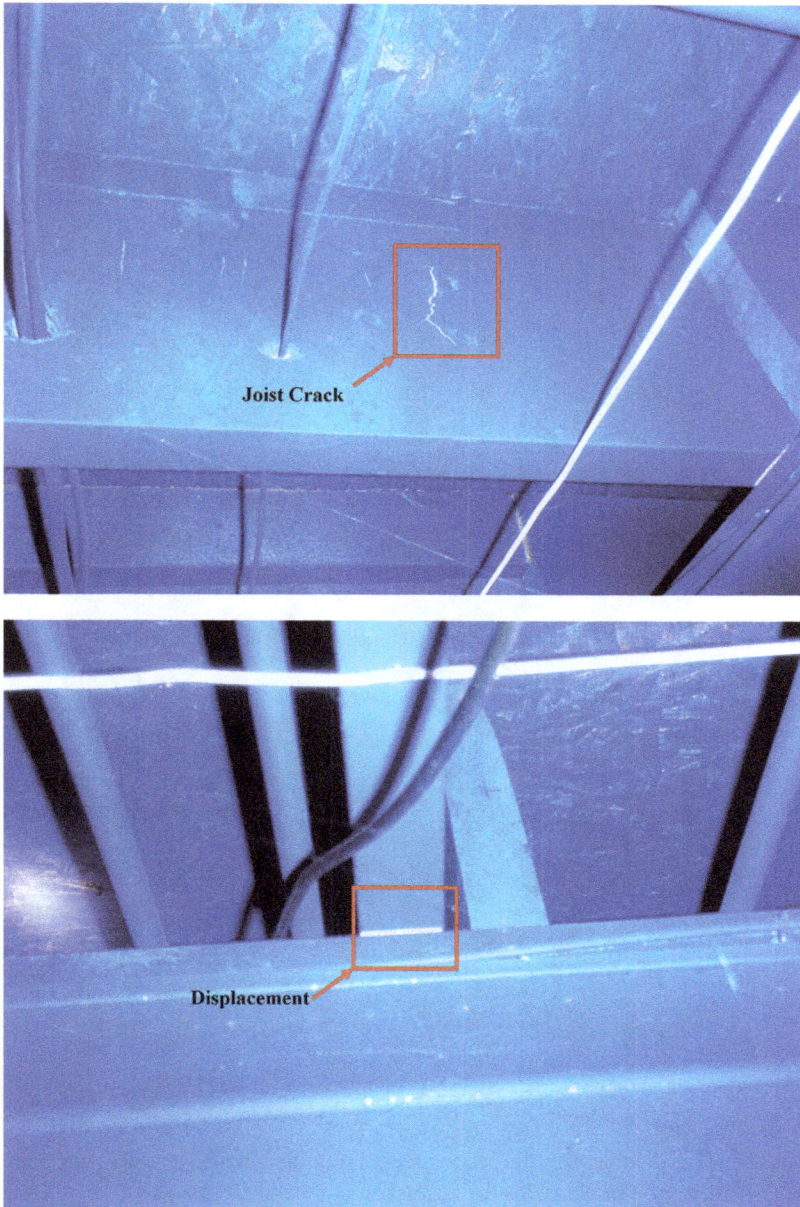

FIGURE 20.11   Explosion damage and movement to basement wood members.

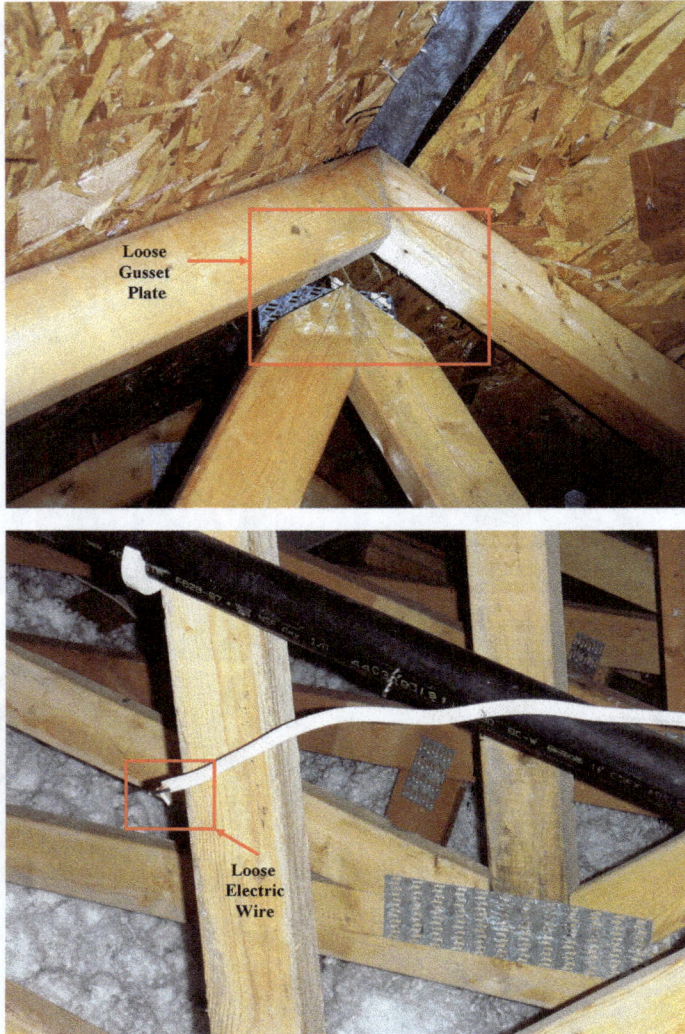

FIGURE 20.12 Explosion damage – dislocated truss members and wire in attics.

The inspection should conclude with the recording (in tabular format) of wall plumb and floor-level measurements following the methodologies discussed in Chapters 14 and 15.

20.3.2.3 Explosion Damage Assessment Methodology – Analysis of Collected Information

The information collected as outlined in Sections 20.3.2.1 and 20.3.2.2 should be utilized to make the following determinations:

- Determine by space and/or elevation pre-explosion versus post-explosion damage. This can be done using crack analysis described in Chapter 15 in this book. For example, newer damage to building materials is nearly always associated with brightness of color at the crack and lack of debris or paint/finishes on cracked surfaces.

- Does the extent of the damage require replacement of the damaged item(s) or can it be repaired? In general, damaged structural members should be replaced as well as items containing significant cracks. Minor cracks in finished surfaces can usually be repaired.

Wall plumb and floor-level measurements taken during the inspection should be analyzed following the methodologies discussed in Chapters 14 and 15 and compared with the recommendations discussed in these same chapters to determine if the damage present would be classified as minor, moderate, or major/significant. An example of actual project wall-level measurements is shown in Table 20.9.

TABLE 20.9
Example of Wall Level Measurements

ID	Location	Vertical Direction Direction Top Sloping	Slope
1	First floor – NE BR – W wall	–	~0.0°
2	First floor – NE BR – N wall	North	~0.2°
3	First floor – NE BR – E wall[a]	West	~0.3°
4	First floor – NW BR – W wall	West	~0.2°
5	First floor – NW BR – N wall	North	~0.4°
6	First floor – bathroom – W wall	West	~0.1°
7	First floor – LR – E wall	West	~0.1°
8	First floor – DR – W wall[b]	–	~0.0°
9	First floor – kitchen – W wall[c]	West	~0.4°
10	First floor – laundry – E wall	West	~0.1°
11	First floor – master BR – E wall	West	~0.3°
12	First floor – master BR – S wall	North	~0.3°
13	First floor – master BR – W wall[d]	–	~0.0°
14	First floor – master BR – N wall	North	~0.3°
15	First floor – master BR bath – W wall	West	~0.1°
16	First floor – master BR bath – N wall	North	~0.3°
17	First floor – hallway – N wall	North	~0.1°
18	Garage – E wall – N end	West	~0.3°
19	Garage – E wall – S end	–	~0.0°
20	Garage – W wall	West	~0.3°

[a] Window sash sloped 0.3° to the west and the wall surface below the drywall crack sloped 0.5° to the west.
[b] Door sloped 0.5° to the west.
[c] Window sash sloped 0.5° to the west.
[d] Sliding door window sloped 0.3° to the west.

Walls that are either bowed or tilted/sloped in or out have likely been impacted by the explosion(s). Walls with tilt/slopes <0.6° with finished surfaces (e.g., drywall or plaster) that are not freshly and/ or significantly cracked are within the range of slopes commonly encountered in residential and light commercial construction and not likely to have been impacted by the explosion(s).

20.3.2.4   Explosion Damage Assessment – Written Report

Based on interview information, site inspection information, analysis, technical literature, and experience, a report of findings can then be prepared. The written report should follow the overall approach outlined in Chapter 1. Particular attention should be spent documenting pre-explosion versus post-explosion damage and determining whether or not the damage requires the item to be removed/replaced or repaired following the methodologies outlined in Chapters 14 and 15. A particularly effective method for summarizing damaged items and recommended remedial action(s) is to use a table format as illustrated in Table 20.10 for each space or elevation where damage was observed.

TABLE 20.10
Example Damage Summary Table

Component(s)	Photograph	Recommendation
Main attic – south gable Garage attic – east gable		Remove and replace
Main attic – SW corner – hole in decking		Repair decking and roof shingles
Main attic – loose electrical wire		Inspect and repair

(Continued)

TABLE 20.10 (*Continued*)
Example Damage Summary Table

Component(s)	Photograph	Recommendation
Main attic – loose soil stack line		Inspect and repair
Loose boards on floor of attics Missing insulation		Repair and replace
Garage attic – loose gusset plate		Repair

This methodology of summarizing inspection results and analysis provides a succinct summary of damage and remedial recommendations to all parties performing future activities on the property.

IMPORTANT POINTS TO REMEMBER

- Blast damage normally occurs as a result of quarrying and mining surface and subsurface operations, whereas explosions typically result from accidents associated with leakage and ignition of fuels such as natural gas, propane, LPG, and gasoline.
- Considerable research has been completed to define "safe" overpressure and ground movement parameters to nearby buildings. As a rule of thumb, if windows were not

broken by the blast, it is unlikely that the structure suffered other significant blast damage.
- Often pre-blast surveys and seismic information are available to the forensic investigator, although the records are often incomplete or inaccurate. These can be used, along with GPS coordinates, to rule in or out the potential of blast damage to a structure.
- Explosion damage assessments are typically associated with nearby buildings since the building where the explosion occurred is often a total loss. The focuses on forensic explosion inspections are as follows:
  1. Determining whether or not the reported damage occurred before or after the explosion.
  2. Taking inventory of the damaged items.
  3. Performing an assessment of which damaged items must be removed and replaced versus those that can be repaired.

REFERENCES

1. National Fire Protection Association (NFPA). "Guide for Fire and Explosion Investigations." 2017, p. 921.
2. Lees, F.P. *Loss Prevention in the Process Industries*. Boston, MA: Butterworth-Heinemann, 1996.
3. Brasie, W.C., and D.W. Simpson. "Guidelines for Estimating Damage Explosion." *Loss Prev* 2 (1968), p. 91.
4. U.S. DOT. "Hazard Analysis of Commercial Space Transportation." U.S. Department of Transportation, Office of Commercial Space Transportation Licensing Programs Division, May 1988.
5. U.S. Air Force. "Explosives Safety Standards." Air Force Regulation 127-100. May 20, 1983.
6. McRae, T.G., et al. *The Effects of Large Scale LNG/water RPT Explosions*. Washington, DC: Lawrence Livermore National Laboratory, April 27, 1984.
7. Fletcher, E.R., D.R. Richmond, and J.T. Yelverton. *Glass Fragment Hazard from Windows Broken by Airblast*. Washington, DC: Defense Nuclear Agency. Report Number DNA 5593T, May 30, 1980.
8. National Fire Protection Association (NFPA). "Guide for Fire and Explosion Investigations." 2004, p. 921.
9. Noon, R.K. *Forensic Engineering Investigation*. New York: CRC Press, 2001.
10. Mainiero, R.J. and H.C. Verakis. "A Century of Bureau of Mines/NIOSH Explosives Research." Accessed January 2012. http://www.cdc.gov/niosh/mining/ pubs/pdfs/acobo.pdf.
11. Duvall, W.I. and D.E. Fogelson. "Review of Criteria for Estimating Damage to Residences from Blasting Vibrations." U.S. Bureau of Mine, Report of Investigations #5968. United States Department of the Interior, Bureau of Mines, 1962.
12. Nicholls, H.R., C.F. Johnson, and W.I. Duvall. "Bulletin 656: Blasting Vibrations and Their Effects on Structures." United States Department of the Interior, Bureau of Mines, 1971, p. 105.
13. Siskind, D.E., V.J. Stachura, M.S. Stagg, and J.W. Kopp. "Structure Response and Damage Produced by Airblast from Surface Mining." Report of Investigations 8485, United States Department of the Interior, Bureau of Mines, 1980, p. 111.
14. Thoenen, J.R., and S.L. Windes. "Bulletin 442: Seismic Effects of Quarry Blasting." United States Department of the Interior, Bureau of Mines, 1942, p. 83.
15. Windes, S.L. "Damage from Air Blast." Report of Investigations 3708. United States Department of the Interior, Bureau of Mines, 1943, p. 50.
16. Ohio Administrative Code. "Chapter 1501:13-9-06 Use of Explosives." February 6, 2009.
17. Telford, T. ed. *Blast Effects on Buildings*, Second Edition. London: Thomas Telford Limited, 2009.
18. Ngo, T., P. Mendis, A. Gupta, and J. Ramsay. "Blast loading and effects on structures: An overview." *eJSE International*. Special Issue: Loading on Structures, 2007, pp. 76–91. Accessed January 2012, http://www.ejse.org/Archives/Fulltext/2007/Special/200707.pdf.

21 Lightning Damage Property Assessments

Stephen E. Petty
EES Group, Inc.

CONTENTS

PURPOSE/OBJECTIVES

The purpose of this chapter is to:

- To be able to determine whether or not lightning struck the subject property.
- Understand typical effects of lightning-strike damage to equipment and materials.
- Provide methodologies to conduct lightning damage assessments for HVAC outdoor units, well-pump motors, and brick masonry.

Following the completion of this chapter, you should be able to:

- Understand typical characteristics of lightning damage to property.
- Be able to understand typical characteristics of lightning damage to HVAC outdoor units, well pumps, and brick masonry.
- Be able to collect necessary site-specific information needed to determine whether or not lightning likely damaged the equipment or property.
- Be able to prepare a report with analysis and documentation regarding lightning strike damage.

21.1 INTRODUCTION

The University of Florida, located in an area with one of the highest rates of lightning strikes in the United States, published the following interesting fact-versus-fiction statements regarding lightning[1]:

Myth #1: Lightning never strikes in the same place twice.
Fact #1: The Empire State Building is struck about 23 times in an average year.

Myth #2: In medieval times, church bells were rung during thunderstorms, because people thought the sound waves from the bells would suppress the lightning.
Fact #2: This belief was discarded when one historian noted that in a 33-year period, 386 church towers were struck and 103 bell ringers killed.

Myth #3: Lightning follows the most direct path to the ground.
Fact #3: It has been known to travel through clear air and strike 10 miles from the storm like a "bolt from the blue!"

As suggested, facts surrounding lightning are often misunderstood as are facts regarding lightning damage to property.

Approximately 20 million cloud-to-ground lightning flashes per year occur in the United States.[2] Cloud-to-cloud lightning flashes occur 5–10 times more frequently, so upward of 100–200 million total lightning flashes occur in the United States each year.

Lightning is simply the discharge that occurs with the buildup of positive and negative charges in a thundercloud. Charge initiation occurs at the mid-levels of the cloud when graupel (soft ice) forms in a process called riming (supercooled liquid precipitation that freezes around an ice crystal). This occurs at temperatures between 14°F and −4°F (−10°C to −20°C). The graupel has negative charges; electrification results from millions of collisions between graupel and ice crystals. The negative charge of the graupel within the cloud induces positive charges within the upper reaches of a cloud. As the graupel nears the earth, it also induces a positive charge on objects on the ground. The anvil upper portion of a cloud, which is typically offset away from the main column of the cloud and is positively charged, also induces an opposite negative charge on the ground location below. Ground-to-cloud lightning occurs when sufficient charge builds up to bridge the gap between: (1) the graupel in the bottom of the cloud and objects on the earth below and (2) the charge in the anvil portion of the cloud and objects on the earth below. These anvil-type strikes can occur from 5 to 10 miles ahead or behind the main storm area.

Typical lightning bolts are ~1-inch wide and ~3–5 miles long, but some bolts have been reported to be upward of 118 miles long. The lightning, in less than a millionth of a second, heats up the air in the channel between 40,000°F and 50,000°F, causing the shock wave that one hears as thunder. The sound produced by load lightning ranges from 100 to 120 decibels, well above the sound levels of normal conversation, which is ~60 decibels. The energized state of the air gives off the light one sees as lightning. In terms of electrical output, a lightning bolt can produce voltages upward of 200 million volts of electricity. Within power lines, power spikes up to 6,000 volts over times of 8μ-seconds are not uncommon. Power spike rise times can be as low as 500 nanoseconds, thus blown fuses or breakers do not provide an indication of lightning-induced power surges and are not protective of downstream equipment.

Lightning damages both property and individuals. Property damage losses from lightning strikes have been estimated to be $4–$5 billion each year.[3] Annual deaths from lightning strikes in the United States, from 1998 to 2010, have ranged from 27 to 51[4] (Figure 21.1).

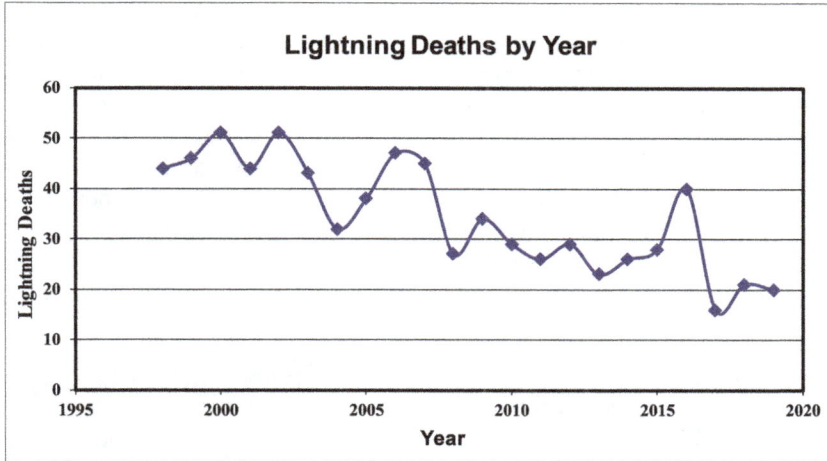

FIGURE 21.1 Annual deaths from lightning in the United States by year.

From 1978 to 2008, the average number of people killed by lightning in the United States was 58, with ~300/year reported injured by lightning.[3] For comparison, annual deaths from other weather-related hazards such as those from tornados and hurricanes were, on average, lower; annual deaths from flash floods were greater than those for lightning.[5]

Thunderstorms, the sources of lightning, occur more frequently in the eastern two-thirds of the United States as illustrated in Figure 21.2.[6]

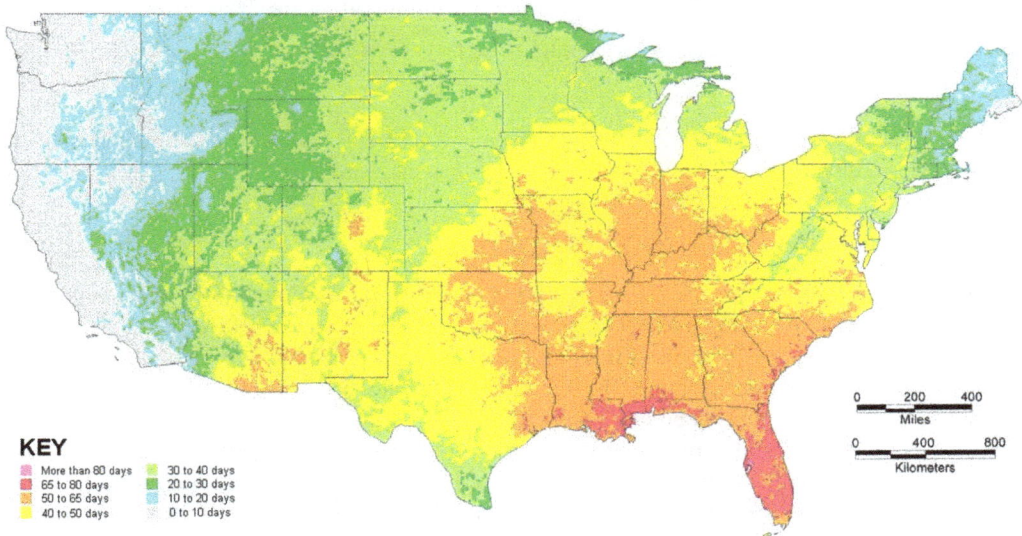

FIGURE 21.2 Number of thunderstorm days per year by location in the United States.

The National Oceanic and Atmospheric Administration (NOAA) provides a simplified map showing essentially the same information.[7] Peak thunderstorm day/year locations are in Florida, along the gulf and southeast Atlantic coast, and throughout the south and Midwest. Throughout much of this area, the number of annual thunderstorm days can range from 50 to 90 days or 13.7%–24.6% of the days per year. In the Midwest, there is a one chance of six (~17%) probability that a thunderstorm (and lightning) will occur on any one day during the year, so it is to be expected that equipment failures are associated with lightning storm events, whether true or not.[8] Thunderstorms will have multiple lightning strikes. Worldwide cloud-to-ground lightning strike data from April 1995 to February 2003 are shown in Figure 21.3.[9]

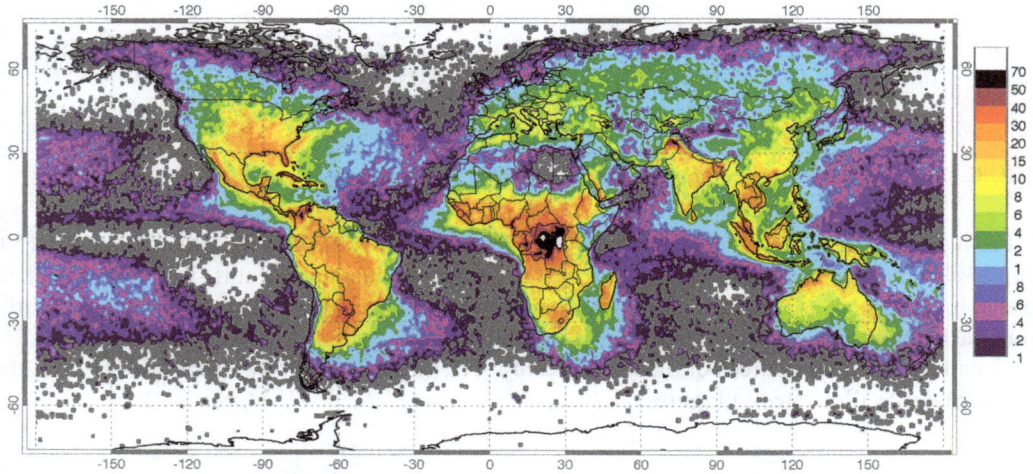

FIGURE 21.3   Worldwide lightning strike intensity. (Courtesy of NASA.[9])

Not surprisingly, the intensity of lightning strikes in the United States, by location, parallels that of thunderstorm intensity.

21.2   LIGHTNING CODES AND STANDARDS

The International Residential Code[10] and the International Building Code[11] (for commercial buildings) are remarkably silent regarding design requirements for protection of property from lightning strikes, but the issue is addressed in guidance documents and by governmental agencies. The National Fire Protection Association (NFPA) has a Technical Committee on Lightning Protection that has updated and continues to update NFPA 780-2020, Standard for the Installation of Lightning Protection Systems. This standard is published in the National Fire Codes. The Lightning Protection Institute (www.lightning.org) publishes Standard LPI-175 (2017 Edition currently – see https://guardianlp.com/media/2019/07/LPI-175-2017.pdf), which contains valuable detailed information on nationally recognized methods for the proper design, installation, and inspection of lightning protection systems. The Underwriting Laboratories (UL) publishes (https://ifs.ul.com/lps/lightningprotectionhome.nsf/HomePage?OpenForm) Standard UL 96A Installation Requirements for Lightning Protection Systems (March 18, 2016 Edition currently), the scope of which is as follows:

1.1   These requirements cover the installation of lightning protection systems on all types of structures other than structures used for the production, handling, or storage of ammunition, explosives, flammable liquids or gases, and other explosive ingredients including dust.

1.2 These requirements apply to lightning protection systems that are complete and cover all parts of a structure. Partial systems are not covered by this standard.

1.3 These requirements shall not apply to adjacent structures.

1.4 Adjacent structures shall be considered separate structures.

1.5 Adjacent structures with lightning protection shall be considered part of the structure if the adjacent structure's lightning protection system complies with this standard and is connected to the lightning protection system of the structure in accordance with Section 10.4.

1.6 Walkways that are attached to a structure shall be considered part of that structure.

1.7 Free-standing walkways shall be considered an adjacent structure under the following conditions:

    a. It is separated by a fire wall and conductive media that is shared by both facilities has an SPD in accordance with Section 13.

    b. It is isolated by a distance of not <6 feet and conductive media that is shared by both facilities has an SPD in accordance with Section 13.1.8.

        This standard does not cover lightning protection for:

        a. electric transmission lines or open air distribution racks,

        b. outdoor substations or switch yards, and

        c. electric generators unenclosed by a building or other enclosed structures.

1.9 Enclosed generators and conventional building structures at or associated with generators or power plants, etc. are covered.

1.10 These requirements do not cover lightning protection components, which are covered by the Standard for Lightning Protection Components, UL 96.

Thus, lightning best practices are covered by standards rather than codes in the United States. The National Lightning Safety Institute (NLSI) publishes a list of worldwide codes and standards associated with lightning protection[12]; these are summarized in Table 21.1.

TABLE 21.1

Worldwide Codes and Standards Regarding Lightning Protection

Entity	Code or Standard Description of Code or Standard	Electronic Link to Code or Standard
API	API RP 2003 Protection Against Ignitions Arising out of Static, Lightning, and Stray Currents Revision/Edition: 8 September 2015. American Petroleum Institute (2015)	Purchase at: *https://www.api.org/~/media/files/ publications/* whats%20new/2003_e8%20pa.pdf
DOE	DOE M440.1-1 Explosives Safety Manual, Chapter 10, Section 3.0 Department of Energy Electrical Storms and Lightning Protection	N/A.
DOE	DOE/EH-0530 Lightning Safety Department of Energy Office of Nuclear and Facility Safety (1996)	N/A.
DOE	DOE Order 420.1 Facility Safety Issued on 5/20/02 and Canceled 12/22/07 Department of Energy	N/A.
DDESB	DDESB 6055.9 DOD Ammunition and Explosives Safety Standards, Chapter 7, Lightning Protection Department of Defense (2004); Replaced by DESR 6055.09, Edition 1, January 13, 2019.	https://www.denix.osd.mil/ddes/home/ home-documents/desr-6055-09-edition-1/

(Continued)

TABLE 21.1 (*Continued*)

Worldwide Codes and Standards Regarding Lightning Protection

Entity	Code or Standard Description of Code or Standard	Electronic Link to Code or Standard
FAA	FAA-STD-019D Lightning and Surge Protection, Grounding, Bonding, and Shielding Requirements for Facilities and Electronics Equipment (9 August) Federal Aviation Administration (08/09/2002)	N/A
FAA	FAA 6950.19A Practices and Procedures for Lightning Protection, Grounding, Bonding and Shielding Implementation Federal Aviation Administration (07/01/2002)	N/A
LPI	Lightning Protection Institute – LPI-175/2017 Edition - Standard Of Practice for the DESIGN – INSTALLATION – INSPECTION of Lightning Protection Systems	https://guardianlp.com/media/2019/07/LPI-175-2017.pdf
NASA	KSC-STD-E-0012E. Facility Grounding and Lightning Protection Standard National Aeronautics and Space Administration (08/01/2001)	N/A
National Weather Service	NSW Manual 30-4106/NSWM 30-41 Lightning Protection, Grounding, Bonding, Shielding and Surge Protection Requirements National Weather Service (06/02/05)	N/A
NFPA	National Fire Protection Association (NFPA) NFPA 780 – Standard for the Installation of Lightning Protection Systems	Purchase at: https://www.nfpa.org/codes-and-standards/all-codes-and-standards/list-of-codes-and-standards/detail?code=780
NLSI	Lightning Protection for Engineers National Lightning Safety Institute (NLSI), 2006	https://www.scribd.com/doc/92206504/Lightning-Protection-for-Engineers-National-Lightning-Safety-Institute-NLSI-2006-Year
UL	Underwriting Laboratories UL Standard 96A, Standard for Installation Requirements for Lightning Protection Systems, Edition 13, Edition Date: March 18, 2016	Purchase at: *https://standardscatalog.ul.com/standards/en/* standard_96a_13
US Air Force	Air Force Instruction (AFI) 32–1065 Grounding Systems United States Air Force (Current Rev. 06/14/2017)	*https://static.e-publishing.af.mil/production/1/af_a4/publication/* afi32–1065/afi32–1065.pdf
US Air Force	Air Force Manual 91–201 Explosives Safety Standards United States Air Force (03/21/2017)	*https://static.e-publishing.af.mil/production/1/af_se/publication/* afman91–201/afman91–201.pdf
US Army	Department of Army Pamphlet 385-64 Ammunition and Explosive Safety Standards Chapter 12 – Electrical Hazards and Protection – Section IV – Lightning Protection United States Army (05/24/11)	*https://www.wbdg.org/FFC/ARMYCOE/ARMYCRIT/* pam38564.pdf
US Corps	EP 385-1-95a Basic Safety Concepts and Considerations for Munitions and Explosives of Concern (MEC) Response Action Operations United States Corps of Engineers (08/27/04)	*http://asktop.net/wp/download/27/EP%20 385-1-95a%20Basic%20Safety%20 Concepts%20and% 20Considerations%20 for%20Munitions%20and% 20Explosives%20 of%20Concern%20MEC% 20Response%20 Action%20Operations.pdf*

(Continued)

TABLE 21.1 (*Continued*)
Worldwide Codes and Standards Regarding Lightning Protection

Entity	Code or Standard Description of Code or Standard	Electronic Link to Code or Standard
US Military	MIL-STD188-124B Grounding, Bonding and Shielding for Common Long Haul/Tactical Communications Systems Including Ground Based Communications – Electronics Facilities and Equipment United States Military Specifications (Revision 12/18/00 of 02/01/92 Std.)	http://www.tscm.com/MIL-STD-188-124B.PDF
US Military	MIL-HDBK 419A (Volume I Basic Theory) Grounding Bonding and Shielding for Electronic Equipment and Facilities United States Military (12/29/87)	*https://www.wbdg.org/FFC/NAVFAC/ DMMHNAV/* hdbk419a_vol1.pdf
	MIL-HDBK 419A (Volume II, Applications) Grounding Bonding and Shielding for Electronic Equipment and Facilities United States Military (12/29/87)	*https://www.wbdg.org/FFC/NAVFAC/ DMMHNAV/* hdbk419a_vol2.pdf
US Navy	NAVSEA OP5 Ammunition and Explosives Ashore – Chapter 6 – Lightning Protection United States Navy (08/30/07)	N/A
Australian Standard	AS 1768:2007 Australian Standard for Lightning Protection Australia (2007)	*https://www.saiglobal.com/PDFTemp/Previews/* OSH/as/as1000/1700/1768–2007.pdf
British Standard	BS EN 62305:2011 Protection of Structures against Lightning Great Britain (July 30, 2011)	https://343lzp26ts7kiqip8biqf71a-wpengine. netdna-ssl.com/wp-content/uploads/BS-EN-62305-1-2011-Protection-against-lightning-Part-1-General-principles.pdf
Chinese Code	GB 50057-94 (2000) Design Code for Lightning Protection of Buildings China (2000)	N/A
Indian Code	IS 2309 Protection of Buildings and Allied Structures Against Lightning – Code of Practice India (1989, reaffirmed 1/2006)	http://www.doksunpower.com/wp-content/ uploads/2013/06/BIS-2309.pdf
Polish Standard	PN-86/ E-05003/02 Lightning Protection of Structures Poland (2009)	*https://www.google.com/ url?sa=t&rct=j&q=&esrc= s&source* =web&cd=2&ved=2ahUKEwjQ2u248 aTpAhU NPK0KHTw2AlIQFjABegQIAhAB&url= https %3A%2F%2Fbazakonkurencyjnosci. funduszeeuropejskie.gov. pl%2Ffile%2Fdownload% 2F1248914&usg=AOvVaw0Ncm-nN_ yOJowY8VaH3dzk
Russian Code	RD 34.21.122-87 Manual for Installation of the Lightning Protection of Buildings and Structures Russia (1987)	Purchase: RD 34.21.122–87 Manual for Installation of the Lightning Protection of Buildings and Structures
South Africa Standard	SABS 03-2012 The Protection of Structures Against Lightning South Africa (2012 Revision)	https://store.sabs.co.za/pdfpreview.php?hash= a7d1fd1ad4c12fd848 35753890558c9ee68a528& preview=yes

Within the United States, standards regarding lightning have been prepared by the military (e.g., Air Force, Army), governmental agencies (e.g., Department of Defense, Department of Energy, Federal Aviation Administration, National Air and Space Administration, and the National Weather Service) and by industry (e.g., American Petroleum Institute, Lightning Protection Institute, NLSI, and Underwriting laboratories). Other countries, including Australia, China, Great Britain, India, Poland, Russia, and South Africa, have developed codes for protection from lightning.

21.3   LIGHTNING DAMAGE AND FORENSIC INVESTIGATIONS

Forensic investigations for lightning damage stem primarily from claims in the insurance industry for reported failures of heating, ventilation, and air conditioning (HVAC) outdoor units and from failures of well-pump motors. Reports of lightning damage to brick masonry chimneys and to motors and equipment are also sometimes encountered. Experience has shown that failures of HVAC compressors, well pumps, motors, and other electrical equipment are often associated with thunderstorm or lightning events.

Damage from lightning is caused by relatively high voltages that occur quickly enough that it bypasses surge protection equipment. Lightning can send extremely high-voltage surges into an electrical installation. Because the voltages and currents from lightning strikes are so high, arcs can jump at many places, cause mechanical damage, and ignite many kinds of combustibles.[13]

NFPA states the following in their sections on Lightning Strikes and Lightning Damage (Reproduced with permission form NFPA 921.2017, *Guide for Fire and Explosion Investigations,* Copyright© 2016, National Fire Protection Association. This reprinted material is not the complete and official position of the NFPA on the referenced subject, which is represented only by the standard in its entirety.):

> *8.12.8.2 Lightning characteristics*: Typical lightning channels have a core of energy plasma 12.7–19 mm (½ to ¾ inches) in diameter, surrounded by a 102 mm (4 inches) thick channel of superheated ionized air. Lightning return stroke currents average between 30,000 and 45,000 A depending upon location, but can exceed 200,000 A. Potentials can range up to 15,000,000 V.
>
> *8.12.8.3 Lightning strikes*: Lightning may strike any object that generates an upward-extending streamer that successfully connects with the downward extending step leader generated from the base of the cloud. In many cases, this may be the tallest object but could also be any protruding or elevated surface or mass. Lightning threats to a structure consist of the following:
>
> 1. A direct strike to the structure or an item attached to the structure, such as a TV antenna, air-conditioning unit, and so forth, extending up and out from the building roof
> 2. A strike near a structure that couples or channels energy onto energized or non-energized conductors
> 3. A direct strike to incoming conductors connected to the structure
> 4. A strike near overhead conductors that can couple lightning currents onto conductors connected to the structure

The bolt generally follows a conductive path to the ground. At points along the path, the main bolt may divert, for example, from wiring to plumbing, particularly if underground water piping is used as a grounding device for the electrical system of the structure.[13]

> *8.12.8.4 Lightning damage*: Damage by lightning is caused by two characteristic properties; high currents and energy in a lightning strike and extremely high heat energy and temperatures generated in the channel by the electrical discharge. Examples of these effects are shown in "A" through "D" below:

A. A tree can be shattered by lightning current conducted deep into the tree's heart-wood with the heat vaporizing the moisture in the tree into steam – with explosive effects.

B. Copper conductors not designed to carry the thousands of amperes of a lightning strike may be melted, severed, or completely vaporized by the overcurrent effect of a lightning discharge. Lightning currents may also generate overvoltages that trigger power system overcurrents sufficient to sever conductors in one or more locations. Copper and aluminum conductors properly sized and routed in accordance with NFPA 780 will not be damaged by a lightning impulse current up to 200,000 A.

C. Where lightning strikes a steel-reinforced concrete building, the current may follow the steel reinforcing rods as the least resistive path. The high energy may destroy the surrounding concrete with explosive force to get to the reinforcing steel.

D. Lightning can also cause fires by damaging fuel gas systems. Fuel gas appliance connectors have been known to have their flared ends damaged by electrical currents induced by lightning and other forms of electrical discharge. Where gas lines are damaged, fuel gas can leak, and the same arcing that caused the gas line to fail may also cause ignition of the fuel gas.[13]

Experience has shown lightning to damage the top edges of brick masonry chimneys since they are a high point. The damage is caused when moisture in the brick masonry is vaporized, causing the brick to be damaged as if it exploded. Lightning damage to air-conditioning outdoor units is typically manifested as high-voltage damage (blackened and/or melted) to items such as the contactor points and run capacitor (Figure 21.4).

FIGURE 21.4 Explosive lightning damage to HVAC outdoor unit run capacitor.

21.3.1 LIGHTNING DAMAGE ASSESSMENT METHODOLOGY

Most lightning damage assessments encountered will be associated with reported lightning damage to HVAC outdoor units, water well pumps/motors, and commercial motors. It is important that the inspection occurs soon after the reported failure so the equipment can be inspected before it is either discarded or spoliated (from a forensic standpoint). The methodology for completing lightning damage assessments includes the following sequential elements:

1. Determine date(s) of loss.
2. Obtain (and interpret) weather data from sources such as Weather Decision Technologies/ DTN and CoreLogic STRIKEnet for location on date(s) of loss.
3. Interview tenant/owner regarding recollections on date of loss.
4. Inspect premises for signs/evidence of collateral lightning damage.
5. Inspect equipment or property – for equipment; look for signs of damage to outdoor unit components typical of damage by high voltage associated with lightning strike(s). For property such as a brick chimney, look for explosive-like damage and burn marks. If struck, the moisture in the masonry will rapidly turn to steam and cause explosion-like damage.
6. For equipment – document findings (in writing, with measurements and photographs). This should involve a preliminary overview inspection of the equipment; recording model number and serial numbers. Also, record the date of installation if given or written on the unit. Note any burn marks, melting or burning of wires and contact points, and any bulging or exploded components such as capacitors, which typically suffer high-voltage damage. Then follow up the preliminary inspection with a detailed electrical check of the equipment, including the power supply back to the circuit breaker or fuse box. Then perform detailed electrical checks on the equipment itself.
7. For property – document (in writing, with measurements and photographs) the location of the damage, the extent of the damage, the material(s) damaged, and a description of the damage itself.
8. Prepare a written report based on interview information, inspection visual observations, inspection field measurements, and a review of the pertinent literature and experience.

Details regarding these elements follow.

21.3.1.1 Lightning Damage Assessment Methodology – Collection and Review of Weather Records for Lightning Strikes at/near the Date of Loss

Prior to arriving at the site, attempt to determine if lightning struck the area at the reported time of loss; if it did not, then it is likely that the equipment failed for other reasons. Sources of data on thunderstorms and lightning include the NOAA National Climactic Data Center (NCDC), the NOAA Storm Prediction Center (SPC), the Weather Underground (www.weatherunderground. com), and specialized services such as LightningTrax™ Weather Decision Technologies, Inc. (www. weatherforensics.com) and STRIKEnet® by CoreLogic. The public NOAA sources can be used to search for evidence of thunderstorms or lightning at their nearest weather station to determine if thunderstorms were present in the area at the reported time of the lightning loss. The Weather Underground site can further refine the analysis since often they have more historical data online for local weather stations. Weather Underground also provides hour-by-hour weather information such as thunderstorm activity at a specific station, so the times of thunderstorm and lightning activity can be determined within an hour of its occurrence. However, for specific lightning data, commercial sources such as LightningTrax™ and STRIKEnet by CoreLogic can be utilized to obtain site-specific lightning strike data for a nominal fee. The user provides the date of interest and the address or GPS coordinates for the site and a lightning strike. The company can then generate an electronic report for a period of typically 72 hours on or around the date of loss and out to distances of 1, 5, or 10 miles. An example output from LightningTrax™ is shown in Figure 21.5 to illustrate the type of product one might receive.

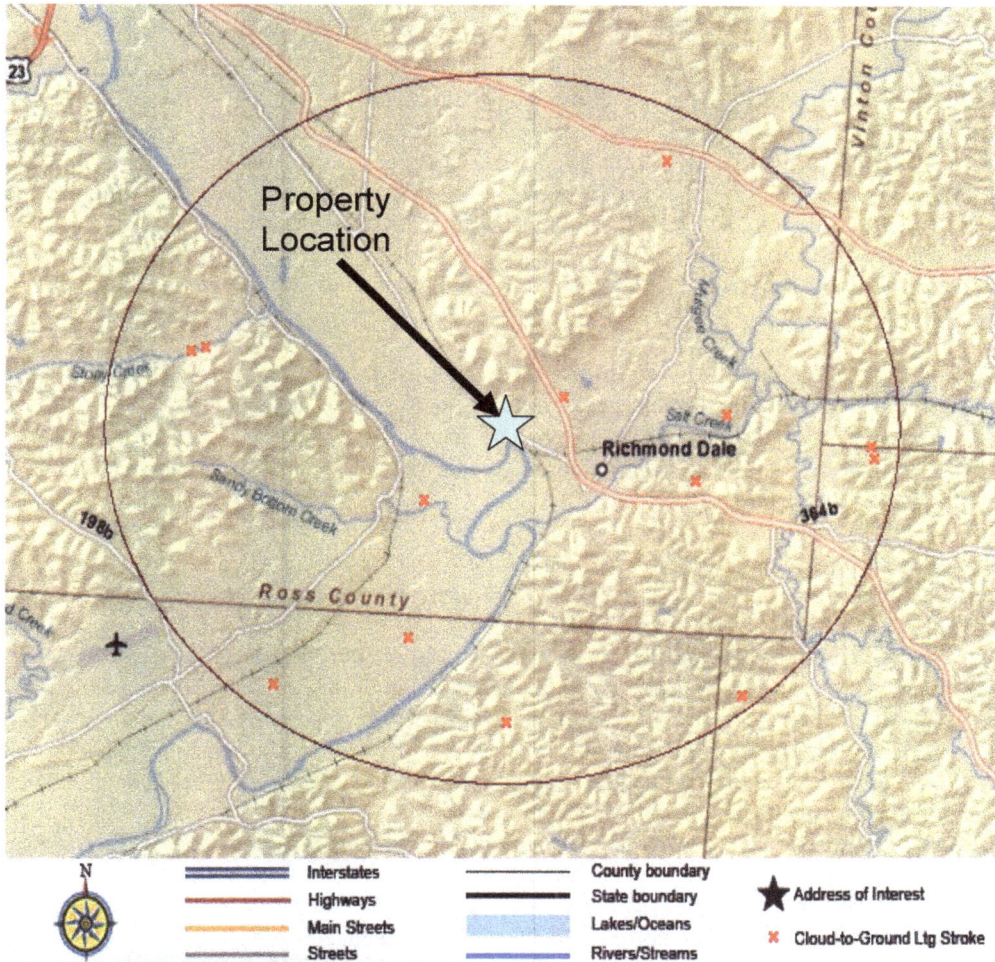

FIGURE 21.5 Example lightning strike map output. (Courtesy of Weather Decision Technologies/DTN, Norman, OK)

The reports also list the times and distance of nearby lightning strikes over the time and distance specified in tabular format. In this particular example, the nearest lightning strike over the period of interest was 0.9 miles away from the address of the property where lightning damage reportedly occurred.

However, these services are not perfect. The commercial lightning services provide the basis for the data they use, along with data on the accuracy (distance) and probability that all lightning strikes are recorded. The detection efficiency of identifying lightning strikes in most of the continental United States is >95%, and the distance accuracy is <250 m (~820′ or 0.155 miles). This means that 19 in 20 lightning strikes will be detected by this technology, and conversely, <1 in 20 lightning strikes will not be recorded by this technology. In addition, the location of the strike should be within 0.155 miles. In this example case, the closest reported strike was 0.9 miles from the business; given the accuracy of the data is within 0.155 miles, it appears that actual lightning strikes would be beyond known lightning strikes for the period of interest, and it is unlikely that lightning struck this particular address (note NFPA 921, Section 9.12.8.5 Lightning Detection Networks states that lightning strikes can be determined to within 500 m–1,640.5 feet or 0.31 miles[13]). Nevertheless, there is a <5% probability that a lightning strike was missed by the technology, so a field inspection may still be warranted.

21.3.1.2   Lightning Damage Assessment Methodology – Interview of Owner or Owner's Representative

Once arriving at the site, the lightning damage inspection process should follow the overall approach outlined in Chapter 1. Specific interview questions for lightning damage should include the following:

- What were the date and time of the lightning strike? If the date is not recalled, can the week and/or time of week be recalled? What time of day did the lightning strike (a.m. or p.m.)? Did it occur during working hours?
- What collateral damage occurred? Were clocks, garage door openers, or other equipment damaged? If so, what? Was there any damage to the building? If so, what and where?
- Is any of the damaged equipment available for inspection? If so, inspect. If not, does someone else have or possibly have the equipment?
- Has anyone worked on the damaged equipment? If so, who are they, what is their contact information, and are they available for an interview?
- If the equipment has been repaired, what specific damage was observed (e.g., blown or bulging capacitor, etc.)?
- Was any neighboring building or equipment from nearby buildings or homes damaged at this same time? If so, whose property and what was damaged.
- What is the age of the damaged equipment; when was it installed?
- What is the history of maintenance and repair activities on the subject equipment?
- Is the nameplate from the manufacturer on the equipment? If so, what are the model and serial numbers of the equipment?

The sum of this information should help to identify the date of the damage so weather data can be researched to determine if lightning did or did not strike the area at the time it was reported to have struck the building and/or the equipment. Lightning will often damage items such as clocks, so this provides collateral evidence indicating that lightning may have struck the building and/or equipment. The age of equipment can often be determined from the serial number on the nameplate. This is important since it is not unusual for equipment to fail of "old age" once it enters the range of known lifetimes for given equipment.

21.3.1.3 Lightning Damage Assessment Methodology – Site Inspection

The site inspection should follow the overall approach outlined in Chapter 1. The intent of the lightning damage inspection is to collect necessary information needed to make the following determinations:

- Determine the date of the lightning damage, age of equipment damaged, and collateral site damage from the lightning.
- Obtain information regarding the age of the equipment (from serial numbers).
- Look for evidence of high-voltage damage to property or equipment. On equipment, look for damage to components that typically would be damaged by a lightning surge.

It is unusual for an electric motor, in even fair condition, to be damaged by a lightning surge since other more delicate electric components connected to the same service would likely be damaged first. This is because the damage threshold for electric motors is between 1,000,000 and 10,000,000 watts/microseconds while that for relays is between 100 and 1,000 watts/microsecond, and for computer control electronic components 0.001 and 0.1 watts/microseconds. This is why, on HVAC equipment, components such as the contactor points and run capacitors are typically damaged before the compressor motor.

- Perform electrical checks, primarily continuity, on the power supply from the breaker through the damaged equipment (is something else wrong – blown fuse).
- Perform equipment-specific electrical checks.

To illustrate this inspection process, abbreviated case study examples for an HVAC outdoor unit and a well pump follow.

21.3.1.3.1 Lightning Damage Assessment Methodology – Site Inspection Example for Damaged HVAC Equipment

Inspect Premises for Evidence of Collateral Lightning Damage:

- No high-voltage damage to exterior surfaces of the building observed.
- High-voltage damage to a time clock present.

Inspect Equipment to Obtain Model and Serial Numbers and Date of Installation from Name Plate (if provided): Using the serial number on the unit (WIAF337760) and information from the manufacturer's (York) webpage, the following information was obtained: Manufactured in Wichita, Kansas in August of 1992. Thus, this particular unit was ~19 years old at the time of the inspection. Based on the life expectancies of HVAC units, which range from ~8 to 19 years with a mean of 13 years, this particular unit was well beyond its typical average life expectancy.

Inspect Equipment – Look for Damage to Components Typically Damaged by High Voltage:

- Circuit breakers supplying power to the unit were not blown closed.
- The points on the contactor were melted (Figure 21.6), and the packing of the run capacitor was blown out the end of the capacitor (Figure 21.7).

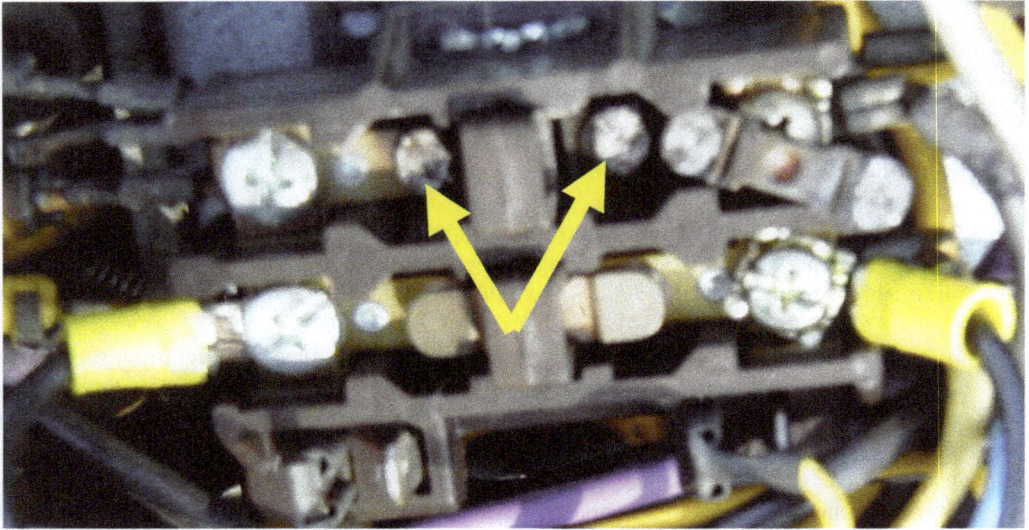

FIGURE 21.6   Melted points on an HVAC contactor.

VS

FIGURE 21.7   Normal and lightning-damaged HVAC outdoor unit run capacitors.

Note that high-voltage lightning strike energy is typically too fast for circuit breakers to close in time to protect equipment. Based on experience with lightning strikes to outdoor units, the components most likely to show damage are the contactor(s) and the run capacitor. Wiring leading into these components may also contain burned insulation, or the wires may be melted. If these components were not damaged, and the unit is at or beyond its normal life, it is much less likely that failure of the compressor was caused by lightning.

In addition, the area around the base of the compressor should be inspected for oil/oil leaks and the compressor surface should be checked for evidence of heat damage (paint thermally degraded). Both are signs of mechanical failure of the compressor.

Inspect Equipment – Perform Electrical Checks on Key Components, including the Compressor: Typical electrical checks would be continuity or resistance (units of ohms), amperage (amps), voltage (volts), and impedance (ohms). Typically, for an HVAC outdoor unit, electrical checks on resistance are completed from the power supply to the compressor using resistance checks. The checks are completed across the fuse or breaker, through the contactor, through the run capacitor, the fan motor, and to the compressor. The fuses, breakers, and contactors can be manually closed to check continuity (resistance in ohms). These checks are illustrated below for a situation where the contactor was not melted and the capacitor was not bulging or blown:

Contactor observations: The sheet metal cover to the controls section of the outdoor unit was removed and the controls were exposed. When first inspected, the contactor was manually engaged to determine if power was present to the unit and if the contactor energized the system. When energized, the power was on and the fan started. This clearly indicated that the contactor was functioning (i.e., not shorted). The contactor was free of melting on the contact points and free of burn marks. No insect debris was found on the contact points. It moved as expected when manually actuated.

Run capacitor observations: The unit was manufactured by General Electric and contained three contact points (yellow – common, brown – fan, and black – compressor). The product ID number was HC98JA062D. No blown-out sides, leaking oil, or burned leads, common when struck by high voltages such as those caused by a lightning strike, were observed. Electrical checks for resistance and capacitance were completed after disconnecting the leads and shorting the terminals. Results of the electrical checks are summarized in Table 21.2.

TABLE 21.2
Summary of Run Capacitor Electrical Checks

Wire Leads Checked	Resistance (Ω)	Resistance Rising with Time	Capacitance (μF)
Brown – Yellow (fan)	>5M	Yes	7.45
Brown – Black	>5M	Yes	6.59
Black – Yellow (compressor)	>5M	Yes	57.4

The capacitor appeared to be item # G32-461 with a rating of 55 and 7.5 μF; actual measurements appeared to be consistent with ratings from the manufacturer. The rule of thumb is to change out the run capacitor if the reading drops by 10% or more.[14]

Fan motor observations: Similar electrical checks for resistance were completed after disconnecting the leads from the fan motor. Results of the electrical checks are summarized in Table 21.3.

TABLE 21.3
Summary of Fan Motor Electrical Checks

Wire Leads Checked	Resistance (Ω)
Brown – Yellow	44.2
Brown – Black	34.7
Black – Yellow	78.8

Since resistance was observed between all contacts,[15] the fan motor appeared to be operating properly.

Compressor observations: Light heat effects were present in the paint on top of the unit. On removing the fan, the inside of the unit and the compressor could be observed. Light amounts of leaf debris were present near the base of the compressor; no compressor oil was observed near the base of the unit (Figure 21.8).

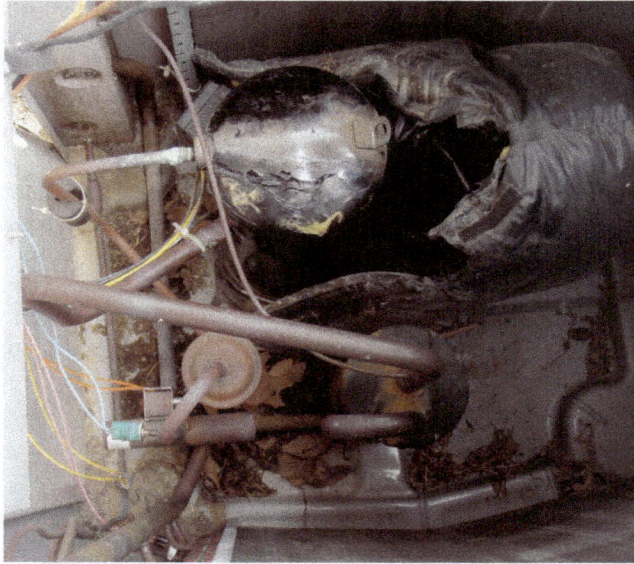

FIGURE 21.8   Heat pump outdoor unit – view from above.

The compressor was a Copeland Scroll compressor. The model number on the label was ZR49K3-PFY-230 and the serial number was 98J873539. The ends of all three wires to the compressor were not blackened, nor were the leads into the compressor. The electrical leads, by color, were black, blue, and yellow (Figure 21.9).

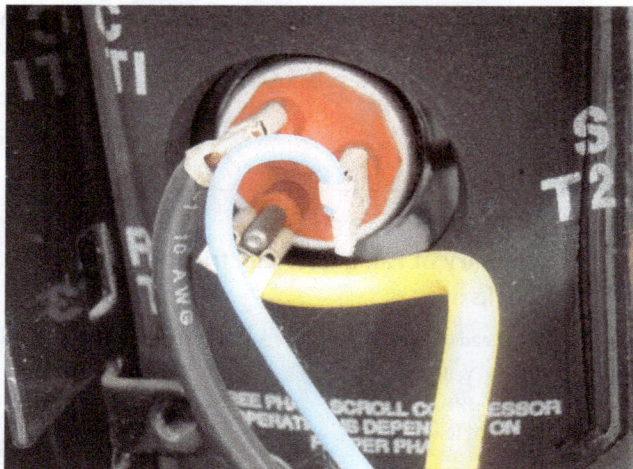

FIGURE 21.9   Heat pump outdoor unit – compressor power lead lugs.

Once the fan and fan housing were pulled up, the cover for the compressor leads was exposed to complete continuity and impedance checks (electrical checks) on the compressor. Results of the electrical checks on the compressor motor are summarized in Table 21.4.

TABLE 21.4
Summary of Compressor Electrical Checks

Wire Leads Checked	Resistance (Ω)
Black – Blue	0.8
Black – Yellow	0.5
Blue – Yellow	1.3
Black to Ground	0.0
Blue to Ground	0.0
Yellow to Ground	0.0

The interpretation of these findings is as follows[16]:

- When a sealed compressor motor has "burned out," this means that the internal wiring of the motor has become irreparably damaged: the compressor motor windings may be burned and shorted together or shorted to the steel shell of the motor, or the windings may have burned and simply become "open" or disconnected. If the motor has burned out in either of these ways, it needs to be replaced.
- When an air-conditioning compressor has "burned out" by shorting of internal components, it will fail to start at all. This failure is detected by disconnecting all power and wiring from the unit and measuring resistance (ohms) between the motor start/common and run/common terminals. If there is zero resistance, the winding is open or broken.
- If you measure the resistance across a compressor winding and the needle on the digital meter is stuck over at infinity, or "OL"/"OVER," that would indicate the compressor winding is open (burned through). The same effect can be observed from simply connecting the meter to absolutely nothing.
- *If the resistance measured across the air-conditioning compressor winding is too close to 0 Ω, it is shorted.* The compressor should blow the fuse or trip the circuit breaker when power is turned back on.
- If there is resistance between the motor terminal and the motor casing, the motor has become shorted to ground internally and the unit needs to be replaced. If there is no resistance between the start and run terminals to common, but there is resistance between the start and run terminals, this means that the internal motor overload protection circuit is open. In this last case, allow the motor to cool and retest it before replacing it.

In this case study, based upon lightning records, the life of this unit, lack of high-voltage damage to the contactor, run capacitor, or wiring, and based on electrical checks of the compressor, the failure of the unit was not likely due to lightning; the motor windings appear to be shorted.

21.3.1.3.2   Lightning Damage Assessment Methodology – Site Inspection
              Example for Damaged Water Well Pump/Motor

See Chapter 23 on Equipment and Installation Failures.

IMPORTANT POINTS TO REMEMBER

- Lightning weather data can be used to determine whether or not lightning struck the location on the reported date of loss.
- Lightning damage normally affects more delicate electrical components such as relays, contactors, run capacitors, and computer electronic boards before impacting less sensitive components such as electrical motors.
- Lightning strikes occur so rapidly that common fuses and circuit breakers cannot react fast enough to protect downstream equipment.
- In general, it is unlikely that commonly reported lightning-damaged equipment such as air-conditioner outdoor unit compressors or well-pump motors were actually damaged by lightning. Most of the time the failures are from other causes.
- Serial numbers can be used to date the equipment and determine if it is at, near, or past its normal useful life.
- Electrical checks to motors and compressors can be used to help determine whether or not lightning was responsible for its reported failure.

REFERENCES

1. University of Florida IFAS Extension. "When Lightning Strikes." Accessed January 31, 2012, http://pasco.ifas.ufl.edu/fcs/Lightning.shtm.
2. NOAA. "Lightning Climatology: Where Does Lightning Usually Strike?" Accessed January 31, 2012. http://www.nssl.noaa.gov/primer/lightning/ltg_climatology.html.
3. NOAA. "Damage and Impacts – Lightning Losses," 2012. http://www.nssl.noaa.gov/primer/lightning/ltg_damage.html.
4. NOAA. "Lightning Victims." Accessed May 8, 2020, https://www.weather.gov/safety/lightning victims.
5. Funk, T. "Lightning: Facts, Fiction, Safety Issues Lightning Formation and Meteorological Considerations." National Weather Service. Accessed January 31, 2012, http://www.crh.noaa.gov/lmk/soo/presentations/Lightning_WFO_LMK.pdf.
6. Eastern US Weather. "Thunderstorm Days per Year by Location." Accessed January 31, 2012, http://www.easternuswx.com/bb/index.php?/topic/152353-thunderstorm-days-per-year/.
7. NOAA. "Thunderstorm Days per Year by Location." Accessed January 31, 2012. http://www.srh.noaa.gov/jetstream/tstorms/tstorms_intro.htm.
8. Noon, R.K. *Forensic Engineering Investigation.* New York: CRC Press, 2001.
9. NASA. "Global Average Annual Occurrence of Lightning." Accessed January 31, 2012, http://www.srh.noaa.gov/jetstream/lightning/hirez_72dpi.htm.
10. International Code Council. "2018 International Residential Code (IRC) – Third Printing." Accessed 8, 2020, https://codes.iccsafe.org/content/IRC2018P3.
11. International Code Council. "2017 International Building Code (IBC – First Printing)." Accessed May 8, 2020. https://codes.iccsafe.org/content/IBC2018?site_type=public.
12. National Lightning Safety Institute (LNSI). "Section 2.3.2: Codes and Standards on Lightning Safety, Updated November 2015." Accessed May 8, 2020, http://lightningsafety.com/nlsi_bus/nlsi_publ.html.
13. National Fire Protection Association, Inc. *NFPA 921–2017: Guide for Fire and Explosion Investigations.* Quincy, MA: NFPA, 2016.
14. HVAC Helpdesk. "Troubleshooting a Capacitor – Troubleshooting Guides." Accessed February 3, 2012, http://www.hvacpartsoutlet.com/troubleshootacapacitor.aspx.
15. IceHouse. "Topic: Heat Pump Electrical Component Check." Accessed February 3, 2012, http://appliancejunk.com/forums/index.php?PHPSESSID=   0aa151be5684a8700fca8b6bd8d30572&topic=2909.msg10999#msg10999.
16. InspectAPedia. "How to Diagnose a Burned-Out Air Conditioning Compressor." Accessed February 3, 2012. http://inspectapedia.com/aircond/aircond15f.htm.

22 Plumbing/Piping/Tubing Failures

Stephen E. Petty and Bryan E. Knepper
EES Group, Inc.

CONTENTS

PURPOSE/OBJECTIVES

The purpose of this chapter is to:

- Explain the differences between the plumbing terms of "tubing," "piping," and "hosing."
- Understand the main types of plumbing failures.
- Understand the methodology for conducting a plumbing failure assessment.
- Following the completion of this chapter, you should be able to:
- Recognize different types of plumbing failures.
- Be able to complete a plumbing failure assessment.

22.1 INTRODUCTION

It is not uncommon for a forensic engineer/scientist to encounter plumbing failures involving tubing, piping, or hosing. These failures generally fall into the following categories:

1. Potable Water Plumbing Failures:
   a. Freeze failures – except for tropical climates.
   b. Installation failures.
   c. Manufacturing defects.
   d. Chemical/corrosion failures.
   e. Mechanical damage (e.g., nail damage).
2. Sanitary Plumbing and Fire Suppression System (FSS) Failures:
   a. Freeze failures (uncommon in sanitary/more common in suppression).
   b. Installation failures.
   c. Corrosion failures.

The purpose of this chapter is to provide an understating of the methodology used when conducting inspections associated with plumbing failures.

22.2 TUBING VS PIPING VS HOSING?

Most new forensic engineers or scientists have not been formally trained on plumbing systems and nomenclature unless they worked earlier in the trades, construction, or grew up in rural areas. Plumbing components consist of "tubing," "piping," or "hosing" along with associated fittings, valves, control items, and other components. The "tube," "pipe," or "hose" consists typically of a hollow section with a known inside diameter (ID), outside diameter (OD) forming a known wall thickness used for the transfer of products including fluids (liquids and gases) and solids (e.g., powders); for most forensic investigations, the fluid is water (potable) or water and solids (sanitary lines for materials such as sewage). The pressure drop for a given length, inside diameter of the "tubing," "piping," or "hosing," and surface roughness determine flow. Key differences among "tubing," "piping," and "hosing" are summarized in Table 22.1.

TABLE 22.1
Typical Differences among the Terms "Tubing," "Piping," and "Hosing"[1-21]

	Differences by Term		
Parameter	**Tubing**	**Piping**	**Hosing**
Rigid or flexible	Rigid	Rigid or flexible	Flexible
Designation based on	Outside diameter typically in inches. Also defined as Birmingham Iron Wire Gage (BWG) with values from 1 (0.012″) to 30 (0.30″)	Nominal Pipe Size (NPS), Nominal Diameter, or Trade Size. Defined as Schedules (10, 20, 30, 40, 60, 80, 100, 120, 140, and 160) and also with the terms Standard Wall Designated (STD), Extra Strong Wall (XS), Extra Heavy Wall (XH), Double Extra Strong Wall (XXS), and Double Extra Heavy Wall (XXH)	Inside diameter

(Continued)

TABLE 22.1 (*Continued*)
Typical Differences among the Terms "Tubing," "Piping," and "Hosing"[1–21]

Parameter	Differences by Term		
	Tubing	Piping	Hosing
Materials of fabrication	Aluminum, brass, carbon steel, copper, plastic, nickel alloy, and stainless steel. Copper tube types K, L, and M are available in annealed (soft) temper copper or drawn (hard) temper copper and DWV. Type K copper has the heaviest wall thickness and is color-coded green. Type L is the intermediate wall thickness tube and is color-coded blue. Type M is the thinnest wall thickness of pressure tube and is color-coded red. DWV (Drain, Waste, Vent), a non-pressure copper tube, is color-coded yellow. ACR (Air-Conditioning, Refrigeration) is color-coded blue OXY, MED OXY/MED OXY/ACR ACR/MED (Medical Gases/Vacuum) is color coded green for K and blue for L. Plastic tubing consists of Polyethylene, Polypropylene, Cross-linked polyethylene (PEX), and Polyvinyl chloride (PVC)	Carbon steel, glass, lined metal, iron, non-ferrous, and plastic (e.g., PVC and CPVC).	Buna-N, butyl, ethylene neoprene, propylene diene rubber (EPDM), and Teflon™
Wall thickness change to handle pressure	OD fixed and ID shrinks or becomes smaller	OD fixed and ID shrinks or becomes smaller	Grows outward – OD increases – even though often referred to as a pipe
Shapes	Round, rectangular, square, oval, elliptical, hexagonal and octagonal	Typically round	Typically round
Typical diameters	~1/4″ to ~12″ for K, L, M; ~1–14/4″ to ~8″ for DWV	Up to 80+″	Up to 5+″
Terminations	Threaded and grooved ends (for quicker connections)	Most commonly beveled, plain and screwed ends	Threaded and grooved ends (for quicker connections)
Partial listing of standards	ASTM B88, ASTM B326, ASTM B280, ASTM B819 PEX (ANSI/NSF 14, ANSI/NSF 61, ASTM F876, ASTM F877, ASTM F1281, ASTM F1807, ASTM F1281)	ANSI B36.10	ASTM D3767, SAE J-517

Tubing systems typically transport fluids (gases and liquids). Type K or green copper tubing is the only one suitable for burial and is often used to bring the water supply into a home. Type K tubing is thicker than the other types available and therefore is able to withstand pressure from backfilled earth in trenches. Types L and M make up the water supply pipes in homes. The International Building Code allows both L and M, as well as type K for use as water supply tubing. However, local codes often specify the use of type L and/or M for use in a home. Type L copper tubing is the most common type as it can be used in more applications than type K and is thicker than type M. Other types of copper plumbing are type DWV or "drain-waste-vent," copper-coded green, which are used in systems that allow air into plumbing systems; ACR or "Air Conditioning, Refrigeration," which is used in HVAC systems and color-coded blue; and

OXY, MED OXY/MED OXY/ACR ACR/MED or "Medical Gases/Vacuum," which are typically used in medical facilities. An excellent source of information on copper tubing is the Copper Development Association, Inc. (https://www.copper.org/resources/).

Piping is defined as tubular vessels used in pipeline and piping systems commonly transporting fluids (gases or liquids) and solids. Piping is specified by "Nominal Pipe Size" (NPS) and Schedule (wall thickness). NPS is a size standard established by the American National Standards Institute (ANSI); once the pipe size reaches an OD of 12″, the NPS reflects the actual OD.

The differences in outside diameter (OD), inside diameter (ID), and wall thickness (T) for ¾″ round tubing and piping are illustrated in Table 22.2.

TABLE 22.2
Illustration of Differences between Tubing and Piping Dimensions for 3/4″ Sized Tubing or Piping (Round)

Type of 3′4″ Tubing or Piping	Dimensions in Inches		
	Outside Dimeter (OD)	Inside Diameter (ID)	Wall Thickness (T)
¾″ Round Copper Tubing			
Type K	0.875	0.745	0.065
Type L	0.875	0.785	0.045
Type M	0.875	0.871	0.032
¾″ Round Steel Piping			
Schedule 5	1.050	0.920	0.065
Schedule 10	1.050	0.884	0.083
Schedule 40	1.050	0.824	0.113
Schedule 80	1.050	0.742	0.154

Interestingly, copper tubing for residential plumbing (or "plumbing tube") generally has a nominal size that is 1/8″ smaller than the actual OD. For copper tubing used in air-conditioning and refrigeration (ACR), the tubing is sized by the actual OD. Sometimes plumbers, engineers, and others mistakenly believe that NPS refers to the ID on smaller pipes, but as illustrated in Table 22.2, this is not the case.

Plumbing failure inspections typically involve the failure of tubing systems containing water. Upon failure, large amounts of water can be released into a structure as illustrated in Figure 22.1.

Flow GPM	Nominal or standard size, inches																	
	1/4			3/8			1/2			3/4			1			1 1/4		
	K	L	M	K	L	M	K	L	M	K	L	M	K	L	M	K	L	M
1	0.138	0.118	N/A	0.036	0.023	0.021	0.010	0.008	0.007	0.002	0.001	0.001	0.000	0.000	0.000	0.000	0.000	0.000
2				0.130	0.084	0.075	0.035	0.030	0.024	0.006	0.005	0.004	0.002	0.001	0.003	0.001	0.000	0.000
3										0.014	0.011	0.009	0.003	0.003	0.001	0.001	0.001	0.001
4							0.125	0.106	0.086	0.023	0.018	0.015	0.006	0.005	0.004	0.002	0.002	0.002
5							0.189	0.161	0.130	0.035	0.027	0.023	0.009	0.007	0.006	0.003	0.003	0.002
10										0.126	0.098	0.084	0.031	0.027	0.023	0.010	0.010	0.009
15													0.065	0.057	0.049	0.022	0.020	0.018
20													0.096	0.084		0.037	0.035	0.031

FIGURE 22.1 Loss of water from K, L, or M tubing. (From Copper Tubing Handbook.[6])

As illustrated in Figure 22.1 for typical pressure drops, the failure of a ¾" tubing will release between 1 and 5 gallons per minute (GPM) from the failed line. Using range of values, a 1–5 GPM leak corresponds to the release of 1,440–7,200 gallons/day, 10,080–50,400 gallons/week, and 302,400–1,512,000 gallons/month depending when the leak is discovered and the water is turned off. If the leak situation is discovered early, damage may be limited to the drywall, floor coverings, and local furnishings. However, if the leak continues for days to weeks or more, the damage will increase potentially resulting in mold formation on surfaces, damage to appliances, and potentially structural damage.

Cross-linked polyethylene (PEX) products were first developed in ~1965 and introduced commercially in the United States and Canada in ~1980. PEX gained popularity over traditional piping/tubing due to its flexibility, ease of installation, no risk of fire during installation from joining, greater water pressures at fixtures due to fewer sharp turns necessary, and due to being less likely to burst from freezing. However, as with any new product, there are several flaws with PEX including degradation in direct sunlight, the potential for perforation during installation or by insects, and chemical reactions between various fittings and "hard water." In 1994, IPEX introduced products and fitting known as KITEC, which by mid-1995 were found to fail under certain conditions.[7–11] One major condition documented included "yellow brass" fittings (~30% zinc) used to connect KITEC, which would leach into the mineral-rich "hard water" and cause a reaction known as dezincification. Newer products contain fittings with ~5%–10% zinc, which mitigates this reaction. KITEC products were recalled in 2005. PEX has been produced using the following four main methods[8]:

- *PEX-A or PE-Xa*: Developed by Engel, PAM, Frankische, and others, uses peroxides to provide a very high level of molecular cross-linking. PEX-A is the most flexible and the most kink-resistant of these forms of polyethylene tubing.
- *PEX-B or PE-Xb*: Developed by DOW Corning/Sioplas. PEX-B uses silane (a catalyst) and steam after extrusion to form a high-density polyethylene. PEX-B is not expandable and has a lowered kink resistance. Do not try to connect PEX-B by shoving its end over a fitting that would cause expansion of the tubing. Doing so would lead to a plumbing failure and leaks.
- *PEX-C or PE-Xc*: PEX-C uses gamma radiation or an electron beam to create the polymer cross-linking.
- *PEX-D or PE-xd*: Developed by AZO (Gustavsberg-Lobonyl). Not currently in production.

KITEC later produced a PEX sandwich with a layer of aluminum inside (PEX-AL-PEX or PE-AL-PE).[9,10] PEX has been approved in all 50 states and Canada including California, which approved its usage in 2009.[12]

PEX IDs typically are 1/4", 3/8", 1/2", 3/4", 1" up to 2" (5, 10, 15, 20, 25, up to 50 mm); the most popular dimensions used are 3/8" and 3/4".[13,14] These products were manufactured in the following colors: black, blue, gray, green, orange, red, yellow, and translucent; red is often seen for the hot water line and blue for the cold water line.[8] PEX products cannot be: (1) stored or installed in areas where it will be exposed to sunlight, either direct or indirect, or (2) used with fittings under slabs, unless protected.[15]

Based on hundreds of inspections, plumbing failures mostly are found to be associated with tubing failures associated with freeze failures from lack of heat and/or insulation, manufacturing defects of fittings, or from installation (failure to tighten connections or overtightening connections). These situations, and others, will be discussed within subsequent sections of this chapter.

22.3   EXAMPLES OF PLUMBING FAILURES

Illustrations of the following plumbing failures are demonstrated in the following sections:

- Plumbing Freeze Failures.
- Plumbing Installation Failures.
- Manufacturing Defect Failures.
- Chemical/Corrosion/Leaching Failures.
- Mechanical Damage Failures.
- Dishwasher and Icemaker Failures.

22.3.1   PLUMBING FREEZE FAILURES

In colder climates, freeze failures of tubing and piping are the most common plumbing failures seen. While the type of failure is known as a "freeze" failure, it is actually a "freeze/thaw" failure. For the tubing or piping to fail, it must first be subjected to subfreezing temperatures. Interior pipes are susceptible to freezing when the outdoor temperature drops below ~20°F. This temperature is critical for two reasons[16]: (1) the temperature of an unheated portion of a home is almost always a few degrees above the outdoor temperature due to heat transfer from heated portions of the home and from the exterior materials' R-values, and (2) water supercools several degrees below freezing before any ice begins to form. As water freezes and expands, the pressures exerted on plumbing are significant; for example, pressure failure limits range from ~200 to ~1,800 psi depending on the type and diameter of the tubing.[6] The "thaw" part of the equation occurs typically several days after the freeze and expansion of the piping when temperatures rise above freezing and the ice blockage thaws. The water service then returns and leaks at the "burst" section of the piping or tubing. Bursts can also occur at weak locations in the piping system, including soldered joints. In this case, the components such as a coupling or elbow separate from the main tubing. The most likely scenario for a freeze/thaw burst to occur is a vacant home that has not been provided sufficient heat over the winter or that has not had the plumbing system winterized properly. It is not uncommon for a forensic expert to be asked to perform a gas usage analysis to determine whether sufficient heat was provided to a home. Examples of freeze/thaw failed tubing and piping are illustrated in Table 22.3.

TABLE 22.3

Illustrations of Freeze Failures to Tubing and Piping

Scenario	Photograph	Comments
Residential home Basement near sill plate and Rim Joist to Hose Bib 1/2″ copper tubing		Classic Split Freeze Failure to Copper Tubing Burst Failure Longitudinal on Tubing and Metal Moved Outwards
Commercial car wash facility 1-1/2″ copper tubing		Classic Split Freeze Failure to Copper Tubing Burst Failure Longitudinal on Tubing and Metal Moved Outwards Outside Temperature Dropped to −1°F
Residential home Second floor bathroom 1/2″ copper tubing		Classic Split Freeze Failure to Copper Tubing Burst Failure Longitudinal on Tubing and Metal Moved Outwards Note Corrosion on Copper Tubing Below Likely from Leak Above
Commercial building Fire suppression system Sixth floor – wall next to stairwell ~2″ schedule 40 carbon steel pipe (note leaks of water from split in pipe)		1st Floor Heated and Used for Offices; Upper 5 Floors Not Heated and Used for Storage Burst Longitudinal on Pipe (classic split failure) Failed section in CMU Wall Failure Outside Temperature Dropped to −10°F

(Continued)

TABLE 22.3 (*Continued*)
Illustrations of Freeze Failures to Tubing and Piping

Scenario	Photograph	Comments
Commercial building – 23,541 square feet three-story hotel Fire suppression system – ~1-1/4″ sock-it tee – carbon steel Attic space - ~12″ below roof decking		The first suppression system was designed to operate dry – however, the line had a low spot in the line that held residual water – which froze. Burst failure from freezing of the water coupled with some corrosion. Record Minimum Temperatures of –3°F & –6°F set at time of Failure. No insulation between roof decking and piping
Commercial car wash facility ~1″ PVC piping	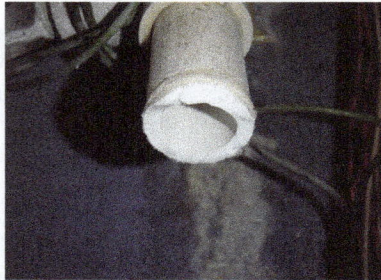	Classic Shatter Failure of Plastic Pipe Outside Temperature Dropped to –1°F
Residential villa of major golf figure 1″ CPVC (~1.3495″, the inside diameter of the pipe was ~1.0870 with a wall thickness of ~0.1150″) Fire suppression system to outdoor deck/grill	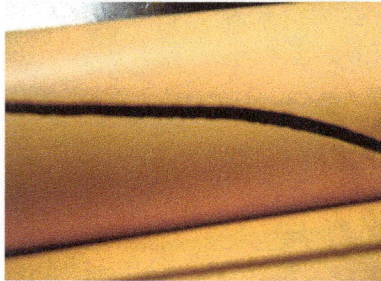	State and National Codes did not require fire suppression system for grill on deck in a residential structure but local codes officials required its installation after initial construction. System had failed twice in the past on low-temperature events. Outside temperature dropped to –7°F for two consecutive days during this failure event. System was a dry system from the outside into the inside ceiling space where a valve separated the wet side from the dry side. Analysis determined that the valve was too close to the exterior wall, allowing the water on the wet side of the valve to freeze
High-rise building Line in plenum space	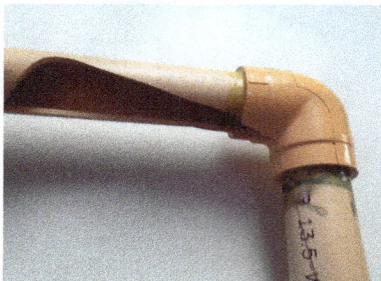	Failure Near Elbow

(Continued)

TABLE 22.3 (*Continued*)
Illustrations of Freeze Failures to Tubing and Piping

Scenario	Photograph	Comments
Residential home ~3/8″ PEX tubing Line to sink		Home Subject to Power Failure at the Time of the Loss Burst Failure Longitudinal on Tubing and PEX Material Bulging Outwards Outside Temperature Dropped to –5°F

As illustrated in Table 22.3, for metal tubing and piping, freeze failures are associated with longitudinal splits. Moreover, for copper tubing, the splits bow outward, whereas for carbon steel, the piping simply splits. Plastic materials (i.e., PVC and CPVC) tend to crack longitudinally or in spiral-type fractures and then shatter when subject to freeze failures.[22] PEX products tend to bulge and then tear or rip when subjected to freeze failures.[23] However, PEX freeze/thaw failures are less common.

22.3.2 PLUMBING INSTALLATION FAILURES

Plumbing installation failures are a second cause for pluming failures. Examples of plumbing installation failures are illustrated in Table 22.4.

TABLE 22.4
Illustrations of Plumbing Installation Failures

Scenario	Photograph	Comments
Residential home Kitchen sink – hot water line 1/4″ Copper tubing separated from copper fitting		Compression fitting failure Nut in compression fitting not tightened sufficiently; ferrule not retained on separated tubing – If sufficiently tightened, the ferrule would bite onto the tubing and could not be removed
Residential home Bathroom sink – hot water line 1/4-inch CTS cross-linked polyethylene (PEX) tubing separated from stainless steel valve		Compression fitting failure Nut in compression fitting not tightened sufficiently; ferrule not retained on separated tubing – if sufficiently tightened, the ferrule would bite onto the tubing and would not pull loose

(*Continued*)

TABLE 22.4 (*Continued*)
Illustrations of Plumbing Installation Failures

Scenario	Photograph	Comments
Shark-bite fitting failure		Not tightened properly
Residential home – rental Bathroom tub – hot water line 1/2″ copper tubing separated from copper fitting		Solder joint failure Corroded soldered surface; a properly soldered joint, if heated and removed would have a silvery coating of solder. This failure typically reflects improper cleaning of surfaces prior to soldering. Mating pipe will often show longitudinal scratches where coupling slipped away from pipe
Elementary school Fire suppression system FireLock EZ couplers (Style 009H) Associated with 2-inch diameter, schedule 10, steel pipe		Coupling failure Failure to tighten bolts with nuts properly Mating pipe will often show longitudinal scratches where coupling slipped away from pipe
Bathroom toilet water supply line fitting Overly hand-tightened		From Pimpanaro et al. – analysis of acetal toilet valve supply nut failure https://www.sciencedirect.com/science/ article/pii/ S2213290216300189

As illustrated in Table 22.4, compression fitting and coupling failures are generally caused by insufficient tightening of the fitting or coupling. Solder failures typically are caused by insufficient cleaning of the tubing/fitting/valve surfaces prior to soldering. Overtightening of components such as couplings, valves, or hose connections can induce cracked threads or produce microfractures in the joining pipe. Under-tightening of components is more common and is typically obvious immediately after turning on a water supply; overtightening typically occurs when homeowners may tighten connections at toilet supply lines or between the supply line and solenoid valve of a dishwasher.

22.3.3 Manufacturing Defects

Due to quality control procedures, manufacturing defects are rare, but they do occur. Examples of manufacturing defects include but are not limited to an incorrect size or defective O-ring within a valve, cracked threads due to mishandling, or improper materials used during the manufacturing process, allowing for premature failures. Typically, manufacturing defects show up in the first 1–3 years of usage although some piping manufacturers provide warranties of their products up to 25 years after the manufactured date. One of the more widely known manufacturing defects in the plumbing industry involved the Watts Regulator Company. A class-action lawsuit was filed against Watts due to a stainless-steel braided hose ("Floodsafe") used for water supply lines bursting catastrophically within homes. This was determined to be from the inner tubing of the hoses being manufactured from a thermoplastic polymer that degraded as it was exposed to hard metals within most public water systems.[24]

22.3.4 Chemical/Corrosion Plumbing Failures

Chemical/corrosion plumbing failures are less commonly encountered by forensic engineers/scientists, but are seen occasionally. Examples of chemical/corrosion failures are illustrated in Table 22.5.

TABLE 22.5
Illustrations of Chemical/Corrosion/Leaching Plumbing Failures

Scenario/Comments	Photograph - Far	Photograph – Close-Up
Residential home Basement – water heater – cold water supply line Brass fitting between 3/4″ Rigid copper tubing and flexible tubing Cause of leak: pit corrosion cracking		
Residential home Basement – main water supply lines Copper tubing – type 'M' Cause of leaks: Pit corrosion inducing perforations due to exposure to hard water		

(Continued)

TABLE 22.5 (Continued)

Illustrations of Chemical/Corrosion/Leaching Plumbing Failures

Scenario/Comments	Photograph - Far	Photograph – Close-Up
Residential home Basement – hot water supply line Nominal ~1/2″ (~5/8″ OD) Polyethylene (PEX) tubing Cause of leak: likely water chemistry		
Residential home Orangeburg sewer line outside front of home Nominal ~6-5/8″ diameter but collapsed Cause of lack of flow: Leaching/degradation of pipe – no longer used after around the mid-1970s "Orangeburg" pipe refers to the brand name of a sewer pipe made by the Orangeburg Manufacturing Co. of Orangeburg, New York. Its generic name is "bituminous fiber pipe" and is manufactured from a combination of cellulose and asbestos fibers impregnated with a bituminous (coal tar) compound		

As illustrated in Table 22.5, corrosion, chemical, and leaching can cause failures of plumbing. Pitting corrosion of metal plumbing is typically observed; literature suggests that such corrosion is the effect of water chemistry and erosion of a surface. Copper plumbing, used since the 1930s, can fail through several mechanisms[21,25–27]:

- *Cold water pitting*: This type of failure is associated with well or other ground waters containing free carbon dioxide in conjunction with dissolved oxygen. Pits develop from the inside of the tube and typically have a blue/green tubercle or hollow mound of corrosion products over the pit.

- *Cold water type I pitting (pH from 7.0 to 7.8)*: The pitting is deep and narrow; in many cases, a carbon film or silica scale forms with this type of pitting. Factors associated with this type of corrosion include the following:
  - Alum coagulation.
  - Deposits within the pipe, including dirt or carbon films.
  - High chlorine residuals.
  - Stagnation early in pipe life.
  - Water softeners.
- *Cold water type III pitting (pH above 8.0)*: This form of pitting is more generalized. It tends to be wide and shallow and results in blue water, by-product releases, or pipe blockage. Factors associated with this type of corrosion include the following:
  - Alkaline water.
  - Alum coagulation.
  - Stagnation early in pipe life.
- *Hot water type II pitting (pH below 7.2)*: The pitting is narrower than in type I and rarely occurs at temperatures below 140°F. Factors associated with this type of corrosion include the following:
  - Alum coagulation.
  - High chlorine residuals.
  - Higher temperatures.
- *Flux corrosion*: A process that also results in pitting corrosion. It is associated with excess usage of solder fluxes and poor workmanship (e.g., insufficient heat), which results in corrosive waxy residuals in the tube. The effect is seen more often in cold water lines than in hot water lines because the hot water tends to dissolve and carry away the flux residuals. Leaks due to flux corrosion typically develop after 1–5 years.
- *Erosion corrosion:* Normally the inside of the tubing or piping will contain a protective oxide film. In situations where excessive localized water velocity or turbulence exists, the film can be stripped away, leaving the area bright and shiny with horseshoe-shaped pits present. Conditions that may cause excessive local velocities (design ranges are from ~4 to ~8 fpm) include: (1) failure to deburr (ream or smooth) the inside edge of a tube after cutting and (2) too many abrupt changes in flow direction (e.g., elbows, tees, bends, sections of too small of tubing or too large of pumps – cavitation).
- *Hot water recirculating systems*: Require special mention. Excessive velocity in such systems is a common cause of erosion corrosion and failure. Installations that use small sizes of tube or too large pumps result in higher than recommended flow rates. Operating temperature and water chemistry must also be taken into consideration.
- *Certain soils for underground copper lines*: While copper tubing and piping do not corrode in most clays, chalks, loams, sands, and gravels some, such as cinder fill containing sulfur, will corrode copper.
- *Pipe electrolysis*: While less common, stray direct current (DC) electricity causes underground pipe to corrode, especially piping with high zinc content. Electrolysis can also happen to galvanized pipe. This process is known as "galvanic action."[28]

Cuprosolvency: Occurs in soft waters with low hardness, low alkalinity, and a pH of 7 or lower. It causes a blue/green color in the water and staining of plumbing fittings or laundry. The general dissolution of copper tube associated with cuprosolvency is a very slow process that thins the tube but does not usually result in failure of the wall of the tube.

PEX plumbing and fitting failures have been observed and are associated with the following conditions[29-35]:

- *Dezincification*: This is the leading cause of failure PEX installation and is caused by the leaching of zinc from brass fittings. In PEX installations, when water flows through the brass (brass is a copper–zinc alloy with zinc contents ranging from ~15% to ~40%) fitting, the zinc leaches from the brass creating a powdery buildup. This buildup can cause a blockage within the fitting. Moreover, when the zinc leaches from the copper, it leaves the copper very porous and mechanically weakened, which then can cause the fitting to leak and possibly rupture. PEX manufacturers have recommended that the zinc content of brass (copper-zinc alloy) be reduced to ~15%–19% of the total alloy, but since copper is more expensive than zinc, this has been resisted.
- *Chlorine in water*: Interestingly PEX piping in Europe also does not appear to fail as much as that in the United States. This is because the chlorine levels to disinfect potable water in the United States are much higher than that used in Europe. Attempts have been made to add antioxidants to PEX, but it is consumed and then failures occur.
- *Heat*: Elevated temperature can thermally degrade PEX. For example, best installation practices recommend that PEX be kept at least 12″ from lighting and other heat sources.
- *Sunlight*: PEX degrades due to the UV radiation from sunlight if exposed to the sun before, during, and after installation.

Coal Tar Impregnated Wood Fiber Pipe (Commonly referred to as "Orangeburg," "Bermico," etc., pipe) was used for sewer piping until the early to mid-1970s when it was displaced with PVC.[36-38] While no longer produced and used, millions of linear feet of the product used remain within building structures and under yards, streets, and roads. Fibre Conduit Company, the name of which changed to the Orangeburg Manufacturing Company in 1948, was the largest producer of this type of pipe, thus the name "Orangeburg" pipe. It was produced in sizes from 2″ to 18″ ID mostly for sewer and drain applications. A perforated version was manufactured for leach fields and for drain tile for farm fields. Joints were made with couplings of similar material and sections simply compressed together; no gaskets or joint sealants were used. It was relatively inexpensive, lightweight (but brittle), and could be cut with a simple carpenter's saw. On the other hand, the pipe was vulnerable to root intrusion and degradation due to leach of the coal tar and decay of the cellulose fibers forming the pipe.

22.3.5 MECHANICAL PLUMBING FAILURES

Mechanical plumbing failures are the least commonly encountered by forensic engineers/scientists, but are also seen occasionally. Examples of mechanical plumbing failures are illustrated in Table 22.6.

TABLE 22.6

Illustrations of Mechanical Damage Plumbing Failures

Scenario/Comments	Photograph – Far	Photograph – Close-Up
Residential home Basement – water heater – cold water supply line brass fitting between 3/4″ rigid copper tubing and flexible tubing Cause of leak: pit corrosion cracking		
Condo Laundry room – Cold water line Nominal ~1/2″ (~5/8″ OD) Polyethylene (PEX) tubing Cause of leak: Mechanical damage – scraping against metal stud		

As illustrated in Table 22.6, pitting corrosion of metal plumbing is typically observed; literature suggests such corrosion is the effect of water chemistry and erosion of a surface. PEX plumbing degradation has been associated with water chemistry, but the specific cause(s) continues to be debated.

22.3.6 DISHWASHER AND ICEMAKER PLUMBING FAILURES

As illustrated in Figures 22.2 and 22.3, failures of dishwasher and icemaker systems are generally associated with the solenoid valves controlling water to these systems.

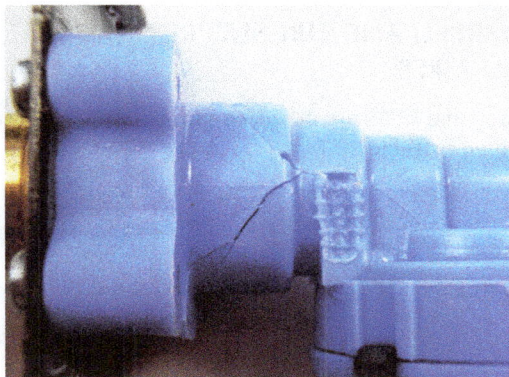

FIGURE 22.2 Crack in dishwasher water solenoid valve.

FIGURE 22.3   Crack in icemaker water control valve.

As illustrated in Figures 22.2 and 22.3, water leaks from dishwashers and icemakers are often associated with cracks in the water control valves. These cracks occur as a result of overtightening attached plumbing lines or torque on the valve associated with these same lines. Sometimes evidence suggests flaws in the material(s) of construction, but this is often very difficult to prove.

22.4   SANITARY PLUMBING AND FIRE SUPPRESSION SYSTEM (FSS) FAILURES

Failures of sanitary plumbing and FSSs are not as common for various reasons. Examples of these types of failures are illustrated in Table 22.7.

TABLE 22.7
Illustrations of Sanitary and Fire Suppression Failures

Scenario/Comments	Photograph	Comments
Residential home Sanitary drain pipe within exterior wall, hidden until Interior finishes removed Cause of leak: Heavy corrosion to pipe due to caustic effects of sewage and pitting corrosion		This failure was inspected to determine whether it was a long-term leak or from a recent freeze/thaw failure. Note that sanitary drain lines are typically gravity drains, which would not contain stagnant water that would be able to freeze
Commercial warehouse dry fire suppression system 4″ nominal pipe Cause of leak: Interior corrosion due to improper drainage and sagging of piping		The 4″ diameter pipe was 50% filled with scaled sections of piping distributed within the system from years of exposure to moisture within the "dry" system. Corrosion = water + iron + oxygen
Commercial warehouse wet fire suppression system Cause of leak: freeze/ thaw failure at main valve; insufficient heat provided to vacant facility		It is not uncommon for building owners to forget that a fire suppression system is not supplied by water from the potable water service lines. Therefore, the FSS has to be shut-off and winterized just like regular water supply systems. However, even if a building is vacant, most jurisdictions require that fire suppression be provided to the building to prevent the spread of fire between properties. This makes it paramount to maintain sufficient heat within a commercial building

One main reason for less common failures is the amount of inspections performed on these types of systems. When a sanitary system is not working properly, it is typically inspected almost immediately by a plumbing professional. This is probably due to the nature of the materials being handled in the lines. In the same respect, FSSs are required by local jurisdictions and state authorities to be properly maintained and inspected periodically. However, even with the amount of publicity these systems receive, failures can still occur. The main types of failure for a sanitary plumbing system are corrosion and lack of maintenance. The main types of failure for an FSS are corrosion and freeze/thaw failures.

22.4.1 SANITARY PLUMBING FAILURES

Prior to the advent of PVC and CPVC plumbing and components, most sanitary lines were made from cast iron or carbon steel. Further, most sanitary piping is hidden within walls and for good reason.

With the exception of basements, sanitary plumbing systems are not normally visible to the occupants of a home. The problem with unseen piping is unseen deficiencies. Cast iron piping installed in the 1920s may not look the same in the 2020s as both exterior and interior moisture and the effects of caustic materials on the inside of the pipes/components take their toll. An extreme example of this type of failure is provided in Table 22.7. The other major failure mode for sanitary plumbing systems is a lack of maintenance. This could include a failure to mitigate minor leaks when they first occur (i.e., leaks at fittings/threaded connections) and from a failure to clean out systems when initially clogged. Another typical issue that causes sewage backups is tree roots at the exterior of the home growing into the sewer lines and thus reducing the capacity of the system. Installation failures are one of the least common failures for sanitary lines as they are typically tested once installed. Installation failures include misthreading at joints, not adhering pipe properly in the case of PVC lines, and sag (or "bellies") in lines that allow for waste to remain stagnant causing clogs.

22.4.2　Fire Suppression System Failures

The two most common failures for FSSs are corrosion failures and freeze/thaw failures. Interestingly, the heaviest corrosion damages typically occur in "dry" suppression systems. These are systems that only activate when a pressure release occurs at a pendant, which induces a pressure drop on the system and then sends water from a centralized location to the affected area(s). A recent study into dry suppression systems revealed that ~73% had significant corrosion issues.[39] The main issue with "dry" systems is that they are never completely dry. This is due to trapped water from hydrostatic testing combined with humid air supplied by air compressors for the system. In general, corrosion requires three elements: water, iron, and oxygen. A small amount of water remaining in a "dry" system combined with the oxygen in the lines is a recipe for disaster.[40] Table 22.7 shows the detrimental effects of an improperly drained "dry" system. The most catastrophic failure for a commercial building, however, is a freeze/thaw burst to the FSS. Once the thaw portion of the freeze/thaw occurs, it can produce untold damage to the interior of a building as the pressures on an FSS are much greater than regular potable water supply systems (typically between 50 and 150 psi). One of the most common causes for an FSS failure is a building owner not recognizing that the water being supplied for fire suppression is on a different supply line than the main potable water system. FSSs are tied into the same lines as the public hydrant systems and are required (in most cases) to remain active, even in vacant facilities or warehouses. However, forensic experts typically see vacant facilities with widespread damages due to the lack of sufficient heat supplied to a facility/warehouse over each winter. Table 22.7 shows one of these such catastrophic failures. Installation failures for FSSs are also rare but can occur typically from mis-sizing of components, placing sprinklers too close to ceilings or objects that may get hot on a regular basis as part of operation, the placement of equipment behind obstacles, and improper testing.

22.5　PLUMBING/PIPING/TUBING FAILURES ASSESSMENT METHODOLOGY – SITE INSPECTION

The site inspection should follow the overall approach outlined in Chapter 1 and the water damage assessment methodologies outline in Chapters 6–10. Additional steps to include in plumbing failure assessments during the site inspection are as follows:

- During the interview, ask the site representative when the leak was discovered (date and time) and their best estimate of the ranges of times when it likely occurred (data, time).
- During the interview, ask the site representative where they first observed evidence of the leak, what the evidence was (e.g., water on floor(s), wall(s) or ceiling(s); staining on

surfaces, noise, steam – hot water, and how the leak was stopped if stopped). In addition, ask them if they know specifically where the leak occurred and if so, to show you the location.

- During the interview, ask the site representative whether any modifications were made to the plumbing system or component and whether any past repairs were made to indicate a historic or recent issue.
- Begin the inspection from the lowest level (e.g., basement, crawlspace, first floor, floor below the known leak location). Trace signs of water damage (e.g., staining on walls, rust on metal surfaces such as nail heads), document and verify with a moisture meter.
- Locate and verify the leaking plumbing item.
- Take detailed measurements and photographs of the leaking component, its location, its location from an exterior wall and nearby construction including insulation.
- Collect details of water-damaged materials (e.g., type, amount, condition, location).
- Take IAQ measurements (temperature, CO, CO_2, and relative humidity), moisture meter measurements, and FLIR measurements to demonstrate wet and/or cold areas.
- Always collect leaking component; photograph location before and after removal of the evidence. Let the owner or site representative know that the component has been removed so the system is not reactivated leading to further potential damage.
- Photograph the collected evidence.
- Once arriving back at the office, log in the evidence using a Chain-of -Custody form (Figure 22.4).

EES GROUP
ENGINEERING & EXPERT SERVICES

Evidence Form

Date of Loss	Claim No.	Storage Location	EES Project No.
2/15/2008	Aug-05	Garage	8-045-S

Description of Evidence	
1	1/2" copper pipe - from cold water side - with valve and flexible tubing
2	1/2" copper pipe - from hot water side - with valve and 3/8 copper tubing

Released ____________

Destroyed ____________

Removed From Scene by Investigator: Stephen Petty

Received By Investigator From: Received by Stephen Petty - taken from home about 2:40 PM
Address: 3967 Fieldcrest Drive, Beaver Creek, OH 45431

Date Received: 2/26/2008 - ~2:40 P.M.

Date Stored: 2/26/2008

How Evidence Received: In person; removed from site

Received By: Stephen Petty

Evidence Transfer			
Date	Time	Samples Relinquished By	Samples Received By
2/26/2008	~2:40 P.M.	Obtained from home by	Stephen Petty - EES Group, Inc.
		Stephen Petty, P.E., C.I.H.	

Client Information: Ohio Mutual Insurance Group
Mr. Tom Pfeifer
P.O. Box 997
Washington Court House, OH 43160
(740) 333-3405

754-220-8844 phone
Engineering & Expert Services
EES Group, Inc.
1701 E. Atlantic Blvd., Suite 5
Pompano Beach, FL 33060
www.eesgroup.us
754-220-8783 fax

FIGURE 22.4 Example of chain-of-custody form.

22.5.1 PLUMBING/PIPING/TUBING FAILURES ASSESSMENT METHODOLOGY – ANALYSIS OF INFORMATION COLLECTED

In preparation of the written report, the following information should be collected and resulting analyses completed depending on the specific plumbing failure:

- As part of the report preparation, check pertinent Codes and Standards, Industry Best Practices, and Manufacturer's Installation Instructions or other documents related to failed component. Compare the failed component installation and location vs these documents to determine if the installation met Codes and Standards, Industry Best Practices, and Manufacturer's installation instructions.
- Consider obtaining local water service records for the past ~15 months to determine increased rate of water usage at the property. Compare any differences with estimated loss of water determined by hole or crack size, pressure drop, and likely time of failure to potentially validate water loss.
- For plumbing freeze/thaw failures, obtain local weather data for the range of days during which time the failure may have occurred. This often provides a date where the temperature reached a minimum where the freeze initially occurred, causing the split or break of the line/component and the subsequent thaw allowing for the release of water into a home or building (Figure 22.5). Check to see if this temperature was an extreme for the year and/or date and whether or not the failure location was in an uninsulated or poorly insulated space. This may also include an energy usage analysis to determine the amount of gas or electricity used within the home during a certain period (verified with the local utility company) and whether that usage was enough to maintain sufficient heat within the property. Further, water usage can be verified with the local water department to determine how long the leak may have occurred.

```
                                        MONTH:     JANUARY
                                        YEAR:      2014
                                        LATITUDE:  40  0 N
                                        LONGITUDE: 82 53 W


      TEMPERATURE IN F:        :PCPN:   SNOW:  WIND        :SUNSHINE: SKY     :PK WND
   ===============================================================================
    1   2   3   4   5  6A  6B    7     8   9   10   11  12  13   14  15   16    17  18
                                          12Z  AVG MX 2MIN
   DY MAX MIN AVG DEP HDD CDD  WTR   SNW DPTH SPD SPD DIR MIN PSBL S-S WX    SPD DR
   ===============================================================================

    1  40  23  32   2  33   0 0.00   0.0   0  4.0 10 180   M    M   8         13  20
    2  30  11  21  -9  44   0 0.26   3.3   1 14.1 29 350   M    M  10 128     38 350
    3  15   1   8 -22  57   0   T     T    3  7.0 22 330   M    M   4         28 330
    4  35  11  23  -7  42   0 0.00   0.0   3  7.8 18 190   M    M   4         23 180
    5  47  27  37   7  28   0 0.35   T     2  7.7 28 290   M    M  10  1      33 290
    6  33  -7  13 -17  52   0 0.02   0.3   T 20.6 32 290   M    M   8 18      38 300
    7  11  -7   2 -28  63   0 0.00   0.0   T 15.0 24 250   M    M   8  8      31 240
    8  26  10  18 -12  47   0 0.00   0.0   T  4.8 13 200   M    M   8         15 200
    9  35  21  28  -1  37   0 0.10   1.2   T  4.5 12 180   M    M   9 18      16 130
   10  48  32  40  11  25   0 0.01   0.1   1  7.5 15 130   M    M   8  1      21 120
   11  53  36  45  16  20   0 0.69   0.0   0 13.6 33 270   M    M  10  1      40 270
   12  41  33  37   8  28   0 0.00   0.0   0  9.2 20 290   M    M   8         23 290
   13  50  40  45  16  20   0 0.23   0.0   0  9.2 17 190   M    M   7  1      24 180
   14  47  35  41  12  24   0 0.01   0.0   0  8.3 26 290   M    M   6 18      32 290
   15  35  23  29   0  36   0   T     T    0 10.5 20 240   M    M   9  8      24 240
   16  31  21  26  -3  39   0 0.04   0.4   0 10.2 20 190   M    M   8  1      24 170
   17  35  13  24  -5  41   0 0.09   0.9   T 10.2 26 270   M    M   9 128     35 270
   18  23  11  17 -12  48   0 0.09   1.5   1  9.0 15 230   M    M   9 18      20 240
   19  33  21  27  -2  38   0 0.01   0.1   2 10.5 22 210   M    M   7 18      29 210
```

FIGURE 22.5 Example of weather. (Data from Columbus, OH from National Weather Service January 2014 - https://w2.weather.gov/climate/getclimate.php?wfo=iln)

- Perform a detailed inspection of the failed component and compare with failure characteristics known to be associated with a tube/pipe/hose/component failure. In some cases, additional literature searches and evaluation under a microscope will be needed to confirm the cause of the failure.

22.5.2 Plumbing/Piping/Tubing Failures Assessment Methodology – Written Report

The written report should follow the overall approach outlined in Chapter 1 as well as the water loss report outlines provided in Chapters 6–10. Particular attention should be spent documenting the following observations and results from the inspection in the body of the report:

- The date, time, location, and plumbing component associated with the plumbing failure.
- Visual inspection observations. The inspection should be completed indoors for all rooms/spaces for each impacted floor, wall, ceiling, and contents. The report should include a delineation of damaged areas, including extent of damage to wall, floor, and ceiling surfaces, including structural support materials such as those formed from concrete, CMU, stone, steel, and wood. Results should be summarized by space as illustrated earlier in Table 17.7.
- Summation and analysis of measurements, photographs, and FLIR measurements taken during the site visit.
- Plumbing failure cause determination. Every effort should be made to provide a specific causation for the plumbing failure. This basis for this causation should be well supported by observations, measurements, and the literature.

The report should also provide guidance on water damage repair(s) and close with any limitations noted.

IMPORTANT POINTS TO REMEMBER

- Different types of tubing (i.e., K, L, and M) and piping (various Schedules) exist in various materials of fabrication; the Forensic Engineer should know how to identify each of these types of plumbing and how it typically can fail. This can take considerable time given the extent of products, materials, and failure mechanisms.
- In colder climates, freeze failures are often encountered. Each type of plumbing has characteristic failure mechanisms that tend to result in similar characteristic visible failure observations.
- Plumbing failures generally lead to significant consequential damages such as water damage and mold formation. These damages should be quantified as part of the plumbing failure site investigation.
- Once the cause of the plumbing failure has been determined, the basis for this determination should be thoroughly documented. Documentation should include site observations, site measurements, photographs possibly including microscopic photographs, and pertinent technical literature.

REFERENCES

1. Commerce Metals. "Tube vs Pipe: The Differences Explained in Plain English." Accessed May 18, 2020, https://www.commercemetals.com/tube-vs-pipe-the-differences-explained-in-plain-english/.
2. Explore the World of Piping. "What is the Difference between Pipe and Tube?" Accessed May 18, 2020, http://www.wermac.org/pipes/pipe_vs_tube.html.

3. espo CRM. "Flow of Fluids Through Valves, Fittings and Pipe, Crane Company" Crane Technical Paper No. 410 (TP-410), 2009.
4. Allied Tube & Conduit. "Tube and Pipe Size Overview." Accessed March 18, 2020, http://www.atc-mechanical.com/tube-pipe-101/tube-pipe-size-overview/.
5. Industrial Wiki. "Piping Tubing and Hoses." Accessed March 18, 2020, https://www.myodesie.com/wiki/index/returnEntry/id/2971.
6. Copper Development Association, Inc. (Copper Alliance). *"Copper Tube Handbook – Industry Standard Guide for the Design and Installation of Copper Piping Systems"*, 2020 Edition. Accessed May 19, 2020, https://www.copper.org/publications/pub_list/pdf/copper_tube_handbook.pdf.
7. American Society for Testing and Materials (ASTM). "ASTM F 1281-17. Standard Specification for Crosslinked Polyethylene/Aluminum/Crosslinked Polyethylene (PEX-AL-PEX) Pressure Pipe," 2017.
8. InspectAPedia. "PEX Tubing & Piping." Accessed May 20, 2020, https://inspectapedia.com/plumbing/PEX_Piping.php.
9. California Real Estate Inspection Association. "Kitec® Water Supply Piping and Brass Fittings: What You Need to Know." Accessed May 20, 2020, https://www.creia.org/kitec-water-supply-piping-and-brass-fittings--what-you-need-to-know.
10. InspectAPedia. "Kitec® Piping & Connector Leak Causes & Lawsuit Settlement." Accessed May 20, 2020, https://inspectapedia.com/plumbing/Kitec_Plumbing_Leaks_Settlement.php.
11. Atlas Care. "Must Know Facts about Kitec® Plumbing Recall, Lawsuit & Repair Costs." October 4, 2019. Accessed May 20, 2020, https://atlascare.ca/blog/must-know-facts-about-kitec-plumbing-lawsuit/.
12. Mader, Robert P. "California Approves PEX for Plumbing – Again." September 2010. Accessed May 25, 2020, https://www.contractormag.com/plumbing/article/20879315/california-approves-pex-for-plumbing-again.
13. Watts. "Watts WaterPEX® Installation Guidelines" 2011.
14. NSF. "Bulletin, Crosslinked Polyethylene (PEX) Flexible Tubing for Hot and Cold Water Applications." Accessed December 19, 2006, http://www.nsf.org/business/newsroom/plumbing98-1/tubing.html.
15. MacNevin, Lance and Camille G. Rubeiz. "PEX pipes for plumbing", *Presented at the 40th ASPE Convention*, Plastic Pipe Institute, Cleveland, OH October 2004.
16. Natural Hazard Mitigation Insights. "Freezing and Bursting Pipes." Institute for Business and Home Safety. Accessed May 25, 2020, https://www.iccsafe.org/cs/PMG/Documents/DIS-FreezeBurstPipe.pdf.
17. NAHB Research Center. "Design Guide: Residential PEX Water Supply Plumbing Systems," November, 2006. Accessed May 20, 2020, https://www.huduser.gov/portal/publications/pex_design_guide.pdf.
18. NAHB Research Center. *"Design Guide: Residential PEX Water Supply Plumbing Systems,"* 2nd Edition, November 2013. Accessed May 20, 2020, https://plasticpipe.org/pdf/pex_designguide_residential_water_supply.pdf.
19. Plastic Pipe and Fittings Association. *"Installation Handbook: Cross-Linked Polyethylene (PEX) Hot- and Cold- Water Distribution Systems"*, Glen Ellyn, IL: Plastic Pipe and Fittings Association, 2006.
20. Plastics Pipe Institute. "Guide to Chlorine Resistance Ratings of PEX Pipes and Tubing for Potable Water Applications." TN-53, 2018. Accessed May 20, 2020, https://plasticpipe.org/pdf/tn-53-pex-chlorine-ratings.pdf.
21. Canadian Copper & Brass Development Association (CCBDA). "Copper Tube & Fittings." CCBDA Publication No. 28E, 2nd Edition 2000. Accessed May 20, 2020, http://en.coppercanada.ca/pdfs/28e.pdf.
22. Cruz, J., B. Davis, P. Gramann, and A. Rois. "A Study of the Freezing Phenomena in PVC and CPVC Pipe Systems." The Madison Group. Accessed May 20, 2020, https://docplayer.net/23613151-A-study-of-the-freezing-phenomena-in-pvc-and-cpvc-pipe-systems.html.
23. Burch, J. "PEX Piping as a Fail-Safe Backup for Pipe Freeze Protection." U.S. Department of Energy, National Renewable Energy Laboratory, 2006.
24. "Watts Water Heater and Floodsafe Connectors Class Action Settlement." Accessed May 25, 2020; https://classactionsreporter.com/settlement/watts-water-heater-and-floodsafe-connectors-class-action-settlement/.
25. Eaton, G.C. "Task Force to Study Pinhole Leaks in Copper Plumbing." December 2004, State of Maryland. Accessed May 20, 2020, https://www.wsscwater.com/files/live/sites/wssc/files/PDFs/Study%20of%20Pinhole%20leaks%20Report_1503042.pdf.
26. Lewis, R.O. "A White Paper Review: History of Use and Performance of Copper Tube for Potable Water Service." Lewis Engineering and Consulting, Inc., August 23, 2016. Accessed May 20, 2020, https://www.wsscwater.com/water-quality--watershed-informa/copper-pipe-pinhole-leaks/copper-pipe-white-paper.html.

27. Coyne, J.M. "Flow Induced Failures of Copper Drinking Water Tube" Master Thesis, Virginia Polytechnic Institute and State University. May 5, 2009, Blacksburg, VA. Accessed May 20, 2020, https://vtechworks.lib.vt.edu/bitstream/handle/10919/32765/JeffCoyneThesis_June_10_09.pdf?sequence=1&isAllowed=y.

28 The Engineering Toolbox. "Piping Materials and Galvanic Corrosion." Accessed May 25, 2020, https://www.engineeringtoolbox.com/galvanic-piping-corrosion-d_906.html.

29. Calkins, C. "PEX Plumbing Failures." Accessed May 20, 2020, https://sienapex.weebly.com/technical-causes-of-pex-plumbing-failures.html.

30. E.R. Services. "Problems with PEX Pipes with Brass Fittings." Accessed May 20, 2020, https://www.erplumbing.com/blog/problems-with-pex-pipes-with-brass-fittings/.

31. Janowiak, R. "Chlorine Failures in PEX Pipe Plague Washington Home." March 28, 2017. Accessed May 20, 2020, https://plumbing.corzan.com/blog/chlorine-failures-in-pex-pipe-plague-washington-home.

32. Super Brothers Plumbing, Heating & Air. "PEX Plumbing Failures." Accessed May 20, 2020, https://www.repipeyourhouse.com/pex-plumbing-failures/.

33. InspectAPedia. "Brass Connector Corrosion, Leaks, Dezincification in PEX Piping Systems." Accessed May 20, 2020, https://inspectapedia.com/plumbing/PEX_Brass_Fitting_Leaks_De_Zincification.php.

34. Zurn Plumbing Products Group. "PEX Plumbing Design and Application Guide." Accessed May 20, 2020, http://0323c7c.netsolhost.com/docs/PEXDesApplGuide.pdf.

35. Zurn Plumbing Products Group. "PEX Plumbing Installation Guide" March 2012. Accessed May 20, 2020, https://api.ferguson.com/dar-step-service/Query?ASSET_ID=1272622&USE_TYPE=INSTALLATION&PRODUCT_ID=1857360.

36. Schladweiler, J.C. "Coal Tar Impregnated Wood Fibre Pipe." Sewerhistory.org. Accessed May 20, 2020, http://www.sewerhistory.org/articles/compon/orangeburg/orangeburg.htmhttp://www.sewerhistory.org/articles/compon/orangeburg/orangeburg.htm.

37. "Orangeburg Pipe: Unwelcome History." Accessed August 30, 2012, http://plumberologist.com/2011/04/orangeburg-pipe-unwelcome-history/.

38. "Orangeburg Sewer Piping (Bituminous Fiber Pipe)." February 2006, http://www.a2gov.org/government/communityservices/planninganddevelopment/building/Documents/building_info_orangeburg.pdf.

39. Potter Corrosion Solutions. "Corrosion in Fire Sprinkler Systems." Accessed May 25, 2020, https://www.nfpa.org/-/media/Files/News-and-Research/Resources/Research-Foundation/Symposia/2015-SUPDET/2015-papers/SUPDET2015Tihen.ashx?la=en40.

40. Hopkins, M.P.E. "An Introduction to Corrosion in Sprinkler Systems: Its Identification and Mitigation." December 7, 2018. Accessed May 25, 2020, https://nfsa.org/2018/12/07/an-introduction-to-corrosion-in-sprinkler-systems-its-identification-and-mitigation/.

23 Equipment Failures and Investigations

Bryan E. Knepper and Stephen E. Petty
EES Group, Inc.

CONTENTS

PURPOSE/OBJECTIVES

The purpose of this chapter is to:

- Discuss various types of common and less commonly observed equipment failures.
- Provide a methodology (differential diagnosis) for completing equipment failure forensic inspections.
- Provide examples of various types of equipment failure investigations.

Following the completion of this chapter, you should be able to:

- Conceptually understand the process for determining how to diagnose an equipment failure.
- Have a basic understanding of the difference between an equipment failure inspection and other types of forensic investigations.
- Be able to complete an equipment failure inspection and provide a written report.

23.1   INTRODUCTION

Equipment failures within a residence, business, or on farms can cause large-scale damage including major water losses, structural failures, and in some cases explosions. Many insurance claims involve a loss that could have a root cause with an equipment failure such as a sump pump not working during a heavy rain storm, the failure of a water well pump following a lightning storm, or a part within a dishwasher or washing machine. For this reason, it is paramount that a forensic investigator be able to determine how and why various equipment fail.

23.2   VARIOUS TYPES OF EQUIPMENT FAILURES

While this is by no means an exhaustive list, the following are examples of equipment failures that will be discussed in this chapter. Note that equipment failure investigations typically follow a general forensic process similar to that provided in the preceding chapters but will add a layer of research and correspondence not typically involved in residential and commercial forensic inspections.

More common failures:

- Sump Pump Failures (23.2.1).
- Well Pump Failures (23.2.2).
- Radiant Heating System Failures/Cast Iron Radiators (23.2.3).
- Dishwasher Solenoid Failures (23.2.4).
- Washing Machine Gasket Failures (23.2.5).

Less common failures: The equipment in these sections will not be discussed in as much detail with examples from actual forensic investigations provided as case studies instead.

- Wind Turbine Failure (23.2.6).
- Boat Hoist Failure (23.2.7).
- Grain Elevator Explosion (23.2.8).
- Grain Bin Re-corrugation Failure (23.2.9).

In general, the base methodology outlined in Chapter 1 is followed completing forensic investigations of equipment failures. However, the added element for equipment investigations is a differential diagnosis of the equipment to determine specifically why it failed. A differential diagnosis is based on an understanding of how the equipment should function vs how it is functioning at the time of the inspection. The difference between these two functioning states normally determines why the equipment is not functioning as desired. Illustrations of this methodology follow.

23.2.1   SUMP PUMPS

In most residential settings, a sump pump, commonly located in the basement, is utilized to remove water from around the foundation. Collected water flows into a sump pump pit where it is then pumped into the subgrade drain lines or to the ground surface where it flows away from the foundation walls by gravity. Four types of sump pumps discussed below are: (1) pedestal, (2) submersible, (3) water powered, and (4) floor sucker[1]:

- Pedestal pumps are upright electric pumps where the motor is located above the water line. These pumps can work well when frequent drainage is needed. Since the motor is above the water line, these pumps are louder than submersible pumps and not well suited for residential settings.
- Submersible pumps are the most common type of sump pumps. The motor sits below the water line making them quieter. The pumps are activated with a float switch. A check valve is required to prevent backflow of water into the sump. Since the pumps are powered by electricity, they will fail during power outages unless backup power is available.
- Water-powered pumps are normally used as a backup for electric pumps and are mounted near the ceiling. These pumps work with suction provided from pressure supplied by a municipal water supply to the home. If the electric pump fails and the water rises to near the top of the sump, a float switch opens a one-way check valve connected to the municipal water supply. The pressure from the municipal water supply creates a suction similar to a straw effect to discharge the water from the sump pit. The system is normally discharged independently or separately from the system connected to the primary sump pump.
- Floor sucker pumps are used in cellars or crawlspace areas that have no sump pit. The pumps are designed to remove water within 1/8″ of the floor and are mostly used as a short-term solution to aid in controlling water accumulation.

Code requirements from Section 1113 of the 2018 International Plumbing Code (IPC) regarding sump pumps are[2] as follows:

1113.1 Pumping system: The sump pump, pit, and discharge piping shall conform to Sections 1113.1.1–1113.1.4.

1113.1.1 Pump capacity and head: The sump pump shall be of a capacity and head appropriate to anticipated use requirements.

1113.1.2 Sump pit: The sump pit shall be not <18 inches (457 mm) in diameter and not <24 inches (610 mm) in depth, unless otherwise *approved*. The pit shall be provided with access and shall be located such that all drainage flows into the pit by gravity. The sump pit shall be constructed of tile, steel, plastic, cast iron, concrete, or other *approved* material, with a removable cover adequate to support anticipated loads in the area of use. The pit floor shall be solid and provide a permanent support for the pump.

1113.1.4 Electrical: Electrical service outlets where required, shall meet the requirements of NFPA 70.

1113.1.5 Piping: Discharge piping shall meet the requirements of Section 1102.1, 1102.3, or 1102.4 and shall include a gate valve and a full flow check valve. Piping and fittings shall be of the same size as, or larger than, the pump discharge tapping.

Exception: In one- and two-story dwellings, only a check valve shall be required, located on the discharge piping from the pump or ejector.

Sump pump failures are a common occurrence in residential homes. Failures can occur as a result of power failures, improper installation, and mechanical failures related to improper/lack of maintenance, wear, or defects during manufacturing,[3] which can result in water backup and flooding. These are further discussed in the following sections.

23.2.1.1 Power Failures

Electrical failures can be difficult to predict or control and are often related to storm events. Inevitably, these storm events bring heavy rains that raise the groundwater level and increase the need for a sump pump. To prepare for this problem, battery powered backup systems are available and are recommended where power outages occur frequently.

A second means of electrical failure includes a branch circuit overload. If too many appliances draw electrical power from the same circuit as the one to which the sump pump is attached, the circuit can be overloaded and the circuit breaker will open, stopping power to the sump pump. A dedicated circuit for a sump pump is recommended.[3]

23.2.1.2 Improper Installation

Based on experience, the main failures involved with the installation of a sump pump are typically the result of one, or a combination, of the following causes: (1) improper housekeeping, (2) improper discharge pipe assembly, (3) undersized pump, and (4) failure or blocking of float switches.

In new construction, debris can easily accumulate in a sump pump. The pit should be kept clean and free of debris. This will prevent debris from entering the pump and jamming the impeller or interfering with the float. Sump pump pits are typically covered with the intent of limiting debris falling into the pit.

As required by code, the discharge piping needs to be of the same size or larger than the discharge tap from the sump pump. If a smaller-sized pipe is used, especially at an elbow, this increases the chances of discharge line blockage. In an actual forensic investigation, a sump pump backup occurred in a home under new construction as a result of debris blocking a 90° fitting that reduced the 1-1/2 inches pump line to a 1-inch discharge line. This situation probably would have been avoided had the debris been cleaned from the sump pump and the same size or larger elbow been used (Figure 23.1).

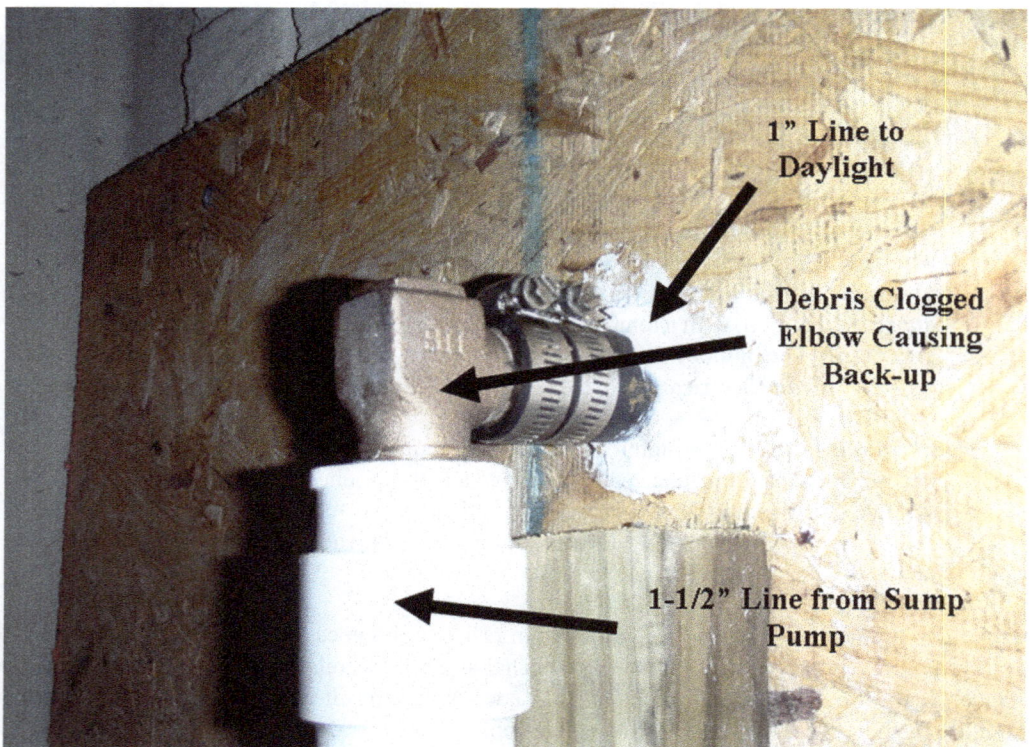

FIGURE 23.1   1-1/2″ discharge line from sump pump reduced to 1″ line at elbow – resulted in debris clogged elbow and water backup.

Sump pump failures can also occur if a sump pump float switch is not positioned in a manner to allow for free movement of the float. An interesting case occurred where a homeowner replaced a sump pump. The basement had recently undergone a major renovation. Within days of installation, the sump pump quit working and caused extensive water damage in the newly finished basement. It was later determined that the sump pump was positioned too close to the sump pump pit wall, thus preventing the float from rising to move the float switch and activate the sump pump motor (Figure 23.2).

FLOAT STUCK AGAINST WALL

FIGURE 23.2 Sump pump float stuck against wall.

When installing a sump pump, it is important to select the proper size for the application. If the capacity of the pump is inadequate, this can cause the pump to either overwork, which can lead to premature failure, or to be overwhelmed and not be able to keep the basement from flooding. Most pump manufacturers provide online pump-sizing calculators to aid in the selection of the correct pump.

23.2.1.3 Mechanical Failures

Mechanical failures to sump pumps occur for a variety of reasons. The failures can be related to defects during the manufacturing process, improper installation practices, and debris clogging or damaging the pump impeller. For example, if the sump pump is installed without a check valve, the water remaining in the pipe head will return to the sump pit. This will cause repetitive short cycling that can burn out the motor or electrical contacts for the float switch.

Mechanical failures can also occur as a result of air lock. Air lock occurs when air builds up between the pump discharge outlet and check valve, thus causing cavitation of the pump. In order to eliminate air lock issues, some sump pump manufacturers recommend drilling a 3/16″ hole between the outlet and check valve to relieve air.[4] When drilling this hole, it is important that the hole be placed at a downward angle so the water that sprays from the hole during its cycle will be directed down into the pit. During one inspection, the relief hole was drilled at an upward angle, which caused water to spray or splash around the cover (Figure 23.3).

FIGURE 23.3 Water spraying from relief hole during cycling.

The owner had thought that the water was a result of the pump malfunctioning.

23.2.2 WELL PUMPS

One of the most common claims for well pump failures is from lightning damage (see Chapter 21 for more information on lightning strikes). While well pumps can fail from lightning strikes, there are several other more likely causes for well pump failures. Lightning damage to a water well pump/motor is highly unlikely since the well pump/motor is located deep below the ground surface, which is the natural grounding point for lightning. The electrical flow from the water pipe into the water well pump motor windings is actually the path of highest resistance,[5] ~30,000,000 ohms vs 20–30 ohms from the well pipe to the ground. Furthermore, most well pump/motor manufacturers build lightning protection into their products. Thus, while failures of water well pumps/motors are often associated with thunderstorm/lightning events, the timing of the failures is likely coincidental with the storm events and is not the result of lightning strikes. More likely, causes for failure are age and/or mechanical wear of the pump/motor system. Typical failure modes include the following:

- Worn bearings (pump or motor) – this is manifested in the inability or difficulty in rotating the motor/pump shaft.
- Water leakage into the motor windings – causes electrical short.
- Worn packing – allows silt to enter the pump/motor system.
- Breakdown or loss of lubricating oil – seal failure or overheating.

Lightning damage to a well pump/motor would be associated with other damage on the same power supply to more delicate electrical components such as high-voltage damage to the internal lightning arrestor, high-voltage damage to the run capacitor (bulged or blown out), high-voltage damage to circuit boards in controllers and/or a motor with the start winding (normally) having normal dielectric resistance readings while the run winding is grounded out. Nevertheless, many forensic investigations occur as a result of claims that lightning caused the failure of a well pump motor.

23.2.3 RADIANT HEAT SYSTEM/RADIATOR/BOILERS

Older homes, typically in cities, were built with boiler systems to provide heat. The most common type of system was a coal-fired boiler connected to radiators in each of the rooms of the home. The first popular cast iron radiator was invented in 1874. By the 1880s, sectional cast iron radiators became popular and manufacturers began to develop ornate radiators in order to capture the market.[6] Prior to these systems, heat by fireplace was the most common home heating system. The radiators were connected to the boiler with cast iron, carbon steel piping, and in some cases copper tubing. While newer forms of heating homes (including furnaces and geothermal systems) have been developed over the last 100 years or so, many older homes still employ the use of boilers and radiators as the main source of heat for the home. Newer natural gas boiler systems have been developed to replace the older coal-fired boilers, but the implementation of the heat (in the form of steam) throughout the home has not changed, and many of the original sectional radiators can be found to this day in homes built typically from the 1880s to the 1950s.

As with many types of older equipment, boilers and radiators will fail over time. There are many reasons for the failures to occur including a lack of maintenance, either slow pressure buildup or pressure loss, water leaks, blockages or scale buildup within the piping delivery system, or a lack of insulation.

23.2.3.1 Improper/Lack of Maintenance

One of the most common reasons for the failure of a boiler/radiant heat system is the lack of proper maintenance. This is partly due to the lack of general knowledge of the public on how the systems work and partly due to the lack of training of new HVAC personnel with the older boiler systems. If the procedures are not followed for an extended period of time, the boiler system will likely have, at a minimum, efficiency issues and as a worst-case scenario, catastrophic failure. Forensic inspections have revealed cases where the boiler-radiant system had not been inspected by a qualified professional for decades, leaving the boiler and radiators in a heavily degraded state. The following is excerpted from the suggested minimum maintenance schedule for a Weil-McLain Boiler[7] commonly used in older boiler-radiant heat systems:

At the beginning of each heating season:

1. Annual service call by a qualified service agency.
2. Check burners and flueways and clean if necessary.
3. Follow flow procedure "To Place in Operation."
4. Visually inspect pilot and burner flames.
5. Visually inspect venting systems for proper function, deterioration, or leakage.
6. Visually inspect base insulation.
7. Check operation of low-water cutoff, if used, and additional safety devices.
8. Check that boiler area is free of combustible materials, gasoline, and other flammable vapors and liquids.
9. Check for and remove any obstruction to flow of combustion or ventilation air.

Daily during heating season:

1. Check that the boiler area is free from combustible materials, gasoline, and other flammable vapors and liquids.
2. Check for and remove any obstruction to flow of combustion or ventilation air.

Periodically during heating season:

1. Check safety relief valve.
2. Test low water cutoff, if used. Blowdown if low-water cutoff is float type.

Monthly during heating season:

1. Check for leaks in boiler and piping. If found, repair at once. Note: continuous use of makeup water can damage boiler sections due to addition of minerals. DO NOT use petroleum-based stop-leak compounds – leakage between the sections will occur.
2. Check any gaskets for leakage. Tighten or replace, if needed. Do not overtighten bolts – damage to the gasket can occur.
3. Visually inspect pilot and burner flames.
4. Visually inspect venting systems for proper function, deterioration, or leakage.
5. Check automatic air vent for leakage. If leaking, remove vent cap and push valve core into wash off sediment that may have accumulated on the valve seat. Release valve, replace cap, and open one turn.

End of each heating season:

1. Follow "Annual Shutdown Procedures."

23.2.3.2   Issues with Pressure

Problems with a lack of pressure typically point to a leak within some portion of the system. If the supply to the system is steam, a leak will prevent the system from building up enough pressure to force the steam through the system into the radiators. The most common areas for steam leaks are at valves and other components built into the system. If the boiler has an on-off water feed, the water entering into the system (although heated) is colder than the boiler water. This temperature differential can cause a reduction in steam production until the newly introduced water heats up. Lower steam production will drop the pressure in that time span. The use of a continuous water feed will help to alleviate this phenomenon. Another cause of pressure loss is an undersized heating element. A heat source that cannot adequately boil the water in the tank will fail to generate enough steam to support the system.

23.2.3.3   Water Leaks

There are several reasons that a boiler-radiant heat system can and will leak, including improper installation (typically newer replacement components), freeze/thaw damages in the piping system where either improper insulation is used or the heating system is turned off during prolonged sub-freezing temperatures (this typically occurs due to a vacancy), and the most common cause: corrosion. Water leaks will cause a loss in steam pressure, which will keep the system from operating properly and in some cases cause extensive damages within a home.

23.2.3.3.1   *Causes of a Burst Home Radiator*

Forensic investigations for the purpose of insurance claims typically derive from the question: "why did a water loss occur?" The two most common causes of a burst home radiator are from (1) freezing water and (2) internal corrosion. As shown in Chapter 22, freezing water inside a piping system can cause substantial water damage to a home. While it is common knowledge that water freezing inside a pipe expands and bursts the pipe, it is less commonly known that the same can happen inside a radiator. Of course, the idea of a steam system "freezing" is antithetical to the process, but if a house is left unheated (i.e., the boiler system turned off for a prolonged period of time) during colder outdoor temperatures, accumulated water within the radiator can freeze. The expansion of the ice inside the radiator causes hairline cracks to develop, compromising its integrity. When pressurized, the radiator no longer can take the pressure and bursts along the hairline cracks (Figure 23.4). While freeze/thaw bursts do occur in radiators, they almost never occur without some corrosion present to the unit, serving to weaken the sections (or fins).

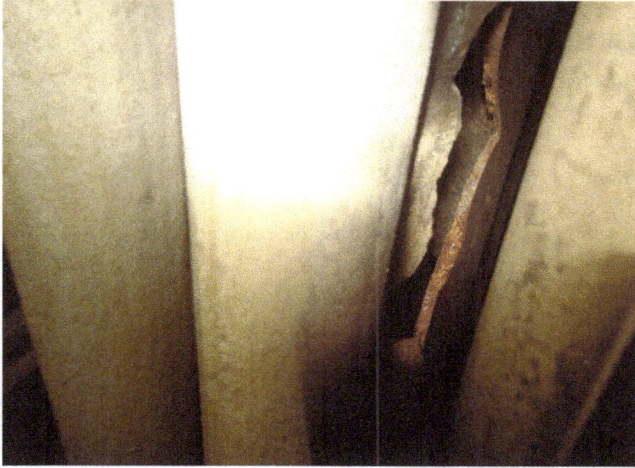

FIGURE 23.4 Freeze/thaw burst of radiator section.

The most common failure of a radiator, however, is internal corrosion, typically in the form of localized pitting. Most water is not pure as it includes dissolved minerals and gases. These contaminants are very corrosive and over time corrode the radiator from the inside out. The result of this corrosion is that the walls can become almost paper thin and perforations can form. When the radiator is pressurized, these sections can blow out, causing heavy water leaks to occur. An example of a crack induced in a radiator fin by pitting corrosion is provided in Figure 23.5.

FIGURE 23.5 Induced crack in radiator fin from pitting corrosion.

23.2.4 DISHWASHER SOLENOID VALVE FAILURES

These types of failures were briefly touched upon in Chapter 22. The two most common failures for a dishwasher are solenoid valve failures and water supply leaks. Piping supply leaks have been detailed already in this textbook, so the concentration here will be on solenoid failures. There are several reasons a solenoid valve will fail/leak including but not limited to manufacturing defects, improper installation, and wear/tear.

23.2.4.1 Manufacturing Defect

Most products made today are quality tested to meet various standards. However, many products are not individually tested, but are sample-tested, meaning that portions of the batch products are tested for quality and to ensure they meet specifications/standards. Customers typically set the quality standards for a product (i.e., if a product is manufactured poorly, people will stop buying that product), but industry standards such as Six Sigma[8] provide acceptable defect levels by end users, which is known as the "acceptance quality level." Critical defects, which are defined as those that could lead to harm of the users, have an acceptable limit of 0%. The AQL for major defects (or defects that likely result in failure of the product) is 2.5%. The AQL for minor defects (or defects not likely to reduce usability but only slightly defer from specified standards) is 4%. Despite these manufacturing quality control efforts, a certain percentage of a product will have manufacturing defects. These defects in regard to a dishwasher, or any other appliance in a home, will typically become apparent within the first year of use. This is one of the reasons that 1-year warranties provided by manufacturers has become a normality. Defects that arise outside of this time period will typically be due to wear/tear or usage error. An example mechanical defect for a solenoid valve would be an induced crack to the threads for the supply line from the manufacturing process or a deficient magnet.

23.2.4.2 Installation Error

As discussed in the previous section, a manufacturing defect will typically reveal itself shortly after initial use and almost always in the first few years of usage. An installation error in regard to a dishwasher solenoid valve will occur at two distinct times: at initial installation, meaning when the water supply is first connected to the valve and at the time of replacement. The most common issue with the installation of a new supply line or after replacing a damaged or inoperable solenoid valve is the overtightening of the supply line onto the valve inlet or the under-tightening of the supply line onto the valve inlet. More uncommon installation errors would be cracking the housing due to lack of care or separating the parts during install.

23.2.4.3 Cyclic Fatigue – "Wear and Tear"

A solenoid is an electromagnet around a core in which a magnetic plunger is located. When a voltage is applied to the solenoid coil, an electromagnetic field is developed around it and the magnetic field pushes the plunger upward in the solenoid coil. The plunger is connected to the valve through a connection that forces the valve to open. The tasks of a solenoid valve are to shut off, release, dose, distribute, or mix fluids. In the case of dishwashers, a solenoid valve is used to control the amount of water that is supplied to the dishwasher tub. As with any major appliance, a dishwasher has an expected service life (e.g., 9–16 years) as illustrated in Table 23.1.[9]

TABLE 23.1

Appliance Life Expectancy's Tables

Appliance	Life Expectancy, Years			Units to Be Replaced in 2001	Units Shipped in 1999
	Low	High	Average		
Major Home Appliances (Excludes Commercial Appliances)					
Microwaves	5	10	8	8,132,300	11,581,085
Ranges, electric	13	20	16	3,227,700	7,016,939
Ranges, gas	15	23	19	1,367,400	3,136,200
Ranges, hoods	9	19	14	2,595,000	3,000,000
Refrigerators, compact	4	12	8	1,030,000	141,283
Refrigerators, standard	10	18	14	6,972,100	9,098,600
Water heaters, electric	6	21	14	3,396,395	4,281,199
Water heaters, gas	5	13	9	4,241,354	4,933,659
Washers	8	16	12	6,607,500	7,508,200
Dryers, electric	11	18	14	3,381,200	4,864,700
Dryers, gas	11	16	13	1,046,800	1,443,000
Dishwashers	9	16	12	3,668,400	5,711,200
Food waste disposers	10	15	13	4,232,600	5,369,400
Freezers	12	20	16	1,472,800	1,987,200
Water softeners					951,498
Compactors	7	12	11	185,000	114,700
Totals				**51,556,549**	**71,138,863**
Comfort Conditioning Appliances					
Fans, ceiling	7	18	13	6,400,000	19,100,000
Air conditioners, room	7	16	12	5,091,100	6,113,600
Air conditioners, unitary	8	19	13	3,214,606	5,353,676
Humidifiers	6	13	10	612,000	9,800,000
Furnaces, electric	9	20	14	375,055	
Furnaces, gas	11	23	17	2,049,335	3,293,646
Furnaces, oil	13	23	18	127,305	125,378
Portable heaters	8	13	11	5,542,900	2,700,000
Heat pumps	6	21	14	918,432	1,293,395
Dehumidifiers	9	13	11	742,500	871,000
Room heaters, vented gas	7	18	13	91,426	35,927
Room heaters, unwanted gas	13	23	18	217,566	467,204
Boilers, gas					200,893
Boilers, oil					149,050
Totals				**25,382,225**	**49,503,769**

Source: "23rd Annual Portrait of the U.S. Appliance Industry," *Appliance Magazine*, September 2000.[9]

While the new purchase of an appliance could be related to updating a kitchen or utility area, many times the replacement is due to a failure of the appliance within one of its major components such as in the case of dishwashers, the solenoid valves or the motor.

23.2.5 WASHING MACHINE GASKET FAILURE

As with any major appliance that uses water, the main failure mode is the water supply to the solenoid or main inlet/valve connection. One of the more atypical failures for washing machines involves a door gasket failure on front-loading units. The gasket is located at the interior portion of the door, which allows for a seal when the door is closed. The gasket is attached to the door opening with a wire that constricts around the protrusion at the opening. As with other appliances, the major modes of failure are from manufacturing defects, installation errors, and wear/tear.

23.2.5.1 Manufacturing Defects

A manufacturing defect to a gasket for a front-loading washing machine would be apparent relatively quickly as the amount of water introduced during a typical washing cycle would be enough to escape through a defective part such as the door seal. The water from a leak would develop quickly beneath the unit, causing damages immediately to the floor below. As indicated in the section on dishwasher solenoids, manufacturing defects do occur; however, this type of defect would be considered a "major defect" due to the amount of damage that could occur in a relatively short amount of time and is thus not as likely to occur.

23.2.5.2 Installation Error

Perhaps the better subtitle heading should be "re-installation" or "repair errors." Moving parts or parts frequently exposed to moving parts typically wear out quicker than stationary parts. In the case of a washing machine door, the gasket is exposed to water throughout the washing cycle and then is moved/displaced during the opening of the doors. This cyclic wear will cause the gasket to wear out over time. This fatigue can become exacerbated if the washing machine is not cared for according to the manufacturers' guidelines for maintenance. As such, the gasket is one of the items that is replaced the most. If the proper size gasket is not used or if the wire to hold it in place is not properly tightened, a leak will occur in a short amount of time. An example of a gasket that was not tightened/installed properly is provided in Figure 23.6.

FIGURE 23.6 Washing machine failure - detached gasket wire.

23.2.5.3 Wear and Tear/Cyclic Fatigue

If other parts are not working properly within a washing machine, such as the motor or rotating drum, an induced vibration may occur. This vibration has been known to cause the wire for the gasket to come loose, thus causing a leak at the tub to door interface. This of course happens over a period of time and washing cycles. Another instance of cyclic fatigue is mildew growing on the rubber material that can cause the material to wear out quicker than expected. Proper maintenance of the washing machine is required for it to reach its assumed life expectancy of 8–16 years (see Figure 23.6).

Case Study – Wind Turbine Failure

Scope of inspection: to determine the cause of a wind turbine failure on a residential property.

System specifications: Ventera V12 Hybrid Wind and Solar System – VT10 Wind Turbine. Note that this unit was manufactured with a governor that allows for the composite blades to pitch in the wind and a bearing plate that allows the entire assembly to rotate on top of the mounted shaft.

Background information: the wind turbine failed (i.e., came off its mount and crashed to the ground) during a reported wind-driven ice storm. The composite blades and governor had been replaced the year before. Because the damage occurred in a storm recently after repairs were made, an investigation was launched to determine if the failure was caused by a manufacturing defect, and installation error, or from the storm itself. Note that rub marks were present to the base of the turbine body at the location where the yaw-bearing plate should have been located. The damages present to the interior shaft were that of sudden shear failure.

Review of weather data: The survival speed for the unit being investigated was listed as 130 MPH. The winds on the date of the loss were up to ~44 MPH. At a height of ~70 feet in the air, the expected wind speeds would be up to ~50 MPH, nowhere near the survival speed. The winds that date, however, were accompanied by ~0.63″ of freezing rain.

Field observations: There were no signs of a manufacturing defect or obvious installation deficiencies. The wind turbine had operated since the repairs for nearly a year.

Review of troubleshooting manual: As provided in the manual for the specific wind turbine, there were two symptoms identified that appeared to have occurred based on evidence collected at the scene: (1) Turbine does not orient to the wind – probable cause: tower and turbine not level OR yaw-bearing binding; (2) blades only start in high wind – probable cause: ice on propeller.

Forensic conclusions: The damage to the wind turbine was concluded to be caused from a binding failure at the yaw-bearing plate to mounting shaft assembly (Figures 23.7 and 23.8).

FIGURE 23.7 Failed wind turbine – far and close views.

FIGURE 23.8   Failed wind turbine – far and close views.

It is possible that the binding of the yaw-bearing plate occurred due to a buildup of ice in the form of freezing rain just prior to the ultimate failure.

Case Study – Boat Hoist Failure

Scope of inspection: To determine the cause and origin of the failure of a boat hoist.

Background information: The boat hoist was ~20 years old and at a dock within freshwater. The water depth was ~4′. The boat hoist was typically used to put a boat in the water around Memorial Day and out of the water in September. During the latest placement of the boat in the water, the owner attempted to lift the hoist back into position but could not get the unit to stop moving upward. The north cable for the hoist snapped causing the support structure to collapse. The owner indicated there had been no modifications or repairs made to the boat hoist during its lifetime and that the cables were oiled/greased prior to each Winter.

Field observations: The broken cable, which consisted of seven (7) wound wire cords with up to seven (7) individual wires in additional windings, was heavily frayed with corrosion to individual wires within the cable consistent with long-term exposure to moisture (Figures 23.9 and 23.10).

FIGURE 23.9   Boat hoist failure – far and close views.

FIGURE 23.10 Boat hoist failure – far and close views.

The wire windings were expanded consistent with the effects of heavy oxidation.

Forensic conclusions: The failure was ultimately due to heavy corrosion to the portion of the cable that caused it to snap when reportedly lifted past its normal resting position. Based on the heavily corroded and frayed section of the broken cable, the cable had been in the condition suitable for failure for an extended period of time (i.e., the cable was subjected to cyclic fatigue exacerbated by the corroded state).

Case Study – Grain Elevator Explosion

Scope of inspection: To determine the cause and origin of damages to the bucket grain elevator at the subject farm.

Background information: The ~115′ high subject grain elevator had been installed between 8 and 10 years prior to the inspection. Two weeks prior to the inspection, an explosion occurred within the north trunk of the grain elevator while loading corn into a semitruck. Several sections of the north trunk had blown out and the south trunk had damages as well. The bearings were routinely greased within the elevator. There were no temperature sensors on the bearings.

Field observations: Several sections of the north and south trunks of the grain elevator contained damages consistent with an internal explosion (Figures 23.11 and 23.12).

FIGURE 23.11 Grain elevator failure – far and close views.

FIGURE 23.12   Grain elevator failure – far and close views.

A strong burning smell was present within the head assembly at the top of the grain elevator. No damages were present to the motor or head assembly. Heavy grain dust and debris were present within the trunks, buckets, and surrounding area.

Forensic conclusions: The damage to the grain elevator was likely due to a confined grain dust explosion. The burning smell originating near the top bearings indicated a potential ignition source for the explosion. The fuel for the explosion likely derived from the buildup of grain dusts within the elevator.

Case Study – Grain Bin Re-corrugation

Scope of inspection: To determine the cause and origin of reported damages to a grain bin.

Background information: The grain bin in question had been installed in 1980 by the Amish and manufactured by GSI (Grain Systems). There were no stiffeners installed on the grain bin. A farmhand noticed that a portion of the corrugated panels on the bin had collapsed downward. The moisture content of the previous harvest had been between ~18% and ~19%. The corn in the bin had only been dried internally from the fan at the base of the bin.

Field observations: The grain bin was ~36′ in diameter and consisted of ten concentric rings of ~9′ 4″ wide × ~2′ 8″ high, corrugated metal sheets with 12 sheets total at the circumference. Several replacement sheets were present on the ninth ring from the base (Figures 23.13 and 23.14).

FIGURE 23.13   Grain bin failure – far and close views.

FIGURE 23.14 Grain bin failure – far and close views.

The lower corrugation of the fifth ring from the base had folded over (or re-corrugated) from the north end of the first base sheet and counterclockwise to the north end of the ninth sheet. The area of damage was free of corrosion. The remainder of the bin appeared undamaged.

Forensic conclusions: The observed re-corrugation of the fifth ring from the base of the grain bin was likely due to an unbalanced loading caused by a higher than normal moisture content of the corn being stored in the bin. While the grain is being dried out with the use of a base-drying system (and entrainment), the upper portion of the grain is still moist and tends to cling to the sidewall metal sheets causing an unbalanced loading, which can lead to re-corrugation.

23.3 METHODOLOGY FOR EQUIPMENT FAILURE FORENSIC INSPECTIONS

The basic steps in performing an equipment failure inspection have been discussed in previous chapters. These same methodologies, analyses, and inspection tools can be used to investigate equipment failures. However, further analysis of collected evidence through the use of microscopes or other nondestructive and destructive testing methods may be implemented.

23.4 METHODOLOGY FOR EQUIPMENT FAILURE FORENSIC INSPECTION REPORTS

The forensic report for equipment failure inspections should follow the same methodology outlined in previous chapters with modifications to adjust for equipment-specific observations and findings. An example of equipment specific methodology is provided below for a well pump failure.

23.4.1 OBTAIN WELL LOGS BEFORE SITE INSPECTION

Obtain well records and nearby well records from the State Department of Natural Resources website. This will provide data on the age, construction, recharge rate, water level, and date the pump/motor was installed into the well. This may also provide data from similar nearby wells. In one case, it was noted that several wells were installed nearby at slightly deeper elevations after the subject well was installed. This clearly caused the local well water table to lower, thus lowering the water level in the well of interest; the well pump/motor drew in more silt and debris, which led to mechanical failure. Information from the installation of the pump was copied down and is reproduced below:

Pump: Red Jacket
Model #: 50CNSW1 – CNS9BC
Hp: 1/2; *Volts*: 208/230
Amps: 5.5
Installation date: 7/17/1996
Static water level: 112'
Drawn down: To 0' at 12 GPM

This information, along with information collected at the site, allows one to date the pump/motor, verify the manufacturer of the unit, and determine if the original unit is the unit still installed.

23.4.2 Inspect Premises for Evidence of Collateral Lightning Damage

No high-voltage damage was observed to the exterior surfaces of the building or other components.

23.4.3 Inspect Equipment to Obtain Model and Serial Numbers and Date of Installation from Name Plate (if Provided)

In this example, the following information was obtained from the manufacturer's nameplate:

- *S/N*: 99A18.
- *Stamp*: 337445920.
- Franklin Electric, Bluffton, IN.
- *Model #*: 2445059004.
- *Hp*: 1/2; *Hz*: 0; *V*: 230; *Amps*: 5.0; *Max. Amps*: 6.0 ; *KW*: 0.37 ; *Ph.*: 1.
- *RPM*: 3,450.
- *KVA code*: R.
- S.F. 1.6.
- Continuous Duty E79319.
- Two-Wire Submersible Motor
- Thermally Protected
- Equipped With Lightning Arrestor

The serial number indicated that the well pump/motor was manufactured in 1999, making the unit ~13 years old at the time of the inspection. Note that it is stated on the pump that it is equipped with a lightning arrestor, suggesting it should have been protected from lightning strike damage.

23.4.4 Inspect Equipment – Look for Damage to Components Typically Damaged by High Voltage

Both the well-head area and wiring, along with the well pump/motor, should be inspected for evidence of high-voltage damage. An example of a well-head casing and well-head wiring is shown in Figure 23.15.

FIGURE 23.15 Well head casing and pump/motor wiring.

As seen, no high-voltage damage was present to the well casing or well pump/motor wiring.

Similarly, once the well pump/motor is removed (Figure 23.16), general observations of the casing and wiring entering the casing should be made with a focus on evidence of high-voltage damage (e.g., burn marks, melted or burned wires).

FIGURE 23.16 Removed well pump/motor.

In this example, no high-voltage damage was present, but heavy iron and soil deposits covered the unit, including the intake area of the pump. The motor section is typically located at the bottom of the unit and the pump section above the motor section.

The well pump/motor is then typically logged in as evidence and stored. Like with most evidence, no destructive testing should occur until all impacted parties are informed and agree on a date and time for destructive evaluation of the well pump/motor. At times, one or more parties either do not respond or miss the agreed-upon inspection. If this occurs, it is assumed that reasonable notice had been provided and destructive testing can be completed.

Destructive testing of the unit should include a sequence of repeated steps: observations should be completed, photographs taken, and then disassembly is completed. For well pump/motors, aside from looking for evidence of high-voltage damage, the following details should be recorded:

- Whether or not the pump and/or motor will rotate freely (if not, this is a sign of motor failure, clogging with debris, failure of seals, or packing, etc.).
- The presence and condition of a lightning arrestor.
- Evidence of debris or soils in sensitive pump and/or motor areas.

As an example situation, actual destructive pump/motor disassembly findings from this case study are summarized below:

The screen at mid-level was removed and exposed the nuts holding the lower motor assembly to the upper pump assembly. The screen was over half clogged with wet, rust-colored deposits. Once the screen was removed, the area behind the screen (pump inlet) was also heavily coated in wet, rust-colored deposits (Figure 23.17).

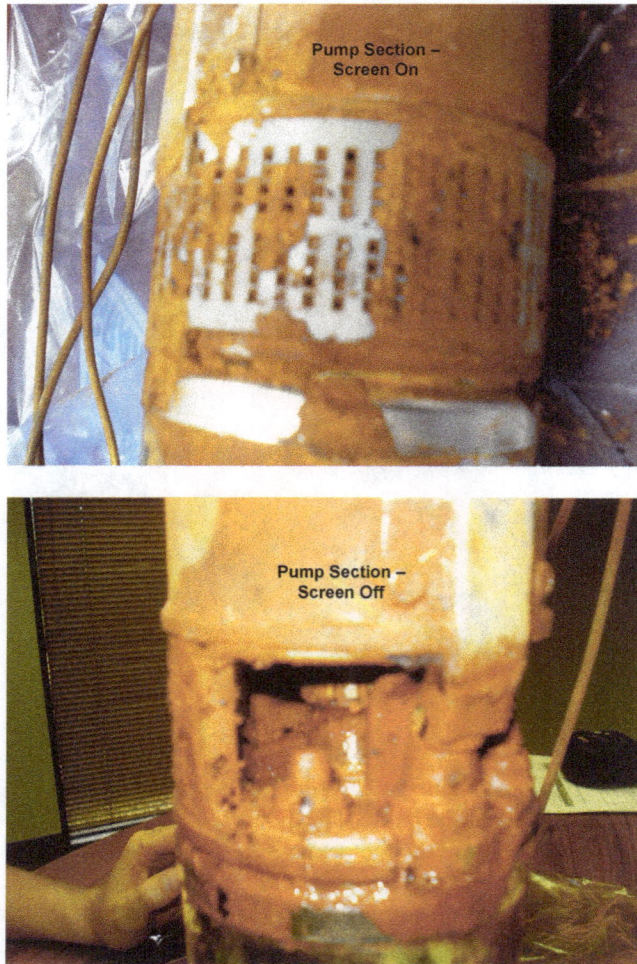

FIGURE 23.17   Well pump screen (a) and assembly behind screen (b).

The nuts and electrical assembly were detached allowing one to pull the pump assembly from the motor assembly (Figure 23.18). No burn marks or melted wires were found on the wire bundle terminal assembly.

FIGURE 23.18 Well pump – motor assembly and drive to pump assembly.

The pump shaft, when manually manipulated by hand, rotated freely but with the feel of some grit on the impeller.

The motor shaft, when manually manipulated by hand, would not rotate. When vice-grips were place on the shaft, the shaft would not turn clockwise under full hand force, but would barely rotate counterclockwise with full hand force. The motor was clearly seized.

The top of the motor case was removed and exposed the motor windings and casing. The windings were rusted and very little oil was present in the case. What little material was drained from the motor case had a water/sediment appearance. Neither the electrical wiring connector nor the arrestor plug appeared to contain burn marks or melted leads consistent with high-voltage damage (Figure 23.19).

FIGURE 23.19 Lack of lubricant near motor windings and lack of high-voltage damage to wire connector or voltage arrestor.

In this particular example, the pump failed due to a combination of factors not related to lightning. These included failure of the seal, which caused a loss of lubricant and allowed water to enter into the motor windings. This failure was likely due to a lack of water at the bottom of the well due to a lowered water table from the installation of several wells into the same aquifer after the installation of this well. The bottom of the well consisted mostly of iron mud and sludge, which likely cause the pump motor to run hot.

IMPORTANT POINTS TO REMEMBER

- Equipment failures are commonly seen during forensic engineering inspections.
- Typical equipment failures are associated with air-conditioning systems, dishwashers, washing machines, sump pumps, and icemakers. Other failures seen in rural areas are well pump failures and farm equipment failures.
- Use of a differential diagnosis approach is the most successful approach to determining equipment failures. This involves an understanding of how the equipment is supposed to work, why it is now not working, and what is the differences between the two conditions.
- Solenoid valve failures are commonly seen in dishwasher failures.
- Sump pump failures occur from several factors including obstructions of the float valve, drain line failures, operating too long with a dry sump, and sump pump motor and pump failures.
- In general, it is unlikely that commonly reported lightning-damaged equipment such as well-pump motors were actually damaged by lightning. Most of the time the failures are from other causes.
- Electrical checks to motors and compressors can be used to help determine whether or not lightning was responsible for its reported failure.

REFERENCES

1. "Sump Pumps." Accessed June 8, 2020, https://www.sumppumpinfo.com/sump-pump-types.html.
2. International Code Council. "2018 International Plumbing Code (IPC)." Accessed June 8, 2020, https://codes.iccsafe.org/content/IPC2018/chapter-11-storm-drainage#IPC2018_Ch11_Sec1113.
3. Roberts, Charles, C, Ph.D. "Sump Pump Failures." Accessed June 8, 2020, http://www.croberts.com/sumppumpfailure.htm.
4. Zoellner Pump Company. "Installation Instructions – Models 49, 50, 70, 98, 130, 141/4140, 145/4145, 150, 160/4160, 180/4180, 191, 371, 372, 373 Series Pumps," 2017. Accessed June 8, 2020, https://www.zoellerpumps.com/content/literature/fm2549_Ec.pdf.
5. Noon, Randall K. *Forensic Engineering Investigation*. New York: CRC Press, 2001.
6. The ACHR News. "An Early History of Comfort Heating." Accessed September 26, 2020, https://www.achrnews.com/articles/87035-an-early-history-of-comfort-heating#:~:text=Nelson%20Bundy%20invented%20the%20first,bankruptcies%20and%20consolidations%20were%20frequent.
7. "Weil-McClain Boiler Manual: EG and P-EG (Series 1) and EGH (Series 2)." Accessed September 26, 2020, https://www.weil-mclain.com/sites/default/files/field-file/eg-peg-series-1-egh-series-2-manual_1.pdf.
8. "Acceptable Quality Level." Accessed September 26, 2020, https://www.whatissixsigma.net/acceptable-quality-level/.
9. "23rd Annual Portrait of the U.S. Appliance Industry", *Appliance Magazine*. September 2000. http://www.corrosioncost.com/pdf/homeappliances.pdf.

24 Serving as an Expert Witness

Thomas E. Schwartz
Holloran White & Schwartz LLP

Stephen E. Petty
EES Group, Inc.

CONTENTS

PURPOSE/OBJECTIVES

The purpose of this chapter is to:

- Provide an overview of the civil litigation process.
- Provide resources regarding how to serve as an expert witness.
- Provide "lessons learned" serving as an expert witness.

Following the completion of this chapter, you should be able to:

- Have an understanding of the civil litigation process.
- Know where to go for information on serving as an expert witness.
- Be aware of the pitfalls that can befall an expert witness.

24.1   INTRODUCTION

D.A. Jim Trotter: Ms. Vito, what is your current profession?
Lisa: I'm an out-of-work hairdresser.
D.A. Jim Trotter: An out-of-work hairdresser. In what way does that qualify you as an expert in automobiles?
Lisa: It doesn't.

In the American movie classic *My Cousin Vinny*, two New Yorkers are charged with murder in a small southern town. The only thing standing between the accused murderers and death row is their less-than-perfect attorney. Just when the trial could not be going any worse, the defendants' attorney (played by Joe Pesci) calls his fiancé to the witness stand to testify as an "expert" in the field of automobiles. When her qualifications come into question, the would-be expert establishes her ability by rattling off obscure facts that could only be known by a true car expert. Her surprise testimony proves that the defendants did not commit the crime, and the innocent men are set free. Thanks to the expert, our justice system worked again!

Although this movie scene takes the usual Hollywood liberties in stretching reality, there are some very important truths to take away. First and foremost, experts play a crucial and oftentimes outcome-determinative role in our legal system. Lawyers, judges, and juries rely on experts to apply their specialized knowledge to the facts of a case and explain how or why an event happened. Another important point to remember is that experts come in many different shapes and sizes. Experts can have varying levels of education and occupational histories. Regardless of how an expert witness came to serve as an expert in a particular case, the only imperative is that they know the subject of your expertise.

This chapter has been included because forensic engineers are frequently called upon to serve as experts in litigation. This chapter is not meant to serve as "Law 101," many books, including several referenced here,[1-8] better cover the topic. The purpose of this chapter is to give a basic understanding of an expert witness' role in the civil litigation process. Tips gained as part of these experiences are also imparted at the end of this chapter.

24.2   THE BASIC ELEMENTS OF A CIVIL CASE

Although expert witnesses are used in both criminal and civil matters (i.e., lawsuits), a forensic engineer or scientist is more likely to be asked to serve in the context of a civil case. Interestingly, the term "suit" in lawsuit was defined in an 1875 Ohio eminent domain case by Chief Justice Marshall of the United States Supreme Court,[7] when speaking for this court, he said, "The term [suit] is certainly a very comprehensive one, and is understood to apply to any proceeding in a court of justice by which an individual pursues that remedy which the law affords."[9]

For simplicity sake, the civil litigation process can be broken down as follows: the plaintiff's complaint, the defendant's answer, the discovery phase, and the actual trial (Figure 24.1).

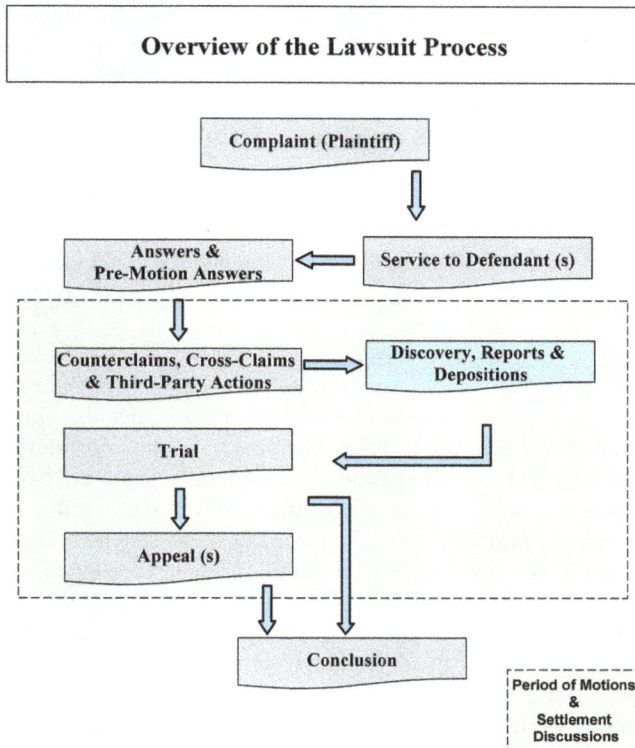

FIGURE 24.1   Illustration of the elements and processes in a typical civil lawsuit.

Generally speaking, an expert's role is limited to discovery and trial.

24.2.1 THE COMPLAINT

To begin a lawsuit, someone files a complaint (called a "petition" in some jurisdictions) in civil court. The complaint is filed by the plaintiff and outlines the allegations against the defendant. The complaint identifies the parties involved in the case, the facts surrounding the events at issue, the nature of the defendant's actions that caused injury to the plaintiff, and the extent of the damages suffered by the plaintiff as a result of the defendant's actions. The complaint essentially gives the "who, what, when, and where" of the case. The complaint is served on the defendant.

24.2.2 DEFENDANT'S ANSWER

After the defendant receives a copy of the complaint, he must file an "answer" with the court. The defendant's answer addresses the allegations in the complaint. The defendant can admit the allegations, deny them, or claim that he does not have enough information to admit or deny them. In answering the complaint, the defendant also lists his affirmative defenses to the case against him. An affirmative defense is a legal reason why the defendant should not be held liable for the plaintiff's injuries. For example, a plaintiff alleges that the general contractor failed to seal a door frame and allowed moisture to enter the home, resulting in mold. The complaint alleges that the mold caused the plaintiff to become sick and that the mold has greatly reduced the value of his home. As an affirmative defense, the general contractor could state that he was not responsible for sealing the door frame pursuant to the construction contract.

In some cases, the defendant may file a motion to dismiss in lieu of filing an answer. A motion to dismiss requests the court to reject the plaintiff's case because the plaintiff's complaint is defective in some way. A motion to dismiss may claim that the plaintiff has failed to provide enough information in the complaint in order for the defendant to know why he is being sued. A motion to dismiss can also claim that the case was not filed in the proper jurisdiction. In most instances, a motion to dismiss essentially argues that even assuming that the allegations in the plaintiff's complaint are true, the plaintiff still loses. After the defendant files his answer, the case then moves onto the next stage of litigation.

If the defendant files a motion to dismiss the complaint and the court grants the motion, the case is over. If the court denies the defendant's motion, the case continues on to the discovery phase.

24.2.3 DISCOVERY

The discovery phase is the most time-consuming part of litigation, but arguably, the most important part. During this time, both the plaintiff and the defendant have an opportunity to learn more about the other party's allegations and the facts of the case. Generally, there are two types of discovery: written and oral. Written discovery is generally in the form of "interrogatories" and "requests for production." An interrogatory is simply a set of written questions sent by one party to the other, and it must be answered within a certain period of time. In our mold case example, the defendant may ask the plaintiff to identify any witnesses who saw the defendant install the door frame. A written request for production asks one party to provide copies of certain relevant documents or other tangible items. The plaintiff may ask for the defendant to provide a copy of the receipt generated from the purchase of the door frame sealant. In response to the request, the defendant must provide the plaintiff a copy of the receipt, let the plaintiff know that he does not have the receipt, or explain to the court why he should not be forced to give the plaintiff the receipt even though he has it in his possession.

Oral discovery includes the taking of witness depositions.[10,11] In preparing for trial, the parties search for information relating to the plaintiff's allegations and the defendant's defenses. A deposition is a time where the parties' attorneys sit down with a knowledgeable witness and ask questions relating to the facts of the case. The parties (the plaintiff and defendant) are usually deposed before any other witnesses. A deposition is a great tool used to learn more about the facts so that a proper strategy can be developed.

The pursuit of information in the discovery process sometimes goes on for a long time if new information continues to become known. In our mold case, let us assume that in his deposition, the plaintiff testifies that he saw the contractor, Joe, installing the door frame. Inevitably, the attorneys are going to want to ask Joe some questions about the door frame installation. In Joe's deposition, he admits to installing the door frame, but he says that Frank provided the sealant and told him what to do. You can probably guess what is coming next – Frank's deposition. When Frank sits down to answer the attorneys' questions, he testifies that he purchased the sealant from Scott's Hardware. The parties can now find out exactly what type of sealant was used on the door frame. As you can see, as more people are deposed and more questions are asked, new areas of information are found.

Inspection of evidence, or site inspections, can be valuable in order to gain an understanding of the condition of the evidence and/or the site and how/why it contributed or did not contribute to the complaint. For example, disassembly of a reportedly failed sump pump, causing water damage and visible mold, can be undertaken to determine whether or not the pump actually failed and if it did fail, why? The results of such an analysis have significant implications on the outcome of the lawsuit. For this example, if the failure occurred with the pump, the pump manufacturer may be implicated. If the failure occurred due to construction debris in the intake of the pump, the installation contractor or the owner may be implicated for the cause of the damage. Precautions should be taken on the storage and destructive testing of collected evidence. The following steps should be taken for the collection and storage of evidence:

- Take notes and photographs of the evidence at the site before and after you remove it.
- Take care in collection of the evidence so as not to spoil the evidence.
- Prepare a chain of custody form for the evidence (Figure 24.2).

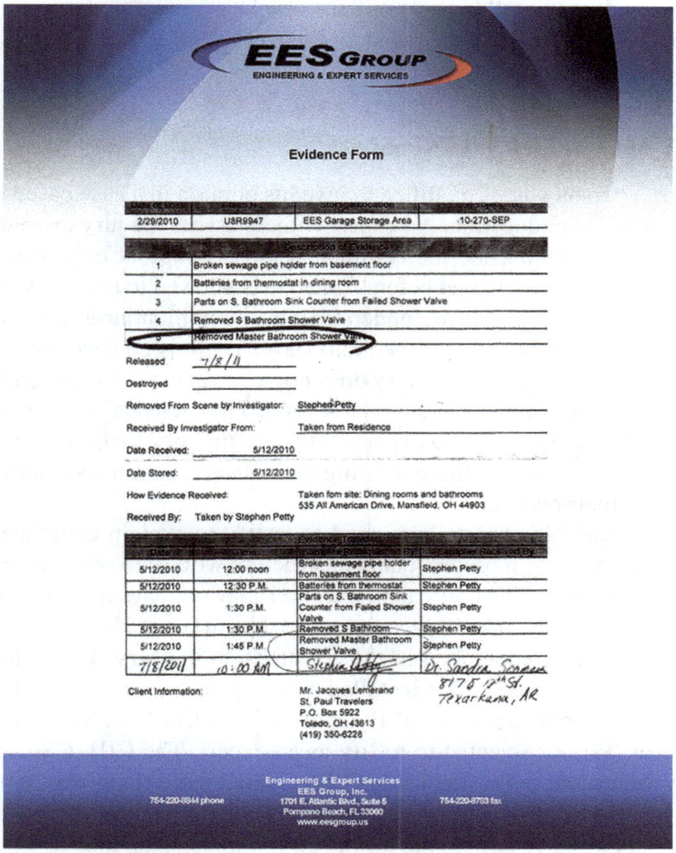

FIGURE 24.2 Example of evidence chain of custody form.

It is not unusual for some time to pass before all parties can be notified and a common date, time, and location agreed upon to meet to examine the evidence. In this case, the evidence was collected and stored on May 12, 2010 and then released to a manufacturer's representative after examination on July 8, 2011.

- Store the evidence in a secure climate-controlled area.
- Do not perform destructive testing on the evidence until all affected parties have been given proper notice of a date for evidence review. Spoilage of evidence without the opportunity for the other party to review it may result in the inability of any party to use such evidence at trial since not all parties had equal access to the equipment or its examination.

24.2.4 TRIAL

Although cases move through the stages of litigation with an eye toward trial, the vast majority of cases never get there. Most civil cases are resolved, by either settlement or dismissal, before a trial becomes necessary. When a case goes to trial, the attorney for each party assembles all of the information gained through the discovery process for presentation to the judge or jury. The plaintiff puts forth his evidence first to explain what the defendant did to cause him harm, why the defendant should be held responsible, and what damages the plaintiff suffered. The defendant then gets his chance to introduce evidence to tell his side of the story. Depending on the complexity of the case, there could be two witnesses called to the stand, or there could be 100. Regardless, the judge or jury determines who is right in the end.

24.3 WHAT IS AN EXPERT WITNESS?

An expert witness is a person who is qualified to give his opinion in a case based on his knowledge, skill, experience, training, or education. An expert witness assists the jury in understanding certain aspects of a case that may be outside the common person's scope of experience and knowledge. There are requirements that must be met before someone is allowed to testify as an expert. First, the subject area must be one that is not easily understood by a common juror. Let us take a simple car accident case. Sam ran a stop sign and crashed into Dave. There is no need for an expert to explain to the jury the purpose of a stop sign. The jury does not need an expert to tell them why it is important to stop at a stop sign before entering an intersection. The "science" of stop signs is commonly understood. An expert is only needed to explain things to the jury that they do not already know. The Federal Courts have six rules for those offering opinions and expert testimony[12,13] and the States and Local Courts have their own rules.

In some situations, multiple experts may need to testify to explain complicated issues. In our mold case, the plaintiff may ask a professional construction worker to serve as an expert regarding the installation of door frames. The construction worker may not have a high school degree, but he may have 30 years of experience installing and sealing door frames. He can explain to the jury the proper method of installation recognized in the construction industry. This information will help the jury determine whether the door was installed properly and if not, what was done wrong. The plaintiff may also ask a Certified Industrial Hygienist (CIH), Certified Safety Professional (CSP), Professional Engineer (PE), or scientist to testify as an expert. The CIH, CSP, PE or scientist can explain to the jury where the mold was located, how much mold there was, how the mold could have gotten there, and the negative impact of the mold. Without the specialized knowledge of both these experts, it would be nearly impossible for the jury to understand how or why the plaintiff was damaged.

An expert's testimony must be reliable. Reliability is judged by the principles and methods the expert applies in coming to their opinion. An expert's specialized skill should allow them to analyze the data in a manner that is consistent with others in that field. Whether or not an expert can be deemed "reliable" is often the subject of much contention in the context of civil litigation. An expert must be able to explain how they came to the opinions and conclusions they offer and their scientific or technical process must be dependable.

24.4 WHERE DOES AN EXPERT WITNESS FIT IN THE CIVIL CASE?

In some situations, an expert is asked to review the facts of a case before the complaint has even been filed. In these circumstances, the plaintiff's attorney is investigating the case to determine whether a complaint should be filed. In our mold case, at the time the homeowner hires an attorney, it may not be clear whether the type of mold growing in the plaintiff's house could be caused by moisture from an unsealed door frame. If the plaintiff files the case without knowing this information, he could eventually lose the case after spending much time and money in the litigation process. In the face of questionable facts, a conscientious attorney will consult an expert before filing the lawsuit. In such a pre-suit consultation, the expert is asked to give his opinion despite not having the benefit of all the facts that will eventually develop in the discovery process. In our case, a CIH or scientist may be asked to visit the plaintiff's home, assess the mold growth, and advise the plaintiff's attorney. The attorney will use the expert's opinion when determining whether to file a lawsuit.

Once a case is commenced, an expert's main role is in the discovery and trial phases of the litigation process. Once the facts of the case have been gathered in written and oral discovery, an expert is asked for their opinion on certain aspects of the case that fall within their area of expertise. The expert conducts a thorough review and reliable analysis. In most circumstances, the expert is asked to provide a written report outlining their opinions and conclusions. In discovery, both the defendant and the plaintiff are given the report, and both sides have a chance to question the expert in a deposition if it is requested by either party. Ultimately, if the case goes to trial, the expert will likely be asked to testify on the witness stand.

24.5   SERVING AS AN EXPERT WITNESS IN A CIVIL CASE: THE STEPS OF THE PROCESS

There are two types of expert witnesses: (1) non-retained and (2) retained. A non-retained expert is an expert who is actually involved in certain factual developments of a case. A "non-retained" expert has personal knowledge of the facts of the case and can apply his specialized skill to assist the jury in understanding those facts. In our mold case, a non-retained expert could be the plaintiff's treating physician who diagnosed his mold-related illness. The treating physician, who found himself in the middle of the litigation simply by doing his job, may have the expertise to explain to the jury the detrimental effects mold had on his patient. On the defense side, one of the plaintiff's doctors may have indicated in a medical record that he did not believe the plaintiff's illness was caused by mold. The defendant's attorney would likely seek to have this non-retained expert testify on the defendant's behalf.

A "retained" expert is an expert who is not involved in the case (and has no knowledge of the case's facts) until they are asked to serve as an expert by one of the parties. A retained expert is provided with information about the case by the party who has hired him. They are then asked to give their opinion on certain issues involving their area of expertise. The sections below focus on the job of a retained expert.

24.5.1   RETAINED EXPERT – INITIAL CONTACT BY AN ATTORNEY

As mentioned above, experts are sometimes consulted before a lawsuit is filed. Most of the expert's work is conducted during the discovery phase of the litigation process, so they are usually contacted by an attorney before or during discovery. If an expert never worked with a particular attorney, the initial contact is quite important. During this conversation, the expert would be able to obtain the following information:

How the attorney came to hear about them as a potential expert. The attorney could have gotten their name and contact information from an industry list, a court filing in another case, or a friend or colleague who had worked with that expert in the past. Once this is known, it is easier for the expert to know how much they need to explain about their particular area of expertise and what the attorney should expect from them in the process.

When the expert is contacted by the attorney for the first time, they are given an explanation of the facts. The attorney should tell them the specific issues where their opinion is needed. Although the attorney is not likely to get into much detail on initial contact, it is important for the expert to ask questions to understand fully the nature of the claim and the parties involved.

It should be disclosed to the attorney if the expert has previously served as an expert witness for one of the parties in the case. If the expert's brother-in-law owns the company being sued, this is obviously something they would want to know prior to entering into an agreement to serve as an expert in any capacity.

It is important to note that a lot can be learned about the attorney during this first contact. If the attorney seems unprepared or confused about his own case, the expert may want to think twice about getting involved. The expert may come across someone who believes that since he is paying for his or her opinion, he gets to dictate the substance of that opinion. This is not how the attorney/expert relationship works. If the attorney is overly suggestive regarding what the expert's opinion must be, this is a definite red flag and an expert should not consult on the case under these circumstances. The great majority of attorneys handle experts very professionally and ethically.

24.5.2 RETAINED EXPERT – AGREEMENT TO CONSULT WITH AN ATTORNEY TO REVIEW THE FACTS OF THE CASE

During this initial contact, the attorney may ask the expert to take a closer look at the case. If the expert agrees, their relationship at this time is defined as a "consulting expert." As a consultant at this early stage, the attorney will provide certain materials to give the expert a better understanding of what happened in the case. Depending on the type of case, a consulting expert may be given photographs of the accident scene, witness statements, maps or geographical descriptions, or the owner's manual for an alleged defective product. The expert may be asked to visit the location of the incident or inspect a particular item. As a consultant, they have only agreed to take a closer look at the case based on what the attorney told them in the initial contact. The expert has made no guarantees as to what their opinion may or may not be. The attorney is under no obligation to disclose the expert's role in the case to the opposing side at this time.

24.5.3 RETAINED EXPERT – AGREEMENT AS TO COMPENSATION FOR YOUR TIME

At the time an expert agrees to consult with the attorney on his case, they should also enter into an agreement as to how the expert will be compensated for his or her time. The expert should make sure to explain exactly how their time is billed. Most importantly, the expert needs to have a clear understanding of what the attorney wants them to do and should communicate to him the amount of time such tasks can take. If both the expert and the attorney understand these points, future problems in the relationship can be avoided. In our mold case, let us assume that the attorney asks the expert to determine the source of the mold. The attorney provides them with 300 photographs and three deposition transcripts for review. The expert estimates that it will take at least 12 hours to review the material and develop an initial determination. Knowing their hourly rate and the approximate time needed, the expert can now provide the attorney with an estimated amount of the bill to be expected. In addition, specific terms and conditions associated with depositions and trial support should be outlined in writing (e.g., rate, cancellation fees, minimum times, travel, and expense costs). For illustration purposes, an example retention agreement is shown in Figures 24.3–24.5.

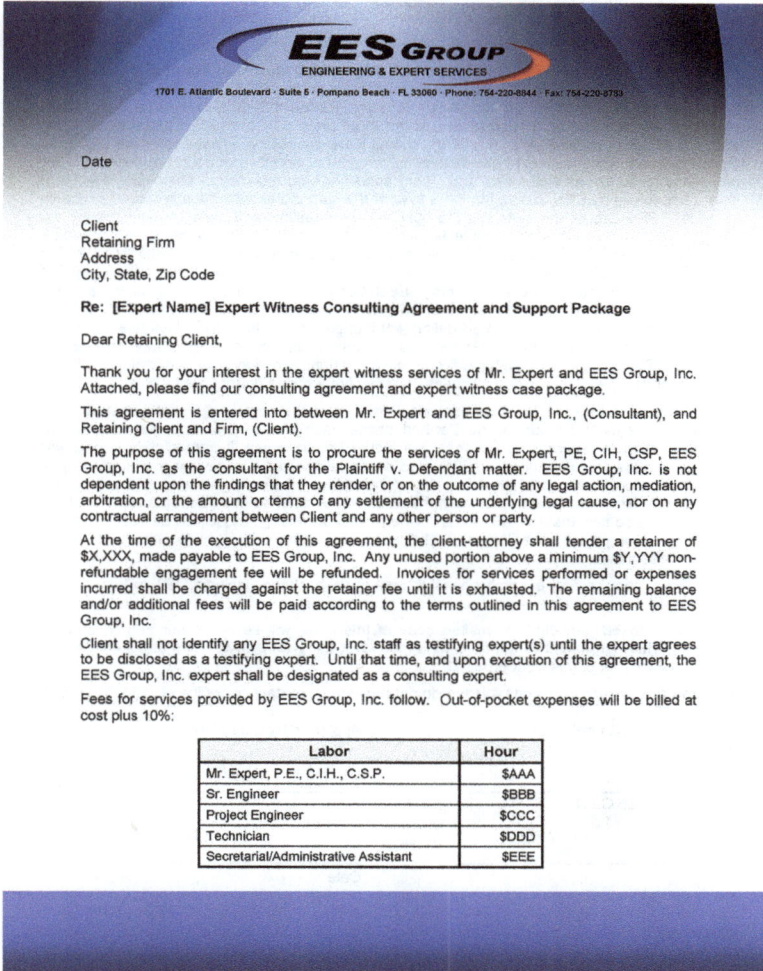

EES GROUP
ENGINEERING & EXPERT SERVICES

1701 E. Atlantic Boulevard · Suite 5 · Pompano Beach · FL 33060 · Phone: 754-220-8844 · Fax: 754-220-8753

Date

Client
Retaining Firm
Address
City, State, Zip Code

Re: [Expert Name] Expert Witness Consulting Agreement and Support Package

Dear Retaining Client,

Thank you for your interest in the expert witness services of Mr. Expert and EES Group, Inc. Attached, please find our consulting agreement and expert witness case package.

This agreement is entered into between Mr. Expert and EES Group, Inc., (Consultant), and Retaining Client and Firm, (Client).

The purpose of this agreement is to procure the services of Mr. Expert, PE, CIH, CSP, EES Group, Inc. as the consultant for the Plaintiff v. Defendant matter. EES Group, Inc. is not dependent upon the findings that they render, or on the outcome of any legal action, mediation, arbitration, or the amount or terms of any settlement of the underlying legal cause, nor on any contractual arrangement between Client and any other person or party.

At the time of the execution of this agreement, the client-attorney shall tender a retainer of $X,XXX, made payable to EES Group, Inc. Any unused portion above a minimum $Y,YYY non-refundable engagement fee will be refunded. Invoices for services performed or expenses incurred shall be charged against the retainer fee until it is exhausted. The remaining balance and/or additional fees will be paid according to the terms outlined in this agreement to EES Group, Inc.

Client shall not identify any EES Group, Inc. staff as testifying expert(s) until the expert agrees to be disclosed as a testifying expert. Until that time, and upon execution of this agreement, the EES Group, Inc. expert shall be designated as a consulting expert.

Fees for services provided by EES Group, Inc. follow. Out-of-pocket expenses will be billed at cost plus 10%:

Labor	Hour
Mr. Expert, P.E., C.I.H., C.S.P.	$AAA
Sr. Engineer	$BBB
Project Engineer	$CCC
Technician	$DDD
Secretarial/Administrative Assistant	$EEE

FIGURE 24.3 Example of EES Group, Inc. retention agreement – p. 1.

For testimony at deposition or trial, Mr. Petty shall be compensated at the rate of $X,XXX per half day (≤ 4 hours) and $Y,YYY per day (> 4 hours; maximum of 8 hours per day) plus travel and expenses if applicable. If continuous time for deposition/trial is to exceed eight hours, testimony will be billed as a second day, at the rates previously defined. The tabulated rates above shall apply to time while travelling, should it be necessary. These rates for testimony shall apply both while waiting to give testimony, whether at an office or court and for time taken for breaks or meals, as well as for time spent actually giving testimony. Deposition payments must be received by EES Group, Inc. prior to or at the start of the deposition in order for experts to testify. EES Group, Inc. reserves the right to cancel testimony if account balances are not current. If less than 48 hours' notice is given for the cancellation of a deposition or trial, a minimum of one-half day fee of $X,XXX will be charged. The client is responsible for payments to EES Group, Inc. as outlined in this contract, regardless of any arrangement the client has with any party or parties he represents. EES Group, Inc. will issue invoices on a semi-monthly/monthly basis, or whatever other interval is deemed appropriate. Invoices are due upon receipt, and shall be considered delinquent if unpaid more than thirty days after their date of issuance. In the event that a bill remains unpaid for sixty or more days after the date of issuance, EES Group, Inc. shall have the unrestricted right to resign from performing additional services for the Client on any and all cases that they are working on for Client's firm.

In the event of a Daubert (or state equivalent) motion of the above-reverenced case(s), EES Group, Inc. wants to be part of the Daubert phase each step of the way. As soon as the Daubert motion is received by Defense, we would like to review it immediately (along with relevant exhibits to the motion). EES also reserves the right to review the draft of your opposition before it is filed at least five days (preferably 7) before it is due. EES will review, provide comments, and return it to you right away. If the defense is permitted a reply brief, we ask that you also sent that to EES. Finally, when the court rules, the decision be sent to EES as well. If these steps are not followed, EES Group, Inc. reserves the right to withdraw from this case or these cases.

EES Group, Inc. carries a comprehensive insurance package, including $2 million general liability insurance. Certificates will be provided upon request.

One signed faxed, e-mailed, or mailed copy of this letter will serve as our authorization to proceed. This agreement is effective for a period of thirty (30) days. The attached General Conditions are part of this work order.

If we can answer any questions, please do not hesitate to contact our office at (754) 220-8844.

Approved by EES Group, Inc.:   Approved by Client Firm:

_____   _____
Mr. Expert, P.E., C.I.H., C.S.P.
Title

_____   _____
Date   Date

Attachment

EEs Group   Engineering & Expert Services, Inc. ©

FIGURE 24.4   Example of EES Group, Inc. retention agreement – p. 2.

Engineering & Expert Services, Inc. (EES Group, Inc.)
General Conditions

1. **PARTIES & WORKSCOPE:** Engineering & Expert Services, Inc. (hereinafter referred to as EES) shall include said company, its affiliates, suppliers, or subcontractors performing the work. "Work" means the specific services to be performed by EES as set forth in EES's proposal, and these General Conditions. Client refers to the person or business entity ordering the work to be done by EES. If the client is ordering the work on behalf of another, the client represents and warrants that the client is the duly authorized agent of said party for the purpose of ordering and directing said work. Unless otherwise stated in writing, the client assumes sole responsibility for determining whether the quantity and nature of the work ordered by the client is adequate and sufficient for client's purposes. The ordering of work from EES constitutes acceptance of the terms of EES's proposal and these *General Conditions*.

2. **SCHEDULING OF WORK:** The services set forth in EES's proposal will be accomplished in a timely, efficient, and professional manner by EES personnel at the prices quoted. If EES is required to delay commencement of the work, or if upon starting the work, EES is required to stop or interrupt the progress of its work as a result of changes in the scope of work requested by the client, to fulfill the requirements of third parties, or other causes beyond the direct reasonable control of EES, additional charges will be applicable and payable by the client. Additional charges will be billed at rates so stated in the proposal.

3. **CONFIDENTIALITY:** EES agrees not to violate the confidentiality of the client through the release or disclosure of contractual agreements, testing results, processing procedures and any other information made available by the client in the conduct of said project, without client's express consent, except where such disclosure is required by a court or governmental agency of competent jurisdiction.

4. **INSURANCE:** EES shall maintain all appropriate Workmen's Compensation and General Liability Insurance.

5. **LIMITED WARRANTY:** Materials supplied by EES are warranted per the manufacturer's written warranty. Installation is warranted for six months, ordinary use, wear or tear, or damage from abuse or accident accepted. NO OTHER WARRANTY IS EXPRESSED OR IMPLIED.

6. **PAYMENT:** The client shall be invoiced at least once each month for work performed during the preceding period. The client agrees to pay each invoice upon its receipt. Past due payments shall bear interest at the rate of 1 1/2% per month on the outstanding balance until paid. In the event of default in payment or any other terms of this Agreement by client, EES may, at its option: (i) terminate this Agreement; or (ii) declare the unpaid balance due and payable, without notice or demand to client, and sue and recover from client said amount, together with all reasonable costs and attorney's fees incurred by EES relating to its enforcement or preservation of its rights hereunder.

7. **TERMINATION:** This Agreement may be terminated by either party upon seven (7) days written notice. In the event of termination, EES will be compensated by the client for all services performed up to and including the termination date.

8. **INDEMNITY:** The client, and if the client is acting as an agent for a principal in ordering work from EES, then also said principal, agrees to indemnify, defend and hold EES, its officers, employees and agents harmless from any and all claims, suits, losses, costs and expenses, including but not limited to, court costs and reasonable attorney's fees arising or alleged to have arisen out of or resulting or alleged to have resulted from the performance of EES's work on or about the project and caused in whole or in part by any negligent, willful or wanton act or omission of the client or the client's principal or any party directly or indirectly employed by the client or the client's principal or anyone for whose costs the client or the client's principal may be liable except to the extent, and only to such degree, as such claim, suit, loss or damage is caused by the sole negligence or willful, or wanton act of EES, its officers, agents, employees or anyone for whose acts EES may be liable. In the event that the client or the client's principal shall bring any suit, cause of action, claim or counterclaim against EES, and to the extent that EES shall prevail upon such suit, cause of action, claim or counterclaim, the person or business initiating such actions shall pay EES the costs expended by EES to answer and/or defend against such suit, cause of action, claim or counterclaim, including reasonable attorney's fees, witness fees, and other related expenses.

9. **NOTICE OF COMMENCEMENT:** This contract constitutes an immediate and continuing request by EES that it be provided with a copy of the Notice of Commencement on this project from the owner and execution of this agreement constitutes acknowledgment by client of this request. Owner shall prepare and record a Notice of Commencement on this project and respond timely to requests for copies of the Notice of Commencement by sub-trades. EES shall not start work until it has received proof that a Notice of Commencement has been recorded by the owner with the County Recorder and shall be entitled to an extension of time for every day of delay caused by a late filing of the Notice of Commencement.

10. **DIRECT PAYMENTS:** Owner shall make no direct payments to subcontractors or suppliers on the project without giving ten (10) day written notice to EES of its intention to do so. If EES disputes or contests this direct payment in writing to the owner within this ten (10) day period, owner will refrain from any such direct payment unless EES is adjudged bankrupt, insolvent or in receivership.

11. **COMPLETE AGREEMENT:** Client and EES mutually agree that the EES written proposal and these *General Conditions* comprise the full and entire agreement between the parties and no other agreement or understanding has been entered into or will be recognized, and that all negotiations, acts, or representations made prior to the ordering of work shall be deemed merged in, integrated and superseded by the EES proposal and these general conditions. This agreement may be amended only by an instrument in writing signed by both parties.

EES Group   Engineering & Expert Services, Inc.©

FIGURE 24.5 Example of EES Group, Inc. retention agreement – p. 3.

It is a good policy to require a retainer in an amount that will cover the expert's initial work on the case. Commencement of work on the case can begin once the retainer fee is received and the retainer agreement is executed.

It is important to note that if the expert is consulting on a plaintiff's case, the attorney they are working with is likely to be working on a contingency fee basis while fronting the cost of the case for his client. Expert consulting fees are included as a cost of the case. Sometimes, the attorney may not have decided to move forward with the case and is only investigating the merits of the case. If the attorney does not have a clear picture of how much time and money your initial work will take, an expert bill for thousands of dollars, for a case that he ultimately does not pursue, could come as quite an unpleasant surprise.

24.5.4   RETAINED EXPERT – CAREFUL REVIEW OF THE FACTS OF THE CASE

After the scope of work has been agreed upon and financial arrangements are in place, it is time to begin organizing and reviewing the materials that have been provided. As the material is reviewed, any questions the expert may have should be noted. If a certain piece of information is needed, the attorney should be asked. Oftentimes, the attorney has only provided materials that he believes will be helpful. Because the attorney is not the expert, he may not know the importance of certain pieces to the puzzle. In our mold case, photographs of the floor joists underneath the door frame may not have been provided. Experience indicates that the floor joists directly beneath a source of moisture help trace the growth pattern of mold. The expert should make sure the attorney is asked for these photographs. He may have held them back because he did not see any mold on the joists and did not think the photos were important. If photos were not taken of the joists, he may send someone to photograph them. Depending on how important the missing information is to the expert's review, he may ask them to visit the site. In summary, the attorney must know what information is needed in order to develop a reliable opinion. While reviewing the materials, taking notes on the pieces of evidence is helpful in the expert witness' analysis.

24.5.5   RETAINED EXPERT – GOOD UNDERSTANDING OF THE ALLEGATIONS

In order for an opinion to be developed that is germane to the case, the expert needs to have a thorough understanding of the allegations in the plaintiff's complaint and the defendant's answer. When it is said that the allegations need to be understood, it does not mean that the expert's opinions should be molded to fit the allegations. As a consultant, the attorney is paying for complete honesty with the facts of his case. With that said, the attorney is also paying for the expert to teach him something about his case that he needs to know. Ultimately, the expert opinion will be used to help prove or disprove certain propositions, so the review of materials pertinent to the case needs to be tailored to answer those questions.

After the review is completed, the expert's finding should be relayed to the attorney. It is crucial that none of the pros or cons of the case that have been discovered are withheld. If the expert's findings are "sugarcoated," both the expert and the attorney will pay the price when the stakes are much higher in a future expert deposition or on the witness stand in trial. Any questions the attorney may have should be answered, and they should be informed if any further analysis would better answer their inquiries. The attorney is likely interested in the pieces of evidence that were most helpful in developing an opinion. The attorney should be advised of both the "good" and the "bad" evidence.

24.5.6   RETAINED EXPERT – AGREEMENT TO BE ENDORSED AS AN EXPERT WITNESS

At this point, the expert has reviewed materials provided by the attorney and the attorney has been advised of the expert's initial thoughts on the case. If the expert's findings do not support the attorney's allegations, the consulting relationship will likely end here. There is no reason for the attorney to continue paying an expert when the expert's opinion does not support the attorney's theory in the case. If the expert's initial opinions are helpful to the case, the attorney may ask if they would agree to be "endorsed" as an expert witness. Becoming an endorsed expert takes the expert's role in the case to the next level. This means that the expert is no longer an anonymous consultant. Eventually, their identity and credentials will be disclosed to the opposing party. Further analysis and review will most likely be done so that a more in-depth and supported opinion can be developed. An endorsed expert will then prepare a written report that outlines their opinions and conclusions. The expert will also be asked to give a deposition where they will be subject to rigorous questioning by the opposing party's attorney. If the case makes its way to trial, the expert will likely need to testify on the witness stand.

24.5.7 RETAINED EXPERT – FORMATION OF OPINIONS

In forming expert opinions, the most important thing is sticking with the principles and methods that are accepted in that particular area of expertise. The reliability of an expert's opinions is measured mainly by the process that the expert takes to get there. The basis of an expert's opinions is going to be the subject of much scrutiny, so thoroughness is needed.

All of the evidence provided or obtained must be taken into consideration. If there is evidence in the case that tends to disprove your position, you should address the issue and account for it in your opinion. In our mold case, it was ultimately found that the moisture from the unsealed door frame caused the growth of dangerous mold. Regardless of how sure the expert is that their conclusions are correct, they should be prepared for the opposing side to retain an expert who differs. The expert needs to consider all possible causes of the mold growth and be able to rule out each opposing theory systematically. Remember, the expert has been asked to opine on the *most likely cause* of the mold and experience and knowledge should be used to explain why the other possible causes are not supported by the evidence.

24.5.8 RETAINED EXPERT – THE ATTORNEY'S EXPERT "DISCLOSURE" OR "ENDORSEMENT"

When an expert agrees to serve as an endorsed expert witness, permission is granted to the attorney to identify them to the opposing party as an expert witness. As a case moves through the discovery process, there are usually court-ordered deadlines for certain key events. Usually, there is a deadline by which the attorneys need to disclose or endorse their experts to the other party. The plaintiff is required to endorse their experts before the defendant because the plaintiff bears the burden of proof on the issues outlined in his complaint. The defendant is then given time to seek out a rebuttal expert witness, if they so choose.

The endorsement is filed with the court and served on the opposing party. Generally, the endorsement contains the expert's contact information, a copy of his curriculum vitae (including his education, employment history, and any authored publications), and a list of all cases in which the expert has been involved (and whether he was on the plaintiff's or defendant's side). If the attorney has asked for the expert to complete a written report outlining their opinions, this report will be made part of the endorsement.

24.5.9 RETAINED EXPERT – PREPARING THE EXPERT REPORT

Depending on the type of case, the expert report could be one page or it could be 500 pages. The report should outline the facts that the expert relied upon in coming to their opinions. All factual evidence should be cited so that the source of the information is readily available. After the relevant facts have been established, the process used in determining the conclusions should be described in detail. Why the process is reliable should be explained by citing pertinent industry literature or other supporting evidence. The expert's conclusions are then outlined and their opinions listed. In some cases, the opinion as to why the opposing side's position is flawed will be asked to be provided in the report. The scope of the final expert report should be discussed with the retaining attorney.

24.5.10 RETAINED EXPERT – DEPOSITION BY OPPOSING ATTORNEY

The process of developing informed opinions is crucial to the role of an expert. All of the hard work an expert puts forth can be rendered meaningless unless the expert's opinions can be defended. In our adversarial legal process, the deposition will be the first time an expert must defend their opinions. A deposition is a time when the attorneys for all of the parties involved in the case have a chance to ask the expert questions. The deposition usually takes place in an office conference room, and the attorney who retained the expert will be present. There will be a court reporter who will transcribe the questions and answers and possibly a videographer who would video record the deposition. Around the table will be the attorneys who represent the opposing parties in the case.

The expert will need to prepare for the deposition and the report that was written by the expert should be reviewed. During the questioning, it is important to listen very closely to how each question is worded. If the question needs to be repeated or reworded to make more sense, the request should be made. A question that is not fully understood should never be answered. Although the order of the questions varies depending on the opposing attorney's style, they will likely first attempt to discredit the expert witness' qualifications to serve as an expert in this particular matter. Questions about the expert's education and experience will be asked, and it will be suggested that the expert is outside his area of expertise. The expert must stay poised and respectful when answering these questions.

When the opinions of the expert are questioned, the answers should be thoroughly explained. If the opposing attorney requests a "yes" or "no" answer to a complicated question, it can be politely explained that such a question deserves more of an explanation. If there is scientific literature that contradicts an expert's opinions, it is very likely they will be confronted during the deposition. If the presented literature is unfamiliar, the expert can ask to take a closer look before answering questions about it. The expert should be careful in giving answers to unreasonable hypothetical questions offered by the opposing lawyer. Usually, peculiar questions are meant to hurt the expert's position.

Upon conclusion of the deposition, the expert should not waive the right to review the deposition transcript; it will help to familiarize themselves with their testimony and to make any minor corrections. However, if major corrections are made, opposing counsel will likely reserve the right to re-depose the expert.

24.5.11 Retained Expert – Potential Expert "Challenge" by Opposing Attorney

At some point prior to trial, it is possible that the ability of the expert to serve as an expert witness will be challenged by the opposing party. This tactic is commonplace in civil litigation and the expert should not take such a challenge personally. When challenged, the opposing attorney argues that the expert's opinions cannot be relied upon because they are not based on sufficient facts and data, the expert failed to use established principles and methods in coming to their conclusions, or the expert lacks the specialized experience and training to develop trusted opinions. The opposing attorney files a motion with the court, which outlines the reasons why the expert should not be allowed to testify in the matter. Such a challenge is often referred to as a "Daubert" or "Frye" challenge (depending on the jurisdiction), named after the judicial opinions that discuss the requirements of an expert witness.

The attorney who retained the expert will file a response to the challenge. The response will explain to the court why the expert is qualified to give an opinion on certain issues in the case and that the principles and methods used in forming their opinions are reliable. Frequently, the challenge will be the subject of a court hearing. If a hearing takes place, it will most likely occur after the expert's deposition has been given. In a Daubert or Frye hearing, questions will be asked much like the ones in the deposition, only this time the expert will be on the witness stand in front of a judge. Unlike the deposition, the hearing will probably not encompass the intimate details of the expert's opinion on the case. Rather, the challenge hearing is focused entirely on whether the opinion of the expert should be allowed. From a preparation standpoint, the hearing should be treated much like the deposition. The expert will need to be especially ready to defend his knowledge, skill, experience, training, and education in relation to the subject matter at issue.

The importance of disclosing prior experience the expert has to the attorney who retained them cannot be over emphasized. If the testimony of an expert witness has been previously stricken, it should be discussed with the attorney. If there is any doubt, it can be guaranteed that the opposing attorney will discuss any previous challenges at the Daubert or Frye hearing. Just because expert testimony was not allowed in one case, it does not mean that testimony cannot be offered in a different case. Regardless, the attorney who retains the expert needs to know the past history of the expert in order to make the best argument on his behalf.

24.5.12 RETAINED EXPERT – TRIAL DEPOSITION

In lieu of an expert's appearance at trial, a trial deposition may be taken. Essentially, the expert will be testifying in the deposition as if speaking on the witness stand in front of the jury. At trial, the deposition testimony will be read to the jury, or if the deposition was videotaped, the video will be played. The main difference between the trial deposition and the earlier deposition is the format. In the trial deposition, the attorney who retained the expert gets to start by asking them questions, and the opposing attorneys follow. There will likely be more objections made during the trial deposition since the attorneys want to preserve these issues for trial. If an objection is made, the expert should stop speaking and wait for instructions as to whether the question should be answered. All of the comments outlined below in the section regarding trial testimony are equally applicable to a trial deposition.

24.5.13 RETAINED EXPERT – TRIAL TESTIMONY

Finally, the trial has arrived. It is understood that preparation is the most important part of the expert's trial experience. When the expert takes the witness stand, it is important for them to remember that the jury does not have a mastery of the facts in the same way as the expert. They should start from square one in explaining to the jury how their opinions were formed. While testifying, the expert becomes the teacher and the jurors are the students. The testimony should be interesting even if that seems impossible given the subject matter. Jurors appreciate demonstrative evidence in the form of charts, graphs, photographs, and other visuals. The better the jury understands the opinions of the expert, the more likely they are to agree with the expert's position.

During cross-examination by the opposing attorney, it is important that a calm demeanor be maintained. His job is to discredit the expert and his opinions in front of the jury. If composure is lost, the other side's case has been helped as this makes the expert look unprofessional. They should be assertive and authoritative, but not argumentative.

Most importantly, when it comes to trial, the expert should be themselves. Many experts take the witness stand and try to act like they think an expert witness is supposed to act. Juries can smell a fake a mile away. If an expert is perceived as disingenuous and insincere, a juror is less likely to believe what they are hearing.

24.6 SOME "TIPS" FOR SERVING AS AN EXPERT WITNESS

The information or "tips" on serving as an expert witness and writing an expert report are voluminous[1–9] and far beyond the ability of this author to know, let alone cover in a few paragraphs. Nevertheless, a few tips, both general and regarding specific situations, based solely on experience as an expert witness, follow.

24.6.1 GENERAL TIPS

1. Know the facts of the case inside and out. As in every aspect of engineering/scientific work, preparation is essential. Read everything – pertinent records, photographs, the complaint and answer, fact witness statements and depositions, expert witness reports, and expert witness depositions. The easiest way to make mistakes and have the opinions of the expert called into question is for the expert not to have a good handle on the facts.

2. When being deposed, the expert should listen carefully and use words carefully. Particularly pay attention to adjectives used in questions that will affect the response to a question asked of an expert. In a legal matter, words matter!

3. An expert's reputation is the most important professional asset as is that of his profession, whether it be an engineer, scientist, or trade expert. An expert witness should never testify in a manner that will compromise his reputation.

4. If there are any questions as to the qualifications or experience of an expert with the subject matter of a case, the case should not be taken. When an expert witness stretches their qualifications and experience, they are making a mistake that will eventually be exposed.

5. Opportunities to work as an expert witness will come to someone who is qualified through education and experience. Advertising and self-promotion are rarely necessary.

6. Before an expert witness agrees to be endorsed, the facts in the case and the position to be supported should be clearly understood. It is essential that the expert maintain thorough communication with the retaining attorney.

7. Trust in the attorney who would like to retain an expert is crucial before agreeing to be endorsed. This is especially true since often the expert witness expertise will be challenged using "Daubert," "Frye," or other local rules.[14] It is important that the expert's contract require their client to make them aware of and participate in all motions challenging their expertise.

8. Be sure that the attorney who would like to retain the expert is organized and prepared before an agreement to be endorsed is made. If the attorney is not organized and prepared, then it is likely that the expert will not appear to be organized and prepared.

24.6.2 Tips Regarding Expert Written Reports

1. *Using references that post-date time item of contention constructed*: In many hail, wind, or structural project reports, information regarding how an improperly installed or designed item should have been installed or designed is provided under the discussion section of the report. At times, these documents, including codes and standards, post-date the time the home or structure was fabricated. The opposing attorney almost always notes this and argues in court that the document either should not be allowed in as evidence or even worse, that the expert's report should be disallowed and the case dismissed under summary judgment because, "how could my client (e.g., a builder) know about this information when it was published after he constructed this home?" For the expert in this situation, they should: (1) use documents in their report that pre-date the construction period of interest or, if not available (2) represent that the document is representative of known best practices that pre-dates the date of the construction period of interest. Under this later situation, the expert witness should be prepared to demonstrate, based on experience, or other evidence that such a representation is true.

2. *Speculation*: A classical approach used by attorneys to discredit an expert witness is to find a portion of the report that is speculative or cannot be supported and use this speculative section to state that this is how that expert conducts themselves and that it calls into question all their other work, including their opinions. Often the point of speculation is minor or not germane to the majority of the report or opinions, but nevertheless can be used to discredit the overall work of the expert and the expert himself.

   An attorney will play into the natural tendency of the engineer or scientist to provide a solution to a problem when such a conclusion cannot be explicitly proven (i.e., not all the "I's" and "t's" were crossed). To illustrate this point, consider the following scenario:
   Situation: Water Claim: Staining on cathedral ceiling.

Engineer: Finds a defect in the soil stack flashing on the roof and concluded that the stain was caused by the defect in the soil stack. No water testing was completed and the attic space above the ceiling was not accessible.

Problem: The water path between the soil stack and the ceiling has not been proven either visually or by water testing. The leak could be from clogged gutters, ice damming, or even a pipe break. Without water testing or visual verification of the staining to the attic decking and drywall above, this conclusion cannot be supported.

Expert problem: Assuming this is only one of several leaks in the home investigated by the expert, and the matter goes to litigation, an opposing attorney could use this "speculation" regarding the cause for this one leak to condemn the expert's entire report even though the rest of the conclusions in the report were rock solid.

3. *Use of broad language*: In an effort to be definitive, engineers and scientists will sometimes rely on overly broad words such as "all," "always," and "never," when the situation is not that precise. Further, such language is often not necessary. Recall that the preponderance of evidence criteria only requires that the situation regarding facts be "more likely than not" (i.e., >50%) or "to a reasonable degree of engineering or scientific certainty." Nowhere is the requirement that an expert deems the situation be 100% true (or false), nor is the evidence likely to support such a definitive answer. Yet experts often write reports stating the event occurred with effectively 100% certainty, which gives great opportunity for the opposing attorney to discredit such an expert's report and opinions during future cross-examination.

4. *Advocacy or use of extreme values*: An expert report should not appear to be, nor take on the position of being, an advocate for either side. Opposing experts, attorneys, judges, and juries will see through the use of such language, information, or data by the expert with time, which in turn will damage an expert witness' credibility. For example, if the data suggest that the humidity in the room varied from 35% RH to 52% RH with a mean of 40% RH, it would be a stretch to suggest that the room conditions were above levels associated with the amplification of mold (>50% RH).

5. *Draft reports, notes on documents, and notes on margins of documents*: As an expert witness, any and typically all documents, notes, draft materials, sketches, etc., are items the opposing attorney is entitled to have, review, and question an expert. Preparation and distribution of draft reports should be avoided. Placing notes in the margins of papers, books, and other discovery documents that will be subject to examination should also be avoided. Frequently, these notes can be the topic of detailed examinations by an opposing attorney when they may be trivial and may not even have been added to the document as a result of the particular case. Oftentimes, why the note was written or even who wrote the note may have been forgotten, yet they are materials that can be used for examination by an opposing attorney.

6. *E-mails and/or electronic correspondence*: In most jurisdictions, an expert's e-mails and electronic correspondence related to a specific case are discoverable. EES Group, Inc. has had a policy for nearly 20 years that all e-mails older than 30 days are to be deleted, saving server space.

7. *Mistakes*: Inevitably, despite efforts to avoid them, mistakes will occur in an expert report. Sometimes these are trivial and sometimes they can be substantial. Normally they will be found when reviewing the report for a future deposition or before trial testimony during trial preparation activities. In all cases, the report should be corrected and the client notified immediately so that they are aware of the situation and can provide the corrected report to the opposing attorney. While finding, correcting, and admitting mistakes are embarrassing, it is far better to find and correct the mistake proactively rather than having it identified at trial by the opposing attorney, who could use it to damage the credibility of

the expert in front of the client, judge, and/or jury. In fact, such admissions, provided they are infrequent, can actually benefit the expert's credibility by the professionalism exhibited in correcting and admitting a mistake proactively.

24.6.3 Tips Regarding Opposing Attorney's Questions in Depositions and Trials

1. *Questions asked in depositions are only designed to obtain the truth*: Not necessarily so. In most cases, the opposing attorney will be asking questions to support their case and to obtain information that may be helpful in motions to exclude at least portions of an expert's testimony. As described earlier in the goals for an attorney completing a deposition, many questions will be asked regarding an expert's qualifications and methodologies in an attempt to use the Rules of Evidence[12,13] against you. It should always be ensured that the calculations, analysis, and opinions of an expert have a basis in the peer-reviewed technical literature.

2. *Repeated or similar questions*: Questions generally repeated by an attorney are not because the earlier answer was not understood, but because the answer is not the answer they desire. Remember, that if a question is asked multiple times, two things are in play: (1) the particular issue being questioned is important and (2) the opposing attorney is driving to obtain a specific answer to the question to support their case. Also, if an expert is asked the same or similar question multiple times and answers the question essentially the same way in every case but the last time when a different answer is provided, the last answer, if it is the one desired by the opposing attorney, will and can be used in motions and at trial despite all the previous answers.

3. *Multiple question questions*: Questions are sometimes asked with multiple questions imbedded within an apparently single question. This type of question should always be recognized and the attorney should be asked which question he would like to have answered. Attempts to answer a question that contains multiple questions are dangerous since the answer can (and will) be applied to the portion of the multiple question that best supports the opposing attorney's case.

4. *Yes/no questions*: As briefly discussed earlier, an attorney will sometimes pressure a deponent to answer a question either "yes" or "no." This situation should raise a red flag, since rarely are complex technical questions posed in this scenario answered that simply. A simple version of such a question might be, "Was/is the air in the room humid, please answer yes or no." The answer is likely both yes and no; that is, yes there is some humidity in the room; now whether or not it is high (yes or no) would depend on what criteria the level were compared with (e.g., human comfort at 60% RH or mold growth amplification beginning at ~50%). In this scenario, it might also be time-dependent. Again, be cautious of answering yes or no questions with a simple yes or no.

5. *Attorneys have to be truthful with the deponent*: While seemingly unethical, it is not illegal for an attorney to misrepresent the truth regarding the testimony of others or to alter documents and present them to you as original documents. Further, an expert's answers to these misrepresentations can, and will, be used to support the opposing attorney's case, including motions against the expert. In an actual case, a document had a label removed on an axis of a chart that affected the apparent values in the chart. The attorney presented this plot and then proceeded to ask a series of questions regarding the altered document. Fortunately, the basis of the original document was recalled and no damage was done in responding to questions regarding this altered document.

IMPORTANT POINTS TO REMEMBER

- The legal process allows someone a formal process to right what he or she perceive as a wrong.
- The person(s) or entity filing the lawsuit is known as the plaintiff and those who the lawsuit is filed against are known as the defendants.
- Expert witnesses are to provide an understanding of complex issues not readily apparent to those with general background knowledge.
- An expert witness is someone who is qualified to give their opinion based on their knowledge, skill, experience, training, or education.
- Experts should conduct themselves, and their work, as they would in their normal professional lives with attention paid to details.

REFERENCES

1. Bronstein, Daniel A. *Law for the Expert Witness*, Fourth Edition. CRC Press: Boca Raton, FL, 2012.
2. Babitsky, Steven, James J. Mangraviti, Jr., and Alex Babitsky. *The A-Z Guide to Expert Witnessing.* SEAK, Inc.: Falmouth, MA, 2006.
3. Speight, James G. *The Scientist or Engineer as an Expert Witness (Chemical Industries).* CRC Press: Boca Raton, FL, 2009.
4. Babitsky, Steven and James J. Mangraviti, Jr. *Writing and Defending Your Expert Report: The Step-by-Step Guide with Models.* SEAK, Inc.: Falmouth, MA, 2002.
5. Babitsky, Steven and James J. Mangraviti, Jr. *How to Excel during Depositions: Techniques for Experts That Work.* SEAK, Inc.: Boca Raton, FL, 1999.
6. Noon, Randall K. *Forensic Engineering Investigation.* CRC Press: New York, 2001.
7. Postol, Lawrence P. "Certifiably credible: The industrial hygienist as expert witness," *The Synergist*, January 2010, pp. 35–37.
8. Dreger, Kurt W. and David L. Dahlstrom. "Make Your Evidence Count: The implications of 'Hearsay' for IHs in the Courtroom." *The Synergist*, pp. 24–26, December 2013.
9. Kohl v. US, 91 U.S. 367, 1875. Accessed May 18, 2020, http://caselaw.lp.findlaw.com/cgi-bin/getcase.pl?court=us&vol=91&invol=367.
10. US Legal. "Discovery Depositions Law and Legal Definition." Accessed May 18, 2020, https://definitions.uslegal.com/d/discovery-depositions/.
11. UpCouncil. "Legal Definition of Deposition: Everything You Need to Know." Accessed May 18, 2020, https://www.upcounsel.com/legal-def-deposition.
12. Anjelica Cappelino, J.D. "Federal Rules of Evidence and Experts: The Ultimate Guide" – updated February 18, 2020. Accessed May 18, 2020, https://www.expertinstitute.com/resources/insights/the-ultimate-guide-to-the-federal-rules-of-evidence-and-expert-witnesses/.
13. Federal Rule 702. "Testimony by Expert Witnesses." Accessed May 18, 2020, https://www.law.cornell.edu/rules/fre/rule_702.
14. Expert Institute. "Daubert v. Frye: A State-by-State Comparison", updated August 9, 2018. Accessed May 18, 2020, https://www.expertinstitute.com/resources/insights/daubert-v-frye-a-state-by-state-comparison/.

Index

Note: **Bold** page numbers refer to **tables** and *Italic* page numbers refer to *figures*.

For Product Safety Concerns and Information please contact our EU
representative GPSR@taylorandfrancis.com
Taylor & Francis Verlag GmbH, Kaufingerstraße 24, 80331 München, Germany